ENVIRONMENTAL TECHNOLOGY

Environmental Technology

Proceedings of the Second European Conference on Environmental Technology, Amsterdam, The Netherlands, June 22–26, 1987

edited by

K.J.A. DE WAAL

TNO Division of Technology for Society
Apeldoorn, The Netherlands

W.J. VAN DEN BRINK

TNO Corporate Communication Department
The Hague, The Netherlands

1987 **MARTINUS NIJHOFF PUBLISHERS**
a member of the KLUWER ACADEMIC PUBLISHERS GROUP
DORDRECHT / BOSTON / LANCASTER

Distributors

for the United States and Canada: Kluwer Academic Publishers, P.O. Box 358, Accord Station, Hingham, MA 02018-0358, USA
for the UK and Ireland: Kluwer Academic Publishers, MTP Press Limited, Falcon House, Queen Square, Lancaster LA1 1RN, UK
for all other countries: Kluwer Academic Publishers Group, Distribution Center, P.O. Box 322, 3300 AH Dordrecht, The Netherlands

Library of Congress Cataloging in Publication Data

```
European Conference on Environmental Technology (2nd :
  1987 : Amsterdam, Netherlands)
  Environmental technology.

  1. Environmental engineering--Europe--Congresses.
2. Environmental policy--Europe--Congresses.  3. Factory
and trade waste--Environmental aspects--Europe--
Congresses.  4. Recycling (Waste, etc.)--Europe--
Congresses.  5. Salvage (Waste, etc.)--Europe--Congresses.
I. Waal, K. J. A. de.  II. Brink, W. J. van den.
III. Title.
TD55.A1E97  1987         628             87-15241
```

ISBN-13:978-94-010-8139-9   e-ISBN-13:978-94-009-3663-8
DOI:10.1007/978-94-009-3663-8

Copyright

© 1987 by Martinus Nijhoff Publishers, Dordrecht.
Softcover reprint of the hardcover 1st edition 1987

All rights reserved. No part of this publication may be reproduced, stored in a retrieval system, or transmitted in any form or by any means, mechanical, photocopying, recording, or otherwise, without the prior written permission of the publishers,
Martinus Nijhoff Publishers, P.O. Box 163, 3300 AD Dordrecht,
The Netherlands.

**Second European
Conference on
Environmental Technology**

in the 'European Year
of the Environment'

**Recommending Committee of the
Second European Conference on Environmental Technology**

G.A. Wagner (chairman)
Chairman of the Supervisory Board of the Royal Dutch Petroleum Company. The Netherlands.

J.C. Blankert
President of the Association of the Mechanical and Electrical Engineering Industries (FME). The Netherlands.

G. Bresser
Chairman of the Association of the Dutch Chemical Industry (VNCI). The Netherlands.

S. Clinton Davis
Member of the Commission of the European Communities. Belgium.

J.M. Dirken
Vice-chancellor of Delft University of Technology. The Netherlands.

M.C. van der Harst
Director General of the Netherlands Ministry of Economic Affairs. The Netherlands.

J.D. Hooglandt
Chairman of the Dutch Steel Producers' Association. The Netherlands.

E.W. ter Horst
President of the Industrial Council for Energy and Environmental Technology. The Netherlands.

W.A. de Jong
President of the TNO Board of Management. The Netherlands.

N.G. Ketting
President of the Board of Managing Directors of the NV SEP Dutch Electricity Generating Board. The Netherlands.

C.J.A. van Lede
President of the Federation of Netherlands Industries (VNO). The Netherlands.

Lord Nathan
Chairman of the Environment Committee of the House of Lords. Great Britain.

S. Patijn
Chairman of the national committee of the European Year of the Environment. The Netherlands.

W.C. Reij
Director General of the Netherlands Ministry of Housing, Physical Planning and the Environment. The Netherlands.

K.C. Sivaramakrishnan
I.A.S. Project Director for the Central Ganges Authority. India.

**Programme Committee of the
Second European Conference on Environmental Technology**

K.J.A. de Waal (chairman)
TNO Division of Technology for Society, Apeldoorn. The Netherlands.
W.J. van den Brink (secretary)
TNO Corporate Communication Department, The Hague. The Netherlands.
F. van den Akker
Ministry of Housing, Physical Planning and the Environment, Leidschendam. The Netherlands.
H.M. de Bliek
Ministry of Economic Affairs, The Hague. The Netherlands.
A.M. Buekens
Free University of Brussels, Brussels. Belgium.
S. Dudenkov
USSR.
D. Huisingh
North Carolina State University, Raleigh, N.C. USA.
P.J. Huiswaard
Association of Consulting Engineers of the Netherlands (ONRI), The Hague. The Netherlands.
G.S. Jonker
Delft University of Technology, Delft. The Netherlands.
F.E. Joyce
ECOTEC Research and Consulting Ltd, Birmingham. Great Britain.
J.M. Junger
Commission of the European Communities, Brussels. Belgium.
K. Müller
National Agency of Environmental Protection, Copenhagen. Denmark.
Ms. F. Petillot
Direction de la Prévention des Pollutions, Neuilly-sur-Seine. France.
H.F. Peze
Gist Brocades B.V., Delft. The Netherlands.
M. Tels
University of Technology Eindhoven, Eindhoven. The Netherlands.
A.W. Veenman
Netherlands Association of Suppliers of Environmental Equipment and Technics (VLM), Zoetermeer. The Netherlands.
L. Verweij
Federation of Netherlands Industries (VNO), BMRO Bureau of Environment and Physical Planning, The Hague. The Netherlands.
E.J. Vles
Association of the Dutch Chemical Industry (VNCI), Leidschendam The Netherlands.
J. Voskamp
Hoogovens Groep B.V., IJmuiden. The Netherlands.

The Programme Committee wants to thank the following persons for their assistance in the preparations of this conference:

Ms. C.A. Arensman
TNO Corporate Communication Department, The Hague. The Netherlands.
G.A. Maas Geesteranus
Ministry of Housing, Physical Planning and the Environment, Leidschendam. The Netherlands.
W.J.C. Melgert
TNO Corporate Communication Department, The Hague. The Netherlands.
S.V. Swolfs
Netherlands Association of Suppliers of Environmental Equipment and Technics (VLM), Zoetermeer. The Netherlands.
J.I. Walpot
TNO Division of Technology for Society, Apeldoorn. The Netherlands.

CONTENTS

Programme Committee and Recommending Committee of the
Second European Conference on Environmental Technology　　VII

Introduction
E.H.T.M. Nijpels　　1

RAW MATERIALS AND ENERGY

Combined-cycle power generation - a promising
alternative for the generation of electric power from
coal.
E. Nitschke　　5

A proposal for the energy infra-structure of Atatürk
Organized Industrial District in Izmir.
A. Durmaz, Y. Ercan, Ö.E. Ataer & M. Sivrioglu　　18

Fluidized bed combustion technology for low grade
lignite utilization in Turkey.
M. Arikol, E. Ekinci, D. Bilge & A. Serpil　　28

Pressurised combustion - a new tool in emission
abatement.
L.J.M.J. Blomen, J.E. Hille & P.F. van den Oosterkamp　　37

Safety assessment and the selection of detergent raw
materials.
N.T. de Oude　　43

Thermodynamic analysis of solar powered absorption
refrigeration systems for comparison of working fluids.
Ö.E. Ataer　　49

PROCESS-INTEGRATED ENVIRONMENTAL TECHNOLOGY

Process integrated environmental technology. A must to
survive.
J. Quakernaat, J.A. Don & F. van den Akker　　55

The development of clean technology phosphate
fertilizer production processes.
G.H.M. Calis											67

Hydrolysis of waste paper.
J.I. Walpot & H. Visscher									69

The effect of processing on the heavy metal content in
compost.
G.R.E.M. van Roosmalen, M.M.G. Senden, T. Brethouwer &
M. Tels											77

Autothermal incineration of waste water sludge in a
fluid bed furnace.
R. Tize											87

Dealing with environmental affairs in the Dutch sugar
industry.
J.A. Don & L. Feenstra									96

Removal of nitrogen compounds from waste water.
L.W.F. Harmsen										110

Cleaner technologies in Denmark - examples and
experiences.
K. Müller											114

The application of the Best Practicable Environmental
Option to pollution control in 1987.
R.G.P. Hawkins										121

Removal of cadmium by anion exchange in a wet
phosphoric acid process.
T.T. Tjioe, P. Weij & G.M. van Rosmalen							145

Cadmium incorporation in calcium sulfate modifications.
G.J. Witkamp & G.M. van Rosmalen							148

Removal of cadmium from phosphoric acid solutions by
selective coprecipitation in a minor amount of
anhydrite.
J.A. Kroon										151

A clean technology phosphoric acid process.
S. van der Sluis & G.M. van Rosmalen							153

TREATMENT OF INDUSTRIAL EMISSIONS / WASTE

Abatement of HCl and HF emissions from waste
incinerators by injection of hydrated lime.
A. Verbeek, D. Schmal & C. van der Harst						157

Residual products from flue gas desulphurization by
spray-dryer method - technical and economic aspects of
their disposal and recovery for utilization.
H. Ludwig, M. Hetschel & H. Fitjer							165

Flue gas treatment according to the BF/UHDE process.
U. Neumann   182

Technical/economical developments of fluegas desulphurization (FGD) installations.
L.A.J. Tol & W.L. Prins   194

The Bischoff flue gas desulphurization process in the power plants of Borssele and Maasvlakte in the Netherlands.
T. Risse & I. Maassen   211

The biological treatment of waste gases from small urban sources (emission of volatile organic compounds).
A.J. Dragt, A. Jol, C. van Lith & S.P.P. Ottengraf   224

Biofiltration - a relatively cheap and effective method of waste gas treatment.
P.G. Paul & F.J. Castelijn   231

Experience with full-scale Biopaq - U.A.S.B.-plants treating various types of effluent.
P.J.F.M. Hack & L.H.A. Habets   238

Wet air oxidation of toxic wastewater.
M.A.G. Vorstman & M. Tels   247

Thermochemical treatment of solid waste obtained from pulp and paper factory with 15 % acetic acid and NaOH for conversion to crudes.
F. Taner, H. Boztepe & Ü. Kimyonsen   256

Ozonation of the aqueous layers obtained from thermochemical treatment of solid waste with 15 % acetic acid and NaOH.
H. Boztepe, F. Taner & Ü. Kimyonsen   263

Chloroff, a non-destructive dechlorination process.
P.F. van den Oosterkamp, L.J.M.J. Blomen, A.S. Laghate & H. ten Doesschate   275

Electrochemical treatment of organohalogens in process waste waters.
D. Schmal, J. van Erkel, A.M.C.P. de Jong & P.J. van Duin   284

Recovery of heavy metals by crystallization in the pellet reactor.
M. Schöller, J.C. van Dijk & D. Wilms   294

Removal of arsenic from waste waters of the lead glass industry.
A. Kaiser, F. Hutter, J. Kappel & H. Schmidt   304

Mercury removal by activated carbon process.
K.-D. Henning, K. Keldenich & K. Knoblauch   314

Fate of some trace elements in combustion and
gasification processes.
W. Mojtahedi & K. Larjava 323

Reduction of the cadmium content of waste gypsum
produced by the DSM-phosphoric acid plant at Pernis.
R. Spijker 334

Investigations of the adsorptive removal and recovery
of halocarbons.
K.-D. Henning, M. Schäfer, W. Bongartz & K. Knoblauch 337

The use of Zoogloae ramigera for removal and recovery
of Cr^{6+} ions from industrial waste waters.
T. Kutsal & Y. Sag 340

Inactivation of sewage sludges with calcium oxide prior
to agricultural uses.
T. Marcinkowski 343

Fly-ashes in soil improvement.
B. Quandt 349

Vegetation establishment on fly ash ponds by means of
hydroseeding.
D. de Vleeschauwer & R. Imler 352

Atmospheric probing with lidar.
G.J. Kunz 355

VT-Biofilter.
J.J.W. Bijl 358

Removal of heavy metals and dioxins in flue gases from
waste to energy plants.
K. Carlsson 361

PRODUCTS (DESIGN)

Designing of products in view of recycling.
W. Jorden 367

Electrodeposited aluminium by the $SIGAL^R$-process. A
superior coating for corrosion protection as an
alternative to cadmium and zinc coatings.
H. de Vries 377

Economic aspects of environmental product design.
G. Huppes 383

Waste prevention in the ecological building project of
Delft University of Technology.
H. Hubers 391

APPLICATION / UTILIZATION OF SALVAGEABLE REFUSE COMPONENTS

Application of some waste materials in hydraulic engineering.
K.W. Pilarczyk, G.J. Laan & H. den Adel — 399

The processing of industrial waste for immobilization and/or recycling applying pozzolanic reactions.
P.D. Rademaker & R.B. Wiegers — 411

Utilization of industrial waste gypsum for the manufacture of gypsum-bonded particleboards in a semi-dry process.
K. Lempfer — 422

The use of concrete and masonry waste as aggregates for concrete production in the Netherlands.
Ch.F. Hendriks — 431

Utilization of residues of desulphurization and denitrification technologies.
H.-J. Pietrzeniuk — 441

Glass-recycling.
C.Q.M. Enneking — 448

Photometric sorting of cullet.
A. Reichert & H. Hoberg — 457

The use of off-gas CO_2 in greenhouses. Removal of NO_x and ethylene.
C.M. van den Bleek, P.J. van den Berg & A.G. Montfoort — 466

Product development needs of waste management.
P. Vilppunen — 476

REMEDIATION TECHNIQUES AND WASTE HANDLING

Review of soil treatment techniques in the Netherlands.
E.R. Soczó, E.J.H. Verhagen & C.W. Versluijs — 481

In situ remedial action techniques for treatment of contaminated soil and groundwater by means of groundwater extraction and infiltration techniques.
E. de Zeeuw — 493

Cleaning soils contaminated with heavy metals.
J.W. Assink & W.H. Rulkens — 502

Steam stripping organic compounds from contaminated waters.
F.H.M.M. Langen, P.G. Paul & R. van Booren — 513

Extractive cleaning of heavy metal contaminated clay soils.
B.J.W. Tuin, M.M.G. Senden & M. Tels — 520

Characterization and remediation of coking plant sites in the Ruhr territory.
H. Koesters & H. Spittank — 529

In situ remedial action techniques of cadmium polluted soil.
L.G.C.M. Urlings, A.T. Blonk, J.A. Woelders & P.R. Massink — 539

Need and availability of cover material for landfill sites in South Wales.
E.M. Bridges, A.T. Evans & D.J. Leech — 549

Development of technology for contaminated dredged material remediation.
H.J. van Veen & A.C. de Waaij — 559

Research on polluted sediment.
P.J.A. Baan — 568

Cone penetration testing in relation to environmental problems.
J.G. de Gijt & G. van Roekel — 577

Re-infiltration of waste tip leachate: an inexpensive alternative.
P.A. de Boks — 587

Nitrate removal from ground water.
J.P. van der Hoek — 593

Nitrates in groundwater.
A.L. Kowal & A. Polik — 604

Indaver n.v. - a new industrial waste treatment plant in Belgium.
L.M.J. Sterckx — 610

Methodological evaluation of containment strategies.
C.C.D.F. van Ree, R. Kabos, F.A. Weststrate & M. Loxham — 614

The use of diaphragm walls of over 50 meter deep to stop spreading of contamination.
H.H. van Breukelen — 617

Leaching of cyanides.
W.P. van Oosterom & L.G.C.M. Urlings — 619

Polymeric flocculants in waste water treatment and in sludge dewatering - technical and economical aspects.
A. Landscheidt & J.M. Reuter — 622

On-site application of trickling filter and rotating biocontactor in treatment of groundwater, polluted with chlorinated hydrocarbons.
A.L.B.M. van Campen, L.G.C.M. Urlings & B. Bethe — 624

Pollution-controlling effect of adsorbing materials on the leaching of pollutants from dredging spoil depots.
A.W. Grinwis & L.G.C.M. Urlings — 627

Characterization and environmental effects of sludges produced by metallurgical factory.
J.R. Dobosz & M. Sebastian — 630

Recovery of chemical wastes - results of a Danish case study and implications for other countries.
K. Christiansen — 635

Waste management: a comprehensive service to industry.
P.H.M. Meyer zu Schlochtern — 638

INCENTIVES / ATTITUDES

Implementation of energysaving and environmental technologies.
R.W. Hommes, J.C. Brezet & L.W. Baas — 643

Innovative strategies in the environmental industry - an international comparison.
D. Drouet — 653

Risk analysis as a rational basis for soil protection and contaminated land clean up.
M. Loxham & C.C.D.F. van Ree — 662

Methods for assessing the risk of environmental contamination.
C.L. van Deelen — 671

Risk analysis and risk policy in the Netherlands.
B.J.M. Ale & M. Seaman — 682

Growth of environmental technology as foreseen by the recent administrative and legislative developments.
A. Müezzinoglu & F. Sengül — 690

Development of appropriate technology and way out for environmental protection of China - a way for the developing countries like China.
R. Hou — 698

Surabaya Indonesia: options in solid waste management.
S.A. Vigil — 707

Application of the UASB-process for anaerobic treatment of municipal wastewater under (sub)tropical conditions.
A.F.M. van Velsen & J.A.W. Maas — 716

Wastewater treatment technology for developing countries.
J.G. Bruins — 729

A waste reduction program and assessment of current status for Illinois.
D.L. Thomas, D.D. Kraybill & G.D. Miller — 740

Towards a network for environmental transfer.
F. Joyce — 750

Research and development of environmental biotechnology in the Netherlands.
K. Visscher & W.H. Rulkens — 755

American nongovernmental organizations' role in promoting waste reduction and clean technology transfer.
L.R. Martin — 763

The waste water sanitation program of Gist-Brocades N.V. location Delft.
P.A. Lourens — 773

Environmental technology and ecologization.
H. Marinov — 777

Optimization of environmental policy by integrated environmental research.
J.S.A. Langerwerf — 784

National Environmental Centre. Outline summary: objectives, structure, function and procedures.
A.W. Veenman — 788

Environmental Care Programme: Exploring the market for environmental technology.
E.J. Vles — 795

Innovation and pollution abatement.
D. Wiersma & T. Pulles — 806

A micro-computer package for calculating the cost of waste management options.
F. Etemad & D.J. Leech — 809

Hazardous waste management on the regional, macroregional and national levels.
R. Szpadt — 812

Siting of industries and status of wetland ecosystem on the Indian coast.
R. Mahalingam & K. Gopinathan — 817

List of addresses of first-named authors — 819

INTRODUCTION

Five billion people, at present the world population, inevitably affect the quality of the environment. The general public in an increasing number of countries is getting more and more concerned about this deterioration in quality. As a result many people cast doubts upon the desirability of the increase in energy consumption, the production of superfluous goods, ever-growing waste flows, harmful emissions of industrial processes, and so on.

Actually, no one can simply ignore these issues. For instance, the authorities could introduce more environmental legislation aiming at a healthy environment; industries could change to cleaner production processes; the public at large should assume an even more conservation-minded attitude rather than confine themselves to shaking a finger at 'the industry'. In short, in all sectors of society there are often numerous ways and means of curbing environmental pollution. Clearly, environmental technology - the development and application of techniques to identify, quantify and reduce environmental problems - can make a substantial contribution here in many situations.

Until now a large number of such new techniques have been developed. Many of these techniques not only appear to add greatly to reducing the burden on the environment, they sometimes also offer interesting economic advantages (savings in raw material and energy, etc.).
However, which solution is chosen from the broad range of technology available depends on a variety of factors. That's why this Second European Conference on Environmental Technology, which takes place in the middle of the European Year of the Environment, aims at presenting the state-of-the-art in this broad field. Another major objective is to show the wide range of possibilities offered. Excursions to industries already applying environmental technology therefore form an essential item of the conference programme. Thus the conference will make it easier for industries, governments, consultants and scientists to make a sound choice.

There are various ways of arranging the subjects that come within the scope of environmental technology. For this Conference (and for these Proceedings) the life-cycle of a product has been taken as a guiding principle.
The opening session of the Conference deals with such general aspects as national environmental policies of western industrialized countries. The views of the EEC and of industry will be discussed, too.

The life-cycle starts with raw materials and the energy required

for production. Subsequently, attention is paid to production processes and emissions.
Quite a different aspect of the product life-cycle is that the product is bound to end up as waste. In this respect, too, environmental technology has come to offer interesting possibilities. More and more products are designed and constructed in such a way that the use of harmful and toxic components is avoided, components are easy to separate, and so on. Also, several interesting techniques have been developed for recycling various components. Where in the past these techniques were not available, one often had to resort to dumping the waste into the ground. Consequently, removing this 'heritage from the past' and present-day waste handling, will constitute the last subject within the framework of the product life-cycle.

Apart from the technical feasibility, another requisite for the successful, large-scale introduction of environmental technology is the willingness to actually try and check environmental pollution. That's why 'incentives and attitudes' features as a separate topic on the conference programme. This topic includes subtopics such as: risk evaluation, national and regional programmes, environmental technology centres, and the promotional aspects of introducing environmental technology. Particular attention is given to the role environmental technology can play in developing countries.

It should be remarked that maintaining the existing competition balance between the various countries is an important prerequisite for the breakthrough of the application of environmental technology in practice. Among others, the EEC could play an important role in this field by taking steps which prevent this balance to be destroyed.

A large number of organizations, institutes and authorities have been involved in the organization of this Conference; they have also played a role in the realization of these Proceedings. Government, trade and industry and the research world have exerted themselves to make this event a successful one. They have done so with enthusiasm, in spite of sometimes conflicting interests. This broad basis guarantees that attention is paid to virtually all the aspects of the extensive field of environmental technology.

If I mention the names of those who contributed to the realization of this event, I might easily forget someone. Therefore, I had rather conclude by saying simply: 'Thanks to all of you who contributed'.

The Hague, 17 June 1987

E.H.T.M. Nijpels
Netherlands Minister for Housing,
Physical Planning and the Environment

RAW MATERIALS AND ENERGY

**Second European
Conference on
Environmental Technology**

in the 'European Year
of the Environment'

COMBINED-CYCLE POWER GENERATION - A PROMISING ALTERNATIVE
FOR THE GENERATION OF ELECTRIC POWER FROM COAL

EBERHARD NITSCHKE, UHDE GMBH

1. INTRODUCTION
The classic concept of generating electric power from a fossil energy source (coal, oil, gas) comprises the following essential process steps (Fig. 1):
- Combustion of coal and generation of thermal energy in the form of hot flue gas.
- Transfer of the heat to a water/steam cycle and generation of high-pressure steam.
- Conversion of the thermal energy contained in the steam to mechanical energy in a steam engine.
- Conversion of the mechanical energy to electric power in a generator.

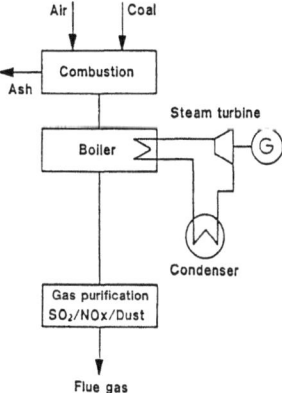

Fig. 1: Block diagram of a conventional power station

This process combination has been used for the generation of electric power for more than 100 years. Needless to say, the efficiency has been increased considerably in the course of time. An important improvement was achieved by introducing a steam turbine. Further improvements resulted from the introduction of intermediate superheating in the steam cycle, the use of bleeding steam for feedwater preheating, as well as from the change to higher pressures and temperatures. Nevertheless, the thermal efficiency of such a power station concept is limited, this limitation resulting particularly from the thermodynamic laws applying to the steam cycle. In theory it is possible to further enhance the efficiency by using even higher steam pressures and temperatures. However, this alternative is not put into practice due to material-related difficulties.

In practice, the net efficiency of a coal-fired power station has actually declined in recent years, this being due to the more stringent regulations governing environmental pollution. The decline began with the change in the type of cooling system from cooling with fresh water to a closed cooling-water cycle and the installation of a dust removal filter and continued with the requirement to install desulphurization and DeNOx units to clean the flue gas thoroughly. This led to high internal power consumption figures which, in turn, resulted in a decrease in the net electric power generation figures. All in all, it can be assumed that a coal-fired power station today achieves a net efficiency of approx. 34% (lignite) to 36% (hard coal) under optimum conditions, taking into consideration the statutory environmental pollution abatement requirements.

Efforts have been made for quite some time to find ways of circumventing the restrictions resulting from the thermodynamic laws.

Of the many ideas proposed, the utilization of a gas turbine has progressed considerably of late. Consequently, a large number of power stations based on the utilization of gas turbines have gone on stream in recent years. Such plants have also been built here in the Netherlands and have been operating successfully for a number of years.

2. COMBINED GAS/STEAM TURBINE POWER STATION

The idea of using gas turbines for the generation of energy was conceived about 100 years ago. But it was not until the fifties that commercial gas turbines were built.

Due to their advantages, i.e. low specific investment costs, small space requirement, quick and unproblematic start-up, the gas turbines were favoured for application in power stations operating at peak load. Their efficiency, however, was limited owing to the high exhaust gas temperatures of the turbine. It was therefore reasonable to utilize the energy of the exhaust gas for the generation of steam. The combination of gas and steam turbines thus offers a good basis for a thermal process with high efficiency. At high process temperatures, the energy is utilized directly in the gas turbine. In the steam cycle, the max. process temperature is relatively low and the waste heat is discharged into the atmosphere at low temperature. Fig. 2 shows the temperature/entropy diagram of such a combined gas and steam turbine cycle. The gas turbine operates in a so-called "topping cycle" and the steam turbine in a "bottom cycle", these cycles having to be adapted to one another in such a manner as to optimize the overall efficiency.

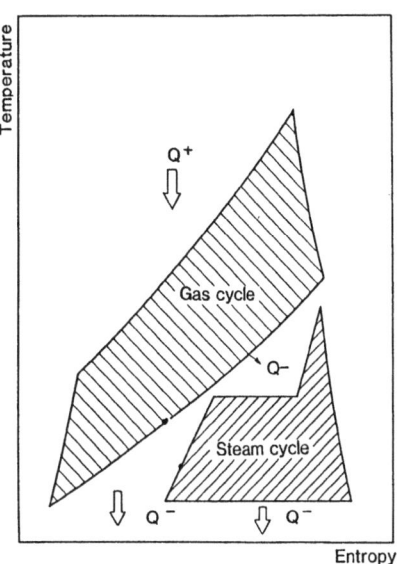

Fig. 2: Temperature/entropy diagram of combined power cycle

The temperature of the exhaust gas from the gas turbine is decisive for the design of the steam cycle. This temperature determines the steam pressure and thus the efficiency of the steam turbine. If the exhaust gas temperature of the gas turbine is raised, the efficiency of the steam cycle is increased but the disadvantages inherent to the gas turbine decrease. To overcome this difficulty, the exhaust gas temperature may be raised by installing an additional firing system. Such an additional firing system can be of particular advantage in combination with gas turbines operating at low inlet temperatures.

The design of a so-called dual-pressure steam cycle permits better utilization of the heat which is available only on a lower level. Such a dual-pressure steam cycle is particularly advantageous if the inlet temperatures of the gas turbine are high.

To summarize, it can be stated that the combined-cycle power station does not only distinguish itself by its considerably higher efficiency, which can be more than 50%, but also has the advantages of low specific investment costs for the gas turbine and a low space requirement. This is balanced by the disadvantage that only high-grade fuel, such as natural gas or oil, with a low evaporation temperature can be used.

3. COMBINED-CYCLE POWER STATION WITH INTEGRATED COAL GASIFICATION

The disadvantage of being unable to use solid matter directly as fuel in a combined-cycle power station can be overcome by means of an upstream process stage where the solid fuel is converted to a combustible gas. A coal gasification process which is suitable for a combined-cycle power station must meet the following requirements:
- Gasification under pressure for optimum adaptation to the pressure level of the gas turbine and high specific gasifier output.
- High carbon conversion rate by optimum adjustment of the gasification parameters.
- Utilization of the waste heat for the generation of high-pressure steam.
- High environmental protection by appropriate process conditions as well as the use of advanced treatment stages.
- High on-stream time, operational flexibility.

In view of the effects of the first oil crisis, intensive research was started all over the world, especially in the USA and in the Federal Republic of Germany, either to improve the available coal gasification processes or to develop new ones.

In 1986, two commercial-scale coal gasification plants went on stream in the Federal Republic of Germany. The plant of Rheinische Braunkohlenwerke AG, Cologne, at Berrenrath is based on the Rheinbraun HTW (high-temperature Winkler) process. The fluidized-bed process used in this plant was applied for the first time in the twenties. However, the plants built at that time only operated at atmospheric pres-

Fig. 3: The Rheinbraun HTW coal gasification plant at Berrenrath (capacity 730 t/d dried lignite)

sure. The objective was to develop the HTW process for gasification under pressure and to improve efficiency at the same time. This process was developed in long-term operation of a pilot plant, which went on stream in 1978. In the HTW plant at Berrenrath, 730 t/d dried lignite are gasified, thus producing 47 000 Nm³ raw gas. After appropriate treatment, this gas is used as synthesis gas for the production of methanol (Fig. 3 shows a photograph of the plant).

A plant for the gasification of hard coal was put into operation by Synthesegasanlage Ruhr GmbH (a subsidiary of Ruhrchemie AG and Ruhrkohle AG) at Oberhausen-Holten in October 1986. This plant is based on the Texaco process which operates on the entrained-bed gasification principle. In the Texaco process, a slurry-type mixture of coal and water is gasified using oxygen as the gasification agent. Pumps are used to feed this coal slurry to the burner of the gasifier and this makes for high reliability. In the plant, approx. 700 t/d hard coal are converted at a pressure of 40 bar and temperatures exceeding 1300 °C to 50 000 Nm³/h synthesis gas. After appropriate treatment, the main portion of the gas produced is used as carbon monoxide-bearing synthesis gas and fed to the oxo plant of Ruhrchemie (Fig. 4 shows the plant). A partstream is processed to pure hydrogen.

UHDE participated in the development of both processes on the engineering side and acted as general contractor for the design, construction and start-up of the plants.

Both processes can also be used for the generation of gas in a combined-cycle power station.

Fig. 4: The SAR coal gasification plant at Oberhausen (capacity 700 t/d hard coal)

When integrating the coal gasification process, it must be taken into consideration that a significant amount of waste heat is obtained, which is utilized for the generation of additional steam. This steam is introduced into the steam cycle and used by the steam turbine.
Both coal gasification plants are equipped with a waste heat recovery system well suited for application in a combined power process.

4. CONCEPT FOR A COMBINED-CYCLE POWER STATION BASED ON LIGNITE

In recent years, UHDE performed a study for an IGCC plant in cooperation with KWU and Rheinbraun.

The concept is based on the Rheinbraun HTW process which, after the successful commissioning of the plant at Berrenrath, can be considered as commercially proven.

The raw lignite is dried and then fed to the gasifiers. The gas leaving the gasifiers is cooled in a waste heat boiler while high-pressure steam (120 bar) is being generated. The temperature of the gas is reduced further in another cooler, the heat being transferred to the treated gas. The residual

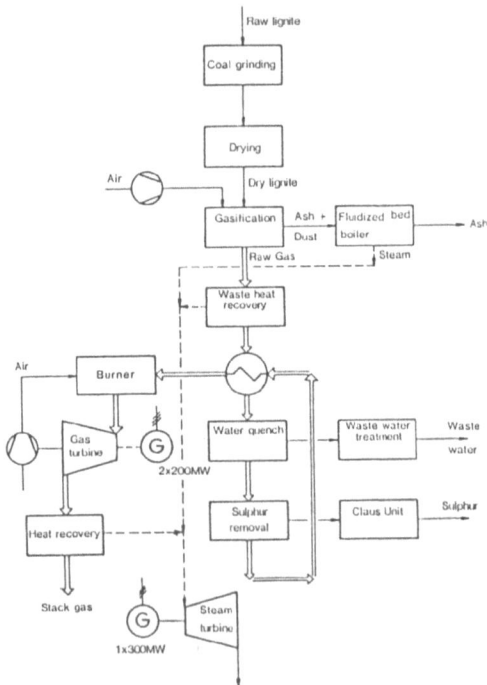

Fig. 5: Block flow diagram of a HTW based 600 MW IGCC plant

dust is removed in a wet scrubbing unit. The dust content obtained is less than 5 mg/m³.

Thereafter, the gas is treated in a H_2S scrubber where a sulphur content of a few ppm can be achieved, if required. Then the gas is reheated in countercurrent with the raw gas, as already mentioned, and burnt in the combustion chamber of the gas turbines. Two gas turbines with an output of approx. 200 MW each are installed in parallel. The gas leaves the turbine at a temperature of approx. 600 °C. In the downstream flue gas waste heat boiler, high-pressure steam of 120 bar is likewise generated which is superheated together with the steam from the waste heat boiler in the gasification section. Low-pressure steam is generated in the low-temperature section. The steam is fed to the steam turbine, which drives a generator with an output of approx. 300 MW. The low-pressure steam required for drying and the medium-pressure steam required for gasification are bled from the turbine. Fig. 5 shows a block diagram of the unit and Fig. 6 an artist's view of such an IGCC plant.

Fig. 6: Combined cycle power station with integrated coal gasification

One special feature of this concept should be mentioned. The ash obtained in the gasifier still contains a certain amount of carbon because the conversion of carbon is limited in a fluidized-bed gasifier. This ash, or rather the residual coke, including the dust from the second cyclone is fed to a fluidized-bed combustor, where high-pressure steam is likewise generated which is additionally fed to the steam turbine.

At an installed gross power of approx. 700 MW, the net power output is approx. 650 MW. The gas turbine/steam turbine out-

put ratio is 1.3, i.e. well above 1. The advantage of a gas
turbine is effectively exploited.

The result of the study shows that, assuming a gas turbine
inlet temperature of 1100 °C, the efficiency is approx. 44%.
This temperature reflects the present state of technology.
According to a competent turbine manufacturer, it can be expected that turbines with an inlet temperature of 1220 °C
will be available within the next few years. This would result in an efficiency of 46%. Compared with a conventional
lignite-fired power station with an efficiency of 34%, this
means a reduction in fuel consumption by approx. 25%. This
fact alone is reason enough for designing and constructing
such plants.

5. ENVIRONMENTAL ASPECTS

Over the years, man has become ever more aware of the fact
that something has to be done to effectively reduce the pollution of the atmosphere. Emission from power stations is of
primary interest in this case. There are stringent requirements in the USA, Japan and the industrial countries of
Western Europe (especially in the Federal Republic of Germany) to limit the emissions of sulphur dioxide, nitrogen
oxides, and dust. For a conventional power station this
means that a desulphurization unit and a DeNOx unit have to
be added and highly efficient dust filters have to be installed. The investment costs for these facilities nowadays
constitute 25% of the total investment costs for a power
station.

In the case of a combined-cycle power station, gaseous waste
and effluents, e.g. the waste water from the gasification
plant, must be taken into consideration. In addition, there
is solid waste in the form of ash and dust.

5.1 Gaseous waste

In a combined-cycle power station, gas purification constitutes an essential part of gas treatment and is relatively
simple. For comparison, here are some figures which make
this clear.

In a conventional 600 MW lignite-fired power station, the
amount of flue gas obtained is 2.4 million m_n^3 per hour,
whereas in a combined-cycle power station, as outlined before, the amount of raw gas to be purified is approx.
360 000 m_n^3 per hour, this being less than 20% of the flue
gas. This does not take into consideration the fact that, in
the combined-cycle power station, the gas to be treated is
under pressure. If the comparison is related to the volume,
the ratio is 100 : 1. It should be mentioned that the sulphur compounds contained in the gas are not present in the
form of SO_2 but in the form of hydrogen sulphide, which is
much easier to remove from the gas and which can be converted directly to elemental sulphur in a Claus unit. The

sulphur dioxide obtained in the flue gas treatment processes presently being used in conventional power stations is converted to gypsum (calcium sulphate). However, it is obvious that this will no longer be possible in the near future due to disposal problems, considering the amounts of gypsum obtained.

The formation of nitrogen oxides in the flue gas of a power station can be traced to two sources. Firstly, nitrogen oxides are formed during the combustion of nitrogen-bearing compounds contained in the coal and, secondly, nitrogen oxides are formed thermally from nitrogen and oxygen at the high combustion temperature. In a combined-cycle power station, the nitrogen compounds contained in the coal are degraded during gasification or completely removed from the gas in the gas treatment unit. The temperature prevailing in the combustion chamber limits the thermal formation of nitrogen oxides.

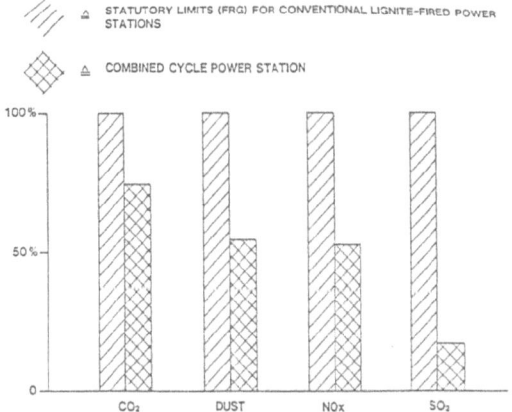

Fig. 7: Emissions for lignite-based power stations

Fig. 7 shows a comparison of the emissions from a conventional power station and a combined-cycle power station. Attention is drawn to the fact that the conventional power station is equipped with a gas treatment unit as required by the present statutory environmental regulations in the Federal Republic of Germany. As shown in the figure, the expected emission from the combined-cycle power station is considerably below the statutory emission limits.

In this connection, the CO_2 emission must also be mentioned which, due to the higher efficiency, is reduced considerably. CO_2 emission can cause a dramatic long-term change in climate.

5.2 Effluents

Waste water is obtained in the coal gasification process. Fig. 8 shows the the basic configuration of a wet purification unit as used in the German plants mentioned earlier. The treatment consists of three steps, i.e. quench, Venturi scrubber and scrubbing column. This configuration permits effective removal of the dust from the gas. The water stream containing solids is fed to a downstream separator (filter or settling tank). In the case of Texaco gasification, a solids-laden stream from the gasifier quench and the ash lock hopper is also fed to the separator, from where sludge containing the solids is withdrawn. The treated water from the separator can be returned to the cycle. Depending on the water-soluble constituents of the coal, a certain amount of waste water must, however, be withdrawn from the cycle in order to prevent salt enrichment in the cycle, as this might cause extensive corrosion. The quantity to be withdrawn is thus primarily determined by the salt content of the coal. It goes without saying that the waste water stream also contains gases which are dissolved in the water, in particular CO_2, H_2S and NH_3, but also water-soluble organic constituents whose point of condensation is below the temperature of the water cycle. The dissolved gases are removed in a downstream stripper and fed to the sour gas treatment unit. A separate treatment step may be required in order to recover the valuable substances, such as ammonia, in concentrated form from the waste water.

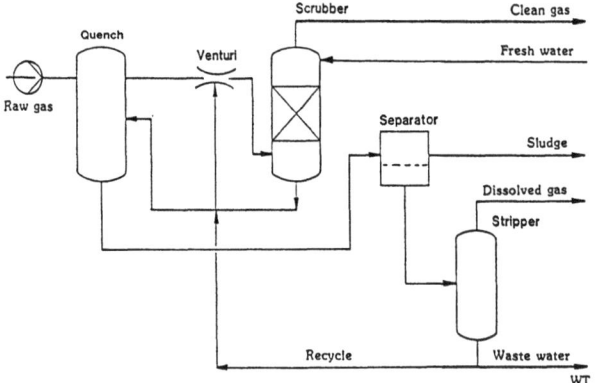

Fig. 8: Dust removal system and water system

The water stream leaving the stripper is of significance with a view to environmental pollution. It contains traces of inorganic and organic substances. The salt content is determined to a large extent by the composition of the mineral constituents of the coal. The content of organic substances depends on the gasification process used. In the case of the

Texaco process, the gasification temperature is very high, so that the gas and, consequently, the waste water will hardly contain any organic constituents at all. The content of aromatic compounds, in particular, is very low. Evidence of this has been obtained in comprehensive test series in the demonstration plant at Oberhausen and the operating results in the commercial plant have likewise confirmed this. But in the HTW process as well, where the gasification temperature is much lower (900-1000 °C), a considerable decrease in hydrocarbons is achieved by the special two-step process in the gasifier. In both cases, the content of aromatic hydrocarbons is in the ppm range only. The waste water can thus be discharged directly into the general waste water system without any further treatment or conveyed to a biological waste water plant, depending on the applicable local regulations.

In this connection, it must be mentioned that the cooling water requirement of a combined-cycle power station is considerably lower than that of a conventional power station. This is not only a consequence of the higher efficiency, which indicates that less heat is dissipated to the environment, i.e. to the atmosphere and to the cooling water. The saving in cooling water is also due to the fact that only part of the total power output is made up by the steam turbines, which again results in a decrease of the cooling water consumption. Referred to the entire plant, about 50 % less cooling water is required.

5.3 Solid waste

The solids leaving the power station consist of ash and dust. Here it is likewise the higher efficiency that accounts for a clear reduction in the specific quantity of ash obtained. However, the decisive factor is the type of ash involved, namely whether this ash can be dumped without causing any problems. In the Texaco process, which operates at temperatures above the ash melting point, vitrious ash is obtained as the extensive leach tests have shown. This ash can be dumped without any difficulty. In special cases, this ash may be used for other purposes. In the HTW process, which operates at temperatures below the ash melting point, there is a risk that, particularly in the case of highly alkaline ash, part of the sulphur contained in the coal is combined in the form of sulphide as a result of the reducing conditions in the gasifier. There is, of course, the risk that these sulphides may be released whenever there is any contact with water. The concept therefore provides for final combustion of the ash and dust. Apart from the additional energy that is generated, sulphide compounds are converted into sulphates in the combustion process in the presence of an excess of oxygen. The ash obtained after combustion can then be dumped under the same conditions as the ash from a corresponding lignite-fired power station.

In summary, it can be said that environmental pollution is drastically reduced by a combined-cycle power station as compared with all other types of fossil-fuel-fired power stations.

6. COMBINED-CYCLE POWER STATION WITH CO-PRODUCTION

The main disadvantage of electric power generation is the fact that electricity cannot be stored. Compared with a chemical plant, the load factor of a power station is relatively low. This particularly applies to power stations which operate in the medium-load range. In this case, there are interesting possibilities for varying the concept of a combined-cycle power station.

These variants are based on the assumption that the gasification section including the gas treatment unit continue to operate at max. load and that the generated gas is used for the production of, for instance, ammonia or methanol. However, it then becomes necessary to provide a further gas treatment stage in order to adapt the generated raw gas to the conditions prevailing in the synthesis unit. Moreover, it is necessary to design the gasification section for a certain excess capacity in order to permit continuous operation of the chemical plant even during max. load operation of the power station, thus reducing the problem of capacity adaptation.

Conversion to methanol is of special interest since it can easily be stored and burnt as additional fuel in the turbine if required for peak-load operation. In this special case, no specific demands are made on the methanol produced and the synthesis unit can be kept relatively simple. Distillation of the methanol is not necessary.

7. ECONOMY

A number of investigations performed by us have shown that the investment costs for a combined-cycle power station are identical with those for a conventional power station. In view of the higher efficiency of a combined-cycle power station, it is to be expected that the operating costs will be appreciatively lower. This, of course, depends on the energy costs, i.e. the price of the coal used as feedstock. The production costs, i.e. the price per kWh, will therefore be lower. An exact calculation will only be possible when comparing alternatives for the same location.

8. SUMMARY AND PERSPECTIVES

The concept of a combined-cycle power station with integrated coal gasification offers a number of advantages. Despite considerably improved efficiency, the plant costs remain in the same order of magnitude as those of a conventional power station. Environmental pollution by the emissions from power stations is drastically reduced. Combined-cycle power stations based on natural gas have become accepted as proven technology, to which the coal gasification technology must then be added for an IGCC plant. The successful commissioning of two gasification plants in the Federal Republic of Germany last year justifies the statement that the integration of a coal gasification plant in such a power station concept is feasible without major difficulties. The combined-cycle power station with integrated coal gasification unit is an important contribution to the improved utilization of the coal energy carrier and, at the same time, a considerable step towards the protection of our environment.

A PROPOSAL FOR THE ENERGY INFRA-STRUCTURE OF ATATÜRK ORGANIZED INDUSTRIAL DISTRICT IN İZMİR

A. DURMAZ, Y. ERCAN, Ö. E. ATAER and M. SİVRİOĞLU

ABSTRACT

Atatürk Organized Industrial District (AOID) is located near İzmir and will comprise 500 small and medium sized factories. The factories require process steam, hot water, hot air, electricity and energy for space heating. The total peak electricity and heat demand is expected to reach to 600 MW within ten years. Production of heat individually by the factories, central production of heat and cogeneration system alternatives for heat and electricity production are investigated technically and economically to meet the energy demand of AOID as well as the heat demand of the 20 000 flats under construction nearby. Pulverized lignite fired cogeneration power plant was found to be the most economical solution which is also favorable from the point of view of environmental protection. Marginal cost of electricity produced by the plant is approximately one-third of the normal purchase price of electricity from the national network. The use of this system conserves the national energy sources by saving considerable amount of fuel during its life time.

I. INTRODUCTION

Rapid industrialization which has been experienced in Turkey during the recent years has led to an ever increasing demand of electricity [1]. Although large projects are beign carried out by the government, additional investment by the private sector is strongly promoted in order to increase the electricity production. For this purpose, a law, which allows new incentives for private investment in electric power has come into act recently. Since the electricity production and consumption centers in Turkey are located far away from each other, the interconnected electricity supply network suffers problems of instability and frequency shift. Therefore, some of the new power plants have to be set up near the major load centers such as large cities and industrial centers.

Major source of energy in Turkey is the low and medium grade lignite which is found in abundance throughout the country. However, the usage of this low quality fuel in the small-sized conventional boilers causes air pollution, which at present is a severe problem in the large cities of Turkey [2]. In order to reduce air pollution, to meet the increasing demand of electricity, to stabilize the electricity network and at the same time to continue its industrialization process, Turkey must direct its efforts towards energy infra-structure projects which utilize its natural resources in the most rational way [3].

In order to reduce the share of energy infra-structure costs paid by small and medium-sized production facilities, to realize the optimal system structure and to eliminate the adverse effects on environment, Turkey has adopted the policy of forming organized industrial districts near large cities. One such district, called "Atatürk Organized Industrial District" is located near İzmir and will comprise 500 factories.

The purpose of this study is to investigate the cogeneration power plant application to industrial districts as a means to solve the above mentioned fundamental problems. AOID was selected as a pilot project and various energy production alternatives were evaluated both technically and economically. The proposed power plant will meet both the electricity and heat demands of the industrial district as well as of the 20000 flats which are located nearby.

2. HEAT AND ELECTRICITY DEMANDS OF AOID AND THE NEARBY FLATS

Determination of the optimal cogeneration system structure and its economical evaluation require detailed information concerning the heat and electricity loads. The magnitudes and properties of these loads, as well as their time variations play an important role in the economical feasibility of the system.

The heat required by the factories of AOID is comprised the loads due to process steam, hot water, hot air and space heating. The information related to the heat and electricity demands was obtained from interviews. Based on this information, the average and peak values of the total heat and electric power demands as well as daily, weekly, monthly and yearly demand variations were determined. The demands for the first, second and third work-shifts of the month January are shown in Figure 1, 2 and 3 [4]. Compared to the first shift and depending on the year, the demands for the second shift are expected to go down by 40 - 68 %. The demands for the third shift are only 10 - 25 % of the first shift demands. Large variations of load from one shift to another present a draw-back for economic operation of the cogeneration plant. If the plant is operated at full capacity during the second and third shifts, the excess electricity can always be sold to Turkish Electric Authority at a reduced price. However if the accompanying excess heat cannot be sold, the operation of the plant will not be as economical as desired. The 20 000 flats which are under construction near AOID seem to present a favorable solution to the problem of excess heat. Space heating and utility hot water heat loads of the flats are expected to change as in Figure 4. The heat loads of the flats during the work shifts of AOID are as in Figure 5 for the month of January. Since the energy demand for heating of the flats increases especially at nights, the excess heat from the cogeneration plant can be utilized to meet this need [5].

3. ALTERNATIVES AND SYSTEM STRUCTURE

Three main alternatives of the system structure were considered to meet the heat and electricity demands of AOID. These alternatives are as follows [4]:

3.1. <u>Production of heat by each factory</u>. In this alternative the required heat energy is produced in boilers individually at each factory. The electricity is purchased from the national interconnected network. The main advantage of this alternative is the relatively low investment cost because a heat distribution network is not required. If this alternative is selected, most of the factories will prefer lignite fired boilers because of the low fuel cost. In this case, since the locally available lignite has undesirable properties, it will be difficult to maintain the emission standarts which are in act [6].

3.2. <u>Central production of heat.</u> In this case the heat required by AOID is produced centrally by a heat plant and distributed to the factories by a heat network which carries process steam and hot water. The electricity is again purchased from the national electricity network. The advantage of this system is the possibility of meeting the requirements set by the emission standards by using advanced combustion and operation techniques and the centrally constructed gas cleaning systems [7], [8]. The investment cost however is higher than the first case.

3.3. **Cogeneration of heat and electricity.** In this alternative heat and electricity are produced together by a cogeneration power plant [9]. The power plant should be operated under full load conditions. Therefore, the plant capacity in this alternative is selected to meet the average load. The peak heat energy demand is met by a stand-by heat plant operating additionally. The surplus heat during the night hours is sold to the nearby 20 000 flats. On the other hand, the surplus electricity is sold to the national electricity network at a reduced price as required by law. The peak electricity demand is met by purchase of additional electricity from the same network at the normal price.

In this alternative three options for the system structure are investigated. In the first option, a cogeneration power plant (70 MW_e plus 130 MW_{th}) is considered. The plant has an extraction and condensing type turbine [9]. Steam is extracted from three points of the turbine. 22 bar and 328 °C extraction is used to feed the process steam network ; the other two extractions are at lower pressures and temperatures and are used for generation of hot water. The hot water supply and return temperatures are 140 °C and 70 °C respectively. Five types of combustion systems, namely pulverized lignite, fluidized bed lignite, pulverized bituminuous coal, fluidized bed bituminuous coal and fuel oil combustions systems, are considered in this option.

The second cogeneration plant option comprises a gas turbine (70 MW_e) and a waste heat boiler (100 MW_{th}), which supplies heat to the process steam and hot water network. Since natural gas is not available in the area, only fuel-oil fired gas turbine is considered [10].

In the third option, a combined power plant is investigated. This plant also comprises a gas turbine (70 MW_e). However the steam generated by the waste heat boiler feeds an extracting and condensing type steam turbine as described in the first cogeneration plant option. Additional electricity is generated by the steam turbine depending on the amount of the steam extracted for the heat network.

The investment costs for all three types of cogeneration system are considerably higher than the first and second alternatives. However, they use the primary energy more rationally and effectively because of the electricity which is generated as a by-product. Effective emission control can be achieved in these options [8].

4. ECONOMICAL EVALUATION OF THE ALTERNATIVES

The total net energy costs for AOID and the 20 000 flats were calculated for the alternatives of Section 3 [4]. In the calculations hourly, daily, monthly and yearly variations of the heat and electricity demands were considered. The investment costs were determined from the price quotations of various manufacturers. Operation costs were determined by assuming that each system is optimally operated to minimize the net cost of energy. In the calculation of the net energy cost, the sale of electricity and heat as well as purchase of electricity are taken into account.

Figure 6 shows the net price of energy (electric plus heat) for the cogeneration power plants with extraction and condensing type steam turbines. The comparison among the plants with different fuel and combustion systems shows pulverized lignite fired cogeneration plant is the most economical one.

The net energy costs for cogeneration systems which employ gas turbines are shown in Figure 7. As seen from the figure, the plant which combines gas and steam turbines, called combined power plant is more economical than the gas turbine with WH-boiler.

The best options from Figure 6 and 7 are taken and compared with first and second alternatives in Figure 8. Figure 8 shows that pulverized lignite fired cogeneration power plant is more economical than the other alternatives. The second best is the combined cycle power plant. The net costs of energy for the first and second alternatives which produce only heat are much higher than the corresponding costs for the cogeneration systems.

In economical evaluations of cogeneration systems, it is also desirable to know the marginal cost of electricity. Figure 9 shows the marginal unit cost of electricity for the pulverized lignite fired system. The results are presented for two types of lignite available near the area. As seen from the curves in this figure, the marginal costs of electricity for both fuels are much less than the cost of electricity from the national network [4].

The use of the cogeneration power plant for the present application also conserves national energy sources. Because the overall system efficiency of a cogeneration plant is higher than those systems which generate heat and electricity seperately. Figure 10 shows the amount of lignite saved by the use of cogeneration plant for the first ten years of operation.

5. CONCLUSIONS

The study presented in this paper shows that the application of cogeneration power plant to the industrial district considered yields a solution that is economically optimal. The resulting system is also favorable from the environmental protection point of view. Among the types of cogeneration power plants that were investigated, the pulverized lignite fired system was found to be the most economical one.

The marginal cost of electricity for this plant is approximately one-third of the normal purchase price of electricity from the national electricity network. This system also conserves the national lignite sources by saving substantial amount of fuel during its operational life time.

REFERENCES

1. Şirin G(ed) : Energy Statistics. Turkish 4. Energy Conference. İzmir, 1986 (In Turkish).
2. Durmaz A : Energy Sources and the Environment. Seminar on Alternative Energy Sources and Environment, Ankara, 1979.
3. Durmaz A, Güngen G, Karabay M, Ercan Y : Application of Energy Conservation Techniques to some Industrial Complexes in Turkey. Symposium on the outlook of energy in the third world, pp.1-30, Tehran, 1983.
4. Durmaz,A, Ercan Y, Ataer E and Sivrioğlu M : Initial Fizibility Study of Cogeneration Power Plant of İzmir Atatürk Organized Industrial District. Vol I-System Structure, Vol II-Economic Analysis, Vol III-Appendices, prepared by Energy-Environmental Systems and Industrial Rehabilitation Research Center, Gazi University Ankara, 1986 (In Turkish).
5. Durmaz A, Ercan Y, Ataer E, Sivrioğlu M : Initial Fizibility Report of Energy

Infra-Structure of İzmir Denizbostanlısı Houses of Construction Credit Bank. Prepared by Energy-Environmental Systems and Industrial Rehabilitation Research Center, Gazi Unversity, Ankara, 1987 (In Turkish).
6. Air Quality Conservation Act. General Directory of Environment, Turkish Prime Ministry, 1986 (In Turkish).
7. Durmaz A : Modeling, Simulation and Optimization of the Combustion Control System of A Steam Generator. Proceedings of AMSE Conference on Modeling and Simulation, Vol.8, pp.80-97, Paris, 1982.
8. NO_x Task Force, Technologies for Controlling NO_x Emissions from Stationary Sources. Economic Commission for Europe, 1986.
9. Schröder K : Grosse Dampfkraftwerke. Planung, Ausführung und Bau. Band I und Band II, Springer-Verlag, 1962.
10. Pfenninger, H. : Combined Steam and Gas Turbine Power Stations. Brown Boveri Rev, Vol.60, No.9 pp.389-397, 1973.

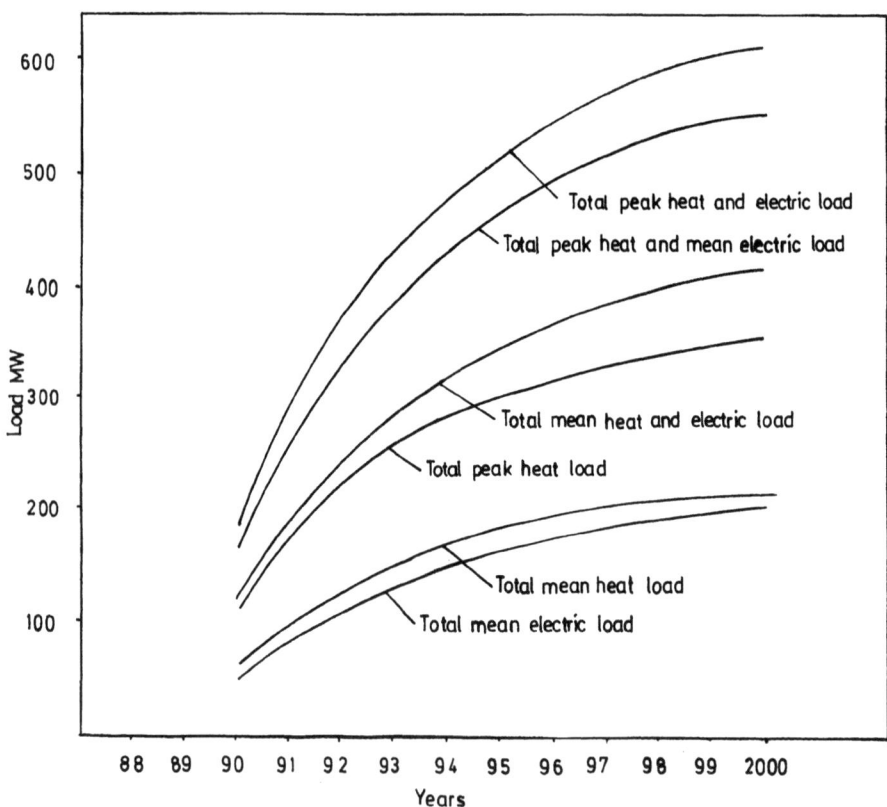

FIGURE 1. Variation of first work-shift load with years.

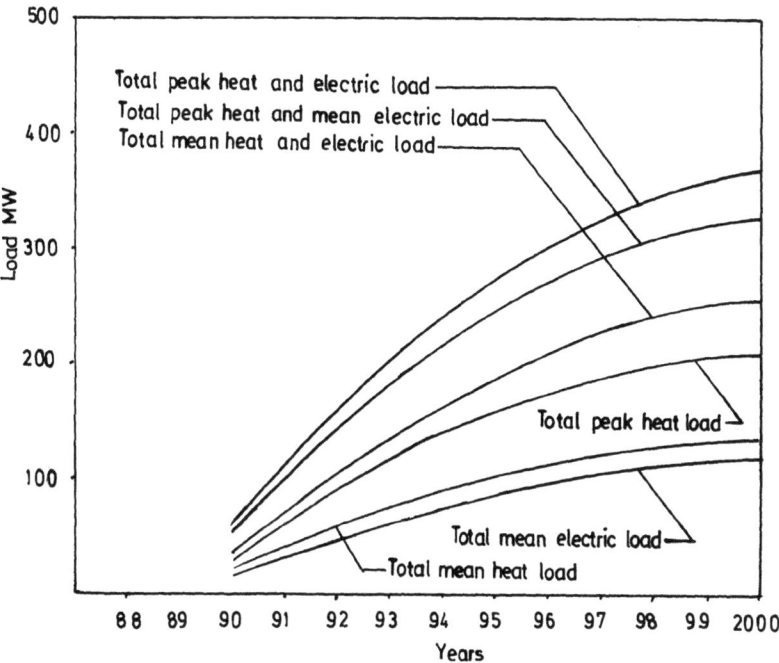

FIGURE 2. Variation of second work-shift load with years.

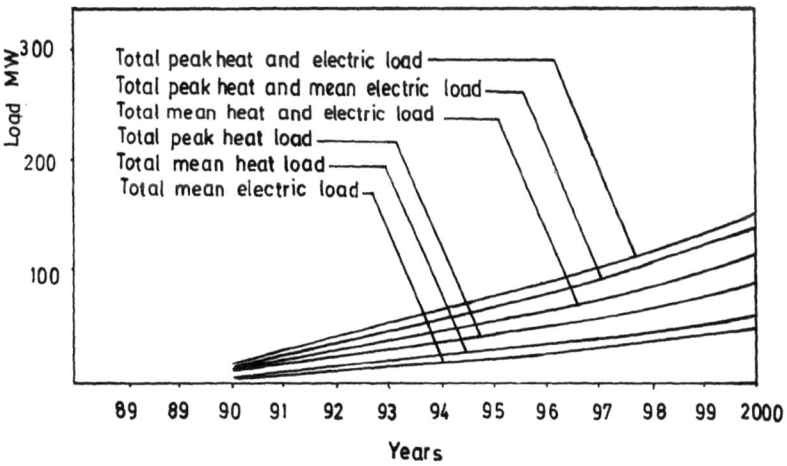

FIGURE 3. Variation of third work-shift load with years.

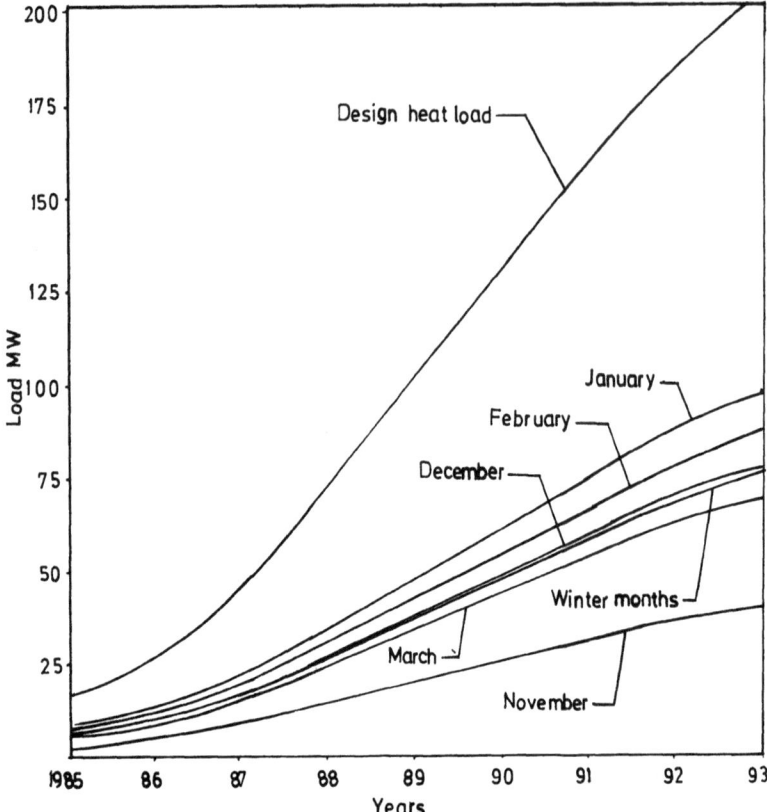

FIGURE 4. Heat demand of the 20 000 flats under construction near AOID.

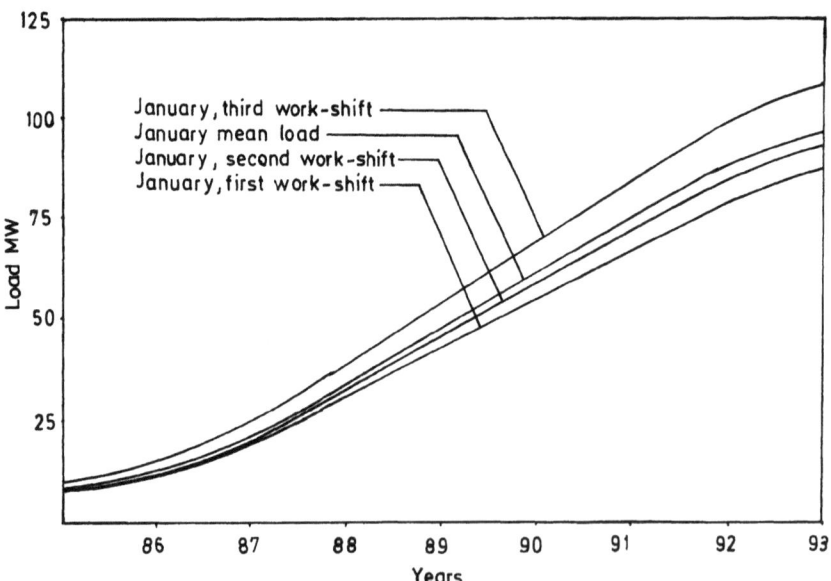

FIGURE 5. Variation of heat load of 20000 flats with years during the work-shifts of AOID.

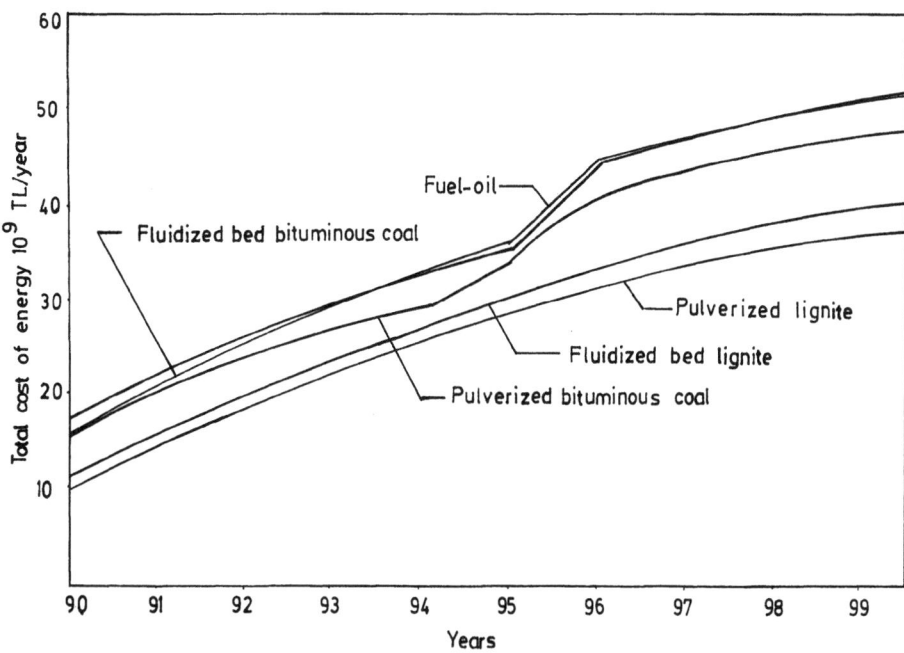

FIGURE 6. Comparison of the variations in annual total cost of energy for different types of cogeneration power plants.

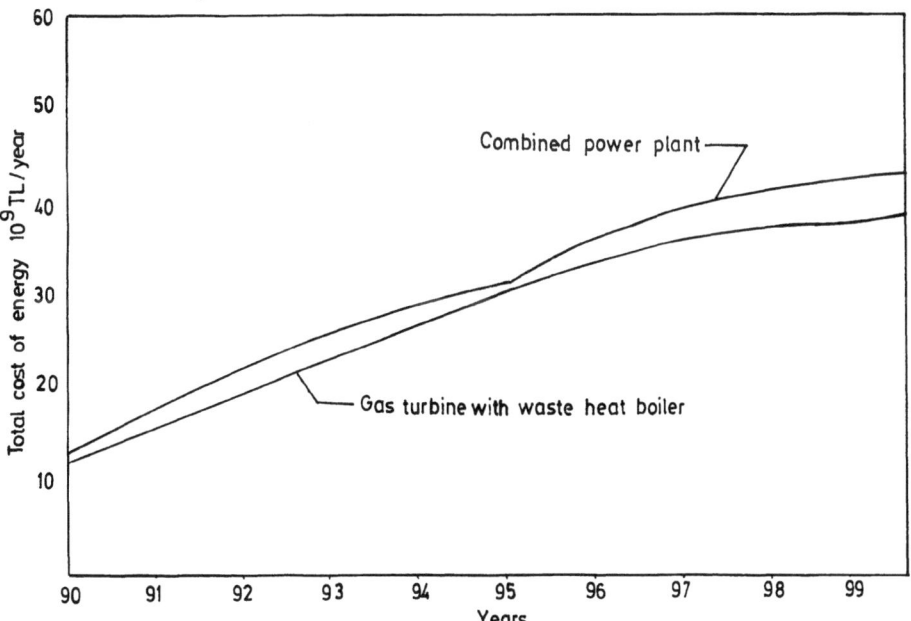

FIGURE 7. Comparison of the variations in annual cost of energy for combined power plant and gas turbine with waste heat boiler.

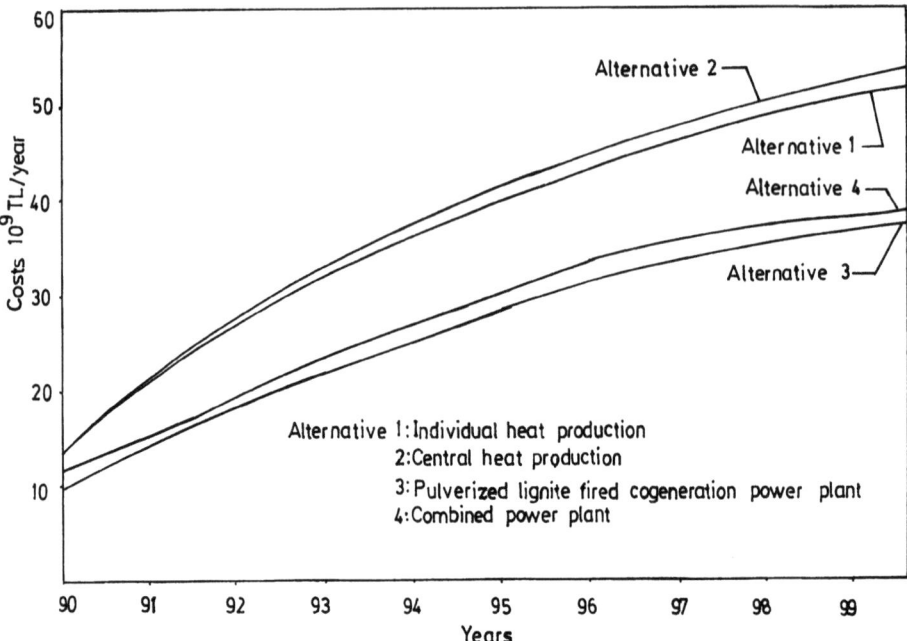

FIGURE 8. Comparison of variations in annual cost for different alternatives

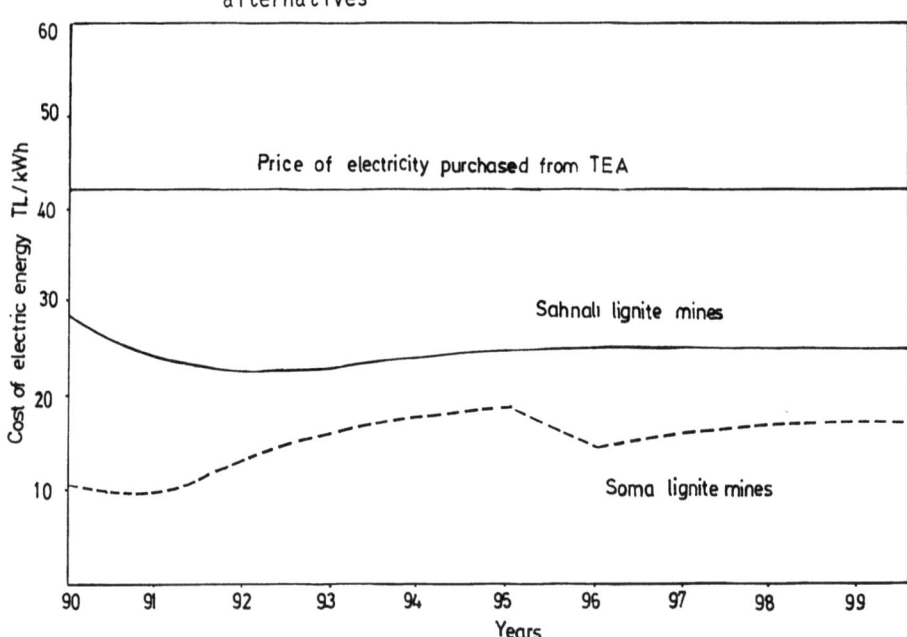

FIGURE 9. Comparison of marginal unit electricity cost for the pulverized lignite fired cogeneration power plant with the purchased electricity cost.

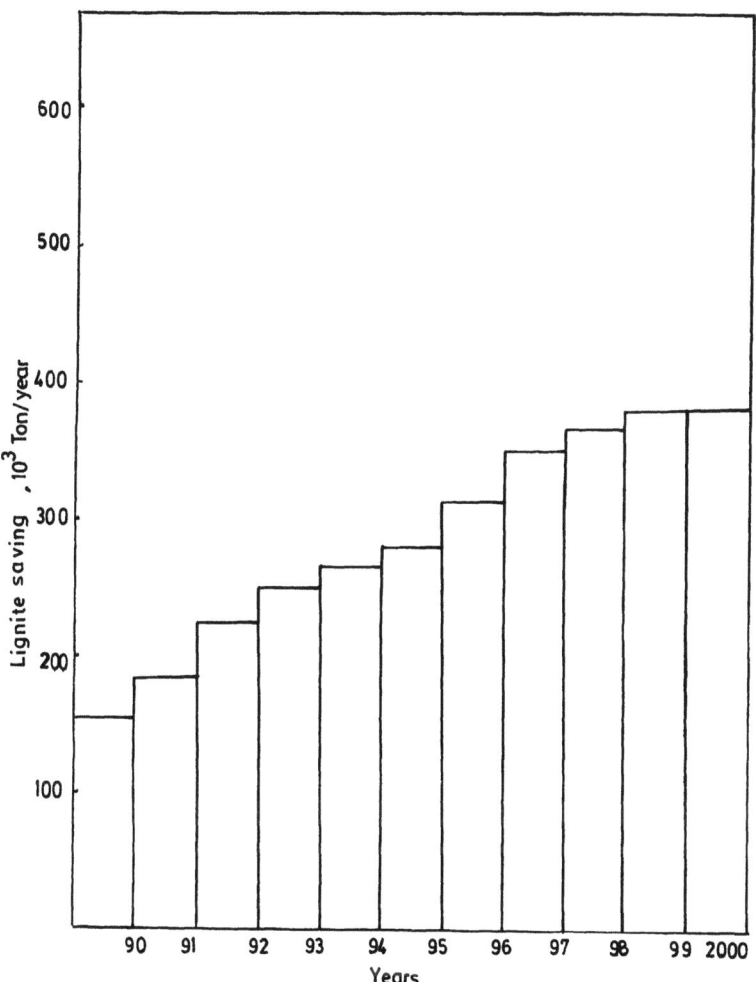

FIGURE 10. Variation of amounts of lignite saved by the use of cogeneration power plant.

FLUIDIZED BED COMBUSTION TECHNOLOGY FOR LOW GRADE LIGNITE UTILIZATION IN TURKEY

M.ARIKOL, E.EKİNCİ, D.BİLGE, A.SERPİL

1. INTRODUCTION

Energy demand is increasing at a far greater rate than the domestic production in Turkey. Since the deficit is being met by importation of oil, the State Planning Organization has already placed special emphasis on better and increased utilization of the country's existing 7.5 billion tons of lignite reserves.

Due to their low calorific values, high ash, sulphur and nitrogen contents and readily changing characteristics, increased utilization of Turkish lignites presents potential environmental problems: Combustion in conventional systems will lead to drastic SO_x, NO_x, CO, unburned hydrocarbon and particulate emissions. Air pollution, already at alarming levels in inland towns of Anatolia, will be aggravated. Furthermore, conventional systems are incapable of burning low grade lignites. High combustion temperatures lead to agglomeration of solid fuel which may cause a reduction in efficiency and increase CO and unburned hydrocarbon emissions. Evaporation of alkali metals which condense on relatively cooler superheater tubes (1) and sensitivity to changes in the characteristics of fuel input are additional problems to be encountered with low grade lignite combustion in conventional systems.

Based on environmental considerations, conversion to cleaner gaseous and liquid fuels prior to utilization may be the best strategy; but the present need for an urgent solution renders direct utilization indispensable for electricity generation, commercial and residential space heating and industrial process heat production.

An outstanding feature of fluidized bed combustion (FBC) technology is easier control of gaseous pollutants such as SO_x and NO_x (2,3,4,5); particulate emissions are also minimal. FB combustors offer numerous other advantages such as adaptability to different fuels, high heat transfer rates (6,7) and combustion efficiencies (8). In view of the present air pollution problems in metropolitan cities and extremely pollutant characteristics of Turkish lignites, adoption of FBC technology in Turkey appears inevitable (9).

2. CHARACTERISTICS AND FBC BEHAVIOUR OF TURKISH LIGNITES

The important characteristics of main Turkish lignite reserves are shown in Table 1 (10,11,12,13). Analysis of this table indicates that volatile matter (VM) content of Turkish lignites is generally higher than 20%, ash and moisture contents may be as high as 70% and 50%, respectively, and calorific value may be as low as 4700 kJ/kg. High sulphur contents up to 14% (14), moderate nitrogen contents and low ash fusion temperatures are also encountered. A very important feature is the drastic fluctuation in lignite properties within a single reserve (Table 2) which will create problems in combustor design and efficient operation.

2.1. Combustibility limit

An important advantage of FB combustors is their ability to utilize low

Table 1. Characteristics of Important Lignite Reserves of Turkey (10,11,12,13)

Name	Total Reserve 10⁶ tons	Ash (%)	Moisture (%)	VM (%)	LCV (KJ/kg)	% daf C	% daf H	% daf S	% daf N	Ash fusion temperature, °C
Zonguldak*	539.2	13.3	9.6	24.4	27839	87.3	5.1	0.5	-	-
Tunçbilek	220.3	24.9	20.8	24.3	15215	74.5	5.5	1.4	2.95	-
Soma	515.0	23.5	15.1	32.1	15591	73.5	4.8	0.9	1.12	1200
Çan	128.3	30.4	21.4	25.5	11704	66.1	5.5	8.4	2.25	1150-1350
Seyitömer	228.6	14.8	31.0	28.4	13627	70.0	5.0	2.0	-	-
Orhaneli	58.5	24.2	26.8	24.6	11202	69.4	5.6	3.2	-	1142
Beypazarı	222.0	34.8	22.1	25.1	10283	70.8	5.8	7.3	2.5	1110
Yatağan	535.1	15.8	37.3	28.2	10617	-	-	3.9	-	-
Saray	143.0	16.8	44.9	19.2	8276	-	-	6.3	-	-
Kangal	176.0	21.0	48.3	19.8	5685	-	-	7.5	-	-
Elbistan	3539.0	23.3	49.5	18.3	4680	66.2	5.5	2.7	-	1195
Seyitömer***	1000.0	68.5	5.0	25.7	12540	13.2	2.0	nm	0.3	-
Göynük-Bolu***	2500.0	32.0	-	-	12720	56	6.75	1.45	1.4	-

* Bituminous Coal

Table 2. Characteristics of Orhaneli lignites

	Sample 1	Sample 2
Moisture (%)	27.71	18.58
Ash (%)	21.78	8.12
VM (%)	29.96	40.42
Fixed carbon (%)	22.55	32.88
LCV (kJ/kg)	13443	19705

calorific value fuels which can not be burned by other combustion systems. Ekinci et al. (15) found that synthetically prepared Avgamasya asphaltite (2800 kJ/kg, 86% ash) sustained FBC at 800°C. FBC of wastes with calorific values of 2340, 3130 and 5000 kJ/kg have also been reported (16,17,18).

LaNauze (19) devised a triangular diagram which can be used to determine combustibility of different fuels in a fluidized bed with 2.4 m/sec air velocity, 20% excess air, 15% carbon loss and 5% heat loss. Turkish lignites listed in Table 1 are plotted on this diagram in Figure 1. It is observed that all of them can be combusted in a fluidized bed under conditions specified in (19) without additional fuel. Ample useful heat can also be recovered in most cases.

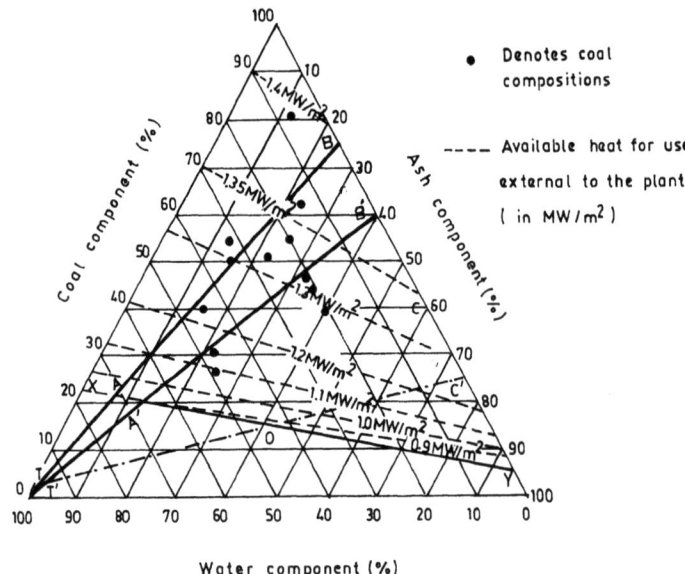

Figure 1. Combustibility of Turkish lignites

2.2. Volatile matter behaviour

Combustion behaviour of Turkish lignites in fluidized beds is different than high grade coals with lower volatile matter. Burning particles tend to spend most of their time on the upper end of active bed during the release of volatile matter (20,21,22,23). Similar observations have also been made by other investigators (24,25,26). This behaviour makes freeboard an active medium for combustion of volatile matter.

2.3. Mixing and segregation

Intensive mixing in fluidized beds is essential for stable and efficient operation with high heat transfer rates. Accumulation of ash in the bed which may form a density or size differential binary system with original bed material, may deteriorate mixing in lignite fired fluidized beds. The segregation patterns mostly are of lignite ash floatsam rich type (17) and in some cases may be of lignite ash jetsam rich type. Ekinci et al. (27) noted that for Çan lignites burning in denser sand, starting with 100% bed material, the fluidized bed became a lignite ash floatsam system after one hour of operation and behaved in this fashion for the remainder of the experiment (8 hours). For lignite ash jetsam rich systems, the tendency of volatile matter to force burning particles to be concentrated at the upper section of the bed may be neglected.

2.4. Agglomeration

For Turkish lignites, two kinds of agglomeration behaviour is observed depending on the segregation systems (27,28). In most cases the lignite is burned in a denser bed material, forming a lignite ash floatsam rich system. A typical differential temperature diagram for such a case is shown in Figure 2. When the bed reaches the defluidization stage, cooling starts steadily above the distributor plate due to the accumulation of a pure jetsam layer. In this type of system there is ample time to control the operation after defluidization to prevent agglomeration. For lignite ash jetsam rich systems, distinct cooling of pure jetsam layer is not observed, but as the extent of segregation increases, differential temperature readings give enough information on the approach of agglomeration.

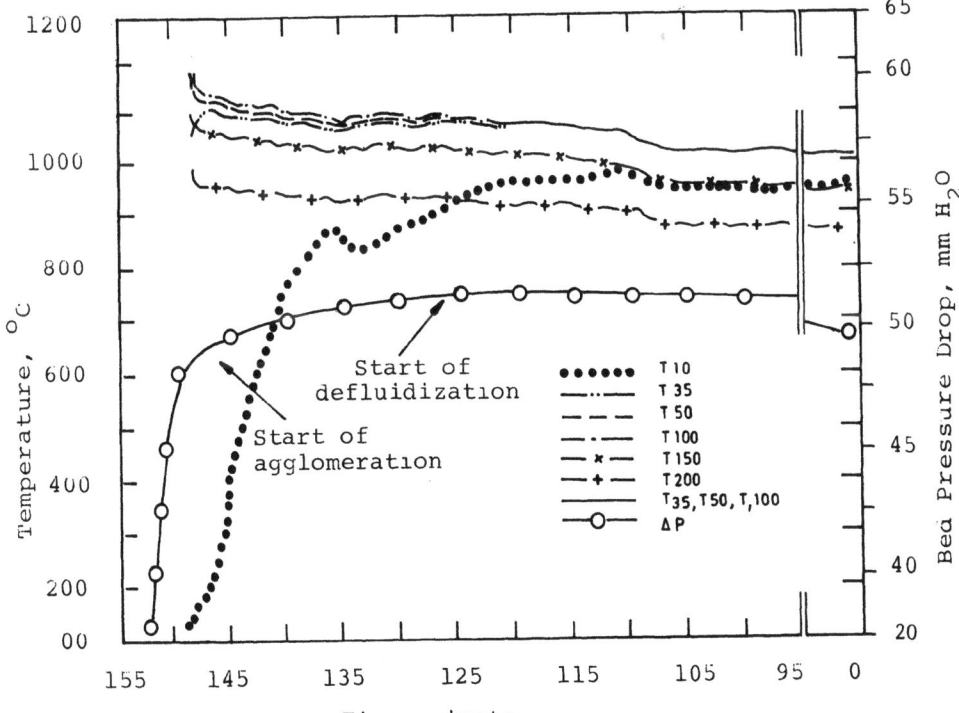

Figure 2. Temperature and pressure drop profiles for a lignite ash floatsam rich system

Table 3. Inherent CaO in coal ash

Coal	CaO content (%)	Reference
Zonguldak	13.30	10
Orhaneli	19.15	43
Beypazarı	9.8	44
Yatağan	12.21	10
Elbistan	14.02	45
Göynük	11.52	46

2.5. SO_2 removal

Due to their low calorific values and high sulphur contents, SO_2 emission during combustion of Turkish lignites is quite high. However, SO_2 removal can be easily achieved by adding a suitable sorbent like limestone or dolomite to the fuel fed to the combustor. It has been reported that (29) calcium and magnesium oxides are better sorbents than limestone or dolomite on equivalent mole bases during FBC. Therefore, depending on the composition of coal ash (Table 3), optimum limestone or dolomite values for different lignites should be less than the calculated theoretical values which must be based on combustible sulphur rather than total sulphur.

3. CRUCIAL DESIGN AND OPERATION FEATURES

Several design aspects of FB combustors have been treated in detail in literature (30,31,32,33,34,35). The general design approach is similar for all fluidized beds, but special considerations are necessary to handle different problems arising from the type of fuel to be employed: A design procedure originally developed by Tolay et al. (36) for a 0.12 MW hot water generator burning lignites with 6300 - 12600 kJ/kg, has also been applied to the design of a waste incinerator (37). Here only features relevant to utilization of low grade Turkish lignites will be discussed.

3.1. Active bed

Turkish lignite reserves with high volatile matter contents and calorific values below 5000 kJ/kg are considerable. FB combustor design for such lignites should take into consideration the fact that during normal operation no heat may be available from the active bed.

3.2. Freeboard

Freeboard should be designed as an active medium for volatile matter combustion, i.e., with secondary air injection provisions and installation of heat transfer tubes at appropriate places. Freeboard can also be utilized as an active medium for SO_2 and NO_x removal. For very low calorific value lignites where no heat can be removed from the active bed, this presents an optimization problem to the designer (38).

3.3. Distributor plate

Design of the distributor plate should enhance mixing in the bed to avoid segregation. This can be achieved by appropriate geometry (39), employment of different fluidization velocities (40) and/or provision for continuous removal of coarse particles or ash (41). Different distributor designs are shown in Figure 3.

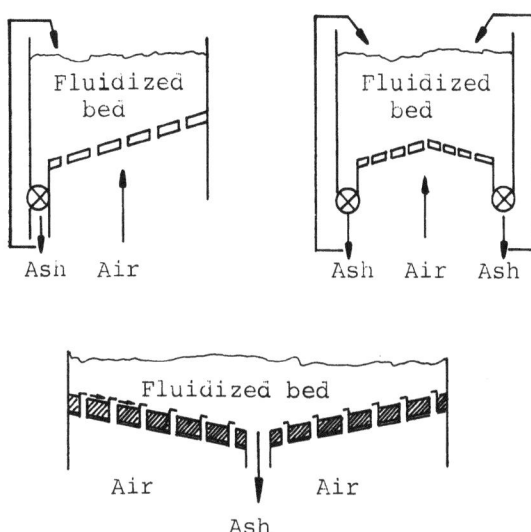

Figure 3. Different distributor plate designs

3.4. Temperature control

Temperature control is essential to avoid segregation and subsequent agglomeration. In view of the information given in sections 2.3 and 2.4, a differential temperature control system monitoring average bed temperature and temperature in the vicinity of distributor plate is recommended (42) to ensure stable operation.

3.5. Heat withdrawal

Heat withdrawal system should be able to account for the fluctuations in the calorific value of fuel input (Table 2). Such a flexible system consisting of a portable heat transfer panel has been developed (38) which can handle large fluctuations in calorific value.

4. ECONOMIC CONSIDERATIONS

Industrial sector is expected to take the lead in implementation of FBC technology in Turkey (47). The most common industrial applications will be atmospheric units in the 3-30 MW range. A cost model has been developed by Serpil et al. (48) to investigate the economic feasibility of such applications. Results based on this model are summarized in Table 4 for a typical lignite.

Analysis of this table indicates that payback periods of industrial FB boilers are most sensitive to competing fuel oil prices and are rather unacceptable under the existing economic conditions. However, the cost model does not account for the benefits incurred by improved environmental pollution control. A new Air Quality Act which imposes strict regulations on polluters has been enacted after the development of this model. Enforcement of this new act will drastically improve the economic feasibility of industrial FB boilers.

5. CONCLUSION

Adoption of FBC technology will render direct utilization of low grade Turkish lignites environmentally acceptable. Hence development of this technology is both urgent and necessary for Turkey. Due to the differences between characteristics and FBC behaviours of Turkish lignites and western bituminous coals, FBC designs in Turkey are expected to be significantly different than their westeren counterparts developed mainly for bituminous coals.

Table 4. Payback periods for an atmospheric FB boiler producing process steam

Capacity (kg steam/hr)	Calorific Value (kJ/kg)	Ash (%)	Sulphur (%)	RCP*	Payback Period (years)
5000	14600	25.0	2.5	0.548	3.96
15000	14600	25.0	2.5	0.548	2.99
50000	14600	25.0	2.5	0.548	2.55
50000	14600	12.5	2.5	0.548	2.42
50000	14600	50.0	2.5	0.548	2.84
50000	14600	25.0	5.0	0.548	3.43
50000	14600	25.0	1.25	0.548	2.24
50000	19250	25.0	2.5	0.548	2.32
50000	10950	25.0	2.5	0.548	2.93
50000	14600	25.0	2.5	0.603	3.53
50000	14600	25.0	2.5	0.493	2.00

* RCP = ratio of coal to fuel oil price, both in $/kcal

REFERENCES

1. Basu P: Design Considerations for Circulating Fluidized Bed Combustors. Jour. Inst. Energy, December 1986, p.179.
2. Ekinci E, Tolay M, Atakül H, Kadıoğlu E: Combustion of a High Sulphur Solid Fuel in a Fluidized Bed Combustor. Archivum Combustionis, 5:2, 161, 1985.
3. Moss G: Mechanism of Sulphur Absorption in a Fluidized Bed of Lime. Inst. of Fuel Fluidized Bed Combustion Symp. Ser. No.1, London, 1975.
4. Pereira FJ: NO_x Formation in a Fluidized Bed Combustor. Ph.D. Thesis, Univ. of Sheffield, 1976.
5. Hammonds GA, Skopp A: Nitrous Oxides Formation and Control in Fluidized Bed Coal Combustion Processes. ASME paper 71-WA/APC 3, December 1971.
6. Gelperin NI, Einstein VG: Heat Transfer in Fluidized Bed. Fluidization, Davidson and Harrison (eds), Academic Press, London, 1971.
7. Botterill JSM: Fluidized Bed Heat Transfer. Academic Press, London, 1971.
8. MacLaren J, Williams DF: Combustion Efficiency, Sulphur Detention in Pilot Plant FBC. J.Inst. of Fuel, 42, 333, 1969.
9. Ekinci E, Öner T, Atakül H, Tolay M: The Air Pollution Capacity of Turkish Lignites. Air Pollution Symp., MMO, Istanbul, 1985 (Tur).
10. The Characteristics of Some Important Turkish Coals. MTA Report, Ankara, 1982 (Tur).
11. Coal Inventory of Turkey. MTA Pub. No. 171, Ankara, 1978.
12. Biron C: The Chemical and Physical Characteristics of Turkish Coals and Their Reserves. International Coal Technologies Seminar, Istanbul, 1982 (Tur).
13. Arıoğlu E, Yüksel A: The Problems of Lignite Mining and Possible Answers in Turkey. Birsen Pub. Co, p.8, 1984 (Tur).
14. Küçükbayrak S: Employment of Different Desulphurization Methods to Some Turkish Lignites. Ph.D. Thesis, İTÜ, 1984 (Tur).
15. Ekinci E, Türkay Ş, Fells I, Kadıoğlu E: Ash Content Limit of Solid Fuels Combustion in a Fluidized Bed. German-Turkish Energy Symp., Izmir, 1981.
16. Copeland GG: Operating Experience with a Variety of Low Grade Fuels in FBC. Proc. of 6th Int. Conf. on FBC, p.579, Georgia, April 1980.
17. Albrecht E: Incineration of Waste in the Fluidized Bed. Thyssen Engineering

GmBH, Am Thyssenhaus 1, Germany, 1986.
18. Knorr F, Novoty P: Operational Application of Fluidized Bed Furnaces in Burning Low Calorific and Waste Fuels in Czechoslovakia. Proc. of 6th. Int. Conf. on FBC, p.822, Georgia, April 1980.
19. LaNauze RD, Dufty GJ, Potter EC, Bradshaw AV: Fluidized Combustion of Coal Washery Waster. Fluidization, Grace and Matson (eds), p.151, Plenum Press, New York, 1980.
20. Urkan K, Arıkol M, Vural H: Volatile Matter Behaviour in FBC. J. of Thermal Sciences and Technology, 9:1, 55, 1986 (Tur).
21. Atımtay A: Combustion of Volatile Matter in Fluidized Beds. Fluidization, Grace and Matson (eds), Plenum Press, New York, 1980.
22. Ekinci E, Tolay M, Kadıoğlu E: Behaviour of Volatile Matter in FBC. Proc. of Combustion Symp., p.63, Bursa, 1983 (Tur).
23. Ekinci E, Yalkın G, Atakül H, Erdem Şenatalar A: Combustion of Volatiles for Some Turkish Coals. Submitted to Jour. Inst. Energy.
24. Stubington JF: The Role of Coal Volatiles in Fluidized Bed Combustion. Jour. Inst. Energy, December 1980, p.191.
25. Pillai KK: A Schematic for Coal Devolatilization in FBC. Jour. Inst. Energy, September 1982, p.132.
26. Pillai KK: Devolatilization and Combustion of Large Coal Particles in a Fluidized Bed. Jour. Inst. Energy, March 1985, p.3.
27. Ekinci E, Yardım MF, Atakül H: Temperature Profile in a Lignite Floatsam Rich Fluidized Bed. Submitted to Fuel.
28. Ekinci E, Tolay M, Atakül H: Predicting Segregation Tendencies in a FB Using Temperature Profiles. Submitted to Powder Technology.
29. Ekinci E, Pogson B, Fells I: SO_2 Capture by the Inorganic Matrix of a Low Grade Fuel in a FBC. Jour. Inst. Energy, September 1984, p.368.
30. Türkay Ş, Atakül H, Ekinci E: Design of a Fluidized Bed Burning Low Grade Fuels: 1st National Design Congress, Istanbul, 1982 (Tur).
31. Highley J, Kaye WG: Fluidized Bed Industrial Boilers and Furnaces. Fluidized Bed, JR Howard (ed), Vol 3, p.77, Applied Science, New York, 1983.
32. Fitzgerald TJ: Fundamentals of Fluidized Bed Hydrodynamics as Applied to FBC System Design. DOE/WVO Conf. on FBC System Design and Operation, West Virginia, October 1980.
33. Newby RA, Keairns DL, Ahmed MM: The Selection of Design and Operating Conditions for Industrial AFBC to Meet Environmental Constraints. DOE/WVO Conf. on FBC System Design and Operation, West Virginia, October 1980.
34. Shang JV, Notestein JE, Mer JS: An Overview of FBC Design Practice. DOE Morgantown Energy Technology Center Internal Report, West Virginia, 1980.
35. Molayem, Bardakçı T, Hall AV, Hewitt DR: Experimental Validation of MIT's AFBC Design Model. 7th Int. Conf. on FBC, Philadelphia, October 1982.
36. Tolay M, Atakül H, Ekinci E: Fluidized Bed Combustion Boiler Design. J. of Thermal Sciences and Technology, 6:3, 37, 1983 (Tur).
37. Ekinci E, Tolay M, Atakül H: FBC Design for Waste Incineration. 2nd Nat. Machine Design and Development Congress, Ankara, 1986 (Tur).
38. Ekinci E, Tolay M, Yardım MF, Özil E: The Development of a FBC System for the Utilization of Turkish Lignites and Oil Shales. Proc. of 8th Int. Conf. on Coal Slurry Fuels Preparation and Utilization, Florida, 1986.
39. Howard JR: A Technology for Helping to Alleviate the Energy Problem: Fluidized Bed Combustion and Heat Transfer. Energy for Industry, O'Callaghan (ed), Pergamon Press, 1979.
40. Sasaki K, Takeuchi Y, Ishiguro S: Newly Developed FB Incinerator for City Garbage. IHI Engineering Review, No.1, p.94, January 1977.
41. Bailie RC: Solid Waste Incineration in Fluidized Beds. Industrial Water

Engineering, p.22, November 1970.
42. Atakül H: The Behaviour of Çan Lignites in a Fluidized Bed Combustion Process. Ph.D. Thesis, İTÜ, 1986 (Tur).
43. Engin O et al: Feasibility Studies of Bursa Orhaneli Coal Seams. MTA Report, Ankara, 1976.
44. Şakir V: Removal of Sulphur and Ash from Beypazarı Lignite by Physical Means. M.S. Thesis, METU, 1979.
45. Bilgin Y et al: Feasibility Study on Elbistan D Section. MTA Report, Ankara 1982.
46. Tolay M: Segregation Forms Leading to Agglomeration in a FBC. Submitted Ph.D. Thesis, ITU.
47. Arıkol M et al: Coal Transportation and Utilization Technologies. MAE Publication, Gebze 1984.
48. Serpil A, Durmuş AH, Bilge D: Economic Feasibility of Industrial Fluidized Bed Boilers. Accepted by Marmara Review.

PRESSURISED COMBUSTION - A NEW TOOL IN EMISSION ABATEMENT

L.J.M.J. BLOMEN
J.E. HILLE
P.F. van den OOSTERKAMP

1. INTRODUCTION

A couple of years ago, an investigation on pressurised combustion was initiated in our company to tackle two drawbacks of present technology. Pollution abatement from flue gases is costly, mainly because of the big volumetric flow which has to be manipulated. At the same time, combustion to supply heat is rather inefficient for many processes. This inefficiency is usually eliminated by the production of steam or other hot utilities. However, this only makes sense when these hot utilities can be used in other processes. It is obvious that from process point of view, utility production should be minimised in a process plant.

Pressurised combustion can be a very attractive solution to these problems, as will be shown. KTI (Kinetics Technology International, Zoetermeer, The Netherlands, executed this investigation with financial support from the Dutch Energy Foundation "PEO" (Project Management Office for Energy Research).

2. PRESSURISED COMBUSTION AS FUEL SAVING TOOL

A process is a system which converts feed streams into product streams. Both groups of streams are usually at ambient or near ambient temperatures. During processing, the streams are heated and cooled. Frequently, also heat of reaction has to be supplied. When we combine all streams which require heat into one with a varying specific heat, this "stream" describes the amount and level of the process energy requirement. It is called "COLD STREAM" in fig. 1. Same can be done for all streams which need cooling. The resultant "stream" is called the "HOT STREAM" in fig. 1. These curves show that all heat required by the process below T_1 can be supplied by the "HOT STREAM". The duty "a" above T_1 has to be supplied by an external source. Dependent upon the level of T_1 and T_{max}, either an external hot stream could be used (if available) or this duty has to be supplied by combustion. The duty available from combustion can be visualised in a similar graph as the cooling down of the combustion gases. These gases cool down from the adiabatic flame temperature (T_{AD}) to the minimum temperature (T_{MIN}) which avoids condensation on the heat transfer surface (see curve T_{AD} — T_{MIN} in fig. 2). This releases the duty "c".

In order to supply the duty "a", this duty should be available in the flue gases above $T_1 + \Delta$ (Δ is the minimum approach).

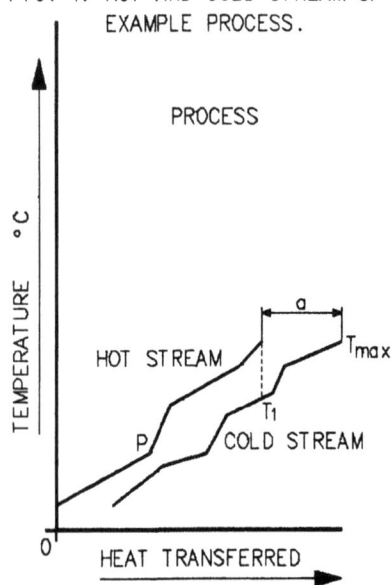

FIG. 1: HOT AND COLD STREAM OF EXAMPLE PROCESS.

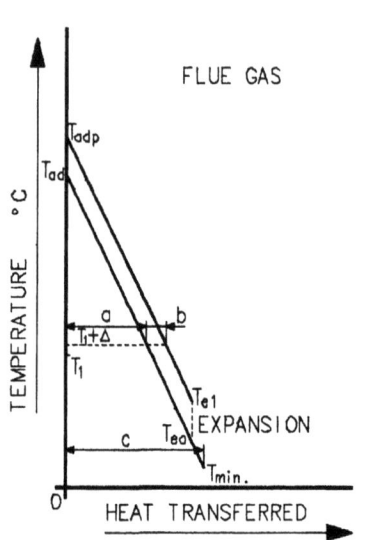

FIG. 2: FLUE GAS CURVES OF NON PRESSURISED AND PRESSURISED COMBUSTION.

Therefore, we have an excess duty available of c - a which cannot be used for this process.

In case we apply pressurised combustion, additional duty is available in the combustion air and the fuel. This is caused by the compression. The resultant adiabatic flame temperature increases to T_{ADP}. The required compression power for the fuel and combustion air is recovered by expanding the flue gases. Careful selection of the expander inlet temperature T_{EI} will make it possible to balance the power requirements. This is represented by the dotted line $T_{EI} - T_{EO}$ in fig. 2. The resultant flue gas curve is $T_{ADP} - T_{EI} - T_{EO} - T_{MIN}$. The available duty for the process is increased till "a+b", if the total releasable duty remains "c". Since only a duty "a" is required, the fuel amount can be reduced by a factor "1 - a/(a+b)". For several values of $T_1 + \Delta$, this saving has been calculated.

$T_1 + \Delta$ (°C)	Fuel saving (%)
200	0
400	10
600	19
800	19

Note: These savings are valid for one particular set of conditions. They can be influenced by changes in fuel type, excess air, combustion pressure, compressor efficiency and turbine efficiency.

Above 600°C, the savings are leveling off, due to the mechanical limitation on the temperature of the air leaving the compresssor. This maximum was taken as 400°C.

To illustrate the above, a practical example has been worked out. For this example, a hydrogen plant was selected. The block diagram of this plant is shown in fig. 3.

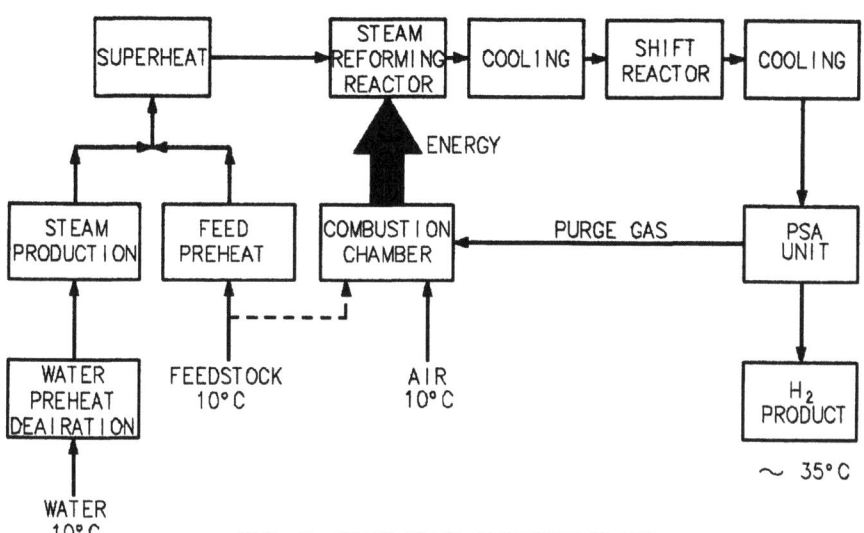

FIG. 3 SIMPLIFIED HYDROGEN PLANT

It contains all the main steps required for hydrogen production. The assumed feedstock was 100% CH_4 at a pressure of about 20 bar. The water has to be demineralised at approx. 25 bar. Other main parameters are

Reformer outlet temperature	850°C
Shift outlet temperature	420°C
PSA efficiency	85%
Hydrogen purity	99.9%
Hydrogen pressure	17 bar

Fig. 4 depicts the hot and cold stream of this process, excluding combustion of the purge gas. In fig. 5, atmospheric combustion and pressurised combustion of the purge gas are incorporated. The curve A-C-D represents the hot stream. The line A-B represents the contribution of the purge gas combustion to the hot stream (this figure excludes firing of external fuel or feed). When we take a closer look at the figure, a problem situation becomes apparent. The curve M-N represents the reforming reactor which should receive its duty from the combustion chamber. However, the temperature level of the combustion gases does not allow this: M-N cuts through A-B. Consequently, although the hot stream has sufficient heat content at a sufficiently high level, this operation mode is not possible without additional firing. The picture changes completely when pressurised combustion is applied.

The hot stream is now represented by the curve E-K-D. The contribution of the flue gases to the hot stream is shown as the curve E-F-G-H-B. In this case, a two stage expansion was selected. The expansion F-G drives the air compressor and H-B drives the purge gas compressor. The approach at the inlet to the reforming reactor is approx. 200°C which is quite sufficient when applying pressurised combustion. These curves show that, once applying pressurised combustion, a hydrogen plant could theoretically be operated on purge gas only.

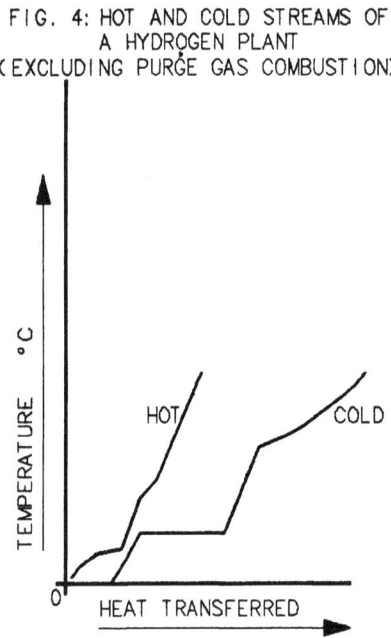

FIG. 4: HOT AND COLD STREAMS OF A HYDROGEN PLANT (EXCLUDING PURGE GAS COMBUSTION)

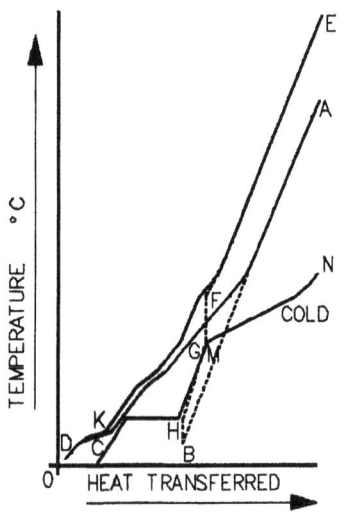

FIG. 5: HOT AND COLD STREAMS OF A HYDROGEN PLANT INCLUDING PURGE GAS COMBUSTION
A-C-D NON PRESSURISED COMBUSTION
A-K-D NON PRESSURISED COMBUSTION

In one of our studies, we did a detailed investigation of a hydrogen plant also including economic analyses. From this study we present some data in tables 1 and 2.

TABLE 1. Simple pay-out time (in years) as function of the steam prices (Dfl/ton) and the combustion pressure (bar).

	Import steam → 15	30	45
	Export steam → 10.7	21.3	32
Combustion pressure → 5	1.5	2.6	9.1
10	1.9	3.7	44.2
15	2.2	6.0	*

* Operating cost for this case is higher than the operating cost of the atmospheric reference case.

Simple pay-out time is defined as additional investment (compared to an atmospheric combustion plant) devided by the annual saving in utility cost.

Since at present, due to large energy saving programs in industry, export steam is an undesirable product and will more and more become so, we also calculated the above table for the case that export steam has no value at all (see table 2).

TABLE 2. Simple pay-out time (in years) as function of the import steam prices (Dfl/ton) and the combustion pressure (bar). Value of export steam is 0 Dfl/ton.

	Import steam → 15	30	45
Combustion pressure → 5	0.4	0.9	1.4
10	0.5	1.1	1.7
15	0.7	1.5	2.3

This table clearly shows that there are circumstances where pressurised combustion can be very beneficial.

It should be noted that investment prices for the estimates in the tables were all based on our standard practice of tailor-made plants. The small size of the units, required for a pressurised combustion hydrogen plant, however, makes them perfectly suitable for production of standard reformer modules. In that case, the investment for a pressurised combustion hydrogen plant will drop to a level almost equal to or even below the investment level of a hydrogen plant with atmospheric combustion. Consequently, all the savings on utilities are direct profits. This makes this concept really attractive.

3. PRESSURISED COMBUSTION AS A POLLUTION ABATEMENT TOOL

There are several ways in which pressurised combustion can affect pollution.
- As discussed in chapter 2, pressurised combustion can save fuel. This inherently reduces pollution, because less fuel is fired and less flue gas is being produced.
- Once the pressure in the flue gas is increased, the partial pressure of contaminants is increased as well. Most contaminant removal processes will operate at a better efficiency when the entering partial pressure is higher.
- A higher pressure level of the flue gases will drastically decrease their volumetric flow. This, in turn, will reduce the size of the treating equipment, even without increasing the absolute pressure drop.
- When operating at a higher pressure, it is economically justified to increase the operating pressure drop. This again decreases equipment size and/or allows the application of cleaning processes which require higher pressure drop. Abatement processes such as filtering and fixed or fluid bed cleaning are becoming suitable for flue gas cleaning.

- Pressure will have an influence on the combustion process and therefore affect the production of polluting substances. Because of the high pressure, the volume of the flame will be reduced which has a decreasing effect on thermal NO_x formation. SO_x formation will not be influenced by pressurised combustion. Because of the high intensity in the flame it is expected that the emission of unburnt hydrocarbons and carbon monoxide will decrease to some extent, especially if sufficient care is taken in burner and combustion chamber design. The same will apply for soot formation.

4. CONCLUSIONS

Pressurised combustion will decrease fuel consumption, especially in those cases where the temperature level of the required duty is high. Consequently, air pollution will decrease proportionally.

Pollution can be more efficiently removed from flue gases under pressure. At the same time, the equipment to remove it can be reduced in size because of the volumetric flowrate.

Pressurisation of the combustion will have a favourable effect on the production of pollutants.

Combustion equipment will reduce in size as a result of pressurised combustion.

SAFETY ASSESSMENT AND THE SELECTION OF DETERGENT RAW MATERIALS

N.T. de Oude
Procter & Gamble European Technical Center,
Temselaan 100, 1820 Strombeek-Bever. Belgium. (*)

INTRODUCTION
   The detergent industry formulates detergents which meet social demands, up to the limit of existing technological knowledge. This definition puts the demands of society squarely in the driving seat. Importantly, it does not put economics first. The reason is that most detergent manufacturers purchase their raw materials from the chemical industry, rather than produce them in-house. Hence, they can easily shift to those materials, required to formulate the products that meet social demands best.
   Economics are, of course, important for the consumer. It is the consumer who, in the end, pays for the total cost of the product. He has virtually no other option, because detergents are part of the essential necessities of life.
   What, then, are the social demands mentioned in the opening sentence? Society is looking for physical as well as psychological rewards. The physical rewards are, of course, cleanliness, i.e. hygiene. Most books on the history of laundering describe this aspect in detail. The importance of hygiene for public health has also been well documented.
   The demand of society for psychological rewards is less well documented. This reward is meeting man's needs for esthetics. The human desire for an esthetically pleasing environment has always been with us and can be seen all around us. Many professions are evidence of, and clearly show, the importance of esthetics in human culture. To name but a few examples, let us take the industrial designer, the interior decorator, and the numerous members of the world of arts and crafts.
   Esthetics are rooted most strongly in the Renaissance. In Northern Europe, the Renaissance started as early as the late 12th century. An example of this, is the art of chivalry, beautifully described by Chrétien de Troyes in his novels "Lancelot" and "Perceval", in which courtesy was the aim of the knight. He established the symbol of the ideal knight as one who measures his abilities in joust, rather than in brutal combat, and who protects women and puts himself in the service of God.
   There was a simple way to refer to people who practiced this lifestyle: they wash their hands after dinner. Washing before dinner was already accepted as desirable for cleanliness, but washing after dinner was good manners, and day-to-day evidence of chivalry.

(*) For: Nederlandse Vereniging van Zeepfabrikanten (Netherlands
    Association of Detergent Manufacturers), Catharijnesingel 53,
    3511 GC Utrecht. The Netherlands.

This conference is about environmental technology, not about history. The effort required in the past to get laundry clean is known from romantic engravings of girls -nearly always attractive- washing clothes, always in nice weather at the banks of a picturesque river, conveniently located just around the corner. There was the option to send the linnen to a commercial laundry, such as has been preserved in the Arnhem Open Air Museum. Wealthy families would send their linnen there twice a year.

The inadequate degree of cleanliness offended people living in these times. It stimulated the development of perfumes and powders in order to satisfy their esthetical needs. However, there was no satisfactory way to meet the need for hygiene.

Today, we live in a world where both of these needs are satisfied. Further, the combination of modern fabrics, washing machines, and detergents has eliminated the drudgery involved in performing the necessary tasks needed to achieve these needs. Society has now added two more requirements:
- safety to the consumer,
- safety to the environment.

Much has been done to safeguard the health of the consumer. Accidents with laundry detergents are rare; if they happen, they do not result in irreversible injury.

ENVIRONMENTAL SAFETY

Safety is, of course, a relative concept. Man has always impacted in some way upon his environment. There are no cedars left in the Lebanon and the Greek hills are still devoid of the vegetation that was exploited some 20 centuries ago. What has changed to justify this conference?

First of all: the population has mushroomed. In that 12th century, when the Renaissance started, Europe had approximately 50 million inhabitants. Today Europe has more than 10 times that number. Our impact on the environment has increased more than proportionally because the resilience of our environment is now being overstretched. The first inhabitant of Broekzele, the city where I live and that is now called Brussels, could use the 10 ft wide Senne river for all his needs without really impacting on it. Today, with more than 1 million people living in that area and discharging their sewage without any treatment into that same creek, it is an open sewer.

Secondly, we now all enjoy a high standard of living, well above even that of the elite of any previous era. In order to achieve and maintain that standard of living we need more, and a larger variety of all products, resulting in more waste. It is no longer possible to continue to have that many people, all living at an acceptable standard without an impact on the environment. This impact has been aggrevated by the lack of a simultaneous development of the means to safely dispose of this waste in the environment. How then can we control that impact and preserve the environment in a condition that we, as society, consider desirable?

This question addresses two subjects: risk <u>assessment</u> and risk <u>management</u>. Risk assessment is a scientific undertaking; this subject will be covered extensively in the remainder of this talk. Risk management is a social subject; the acceptance of risk changes with time and with perceived benefits. For example, fairly toxic compounds are accepted if the social benefit is considered important, as is the case for drugs. The acceptance of risk will be low if the social benefit is considered to be small or even insignificant. There is an ongoing debate about the social benefit of detergents, because they provide the benefit of hygiene so well, that it is taken for granted. Consequently, it means that the detergent industry needs to communicate this benefit better than it has done before. It also means that the industry must continue to put very high demands on the environmental safety of its products.

The detergent industry has been making risk assessments for many years and has continually been improving its capability to do so. It has upgraded its methods as science has progressed and has actually contributed importantly to the development of the environmental sciences. In selecting the raw materials it wants to use, it considers three questions that are discussed below.

Fate

The "fate" of a detergent ingredient is studied in order to determine how much of what ends up where. Most detergent ingredients are biodegradable. About a quarter of a century ago the first test was developed to determine the loss of functionality (so-called primary biodegradation) of anionic surface active agents. Primary biodegradation is a meaningful parameter, because loss of functionality generally parallels loss of aquatic toxicity. However, that is not sufficient and additional tests were developed to determine the ultimate biodegradability, that is the conversion of organic matter into biomass and energy.

Biodegradability was, and to a large extent still is, expressed in a percentage figure. The more meaningful parameter of rate of biodegradation is still not used extensively. A rate figure is more meaningful because the time available for biodegradation varies: it may only be a day or so between a discharge to a river and the intake of drinking water, or it may be many months between successive applications of sewage sludge to arable land.

Although biodegradation testing has come a long way, we still need to go further in our understanding of this process in the real environment. It remains difficult to test slow biodegradation processes: where laboratory tests may show little or no disappearance, the environment often finds means of utilizing the energy sources that organic compounds represent.

Other environmental processes may be important for compounds that do not biodegrade very rapidly. Examples are hydrolysis, photodegradation and catalytic processes. Very few of these, and hardly any combination of these processes, have been investigated. Nature, on the other hand, does not run standardized test methods where only one process is allowed to operate at a time. There is, however, one comforting thought in all this: ignoring these potential removal mechanisms in the safety assessment of detergent ingredients provides an additional margin of safety.

The result of a fate-evaluation is the prediction of the environmental concentration that will occur in the steady state condition of general use of a compound in detergents at volumes that can be estimated with good accuracy from known market volumes.

Effects

In parallel to the above, effect concentrations have also been determined for many years. Such a test usually involves the determination of that concentration at which no toxic effect is observed on relevant organisms.

Early tests (around 1950) were limited to determining acute toxicities to organisms (mainly fish) that were easy to keep in the laboratory. Today, subacute and chronic tests are available to be used when relevant. Examples are:
- reproductive success of daphnia,
- oyster shell deposition,
- inhibition of glucose uptake by bacteria,
- algal growth inhibition,
- emergence of midges,

and many others.

More work is still needed, in particular in establishing the toxicity in the real world. It is not too difficult to develop laboratory tests that show some effect, but it is quite a different challenge to interpret these effects. For example: a few young offspring in a laboratory chronic fish test may show deformations. What is the relevance of this finding when we know that in the laboratory more than 90 % of the brood makes it to the age where deformations can be observed, while in nature less than 10 % does so? The company that I work for has just authorized the expenditure of $ 1 million to construct a small, open air river to study that real world.

Thus far, literature shows that laboratory tests are more conservative than the real world experience is.

Safety Margin

The third and final step is, in principle, easy. It consists of comparing the predicted environmental concentrations with the no-observed effect concentrations and seeing if there is a safety margin. One practical problem is the availability of published statistics, a necessary prerequisite to predicting environmental concentrations. This includes the incidence of sewage treatment, the performance of the plants and the dilution ratio upon discharge. In some countries these data do not exist, in others they are scattered over regional institutes, but in no European country are they available for use in environmental safety assessments, as they are in the U.S.A.

In addition to the above, we voluntarily go out into the field and verify our predictions of environnmental concentrations. Thus, sensitive and specific analytical methods are developed and used, once a compound has been introduced into a detergent, to analyse its actual concentration in untreated and treated sewage, rivers, and other relevant environmental compartments.

With a quarter of a century behind us we can claim that the system appears to work well. There have been few, if any, major errors in the risk _assessment_, carried out by the detergent industry. There have been and still are discussions about the risk _management_. These are primarily the result of the changing demands and expectations of society. It is to be expected that such changes will continue to occur; the detergent industry responds by listening to society, by seeking clarification in discussions and by changing its products. A few of such changes are described in the next section.

ENVIRONMENTALLY RELEVANT DETERGENT CHANGES

Currently, the most intense discussion is about sodium triphosphate, $Na_5P_3O_{10}$. Phosphate plays a major role in lake eutrophication (the excessive growth of algae) and the removal of these compounds from detergents is considered, by some, as the key to solving this problem. The detergent as well as the entire chemical industry has made major efforts to develop a substitute for triphosphate, which plays a major role in cleaning fabrics. Several hundreds of patents have been published, but few of the compounds patented proved effective, safe and economical.

Because of the problem of finding a good substitute, the detergent industry also engaged in the social debate by asking for proof of a meaningful, beneficial effect on the environment that substitution of detergent phosphate would provide. This debate is still continuing. Switzerland has banned detergent phosphate. Germany and Austria prefer to maintain the current reduced phosphate levels because they recognize the toxicological safety of phosphate. The Netherlands are still reviewing the case.

Simultaneously, the detergent industry has continued working to ease the social debate by reducing the phosphate content of detergents as far as possible without impairing the product quality and by continuing the development of substitutes. About 10 years ago, patents appeared that described the use of zeolite A, a grade of sodium aluminiumsilicate, as a partial phosphate substitute. Further developments led to detergents that had been substantially reformulated and with zeolite as the full replacement of phosphate. Such detergents have now appeared on the market.

Bleach activators constitute another example. Detergents contain sodium perborate, $NaBO_3.4H_2O$. This material liberates oxygen in the wash and acts as a bleaching agent at temperatures above approximately 80°. A bleach activator forms organic peroxides with sodium perborate at much lower temperatures; these peroxides decompose, providing bleaching at these lower temperatures. The net effect is a saving of energy.

This technology has been available for a long time and in the past several detergent manufacturers have tried to market products with this benefit. However, the consumer was not interested in energy savings and a product will not sell if the consumer is not interested. All that changed when society became more conscious about energy: products with bleach activators are now well established in most European countries.

Not a replacement, but a new concept was the addition of enzymes, introduced in the late sixties. They provide improved hygiene of the fabrics and reduce the need for the bleaching of enzyme-sensitive stains. Being proteins, they decompose readily and have no toxic effects when they enter the environment after use in the washing machine.

CONCLUSION

This has _not_ been a discussion about replacing dangerous chemicals by safe ones. The emphasis here has been on the development of new compounds that meet social demands for hygiene and safety. That is because detergent manufacturers continue to improve the environmental attributes of the compounds they use. This will result in steadily increasing safety margins. It will also make sure that its products are not only _safe_ (risk assessment), but are also _seen to be safe_ (risk management). Progressing on both fronts simultaneously, and using the best science available in doing so, has resulted in environmental technology becoming an integral part of the development of any new detergent.

THERMODYNAMIC ANALYSIS OF SOLAR POWERED ABSORPTION REFRIGERATION SYSTEMS FOR COMPARISON OF WORKING FLUIDS

Ö.E. ATAER

1. INTRODUCTION

The absorption cooling is widely known as a prospective candidate for efficient and economic use of solar energy. This has led to an intensive search for refrigerant - absorbent combinations suitable for solar energy. However unless a thermodynamic analysis is made, the suitability of a refrigerant/absorbent combination for an application can not be established. In this study he thermodynamic analysis of NH_3/H_2O, R21/DMF (dimethylformamide) and R22/DEGDME (diethylene glycol dimethyl ether) are made in the range suitable for the use of solar energy. The cycle used in the analysis shown in Figure 1, includes generator, condenser, absorber, evaporator and two heat exchangers.

2. THEORY

Coefficient of performance (COP) and the circulation ratio (f) are major thermodynamic parameters affecting the performance of a absorption cooling system. COP of the cycle is defined as $COP = Q_e/Q_g$ where Q_g is the energy input to the system at the generator and Q_e is the cooling effect of the cycle. The circulation ratio is defined as the mass flow rate of rich solution pumped from low pressure to high pressure, to the refrigerant mass flow rate at the generator outlet. These two thermodynamic properties are used to make comparative studies of the refrigerant and absorbent combinations indicated above. To calculate the COP and f of the system, working characteristics at different parts of the cycle should be known. Some of the relations used in the analysis are taken from the references below and some are obtained by curve fitting from the tabulated data.

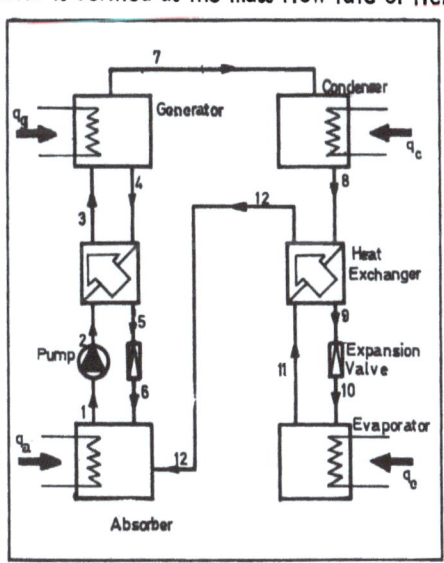

FIGURE 1. Schematic diagram of the cycle.

3. RESULTS

The results of the analysis given in Figure 2 to 4, shows that the COP of the absorption refrigeration cycle decreases by decreasing evaporator (T_e) and generator temperatures (T_g).

On the other hand f of the cycle decreases by increasing evaporator temperature while showing similar trend as COP in variations on generator temperature. The results given are for 20 °C absorber temperature (T_a). Similar variations are observed at different absorber temperatures, but when the absorber temperature increases the system can work at high generator temperatures. The R22/DEGDME combination has better performance than the others, but low critical point of R22 (369.01 K) limits the system to work at high generator temperatures. In the analyses it is assumed that the condenser temperature is equal to the absorber temperature, and the losses in the heat exchangers and the pressure losses in the system are neglected.

REFERENCES

1. Ziegler B and Trepp Ch : Equation of State for Ammonia/Water Mixtures. Int. J. Refrig., Vol.7 Number 2, March 1984.
2. Bourseau P and Bugarel R : Absorption-Diffusion of The Performances of NH_3-H_2O and NH_3-NaSCN. Int.J. Refrig., Vol.9, July 1986.
3. Ando E and Takeshita I : Residential Gas-Fired Absorption Heat Pump Based on R22/DEGDME Pair Part I : Thermodynamic Properties of R22/DEGDME Pair. Int. J.Refrig., Vol.7, Number 3, May 1984.
4. Cleland A C : Computer Subroutines for Rapid Evaluation of Refrigerant Thermodynamic Properties. Int.J. Refrig., Vol.9, Nowember 1986.
5. Badarinarayada K Srinivasa S and Murthy K : Thermodynamic Analysis of R21-DMF Vapour Absorption Refrigeration Systems For Solar Energy Applications. Int.J. Refrig., Vol.5, Number 2, March 1982.

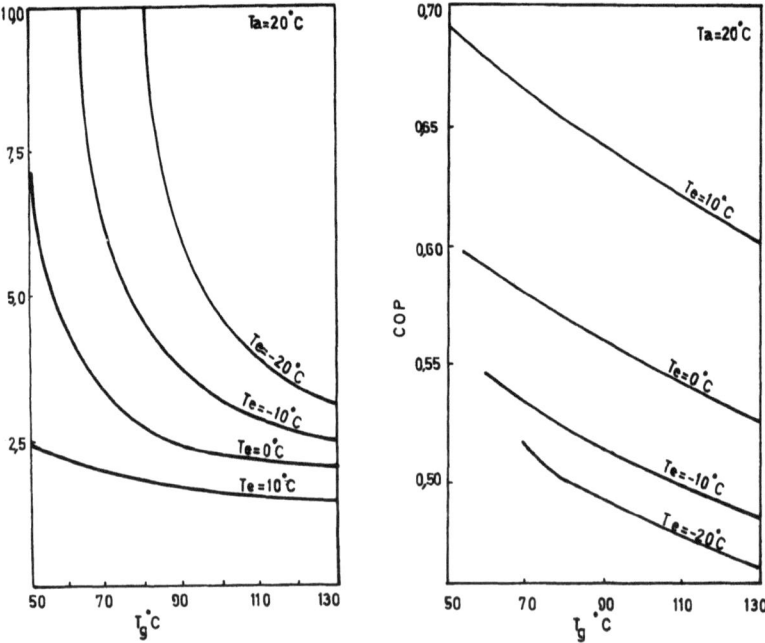

FIGURE 2. Variation of COP and circulation ratio of NH_3/H_2O absorption refrigeration system with operating temperatures.

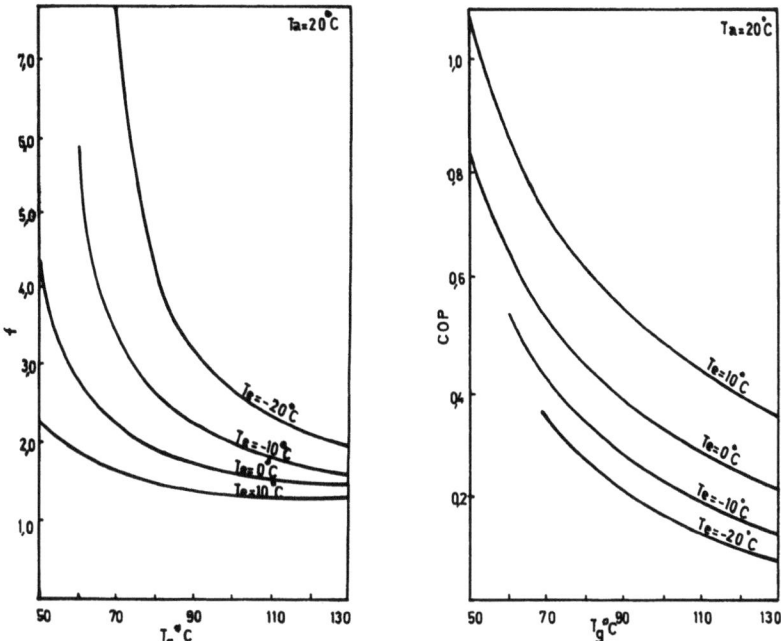

FIGURE 3. Variation of COP and circulation ratio of R21/DMF absorption refrigeration system with operating temperatures.

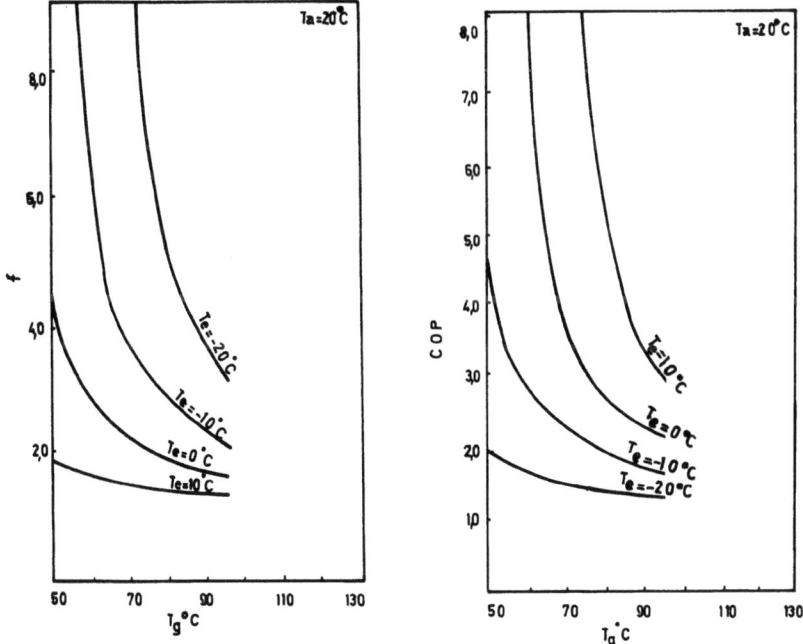

FIGURE 4. Variation of COP and circulation ratio of R22/DEGDME absorption refrigeration system with operating temperatures.

PROCESS-INTEGRATED ENVIRONMENTAL TECHNOLOGY

Second European
Conference on
Environmental Technology

in the 'European Year
of the Environment'

PROCESS INTEGRATED ENVIRONMENTAL TECHNOLOGY
A MUST TO SURVIVE

J. Quakernaat and J.A. Don
Netherlands Organization for Applied Scientific Research (TNO)
Division of Technology for Society
Department of Environmental Technology
The Netherlands

F. van den Akker
Ministry of Housing, Physical Planning and Environment
Directorate General for Environmental Protection
Department for Clean Technologies
The Netherlands

Summary

From experience in the Netherlands it seems that for many production processes environmental benefits with regard to the problem of pollutants can in principle be attained to such an extent as to do justice to the ever increasingly severe environmental demands, as well as to economic feasibility.

As for suitable solutions to this problem, also found to exist in many other countries, a choice will have to be made - on the basis of a proper insight - from the combined potentials of good housekeeping, add-on technology and process integrated environmental technology.

In the somewhat longer run, only another attitude and way of thinking about production activities e.g. the importance of process integrated environmental technology, will offer really conclusive, permanent solutions. In order to overcome the problems linked to a substantial penetration of this option, the government and industrial community will more than ever have to get round to operational participation on the basis of awareness of the problem and the importance of the transfer of knowledge.

I. INTRODUCTION TO THE PROBLEM
Environmental stresses

It is well known that the environment has come under great strain in many places in the world. Pollution of water, soil and air has emerged in conjunction with attacks on the environment and disturbances of biological balances. The primary causes of these developments are obvious; firstly the growth in population, secondly a steadily increasing prosperity based on economic growth. This has led to considerable stresses in the environmental field and in society [1].

The strategies for relieving stresses would appear to be very simple ones: create less damage and less nuisance by introducing less pollutants into the environment, and ensure that other loads such as noise, radiation and heat emission are reduced. Unfortunately, it is very hard to satisfy these demands promptly. This will require a long time, and a large number of conditions will have to be met beforehand. This may be illustrated by examples of problems caused by the increase in pollutants.

The ever increasing extraction of materials from the earth causes considerable dispersion of matter. Besides the relatively concentrated releases of pollutants from so-called point sources, highly dispersed emissions also occur. The latter sources release small quantities of matter, such as wastage and spillage (diffuse emissions).

The spread of natural and xenobiotic materials to the various outer and inner environments creates two significant effects. On the one hand there is an increase in the number of compounds, and on the other hand there is an increase in the concentrations of materials in the environment. When these concentrations exceed the carrying capacity, they produce disruptive and degrading consequences for biological and related systems.

Accumulating diffuse load

Viewed in the long term, the majority of pollutants will go through one form or other of spatial dispersion, called diffuse load. A great many materials transfer in the process from one environmental compartment to another, whereby the soil (above and below the water line) or the sea, often acts as end station. This results in a continuous accumulation of a great variety of materials in the soil/groundwater complex and in the oceans.

Such an accumulation process will in the long run constitute a great danger to the environment, and therefore to the health of mankind. Soil becoming filthier and filthier and less usable will in the long run grow into a permanent stock of environment threatening micropollutants.

Process of becoming aware

In many countries, and especially in Europe, causes and results of changes to the environment have been recognized, and have been discussed at social, scientific and political levels. This has resulted, in these countries, in a comprehensive package of national and international environmental lawmaking and in concrete technical environmental measures.

In the Netherlands the awareness of environmental problems has been developing - although very tentatively - since the sixties. During the last 15 years a great deal of experience has been gained from the fight against air, water and soil contamination, as well as from examining the problems of immaterial loads. Various solutions have been field-tested. However, the level of awareness of these developments is still far too low and must be greatly increased before the environment may become clean again.

No doubt this also holds good for other countries. The fact that the Netherlands is very active in this respect is not surprising. Being naturally positioned at the delta of Rhine and Meuse has made the Netherlands "Gateway and discharge port for Europe", with all the consequences involved. From a comparison of the relative figures for a number of environmental load factors (Table 1) it is clear that the Netherlands had to take the environmental problems seriously. It also shows that other countries too will be confronted - sooner or later - with exactly the same problems.

Table 1. Environmental load factors for a number of countries (densities per km^2; report 1984).

	population	industrial production ($ 1000)	energy consumption (toe)*	livestock	transport (cars)
Netherlands	334	568	1595	334	92
Belgium	321	593	1348	248	84
Japan	298	446	892	30	47
German Federal Republic	248	713	986	145	72
United Kingdom	229	229	833	169	59
France	97	174	307	84	28
U.S.A.	23	42	81	12	12
New Zealand	11	5	38	226	4

* tons of oil equivalent.

Interim report

It is now possible in the Netherlands to give an interim report and to indicate what environmental thinking has produced thus far, and what is expected to happen. In order to cope with the problem of pollutants and immaterial loads, the Netherlands will be actually working towards the development of the philosophy of process integrated environmental technology, sometimes called "clean technology". This development is also being worked out in other countries [2].

2. ABATEMENT OF EFFECTS
General
Towards the end of the sixties the Netherlands gradually introduced technological measures which immediately resulted in a substantial drop in emissions of pollutants [3] from the production consumption cycle of extracting, transporting, processing, applying and waste processing of natural resources, better known as the production-consumption cycle (Figure 1).

Problem recognition and problem analysis
This means, for both government and the industrial community, that they not only have to identify the environmental problems in all their components, but that they also have to acknowledge them! The fact that this acknowledgment of the problems is not yet being given can be illustrated by the reluctant acceptance of environmental problems discussed within the scope of the European Community.

From an environmental-technical point of view the following questions are of current interest:
- What level of the environmental quality is indistinguishable from the original?
- What are the main issues concerning pollutants (dispersal through the various environmental compartments, persistence, eventual wished and unwished accumulation, for example in organisms, reactions with other materials, etc.)?
- What disturbing effects can be expected - in the short and long term - in the environment (dose-effect relationships with regard to relevant environmental constituents/organisms, synergistic and antagonistic effects, etc.)?
- What are the consequences of the short- and long-term effects (permanent damage, required curative measures, etc.)?

Research into analytical field methods and effects will be essential to answer the above questions. The relationship between emissions and the development of long-term environmental demands of the desired technology is of vital importance for drawing up a number of potential plans of action regarding the search for solutions.

Good housekeeping
After the problem has been evaluated, industries have to choose a method of intervention. A large number of good housekeeping measures and adjustments are available. In many instances relatively simple interventions lead to a spectacular drop in environmental emissions. Generally the expenses can be followed and the recoupment times are acceptable to most industries. Cur-

rently such measures are still of very great importance. Not all good housekeeping measures are widely known. It is essential to continue to bring them to the attention of everybody.

Whereas good housekeeping starts for example, with preventing the development of leaking reservoirs of the worst type, and extends to other matters such as noise-abatement, reduction of bad smells or encouraging recycling, it could grow into an approach in which environmental accountancy and auditing would figure.

Add-on technologies

A further step is the installation in the production processes of extra equipment, such as dust and grease traps, washers, filters, purification techniques and the like. In this way emissions of undesirable materials to the air, water and soil are being reduced (the so-called curative (measures). These add-on technologies usually have great benefit for the environment but usually require more preparation time than do good housekeeping techniques. They may also involve substantial increased expenses. Financial recovery also tend to be less favourable.

Legally-enforced environmental standards tend to stimulate many industries in installing add-on technologies. The undesirable emissions from a large number of production processes are being overcome in this way. Nevertheless in every case there remains the problem of the "rest" of the "rest".

A striking disadvantage of a large number of add-on technologies is the difficulty of sustaining the original production process, because of undesirable pollutants remaining in the system. People thinking along economic lines often disagree about this respect. They are, often satisfied with the improved production outline, and they will maintain it in every manner imaginable. This approach tends to consider the add-on measures as unavoidable expenses, required to meet a legally or socially acceptable standard. The measures are therefore included in cost estimates, along with insurance and other overheads.

As society becomes more sensitive to pollutants pressure is maintained and increased on government. The problem of the diffuse load is an example of an aspect which will become obvious and more of a problem in the near future. Even stricter standards will therefore certainly be set for a number of materials. These standards will require additional expenses for industry and may threaten the profitability of some processes. Such a development is naturally undesirable. A dilemma will be created: on the one hand society cannot make do without, on the other hand society cannot accept a decline in the environment.

GOVERNMENT AND THE INDUSTRIAL COMMUNITY ACTUALLY SHARE THIS PROBLEM.
Between them they must decide where and how we set the balance between preservation of the environment and industrial production.

III. SUPPRESSION OF CAUSES OF POLLUTION

General

The solution to the above dilemma will have to be sought in a completely different approach towards the production cycle (Figure 1). First of all we have to decide which elements in the production process must be altered in order to arrive at an environmentally sound solution. Then we must develop methods to improve the process and choose those which are economically acceptable.

Process integrated technologies

Clean technology is the centre of interest in the Netherlands [4] and a start has been made with the application of what is called process integrated environmental technology also called "clean technology". This preventive, more fundamental approach is based on screening the complete production process to improve the overall environmental merit. See Figure 2 for a coherent survey of the relevant components in this respect.

At first sight clean technologies appear to increase costs. However, we have experienced that it often emerges that these costs, viewed in the long term, turn out to be much lower than the costs of add-on techniques. The long term costs are closely linked to the potentials for energy recovery, economics in raw materials and reductions in the costs of removing of waste materials.

It is important to understand that there is a fundamental difference in the time required between add-on technologies and clean technologies. Clean technologies require a much longer time to develop. Setting up a completely new production system may require 10 to 15 years; whereas add-on technologies usually only require 3 to 5 years. The progress of a clean technology will therefore be a highly time-consuming one, and may involve a great deal of risk. The costs may be high. Recovery of these costs may take a long time. The end result is however a totally new production process with considerable competitive force. The process often becomes a saleable product.

Partnership

Government and the industrial community must become partners in the search for solutions to environmental problems. The government has to develop methods for contributing to the costs where these are beyond the resources of the industry concerned. Government must also be active in convincing industry of the long-term merits of clean technology rather than the short-term merits of add-on technologies.

Government and industry share the same objective in process integrated environmental technology. They therefore have to share their knowledge and cooperate in developing strategies which satisfy their different interests.

Only on this basis is it possible to give attention to long-term thinking, whereby everybody will still be answerable for the consequences of their actions.

IV. SOME OBSTACLES
General
It is tempting to assume that the focal point of the environmental--technical evolution is going to shift automatically from add-on technologies to clean technologies. There are however too many differences in attitude for this to happen soon.

The environmental measures that have been achieved so far only came about after the seriousness of environmental pollution had been appreciated and understood, and the political will developed to deal with it. The dilemma of the desire for a better environment, pressing for the government to set high standards, and on the other hand industry's concern for economic feasibility and for limiting measure in question, has lead to prolonged debate and delays. This resulted in the growth of add-on technologies, which have made very large contributions to short-term improvement of the environment. Add-on technologies actually represent symptomatic treatment and are a long way removed from dealing with the cause of the problem. Clean technologies should be able to treat the disease, not merely relieve the symptoms. The add-on technology may however serve as a starting-point and encourage consideration of clean technology within businesses.

Main aspects
Current environmental measures may be considered to the result of interplay of forces between the politically acceptable environmental load an the economically feasible methods of limitation. This interplay of forces, to be characterized as conflict, has in the Netherlands led to conventional, typically short-term solutions of the add-on type.

This trend was, and is being reinforced by a number of factors. Firstly there was in many instances a long period of uncertainty about the actual effects of government imposed rules. This had to be resolved before industry was prepared to commit to any investment for environmental measures. In many cases this uncertainty about the managing and setting of standards led to rather conservative solutions. There were many well-directed and advanced development projects introduced to production processes but it was noticeable that in many cases people were satisfied with a "knitting together" of conventional add-on techniques to create what was presented as a maximal solution rather than directing attention to a better long time solution.

Secondly, there is the problem of a conflict of the generally accepted environmental quality standards and geographic factors in which the location of a process may strongly influence the level of limitation. Local economic factors determining the location of a business often encourage a variety of solutions of differing complexity. Solving all problems connected with the specific location generally leads to application of add-on technologies, rather than to a concern with the somewhat riskier innovations of clean technologies.

Thirdly, there is the problem of the "pollution of the week". The ever increasing stream of cases of environmental pollution throughout the world, stimulates public concern about environmental pollution. Calls for increasingly severe standards for an ever greater variety of pollutants presents those who work in environmental protection sector with (more) new problems. Those cases which featuring a particular chemical constituent lead to a level of concern which government and industry have to take seriously. This tends, at least in the Netherlands, to an accumulation of add-on techniques, modified or not, aimed at that specific chemical rather than the whole process.

Fourthly there are a number of commercial effects which encourage a preference for add-on technologies. Pressures of time play a leading part in the choice of measures. This usually means that the shorter, more direct route is preferred to a probably better but longer route. The attitude giving use to such actions also results in insufficient attention being given to the improvement achieved after the measures have been introduced. This retards the development of valuable and commercially attractive clean technologies. It must be said that a great deal of research is still required in process integrated environmental technology. The fact is that (anno 1987) there are still very few true "clean technologies" available.

A fifth factor, is the tendency to consider only one aspect: a compartment of an environmental issue and to forget the rest. An appreciation of the whole process is an important part of the philosophy of clean technology. It requires an evaluation of the exchange of pollutants between the various environmental compartments. The compartmental approach is very short figured and tends to encourage the apparently relatively simple philosophy of the add-on solution.

A fresh attitude
The possibilities for new add-on technologies is highly dependent on the stages of the life cycle of the actual production process. Incorporation of clean technology from the start takes the component of "permanent environmental merit" into account. In every day practice this involves a different basic attitude. First of all, the production process within the company has to be reanalysed, starting with the following guidelines:
- the use of raw materials and energy consumption should be minimized;

- the production process should prevent, or reduce as much as possible emissions to air, water and soil, as well as nuisance in the form of noise or radiation;
- the product specification will be chosen so that the product causes minimal environmental problems when in use and after;
- failure or damage of the product, apparatus or system may be repaired in a simple way;
- the product, apparatus or system will be produced so that it may be utilized for other purposes after use.

This is a general list of demands that a company must accept, preferably in the early stages of planning of a new process if it is to introduce process integrated environmental technology. Such an approach is a prerequisite to a permanent solution. This does not, however, mean that the options thus obtained can be introduced quickly in a way which is economically acceptable. There are a number of factors which influence the economics of clean technology prominently in this respect. Those that should be mentioned are:
- time required to develop and introduce a new production process or product;
- initial expenses involved in planning and feasibility studies made in order to transform the entire thinking process, research, and adjustments to current schemes;
- discussionand review with experts in other non production fields;
- costs involved in redirective current development plans a.o. replacement and maintenance;
- time taken for the customers to adjust to and to accept the new process and the resulting product;
- uncertainty over market, trends etc. at home and abroad;
- unfamiliarity of the company with regard to possible solutions;
- inaccessibility of data concerning potential solutions.

Only in a careful, integrated evaluation within the triangle of required merit, technical solution and economic requirements will produce clean technologies which are accepted and therefore used willingly by all parties.

V. CONCLUSION

The need for a clean environment is a cause which cannot be stressed often enough. For some time people from all walks of life have been arrived at the conclusion that environmental thinking ought to be an integrated component of industrial thinking [4]. Checking the release of pollutants to the air, water and soil, as well as limiting the immaterial environmental load of noise, radiation and heat emission have become matters of concern to a large number of industries. Choice of methods may vary. The question is not really whether attention is given to preventing environmental disruptions, but rather how people go about it.

As for optimal solutions, during the next few years we must aim for a sensible balance of add-on technologies and process integrated environmental technology. For each production process all the posibilities have to be considered in order to reach short and long-term solutions which meet the demands of society in terms of environmental effect as well as in terms of the economics of the process and its products.

It is of the greatest importance to use the clean technology philosphy as a guiding principle. In reality this means that an attitude is required which focusses on the systematic integration of all possibilities for obtaining a solution. Evaluation of the time required to implement the options is important. If the clean technology options are not easily available than add-on methods may have to be accepted.

Integration of clean technology requires a knowledge of the effects in all environmental compartments (cause and effect chain), drawn from a variety of disciplines (revelatory research) and the relationship between production process and environment (environmental merit). Clean technology often/usually requires that long standing and familiar practices are discarded or modified. There is therefore a need for government to set standards for timing as well as for performance. This requires both industry and government to be aware of the issues and to share a commitment to the goals. The development of standards through debate and conflic is to slow a method.

Government and industry serve the same purpose in development of process integrated environmental technology. Consequently, they have to share their knowledge and to arrive, on the basis of transfer of knowledge, at strategies and guidelines whereby joint interests will be revealed. On this basis only will it be possible to give real attention to long-term thinking and to have everybody answerable for the consequences of their actions. Only along this road does there appear to be a possibility of usefully combining the environmental interests and the interests of industrial production. Process integrated environmental technology is a must to survive!

LIST OF REFERENCES

[1] J. Seymour en H. Girardet, "Far from paradise, the story of man's impact on the environment", British Broadcasting Corporation 1986.

[2] J. Naisbitt and P. Aburdene, "Re-inventing the corporation", Warner Books, 1986.

[3] F. van den Akker, "Recycling Technologies in the Netherlands; an example for Europe", New Technologies - Clean Industry, Dortmund 1985, Institute for European Environmental Policy.

[4] J.A. Don et al, "Dealing with environmental affairs in the Dutch sugar industry", Proceedings of this conference, 1987.

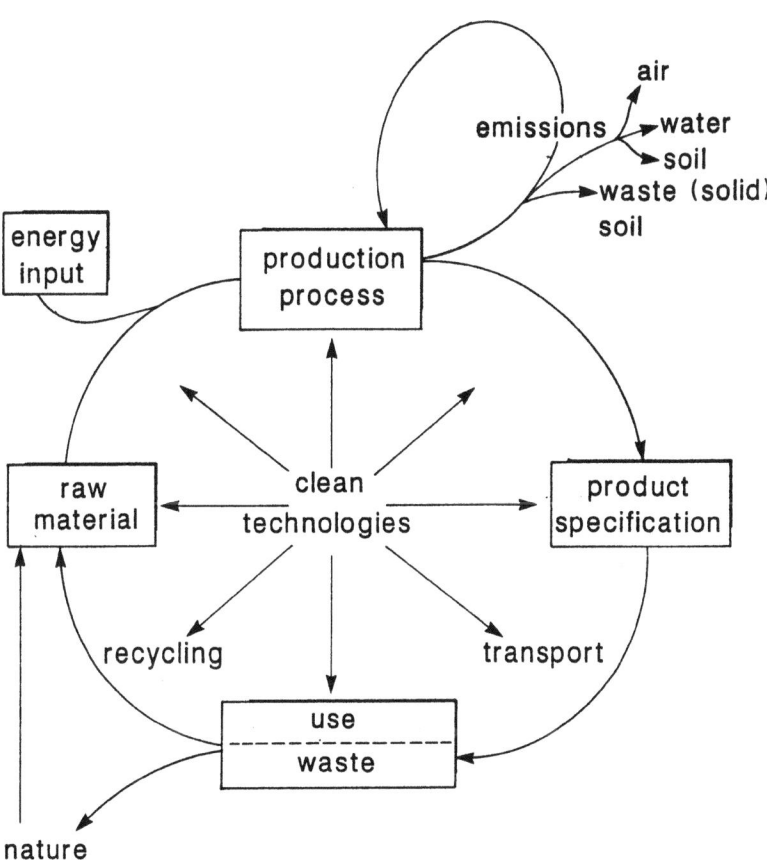

Figure 1 Life-cycle of a product and clean technologies

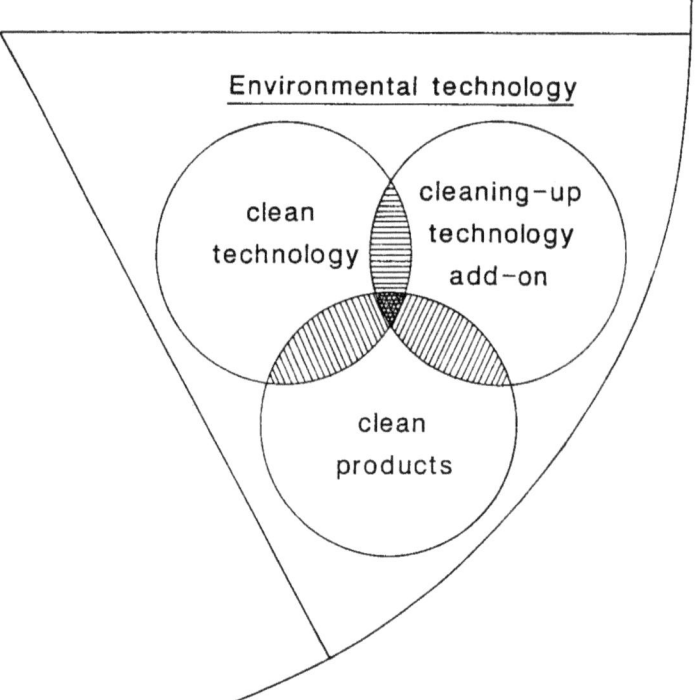

Figure 2 Definition of clean technologies

THE DEVELOPMENT OF CLEAN TECHNOLOGY PHOSPHATE FERTILIZER PRODUCTION PROCESSES

G.H.M. CALIS, DSM Research BV (PT/KG), P.O. Box 18, 6160 MD Geleen, The Netherlands

About 85 % of the world phosphate fertilizer production is based on the use of phosphoric acid, obtained via wet phosphoric acid (WPA) processes. In WPA processes phosphate rock is digested using sulphuric acid, yielding a slurry of phosphoric acid and a calcium sulphate precipitate.
Both the acid and the precipitate are contaminated with impurities from the phosphate rock. Some phosphates have cadmium concentrations leading to unacceptable levels of cadmium in phophoric acid and its byproduct calcium-sulphate. Since phosphate ores vary widely in quality and composition it is often impossible to use ores with low cadmium concentrations, without a serious negative impact on process economy.

To avoid these conflicting interests DSM started a researchprogramme in 1982 for the development of fertilizer production processes, yielding products and byproducts with low cadmium concentrations. This research is being supported by several Dutch ministries, a.o. the Dutch Ministry of Housing, Physical Planning and Environment. This paper deals with the results and conclusions of the labscale experiments, carried out in a common effort of the Delft University of Technology and DSM Research BV.

Investigations on the removal of cadmium from phosphoric acid solutions have resulted in two possible process variants, based on the ion exchange principle and the coprecipitation principle, respectively:

A. Anion-exchange

After addition of complexing halide-ions (I^-, Br^-) to the phosphoric acid solution cadmium is captured in a basic anion exchanger as a negatively charged cadmium complex.
After loading with cadmium the resin is regenerated with diluted phosphoric acid, which can be recycled into the phosphoric acid process.

B. Coprecipitation in calcium sulphate anhydrite (AH)

Sulphate and calcium ions are added to the phosphoric acid solution under such conditions that a cadmium rich AH precipitate is formed. After separation of the precipitate the cadmium content of the solution is reduced considerably. The precipitate can be recrystallized.

Having developed procedures for the removal of cadmium from phosphoric acid solutions, the problem of avoiding the incorporation of cadmium in the bulk of the calcium sulphate byproduct remains to be solved. For this purpose three process variants have been investigated:

1. Addition of chloride in the digestion/recrystallization stage in current WPA processes. This variant is based on the formation of cadmiumchloride complexes, reducing the incorporation of cadmium in the precipitate considerably. It should be combined with cadmium removal from phosphoric acid by anion-exchange.
2. Digestion of phosphate rock in recycled phosphoric acid. In this variant cadmium is removed from the resulting solution by anion-exchange, followed by addition of sulphuric acid to crystallize clean calcium sulphate hemihydrate.
3. Application of either anion-exchange or coprecipitation removing cadmium from recycling phosphoric acid.
This method causes calcium sulphate precipitation at lower cadmium concentrations, yielding a lower cadmium incorporation.

The details of our investigations will be presented in five poster presentations during the conference:
I. Removal of cadmium by anion exchange in a wet phosphoric acid process (T. Tjioe, Delft).
II. A clean technology phosphoric acid process (S. v.d. Sluis, Delft).
III. Cadmium incorporation in calcium sulphate modifications (G.J. Witkamp, Delft)
IV. The reduction of the cadmium content of waste gypsum, produced by the UKF-phosphoric acid plant at Pernis (R. Spijker, DSM Research).
V. Removal of cadmium from phosphoric acid solutions by selective coprecipitation in a minor amount of anhydrite (H. Kroon, DSM Research).

Our results show that production of phosphoric acid and byproduct calcium sulphate with low cadmium concentrations is in principle possible. These results are being incorporated in a feasibility study, investigating the possibilities of the further development of commercial clean phosphoric acid technology.

HYDROLYSIS OF WASTE PAPER

J.I. WALPOT, H. VISSCHER

ORGANIZATION FOR APPLIED SCIENTIFIC RESEARCH TNO

1. ABSTRACT
There are several methods available to convert vegetable cellulose-containing biomass into raw materials for fermentation processes. Two of these methods - acid hydrolysis and enzymatic hydrolysis - will be compared as to their economic merits. For this purpose, use has been made of paper from domestic waste separation installations.
The conclusion can be drawn that the enzymatic route is still considerably more expensive than the acid-hydrolosis route. The most important cause of this is the high cost of enzymes. The cost price of the acid-hydrolysis process per ton of glucose, is of the order of the present market price of melasse (Dfl. 440.- per ton of sugar in the present Dutch situation).
Further reseach into the acid-hydrolysis process is recommended, coupled to fermentation processes for the production of materials like ethanol, butanol, aceton. Paper from domestic waste separation installations, used paper of low quality or other kinds of cellulose-containing waste can be used as raw material for the hydrolysis process. Further research into the possibilities of cheap cellulose production is also recommended.

2. INTRODUCTION
The literature often suggests vegetable biomass, as a substitute for mineral oil, as energy source and raw material for the chemical industry. Various conversion methods have been researched or are the subject of research. For many years research has been carried out in order to find a method to produce glucose, with reference to cellulose, which is both technically and economically feasible.
Generally speaking, a distinction can be made between enzymatic and acid hydrolysis. For both kinds of cellulose hydrolysis an economic comparison is presented, based on recent experimental research into enzymatic paper hydrolysis [1]. Considerations and calculations are based on paper from domestic waste separation installations as the cellulose source, and that because:
(1) Paper is one of the fractions that is released from the processing of domestic waste by means of mechanical separators. This paper has proved to be unmarketable within the paper industry due to impurities (plastics,

sand, and organics). If a useful fermentation raw material can be made from it, via hydrolysis, this would broaden the field of application of domestic waste fractions, thus improving the feasibility of domestic waste separation.

(2) Since paper from domestic waste separation contains impurities (plastics, cloth, soluble organic materials, and anorganic fines), such paper can therefore be regarded as the 'worst sort'. If a hydrolysis process for this is feasible, there is a good chance that it will also be practicable for other cellulose-containing materials.

It should be noted that the nature of the raw material can also be of importance with regard to the components (such as lignin; in the cases of wood and straw, delignification would be necessary).

A cost-benefit analysis has been made for the two abovementioned hydrolysis routes, so that:

(1) it can be indicated where the largest price differences occur;

(2) the glucose prices (cost prices) for the two processes can be compared as a function of external factors (like the melasse price);

(3) an indication can be given of issues with the highest priority, in case of further investigation.

Finally, conclusions and recommendations have been made with regard to the costs expected for the hydrolysis of paper from domestic waste separation installations for either route.

3. PRINCIPLE OF CELLULOSE HYDROLYSIS

On a molecular scale cellulose consists of long series of glucose molecules linked together. Through the addition of water (at the hydrolysis of cellulose) glucose is released. A catalyst helps to accelerate the process. The cellulose enzyme can act as a catalyst (enzymatic hydrolysis). When an acid is used as a catalyst, we speak of acid hydrolysis.

The mutual advantages and disadvantages of the two routes are sometimes very obvious:

(a) Enzymes attach themselves to the surface of cellulose fibers whereby the reaction velocity depends on the free available surface; acid hydrolysis is insensitive to this.

(b) Enzymes work best at an optimum temperature, while acid hydrolysis proceeds more quickly as the temperature rises.

(c) Enzymatic hydrolysis takes place at normal pressure and at pH of abt. 5, in contrast to acid hydrolysis whereby the pH has to be very low and the pressure high.

(d) Enzymes work very specifically, in the case of enzymatic cellulose hydrolysis, glucose is the only end product; acid hydrolysis, however, leads to chain reactions (resulting, among other things, in tarry by-products), making cleaning costs higher.

4. STARTING POINTS

As regards enzymatic hydrolysis the process described by Walpot in 'Conservation & Recycling' [1] was taken as a starting point (see Fig. 1). The process route presented by H.E. Grethlein [2] and Franzidis, Porteous and Anderson [3] was chosen for acid hydrolysis: hydrolysis with diluted sulphuric acid in a plug flow isothermal reactor at 230°C and a residence time of 0.19 min. For the sake of comparison this process was adapted in order to process paper fraction from domestic waste seperation installations (see Fig. 2).

Considerations

(1) For both the enzymatic and acid hydrolysis process of paper we started from the same intake of 12½ tons/hour for 8,000 hours/year.

(2) As a starting point for a proper cost comparison, a 10% hydrolysate solution was taken. In principle, further concentration and crystallization should be the same for both processes. Mention should be made of the fact that with the acid process the hydrolysate is released at a higher temperature than with the enzymatic process which probably results in a lower energy requirement for the concentration process.

(3) The furfural recuperation (about 200 kg/hour) taking place with acid hydrolysis is not taken into account.

(4) Enzyme concentration is 1% with respect to cellulose concentration, and enzymes are recycled for 50%.

(5) The glucose concentration after the acid reactor is a 10% solution, and after the enzyme reactor a 2% solution, which has to be concentrated to 10%.

(6) In the cost comparison an output efficiency of 500 kg glucose per ton of paper for the enzymatic route is assumed, while the acid route has an output of 37% (that is, 370 kg glucose per ton of paper). The annual output in glucose is therefore 50,000 tons (enzymatic route) and 37,000 tons (acid route).

(7) Prices of chemicals:
50% sulphuric acid (delivery per 8 m^3): Dfl. 250.-/ton
Quicklime (unslaked lime)   : Dfl. 160.-/ton
Enzymes   : Dfl. 20.-/kg

Additional provisions must be made to manufacture lime slurry. These provisions are considerably more expensive than the costs involved in the storage of 33% caustic soda.

5. RESULTS

Table 1 shows investment costs for both processes while annual processing costs are given in table 2.

Table 1. Investments based on 1986 prices.

Item	Acid hydrolysis [Dfl.]	Enzymatic hydrolysis [Dfl.]
Unloading/storage of materials	4,550,000	4,550,000
Pretreatment	1,472,000	1,760,000
Dewatering and hydrolysis	2,715,000	3,332,000
Coarse and microfiltration	-	2,148,000
Ultrafiltration	-	3,568,000
Hyperfiltration	-	3,829,000
Neutralization and centrifugation	1,970,000	-
Unforeseen, assembling, and electric installation (50% of abovementioned values)	5,353,500	9,593,000
Total	16,060,500	28,780,000

Table 2. Annual processing costs.

Item	Acid hydrolysis [Dfl.]	Enzymatic hydrolysis [Dfl.]
Interest + depreciation	4,830,000	8,634,000
Maintenance (5%)	805,000	1,440,000
Replacement membrane	-	1,000,000
Manpower: 4x5x80,000 + 3x80,000	1,840,000	1,840,000
Electricity needed for enzymatic hydrolysis 1250 kW		2,000,000
acid hydrolysis 915 kW	1,465,000	
Enzyme costs		10,000,000
Chemicals	880,000	-
Chemicals/general	50,000	50,000
Total	9,870,000	24,964,000
Annual production	37,000 tons	50,000 tons
Costs per ton of glucose	270.--	500.--

The largest differences are found in the items: Investments, maintenance, membrane replacement, enzyme costs, chemical costs (which are much higher for the acid hydrolysis process than for the enzymatic hydrolysis process).

6. DISCUSSION

The difference in cost price per kg glucose between enzymatic and acid hydrolysis is considerable. Grethlein [2] reaches the same relative differences in cost prices between the two routes. The raw material melasse has been chosen for comparison with the market price of glucose solutions from conventional raw materials. The minimum/maximum price of melasse is about Dfl. 200/350.- per ton melasse, with a 50% sugar content. For the sake of comparison we have priced the sugars from melasse at Dfl. 440.- per ton sugar as 50% melasse solution.

Assuming that in the Dutch situation, a domestic waste separation installation in combination with acid hydrolysis and a fermentation step, is built as one plant, the cost price of the glucose production via the acid-hydrolysis process is probably competitive. Even if the transportation costs for the 10% glucose solution are calculated at Dfl. 10.- per ton, the cost price of glucose remains competitive with the cost price of melasse (for the Dutch situation). Should a concentration step to 50% or 100% (pure glucose) be necessary, then the cost price per kg glucose is higher than the market price of melasse, but possibly still competitive considering the market price of sugars from melasse (due to cheaper transportation costs).

7. CONCLUDING REMARKS

- The 10% glucose concentration of the hydrolysate produced by hydrolysis is economically feasible for fermentation processes. Further research is recommended with regard to the glucose solution obtained via the most advantageous route.

- Furfural recuperation can be interesting, but is strongly dependent on the quality, quantities and marketing possibilities. Research into the feasibility of furfural recuperation is recommended.

- The 10% glucose solution seems suitable as fermentation raw material for the preparation of other chemicals, such as ethanol, aceton, butanol, gluconic acid. Further research into the market requirements is recommended.

- The hydrolysis process can possibly serve as a basis for the processing of a number of cellulose-containing wastes. Within this framework it is logical to mention paper from domestic waste separation installations. However, used paper of low quality, straw, roadside grass, agricultural wastes, and possibly even manure, can also be considered as potential raw material. Research into the feasibility of the abovementioned waste processing (possibilities) via hydrolysis seems appropriate.

- On the basis of the economic evaluation a feasibility study of the following process route should be considered:
* separating paper fraction from domestic waste
* acid hydrolysis
* quality optimization of hydrolysate before fermentation
* fermentation into ethanol or other products
followed by investigation into the attractiveness of raw materials for acid hydrolysis, in order to produce basic materials for fermentation.

- It is recommended to look for cheaper enzymes (production methods) because of the numerous advantages of the enzymatic route.

REFERENCES
1. Walpot, J.I.: Enzymatic Hydrolysis of Waste Paper. Conservation & Recycling, Vol. 9, No. 1, pp. 127-136 (1986).
2. Grethlein, H.E.: Comparison of the Economics of Acid and Enzymatic Hydrolysis of Newsprint. Biotechnol. & Bioeng, Vol. XX, pp. 503-525 (1978).
3. Franzidis, J.P., Porteous, A., Anderson, J.: The Acid Hydrolysis of Cellulose in Refuse in a Continuous Reactor. Conservation & Recycling, Vol. 5, No. 4, pp. 215-225 (1983).

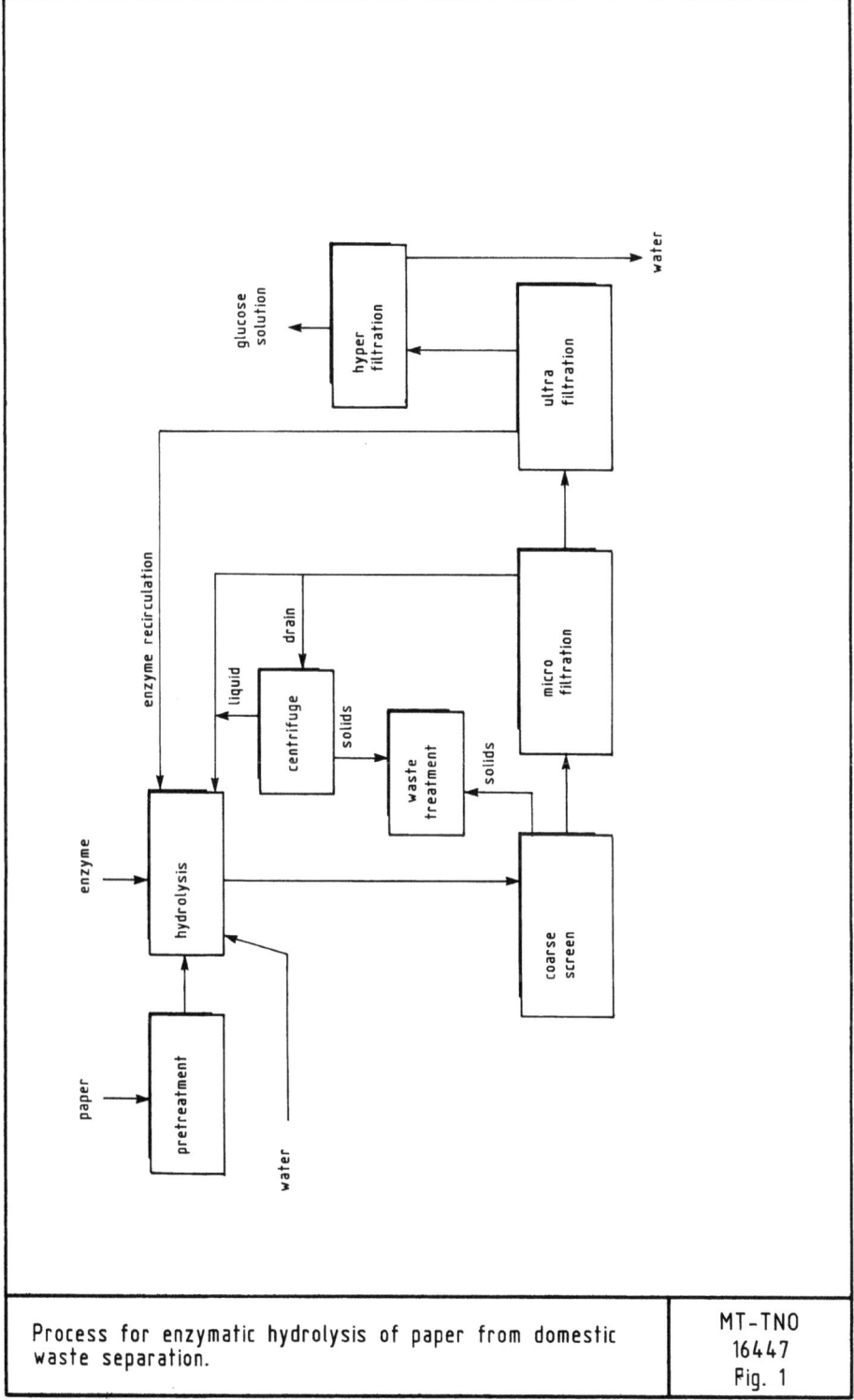

Process for enzymatic hydrolysis of paper from domestic waste separation.

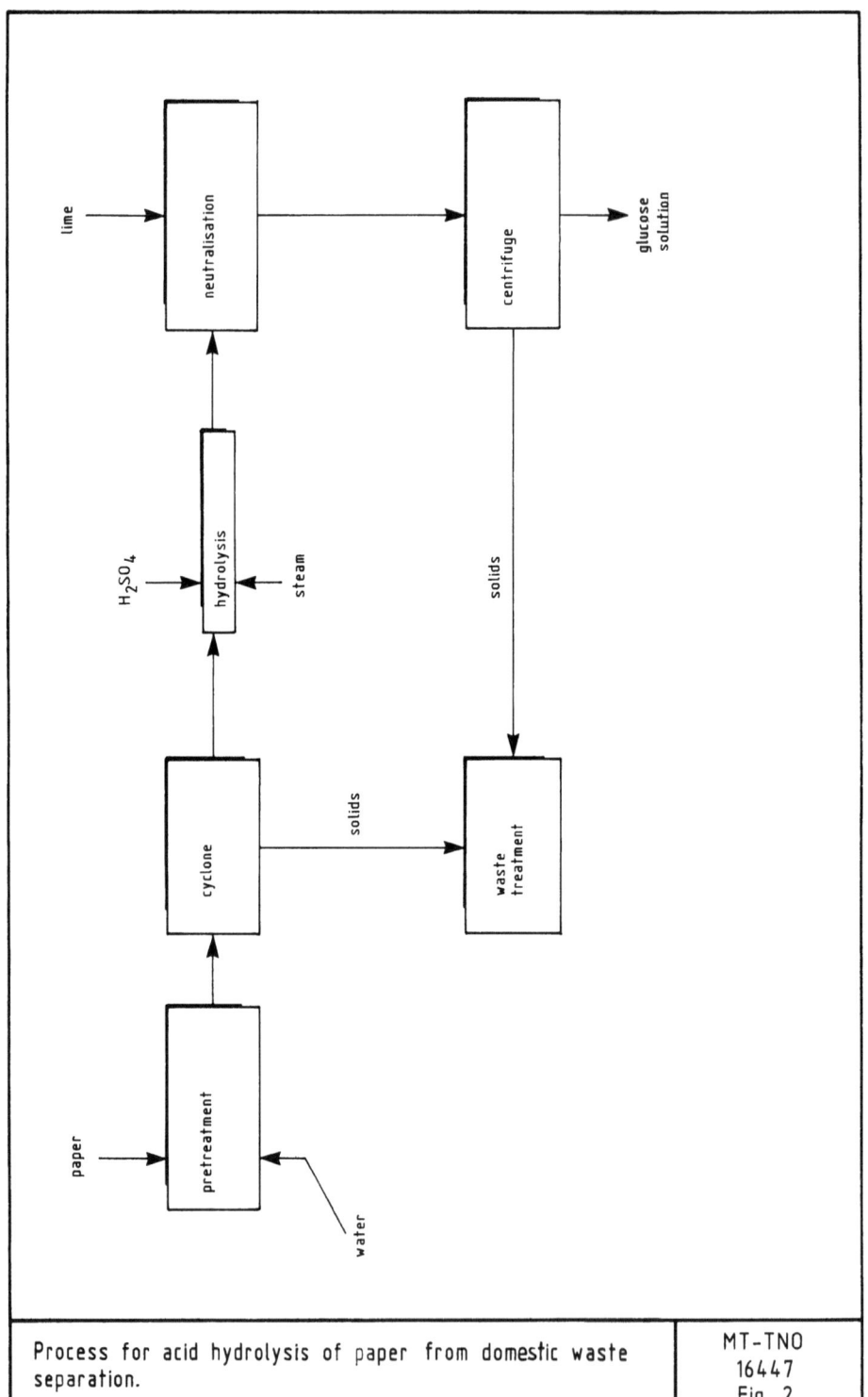

Process for acid hydrolysis of paper from domestic waste separation.

MT-TNO 16447 Fig. 2

The effect of processing on the heavy metal content in compost.

G.R.E.M.v.Roosmalen*, M.M.G.Senden**, T.Brethouwer***, M.Tels****

*	"Waste Magagement", KU Tilburg-TU Eindhoven, P.O.Box 513, BG 3.16, 5600 MB, Eindhoven (The Netherlands).
**	Present address: "Koninklijk Shell Laboratorium", Amsterdam (The Netherlands).
***	N.V.Vuilafvoer Maatschappij VAM, Amsterdam/Wageningen (The Netherlands)
****	University of Technology, Eindhoven (The Netherlands).

Abstract

Mixed collected, mechanically separated domestic waste was composted, applying four different techniques. For each technique applied, the process was monitored in time and the influence of the total time of composting on the heavy metal concentration in the final product was studied.
Waste was composted in a "simulator" (total time of composting (Te) 19 days), on a compost pile with small dimensions and forced aeration (Te=41 days), on a compost pile with small dimensions which was aerated by a turning every 14 days (Te=70 days), and on a compost pile which was aerated by only two intermediate turnings (Te=166 days). For each experiment particle sizes, and heavy metal contents, dry weight, organic content, total organic carbon, and pH, in the total composting material (0-70 mm) and in the small fraction (0-6 mm), were monitored in time.
The particle size analysis showed that the efficiency of the composting (percentage 0-6 mm of the final product), was not influenced by the total time of composting.
The observed heavy metal concentrations indicated a relationship between the total time of composting and the heavy metal concentration of the final product.
Shorter times of composting tend to result in lower Cu, Pb, and Zn concentrations in the mature compost.

1. Introduction

The central production of compost from the organic fraction of domestic waste has a tradition for many decades in the Netherlands. At this moment five plants produce a total amount of 60.000 tons of compost each year as part of their waste treatment activities.
Due to the disadvantageous change in the overall chemical composition of domestic waste over the last decades the heavy metal contents in compost have increased to such a level that its dosage has to be limited which prohibits the use of compost as a successfull soil improver.

From experiments in the recent past (ref. 1,2) it may be concluded that the materials to be composted, i.e. vegetables, fruit- and garden wastes are relatively free of heavy metal contamination. These materials, however, are normally mixed with other, not compostable waste components during storage and collection. This apparently leads to high heavy metal concentrations in the compost after processing.

At the waste treatment plant of the VAM ("Vuil Afvoer Maatschappij"), at Mierlo the organic waste fraction to be composted is obtained from the mixed collected domestic waste by sieving at 70 mm. Ferrous particles are removed by magnetic separation. Obviously the fraction to be composted will contain a certain amount of non-compostables (0-70 mm, non-magnetic) that may prove to be heavy metal sources.

During composting the organic particles become smaller. Non-compostables larger than 6 mm are removed from the final product by sieving.

During composting a number of processes may contribute to the increase of the heavy metal concentration in the dry basis (ref. 3,4):
- loss of dry weight (via CO_2 emission).
- leaching of heavy metals from non-compostables into the organic fraction.
- some non-compostables (6-70 mm) are subject to mechanical degradation during composting and are eventually reduced to a size 0-6 mm. These fragmentized non-compostables are hardly removed from the compost in the final sieving step and will be present in the compost as polluting particles.

The second and third process, which contaminate the organic waste, are believed to be time dependent. A decrease in composting time may well result in lower concentration levels of the heavy metals after composting.

In March 1986 experiments were started at the VAM at Mierlo to study the influence of composting time on the concentration levels in the final compost. The study has been carried out in close cooperation by Waste Management (University of Technology Eindhoven and University of Tilburg) and VAM.

Mechanically separated organic waste was composted applying four different techniques. These techniques were chosen in such a way that for achieving the same level of compost maturity the total times of composting were different. For that purpose the methods of aeration and the dimensions of the composting badges were varied.

For each experiment heavy metal contents, dry weight, organic content, total organic carbon (TOC) and pH were monitored in time in the total composting material (0-70 mm) and in the small fraction (0-6 mm). The particle size of the composting material was monitored in time to determine the influence of time of composting on the efficiency (percentage small fraction in the final product) of the process.

2. Materials and Methods

The organic waste to be composted was separated from domestic waste with a sieve (meshwidth 70 mm). Ferrous particles were removed from the fine fraction with a magnet. This waste was equally divided over the composting methods in the following way:

The waste was collected in a container which was transported to the composting side. Alternately one container was transported to the side of composting with method B, one to the side of composting with method

C, and two to the side of composting with method D. When sufficient waste was transported, the waste of each method was mixed and put into compost piles. Next, from each pile ~40 kg. of waste were removed, mixed, and composted in a "simulator" (method A).

In table 1 for each method of composting the used techniques, the methods of aeration, the dimensions of the waste, and the quantities of waste to be composted, are mentioned. The quantities as well as the dimensions of the piles are indicative. The width of the piles is the width at the base.

	A	B	C	D
technique	simulator (cylindric vessel)	pile	pile	pile
aeration	forced aeration	forced aeration	turnings (at day 13, 27,41,55)	turnings (at day 33,83)
dimensions (LxWxH)	0,3 m3	10x4x2	10x4x2	10x5x4
quantity (kg.)	115	45000	35000	150000

Table 1: Survey of the applied methods of composting.

2.1. Sampling procedure

Before the composting started, 8 samples (~25 kg each) were taken from the organic waste to be composted. From these samples 4 kg were dried, reduced in size, and analyzed (0-70 mm). The remaining of the sample was sieved with a small drum-sieve. The wet weight percentages 0-6 mm, 6-18 mm and 18-70 mm were determined. From the small fraction 4 kg were dried, reduced in size, and analyzed (0-6 mm).

At regular times during the composting, a cross-section was removed from the piles and mixed. From this mixed cross-section 0,5 m3 was separated. 4 kg of this waste were dried, reduced in size and analyzed. Next, the particle size from 100 kg of the remainder was determined. Of the small fraction, 4 kg were dried, reduced in size and analyzed. After this sampling-procedure the piles were turned and put up again.

At regular times, the composting waste in the simulator was mixed. 1 kg. waste was separated, dried, reduced in size and analyzed. 2-3 kg. were sieved with a hand-sieve. The weight percentages 0-6,8 mm, 6,8-13,6 mm and 13,6-70 mm were determined. The small fraction (0-6,8 mm) was dried, reduced in size and analyzed.

For each method of composting the time of sampling (days after the start of the composting) and the number of analyses of the small fraction (0-6 mm, simulator: 0-6,8 mm) is given in table 2.

```
starting material  0(8)
method A           4(1), 8(1), 12(1), 19(1)
method B           13(1), 27(1), 41(3)
method C           13(1), 27(1), 41(1), 55(1), 70(2)
method D           28(1), 55(1), 90(2), 119(1), 166(3)
```

Table 2: times of sampling (days after the start of the composting), and corresponding number of analyses of the small fraction (between brackets).

2.2. Analysis (ref. 5)

The dry matter content was determined by measuring the loss of weight of fresh samples in 24 hours at 105 °C. The organic content was determined by measuring the loss of weight in 4 hours at 500 °C of previously dried samples. The TOC was measured in water extracts from dried samples (105 °C, 24 h) with a TOC analyzer. The pH of these water extracts was measured. The heavy metals were determined by ICP (inductive coupled plasma) after H_2SO_4/HNO_3 extraction (Cr, Cu, Ni, Zn) or by AAS (atomic absorption spectrofotometry) after HCL extraction (Cd, Pb).

3. Results and discussion

The characteristics of the starting material are given in table 3. The mean values, standard deviations, and number of analysis of the dry matter content, the organic content, the TOC, and the pH of the small fraction (0-6 mm) and of the total waste (0-70 mm) are mentioned.

	0-6 mm			0-70 mm		
	mean	s.d.	n	mean	s.d.	n
d.m.	59,6	2	8	48,4	2,2	8
o.c.	35,4	4,8	8	57,8	3,5	8
TOC	22100	4450	7	30700	7950	7
pH	7,1	0,25	8	6,6	0,3	8

s.d. = standard deviation
n = number of analysis
d.m. = dry matter in % wet weight
o.c. = organic content in % d.m.
TOC = total organic carbon in mgC/kg d.m.

Table 3: characteristics of the starting material

The values for dry matter content and pH in the small fraction are higher then those for the total waste (0-70 mm). In the small fraction the organic content and TOC, which are strongly related, are considerably lower. This is caused by the presence of particles with a high dry matter content and a low organic content in the small fraction, like sand and small stones.

During composting the conversion of organics in CO2 and H2O causes a decrease in organic content. This conversion also leads to a decrease in TOC. In the beginning of the composting process the pH will dropp because lower fatty acids are formed. After a while the pH will increase again because NH3 is released.

The dry matter content of the waste will increase due to water evaporation caused by the high temperature. The absolute quantity of dry matter will decrease because of the conversion of organics.

In these experiments it was assumed that the compost was mature at an organic content of 30% dry matter, a TOC lower then 15000 mg C/kg d.m. and and alkalic or neutral pH.

Table 4 gives the total times of composting for each applied technique and for the fractions 0-6 mm and 0-70 mm of the final product the percentage dry matter, the organic content, the TOC, and the pH.

Method	A		B		C		D	
	0-6	0-70	0-6	0-70	0-6	0-70	0-6	0-70
d.m.	69,3	68,6	78,4	66,2	72,0	72,0	70,0	64,8
o.c.	29,4	40,0	29,0	40,3	30,7	45,8	27,8	36,6
TOC	12200	14100	11700	12900	15000	18200	11900	10500
pH	7,09	7,95	7,53	7,22	6,96	7,2	7,40	7,75
Te	19		41		70		166	

d.m. in % w.w. TOC in mg C/kg d.m.
o.c. in % d.m. Te in days

Table 4: characteristics of the final product and total composting times (Te)

Table 4 shows that, according to our definition of maturity, the compost 0-6 mm was mature in all cases. A more extensive presentation of the despatch of the composting is given elsewhere (ref. 6).

3.1. Particle size determination

During composting the particle size of the composting organic waste decreases. This decrease of particle size is caused by the maturing of the composting waste and by the mechanic treatment of the waste (turnings). Both processes lead to an increase of the percentage small fraction (0-6 mm) in the total waste (0-70 mm). The increase in time of the small fraction is shown in figure 1.

The wet weight percentage of the small fraction of the starting material is 19,5%. The mature compost had a percentage small fraction between 29,4% (method D) and 43,2% (method C). As can be seen from figure 1 the velocity of increase of the percentage small fraction is related with the total time of composting. The methods with the lower times of composting show a faster increase of small fraction. The efficiency of the composting (the percentage small fraction in the final product) is not reduced by shorter times of composting.

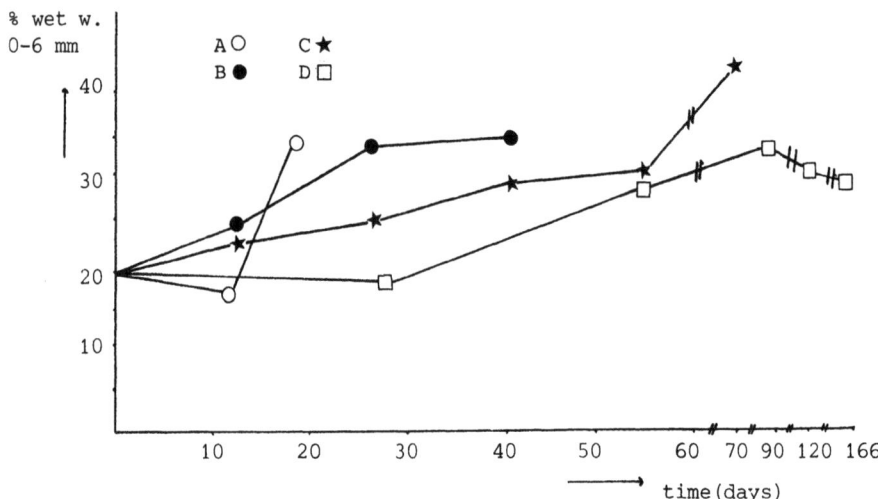

Figure 1: Small fraction in wet weight percentage as function of the time of composting (days).

3.2. Heavy metal concentrations

Of the starting material, the mean values of the heavy metal concentrations, the standard deviations, and the number of analysis of the small fraction (0-6 mm) and the total waste (0-70 mm) are presented in tabel 5.

Due to the inhomogenity of the waste the values for the heavy metal concentrations have a large deviation (up to 30%). It is obvious that the data on individual heavy metal contents should be interpreted with care.

In this paper most attention is payed to the heavy metal concentration in the small fraction. In general the heavy metal concentrations in the total waste (0-70 mm) follow the same pattern. More detailed information is described elsewhere (ref. 6).

	0-6 mm			0-70 mm		
	mean	s.d.	n	mean	s.d.	n
Cd	1,34	0,4	8	1,1	0,6	4
Cr	49	11	8	31	10	4
Cu	68	29	8	53	15	3
Ni	20	9	8	12,5	1,6	4
Pb	358	85	6	293	114	3
Zn	432	160	6	352	101	4

s.d. = standard deviation   n = number of analyses

Table 5: Heavy metal concentrations in mg/kg dry weight, standard deviations and number of analysis of the starting material (0-6 mm, 0-70 mm).

In figure 2 the Cu-concentration in the small fraction is given as function of the time of sampling divided by the total time of composting (T/Te) per technique of composting. During composting dry matter is converted in CO2 and H2O. This conversion increases the Cu-concentration in the composting waste. The data of the Cu-concentration in figure 2 are corrected for the expected increase due to loss of dry matter. In figure 2 observed increases in Cu-concentration can only be caused by the contamination mechnisms leaching and/or mechanical degradation to a size smaller then 6 mm of non-compostables.

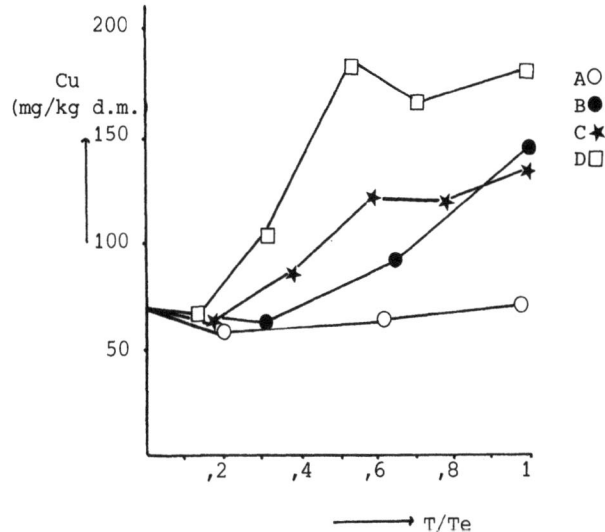

Fig. 2: Cu-concentration in the small fraction as function of T/Te (data are corrected for loss of dry matter).

No significant increase of the Cu concentration was measured in the composting waste in the simulator (method A). The waste composted with methods B and C showed an increase of Cu content to end values of roughly 140 mg/kg d.m.. The Cu content of the waste composted with method D increased to 180 mg/kg d.m.. For the increase in time of the Cu concentration of the different methods of composting the following tendency seems to be present: A < B = C < D.

Figure 3 shows the Pb-concentration in the small fraction as function of T/Te per technique of composting in the small fraction. The data are corrected for the expected increase of the Pb-concentration due to loss of dry matter.

No significant increase of the Pb concentration was measured in the waste composted with methods A and C. The waste composted with method B showed a fast and high increase of the Pb-concentration in time to 762 mg/kg d.m. at day 41. This high increase of the Pb-concentration was not observed in the 'total waste' (0-70 mm). In the 'total waste' (0-70 mm) of method B the Pb-concentration at day 41 was ± 450 mg/kg d.m.

The strange despatch of the Pb-concentration in the small fraction of this method of composting could not be explained.
The waste composted with method D showed a significant increase in Pb-concentration in the small fraction to 579 mg/kg d.m. at day 166.

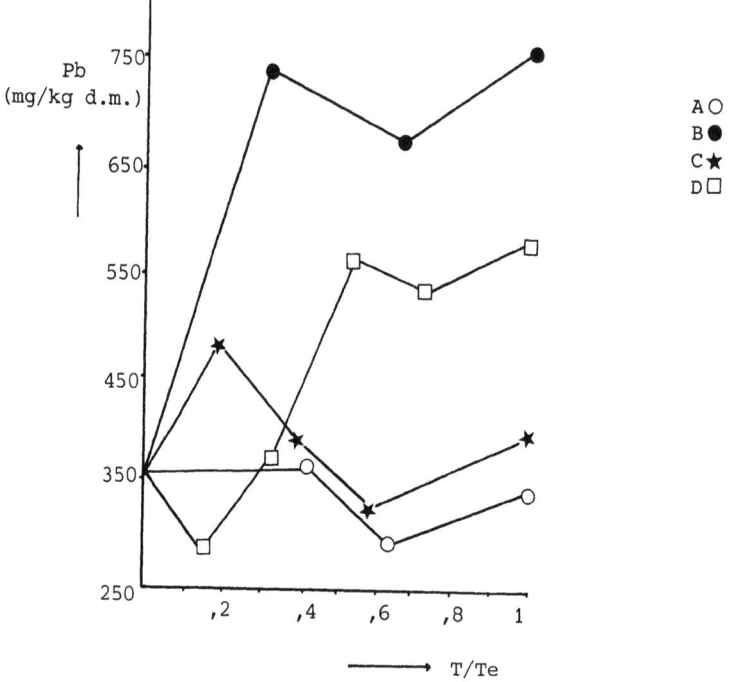

Figuur 3: Pb-concentration in the small fraction as function of T/Te (data are corrected for loss of dry matter).

For the increase in time of the Pb-concentration of the different methods of composting the following tendency seems to be present:
A = C < D < B.

Figure 4 shows the Zn-concentration in the small fraction as function of T/Te per technique of composting. Once more, the data are corrected for the expected increase of the Pb concentration due to loss of dry matter. No significant increase of the Zn-concentration was measured in the waste composted with method A.
The waste composted with methods B and C showed an increase of the Zn-concentration in the small fraction to respectivily 567 mg/kg d.m. and 521 mg/kg d.m.
The Zn-concentration of the waste composted with method D increased to 732 mg/kg d.m. at day 166.

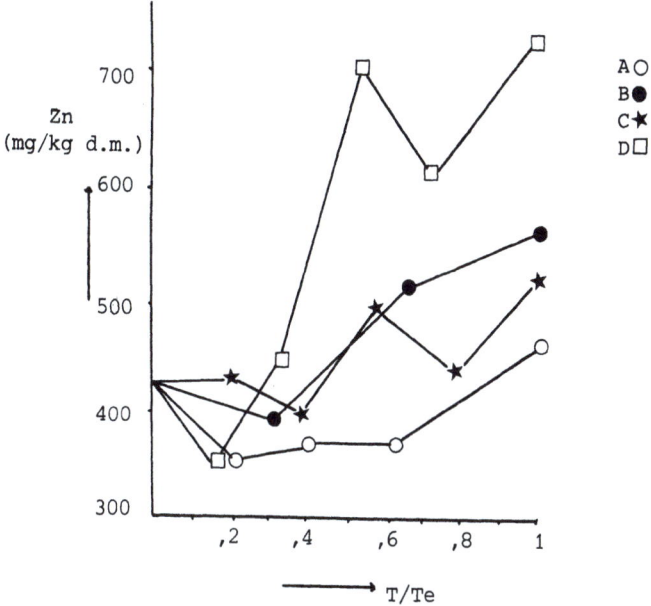

Fig. 4: Zn-concentration in the small fraction as function of T/Te (data are corrected for loss of dry matter).

For the increase in time of the Zn-concentration in the small fraction the following tendency seems to be present A < B = C < D.

For all used techniques of composting the concentrations of Cd, Cr and Ni in the small fraction showed no significant increase in time.

3.3. Summary

Table 6 gives the heavy metal concentrations of the final product for each technique of composting, the mean values of the compost normally produced at Mierlo, the total times of composting and the number of analyses.

method	A	B	C	D	Mierlo (ref.4)
Cd	1,44	1,35	1,89	1,2	1,89
Cr	47	62	44	58	42,8
Cu	76	160	158	204	230
Ni	14,6	30	22	28	22,2
Pb	345	838	425	649	641
Zn	461	624	557	820	751
Te	19	41	70	166	+120
n	1	3	2	3	28

Table 6: Heavy metal concentrations of the final products and of the compost normally produced in Mierlo (in mg/kg d.m.), the total times of composting (days), and the number of analyses.

The heavy metal concentrations (all of the final product 0-6 mm) are not corrected for loss of dry weight.

The data in table 6 show a relationship between the total time of composting and the heavy metal concentration in the final product. Except the waste composted with method B - especially the Pb-concentration in the compost produced with this method - shorter times of composting tend to result in lower Cu, Pb and Zn concentrations in the mature compost.

4. Conclusions

More advanced techniques of composting (for instance use of forced aeration) result in a decrease of the total time of composting. When the waste is processed according to the system used by the VAM at Mierlo, where mature compost is purified from non-compostables by sieving at 6 mm., shorter times of composting may lead to lower Cu, Pb and Zn concentrations in the final product.
Shorter times of composting do not reduce the efficiency of the composting process i.e. the percentage small fraction in the mature compost is not influenced by the time of composting.

Literature
1) J.W.A. Lustenhouwer, Zware metalen in compost uit huishoudelijke afvalstoffen, IVAM, Universiteit van Amsterdam, 1985.
2) G.R.E.M. van Roosmalen, Zware metalen in huishoudelijk afval en compost, Waste Management KUB-TUE, 1985.
3) G.R.E.M. van Roosmalen, J.W.A. Lustenhouwer, J. Oosthoek, M.M.G. Senden. Heavy metal sources and contamination mechanisms in compost production. Proceedings of MER 3, Antwerpen, 1986.
4) G.R.E.M. van Roosmalen. De invloed van enige afvalcomponenten op het gehalte aan zware metalen in huishoudelijke afvalcompost, Waste Management KUB-TUE, 1985.
5) W. Hamminga, Voorschriften voor de bepaling van zware metalen, droge stof, organische stof, pH en totaal organische koolstof in compost, VAM Wageningen, 1986.
6) G.R.E.M. van Roosmalen, Bepaling van de invloed van de composteringstijd op de effectiviteit van de mechanische nabewerking en samenstelling van de compost, Waste Management KUB-TUE, 1985.

AUTOTHERMAL INCINERATION OF WASTE WATER SLUDGE IN A FLUID BED FURNACE

Ronald Tize.

1. SLUDGE TREATMENT IN BELGIUM - SURVEY.

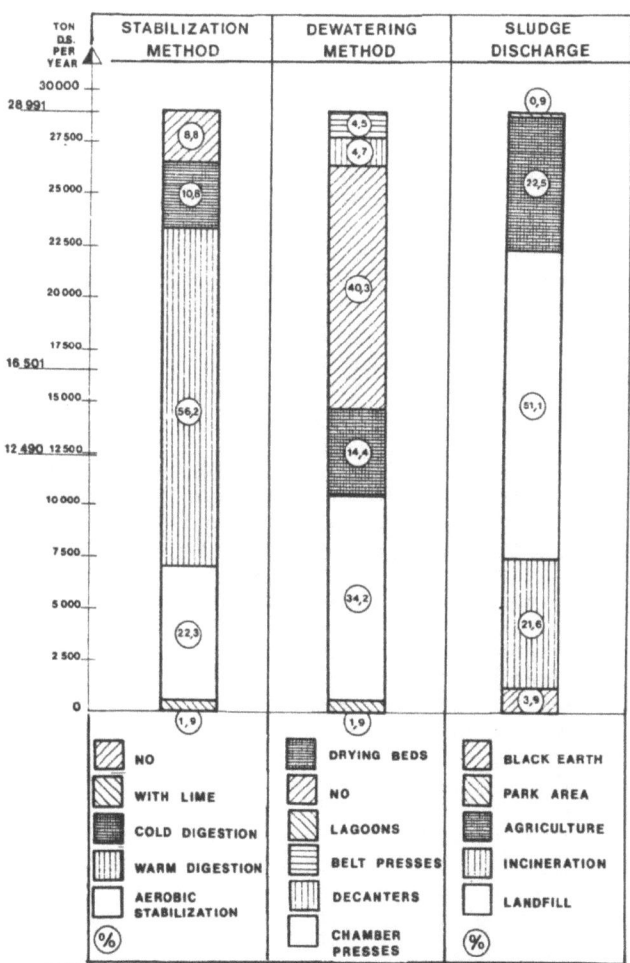

TREATMENT METHODS FOR SLUDGE IN BELGIUM

2. DESCRIPTION OF THE PLANT IN BRUGES.
2.1. Design considerations
2.1.1. Basic data

	Design	Operation
Connected population	305.000 I.E.	+/- 200.000 I.E.
. domestic	189.100 I.E.	-
. industrial	115.900 I.E.	-
Produced sludge		
. mixture of prim. + sec.	90 g/I.E. day	86,4 g/I.E. day
Characteristics of the sludge :		
. DS content		
- from thickener	4 %	4 - 9 % DS (7 %)
- from decanter	24 %	21 - 31 % DS (28 %)
. organic DS content	62,8 - 69 %	58 - 68 % DS (65 %)
. calorific value ODS	5000 kcal/kg ODS	5000 - 5800 kcal/kg ODS

From the preceding table two conclusions can be drawn immediately.
1. There are differences between the predicted and the final values, which can be of importance.
2. The predicted values are not constants, but VARIABLES.
Experience with this plant has shown that these VARIABLES can VARY RELATIVELY FAST.

   2.1.2. Object pursued by the operator. The plant to be built should be **reliable** and easily adaptable to the parameters of the sludge that needs to be incinerated.
This means that it should have **enough reserves** to process a relatively wide variety of sludge qualities.
It should allow flexible operation and **fast reaction** to relatively sudden changes in sludge qualities.
It should **allow for** autothermal incineration.
An important saving can be made by cutting down on fuel consumption. Saving expenses on auxiliary fuel will entail a considerable cut in operating costs. Auxiliary fuel can amount to 30 % of these costs.

   2.1.3. **How was this object realized ?** In order to obtain a reasonable saving, the shortage of energy, necessary to allow autothermal incineration, must be made up for within the system itself. There are 2 technical possibilities.
1. Raising the air pre-heating temperature to a maximum and thus feeding extra energy into the bed.
2. Increasing the dry substance content before feeding it into the furnace, which means an increase in the calorific value of the sludge supplied.

Both recoveries should of course be made out of the heat from the incineration reaction.
Thermal calculations and technical considerations proved that this energy recovery from flue gases is possible and the recovery had to be done on two levels.
- On a high level - flue gas temperature = 850-900 degrees Celsius
- On a low level - flue gas temperature = 300-320 degrees Celsius
With this modular structure it is possible to switch on or off the energy recovery for drying part of the sludge.

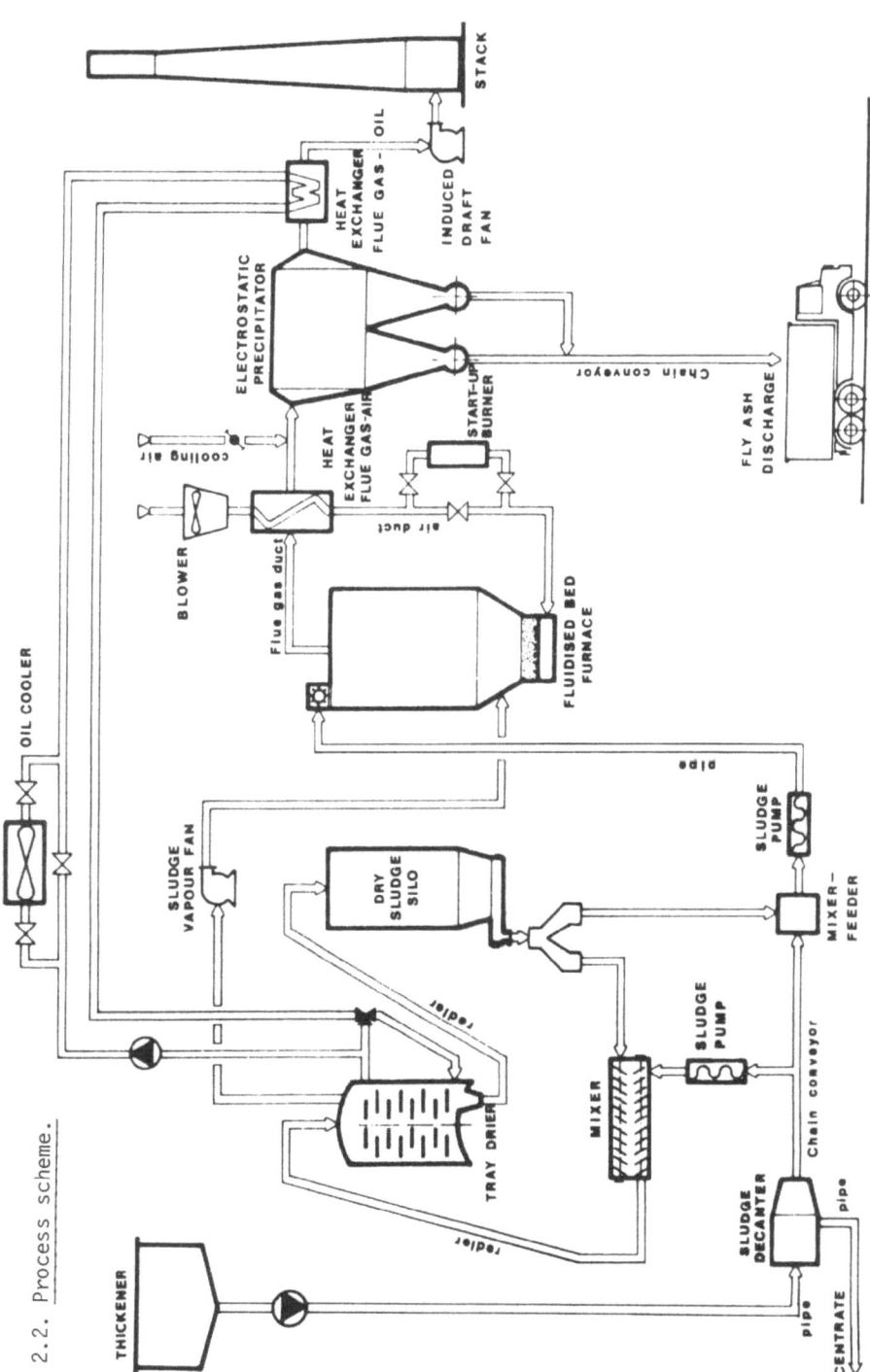

2.2. Process scheme.

2.3. Description of the different elements.
   2.3.1. Centrifuges.
- KHD cylindrical centrifuge
- capacity - 2 x 30 m3/h
The difference in speed between the bowl and the screw is manually adjustable.
   2.3.2. Furnace.
      2.3.2.1. Dimensions.

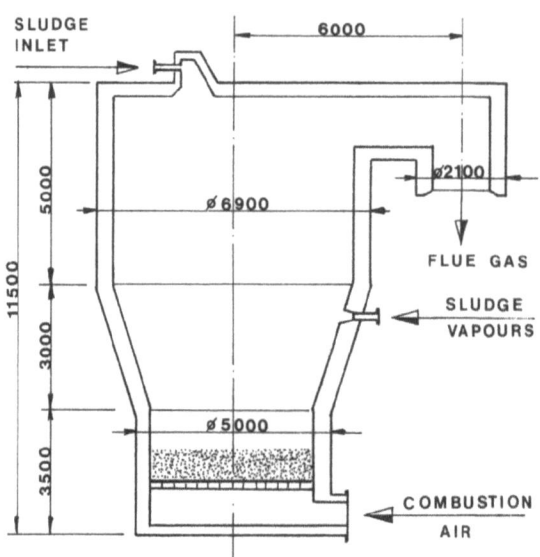

The furnace is one of the three largest ever built in Europe. Because of the high air temperature of 650 degrees Celsius special construction demands were made upon the windbox, the air distribution plate and their disposition with sealings (in connection with differential expansion with respect to the cold furnace wall).

2.3.2.2. Air distribution plate.
The air distribution plate was specially designed to enable
protection against all mechanical forces upto 700 degrees Celsius.
Therefore, a material had to be chosen that does not suffer from
crystalization of the σ-phase (between 600 and 920 degrees Celsius).
The air distribution plate has to guarantee an equal distribution
of the air over the sand bed. This is facilitated by making use
of injectors with a high pressure loss.
Systems with an air distribution by means of pipes instead of a
distribution plate cannot guarantee an equal distribution over the
injectors.

2.3.2.3. Injectors.
The choice of material here is also of importance, a high quality
heat resistant alloy being selected.
The outlet gaps for the air are constructed in such a way that sand
does not enter the windbox when the air suppressor is not running.

2.3.3. Drier

MULTI-TRAY DRIER

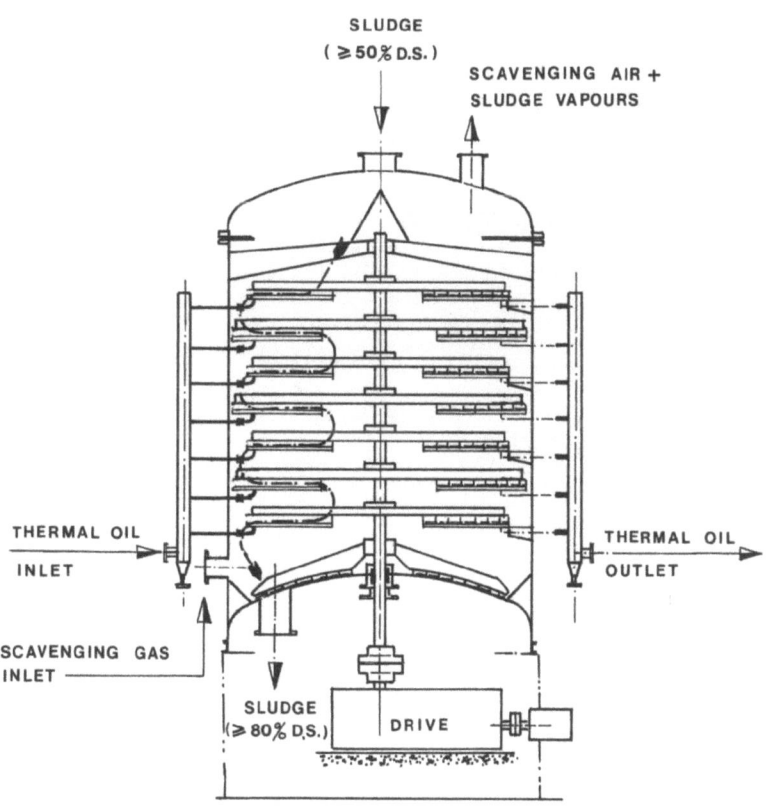

Casing : - The steel casing consists of a cylindrical vessel with
dished bottoms.
Trays : - The trays are jacketed and the heat carrying medium is
circulated in the jacket.
- The upper plate of each tray is constructed in a thicker
material in order to prevent against wear.
Central mechanism :
The central shaft is supported by a combined thrust bearing and
spherical roller bearing with common focus. This enables perfect
vertical alignment of the central shaft. The bearing on top is a
bush bearing which allows a free expansion of the shaft.
At each tray 4 raking arms are provided moving on top
of it and bolted to the shaft. The sludge is moved on the trays
and transported radially from one tray to another.
Sludge inlet :
The mixed sludge (incoming sludge together with recycled dried
sludge to a mixture of about 60 to 70 % DS) is fed in the middle
of the dryer on the upper tray.
Sludge outlet :
The dried sludge is discharged over the complete circumference of
the underneath tray and dropped on the bottom head of the dryer.
A special scraper, fixed at the central shaft, moves the sludge to
a discharge opening.
Connections for steam and condensates or thermal oil :
Feeding and discharging of thermal oil or steam is done for both
by two parallel connections at each tray.

2.3.4. Heat exchanger flue gas - air.
This heat exchanger is of a special design for dust laden gases.
The heated air flows in counter currents through pipes, which have a
special form to avoid clogging.

2.3.5. Heat exchanger flue gas - oil.
Thermal oil flows through the pipes in counter currents with the
flue gases. Three packages in series heat the oil from 187 to
243 °C.

3. OPERATIONAL EXPERIENCES.

3.1. Start-up.

The plant in Bruges operates continuously for periods of 5 days, on an intermittent basis only due to lack of sludge from the waste water treatment plant. As the system is autothermal once in operation, fuel is needed to bring the furnace to temperature during the start-up. A cold start-up takes about 7 hours and uses about 1,000 liter of fuel. A start-up after a shutdown over a weekend can be done in 3 hours and uses about 400 liter of fuel.

3.2. Decanters.

After optimalization, the use of polyelectrolytes (organic - cationic) was lowered to 2.5 kg per ton of dry solids.
The dry solids content of the centrate is about 2 kg of dry solids per m3 (dissolved salts not included).

3.3. Furnace.

The furnace operates without any problem.

3.3.1. Use of sand.
- Only 0.5 ton of sand per 200 T of DS.
- No accumulation of minerals in the bed.
- No slagging in the bed.

3.3.2. Bed height.
The height of the bed was reduced in order to minimize the power consumption. Elutriation is still perfect.

3.3.3. Combustion process.
The difference between the temperature in the bed and in the freeboard is much smaller than in most conventional furnaces. This proves that incineration of the sludge takes place in the bed an not above it in the post combustion chamber. The high combustion air pre-heating temperature ensures a stable combustion process.

3.4. Drier.

It is important to move the sludge in thin layers over the plates. A shorter residence time is of less importance. A good drying process needs good heat transmission values and thus needs a regular contact between all sludge particles and the heating surface.
A tight adjustment of the scrapers above the trays and an adequate number of scrapers moving the sludge backwards on the plates is very important. Thin layers and a good rolling movement of the sludge will produce a crumbly structured sludge. In this way formation of dust in the vapours is minimized.

4. ENVIRONMENTAL ASPECTS.
4.1. Ashes.
4.1.1. Volume reduction.
A tremendous volume (weight) reduction is reached (92 % for sludge coming from the decanter).

Basis : 1 kg of sludge from thickener
       65 % organic dry solids in total dry solids.

	after thickening	after decanter	after predrying	out of furnace
ODS	0.039 kg	0.039 kg	0.039 kg	-
MDS	0.021 kg	0.021 kg	0.021 kg	0.021 kg
H2O	0.94 kg	0.20 kg	0.122 kg	-
TOTAL	1 kg	0.26 kg	0.182 kg	0.021 kg
% DS	6	23	33	100
weight red. from thick.	0 %	74 %	81.8 %	97.9 %
weight red. from mech. dewatering		0 %	30 %	91.9 %

4.1.2. The ashes are inert.
The ashes are discharged out of the electrostactic precipitator in the form of fly ash. The content of organics in the fly ash is less than 1 %.

4.2. Flue gases
Flue gas composition compared to TA Luft (Germany).

	average	TA Luft 1986 (in project)
dust	< 50 mg/Nm3	30 mg/Nm3
Cl	21.2 mg Cl/Nm3	50 mg/Nm3
F	1.5 mg F/Nm3	2 mg/Nm3
CO	20.7 mg/Nm3	100 mg/Nm3

Except for dust, the flue gases already meet the TA Luft standards (in project).

4.3. Heavy metals
Concerning heavy metals, two aspects are of importance :
1) The heavy metals are concentrated in the fly ash for a factor of about 3.
2) The heavy metals are contained in the fly ash.

Because of the basic character of the fly ash, the danger of leaching heavy metals with drain water in a landfill situation is almost negligible.

5. POSSIBILITIES OF THE SYSTEM

The system is built up in a modular way.
1) The furnace.
2) Predrying of the sludge.
3) Storage of dried sludge (can be used an an auxiliary fuel).

In function of the quality of the sludge that has to be incinerated, the fraction of dried sludge can be regulated and consequently the calorific value of the sludge fed into the furnace can be controlled. In practice, the system works autothermally for the following cases :

active sludge
System 1 : thickener - chamber filter press - furnace (if calorific value of organic dry solids is in the normal range).
System 2 : thickener - decanter - furnace - dryer.
System 3 : thickener - filter belt press - dryer - furnace.

digested sludge
System 4 : thickener - chamber filter press - dryer - furnace.
These 4 systems allow the incineration of sludge without the use of additional fuel.

6. COSTS OF SLUDGE INCINERATION.

The overall cost for sludge incineration following the "SEGHERS ZEROFUEL COMBUSTOR" system varies between 400 and 650 Dutch guilders per treated ton of dry solids, depending on the quality of the sludge and its dry solids content.
This cost includes the following items :
(the division in % is given for the plant of Bruges)

- Depreciation of investment costs	
. civil work (25 years - 10 %)	3.2 %
. electromechanical (10 years - 10 %)	47.5 %
- Personnel	23.3 %
- Chemicals	8.1 %
- Maintenance products + spare parts	5.8 %
- Electrical power consumption	6.5 %
- Fuel for start-up (discontinuous operation)	0.4 %
- Sand	0.1 %
- Transport of fly ash to disposal	3.5 %
- Dumping costs	1.6 %
	100 %

7. CONCLUSION

The system is economically justified and the costs are comparable with those for mechanical dewatering, transport and dumping. The described system is environmentally acceptable, only inert ashes and flue gases with water vapour leave the plant. All aromates are destroyed and no waste water is produced.

DEALING WITH ENVIRONMENTAL AFFAIRS IN THE DUTCH SUGAR INDUSTRY

J.A. Don, L. Feenstra
Dutch Organization for Applied Scientific Research,
Division of Technology for Society
Department of Environmental Technology
P.O.Box 342
7300 AH Apeldoorn

SUMMARY

Various environmental problems occurring in the Dutch sugar industry are topical.

Dealing with water pollution and odour emission is described in this article. Characteristic are the conscious problem analysis beforehand and the emphasis on process integrated solutions. This method of approach results in the end in much more effective solutions than the application of add-on cleaning techniques as such. Thinking in the long(er) term with regard to the abatement of environmental load of the Dutch sugar industry has been reached through intensive consultation between government and industry. The combined effort on the optimum abatement plans can be a model for the solution of other environmental problems, also in other industrial branches.

I. INTRODUCTION

Since the beginning of the sixties an increasing insight into the consequences of activities which pollute the environment, both in the short and the long term, has led to an increasing consciousness for environmental questions. As a result a large number of measures have been taken and are still being taken in order to reduce the emissions of the industry in the short term. Besides this, the philosophy that prevention is better than cure is now leading us to a situation in which process-integrated environmental solutions are actually being applied, and an inherently clean technology which can be applied in the longer term is being developed [1].

The Dutch sugar industry was confronted at a very early stage with the necessity to implement environmental measures. As early as the late fifties the waste water question caused to reflect on possible measures. Around 1970 one also began to pay attention to the reduction of noise production in certain parts of the process. Today one of the current matters is dealing with odour emissions. In this publication a further explanation of the approach to two environmental problems will be given. The developments relating to water pollution and odour in the sugar industry serve as an illustration of a strategy for a generally applicable solution.

Today there are 9 sugar factories in Holland, with a total slicing capacity of approx. 80,000 tons of sugar beet per day. The sugar beets are processed in the autumn during a campaign which lasts approx. 90 days. In the sugar manufacturing process the sugar beets are first washed, sliced and then extracted. The juice which is produced in this way is then purified by adding limemilk and CO_2. Sugar is obtained by evaporation of purified juice and by crystallization.

The very simplified process scheme given in fig. 1 shows that sugar beets are not converted into sugar in a closed system. Pulp, limecake and molasses are by-products which can be sold on the market. However, the muddy water, the condensate and the emissions at many stages of the process eventually have to be discharged. These are the most important streams of waste products for the environmental questions in the sugar industry.

II. MANAGING THE WASTE WATER PROBLEM

Recognition and analysis of the problem

The waste water streams from a sugar factory are:
- the water for transporting and washing the beet (approx. 7500 m³ per 1000 ton beet);
- the condensor water (approx. 5000 m³ per 1000 ton beet);
- the condensate surplus (approx. 350 m³ per 1000 ton beet).
- various smaller waste water streams.

Until approximately 1960 this waste water was discharged untreated. At that time the water pollution of a sugar factory amounted to approximately 75 p.e. for each ton of processed beet per day. This pollution was caused mainly by the loss of sugar from the process. When the government began with the preparations around 1960 for an act in surface water pollution which would result in discharges being limited and taxed financially, developments were started up in the sugar industry to deal with the waste water problem. A closed recirculation system was chosen as the set-up for this purpose, in which the water would have to be purified during the campaign. In order to limit the purification measures the sources of sugar loss had to be tracked down.

Measurements made around 1960 showed that approximately 0.5% of the weight of the beet was lost in the waste water in the form of sugar. Of this approximately 0.2% arrived in the muddy water. The latter loss occurs as a result of damage to the beet when washing, which results in sugar dissolving in the washing water. Due to rapid ecological degradation of the dissolved sugar a quantitative approach of the sugar problem was difficult. Therefore, first of all, the analysis methods were perfected. As a result, a good insight could be obtained into the sources of sugar loss and water pollution with the aid of extensive measurements.

Good housekeeping

Due to good housekeeping and the implementation of a programme designed to avoid unnecessary losses the figures for losses via the waste water have dropped from 0.5% to 0.1-0.2%.

Add-on technology

On 1 January 1971 the Act on Surface Water Pollution came into effect in the Netherlands. Despite the fact that the pollution had dropped by approximately 50% between 1960 and 1970 as a result of remedial measures the remaining pollution nevertheless involved a considerable sum in levies. The sugar industry therefore decided to introduce the formerly set-up recirculation system and to clean the waste water. After separating the mud in a mud thickener and subsequently on mud fields, the waste water was purified in the first instance in aerated lagoons. Figure 2 shows a breakdown of the treatment of the muddy water.

In 1970 research into the best method of water purification was far from completed. There were therefore frequent problems in the early years. The mud field in particular caused serious odour problems as a result. By optimalizing the design and the processes for the waste water system a considerable improvement in the situation has arisen. One of the important factors was that the time that the water remained in the washing water circuit was limited to approximately 2 hours, and the time in the purification circuit to 2-3 days.

Process-integrated environmental technology

The aerated lagoons which were applied had important disadvantages, such as a relatively large ground surface area, high energy costs of surface aerators and, to a lesser degree, noise problems.

These aspects led to initiatives to develop new purification techniques. Anaerobic purification in particular seemed to offer perspective as a pre-purification stage, with a yield of 70 to 80%. In anaerobic purification no energy for aeration is required, while furthermore methane is produced which can be used at other places in the process. In 1978, after several years of research with increasingly larger apparatus, the first anaerobic reactor was put into operation on a practical scale. This technique has led to a further reduction of discharge onto surface water, which had already been reduced considerably [3]. As no nitrogen is removed from the waste water in anaerobic purification, the waste water effluent undergoes another final purification by means of an aerobic active sludge installation.

Not only the muddy water, but also the condensor water (see fig. 1) had to be integrated in the long term in a closed system in all factories. The government wanted to limit the discharges of hot water from the mixing condensors of the boiling vapour vacuum system. For this purpose cooling towers and cooling lagoons were installed. In some stages of the process it appeared that it was possible to bring about a certain part of the cooling by means of heat transfer to other process streams.

In the present situation the water circuit can be considered to be virtually closed. As a result of the combinations with the process the original add-on technology is totally integrated in the production process. As a result of this it is difficult to distinguish between a production step and an environmental measure : the stage of process-integrated environmental technology has been reached.

Evaluation of the approach

When solving the waste water problem the approach described above [1] was followed, which consisted of:
- recognition and analysis of the problem;
- good housekeeping;
- add-on technology;
- process-integrated environmental technology.

The policy of the central government was one of the factors which stimulated the progress through these phases. In particular, the preparaion and implementation of the Act on Surface Water Pollution in 1970 was of essential importance. Figure 3 shows that the water pollution of the sugar industry in the period between 1960 and 1970 was halved. After that, however, the emission shows a drop which is considerably faster: in two years' time the discharge was reduced to 30% of the 1970 level.

At the moment the waste water problem is fully under control. It took a period of 20 years, however, to reach this situation. The need to develop new technology can be considered to be the most important factor for this lengthy period of time.

As a result of the approach to the water discharges sugar factories have become more complicated. In the event of defects certain parts of the waste water system are even critical for the total production process, such as the mud thickener and the condensor water cooling.

III. DEALING WITH ODOUR EMISSION
Recognition of the problem

Originally the odour from sugar factories came mainly from the inland waters (canals, rivers), in which microbiological decomposition processes occurred as a result of the discharge of uncleaned water. This is one of the aspects responsible for the investigation into possibilities for reducing the discharges onto surface water. After 1960 this source of odour decreased considerably. In the sixties there was also an increasing public consciousness with regard to odour as an environmental problem. Odour from sugar factories was increasingly considered as pollution, and not as a kind of natural phenomenon which was simply part of the autumn. Furthermore, after 1970 the odour from the sugar industry came into publicity several times as a result of incidental odour waves due to defects in the newly installed water purification facilities (mud fields, aerated lagoons).

The shift in the public acceptance of environmental pollution also occurred in other places and for other industries. For example, the World Health Organization also began to consider odour problems as a factor which could influence human health (in this case: welfare). In the densely populated area of the Netherlands TNO has carried out much work with the support of the central government in order to make a quantitative approach to odour problems possible [4].

As a result of extensive discussions between the authorities and the sugar industry the decision was made in 1983 to carry out an extensive investigation into the most important causes of odour in the area. The obligation to carry out this investigation was included in the licence conditions. The entire Dutch sugar industry gave TNO an order to carry out this investigation. The order also included the request to draw up possible abatement plans and to find out what the consequences would be. An extensive publication of the technical results of the investigation will soon be issued [5].

Results of the investigation survey

The investigation began with a thorough analysis of the processing situation, as it was realized from the beginning that the most attractive abatement methods would have to be found in the process itself. For each factory the contribution from approximately 30 sources to the odour in the area was investigated. The influence of the seasons, of beet quality and of various process variables was also investigated. In order to be able to obtain all the necessary details new odour sampling techniques had to be developed and applied. In the end a reasonably sound survey of the situation in the 9 factories was obtained (see figure 4). It transpired that similar sources on different locations led to virtually the same emission. The total emission per ton of processed beet was in the order of size of 10 - 100 million o.u./ton.

Analysis of the collected data also showed some striking differences. In several cases the deviations which were found provided points of departure for investigation into the possibility to reduce the odour emission by means of process modification. Some examples of this are:

1. The extraction in Roosendaal and Dinteloord. The odour emission of the extraction towers at these locations appeared to be some 10 times higher than that of the other factories. It was determined that microbiological conversion of SO_2 into H_2S was the most important cause of this. SO_2 is added to the extraction water for process-technical reasons. On the basis of an evaluation of the possible alternative forms of process with a low odour emission the SO_2 dosing in Roosendaal has in the meantime been replaced by sulphuric acid dosing. Measurements have proved that as a result of this measure the odour emission of this source has been reduced by more than 95%.

2. The evaporation process in Breda. At this factory the evaporation capacity of the installation is relatively low compared to the beet processing capacity, as a result of which the so-called steam boxes were virtually constantly blown off. While this critical point in the installation was known, the preliminary investigation made it clear that the total odour emission for the factory, not counting the pulp drying plant, is increased by this situation by some 40%.

3. The pulp dryer in Dinteloord has an extremely high odour emission. This must be attributed to the fact that at this factory molasses is added to the pulp before drying. Measurements showed that leaving this dosing out results in a decrease of this particular odour emission by a factor of 5.

4. The odour emission of the crystallization (including the condensor water cooling circuit) is rather high at all locations, with the exception of Roosendaal and Sas van Gent -2. Further investigations into the causes of this are under preparation.

In order to make preparations for possible abatement measures dispersion calculations have been made in order to find out which odour problems occur in the surroundings of the sugar factories as a result of the various emissions. In this respect it is important that the waste gases of the pulp dryer in most of the factories are discharged via a chimney which is approx. 50 m high. As a result of the high temperatures of these waste gases there is also a plume rise of more than 100 metres. The validity of the models for calculating this plume rise was checked in one factory with the aid of SF6 measurements. It appeared there that the influence of the high moisture content of the waste gases and of the surrounding built up area was limited. The conclusion of the dispersion calculations is that in general the emission from the pulp dryer does not make a significant contribution to the odour problems in the area. This rather surprising conclusion is of vital importance for the abatement plans. The priorities here clearly lie differently than was thought; in 1983 it was still assumed that 90% of the odour problems of a sugar factory could be abated by treating the waste gases from the pulp dryer.

Initiative for plans for further abatement

On the basis of the details which are now available and based on existing technology a list has been made of the measures which can now be taken in order to reduce the odour problems in the surrounding area. For each location a range of measures has been drawn up, in the order of their cost-effectiveness.

When drawing up the abatement plans which are technically possible at this moment, it appeared that the treatment of the ventilation of the juice purification building, was rather costly. For this reason efforts were made to find out whether these costs could be reduced by means of more direct air-extraction at a limited number of the many tens of emission places in this building. On the basis of a provisional odour inventory in one factory it appears that a reduction of approx. 80% of the waste gases which are to be cleaned is possible as a result of measures such as improving the closing of the processing apparatus and local gas extraction. For the abatement plan it is assumed that the reduced amount of waste gases could be cleaned by means of biofiltration.

The other sources which would have priority in abating the problems are the carbonatation - part of the juice purification process - and the cooling water circuit. The add-on technology which is technically possible here would, however, entail relatively large financial consequences (see table 1). This is particularly due to the fact that condensation would result in the substances which are now discharged in the air ending up in a waste water stream. This would lead to considerable purification costs. Further thoughts on process-integrated abatement possibilities are therefore now under discussion. Points of attention are:
- effective use of heat released in condensation;
- influencing emissions by changing the process operations and/or the installations;
- prevention of emissions by adapting the process;
- concentration of the emissions to fewer sources and less waste gas.

The optimum abatement possibilities would be found if the compounds now released could be processed in one of the by-product streams.

Table 1: Breakdown of the percentual distribution of the consequences of odour abatement in the sugar industry [5].

Process part	Measure	% annual costs
Extraction	Adjustment of process	0.6
Juice Purification	- Limitation of building ventilation and application of biotration. - Condensation of waste gases carbonatation + incineration or biofiltration + water purification.	6 27
Crystallization	- Indirect condensation waste gas + water purification	59
Pulp drying	- Increasing height of chimney at two locations	7

Evaluation of the approach to the odour problem

When approaching the odour problem the four points mentioned above [1] are of essential importance (see also Evaluation of the approach to the waste water problem). As a result of the far more complex points of departure here compared with the situation for the application of water purification and recirculation, the approach for various sources is not yet at the same stage. For the carbonatation and the cooling water circuit there is still a lot of information missing on the connection between process and emission, while for the extraction the process modification in Roosendaal has already been implemented.

Striking conclusions are that the most suspected source appeared to hardly contribute at all to the odour in the surrounding area, and that the situation varies considerably among the sugar factories. The problem analysis in the various factories has already led to important improvements which require only low costs.

The authorities did not only prescribe the current investigation in 1983 for the licence, but every year they also participate in discussions on the significance of the results obtained and the priorities in the further course of the investigation. In this way, as a result of the complexity of the situation, we are confronted with a long term development, despite intensive research. Particularly responsible for this is the approach which is directed at the process-integrated solution. The approach which was started in 1983 must be concentrated in the coming years on a number of now known bottlenecks. While there is a long period of time needed to reach a complete control of the odour problem one must realize that important improvements have already been reached during the past years. Furthermore, the alternative which is often chosen to this joint problem approach of authorities and industry is also no short term matter: the legal consequences of a conflict between the authorities and industry can also take up some 5 to 10 years (see also [1]).

IV. PROCESS-INTEGRATED ENVIRONMENTAL TECHNOLOGY

In the sugar industry the approach to environmental problems presented above [1] was implemented intentionally for both the waste water and the odour problems. After a phase of problem recognition extensive research was carried out into the causes of the environmental pollution and abatement possibilities. Besides measures which can be characterized as good housekeeping and add-on technology, process-integrated measures have also been taken. The development of new technology has been and is being actively taken up.

Autonomous industrial developments and solving (newly recognized) environmental problems have influenced each other. As examples one can mention the transition to more continuous process operations instead of batch production, process automation and, as a result thereof, more closed process apparatus, and scale enlargement as a result of concentration of the processing in fewer factories. All this has resulted in the fact that for various parts of the production process no clear distinction can now be made between a production step and an environmental facility.

With respect to both waste water and odour the abatement has been concentrated in the first instance on the environmental problem itself. In both cases it appeared that this also effected other environmental compartments. For example, water treatment led to the production of sludge, and sound and odour emission, while the reduction of the odour emission is now in turn threatening to make further measures necessary for the water purification.

While historical developments can provide a good explanation for these problems, it will be increasingly necessary to strive towards a total-environmental-approach. Setting priorities will be an important factor in doing so.

One must not only look beyond the boundaries of the environmental compartments, but also over the process boundaries. In this respect one can state with regard to the sugar industry that attention is being paid to the development of sugar beet breeds with a more rounded shape. As a result of this the tare weight and the damage when washing – and as a result the water pollution – can be considerably decreased. One must also look beyond the process boundaries from the product side. For the sugar industry one can mention as an example the fact that much work has been done recently in order to sell mechanically dried pulp instead of thermally dried pulp. This would lead to the elimination of the entire pulp dryer emission. The clearly limited conservability of the pressed product is, however, still a bottleneck for a good acceptance by the present buyers of dried product.

The situation within the sugar industry does not differ fundamentally from the environmental problems within other sectors of industry. There too vision will be necessary for an effective approach. The correct appreciation of all environmental aspects, in the short and the long term, and of the joint interests of government and industry, is essential. The process of becoming aware of the complex relationship between environment and industry will in our opinion lead more and more to the application of process-integrated environmental technology.

Acknowledgments

We offer our thanks to:
Mr. B.C. Huisman of the Suiker Unie, and
Messrs. H.J. Peters and E. Wind of CSM Suiker b.v.
for their permission to publish the details given above and for the fruitful discussions which have considerably contributed to this article. We would also like to thank the Dutch Ministry of Housing, Physical Planning and the Environment for financially supporting the publication of this paper.

LITERATURE

[1] J. Quakernaat, J.A. Don, F. van den Akker: Process Integrated Environmental Technology, A Must to Survive, Proceedings of this Conference.

[2] R. de Vletter: Twenty years of experience and research concerning waste water control at CSM 1960-1980, Proc. 16th General Assembly C.I.T.S., Amsterdam, 1979.

[3] K.C. Pette: Anaerobic Waste Water Treatment at CSM Sugar Factories, La Sucrerie Belge, 99 (1980) 473.

[4] J.A. Don: Odour Measurement and Control, Filtration & Separation, May/June (1986), 166.

[5] B.C. Huisman, H.J. Peters: Odour Emission and Control in the Dutch Sugar Industry, to be presented at 24th General Assembly C.I.T.S., Ferrara (Italy), June 1987.

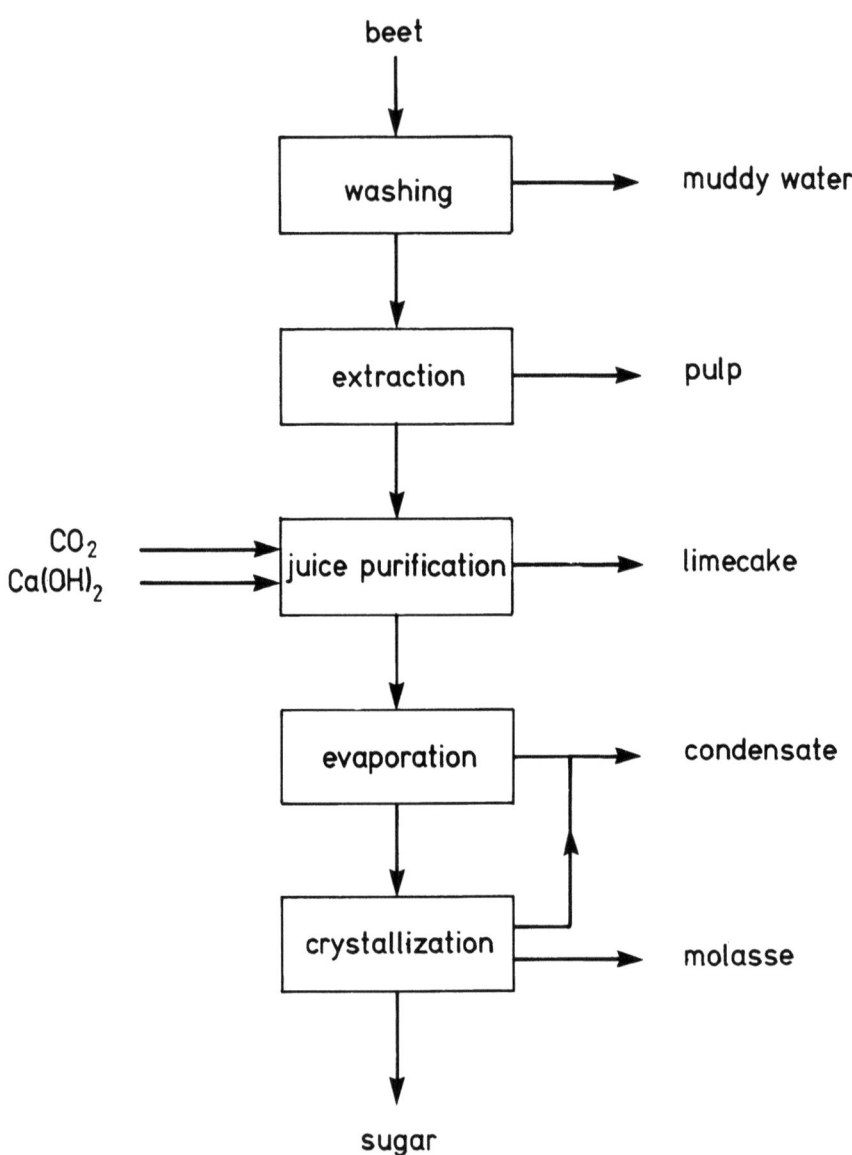

Figure 1 Simplified scheme of the sugar process

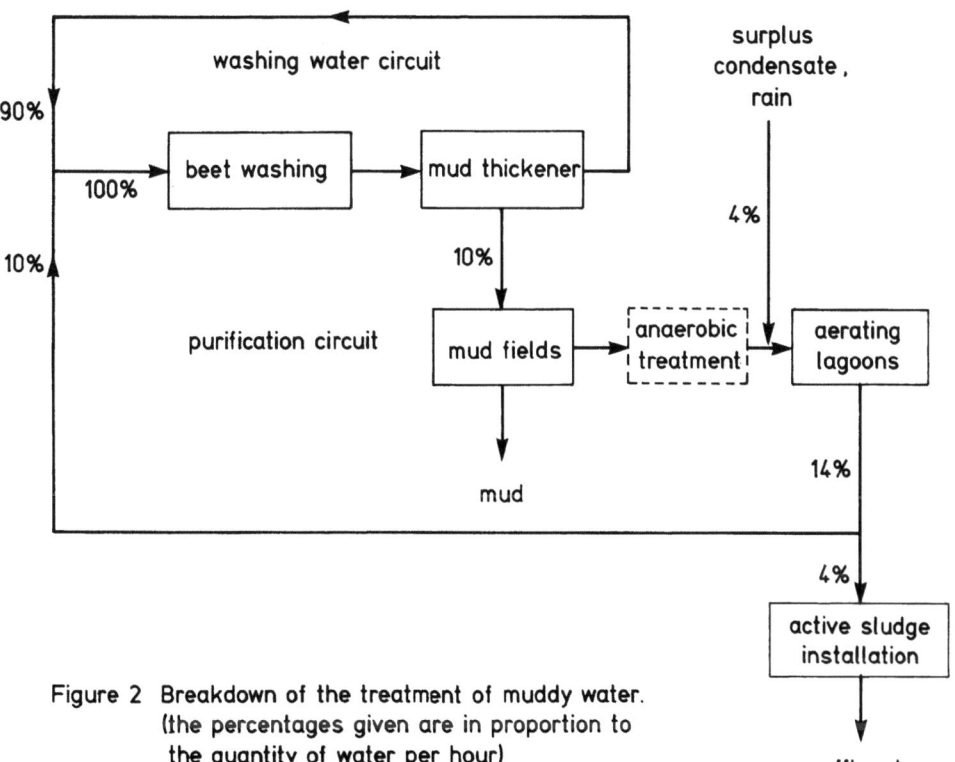

Figure 2 Breakdown of the treatment of muddy water.
(the percentages given are in proportion to
the quantity of water per hour)

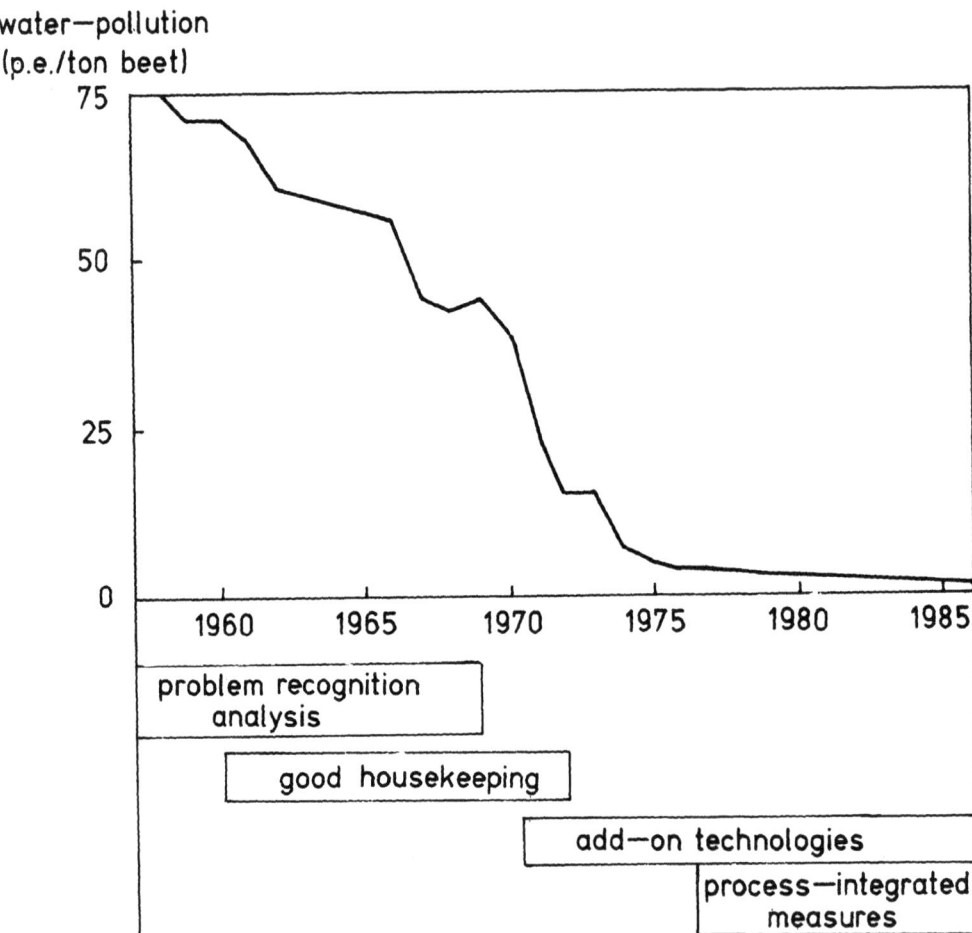

Figure 3 The evolution of the waste water situation at the sugar factories in the Netherlands

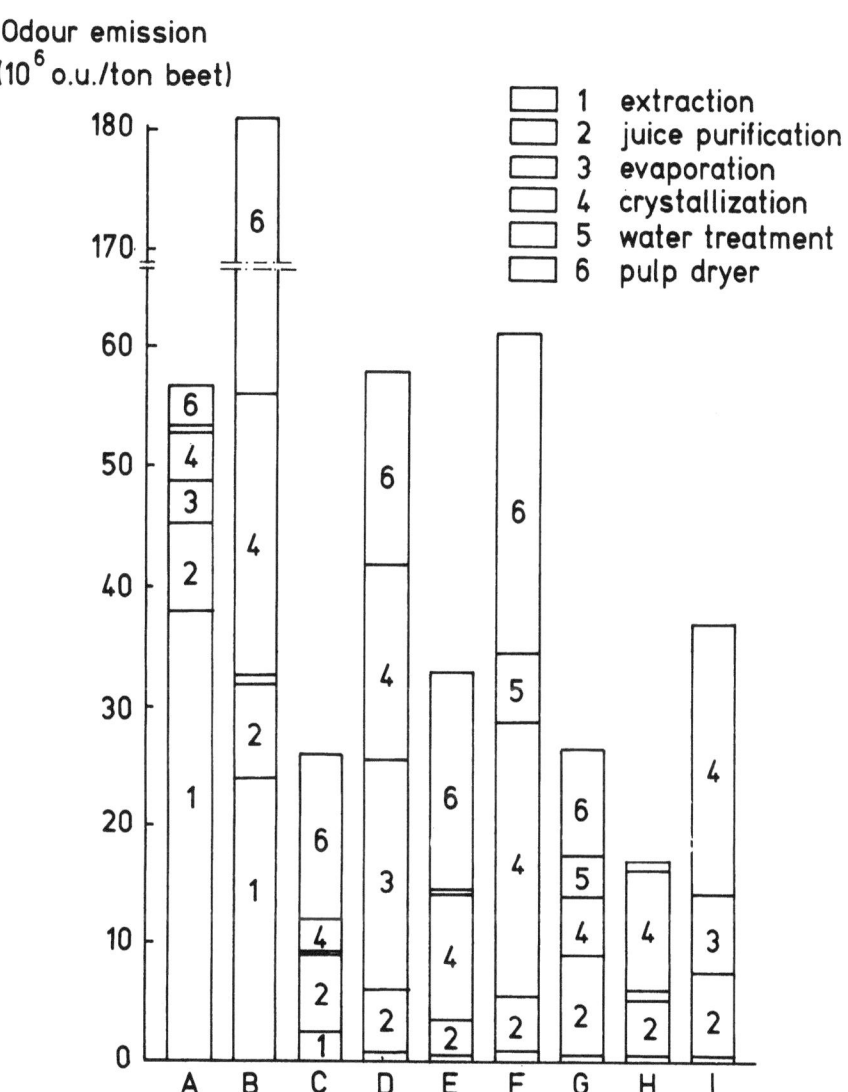

Figure 4 Odour emissions at the Dutch Sugar Factories

(A = Roosendaal, B = Dinteloord, C = Puttershoek, D = Breda,
E = Vierverlaten, F = Groningen, G = Sas van Gent —1,
H = Halfweg, F = Sas van Gent —2)

REMOVAL OF NITROGEN COMPOUNDS FROM WASTE WATER

ING. L.W.F. HARMSEN, Gist-brocades n.v., P.O. Box 1, 2600 MA DELFT

Gist-brocades has developed two processes to remove ammonia and pyridine from strongly polluted waste water streams. Whether or not slightly modified, they are claimed to be suitable to provide technically and economically interesting solutions of many problems related to nitrogenous effluents.
The philosophy behind the way in which Gist-brocades deals with environmental problems always comprises the following points:
1. Where possible, process-integrated solutions are applied (which means the elimination of the environmental problem by adjusting the process operations as such).
2. If that option is not possible, remaining materials are purified for re-use.
3. If also re-use is not possible the installation of additional equipment is necessary.
Waste water problems at the Gist-brocades principal establishment in Delft are likewise approached along these lines. At that site both biotechnological and chemical processes are carried out on a large scale. When combined such processes cause unusually complicated problems in waste water purification. In the past 15 years Gist-brocades has succeeded in designing solutions which have now largely become practicable. Late 1986 it appeared possible to reduce the pollution load, equivalent to aproximately 800,000 population equivalents (PE's) to about 150,000 PE's. As much as 70 percent of the original pollution load is eliminated by measures falling under the first two categories. The most recent acquisition in waste water treatment consists of a biological purification plant based on the well-known sludge on carrier technique.

WASTE WATER COMPOSITION AND METHODS OF PURIFICATION
The two waste water streams produced in the manufacture of semi-synthetic antibiotics by chemical conversion of crude penicillin used to be discharged in unpurified form. Roughly speaking, the streams are composed as follows:
- one containg about 1.5 percent ammonia (3 to 5 m^3/hr) and small amounts of volatile organic compounds several of which are difficult to break down
- another one containing about 1.5 percent pyridine (1 to 2 m^3/hr) and saturated with toluene.
Also present in these two streams are high-boiling organic compounds, inorganic salts and traces of hydrogen sulphide. Nitrogen compounds are hardly removed, if at all, by anaerobic purification. Moreover, in the stream containing ammonia there is a slight concentration of methylene chloride, a compound that is toxic to the micro-organisms in the methane column of anaerobic purification. As a consequence, separate processes had to be developed to purify these waste water streams with their overall burden of about 60,000 PE's.

The main point in our philosophy was that we should develop purification processes permitting the recovery of the discharged compounds. The measures had to satify the following limiting conditions:
- suitability for re-use; in other words, the processes had to be avoided
- environmental safety, notably emissions into the atmosphere had to be avoided
- suitability of the remaining waste water for treatment in the final biological purification processes
- fulfilling standards of business economics
- low energy consumption.

The following review presents a picture of the realized process operations.

AMMONIA RECOVERY

The waste water stream that contains ammonia is collected, made alkaline with sodium hydroxide and fed into an atmospherically operating steam stripper. NH_3 and all other volatile organic pollutants thus pass over the column top.

The stripper is provided with a rectification section, part of the sodium hydroxide is dosed into th column. It is possible to ensure that the NH_3 vapour contains little water and no hydrogen sulphide whatsoever. The stripped waste water stream is free of methylene chloride and can be readily submitted to the biological purification stage where the remaining COD-compounds are eliminated.

The top vapour of the stripper is cooled to about 10°C. This is accompanied by condensation of an aqueous mixture of organic compounds. The condensate is separated into two layers: the lower layer (ammonia-water), which is refluxed to the column, and the upper (organic) layer, which is collected and destroyed by incineration.

Cooled to 10°C, the NH_3 vapour leaving the top condenser is still too much polluted. To achieve a high degree of purity, use is made of a purification process developed by Gist-brocades on the basis of extractive rectification at low temperature. This finding makes it possible to recover with a relatively low energy consumption extremely pure NH_3 (with a content of 99.5 to 99.9 percent) from considerably polluted waste water streams. Finally, the NH_3 is absorbed in water and re-used as 25% ammonia but it is also possible of producing liquid ammonia.

PYRIDINE

In the traditional processes for pyridine recovery from diluted aqueous solutions, the pyridine-water azeotrope (50% pyridine and 50% water) is stripped by the steam of direct heating and pyridine is dried through second distillation. For that purpose, benzene is added to wet pyridine, after which dry pyridine is obtained by distilling the benzene-water azeotrope. This process is costly in terms of energy consumption and another disadvantage is the need to use benzene. There are also rather serious objections, however, to drying methods which involve, for example, the use of concentrated sodium hydroxide.

Gist-brocades has developed an alternative process that required little energy while having no adverse impact on the environment. It comprises the following steps:

The acid waste water stream is passed through an air stripper to remove toluene. Pyridine, which is a weak organic base, remains behind as a salt bound in the liquid phase. The air, which is saturated with toluene, is purified using a compost biofilter. The toluene present in that filter is broken down microbiologically. The compost itself is not converted and can be used for a prolonged period without having to be replaced.

Next, the waste water stream is made slightly alkaline and extracted with suitable solvent. Extraction takes place in a simple countercurrent column extractor which Gist-brocades has designed to allow a fully closed process and to prevent any emissions into the atmosphere. By choosing the right solvent pyridine is recovered in this step in a high yield (over 90%) from the waste water and is taken up in the extract.
The solution of pyridine, obtained by extraction, is distilled in a distillation column with indirect heating. In this step, more than 96% of the solvent is recovered and fed back to the extractor.
The crude pyridine remaining after distillation of the solvent is virtually anhydrous. Because of this, the end stage of purificatioan consists of fractional distillation. To begin with, the remaining solvent is distilled and led back to the extractor. The residue of the distillation of pure pyridine is mainly made up by high-boiling organic compounds and is destroyed by incineration.
Finally, the extracted aqeous layer is freed from the solvent (which is again recirculated) and is then capable of being treated in biological waste water purification.
Apart from the last, fractional distillation of crude pyridine, the entire process is continuously. Pyridine thus recoverd is comparable in quality to fresh pyridine and fully qualified for re-use.

EXPERIENCES
Prior to design and construction of these plants, there were intensive studies on both laboratory and pilot plant scale. Attention was not only paid to primary process development features such as conditions of operation, yield and quality, but also to:
- problems due to pollution and erosion
- minimizing emission and safetely problems
- effects of changes in the compositions of the waste water streams offered for treatment (for example foam or emulsion formation)
- the possibility that certain contaminants may gradually accumulate in the recoverd products, thus giving rise to quality problems.
The ammonia recovery plant was started-up in January 1985, and the pyridine recovery plant has been in use since September 1983. Although intensive studies had been made first, there appeared to be come starting-up troubles but all initial difficulties have been solved and the the recovery processes now proceed smoothly.
In all, approximately 50,000 PE are eliminated by these projects, some 40,000 PE as nitrogen compounds and 10,000 PE as carbon compounds. The remaining pollution load of about 10,000 PE in the aqueous effluents is suitable for anaerobic purification.
Allowing for government grants, the outcome of a cost benefit analysis shows pay-off period of 4-5 years. Purification costs per PE-removed have been calculated at about 25 guilders for the ammonia recovery process. A slightly positive result is obtained for the pyridine process. Related to the absence of discharge taxes between 45 and 50 guilders per PE these are attractive results.
Our company has carefully analysed and also partially tested alternative processes to purify nitrogenous waste water, such as stripping and incineration of impure ammonia and pyridine fractions, or microbiological purification by nitrification and denitrification. The selected recovery routes were chosen for both technical and economic reasons.

POSSIBLE USES OUTSIDE GIST-BROCADES
The extraction process described for pyridine recovery lends itself in

principle also for the recovery of many other organic components from waste water streams. Through variations in process conditions, the process can be utilized for the selective recovery of widely differing types of compounds from strongly pollluted effluents. These variations may refer to:
- temperature and pH during air-stripping to remove first the volatile pollutants
- the choice of the solvent (low boiling, malodorous and inflammable solvents can also be processed provided that the installation is properly closed)
- extraction temperature
- addition of de-emulsifiers.

With regard to the applications of this process, the following compounds may, for example, be mentioned:
- organic bases (aliphatic and aromatic amines, nitrogen heterocycles, etc)
- organic acids (carboxylic acids, sulphonic acids, phosphonic acids, phenols, organic sulphur compounds, etc.)
- neutral, apolar compounds such as pesticides and PCB's.

In view of the above, the extraction process which I have described may be very suitable for the recovery or removal of adjuvants and by-products from effluents in the manufacture of fine chemicals, pharmaceuticals, dyes, insecticides, etc.

The cold extractive rectification, used in NH_3 purification, can be successfully applied in the removal of a large range of pollutants from NH_3 vapour. Since there is a great variety of solvents available for addition to the cold rectification column, the method allows the removal of nearly any pollutant with a boiling point above 20°C from NH_3 vapour. In addition to applications in the chemical industry, there seem to be prospects for processing the excess slurry produced by intensive cattle husbandry. Both in The Netherlands and aborad much study is being made of systems for the anaerobic fermentatin of such waste. The main problem lies in the removal of potentially extremely large amounts of NH_3, and the process described could be applied to good effect in the recovery of NH_3 with the desired degree of purity from a number of central locations for re-use.

The process could furthermore be utilized in the treatment of industrial effluents with relatively low NH_3 contents (larger than or equal to 2 to 3 grams NH_3/l).

COMMERCIALIZATION

In September 1984 Gist-brocades set up the Environmental Business Group whose task consits of providing others with know-how and process knowledge in the field of environmental technology. In addition to the Biothane UASB up-flow technology, taken over from CSM, the anaerobic fluid bed technology (involving the application of sludge of carrier) is offered for use elsewhere. The physical processes described in my lecture for the removal of nitrogen compounds are logical supplements to this know-how package. Gist-brocades with its wide experience in dealing with environmental problems holds a unique position in making recommendations for an optimal solution of specific problems. Know-how is transferred through licencees or special agreements. Gist-brocades will gladly offer advice to those who are interested.

CLEANER TECHNOLOGIES IN DENMARK - EXAMPLES AND EXPERIENCES.

KLAUS MØLLER

1. INTRODUCTION.
Much has been done in Denmark in the past 15 years with regard to solving environmental problems. The development of administrative and legislative systems as well as the rapid growth of the Danish National Agency of Environmental Protection have led to that position Denmark has obtained in Europe: Denmark must be regarded as one of the leading nations concerning environmental protection.
In spite of all our activities and the public awareness it must be stated that there are still unsolved environmental problems - and new environmental threats can be foreseen. Just take the problems of groundwater contamination, the problem of growing waste quantities, the lack or the impossibility of establishing waste deposits or incineration plants, and - last not least - the environmental threat of as well more complicated as still unknown chemical and biological wastes or new materials and products.
All this leads to the conclusion that it is not sufficient to deal with waste treatment technologies, with remedial action techniques or waste handling and abatement methods: there is a need for a preventive policy, for a cleaner technology policy which avoids waste problems as far as possible at source.

2. BACKGROUND.
There are mainly two laws which give the background for the administrative experiences of implementing cleaner technologies en Denmark.
2.1. Environmental Investment Act.
The revised "Environmental Investment Act" from 1980 - 1986 which succeeded two former acts from 1975 and 1978 gave i.a. the possibility of subsidizing "new and less polluting technology" in order
- to create an incitament for starting up cleaner technology projects and
- to enable the environmental administration to influence on future choice of technology within industry and
- to direct the technological development to environmetal benefits.

"New and less polluting technology" was defined as those technologies which had not been used before in the production process of defined or according products, what included a partial or total change of production processes or the implementation of more advanced cleaning (end-of-line) technology. All measurements had to result in an essential reduction of the environmental pollution.

2.2. Act on Recycling.

The first "Act on Recycling" from 1978 did not include the term "less polluting technology" – even though the close relation between recycling and cleaner technology was known at that time. The act was mainly directed towards the packaging (paper and drinks) industry.

The revised "Act on recycling" from 1984 included all waste types and gave the possibility of subsidizing projects for the development of "less polluting technology". This term was not defined more clearly, but practically the traditional ways of limiting environmental pollution (end-of-line measurements) were excluded in order to promote the development of new and integrated cleaner technologies.

3. DEFINITION AND MEANS.
3.1. Definition.

There is a lot of diverging definitions of the term "cleaner technology" - and this will always be an academic discussion, whether you talk about low- and non waste technology, cleaner technology, less polluting technology or environmental technology for just to name some of the existing terms. In many cases these terms and their administration must be seen on the background of historical, administrative or juridical developments which hinder a more technical justified change or adoption of the term in the administrative systems.

The following figure illustrates a model for describing the relation between cleaner technology and recycling:

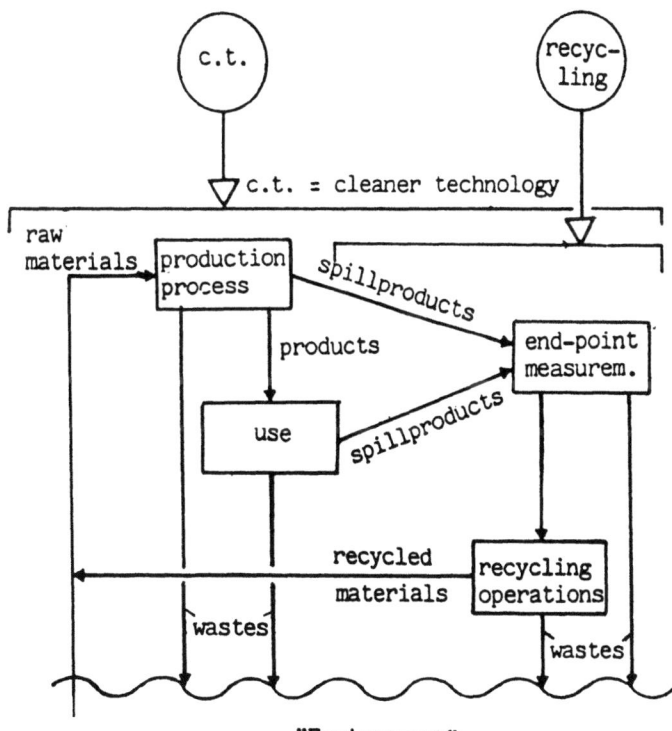

Cleaner technologies can be regarded as those production measurements which reduce the quantity and the hazard of all types of emissions within the production cycle by preventive measurements.

It is obvious that the term recycling just covers those production and product wastes which have occured (and which are inevitable to a certain extent). Also the "end-point measurements" like for example collection, filtering, waste treatment, waste handling etc. deal with wastes which have been produced. Of course it is necessary to develop these technologies and measurements in order to reduce pollution, but in many cases there will be a transformation from one waste type or medium to another, and therefore they cannot be regarded as preventive methods - they are just treatment methods!

3.2. Means.

In our view cleaner technologies should just be regarded as those measurements which from an environmental point of view try to avoid or minimize wastes or spill products at source by regarding the entire production/product cycle.

This can be achieved by the following four measurements:
  1) Choice of other or alternative raw materials
  2) Development of new production processes or methods
  3) Development of new or alternative product design
  4) Combinations of 1 - 3

Ad 1) It is a basic principle that those raw materials which contain fewer pollutants result in fewer environmental polluting emissions. But also the choice of alternative materials with the same function in the production process or in the product will make a given technology cleaner. These substitution processes can be divided into physical, quantitative or functional substitutions.

Ad 2) The development of new production processes or methods is a very important part of a cleaner technology conceptualization. Special attention has to be paid to the fact that there are not only physical, chemical or other machinery-oriented possibilities for the development of new production processes or methods, but also f.ex. organizational ones as the introduction of new wage systems.

Ad 3) New or alternative product design include new approaches with regard to material, function and construction.

Ad 4) In many cases there will be a combination of the above named principles because of the interdependance of raw material, production process and product design.

4. EXAMPLES AND EXPERIENCES.

The previous discussed "Environmental Investment Act" did not result in that number of projects which was supposed. There was only granted subsidies to 18 projects out of 70 applications with a total amount of 16,3 mio. Danish Kroner. One of the reasons may have been the limited prerequisites for becoming subsidized, but mostly it must be stated that the applications did not fullfil the requirement of introducing new technologies as required.

The subsidized projects mainly fell into the food industry branch (11 out of 18). All projects delt with cleaning or recycling of water effluent or the establishment of closed process water systems.

It must be stated that the purpose of the Environmental Investment Act - even if this act also had other intentions, which cannot be discussed here - has not been fulfilled.

Since the introduction of the "Recycling Act" there has been granted subsidies to 14 projects with an amount of 12,8 mio. Danish Kroner (October 1984 - june 1986). Examples for projects are
- Development of less polluting technologies in the metal plating industry
- Substitution of PVC
- Development of new textile-colouring-systems
- Process-water free metal plating
- New de-inking processes
- Reduction of mercury in batteries
- Regeneration of chrome-containing baths
- Evaluation of existing cataloques on cleaner technologies
- Substitution of cadmium in plastic products
- Minimization of spills in industrial painting processes

Because of the limited number of projects any further evaluation shall not be given. Instead of that some general experiences with regard to barriers can be given:
1) The introduction of new programmes for the implementation of cleaner technologies is to a very high degree dependent on information activities which have to be directed towards specific industries or target groups, especially the smaller ones.
2) It must be pointed out that cleaner technology also encompasses changes of input raw materials as well as changes of product design.
3) The willingness of industry of taking risks with regard to development of unproven cleaner technologies is very limited.
4) New and lower emission standards and other environmental regulations enforce the development and implementation of cleaner technologies.
5) There must be a high percentage of subsidy for a given project in order to reach a satisfying participation of industry.
6) It is apparent that the conceptualization of cleaner technologies is very complex, and that the development of new production cycles, processes and products takes a long time.
7) The promotion of cleaner technology programmes makes an active participation of the administrative system inevitable.

5. THE NEW CLEANER TECHNOLOGY DEVELOPMENT PROGRAM.

On the background of the past experiences and forced by the need to accelerate the preventive methods of pollution control the Danish Minister of the Environment has started a new cleaner technology development program for the years 1987 - 1989 which must be regarded as a supplement to the cleaner technology part of the existing Recycling Act, thus emphasizing the importance of cleaner technology.

The program is defined as a development program which shall generate, develop and test experiences in a _limited_ number of industries in order to avoid unspecified actions and to ensure a coordinated policy.

5.1. Aims.

The specified aims of this program are:
- to collect and systematize knowledge on cleaner technologies
- to define the barriers for a better use of cleaner technologies
- to inform industries and the public in a better way on the possibilities and consequenses of cleaner technologies
- to generate a better basis for a dynamic revision of the NAEP's instructions to both industries and local authorities
- to get knowledge of the best means how to improve the implementation of cleaner technologies
- to start research activities in or to get better experiences with the industrial implementation of clean technologies and
- to coordinate and make the clean technology policy more effective.

5.2. Structure and elements.

The new development program consists of two main strings:
  a) to create and spread knowledge on cleaner technologies
  b) to implement cleaner technologies within Danish industries

Industries as well as the public have a great knowledge concerning raw materials, processes and products - but this knowledge is not gathered or treated in a systematic or priority based way. Therefore there is a need for the
- development of methods like production process analyses, input-output functions or models for an evaluation of production cycles from an environmental point of view
- establishing of a data- and information system which includes the development of formats and hard- or software systems
- transformation and adoption of national and international knowledge.

The other basic structure principle of the new cleaner technology development program is the implementation of cleaner technologies within Danish industries. This must become effective by an active decision process including the public with its different possibilities of approval conditions, consulting activities and subsidies.

In order to fulfill these aims the program consists of nine elements:
1) Collection and transformation of knowledge - national and international
2) Development and establishment of a data- and informationsystem within a data-based network
3) Branch-/industry-related research
4) Technology-based research
5) Research- and development projects
6) Demonstration projects (full size scala)
7) Plant investment activities
8) Information and education activities
9) Consulting services

These elements are interdependent and shall not be used as single elements.

Ad 1: The basic precondition for determining possibilities for the implementation of cleaner technologies is the knowledge on the actual status quo of technology and the latest research developments. This knowledge has to be gathered and transmitted to the Danish information system.

Ad 2: It is important to establish a central information system which has to be able to follow the technological development. This system shall be used as well by industries as the public administration, and therefore it must fullfil the specific needs of these two groups. At the same time it must be compatible with existing data-based systems.

Ad 3: The branch-related knowledge is the basis for all further information, consulting and r & d-projects. Some of the branches which are proposed are the furniture-, galvanic-, graphic-, foam plast- or the food industry.

Ad 4: Also the analysis of certain technologies will be necessary in order to ensure environmental advantages and experiences as well as to transform technologies from one branch to another. Examples are the substitution of solvents within different industries or new methods for treating surfaces.

Ad 5: R & d-projects are the logical consequence of the above named elements. Industry, consultants, technological institutes and universities should participate in r & d-projects which should result in laboratory- or pilot plants.

Ad 6: The next step should be the implementation of cleaner technologies by so-called demonstration plants which should be subsidized in order
  - to faciliate the implementation of similar plants
  - to be used for educational purposes
  - to show the conditions - both technological or economical - of how to use certain technologies best and
  - to get an information advantage compared to other nations.

Ad 7: Experiences with former subsidy policies show the necessity of granting investment subsidies, especially when it is environmentally desireable to replace traditional and not depreciated end-of-line equipment by cleaner technologies.

Ad 8: Insufficient knowledge and lacking information - both on the possibilities for participating in cleaner technology programs as on the advantages on known cleaner technologies - must be regarded as those main barriers which hinder an implementation of cleaner technologies. Therefore a variety of different information - and PR-activities is foreseen in order to activate industries and the public.

Ad 9: On the background of the structure of Danish industries, where small- and medium sized companies are predominant, the establishment of an active consulting system is inevitable in order to reach and transmit knowledge to industries.

5.3. Status quo.

The financial means for this program - 50 mio. Danish Kroner for a 3-year program - were granted on February 18th, 1987 and consequently the NAEP is preparing the first phases of the program which mainly will consist of starting the information - and data-based parts of the program.

Because of the short time of the program's existence, no further experiences can be discussed af present.

6. CONCLUSIONS.
A) Experiences with former Danish experiences with the implementation of cleaner technologies showed the necessity of a new and coordinated program.
B) This program tries to overcome economic, financial, personal, information and psychological barriers with regard to the implementation of cleaner technologies which have been stated in former programs.
C) It is inevitable also to include the national administration system in the program in order to enforce the best application of existing or new developed cleaner technologies.
D) It is necessary to define exact aims, target groups, measurements and steps in order to coordinate and fulfill the policies of a program.

THE APPLICATION OF THE BEST PRACTICABLE ENVIRONMENTAL OPTION TO
POLLUTION CONTROL IN 1987

R.G.P. HAWKINS MA BARRISTER, FRSA, FRGS, MInstWM
General Counsel, Cleanaway Limited

"Als er een schaap over de dam gaat, volgen er meer"

                   Dutch Proverb

Summary

A BPEO does not exist as such. It is a combination of scientific
judgement, economic forces, public perception and applied political
policy. There are no easy solutions; it is not a new Jerusalem.
The concept of BPEO is as full of riddles and uneven applicability
as the polluter pays principle. Sometimes that coat is more full of
holes than the cloth itself. The BPEO principle is often as difficult
to apply as the distinction between hazardous and non-hazardous.

Nevertheless, if the given checklist of questions are adressed for
each potential environmental problem, BPEO can be of credible use to
the ecosystem manager and practising engineering environmentalist.

There is the need for a given checklist at
: international policy level, e.g. transfrontier movements
: national policy level e.g. SO_2 scrubbing
: pracitical pollution control level e.g. vehicle movement night
  noise ban.

All or none may be achievable, but at least the application of the
BPEO concept should be applied.

PART A.

Historical Backcloth and Some Definitions

A.1 Best Practicable Means (BPM)

There has never been, and never will be, a commonalty between all nations or indeed, within one nation, as to the extent to which pollution should be controlled. It would however be a churlish environmentalist who disagreed with the statement that in terms of economics pollution should be abated to the point where the extra benefit to society just equalled the extra cost to society of that abatement. Yet when in 1306 a Royal Proclamation prohibited London's artificers from using sea coal in their furnaces, the offender who was executed, would not probably have agreed with even that broad concept.

Pollution of the air by smoke has been a problem ever since mankind began using fire for heating, cooking and metalworking. The artificer who met his Maker in 1306 did so because he wasn't using the Best Practicable Means (BPM) although it is not clear from contemporary chronicles whether this artificer had had explained to him whether there were available to him better other than the ones he used, let alone the best.

First since BPM is the direct precedent of the Best Practicable Environmental Option, let us examine as precisely as possible what we mean by BPM.

The purpose of the employment of the BPM in pollution control, is to reduce, and when necessary eliminate, hazards to human health and safety, taking into account both the magnitude and the certainty of the risks, including the susceptibilities of critical groups of members of the public together with the resulting costs to the community. The deployment of BPM should reduce damage to amenity, property and plant and animal life to a minimum compatible with the wider public interest; this will take into account such factors as economics, employment and trade and, above all, the prevention of irreversible damage to the natural environment. Before we look at the history of the BPM concept let us briefly put the first two words under the etymological microscope.

By "Best" we cannot mean the absolute superlative since such an optimum view will always have a strong subjective element. Indeed there must always be a scientific element, an economic appraisal and the inclusion of a strong degree of public perception however fashionable or transient.

The word "Practicable" connotes reasonably practical having regard among other factors, to local conditions and circumstances, to the current state of technical knowledge and to the financial implications. In practice, it will include inter alia the means to be employed which will include the design, installation, maintenance and manner and periods of operation of plant and machinery together with the design, construction and maintenance of buildings. Whilst BPM was only fairly sketchily described in the Alkali etc. Works Regulation Act 1906 and the Health and Safety at Work etc. Act 1974, other legislation including the Clean Air Act 1956 and the Control of Pollution Act 1974 defined it more fully. (The reader may like to refer to Appendix I which shows all the relevant pollution control legislation in the United Kingdom together with its considerable dependence on land use and development control legislation).

BPM has been particularly developed in England in the context of the Alkali inspectorate. Air pollution from non combustion processes was not regarded as a problem until the industrial revolution throughout Europe at the beginning of the nineteenth century. Alkali works, mostly producing sodium carbonate from salt were commercially viable from the 1820s but the process produced large volumes of hydrogen chloride gas and an unpleasant smell. In England in 1836 a tower containing neutralising soaked branches (a process which was an ancestor of the present gas scrubber) was patented. These towers however allowed up to half the gases to escape.

In 1862 a Royal Commission was set up in the United Kingdom to examine the problem, hitherto parliament and local authorities had attempted to deal with such nuisances by simply banning them, without making any constructive suggestions on how to abate them. The Royal Commission and the Alkaline Act of 1863 which implemented the Commission's recommendations took on a new approach. The Commission recommended that if at least 95% of the hydrochloric acid gas evolved from Alkali works were arrested, the remainder, after adequate dilution, could be allowed to pass into the air. The Alkali Act required that relevant works should use the best practicable means to reduce to the minimum the discharge of noxious or offensive gases. Thus BPM as laid down by the new Her Majesty's Inspectorate of Pollution Air Inspectors, will therefore specify chimney emissions for plant design and operating and maintenance practices so far as these are relevant to air pollution; they will also include general instructions on good housekeeping and may also specify requirements for monitoring and recording emissions.

The Inspectorate doesnot carry out research into air pollution control problems although they occasionally sponsor it; the research is normally carried out by the industry concerned with the Inspectorate under the contaminator (not the polluter) pays concept. Eventually the chief inspector is in a position to produce notes on BPM and to set emission standards. These presumptive standards specify emission levels that are considered to be currently achieveable having regard to the technology available, the nature and effects of the pollutants concerned and the cost to industry. New standards are not usually applied immediately to existing plant. Industry cannot be expected to re-equip their plant with the latest control equipment too frequently, particularly when this forms an integral part of process design. Existing equipment must be allowed a reasonable economic life.

The use of BPM over the years has however, highlighted another problem which although solving one pollution area may create another. For example dust that would otherwise be emitted to the air may be removed by the use of water sprays. Resulting liquid or slurry has itself to be disposed of, possibly leading to a worse pollution problem than if the dust had been widely dispersed through a tall chimney. Again the washing of flue gases may reduce the emission of sulphur oxides from power stations, where the process cools the plume but it will inhibit dispersion. Thus in this case the reduction of widespread emissions is obtained at the expense of higher ground level concentrations near the station, possibly in a more harmful state. Another possibility is that solving an air pollution problem may create noise as large fans have to be used to extract the filth from the air of a large workshop. Such fans may use a great deal of power whose generation at the Power Station could create as much dust as the fans extract in the factory. These examples stress the need to look comprehensively at all forms of pollution arising from a particular process thus arriving at the concept of BPEO.

A.2 Best Practicable Environmental Option.

We have seen above that traditional pollution control measures often only contain pollution problems temporarily and, in many cases, merely convert one form of pollution to another. Pollution control itself also requires energy and materials, the reduction of which in turn often leads to the generation of pollution elsewhere. Effective environmental management requires considering the mosaic as a whole and not the pieces individually; only then can an image appear from which to draw valued conclusions and plan realistically. The concept of BPEO was introduced by the Royal Commission on Environmental Pollution (RCEP) in 1976. BPEO means taking account of the total pollution from a process product or waste and the technical possibilities of dealing with it so that pollution control strategies are overall the most efficient and beneficial to society and the environment. This sounds sensible in principle but like many aspects of pollution control, is not necessarily so easy to put into practice. Indeed the 1976 Fifth Report of the RCEP entitled "Air Pollution Control: An Integrated Approach" noted that

: because of the connections that exist between different forms of industrial pollution it makes little sense to look at one aspect of control and isolation yet
: there was virtually no co-operation between the controlling authorities so that the method of disposal causing least environmental damage overall was identified and indeed
: in sectors other than air pollution, the requirement is for safe management and engineering of waste; there is no power to acquire the adoption of means to eliminate or minimise waste arisings
: in these other sectors the controllers (neither those on the ground or the higher officers) didnot have the expert knowledge of the waste producing processes or the available technology in dealing with wastes that the Alkali Inspectors had and finally
: the most efficient way of getting such controllers would be through a combined inspectorate applying the principle of BPEO

Alas! the Royal Commission didnot think through the problem (and this was not the first time) to the means of achieving that which they proposed. There are clear legal and political difficulties in giving a Central Authority Combined Inspectorate, the power to implement a policy of BPEO when the controls of river pollution, waste management and air pollution from premises not registered under the Alkali Acts rests, with local authorities. However, that it possibly is not practicable to set up a mechanism for BPEO didnot deter the Royal Commission! It was an idea whose time had arrived and they were determined to crystallise the concept into practical operational disciplines and enforceable legislation.

Over twelve years ago therefore the Royal Commission defined the BPEO concept:

> The combined Her Majesty's Pollution Inspectorate would ensure an integrated approach to difficult industrial pollution problems at source, and would seek the optimum

environmental improvement within the concept of BPM,
employing the Alkali Inspectorate approach to reduce or
modify the waste produced. This was in effect, an expansion
of the concept of BPM into an overall BPEO. The Inspectorate
will be instrumental, in consultation with other bodies
involved, in deciding how different sectors should be used
to minimise environmental damage overall.

In the United Kingdom and indeed in the Commission of the European Communities, the European Parliament and the European Community itself that concept lay dormant for some ten years. Let us now ascertain the position in the Spring of 1987.

PART B
BPEO - Development of a Concept
B.1 BPEO in practice.
If any country or sovereign state decides that the principle of BPEO must be applied throughout its protection of environment policy, then there are various types of responses which could help to achieve this

Administrative responses:
these would include rearranging the structure of the pollution control authorities, encouraging greater co-operation between them, and providing official advice on the subject.

Legal measures:
would include legislation which makes the adoption of or consideration of BPEO mandatory.

Voluntary responses:
which would include internal measures taken by some industries to seek the BPEO for their own processes and waste products.

However in order to ascertain whether the superimposition of the concept of BPEO would work in these three areas it would be necessary to review the contemporary pollution control arrangements. In all countries pollution control is developed in a piece-meal and uncoordinated way. Indeed most countries mirror the UK development and in Table 1 below you can see the main legal controls that deal with each medium (air, water, land) largely on a separate basis and with different authorities operating in different levels.

TABLE ONE

Present UK Key Pollution Control Arrangements

(also please see Appendix I)

Type of Pollution	Legislation	Level of Primary Responsibility for Pollution Control.	Enforcement Agency
Air Pollution (registered works) must carry out Best Practicable Means	Alkali Act, Health and Safety at Work	central	Her Majesty's Inspectorate of Pollution Controlled by Department of Environment
Air and Noise pollution (unscheduled works)	Control of Pollution Act Part III, Clean Air Acts, Nuisance Provisions of Public Health Act.	district	Environmental Health Department of the local authority
Water Pollution	Control of Pollution Act Part II	regional	Water Authority and Her Majesty's Inspectorate of Pollution.
Marine pollution	Food and Environment Protection Act 1985	central	Ministry of Agriculture Fisheries and Food
Waste disposal to land	Control of Pollution Act Part I	county	Waste Disposal Authority
Land Use and Planning	Town and Country Planning Acts 1971 etc.	district	Waste and Minerals County

Source: Author and Centre for Environmental Technology, Imperial College, London.

B.2 Individual Applicability

B2.1 Waste on Land

Whilst many local or national waste management authorities might relish the power of using BPEO as an effective tool to control the movement of waste materials, then they must also be prepared to accept waste where a particular site represents the BPEO although this may be politically and psephologically unpopular. There would of course be considerable legal and technical argument as to what was the BPEO; this would ultimately have to be resolved on appeal unless of course the controversial step was taken of confirming power on a national or regional inspectorate of pollution to determine in a binding way the BPEO for a particular waste stream. The environmental bruising and unhapppiness caused by hazardous wastes over the last ten years in the twelve countries of the European Community do not justify such a totalitarian course of action without the necessary scientific dispassionate argument.

B2.2 Discharge to Inland Waters

In many European countries control in this context is by means of consents with the Water Authority under a duty not to withhold consent unreasonably. It seems highly unlikely that any Water Authority would regard their powers as encouraging them to embark upon a wide ranging BPEO exercise. Water Authorities are concerned with outlets and the effluent they discharge (which may be the combined result of several processes) and not the process and plant producing the effluent. Whereas the concept of BPM in relation to atmospheric discharges from registered processes allows standards to be tightened as technology allows (which would appear to be a vital ingredient of BPEO) it is more difficult to see how the control exercise by Water Authorities over discharges could be tightened in the same way. Whilst consents can be reviewed every two years, it may not be possible to alter consent conditions without the payment of compensation. Greenfield developments, having conducted their own environmental impact assessment on their wastes they have been discharging to nearby water courses after some purification, would not like to be instructed four years after it had been perfected, that their whole system of drainage and effluent managements has to be changed.

B.2.3. Discharges at Sea

In one area of the protection of the environment in the United Kingdom and perhaps in some of the other eleven countries of the EC, particularly those with access to the Mediterranean, a form of BPEO is already in force. This is the procedure in relation to the licencing of the deposit of substances at sea under the Food and Environmental Protection Act 1985.

This legislation allows consideration of a wide range of options in considering licence applications and in particular the practical availability of other methods of dealing with the substance or article in question under Section 8 (ii). In practice applicants for such licences are effectively required to demonstrate that what they are proposing is, de facto the BPEO.

B.2.4 Air Pollution

If there is to be a trans media approach in order to ascertain the BPEO for any particular effluvia from a factory chimney, an integrated approach will have to be adopted. At present in the EC it is largely up to the initiative of a particular manufacturer whether the air and water pollution agencies meet to consider jointly, potential problems. The general picture is one of independent consultations between a production director and the various agencies at different stages in the process of designing a new plant. Such an approach is unlikely to apply to all factories; only certain classes of factories will have to be scheduled in order to come under the BPEO scrutiny. It is here perhaps that the integration of the Environmental Impact Assessment Directive into local legislation may assist the BPEO approach. This will institute a form of extra control afforded by prior approval of the planning authorities and once again such a decision will depend on the calibre of the scientific and technical knowledge of the particular authority. We return to this subject in B.2.5.

B.2.5 Planning and Development Control

Thus, the planning system can provide a form for the consideration of BPEO in the case of proposed major developments with serious environmental implications, although such major enquiries will obviously be reserved only for the most important projects.

The EEC Directive on Environmental Assessment (85/337/EEC) is obviously of great importance since it will require the developer to indicate not only the various impacts of the proposed development, but also the main alternatives studied by the developer and an indication of the reasons for his choice. Such a process could provide a clear starting point for consideration and questions of BPEO. In theory, in any country there seems no reason why a local planning authority should not take account of the BPEO in deciding whether permission should be given for the development on a particular site. Certainly in relation to problems such as noise or smell, the BPEO may simply consist of the spatial separation of inconsistent uses rather than expensive control measures.

However, there are problems in using the planning system in this way; a potentially environmentally damaging activity for example, such as some agricultural activities, may not require planning permission. It is not clear to what extent planning conditions can be used where they may duplicate controls provided by other legislation; this varies from country to country, the practice is frowned on in the U.K. Again, it is still not clear to what extent existence of a preferrable alternative site may be a material consideration in refusing permission. Certainly this is not a valid ground in Germany or Italy in 1987, although submissions of this nature have always appeared on behalf of protestors against a potential industrial development application in the United Kingdom.

BPEO therefore, in practical application faces a number of difficulties in child birth, not at least the differences between the enforcement concepts used by the different control authorities and the levels at which they act. If BPEO is imposed then this may encourage pollution transfer rather than pollution control. Let us now focus on what we mean by pollution transfer.

C   Pollution Transfer

C.1   The Principle Reviewed

The First Law of Thermodynamics states that energy is neither created or destroyed; in other words energy and matter may be transformed but not destroyed. If production of a pollutant cannot be avoided, pollution control techniques aim at transforming a pollutant into a substance which has either

: less polluting potential than the original substance or

: is easier to manage or engineer safely

The aim of a treatment plant or control equipment is to process any waste stream into
- substances which can safely be discharged into the environment or recycled, and
- polluting substances which can be concentrated and managed safely with or without further treatment.

Control measures to reduce an air pollutant for example by scrubbing and washing, will give rise to a liquid waste; likewise engineering solid waste to land may result in polluted ground water. Thus pollution is being transferred from one media to another; this is sometimes termed cross media pollution. One must never forget that the use of smokeless fuels in the city has greatly reduced urban air pollution, but its manufacture often increases the pollution load significantly at process sources.

In order to vitiate this pollution transfer, the selection by the concept of BPEO involves an analysis of the costs and benefits (in the widest sense of those terms) of different options for pollution control in a given situation. The aim is to limit damage to the environment to the greatest extent achieveable for a reasonable and acceptable total combined cost to industry and the public purse. The assessment of the benefit i.e. the pollution reduction, involves considerations of the
- polluting potential of the materials released
- qualities for release
- sensitivity to the pollutant and different sectors of the environment

The smaller the combined effect of these variables then the greater the benefit - or so the advocates of BPEO hope. BPEO may hopefully lead to the decision to discharge a large quantity of a slightly toxic material to a sector of the environment where it is rapidly degraded, in preference to the emission of a much smaller quantity of a somewhat more toxic material to a sector of the environment where it accumulates.

The extent of the benefit from pollution abatement has to be triggered off against its costs. This process must take account of local conditions and the current state of technical knowledge. The level of cost considered acceptable may come to be adjusted in the light of evaluation of benefits. Because of the magnitude of the costs it may be necessary to opt for a degree of pollution control that is less than that which is technically achieveable without of course, abandoning the latter as an ultimate goal. Progress towards this goal can often be facilitated by setting a timetable for increasing levels of.

In selecting BPEO for a given case, some of the relevant factors are those operating at a distance as well as those closer to hand, in the longer term as well as the present and loss of amenity as well as the actual damage. There is of course an accent on timely research to provide an adequate basis of knowledge on the effects of a pollutant and on the efficacy of methods in dealing with it in order to permit confident selection of a BPEO.

Table 2 (please see later) shows how conceptually it might be possible to prevent pollution transfer through an omnipresent all enforced BPEO factor.

C.2 The European Dimension

In order to determine the BPEO before a proposed product manufacture, it is necessary to have some knowledge of the environmental effects of the proposal, and its alternatives. We have already mentioned the 1985 EEC Directive concerning the Environmental Assessment of certain types of project to be implemented incidentally in the U.K. in 1988 largely through the context of the present Town and Country Planning controls. There are two further reasons why this Directive may be significant for BPEO application in the 12 countries of the EC

: it is at the planning stage when details of the production technology, pollution control options and choice of site may still be open to some amendment; and therefore at this stage co-ordination between the planning and pollution control authorities is crucial. A greater awareness of the environmental aspects of the development and consultation between development authorities can facilitate the evaluation of BPEO

: to what extent any sovereign state wishes to enforce purported low and non-waste technologies onto its various industries, is a matter of considerable debate. But if it wishes to enforce such technologies in order to achieve a BPEO, it will have to use some form of an Inspectorate. This Inspectorate's best opportunity of influencing events, either by persuasion or by mandate, will be at the planning stage.

In 1986 the Commission published the Fourth Environmental Action Programme 1987-1991. The document showed that community legislation had tended to concentrate on discharges of a particular pollutant to one medium; there had been no concerted attempt to assess substances on a cross media basis. Paragraph 3.3.3 of the draft EEC Fourth Environmental Action Programme states that an integrated substance orientated approach to chemicals would

: take account of the occurrence of a particular substance from whatever source
: proceed towards the integrated risk assessment which takes into account the different routes through which people in the environment are exposed
: lead to choices regarding the most effective and efficient solution to the problems caused

Subsequently the CEC in 1987 proposed a comprehensive strategy to establish an EC wide systems of controls on cadmium embracing all major sources and all environmental media. There are a total of some twenty Directives and one legislative proposal already at CEC level to control human and animal exposures to cadmium sources such as drinking water, feeding stuffs, consumer products and through other environmental pathways. This legislation is not co-ordinated and the CEC feels that a more systematic approach is needed. Whether countries which have sophisticated forms of pollution systems as opposed to Portugal, Greece and Spain are prepared to accept a superimposition of this cross media approach on their national disciplines, is another matter. The CEC will have to achieve two objectives before they find the doors opening before them

: a scientific justification will be needed to be published for every draft directive and regulation, together with
: a plain man's guide explaining the necessity for such a directive and what each individual article means and the derivative rationale.

Both these aspects have been sorely missed, certainly over the last eight years; it has encouraged alienation from DGX1

D Conclusions

In searching for a particular BPEO it is a truism to state that all the relevant facts must be taken into account; but if these are listed it can be seen that BPEO is not the overall panacea that non practising environmentalists hope that it is. Timely research is needed to provide an adequate basis of knowledge on the effects of a pollutant and of the efficacy of methods of dealing with it. BPEO also involves an analysis of the cost both to industry and the public purse in relation to the benefit - that is the limitation of any pollution - of different options for waste disposal, taking into account the polluting potential and the quantities of waste and sensitivity of different sectors of the environment to the pollutants concerned. Selecting the BPEO therefore requires a careful and informed analysis of options; even then the selection is not always straight forward and will require a degree of subjective judgement which can depend upon political principle. More factors therefore emerge. The need to

1) Weigh and balance different types of costs and benefits.

2) To research and ascertain information on environmental effects and control options.

3) Monitor the efficiency of processes and disposal methods.

4) Examine the public perception and the reflection of that perception in the political climate of the sovereign state in which BPEO is to operate.

At present there is often little incentive, either regulatory or financial to encourage the widespread adoption of BPEO. Indeed exhortations such as that the local authority should give due consideration to the wider principle of BPEO when considering disposal site licence applications and operations and in the charging policies of their own sites are valueless without essential structure to assess the relevant BPEO and then to advise local authorities. For example many waste disposal authorities in Wales consist of two intelligent men and probably a dog. Apart from consulting the present BPEOs which are the Department of Waste Management Papers they will have little time, resources or energy left for examining BPEOs in a local context.

In order to identify those measures to be adopted certain key questions need to be asked in order to obtain the BPEO strategy for a particular environmental problem . No doubt the following checklist is imperfect but it would be an inconvenient rule if nothing could be done, until everything had been achieved. Here is the putative list

: in what areas is BPEO strategy to be adopted
: within what timetable?
: which body or bodies are to formulate the necessary BPEO policy?
: within what framework and under what guidelines is that policy to be formulated?
: how is such policy to be updated and revised in the context of extant National and Community law?
: by what bodies is the policy to be implemented?
: what is to be the relationship between the bodies formulating and implementing BPEO policy, if different?
: what is to be the relationship between bodies implementing BPEO policy in different media?
: what relationship will BPEO policy bear to existing law?
: what ultimate sanctions will back up BPEO policy?
: will the BPEO change if there is a change of government after a general or regional election?

It can therefore, be seen that BPEO at the practical pollution control level must be distinguished between BPEO at a national policy level where it can be applied to strategic national issues for example that of nuclear versus fossil fuel. BPEO can be either a practical mechanism similar to BPM to achieve most efficient pollution control arrangements on the ground, or as an instrument to aid sensible policy making in environmental matters, or indeed as a tool with which to influence CEC environmental policy and legislation. Table 2 shows such uses and over which we may all ponder on future developments and hopefully watch the controlled and sensible use of BPEO increase.

TABLE 2

The Need for BPEO

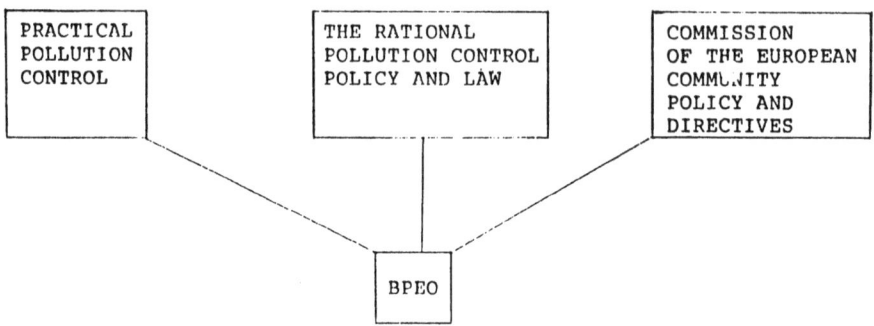

Source: Author and Helen Boutwood Imperial College, London.

Many people feel that guidance on a definition of BPEO and the different circumstances in which it should be used, is urgently needed; even if the definition and interpretation can be clarified, the concept still needs to be capable of being incorporated in the legal and planning framework of each country and there are those who feel that it would be difficult to implement BPEO consistently in the context of current institutional arrangements for controlling pollution. Let us hope that this Paper and the Address complementary to it, will provide acceptable and pollution free fuel for the necessary debate over the next few years.

R.G.P. Hawkins MA
General Counsel
FRSA FRGS MInstWM

Cornard Tye House
Nr Sudbury
Suffolk
CO10 0QA

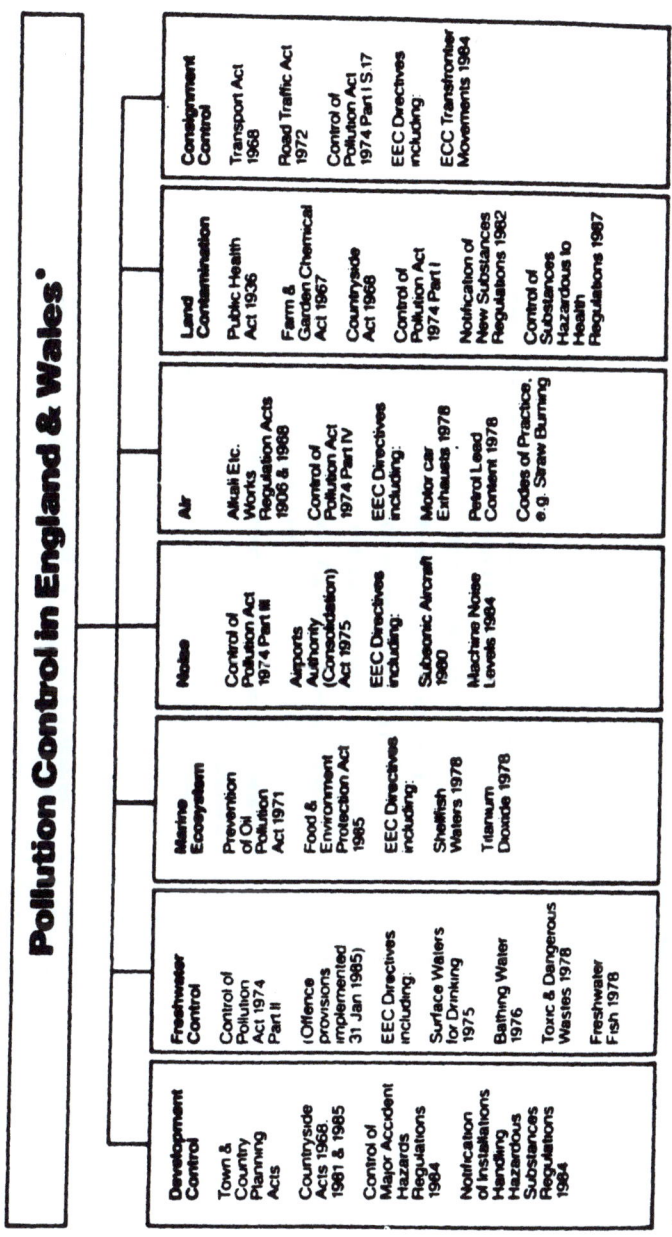

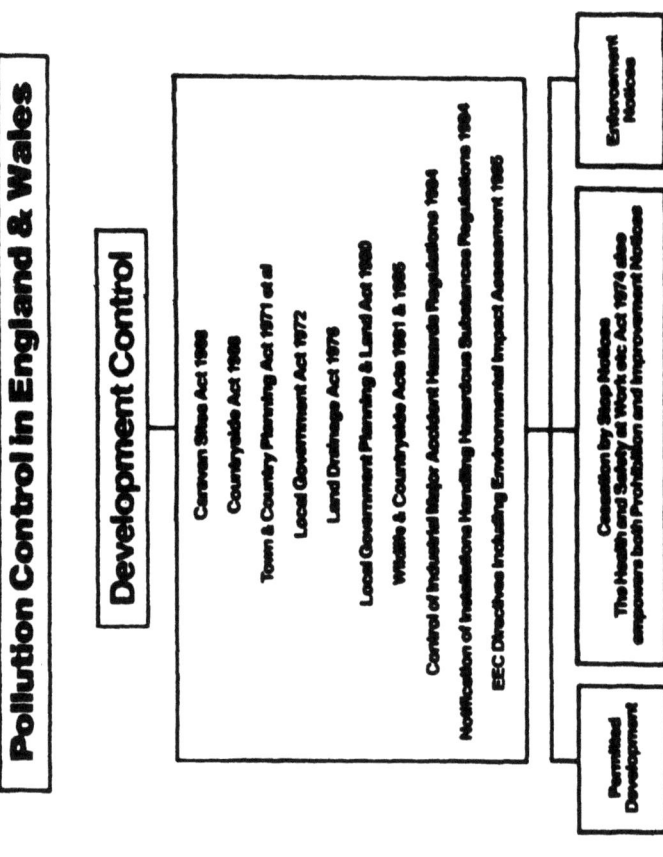

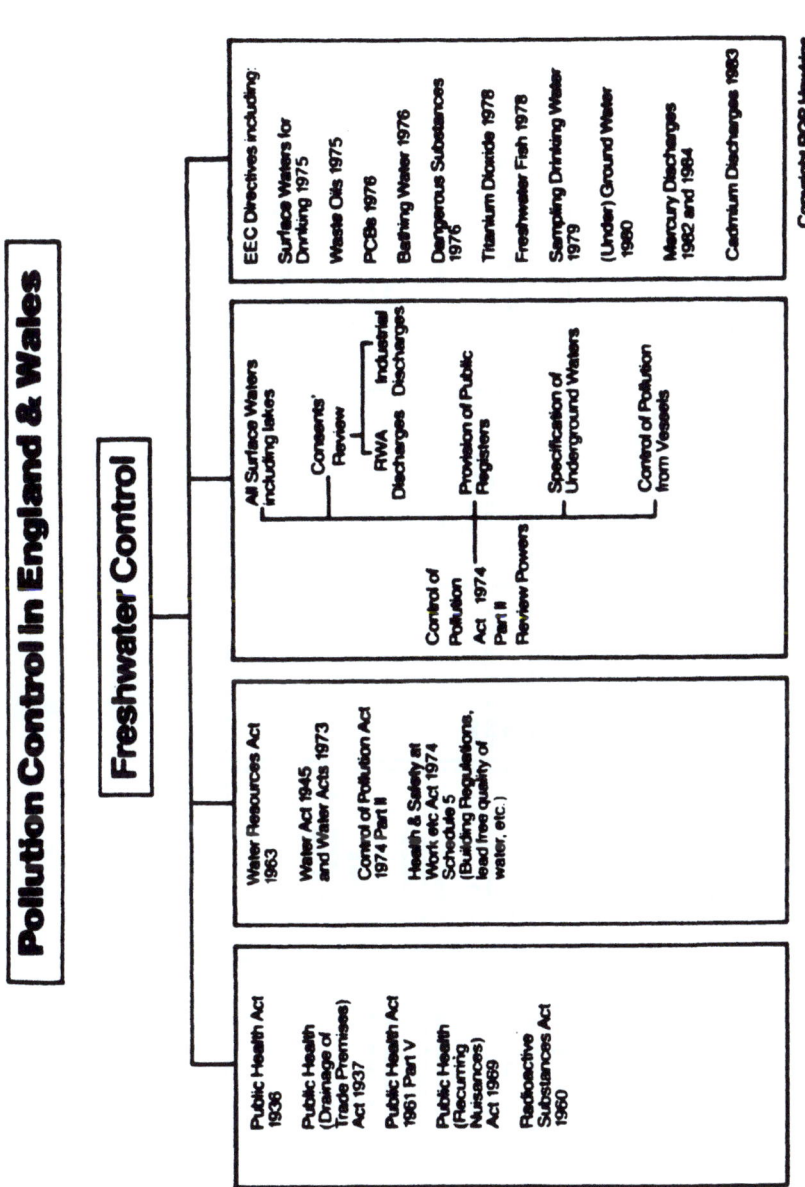

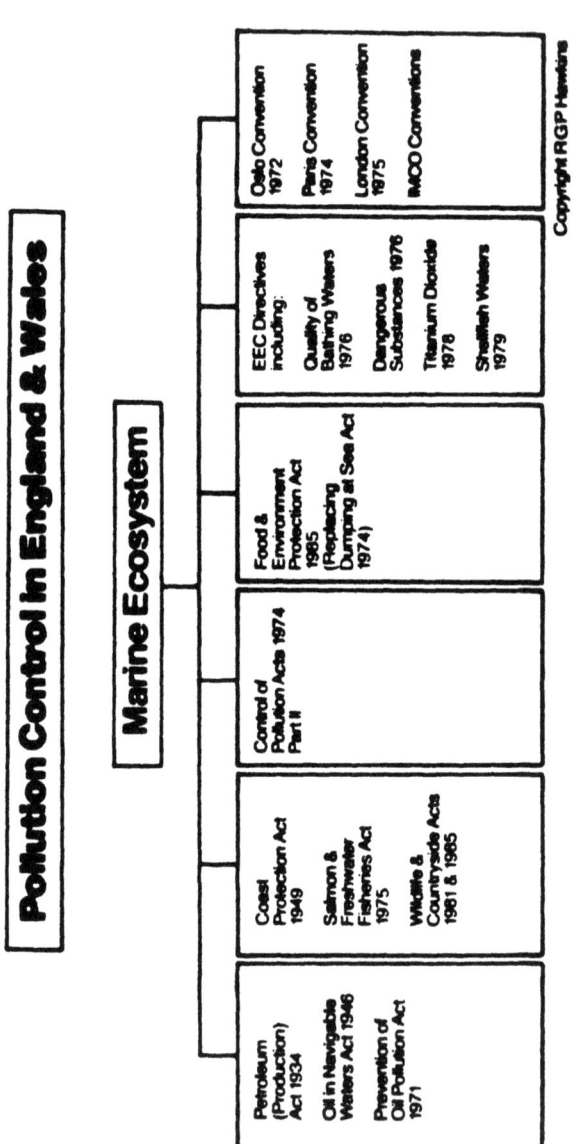

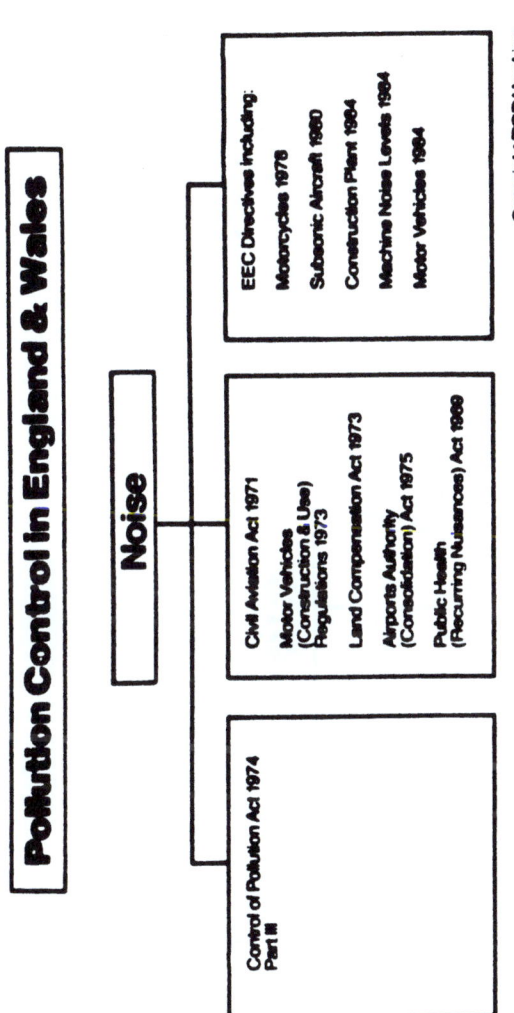

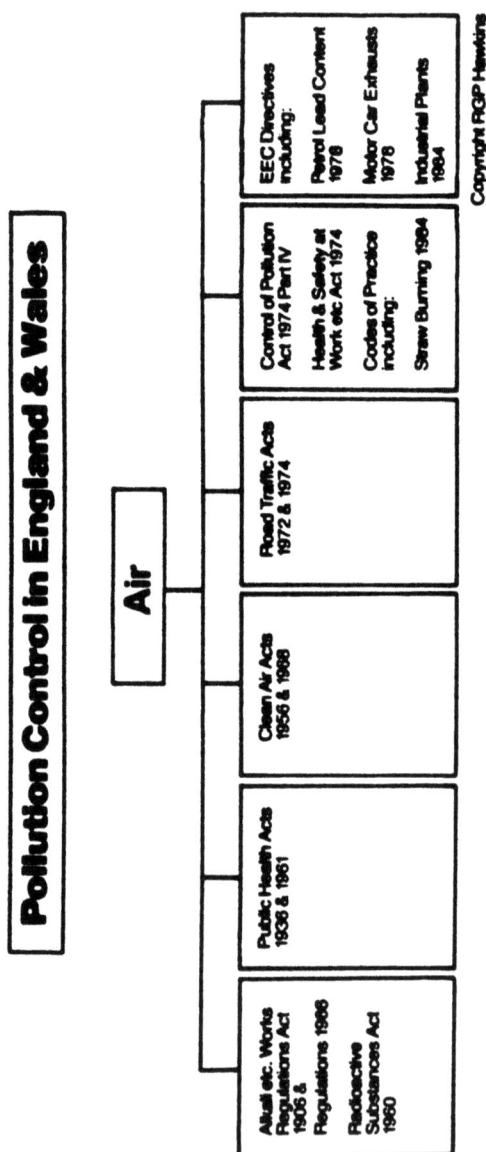

143

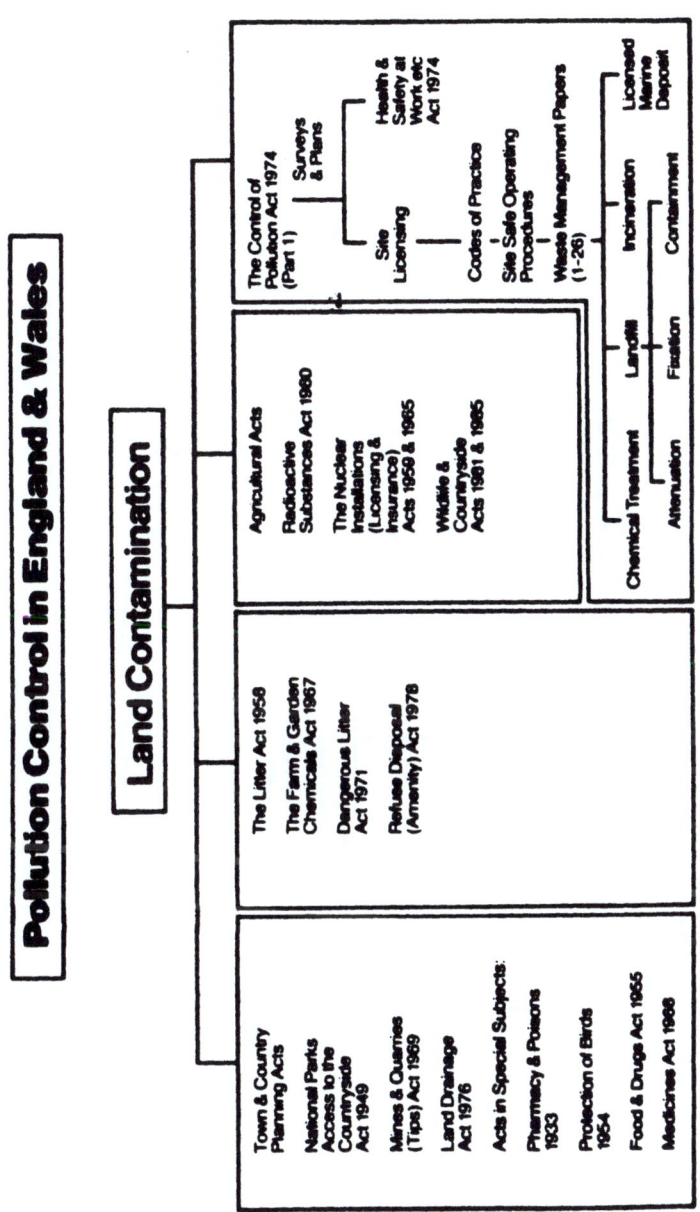

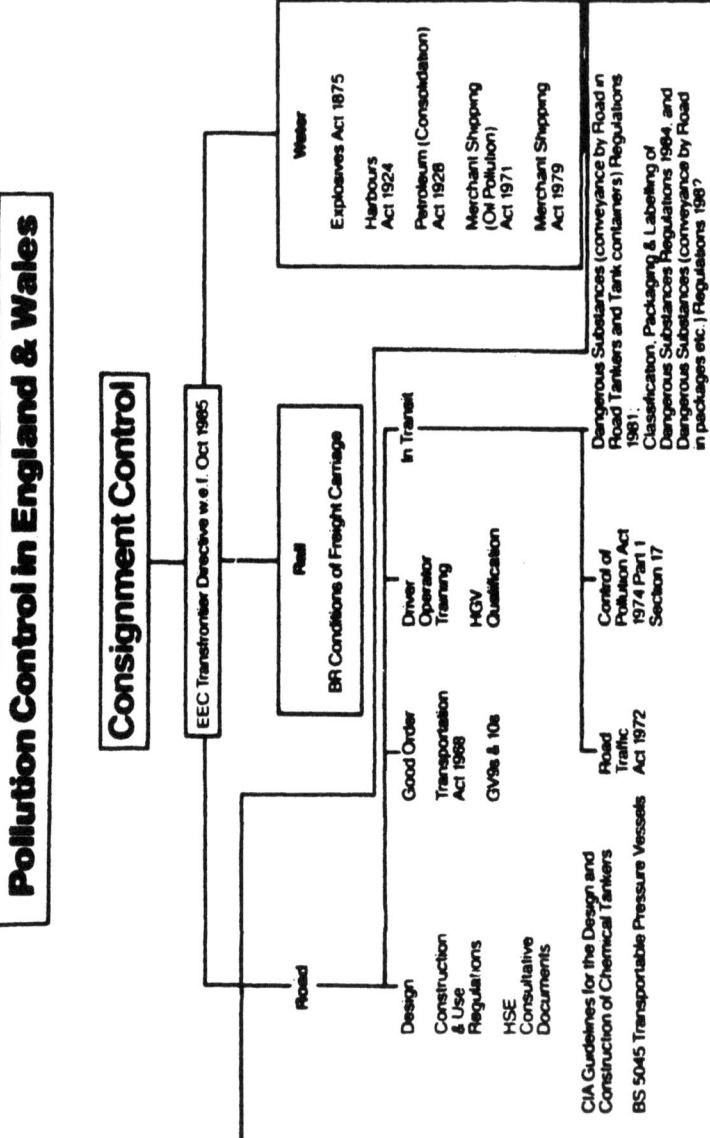

REMOVAL OF CADMIUM BY ANION EXCHANGE IN A WET PHOSPHORIC ACID PROCESS

T.T. Tjioe, P. Weij and G.M. van Rosmalen
Delft University of Technology, De Vries van Heystplantsoen 2, 2628 RZ Delft, The Netherlands

1. INTRODUCTION

Wet-process phosphoric acid is mainly used in the fertilizer production and serves as an intermediate between phosphate ore and phosphate containing fertilizers. The acid contains many of the impurities found in the phosphate ore. Part of these impurities is incorporated in the byproduct $CaSO_4 \cdot 1/2\ H_2O$ (HH) or $CaSO_4 \cdot 2H_2O$ (DH), which is mainly disposed.

At the moment a new wet phosphoric acid process is being developed at our laboratory aiming at the production of phosphoric acid and HH/DH both with a low content of cadmium. In this process Cd has to be removed from a 55-65% w/w H_3PO_4 solution at 90-100°C. In this study we examined the possibilities of removing Cd from the above mentioned solution by anion-exchange in the presence of complexing anions. The removal of Cd can be described with the following overall reactions:

$$Cd^{2+} + nL^- \rightleftarrows CdL_n^{(n-2)-} \qquad (1)$$

$$CdL_n^{(n-2)-} + (n-2)\overline{B^-} \rightleftarrows \overline{CdL_n^{(n-2)-}} + (n-2)B^- \qquad (2)$$

L = complexing anion
B = counter-anion of the ion exchanger (a bar above an ion indicates its presence in the resin sphere)

2. RESULTS

A series of experiments with different halogenides as complexing anions and a strongly basic anion exchanger was performed. Use of chloride resulted in a poor capture of Cd, whereas bromide and iodide both gave a significant reduction of the Cd content in the phosphoric acid solution. Iodide was much more effective than bromide. Subsequently experiments with various types of anion exchangers were performed using iodide as a complexing anion. Three types of anion exchangers were tested:

1) strongly basic $PS-CH_2N(CH_3)_3^+$
2) weakly basic $PS-CH_2N(CH_3)_2$
3) very weakly basic $PS-C_6H_4NH_2$
(PS = cross-linked polystyrene)

In the concentrated acid solution the last two resins are protonated and thus positively charged. From these experiments the following conclusions could be drawn.

Cd is captured in the anion exchanger as CdI_3^- ion and as CdI_4^{2-}. The Cd removal is strongly dependent on the iodide concentration and the phosphoric acid concentration. The stability constants of the CdI_i complexes both in the bulk phosphoric acid and in the resin are larger than in diluted aqueous media. The order of magnitude of the stability constants can be estimated from the distribution of Cd and iodide over both phases.

Iodide is also strongly retained by the anion exchanger. This phenomenon can be used to reduce the iodide content in the phosphoric acid stream after removal of Cd. Separate experiments in Cd-free phosphoric acid were performed to study the iodide removal.

The stability of the anion exchanger and the kinetics and selectivity of the Cd removal are also subject of our investigation. After keeping the resin for 12 weeks in a 70% H_3PO_4 solution at 90°C no loss of capacity could be detected. Cd removal experiments in diluted merchant-grade "black" phosphoric acid from a Nissan H process were performed to study the selectivity. The capacity for Cd is smaller than in chemically pure phosphoric acid indicating competition with other ions. The nature of the competing ions is also studied.

Some other experimental results are shown in fig. 1 and 2.

3. PROPOSED PROCESS SCHEME

Since iodide is expensive, the loss of this anion must be minimized. After the Cd removal step iodide can be removed from the phosphoric acid by a weakly basic anion exchanger. The anion exchanger, which is partly loaded with iodide, can be regeneratedwith a basic solution. The iodide in this solution is captured with a strongly basic resin and the iodide containing resin is transported to the Cd-removal section. The Cd-loaded resin can be regenerated with a diluted phosphoric acid solution. After precipitation of the Cd the diluted phosphoric acid solution is recycled into the phosphoric acid process.

FIGURE 1.
Distribution coefficients of Cd, $[\overline{Cd}]_t/[Cd]_t$, at different [HI] and % H_3PO_4 (strongly basic resin, c.p. phosphoric acid, 90°C).

FIGURE 2.
Elution of Cd from a fixed bed of strongly basic resin with a 1.4% H_3PO_4 solution.
Experimental conditions: 52 g of SB (air-dry weight) containing 780 mg Cd, 90°C. At the start the column contained a 56% H_3PO_4 solution. At m = 0.6 kg of effluent 100 g of 56% H_3PO_4 was introduced into the column to demonstrate the reversibility of the process.

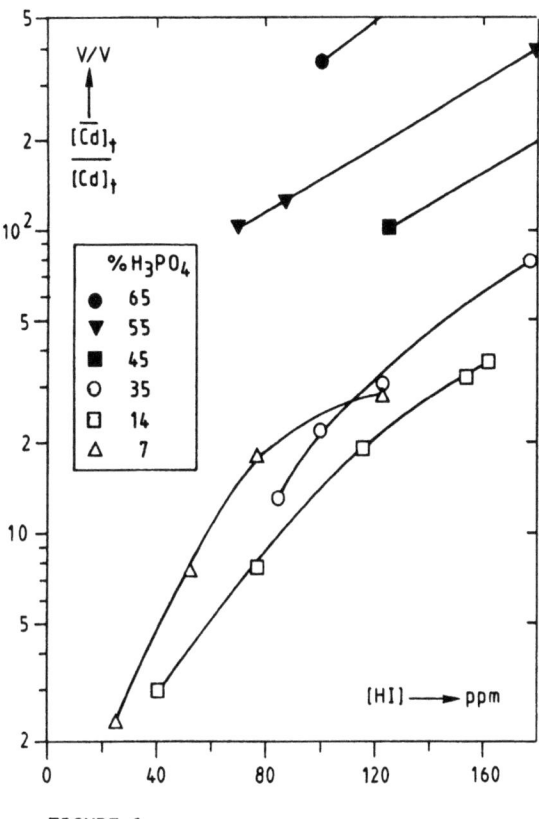

FIGURE 1.

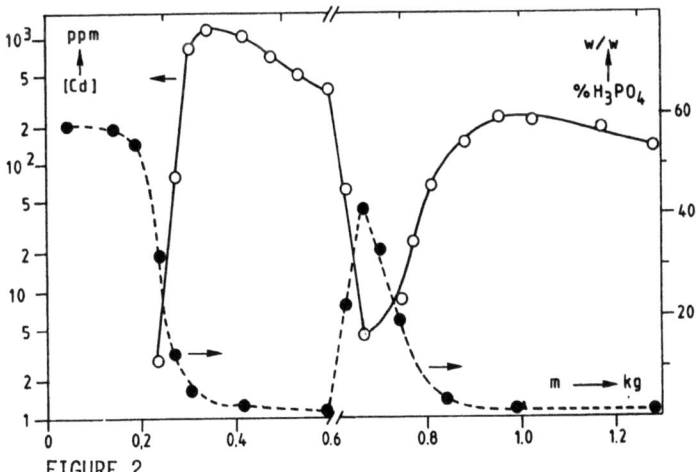

FIGURE 2.

CADMIUM INCORPORATION IN CALCIUM SULFATE MODIFICATIONS

G.J.Witkamp and G.M. van Rosmalen
Delft University of Technology, De Vries van Heystplantsoen 2
2628 RZ Delft, The Netherlands

In various industrial processes like wet phosphoric acid processes, hydrometallurgical zinc production and flue gas desulfurization, huge amounts of various calciumsulfate modifications are formed as a byproduct. These modifications can be dihydrate (DH) or gypsum, hemihydrate (HH) and anhydrite (AH). Many impurities persisting in the crystallizing solution are incorporated into the $CaSO_4$ lattice. Among these, cadmium belongs to the most unwanted, since its toxic character prohibits the disposal of calciumsulfate in the environment. To investigate the degree of Cd-uptake in the three $CaSO_4$ modifications, experiments were performed where $CaSO_4$ was crystallized from a wide variety of solutions under different operating conditions like temperature, calcium over sulfate ratio and supersaturation. Additionally the influence of some cadmium complexing agents on the Cd-uptake was studied.
In the experiments the Cd content of the solution ranged from 30-100 ppm. This content as well as the Cd content of the crystals was measured by ICP. The Cd incorporation is given by a partition coefficient D, defined as:

$$D = [Cd]/[Ca] \text{ (crystal)} / [Cd]/[Ca] \text{ (solution)}$$

Due to the low Cd concentration in solution with respect to the Ca content, the Cd uptake is directly proportional to the Cd concentration in the solution, which makes D here independant of [Cd] (solution).
Four types of experiments together with their results will be described below.

- Continuous crystallization experiments of HH in a 55 w% H_3PO_4 solution at 90°C - [1]

During these experiments a H_2SO_4 solution and a solution of 3.5 w% Ca in phosphoric acid were continuously fed into a 1 liter crystallizer, while unclassified withdrawal of the HH suspension occurred simultaneously. Residence times of 20 to 60 minutes were maintained, and the suspension contained 10 w% HH. The H_2SO_4 concentration in the solution was kept constant at a value between 0.25 and 3 w%, corresponding with Ca concentrations between 1 and 0.1 w%.

For a residence time of 20 minutes the uptake increases with increasing H_2SO_4 content of the solution. A typical D-value is about 10^{-3} for HH at 2 w% H_2SO_4. At residence times longer than 40 minutes minor amounts of AH up to 10% were developed next to the HH, as could be quantified by X-ray diffraction. In this AH a D-value of 10^{-2} was observed at 2 w% H_2SO_4. In AH the replacement of Ca by Cd ions apparently occurs ten times more frequently than in HH.

- Recrystallization experiments of HH into AH - [2]

In these experiments 15 w% HH was suspended in a solution of 55 w% H_3PO_4 and 0 to 2 w% H_2SO_4 at 90°C. Conversion into AH was completed in maximal one day. The observed D-value corresponds roughly with the 10^{-2} value mentioned above.

- Recrystallization experiments of HH into DH -

About 13 w% HH was brought in a solution containing 7 w% seed crystals. Conditions were selected where conversion into DH was completed in a few hours. The H_3PO_4 content was varied between 20 and 35 w%, the H_2SO_4 concentration between 6 and 12 w%, and the temperature between 50 and 70°C. The resulting D-value was approximately 10^{-3}, which is equal to the 10^{-3} from the HH experiments but considerably lower than the 10^{-2} from the AH experiments.

- Suspension growth experiments of DH -

In a suspension, containing less than 1 w% DH in a 10^{-1} M $NaClO_4$ solution at 25°C, well defined, aged seed crystals are allowed to grow to up to 200% of their original volume at a constant supersaturation. A D-value of about $5 \cdot 10^{-4}$ is obtained for supersaturations ranging from zero to 30%. This D-value is almost equal to D-value found in the recrystallization experiments where HH was converted into DH in phosphoric acid. This agreement is surprising, because the crystallizing conditions are very different.

In order to reduce the Cd uptake in the $CaSO_4$, the addition of Cd-complexing agents was applied. Halogenides are known to form complexes with Cd in aqueous solutions [3]. For each of the types of experiments described above, the influence of halogenides on the Cd incorporation was tested.

In general it can be concluded that the effectivity decreases in the order $I^- > Br^- > Cl^-$. For instance, in the 55% H_3PO_4 experiments with HH, 50% reduction in Cd uptake can be achieved by adding $5 \cdot 10^{-3}$ mol I^- or $30 \cdot 10^{-3}$ mol Cl^- per kg solution, which can be seen in the figure below, where the reduction in Cd uptake is plotted versus the concentration of halogenide. For the suspension growth experiments the required amounts of halogenide for a certain percentage reduction are two times larger.

Since at comparable conditions, i.e. 3 M $NaClO_4$ at 25°C, the complexing constants are known, it could be demonstrated that only the free Cd^{2+} ions in the solution contribute to the measured incorporation. In none of the experiments uptake of halogenides could be detected.

INFLUENCE OF HALOGENIDES ON CADMIUM UPTAKE

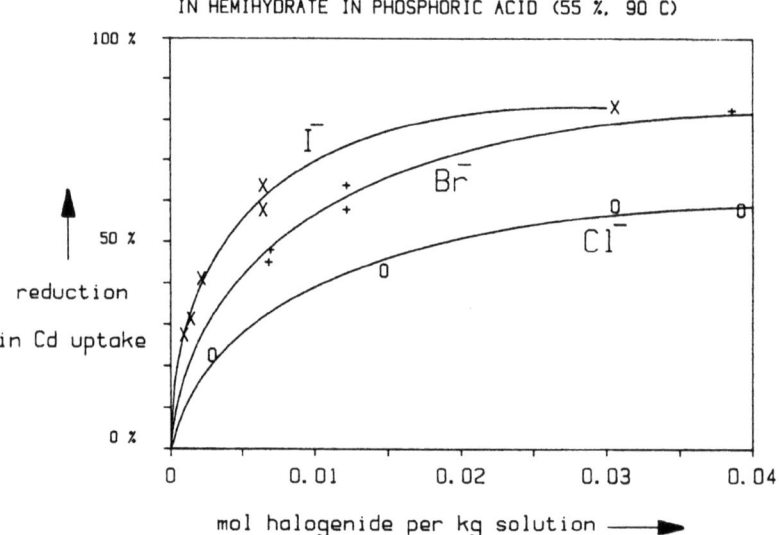

REFERENCES
[1] S. van der Sluis, G.J. Witkamp and G.M. van Rosmalen, J.Crystal Growth 79 (1986) 620
[2] G. J. Witkamp, S.P.J. Schuit and G.M. van Rosmalen, condensed papers of the Second International Symposium on Phosphogypsum, University of Miami, 1986 p. 106.
[3] I.M. Kolthoff and P.J. Elving, "Treatise on analytical chemistry", Vol. 3 Part 2, Interscience N.Y. (1961).

REMOVAL OF CADMIUM FROM PHOSPHORIC ACID SOLUTIONS BY SELECTIVE COPRECIPITATION IN A MINOR AMOUNT OF ANHYDRITE

J.A. KROON, DSM Research BV (PT/KG), P.O. Box 18, 6160 MD Geleen, The Netherlands

To reduce the environmental load of Cd from phosphate fertilizer processes its level in products and calciumsulphate byproducts (as hemihydrate (HH) or dihydrate (DH)) has to be lowered.

It is well known that Cd coprecipitates with calciumsulphate. Experimental evidence shows the affinity for Cd incorporation in calciumsulphate anhydrite (AH) to be much higher than that for HH or DH. Removal of a substantial part of Cd incorporated in a minor amount of AH-precipitate can be obtained especially when operating at very low Ca-concentrations in solution (corresponding to high sulphate-concentrations). Process conditions that allow formation of AH at acceptably short residence times lie in or near region III of figure 1. Working up the separated AH is subject of future study.

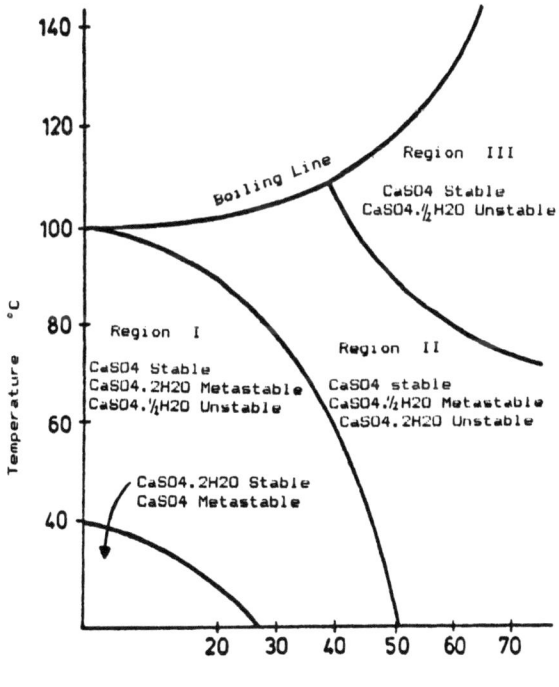

Figure 1. Stability diagram. $CaSO_4$-phases in the $CaSO_4$-H_3PO_4-H_2SO_4-H_2O system (H_3PO_4 expressed as P_2O_5).

The removal technique can be applied in existing nitric phosphate processes and wet phosphoric acid processes after the Ca-removal step. However that would still leave Cd in the calciumsulphate byproducts of phosphoric acid processes. Therefore a new integrated process is designed for the production of clean phosphoric acid, whereby HH low in Cd-content can be produced (figure 2). The conditions for formation of AH are made by introducing all concentrated sulphuric acid into the recycle stream together with a source of calcium. After separation of AH the purified stream is used for digestion of the phosphate rock.

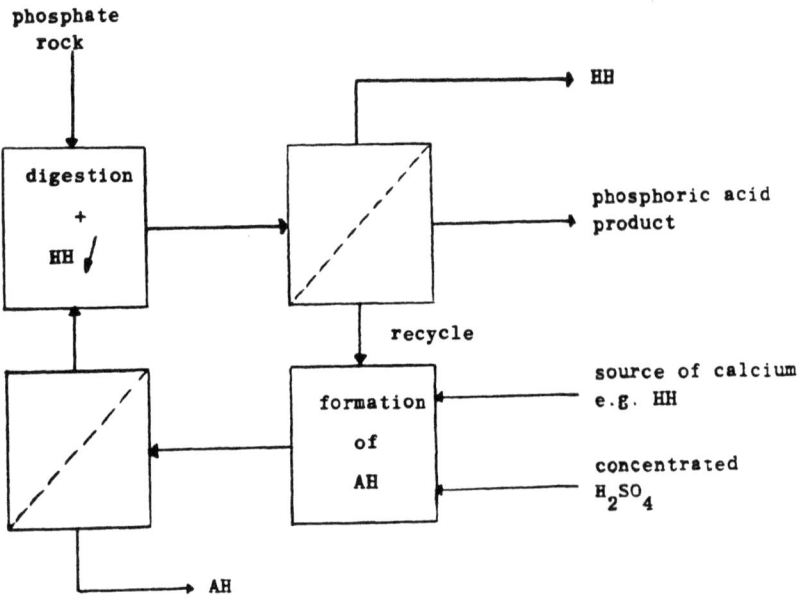

Figure 2. A variant of the new integrated wet phosphoric acid process, producing clean phosphoric acid (40 % P_2O_5, 1 % H_2SO_4, 7 ppm Cd) and HH (1 ppm Cd) at 90 °C and a recycle/product ratio of 4. About 5 % of the calcium sulphate formed is AH containing 200 ppm Cd. The Cd-content of the phosphate rock is 19 ppm.

The anhydrite method adds no foreign operations or materials to the traditional processes, so it is expected to be economically attractive for the removal of Cd.

Possibly also other contaminents are incorporated in AH and removed. This will be subject of future investigations.

A CLEAN TECHNOLOGY PHOSPHORIC ACID PROCESS

S. VAN DER SLUIS AND G.M. VAN ROSMALEN
Department of Chemical Engineering
Delft University of Technology, De Vries van Heystplantsoen 2,
2628 RZ Delft, The Netherlands

Phosphoric acid for use in fertiliser applications is mainly produced by a 'wet-process', i.e. by digestion of phosphate ore with sulphuric acid. In such wet-processes, however, impurities like Cd and Ra, originating from the phosphate ore are distributed between the phosphoric acid and the byproduct, a calcium sulphate modification.
The use of the phosphoric acid as well as the use or disposal of the byproduct is limited by its impurity content.
The aim of the newly developed process is the direct production of concentrated phosphoric acid with a low Cd content, as well as the production of a major amount of calcium sulphate as hemihydrate (HH) with a low phosphate, Cd and Ra content, in a commercially feasible way. For this purpose the phosphate ore is first completely digested in recycled phosphoric acid, containing about 40 w % P_2O_5 and 1.8 w % H_2SO_4. After filtration of the insoluble ore residue, together with a minor amount of calcium sulphate hemihydrate, precipitated in the digester, a clear mono-calcium-dihydrogen-phosphate (MCDHP) solution is obtained. From this solution the Cd-ions can be removed by e.g. ion-exchange. Thereafter the remaining calcium ions are removed by adding concentrated sulphuric acid to the MCDHP-solution in the crystallizer at 90 °C. In this way a clean calcium sulphate hemihydrate can be obtained.
In order to optimize the individual process steps, each step had to be investigated separately.
By performing HH crystallization experiments a linear relationship was found between the molar phosphate over sulphate ratio in the crystals and in the solution. The phosphate content of the crystals decreased until a value below 0.1 w % P_2O_5 with increasing sulphate concentration in the crystallizer. The Cd incorporation was also measured as function of the operating conditions and appears to increase significantly when raising the sulphate concentration above 2 w % H_2SO_4.
Filtration studies showed that the HH crystals obtained during digestion of the ore, are difficult to filtrate, while the HH crystals developed in the crystallizer, filter quite well. A maximum filtration rate was reached, if a sulphate content of about 1.8 w % H_2SO_4 was maintained in the crystallizer.
A study of the fluor removal delivered an expression for the fluor distribution coefficient as a function of the operating conditions, prevailing in the various process steps.
Preliminary results from a continuously operated lab-scale plant, in which part of the process is studied, were in agreement with the results obtained from the separate studies.

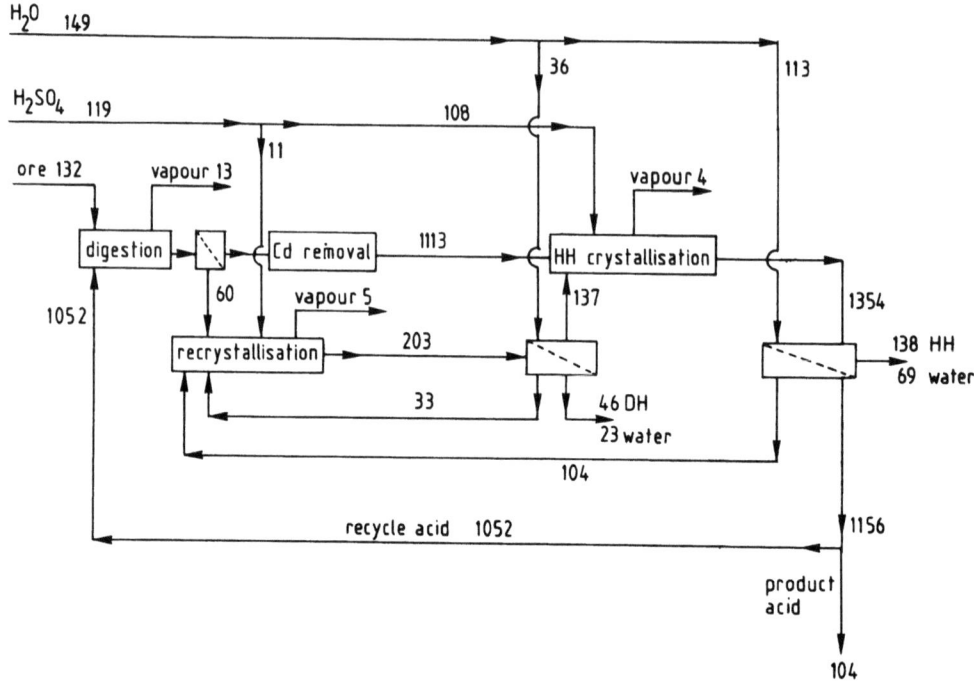

A preliminary proces flowsheet and a mass balance (see figure) show, in combination with our experimental results, that concentrated phosphoric acid (40 w % P_2O_5) with less than 5 ppm Cd as well as clean calcium sulphate hemihydrate with less than 1 ppm Cd and less than 0.2 % P_2O_5 can be produced with our new process.

TREATMENT OF INDUSTRIAL EMISSIONS/WASTE

**Second European
Conference on
Environmental Technology**

in the 'European Year
of the Environment'

ABATEMENT OF HCL AND HF EMISSIONS FROM WASTE INCINERATORS BY INJECTION OF HYDRATED LIME

A. Verbeek, D. Schmal, C. van der Harst,
TNO Division of Technology for Society.

SUMMARY
Investigations were carried out to optimize the design of a dry hydrated lime injection system to reduce the emissions of HCl and HF from waste incinerators. At laboratory scale the chemical aspects of HCl and HF reactions with lime were examined. In a cold model the optimum configuration of the injection sites was determined. With the results obtained an existing lime injection system of an incinerator was modified and the HCl removal efficiencies are measured.

INTRODUCTION
At the incineration of municipal waste, acid gases are emitted which are harmful for the environment. Two of the more important ones are hydrochloric acid (HCl), generated in the combustion of PVC etc., and hydrofluoric acid (HF) from aerosol spray tins etc. Their concentrations are such (HCl: 500-1.000 mg/Nm^3, HF: 1-10 mg/Nm^3) that it has become necessary to restrict emissions to an acceptable level.
In the Netherlands discrimination is made between new installations and existing ones. Requirements for new installations are more strict (HCl<50 mg/Nm^3, HF<2 mg/Nm^3) and can presently only be fullfilled by using separate dry, semi-dry or wet scrubber systems, installed between the incinerator and the chimney. The cost of these systems are high (Dfl. 20-40 per ton of waste [1,2]).
For existing incinerators removal requirements are mostly less strict. In this case a much simpler and cheaper method of pneumatic injection of hydrated lime injection in the incinerator itself can be applied. The feasibility of this method was demonstrated already some years ago using a system developed by "trial and error" [3]. Using a chemically reactive hydrated lime at the rate of 2 kg/ton of waste a HCl removal efficiency of about 30% could be obtained by injection above the incinerator grate. The costs of this system amounted to about Dfl. 1 per ton of waste [3].

With the aim to develop design rules for other installations and, if possible, to improve removal efficiency, a project was started to study the various aspects of the process of lime injection more fundamentally.
The project consisted of three parts:
- Laboratory investigations on the chemical aspects of HCl and HF removal by lime injection under the conditions occuring in waste incinerators, with the purpose to find the optimum conditions (site of injection, temperature, sorbent type) for injection.
- Investigations on the fluid dynamic aspects of the pneumatic injection using a cold incinerator model, with the purpose to find the conditions for fast and complete mixing of the injected lime with the incinerator flue gas.

- Testing the results of the above investigations in an actual waste incinerator in the city of The Hague.

The investigations have been carried out by TNO, in cooperation with Nekami (designer of the injection system, supplier of sorbents) and the Municipality of The Hague (owner of the massburning installation). The project was partly funded by the Dutch Ministry of Housing, Physical Planning and Environment and the Ministry of Economic Affairs in the framework of the Clean Technology Program.

In the following some typical examples of results will be discussed.

CHEMICAL ASPECTS OF HCL AND HF REMOVAL BY REACTION WITH LIME
Conditions

The chemical reactions occuring between the injected lime and the flue gases are dependent on flue gas composition, temperature and type of sorbent used. In Table 1 the average composition of an incinerator flue gas is given.

Table 1 composition of wet flue gas:

Compound	Concentration
N_2	73 vol %
O_2	10 vol %
CO_2	7 vol %
H_2O	10 vol %
HCl	500-1000 mg/Nm^3
SO_2	100- 200 mg/Nm^3
HF	1- 10 mg/Nm^3

The temperature, as far as it concerns the reaction with lime, decreases from about 1000 - 1200° C, just above the grate, to about 280° C in the electrofilter (see Figure 1). The maximum residence time between lime and flue gas (injection at or near the grate) is about 4 s.

The hydrated lime used for the injection is a Nekami-product called "Edelwit" consisting for 94-97% of $Ca(OH)_2$ with a mass median diameter of 8 μm and a specific surface area of 17 m^2/g (BET).

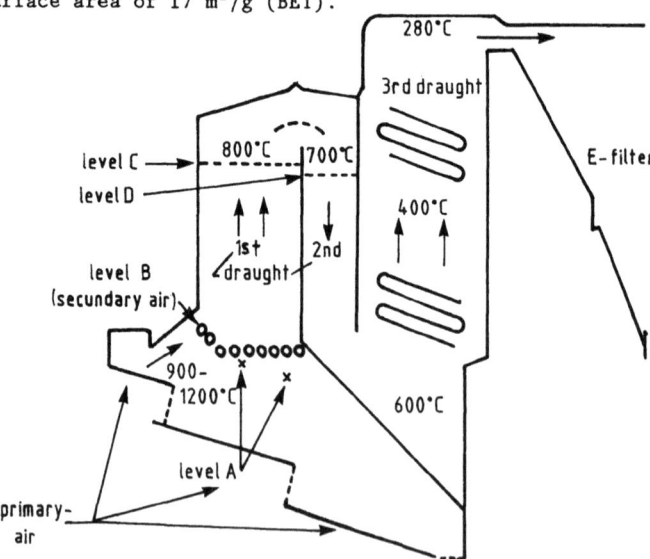

Figure 1: The Hague massburning installation; schematic cross section (capacity 12.5 ton municipal waste/hour).

Chemical reactions

When hydrated lime is injected at temperatures above 400-500° C (first and second draught, see Figure 1) the following reactions occur in order of decreasing temperature:

$Ca(OH)_2$	$\rightarrow CaO + H_2O$	$T > 400°\ C$
$CaO + SO_2 + \frac{1}{2}O_2$	$\rightarrow CaSO_4$	$T < 1000°\ C$
$CaO + 2\ HCl$	$\rightarrow CaCl_2 + H_2O$	$T < 700 - 750°\ C$
$CaO + CO_2$	$\rightarrow CaCO_3$	$T < 700°\ C$
$CaO + 2\ HF$	$\rightarrow CaF_2 + H_2O$	$T < 600 - 700°\ C$
$CaO + H_2O$	$\rightarrow Ca(OH)_2$	$T < 400°\ C$

Taking into account the composition of the flue gas, it will be clear that a number of competing reactions can occur simultaneously.
The temperature given for the various reactions have been calculated from thermochemical data, using the gas concentrations in Table 1.
The equilibrium concentrations at the lowest contact temperature (270° C in the electrofilter where the lime is removed from the gas) are 0.1 mg/Nm^3 for HCL and 1 µg/Nm^3 for HF, making high removal efficiencies possible from a theoretical point of view.

Experimental

Because literature on the reactions occuring is scarce and restricted to low temperature experiences (non-integrated scrubber systems), laboratory tests were carried out to get insight in the chemistry of the process at temperatures above 200 to 300° C. Techniques used were thermogravimetry (TG) for measurements on the chemical reactions given above, scanning electron microscopy (SEM) in combination with energy dispersive analysis (EDS) for structure investigations (sintering etc.) and the BET method for measuring specific surface area of the sorbents used (a measure for the chemical reactivity).

Results

In Figure 2, as an example, the degree of conversion for the various products measured at 300° C is given as a function of "pre-treatment" temperature. It is a simplified simulation of what happens upon the injection of lime at this pre-treatment temperature and the reaction at lower temperatures during the contact between lime and flue gas.
At low temperature injection the sum of the degrees of conversion for the most important products amount to about 100%, leading to the conclusion that pore plugging at these temperatures does not occur. When injecting at higher temperatures conversion decreases due to sintering effects (the remaining part stays in the CaO form). Sintering effects are illustrated in Figures 3 and 4 showing SEM pictures of the same hydrated lime after treatment at two different temperatures.
BET measurements confirm that sintering is responsible for this decrease in total degree of conversion; at temperatures above 400° C, where $Ca(OH)_2$ decomposes, the BET surface area (after an initial increase to 30 m^2/g due to decomposition) decreases more or less linearly with temperature to about 1 m^2/g at 1100° C.
The degree of conversion of the desired reaction ($CaCl_2$ formation) found in the laboratory experiments is about 15-20% (at a HCl concentration of 400 mg/Nm^3). In a first approximation it is proportional to HCl concentration. Therefore the degree of conversion in practice will be in the range of 20-40%, leading to the conclusion that, in order to obtain high removal efficiencies, stoichiometric ratios of 2-5 should be required. These ratios correspond to those applied in practice in separate dry scrubber systems [2].

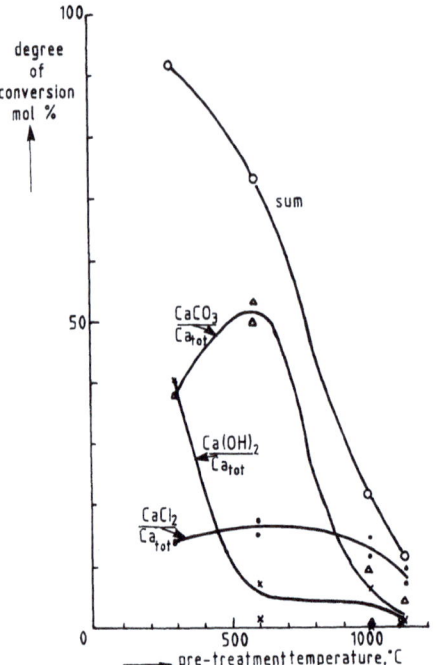

Figure 2: Conversion at 300° C for various pre-treatment temperatures.

Very important from a practical point of view is that the degree of conversion of CaO into $CaCl_2$ is hardly dependent on pre-treatment temperature (below 1000° C). This means that the choice of the injection site for HCl removal is not very critical. Of course the injection site must be chosen such that - on the one hand - temperature is not too high and - on the other hand - residence time of the lime in the flue gas is long enough for the desired reaction to take place.

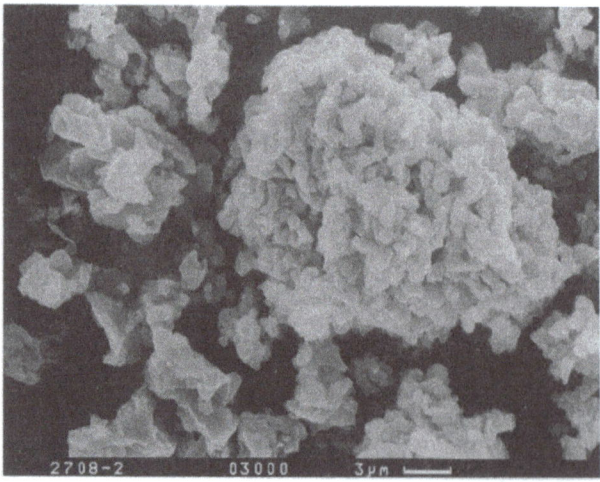

Figure 3: SEM picture of hydrated lime after treatment at 570 °C

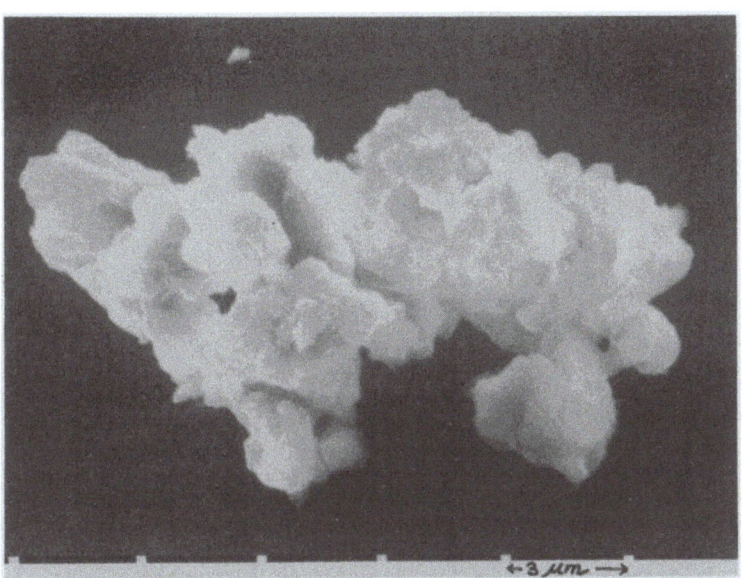

Fig. 4: SEM picture of hydrated lime after treatment at 1125° C.

INVESTIGATIONS ON THE FLUID DYNAMIC ASPECTS OF LIME INJECTION
Incinerator model and measuring techniques

For these investigations a "cold" incinerator model (scale 1:15, see Figure 5) of the first and second draught was constructed.

From theoretical calculations on the trajectories of the lime particles it could be concluded that most of them, due to their small size, follow the streamlines of air flow. Therefore lime injection was simulated with a tracer gas (isobutylene) for which the concentrations were measured downstream of the injection sites chosen.

Flow was vizualized by using a "smoke generator" and by making photographs of the smoke plumes in the transparent plastic model.

Averaged velocities were measured with hot-wire anemometers.

Injection sites

On the basis of the investigations on the reaction chemistry, the flow profile of the flue gases and the accessibility in the real incinerator, three levels were chosen for injection of tracergas in the model (see Figure 1: levels B, C and D). For comparison experiments were also performed on the current injection site (level A in Figure 1). Variables investigated were the injection flow velocity and configuration of the nozzles at one level. At level B apertures are used which are already present in the incinerator for secondary air injection.

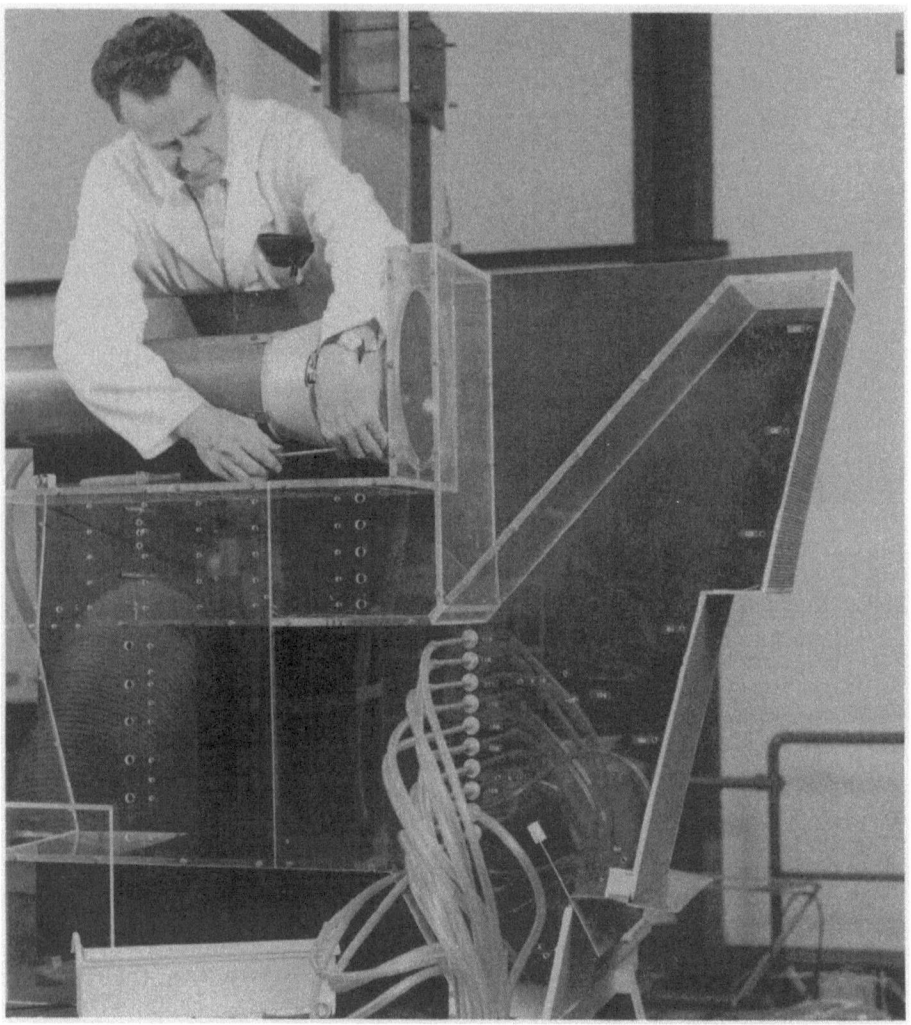

Fig. 5: Picture of the "cold" incinerator model. The model is canted (grate at right side).

Results

As an example of the results found a normalized tracer concentration profile is given in figure 6, measured about 0.5 m (in reality 7,5 m) downstream of the injection points at level A. It can be seen that the tracer gas is well mixed in a relatively short distance, showing that the empirically selected injection site A was chosen well from a point of view of mixing of lime with flue gas.

Best results for mixing of lime with flue gas were found at injection on the current level A and level C, both situated in the first draught.

On basis of these results it was decided to do injection tests in the waste incinerator on level C and also D for comparison.

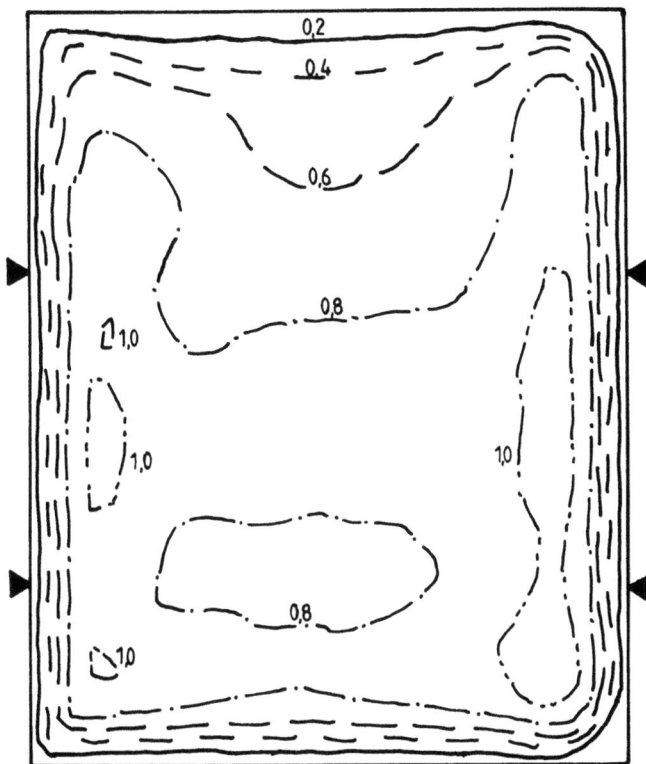

Fig. 6: Isoconcentration lines (C measured/C max.); injection at four sites (>) on level A (see figure 1).

EXPERIMENTS IN THE INCINERATOR
Field tests
The field tests have been carried out in one of the massburning installations of the Municipality of The Hague.
First injection tests were performed on level C and D and the results were compared with the performance on level A (initial injection level). On level D a lower HCl removal efficiency was found. This is explained by the worse mixing conditions on that level, as found in the previously described "cold-model" tests. SEM pictures of lime particles found in the flyash showed that the injected lime was not subjected to sintering (temperature about 600° C). Injection on level C gave better results which are comparable with those at the initial injection level A. SEM pictures of the injected lime showed a slight sintering effect (temperature about 800° C), while lime injected on level A was seriously sintered. Therefore level C was chosen as the optimum injection point.
Subsequently tests were done with the injection of various quantities of lime on level C. This optimalisation process has not yet been finished, but first results show a HCl capture of about 40% at a lime flow of 3 kg per ton of waste.

Operating experience
The injected lime may give rise to fouling on the tube-banks. This depends largely on the configuration of the incinerator (first draughts empty or filled with tube-banks, metal temperature of tubes) and proper distribution of the lime over the injection nozzles. Normally the operating time between downtime periods for cleaning will be slightly shorter when injecting lime. The design of the pneumatic conveying system of the lime to the injection nozzles is critical to prevent blockage in the dosing lines. Inspection of the critical points (splitters, nozzles) once per shift is recommended.

CONCLUSIONS
- From the laboratory work a good insight into the chemical aspects of HCl removal by reaction with lime has been gained. The design of a lime injection system can be supported now by a scientific explanation of the occuring phenomena and is not merely based on experience.
- Normally "cold model" work is required to find the optimum configuration of an injection site. Because the dimensions and consequently the flow pattern inside incinerators are different, no general design rules can be given.
- Lime injection downstream the combustion chamber is a relatively cheap and easily applicable process to reduce the HCl and HF emissions from existing massburning installations. The removal efficiency is lower than can be reached with a separate scrubber system, mainly because the quantities of lime used are much lower (2-4 kg per ton of waste versus 10-40 kg when using separate dry scrubbing).

OTHER INVESTIGATIONS/APPLICATIONS
- Various other sorbents are and will be investigated on their merits for the purpose of improved HCl removal (Nekami, TNO).
- For another massburning plant in the Netherlands (AVR near Rotterdam) investigations were performed by TNO and Nekami for selective removal of HF from incinerator flue gas. Conditions could be found where more than 70% removal was obtained at a lime consumption of about 4 kg per ton of waste.
- Based on the knowledge obtained Nekami designed a lime injection system for the Amsterdam massburning plant (4 incinerators each having a capacity of 16 ton waste/hour).
- Tests for HCl removal in other incineration plants are underway.
- Tests with higher quantities of lime are recommended to improve the removal efficiency. However attention should be paid to the higher fouling tendency.

REFERENCES
1. Westerhuis, A.
   Clean incineration possible, but expensive.
   Pt-Procestechniek 39 (1984) 11, 60-61 (in Dutch).
2. Reimer, H.
   Anwasserlose Rauchgasreinigung in Abfallverbrennungsanlagen, Stand der Entwicklung, Perspectiven.
   Staub-Reinhaltung Luft 43 (1983) 1,28-40.
3. Nocker de, A.N.; D.J. Wiersma,
   Flue gas cleaning in waste incineration.
   Pt-Procestechniek 30 (1984) 1, 21-26 (in Dutch).

Residual products from flue gas desulphurization by spray-dryer method - technical and economic aspects of their disposal and recovery for utilization

by H. Ludwig, M. Hetschel and H. Fitjer
FICHTNER, Consulting Engineers, Stuttgart, Federal Republic of Germany

For the desulphurization of flue gases produced by firing plants in electric power generation and in industry, the spray-dryer process has lately proven to be a competitive alternative to the wet process, particularly for smaller and intermediate-sized plants. In the course of flue gas cleaning by this process, dry products are left as residues, these comprising - according to how the process is controlled - a mixture of calcium salts (e. g. calcium chloride, calcium sulphate and calcium sulphite), excess slaked lime reactant and varying proportions of fly ash.

Apart from the degree of utilization of the slaked lime reactant, i. e. the amount of internal recycling of residues and thus the stoichiometry of the process, the following factors determine the composition of the residue

- for the chloride content, the amount of hydrogen chloride in the flue gases
- for the sulphite/sulphate content, the amount of sulphur dioxide in the flue gases and the ratio of sulphur dioxide to sulphur trioxide
- for the content of fly ash, the degree of particulate removal from the flue gases prior to their entry into the reaction zone of the spray absorber.

The composition of the residual product from a spray-dryer plant used for the desulphurization of flue gas from a slag-tap furnace in a power station fueled with bituminous coal is shown in Fig. 1.

SULPHUR CONTENT OF COAL = 0,9 WEIGHT % S
COMPOSITION OF FLUEGAS TO FGD:
　　CHLORIDE　=　47 - 188 mg HCl/Nm³ DRY GAS
　　FLY ASH　=　100 - 300 mg/Nm³ DRY GAS

Fig. 1: Residual product from spray dryer FGD-anticipated composition for firing with high-grade bituminous coal

COMPONENT	QUANTITY (WEIGHT %)
MOISTURE	< 3
CALCIUMSULPHITE CaSO$_3$	61±6
CALCIUMSULPHATE CaSO$_4$	5-20
CALCIUMHYDROXIDE Ca(OH)$_2$	5-20
CALCIUMCARBONATE CaCO$_3$	2-10
CALCIUMCHLORIDE	8,5±6
MAGNESIUM Mg	< 1
INERTS (SiO$_2$, FeO$_3$, Al$_2$O$_3$)	3-6
FLY ASH	~ 4

Prior to entering the desulphurization plant, particulates are removed from the flue gases by means of electrostatic precipitators down to a level of 100 - 300 mg/Nm³ (dry). There are two basic possibilities for disposal of the spray-dryer product, these being

- recovery for use
- landfill (<u>Fig. 2</u>).

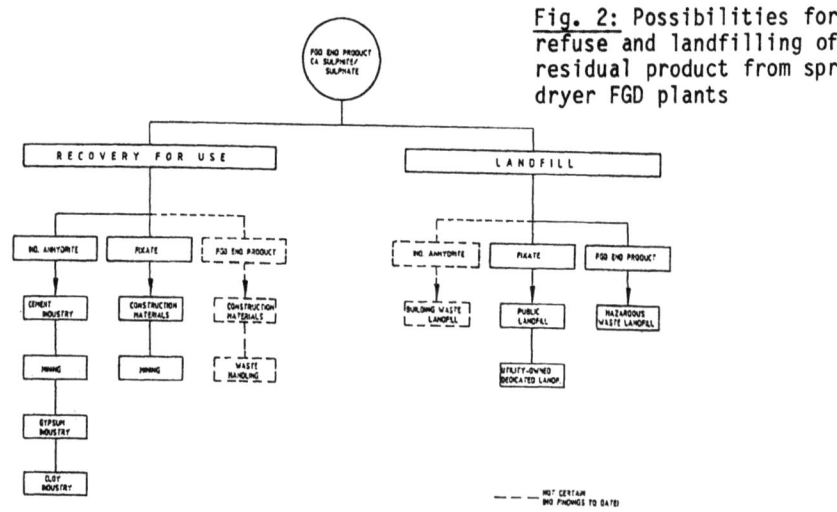

Fig. 2: Possibilities for refuse and landfilling of residual product from spray dryer FGD plants

Recovery for use:
In general, for the recovery option, the material can be used as follows

- the final product itself without modification
- following fixation, i. e. the admixture of substances which consolidate it and thus improve its leaching properties and structural strength
- as industrial-quality anhydrite following oxidation of the calcium sulphite in the product to calcium sulphate.

The prinicpal target market for the recovered material is the construction industry and companies producing raw construction materials. A major obstacle for its acceptance in these industrial sectors is, however, the chloride content of the FGD residual product, which when used directly in structures or in the production of contruction materials or components, can result in reduced product durability or give rise to corrosion problems. Consequently, its theoretically possible direct use in the construction industry is restricted to road works and below-ground civil engineering projects, in so far as the leaching properties of the product can be enhanced, for instance by a fixation process. This fixated product can be then employed as

- coarse beds and filling material in road and dam construction
- as construction aggregate for low-quality concrete blocks, such as paving setts in precast concrete building block works
- as a substitute for bedding courses bound with cement.

Also under laboratory-scale investigation is the use of the residual product with the admixture of sand, lime and fly ash for the manufacture of lime-cemented sandstone as well as the production of artificial gravel by pelletizing the fixated products.

In general, these application possibilities are conceivable. The necessary condition for the use of such materials, possibly as subgrade, as frost-resisting layers or as bedding courses in road building is, however, that the relevant technical standards and guidelines (i. e. TVE, TVT) be complied with. At present, there are no such proofs or suitability certificates available for materials derived wholly or partly from residual products. In the mining industry, the use of fixated product for the stowing of worked-out seams is a possibility. However, up to now there is no proven concept for the utilization of fixated product either in the construction materials industry or in mining.

A potentially interesting possibility is the utilization of the residual product in the cement industry in the form of industrial anhydrite, which is generated by high-temperature oxidation of the residual product. Gypsium and anhydrite are admixed to the burnt cement clinker prior to grinding, to act as a setting and hardening control agent. However, according to DIN 1164, the permissible maximum content of chloride in cement is 0.1 %. Thus the chloride content of the residual product, which lies in a range of 3 - 5 %, must be reduced to meet the requirements of the cement industry if it is to be used in this way. But for the initial "Residual product from spray-dryer FGD" it is hardly possible to obtain on an industrial scale residual chloride contents of less than 0.1 % in the industrial anhydrite so produced, which means that utilization in the cement industry will only represent a possibility if the high-chloride industrial anhydrite can be blended with low-chloride natural anhydrite or gypsum.

Landfill:
Due to its composition and leaching properties the untreated residual product must be disposed of in a landfill to class 4 or 5 according to the draft Guidelines of North Rhine-Westphalia for Water and Waste (Figure 3)

LANDFILL CLASS	TYPE	CRITERIA FOR: 1. TYPE OF WASTE 2. SITING AND 3. LANDFILL EQUIPMENT	MAXIMUM PERMISSIBLE LEACHATE VALUES (EXCL. HEAVY METALS)		
			COND. ms/m	pH	OTHER mg/l
1	DRY PILE	1. NON-CONTAMINATED MINERALS 2. OUTSIDE OF CATCHMENT AREAS FOR DRINKING WATER AND MINERAL SPRINGS	< 100	5,5-10	CSB < 20 LEACHATE ≈ DRINKING WATER QUALITY
2	MINERALS LANDFILL	1. BUILDING RUBBLE AND INERTS, POLLUTED 2. SAME AS LANDFILL CLASS 1 3. LANDFILL BASE ABOVE HIGHEST ANTICIPATED GROUNDWATER LEVEL	< 250	5,5-12	CSB < 50 NO_3 < 100 NH_3 < 5
3	LANDFILL FOR RESIDENT. WASTE	1. DOMESTIC AND TRADE WASTES OF SIMILAR NATURE 2. SAME AS LANDFILL CLASS 1 -SUBGRADE PERMEABILITY <10^9 m/s -LEACHATE COLLECTION AND TREATMENT		5,5-12	
4	LANDFILL FOR TRADE AND INDUST. WASTE	1. WASTES FROM TRADE AND INDUSTRY 2. SAME AS LANDFILL CLASS 3 3. LEACHATE COLLECTION AND TREATMENT	SPECIAL MEASURES FOR LEACHATE TREATMENT		
5	LANDFILL FOR SPECIAL- CATEGORY WASTES	1. SPECIAL-CATEGORY WASTES FROM TRADE AND INDUSTRY 2. SAME AS LANDFILL CLASS 3. IMPERMEABLE NATURAL SUBGRADE SEVERAL m THICK 3. LEACHATE COLL. AND TREATMENT			
6	UNDER- GROUND DISPOSAL	1. TOXIC WASTES UNSUITABLE FOR DISPOSAL IN LANDFILL CLASSES 1 TO 5	NO INFLUENCE ON WATER RESOURCES		

Fig. 3: Classification of landfills in the draft guidelines of the establishment for water and waste, state of North Rhine-Westphalia, issued december 1984

A major reason for this is the high salt content in the leachate from the product, as well as its enhanced content particularly of amphoteric heavy metals and chromates. By admixing fly ash and/or cement, though, a fixated product is obtained which from the point of view of its permeability and structural strength is suitable for emplacement in a dedicated landfill (i. e. for FGD residual product alone), and also exhibits reduced leachability of the neutral salts and heavy metals as well as of the sulphite content.

When comparing the various possibilities for the disposal of spray-dryer FGD residual product, both the manufacture of industrial anhydrite and landfill of the fixated product appear at present to be the safest practicable methods.

For this reason, as part of the investigations of residual product disposal for a spray-dryer plant due to enter service at the end of 1987, the technique of anhydrite manufacture with the various process possibilities for reducing the chloride content and the associated differing specific costs of the process were analysed in more detail on behalf of an electric power utility. This method of utilization was compared to emplacement of the residual product after fixating with additions of fly ash and/or cement in a dedicated landfill. For equipping and utilizing this landfill, the following were investigated more closely:

- the placement of the fixated product
  . in a clay pit without additional liner
  . in a pit with additional synthetic liner
- the utilization of the landfill by the power station operator alone and
- the utilization of the dedicated landfill by more than one power station operator.

The power generating utility operates three power station units with a total installed electric capacity of 300 MW, of which two units are equipped with a two-line spray-dryer/DeNO$_x$ plant. At present the plant is under contruction and will enter service at the end of 1987.

A further unit of 100 MW is also to be retrofitted with a flue gas desulphurization/DeNO$_x$ plant. For complete desulphurization of all units, a flue gas flow of 3 x 375 000 Nm³/h (moist) with a mean content of SO$_2$ of around 1500 mg/Nm³, a mean hydrogen chloride content of 130 mg/Nm³ and a mean dust content of 150 mg/Nm³ is to be treated. Following commissioning of the first two units the amount of residual product will be about 17 600 t/a. After commissioning the additional flue gas cleaning plant of the third unit, the amount of residual product will increase to about 20 000 t/a.

Industrial anhydrite:
In order to dispose of the residual product of the spray-dryer plant in the form of industrial anhydrite, it is oxidized in a fluidized bed reactor at a temperature of 850 °C under the addition of air. The calcium sulphite in the product is oxidized to calcium sulphate, whereas existing calcium hydroxide and calcium carbonate are converted to calcium oxide. At the same time, bound hydration water is set free. The product so obtained contains up to 70 to 80 % of water-free calcium sulphate. The chlorides of potassium, sodium and calcium present in the residual product form eutectic mixtures at the operating temperature of the fluidized bed reactor, so that one part of the product becomes molten and the rest agglomerates with other particles to form pellets. For this pelletization, however, a chloride content of the residual product of about 1.5 to 2 % is necessary. Wet or dry dechlorination may be adopted for reducing the chloride content. For wet de-chlorination (Fig. 4), the recirculation product mixed with water is passed to a vacuum belt or drum filter and rinsed in counterflow with process water.

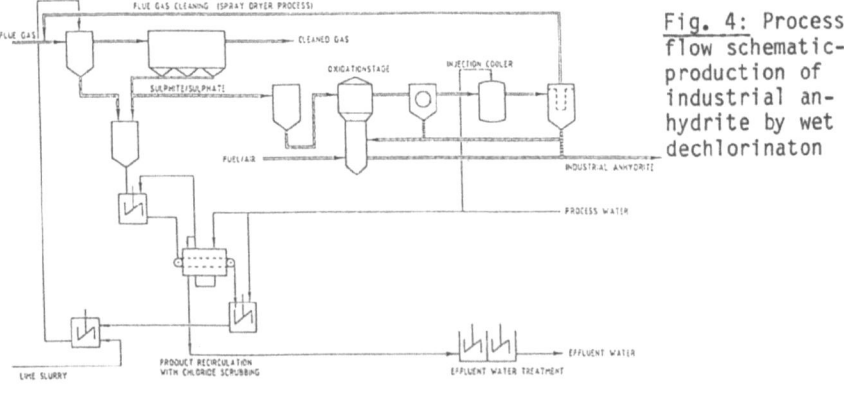

Fig. 4: Process flow schematic-production of industrial anhydrite by wet dechlorinaton

This results in highly saline alkaline effluent water, which must be
treated and diposed of. In the process variant shown in Fig. 4, this
takes place via a physical-chemical waste water treatment stage with
neutralization, flocculation, precipitation and sedimentation.
The scrubbed, dewatered sulphite/sulphate mixture is run into a
dissolving tank were it is mixed with further process water and then
injected into the spray-dryer reactor together with lime slurry. Part of
the recycling product is removed from the circuit and enters the oxidation reactor via a buffer storage silo. The hot off-gas from the oxidation stage is routed via a cyclone collector for preliminary particulate
removal, an evaporative cooler and then a fabric filter for residual
particulate removal, before being returned to the main flue gas flow
ahead of the spray-dryer reactor.

Should it be required to avoid the production of effluent water, the
injection cooler is replaced by an evaporation reactor into which
the effluent water is sprayed (Fig. 5).

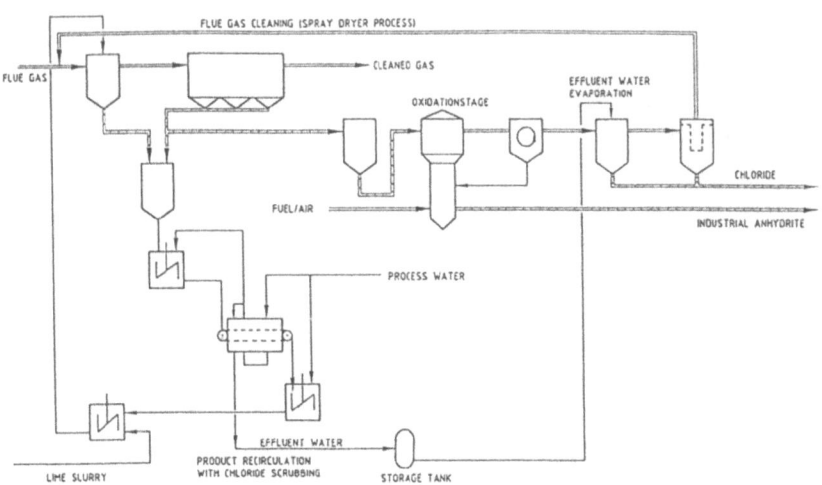

Fig. 5: Process flow schematic-production of industrial anhydrite by
wet dechlorination, no effluent water

A further process variant is dry dechlorination (Fig. 6).

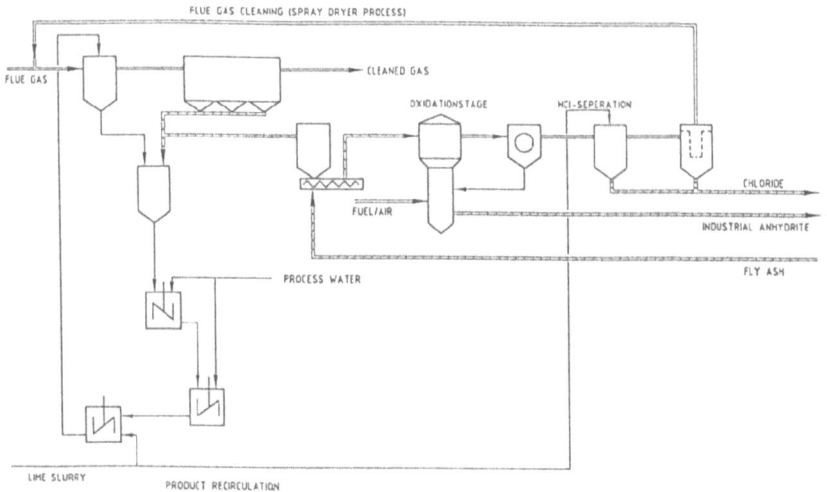

Fig. 6: Process flow schematic-production of industrial anhydrite by dry dechlorination

The reaction product taken from the recycling circuit is mixed, after intermediate storage, with 15 - 20 % by weight of fly ash and supplied to the oxidation stage. At the operating temperature in this stage of about 850 °C, catalytic decomposition of the calcium chloride to hydrogen chloride (HCl) takes place, this exiting the fluidized bed reactor together with the process off-gas. In order to separate the HCl, there is an additional spray-drying stage with lime slurry following the cyclone separator.

For the usual sulphur content of German high-grade bituminious coals of about 1 % by weight and chloride content of a maximum of 0.2 % by weight, the chloride content of the residual product can be reduced from 2.5 - 4 % to 1 - 2 % by dechlorination.

If the three process alternatives for a plant capacity of 18 000 t/a (corresponding to 4 t/h) of residual product are compared, it is apparent that the production of industrial anhydrite by wet dechlorination with total costs of 158 DM/t is the most cost effective process option (Fig. 7).

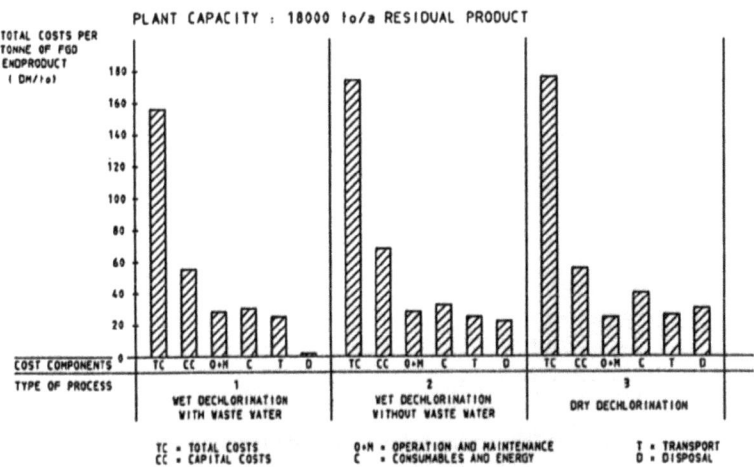

Fig. 7: Residual product from spray-dryer FGD - production of technical anhydrite Survey of comperative costs for different process alternatives

Wet dechlorination without the production of effluent water has a total cost of 170 DM/t, whereas dry dechlorination costs 175 DM/t.

Formation and landfilling of fixated product:
The end product of the spray-drying reaction with a composition as shown in Fig. 1 starts to consolidate when fly ash and about 30 - 40 % of water are added. This so-called pozzolanic reaction influences water permeability, structural strength as well as the leaching properties of the residual product. Whereas the pozzolanic characteristics of the end product without any appreciable content of fly ash are not particularly marked, permeability, i. e. the permeation coefficient K,

is very greatly reduced by the addition of large amounts of fly ash or fly ash + cement (Fig. 8).

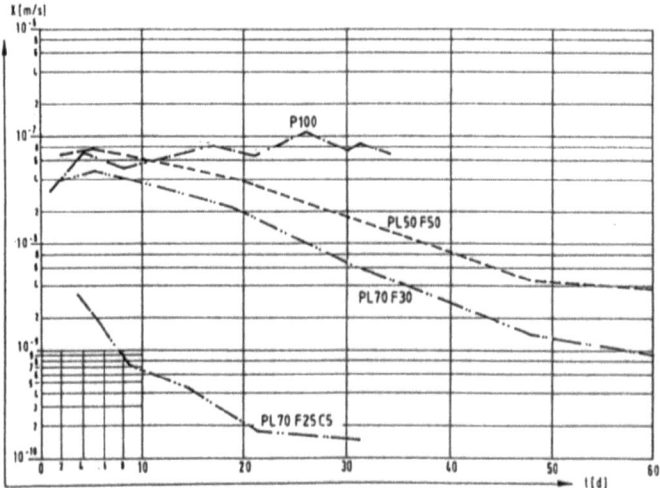

Fig. 8: Residual product from spray-dryer FGD-dependency of permeation-coefficient K on type of admixtures to the product

The leaching properties of the product too are modified by the pozzolanic reaction. Thus, for example leaching out of sulphate and sulphite is considerably reduced by the addition of 50 % of fly ash to the pure end product, whereas the chloride content of the leachate and also the conductivity are influenced hardly at all (Fig. 9). However, the graphs shown here have been plotted from leaching of the crushed material according to German standard DEV S 4. When considering these it is to be borne in mind that this manner of leaching does not correspond to the conditions of emplacement of the materials which, after curing, are in a monolithic form in the landfill. The concentration from leaching of uncrushed material of this type is considerably lower.

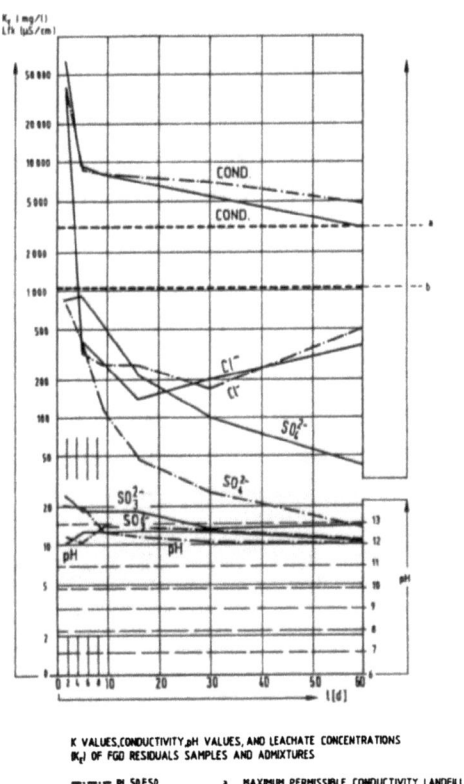

Fig. 9: Residual product from spray-dryer FGD-Leachate concentrations of untreated product and fixate

With such fixated products the residual contents of heavy metals are very low, and are well within the limits for landfills of Class 2. As, however, the salt content of the leachate will sometimes exceed the limit of 2500 µs/cm for landfills of Class 2, for the following determination of the costs of dumping fixated product, an additional synthetic liner was taken into account for the landfill, in order to ensure that there will be no

increase in salinity of the groundwater. Furthermore, the addition of cement as well as of fly ash is planned.
In practice cement addition is provided particularly when placing the base layers of the landfill, whereas for normal dumping of the end product, it is simply mixed with fly ash.

The design of a mixing plant for fixated product is shown in Fig. 10.

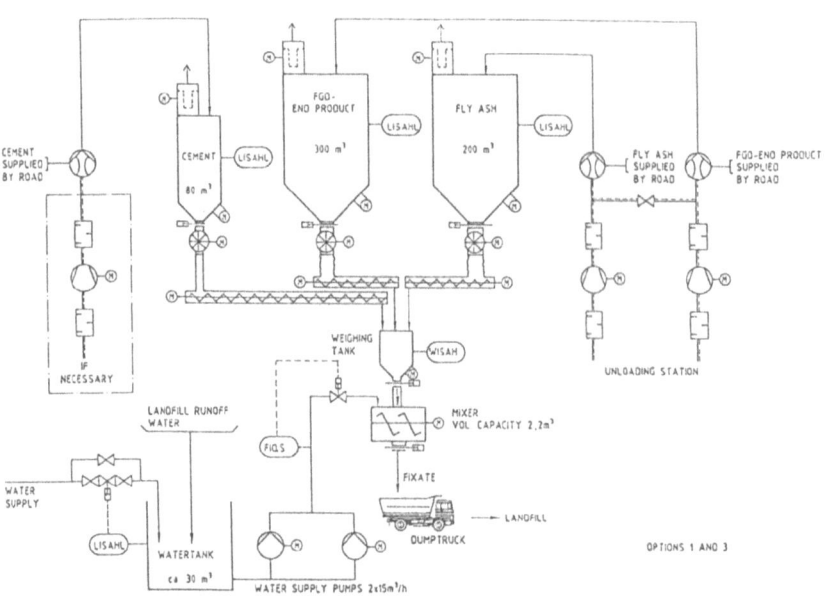

Fig. 10: Process flow schematic-batch operation mixing plant for fixation of product

The plant is batch-operated. The proportions of FGD end product, fly ash and cement are weighed out in the weighing tank for one charge of fixated product and emptied into the mixer. After completely emptying the weighing tank, the mixer commences operation and, at the same time, the appropriate amount of water is metered in. Following a

mixing time of about 1 - 2 minutes, the completed mixture is discharged directly onto a dumper truck and this drives onto the landfill.
On the landfill the material is distributed to form a layer 20 - 30 cm thick by means of a grader, and is compacted by the same machine.
As the fixated product still contains substances which can be leached out, it is expected that the subgrade of the landfill be sealed off from the groundwater.
This seal can either be a natural feature, for instance clay strata, or it can be placed artificially. For the investigation described for the dedicated stabilized landfill, a disused clay pit is selected for which it can be expected that there are clay layers under the bottom of the pit of sufficient thickness and density to act as natural seals.

The dual-liner system additionally considered in the cost analysis can be in the form of an artificially emplaced additional mineral layer or in the form of plastic sheeting. After attaining its final height, the landfill impoundment is provided section-by-section by a surface seal to limit infiltration of seepage water. The rain water running onto the surface seal is collected and channeled off in a controlled way. A cover layer is placed over the surface seal to permit revegetation. At the base of the landfill impoundment, that is immediately above the natural seal or double liner, a leachate control system is installed which collects and drains the infiltrated seepage water. Seepage water flows along the surface of the liner to the collection lines and is then drained in a controlled way via a main header into the seepage water collection pond. If, following analysis, the seepage water is found to be safe it is discharged or if not it is treated on site, e. g. neutralized or transported for processing to an appropriate plant. As the landfill is being filled, the seepage water may also be used as process or mixing water for the fixated product.

Costs of the dedicated stabilized landfill:
The costs of a landfill with a capacity of 1.5 million m³ are investigated for an annual deposition of about 18 000 t of fixated product, corresponding to an end product quantity of 41 000 t/a. The breakdown of the overall investment costs over the individual costs components such as seepage water control system, surface covering and revegetation,

mixing plant, landfill equipment, additional dual-liner etc. is shown in Fig. 11.

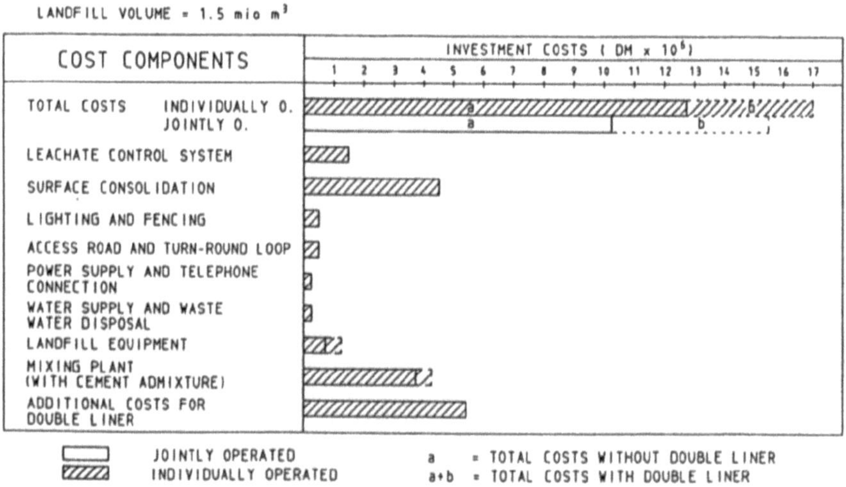

Fig. 11: Residual product form spray-dryer FGD - distribution of investment costs for an individually or jointly operated landfill for fixated residual product (excl. land costs)

The costs of land were not taken into account. Apart from sole operation of a dedicated landfill by one power station operator, deposition of fixated product from flue gas cleaning plants from several operators would also be appropriate for such a landfill.

Costs of common utilization by three operators is compared in the following to utilization of a dedicated landfill by just one power station operator.

Purchasing of fly ash and the use of cement results in the cost situation shown in Fig. 12 as a function of annual cost consumption.

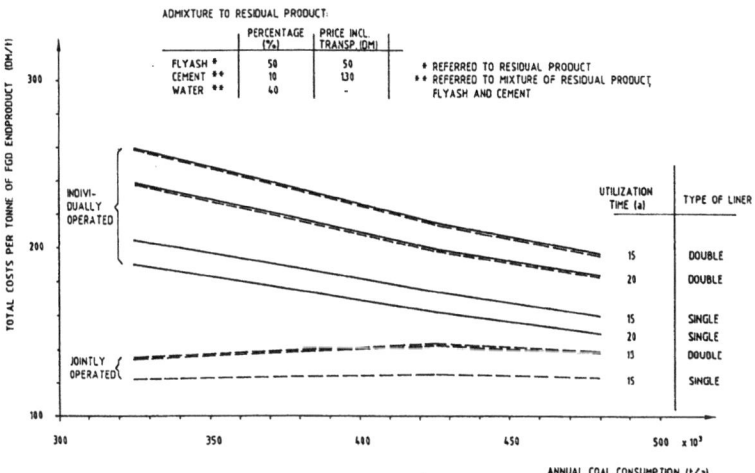

Fig. 12: Residual product from spray-dryer FGD-total costs for individually or jointly operated landfill for fixated residual product

The specific costs for end product disposal in DM/t are thereby very dependent on the utilization period of the landfill as well as the manner of base sealing. For the dedicated landfill for just one operator, these are between 160 and 260 DM/t of end product, whilst for a common landfill for three operators, these costs can be reduced to 120 - 140 DM/t. The disposal costs can also be considerably reduced by the reduction of the costs for the additives (Fig. 13).

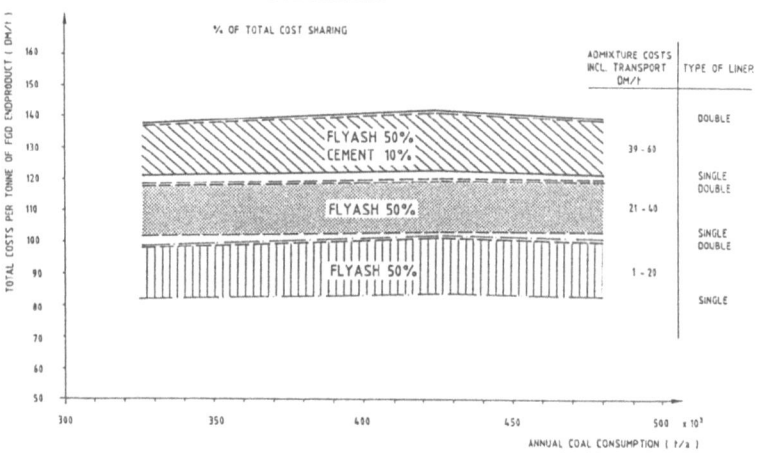

Fig. 13: Residual product from spray-dryer FGD - total cost for jointly operated landfill for fixated residual product - 15 years utilization

Should the type of firing make necessary purchases of fly ash and if additionally cement is used, the disposal costs for a multi-user landfill will be 120 - 140 DM/t of end product, according to the manner of base sealing. If it is decided not to mix in cement, the costs can be reduced to 100 - 120 DM/t, and if fly ash from the operator's own power plant can be used and if only the added water remains as a cost factor, the costs of disposal are reduced to 80 - 100 DM/t. A further reduction in diposal costs can also, naturally, be attained by increasing the number of participants, provided that the available landfill capacity is adequate for a sufficiently long utilization period of at least 10 - 15 years.

Comparison of the disposal costs for the production of
industrial anhydrite and the operation of a stabilized landfill
Under the conditions investigated the disposal of the spray-dryer

end product following fixation is more expensive than the manufacture of industrial anhydrite (Fig. 14).

For a landfill utilization model with just three operators, the disposal costs drop below the costs of the most favourable option for industrial anhydrite even with the provision of a dual-liner system.
By increasing the number of operators, the disposal costs for dumping fixated product may be reduced even further. A possible credit for the product should this be sold is not taken into account in the manufacture of industrial anhydrite. However, the disposal situation of gypsum from wet desulphurization processes indicates that even a credit of this nature will not appreciably reduce the manufacturing costs of the reclaimed product.
Where suitable dumping sites in the form of abandoned clay pits or pit workings suitable for stowing are available, and official permits for the emplacement of fixated product in such disposal sites are obtainable, more favourable costs for the disposal of the FGD end product may be expected for common operation of such dedicated landfills by several operators of spray-dryer plants in comparison to the production of industrial anhydrite. Thereby the costs for stowing in mines will be of the same order of magnitude as the disposal of fixated product without the provision of additional base sealing liners.

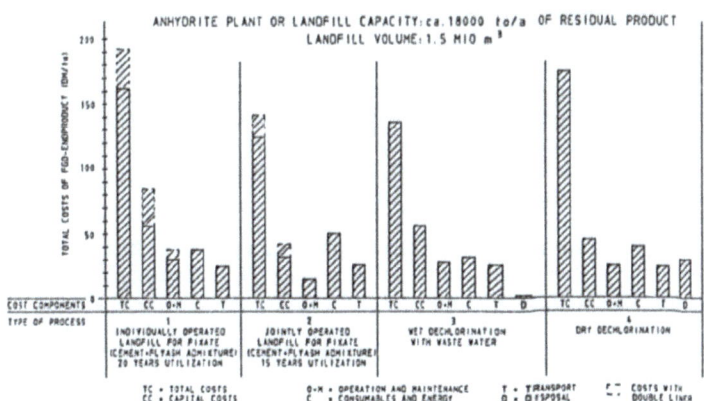

Fig. 14: Anhydrite plant of landfill capacity: ca. 18 000 to/a of residual product; landfill volume: 1.5 million m³

Treatment of industrial emissions and waste

"Flue gas treatment according to the BF/UHDE process"

by Dr.-Ing. Ulrich Neumann

Flue gas treatment according to the BF/UHDE process

The Bergbau-Forschung/UHDE process is a dry process for flue gas treatment with the aid of activated coke. The process takes place at the cold end of the boiler of a power station, i.e. downstream of the air preheater and electrostatic precipitator, and can be used for either desulphurization or NOx removal, or both in a simultaneous process. This paper describes the simultaneous process and the removal of NOx only.

1. Process fundamentals

In the BF/UHDE process, flue gas treatment is performed by means of adsorption and catalytic reduction on activated coke based on hard coal, which is in the form of extruded cylindrical granules of approx. 5 mm diameter and 5 mm length.

1.1 Principles of the adsorption process

The activated coke thus produced has the following specific characteristics:

- high adsorption capacity,
- fast kinetics of adsorption and catalysis,
- low reactivity as regards oxygen,
- no loss of reactivity due to regeneration,
- high mechanical strength.

When flue gas passes through a layer of this activated coke at a temperature of 90-150 °C, the following reactions take place:

Desulphurization: SO_2 is adsorbed by the activated coke and converted to sulphuric acid with the aid of oxygen and water vapour:

SO_2, gas $\longrightarrow$ SO_2, ads
½ O_2, gas $\longrightarrow$ O ads $\longrightarrow$ H_2SO_4, ads
H_2O, gas $\longrightarrow$ H_2O_{ads}

Under the conditions prevailing in commercial plants, the SO_2 load of the activated coke is between 10 and 15 % by weight, and 70-80 % sulphuric acid forms on the coke surface.

Removal of residual SO2: In the adsorption process described, the SO_2 is usually not completely converted to sulphuric acid. Since ammonia is admixed to the flue gas for the removal of NOx, as described later on, this ammonia reacts with the residual SO_2 and causes the formation of ammonium hydrogen sulphate or ammonium sulphate according to the following formulae:

$$NH_3 + SO_2 + H_2O + \tfrac{1}{2} O_2 \longrightarrow NH_4\ HSO_4$$
$$NH_4\ HSO_4 + NH_3 \longrightarrow (NH_4)_2\ SO_4$$

NOx removal: The NOx in the flue gas obtained from boilers of power stations consists of approx. 95 % NO and approx. 5 % NO_2.
NO_2 reacts with carbon according to the following formula:

$$2\ NO_2 + 2\ C \longrightarrow 2\ CO_2 + N_2.$$

With the activated coke acting as a catalyst, the NO is converted as follows when ammonia is added:

$$6\ NO + 4\ NH_3 \longrightarrow 5\ N_2 + 6\ H_2O.$$

Separation of halogen compounds, heavy metals and dust: Halogen compounds (i.e. hydrogen chloride/hydrogen fluoride), heavy metals and dust will likewise be removed to a considerable extent from the flue gas by adsorption and mechanical separation in the activated coke bed.

1.2 Principles of regeneration

The regeneration of the laden activated coke takes place at a temperature of 400-450 °C. During the heating process, the adsorbed sulphuric acid reacts with the carbon of the activated coke and sulphur dioxide and superficial carbon oxides are obtained as intermediate products:

$$2\ H_2SO_4 + 2\ C \longrightarrow 2\ SO_2 + 2\ H_2O + 2\ C....O.$$

These superficial oxides disproportionate as follows:
$$2\ C....O \longrightarrow C + CO_2.$$

The ammonium salts decompose, while the NH_3 obtained reacts with the superficial oxides:

$$2\ NH_3 + 3\ C....O \longrightarrow N_2 + 3\ H_2O + 3\ C.$$

2. Process configuration

2.1 Simultaneous process: (see fig. 1)

The unit for simultaneous removal of SO_2 and NOx according to the BF/UHDE process is at the cold end of the boiler, i.e. downstream of the air preheater and the electrostatic precipitator.

Flue gas conditioning: The flue gas to be treated is withdrawn by means of a blower and compressed to the required pressure. The flue gas stream is conveyed into the flue gas cooler where the gas is cooled to the optimum adsorption temperature of approx. 120 °C.

Adsorption: The flue gas to be treated enters the bottom section of the adsorber, which is divided into several individual chambers of approx. 2 m width, and passes through the first bed of activated coke. The flue gas is then collected in the gap between the chambers and flows upwards to the second stage. While the flue gas passes through the first bed of activated coke, the SO_2 is adsorbed together with a portion of the water vapour and the oxygen contained in the flue gas and deposits on the activated coke in the form of sulphuric acid. At the same time, the gaseous chlorine and fluorine compounds entrained in the flue gas are adsorbed and the NO_2 portion of the NOx undergoes a reaction with the activated coke, thus forming CO_2 and N_2. In addition, the major portion of the dust and the heavy metals still entrained in the flue gas is retained mechanically in the bed of activated coke.

Prior to entering the second stage, ammonia is injected into the flue gas stream in a mixing chamber. Prior to injection, this ammonia is mixed with treated flue gas at a ratio of approx. 1:25, in order to guarantee an optimum NH_3 distribution in the gas.

In the second stage, the catalytic property of the activated coke is used for the decomposition of the NO molecules contained in the flue gas. In the first step, NH_3 is adsorbed which then reacts with the NO on the surface of the activated coke, forming N_2 and H_2O.

In addition, residual SO_2 is removed by the formation of ammonium sulphate or ammonium hydrogen sulphate.

The adsorbers are designed as moving-bed reactors. The adsorption takes place with the aid of the activated coke specially prepared for this purpose. This activated coke is particularly resistant to abrasion. The activated coke, from which the SO_2 has been removed in the desorber, is transported into antechambers which are arranged above the adsorber, from where it flows into the second stage of the adsorber. In the individual chambers, the activated coke gradually (0.1 to 0.2 m/h) slides down between louvre-type plates which are designed such that the flue gas can flow through the sections without the activated coke being entrained. The residence time of the activated coke in the second stage of the adsorber is approx. 100 to 200 hrs. The activated coke then moves from the second stage into the first stage which is arranged underneath the second stage. The laden activated coke from the first stage is withdrawn by belt conveyors and transported to the desorption unit.

The activated coke throughput is controlled as a function of the SO_2 load of the crude gas and the flue gas rate by varying the speed of the belt conveyors arranged underneath the adsorbers.

Desorption: The laden activated coke from the adsorber is transported to an intermediate bin upstream of the desorber, from where it is conveyed to a classifying section, where the major part of the flue dust entrained is removed by blowing flue gas through the activated coke. The dust-laden flue gas that has been used for classification is treated in a hose filter and then returned to the classification section by means of a blower. Further separation of the residual dust and activated coke particles is performed by means of a vibrating screen arranged below the desorber.

The activated coke, from which most of the dust has been removed, is fed from the intermediate bin to the desorption section of the desorber, from where it slides downwards through tubes surrounded by hot gas at a temperature of 300 °C to 600 °C. The coke is thus heated to approx. 400 °C to 450°C, at which temperature desorption of the SO_2 takes place according to the reactions described earlier.

On leaving the tubular desorber, the activated coke is stored in a large intermediate bin (residence time 2-3 hrs) where the residual SO_2 is liberated.

From the bin, the coke slides downwards again through the desorber tubes into the cooling section of the desorber, where it is indirectly cooled by means of air to a temperature of approx. 100 °C. It is withdrawn from the cooling section of the desorber by means of a vibrating chute which feeds the coke to a vibrating screen.

This vibrating screen serves to separate the coke into two fractions:

- The coke with a grain size above 3.5 mm is returned to the adsorber.
- The smaller fractions are used as fuel.

The air from the cooling section of the desorber, which has been heated to 250 °C, is withdrawn by a blower and a portion of it is conveyed to a combustion chamber as combustion air. This combustion chamber serves to produce the flue gas for heating the activated coke. After the flue gas has passed through the desorber, a portion of it is returned to upstream of the desorber for temperature adjustment. The surplus flue gas is admixed to the boiler flue gas upstream of the flue gas treatment section.

The SO_2-rich gas from the desorber is withdrawn from the intermediate bin by means of a SO_2-rich flue gas blower and is conveyed via the dehalogenization section, where HCl and HF are removed, to a further processing unit.

Further processing of SO2-rich gas: (See fig. 2)
The SO_2-rich gas obtained in the desorption process contains approx. 25-30 % by vol. SO_2, 50-60% by vol. water vapour, as well as carbon dioxide and nitrogen.

The simplest and cheapest method of processing this gas is by using an existing chemical plant located in the vicinity (e.g. sulphuric acid or Claus plant). However, in the majority of cases, these facilities are not available, so that the SO_2-rich gas has to be further processed in an integral processing unit.
Suitable final products are sulphur, sulphuric acid or liquid SO_2.

Further processing to **sulphur** is effected in a modified Claus plant. The sulphur produced is of commercial quality will a purity degree of 99.9 %.

In the last stage of the **sulphuric acid** unit, the SO_2-rich gas is scrubbed to produce marketable 96-98 % sulphuric acid which is stored in tanks.

The SO_2-rich gas is first cooled and dried. Thereafter, it is compressed, liquefied by cooling and fed to an intermediate storage tank for the **liquid SO2**.

2.2 BF process for the removal of nitrogen oxides only:
(See fig. 3)

In the BF process for the removal of NOx only, the activated coke acts - as described earlier - as a catalyst for reducing NOx to N_2 and H_2O by means of ammonia. The unit is arranged at the cold end of the boiler downstream of a wet scrubber or spray absorption tower for the removal of SO_2.

Reaction section: The desulphurized flue gas leaves the desulphurization unit at a temperature of 50-60 °C and is then heated to approx. 100 °C in a steam preheater or by another heating system.

The ammonia, which has been diluted with treated flue gas, is admixed to the flue gas on the discharge side of the blower. Prior to entering the reactor, the flue gas and the diluted ammonia are thoroughly mixed in a static mixer.

In the reactor, the catalytic qualities of the activated coke are utilized for decomposing the NOx molecules in the flue gas. In the first stage, NH_3 is adsorbed and reacts with the NOx on the surface of the activated coke, thus forming N_2 and H_2O.

The residual SO_2 of the flue gas likewise reacts with ammonia to ammonium sulphate or ammonium hydrogen sulphate, which is removed in the first narrow activated coke bed.

The Desox stage will be designed as moving bed reactor, the Denox stage as a fixed bed reactor.

Desorption section: The laden activated coke from the Desox stage is conveyed to the desorption section as within the simultaneous process. The exhaust gas obtained is returned to upstream of the desulphurization unit.

Pilot plant: The major advantage of this process is that it allows NOx removal to take place at temperatures as low as 100 °C, thereby obviating the need for an expensive flue gas heating stage, such as is necessary using the conventional plate catalyst process. Individual systems have already proven successful for SO_2 and NOx removal in the Bergbauforschung/Uhde Simultaneous Process.

Uhde and STEAG, Essen, have constructed a pilot plant based on activated coke at the STEAG power plant in Voerde in the Ruhr area with a view to optimizing the operating efficiency of the system. It has been run jointly by the two companies since December 1986 and has a throughput of 10.000 m^3/h. The initial test results have been highly promising.

3. **Characteristics of the BF/UHDE process**

   The main characteristics of the BF/UHDE process are summarized as follows:

   - The possibility of selecting various final products permits optimum adaptation to the location and thus long-term sales guarantees.

   - The adsorption and desorption sections are virtually self-sufficient units which can temporarily be operated independently of each other. Thus, high on-stream factors of the entire system are achieved.

   - As the flue gas is only cooled to 120 °C, the acid temperature does not normally drop below the dew point so that no corrosion problems occur along the flue gas route.

   - Due to the large amount of activated coke in the adsorbers and the high buffer capacity thus achieved, the flue gas treatment unit is not susceptible to load fluctuations of the power station.

   - The BF/UHDE process enables the simultaneous removal of SO_2, SO_3, HCl, HF, Hg, fly ash and NOx. High separation efficiencies are achieved. SO_2 and NOx contents of less than 200 mg/m^3 (Vn) in the treated gas can be obtained without any problems.

   - The process does not produce any waste water. Thus, the problem is not shifted from air to water.

   - As the flue gas is not cooled to below 120 °C, reheating is not necessary. This increases the efficiency of the entire plant.

   - The final products, i.e. sulphur, sulphuric acid or liquid SO_2, can easily be used by the chemical industries. Therefore, these products do not cause any additional costs for dumping.

4. State of development

The first significant findings about desulphurization and later on also about the reduction of NOx in flue gases were obtained between 1973 and 1979 during more than 16 000 operating hours in the prototype pilot plant of Bergbau-Forschung on the site of the Kellermann power station at Lünen, Fed. Rep. of Germany. The flue gas throughput amounts to 130 000 m³/h (Vn).

The Japanese company Sumitomo Heavy Industries has been operating a plant using the simultaneous process for the treatment of 300 000 m3/h (Vn) flue gas, using activated coke of the Bergwerksverband, Essen.

Mitsui Mining Company, Tokyo, with whom Uhde has concluded a mutual cooperation agreement, has been operating a pilot plant at Tochigi with a throughput of 1000 m³/h (Vn) flue gas since 1982 as well as a commercial-scale flue gas treatment plant at Omuta, which was designed on the basis of the results obtained in the pilot plant. This plant, handling a flue gas throughput of 30 000 m³/h (Vn), has been working successfully ever since it was commissioned in October 1984, applying the simultaneous process described earlier.

In 1987, the first commercial plant designed and constructed by Uhde for Energieversorgung Oberfranken will go on stream in order to treat the flue gases of Arzberg power station by the simultaneous process. The throughput will be 1 100 000 m³/h (Vn) with a SO_2 load of 4000 mg/m³ (Vn) and a NOx load of 500 mg/m³ (Vn).

Begin of this year Uhde got a further order for the simultaneous process. The plant will be erected at the Francfort Works of Hoechst AG for the flue gas cleaning of a trivalent industrial boiler.

5. Summary

The BF/UHDE process is a dry process for the simultaneous removal of SO_2 and NOx from flue gases by means of activated coke. SO_2 is adsorbed in the form of sulphuric acid and NOx is decomposed to N_2 and H_2O by adding ammonia. These processes take place in an adsorber, in which the activated coke is transported from top to bottom in the form of a moving bed, while flue gas flows diagonally through the layers. Desorption of the laden coke is performed thermally by indirect heating with flue gas generated in a combustion chamber. The SO_2-rich gas thus obtained can be further processed to sulphur, sulphuric acid or liquid SO_2.

In the BF process for the removal of NOx downstream of a desulphurization unit operating by the wet process, the NOx in the reheated flue gases is removed in a reactor, through which the activated coke is conveyed in the form of a moving bed just as is the case in the simultaneous process. The SO_2-rich gas obtained in the desorber is not processed further, but is returned to upstream of the desulphurization unit.

Having been thoroughly tested in pilot plants, the efficiency and reliability of the BF/UHDE process have now reached a very high standard.

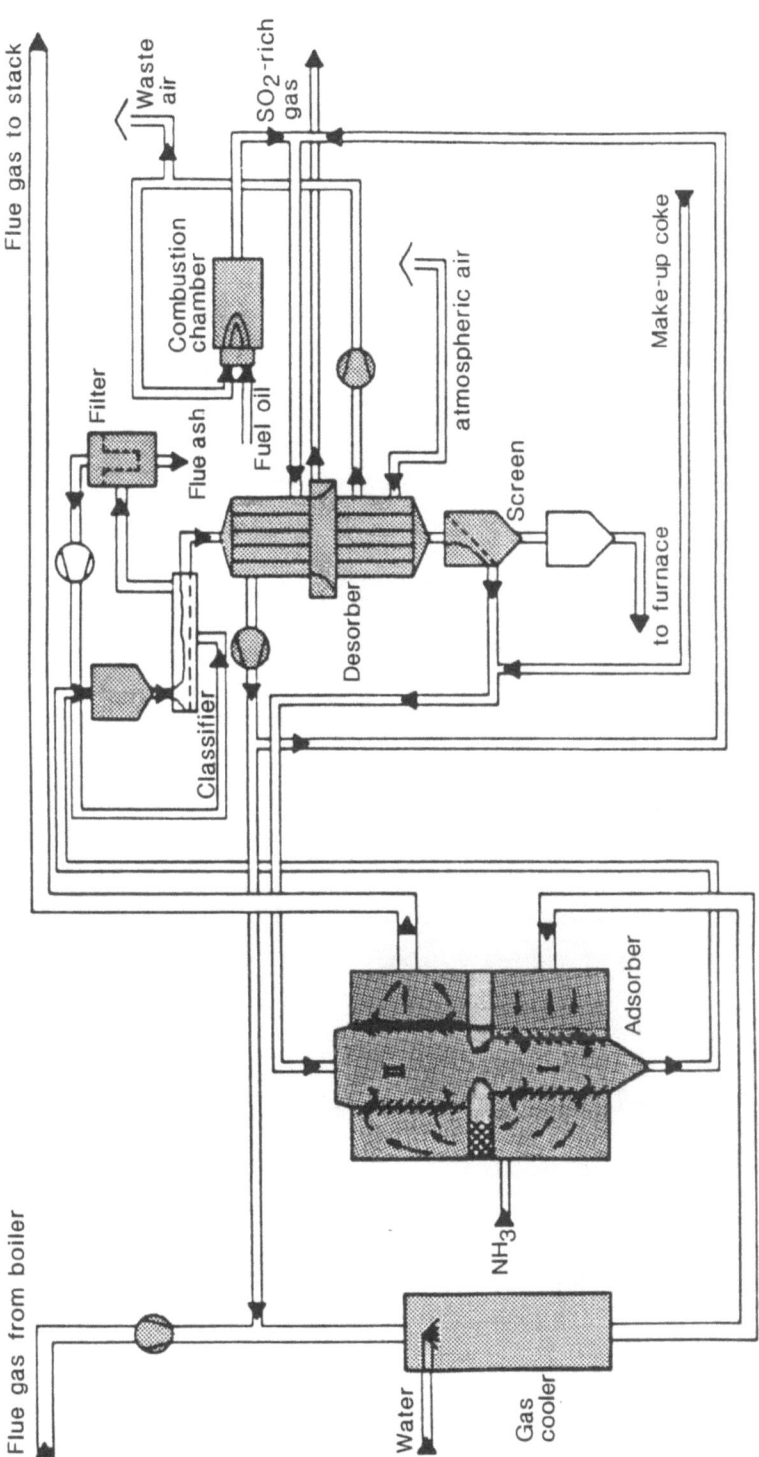

Fig. 1.
BF/Uhde process
Simultaneous flue gas treatment

Processing of SO_2-rich gas

1. Drying tower
2. Reduction burner
3. Waste heat boiler
4. Claus 3-stage catalyst bed
5. Heat exchanger
6. Condenser
7. Sulphur pit
8. Cooler
9. Refrigeration set
10. Drier
11. Blower
12. SO_3 catalytic unit
13. Absorber
14. Pump
15. Storage tank
16. Compressor
17. CO_2 stripper

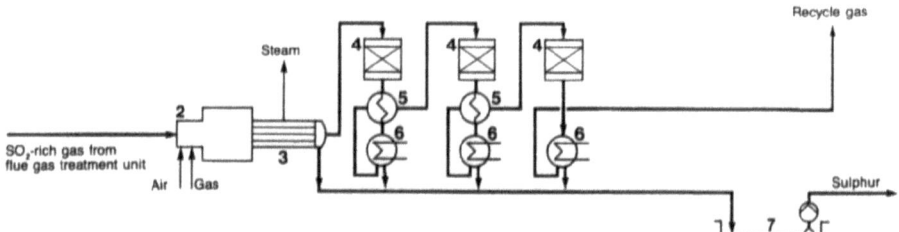

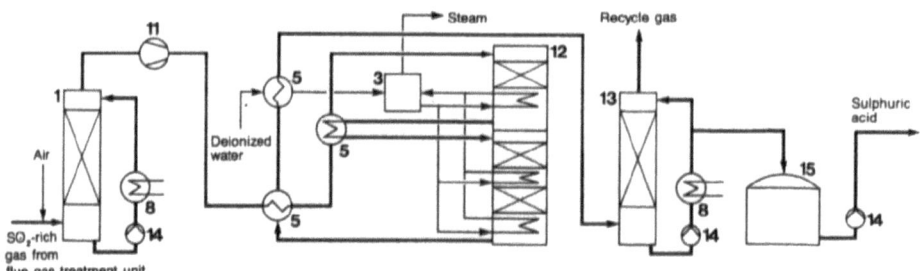

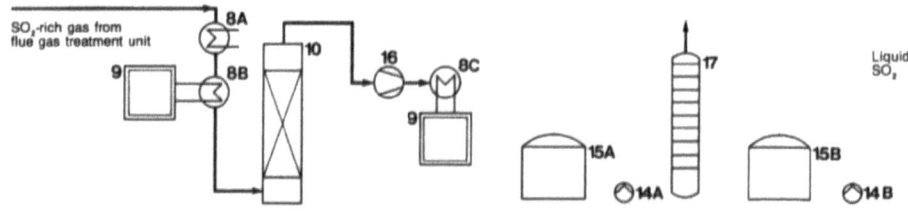

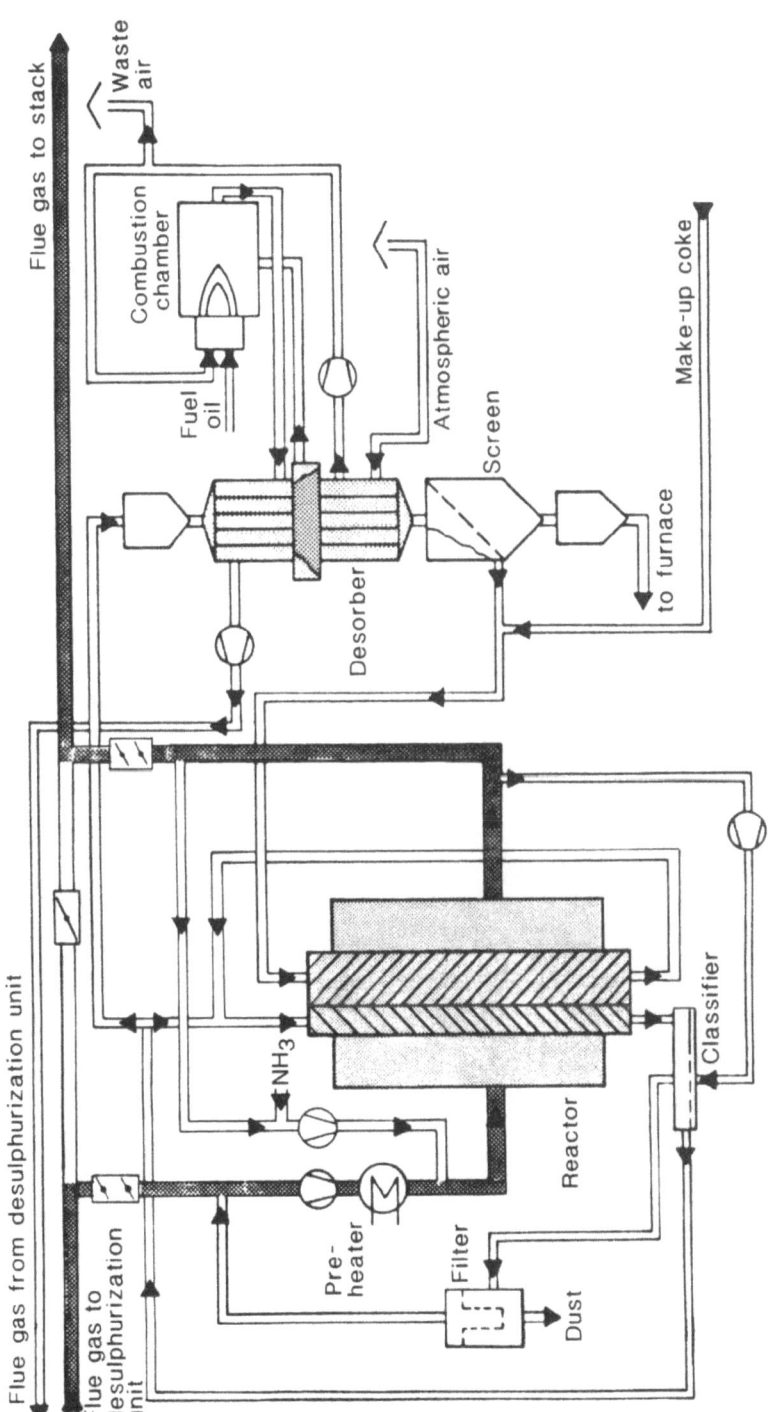

Fig. 3.
BF/Uhde process for the removal of nitrogen oxides only

TECHNICAL/ECONOMICAL DEVELOPMENTS OF FLUEGAS DESULPHURIZATION (FGD) INSTALLATIONS

L.A.J. Tol and W.L. Prins *

Introduction.

Coalfiring of powerplants in the Netherlands is applied again since 1978. Fluegas desulphurization was decided to be necessary for the first coal fired plant in 1982. In 1984 the legislation for SO_2 in the Netherlands was decided as follows:

Figure 1.
Legislation in the Netherlands

Period:	1982	after 1990
SO_2:	50% of fluegases with 90% SO_2 removal	max. 400 mg/Nm^3
NOx:	max. 1000 mg/Nm^3	max. 800 mg/Nm^3

Only those powerstations that will be in operation after 1994 will have to fullfill these SO_2 requirements.
It was also decided that the coal firing share in the firing of powerstations would have to increase to 30% in the 1990's. This results in retrofit-installations at existing and/or converted power plants.
For new powerstations also fgd will be necessary.

* Ir. L.A.J. Tol is Manager APC with ESTS B.V. (Hoogovens Groep)
  Dr. Ir. W.L. Prins is Manager Process Engineering APC with ESTS B.V. (Hoogovens Groep).

In the early 1980's a large number of fgd processes were available with
a variety of end products such as gypsum, sulphur, sulphuric acid,
ammoniumsulfate and fly ash-sulfate-sulfite mixtures.

For coal fired powerstations in the Netherlands a lime(stone)-gypsum
fgd process appears to be the best process with respect to cost,
reliability (process and equipment) and the possibility to market the
gyp.um in the building industry in the Netherlands.

Technical developments in fgd installations.

Fgd in the Netherlands was introduced only shortly after its introduction in West-Germany. Therefore the design know-how and expericence for the first installation were obtained from the USA and Japan. Now approx. 5 years experience in design/operation of fgd installations is available in the Netherlands and it is surprising to see how technological developments took place resulting in improved and cheaper installations.

We appreciate this opportunity to report on the developments by comparing the following fgd installations:

Figure 2:

FGD-installations for this review:

	PGEM Nijmegen G13-fgd-1	PNEM Geertruidenberg Amer 8	PGEM Nijmegen G13-fgd-2
MW el	645	645	645
%of fluegas treated	50%	100%	50%
startup date	01-04-85	01-03-88	01-12-88
fluegas flow(Nm^3/hr)	880560	1926000	1245200
S-content in coal(%wt)	0.5-1,5	0.5-1.5	0.5-1.5
SO_2 removal(%)	90	88	89
fluegastemp.outlet(°C)	80	60	50
gypsum prod.(t/hr)	8	15	10
wastewater (m^3/hr)	27	12	9,5

See the following photographs showing the PGEM and PNEM fgd installations.

PGEM, Nijmegen, G13 fgd-1

PGEM, Nijmegen, G13 fgd-1

Model of PNEM, Geertuidenberg, Amer 8

The period between the design of these fgd installations is approx. 2 years. In the design the following items reflect the changes by operating experience, changing legislation and technological developments:

Figure 4: Improvements in fgd design:

- internal oxidation, no prescrubber
- absorber design
- fluegas reheat systems
- dewatering techniques
- waste water treatment

In the next sections we will discuss these improvements and we will conclude with the effects on costs saving.

Internal oxidation, no prescrubber.

In figure 5 the arrangement of the PGEM Nijmegen, G13-fgd-1, is shown as designed and operated since 1985.
The prescrubber was installed to remove fluorides from the fluegases. In addition chlorides and flyash are removed. The gypsum processing industry experienced that a relatively small amount of insoluble calcium fluoride in the gypsum by omitting the prescrubber was quite allowed. Therefore a prescrubber is no longer necessary for coals fired in the Netherlands.

Introduction of the oxidation air in the absorber recycle tank itself instead of application of (separate) external oxidation reactors resulted from extensive test work and final demonstration on 300MW-scale in 1980 at Monticello followed by a 1400 MW installation at TVA. Nowadays several internal oxidation installations on the basis of lime and limestone are in operation in West-Germany and Austria.
Removal of the prescrubber and conversion from external to internal oxidation appeared to be so advantageous that it was decided in 1987 to apply this for PGEM Nijmegen, G13-fgd-1 as shown in fig. 5.

Figure 5: <u>Retrofit of PGEM Nijmegen, G13-fgd-1.</u>
Installation 1985-1987:

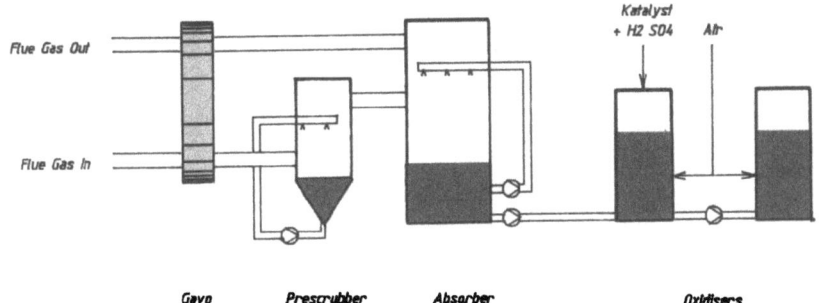

Installation after retrofit in 1987:

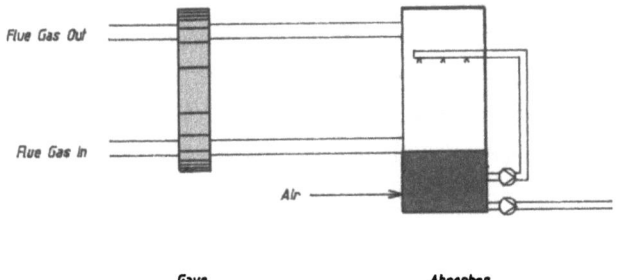

The removal of the prescrubber and modification of the absorber to internal oxidation had the following consequences:
- gypsum quality similar (+ 2% residual limestone)
- no H_2SO_4 required for the external oxidisers for pH control
- no katalyst required for optimum gypsum crystal form.
  The conditions (e.q. residence time) for crystallisation in the recycle tank are more favourable for crystal growth.
- Simpler operation and maintenance. No longer an installation having the corrosive flyash slurry (pH=1) from prescrubber.
- More space available. Removal of oxidisers and prescrubber gave a better design of the second fgd at powerstation G13.
- Lower power consumption by removal of pressuredrop over the prescrubber and elimination of prescrubber recycle pumps.

Absorber Design

In the absorber design we mention three features that have been optimized during the past years: slurry spray nozzles, mist eliminiation and absorber outlet design.

Slurry spray nozzles.

In the 1960's and 1970's the early sulfite based absorbers in the USA suffered (also because of poor pH control) from clogging in the piping and slurry spray nozzles in the absorbers.
Therefore the hollow cone ramp bottom nozzles ("empty nozzles") were chosen in order to avoid fouling and clogging. In the present (limestone based) gypsum fgd installations having "open" spray towers no fouling or clogging problems are experienced any more.
From test work and several years experience in the fgd installation at Matsushima (Japan) it appeared that the socalled spiral (full cone) nozzle is superior for this application.

Figure 6: Slurry spray nozzle development.

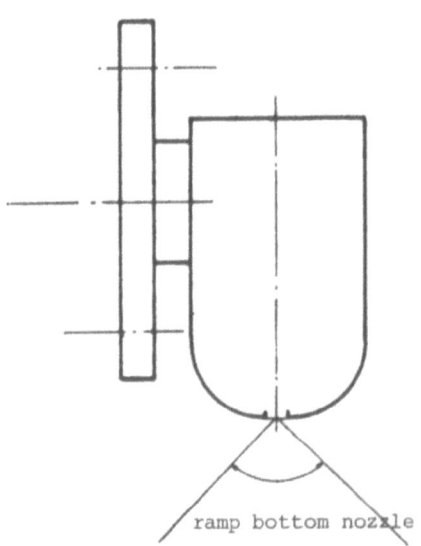
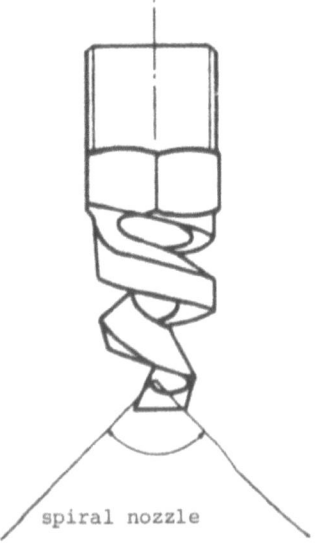

Droplet size distribution can be very well controlled and a much lower pressure drop for the slurry is neccessary (roughly 50%) for the spiral nozzle. This saves approx. 500 kW for a 600 MW installation. Moreover the fluegas pressuredrop over the spray tower can be lower. (saving energy in the boosterfan).

Mist Eliminator.

The most important alternatives to be considered for a certain location are the following.

Figure 7: Mist Eliminator alternatives.

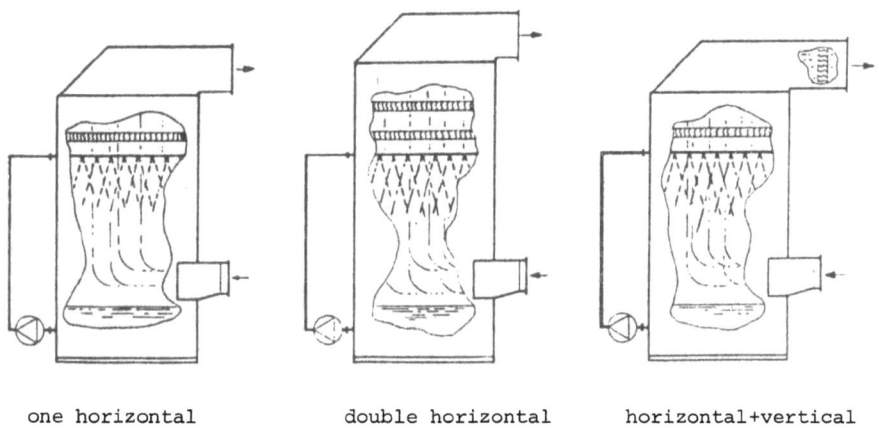

one horizontal   double horizontal   horizontal+vertical

For the first fgd in the Netherlands (PGEM Nijmegen G13 fgd-1) a fluegas reheat by means of a so called GAVO was neccessary.
In order to protect the GAVO from fouling the mist in the absorber outlet had to be minimised to < 200 mg/Nm^3.
Therefore a double demister was installed (horizontal type) + a recycle fan to recycle reheated gas of $> 80°$ C to the absorber outlet in order to raise the absorber outlet gastemperature with $3°C$.
For the fgd at PNEM, Geertruidenberg, Amer 8 an alternative for no GAVO reheat was optimal (see below) and therefore a single horizontal misteliminator was chosen. Besides saving investment cost also energy consumption by the boosterfan is lowered with approx. 100 kW for the 645 MW fgd installation.

Absorber outlet design.

Value engineering exercises for the construction and other aspects of the outlet of the absorber gave the following absorber outlet forms:

Figure 8. Absorber outlet design.

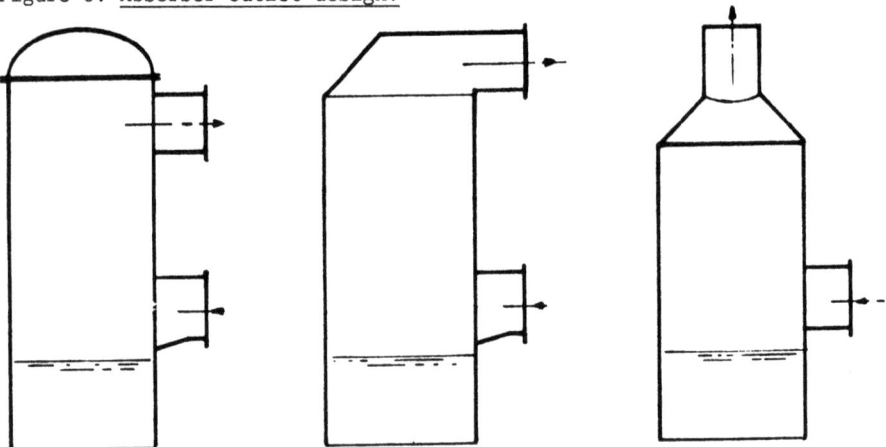

The left outlet was general practise in the USA installations.
The middle outlet was chosen inorder to optimise internal construction and to minimise area to be rubber lined.
The vertical outlet of figure 8 was chosen in order to fit the special ductwork configuration in the design of PGEM, Nijmegen, G13-fgd-2.

Reheat

The cleaned fluegases, leaving the absorber have been cooled down to approx. 50°C, they are water saturated and below the acid dewpoint. Reheat is applied for plume rise and/or chimney protection.
In the early 1980's plume rise was considered to require a stack inlet temperature of > 80°C. Today however 60°C is established to be sufficient. Chimney protection requires at least 5°C reheat.
In the first fgd in the Netherlands a so called GAVO (gas-gas heater) was installed (see fig. 9).
This is a reheat system especially suitable for reheat ΔT's in the range of 20-40°C. Advantages are reliability and proven design, however the system gives a relatively high fluegas pressuredrop and more complex ducting.
For future designs other systems are becoming attractive for reheat temperatures in the range of Δ T=5-15°C. such as a warm water (recuperative) heatexchanger, steam reheat, mixing gasturbine exhaust gas etc.
For the PNEM, Geertuidenberg Amer 8 fgd installation this last system is under construction (fig. 9).
This system has no heat exchanger surface or other obstacles in the fluegasstream, investment is considerably low compared to the alternatives. An advantage is also that cost are primarily variable cost. At mid load situations cost will decrease.

Figure 9: Reheat systems.

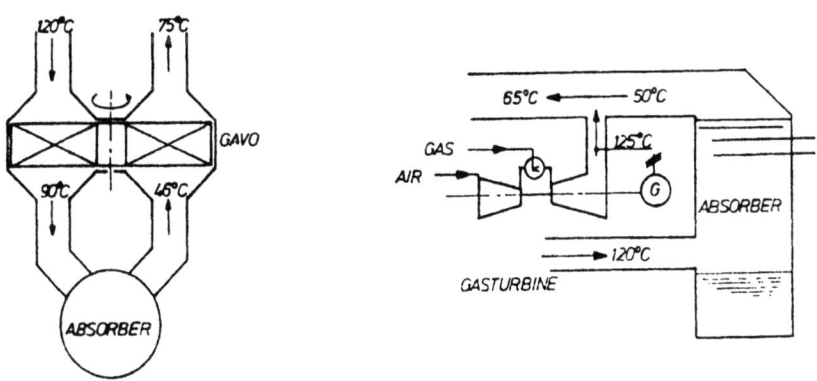

GAVO reheat   turbine gas reheat

Pressure drop reduction

As the fluegas pressuredrop contributes significant to the total electric power consumption of the boosterfan in the fgd unit, it is important to design the absorber and ducting for minimal pressuredrop, in order to keep the power consumption of the boosterfan as low as possible.
The layout of equipment and ducting differs from site to site, depending on the available plot area. This means that each unit requires a tailor-made design and layout in order to obtain the lowest possible pressuredrop.
Therefore, model tests are required for establishing the optimum configuration of equipment and ducting in a particular installation. Especially bends in ducting are important.

Dewatering

The gypsum slurry from the absorber (with internal oxidation) contains 10% solids (gypsum). For further processing in wallboard plants or plants for gypsum blocks or plaster this gypsum slurry has to be dewatered in the fgd installation to gypsum powder having max.10% (wt) water.
Also a water wash is applied to wash out the chlorides from the gypsum.

In the first European fgd installations batch wise operating centrifuges were applied as in Japan.

Since fgd gypsum has a typical grainsize-distribution and morphology several dewatering techniques had to be tested for its suitability to obtain max. 10% H_2O and max. 100 ppm chloride in the gypsum.

Below a survey is given on the different techniques tested.

Figure 10:   Survey of dewatering techniques tested.

laboratorium scale test :   drumfilter
precoat drumfilter
decanter centrifuge
worm screen centrifuge
vacuum belt filter
pusher centrifuge

pilotscale test   :   vacuum belt filter
worm screen centrifuge
pusher centrifuge

full scale (300 MW) test: vacuum belt filter
worm screen centrifuge.

In the lab. scale tests it was possible to establish the limits for dewatering and washing to be sufficient for the mentioned equipment.
In the pilot testing (approx. 1 ton/hr) the high settling velocity of the gypsum with respect to impurities and fines gave segregation in the cake formation which could however easily be overcome for the mentioned belt filter and worm screen centrifuge.
In full scale tests vacuum belt filter and worm screen centrifuge both gave satisfactory results.
Dependent on the aspects as requirements for moisture content, washing efficiency, space availability, investment, powerconsumption, maintenance possibilities one of these systems may be preferred over the other for a certain location.
In figure 11 the difference is shown between the first fgd installation at Nijmegen, having a thickener, centrifuge feedtank and batch centrifuges and the second fgd installation at Nijmegen having hydrocyclones and worm screen centrifuges.

Figure 11: Development in dewatering.

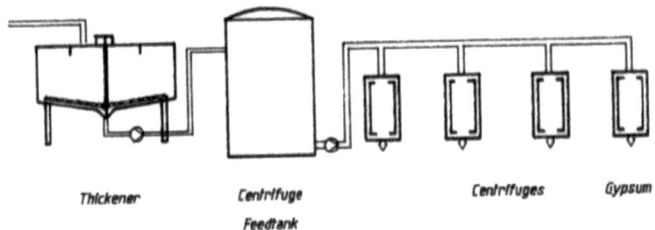

pgem g13-fgd-1

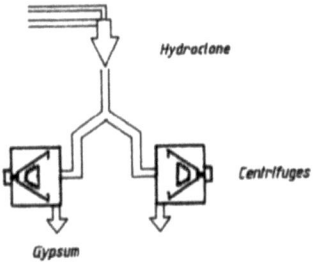

PGEM, G13‑fgd-2

Less equipment (lower investment), less space requirement, less power consumption and simpler operation are the result of this development in dewatering technology.

Waste water treatment.

In a limestone-gypsum fgd installation the chlorides from the coal are absorbed in the water and tend to accumulate as soluble chlorides in the slurry water. In order to restrict the accumulation to acceptable concentrations (e.q. 30.000 ppm Cl in water) a waste water bleedflow is required.
The criteria have been changed as follows:

Waste water treatment effluent criteria

1982:

PH : 5-7
Suspended solids: $<$ 50ppm

1985:

PH : 5-7
suspended solids: $<$ 20 ppm
As : $<$ 50 ppb
Cd : $<$ 10 "
Cr : $<$ 200 "
Cu : $<$ 50 "
Hg : $<$ 10 "
Ni : $<$ 200 "
Pb : $<$ 100 "
Zn : $<$ 200 "

The original neutralisation and clarification was therefore developed further to better heavy metals removal. By means of coprecipitation of metalhydroxides, basic metal carbonates and metal sulfides heavy metals are now removed. In the following scheme a typical wastewater treatment plant is presented.

Figure 12: Wastewater treatment installation.

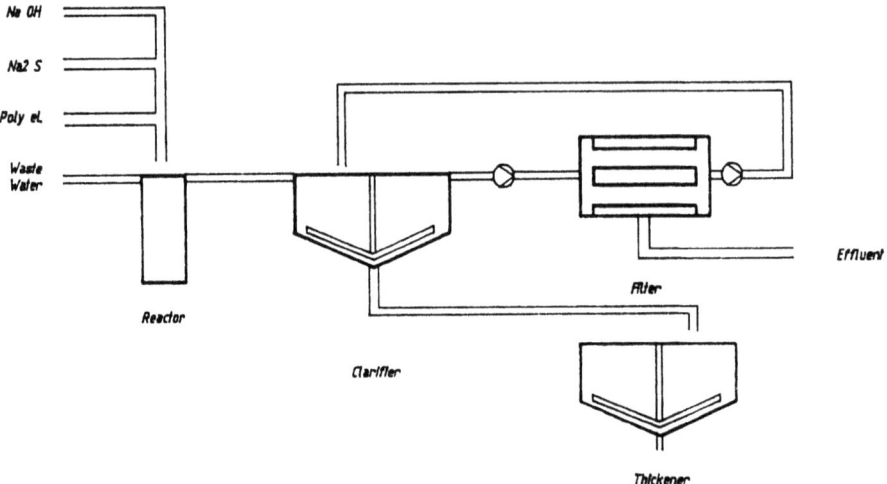

In the neutralisation reactor the pH of the wastewater is raised to PH=9.5. This is sufficient to precipitate the $Mg(OH)_2$. Also Na_2S is added. In the reactor clarifier also polyelectrolyte is added to improve the settling of solids. The overflow (with < 20mg/l suspended solids) is fed to a crossflow microfiltration unit. The underflow is pumped to a thickener.
In the microcrossflow filtration unit suspended solids are further removed to < 1 mg/l.

Costreduction

Cost comparisons between several installations can be made on the basis of the OECD Environmental Monograph "Understanding Pollution Abatement Cost Estimates" of March 1986.

YEAR OF DESIGN	1980	1985	1986/87
Installation *	nr. 1	nr. 2	nr. 3
MWe	2x312	625	312
Capital investment in Dfl/kW	350	190	230
Fixed cost (operating labor maintenance overhead and adm.)	0.22	0.12	0.15
Variable cost (raw materials utilities miscellaneous)	0.21	0.14	0.18
Capital cost + insurance/taxes	0.53	0.29	0.34
Operating cost in Ncts/kWh	0.96	0.55	0.67

* Operating hours: 6000

It is evident that the cost per kWh is the lowest for the biggest plant(s) but there is a clear tendency that these cost are decreasing throughout the years.

The cost per ton SO_2 reduction is also used as an indicator for evaluation of installations.

For these three Dutch plants and a 350 MWe Danish plant (1985) a removal difference of 2700 mg SO_2/Nm^3 was taken into account in order to obtain for each plant the following figures:

	Year	Actual Flue gas flow in $10.^6 Nm_3/hr$	Cost Ngld/tSO_2 removed	MWe
Plant 1	80	0,9	1000÷1250	360
Plant 2	85	2,1	370÷400	625
Plant 3	86	1,2	450÷500	330
Danish Plant	85	1,3	500÷550	350

Conclusion:

During the past 5 years the technical developments resulted in:

- considerable cost reduction
- simpler process, operation and maintenance
- optimisation per location.

It is expected that the development in this field will continue in the next years. Specially since during the coming years no retrofit type installations will be build but installations integrated in new powerplants. Also in the field of denitrification at powerplants developments are expected.

The Bischoff Flue Gas Desulfurization Process in the Power
Plants of Borssele and Maasvlakte in the Netherlands

Dipl.-Phys. Theo Risse; Dipl.-Ing. Ingrid Maassen

The Bischoff Process for Flue Gas Desulfurization

In the sixties, the company of Bischoff, Essen, already began
to develop its own flue gas desulfurization process on the basis of wet absorption with lime or limestone.

Field experience with numerous experimental, pilot and demonstration plants led to the concept of a counter-current scrubber integrating all process steps like absorption, oxidation,
crystallization and mist elimination. This concept allows a
simple plant technology of low cost and space requirement.
High-quality gypsum is produced as desulfurization product.

Figure 1. FGD-plants in Wilhelmshaven

This concept was realized when designing and constructing the
flue gas desulfurization plant 2 ("FGD 2") in the 770 MW Wilhelmshaven power station. This plant is designed for a flue
gas flow of 1 500 000 m3/h (stand.) corresponding to 54 % of
the total boiler flue gases. It was the first single-FGD plant

in Europe of this dimension. The plant was commissioned in 1982 and has been running perfectly since that time.

In 1985, the flue gas desulfurization plant for the 450 MW unit 5 of the Leiningerwerk power station in Zolling near Munich for a flue gas flow of 1 500 000 m^3/h (stand.) followed. This plant was designed according to the concept of "FGD 2" and since its commissioning it has been in operation with good results.

Due to authority regulations demanding the desulfurization of 100 % of power plant flue gases and a maximum sulfur oxide emission of 400 mg/m3 (stand. dry) an additional FGD-plant for 1 300 000 m^3/h (stand.) was designed and constructed. A special feature of this so-called "FGD 1a" is the arrangement of the FGD-fan on the cold clean gas side. "FGD 1a" was commissioned in 1985 and no problems have occured up to now. Figure 1 shows "FGD 1a" and "FGD 2" in Wilhelmshaven.

Experiences gained with the operation of these FGD-plants are the basis for further plants of the Bischoff design including two large contracts in the Netherlands.

Bischoff Plants in the Netherlands

In the Dutch power stations of PZEM in Borssele and GEB in Maasvlakte, the boiler flue gases are desulfurized according to the Bischoff process.

The flue gas desulfurization plant in the 400 MW power plant of Borssele is a single-stream plant designed for a flue gas flow of 1 340 000 m^3/h (stand.) corresponding to 100 % of the boiler flue gases. The sulfur dioxide absorber in the Borssele power station is the largest scrubbing tower presently under construction in the Netherlands. In Summer 1987 the plant will be commissioned. Figure 2 shows the scrubber in the final erection phase.

Figure 2. Scrubber in the Borssele Power Station

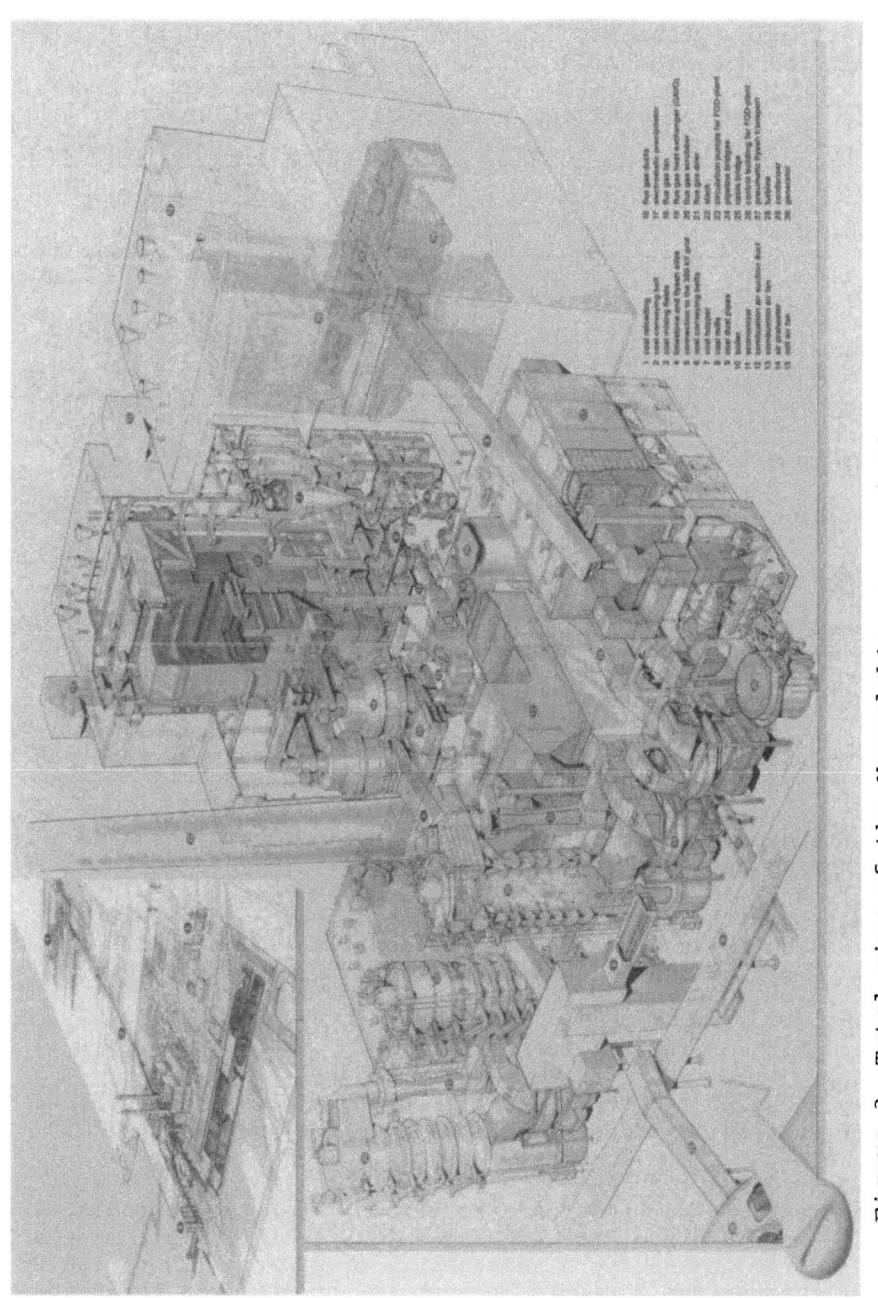

Figure 3. Total view of the Maasvlakte power station

The Maasvlakte power station has a capacity of 2 x 540 MW. Each unit will be equipped with two absorbers. The four-stream FGD-plant is designed for a flue gas flow of 4 x 900 000 m^3/h (stand.). Figure 3 shows the integration of the flue gas desulfurization plant into the power station. The plant will be commissioned by the middle of 1987, resp. early 1988.

Table 1 summarizes the design data of the FGD-plants.

Power station		Borssele	Maasvlakte
Power	MW	400	2 x 540
Number of scrubbing towers		1	4
Flue gas flow	m$_0^3$/h	1 340 000	4 x 900 000
SO$_2$-content of raw gas	g/m$_0^3$	max. 3,5	max. 3,4
Desulfurization efficiency	%	≥ 88	≥ 88
Additive		lime/limestone	lime/limestone
Desulfurization product		gypsum	gypsum
Product quality	%	≥ 95	≥ 95

Table 1. Design data of the Borssele and Maasvlakte flue gas desulfurization plants

For the construction of the FGD-plants in the Netherlands Bischoff cooperates with Dutch companies. In case of the Borssele FGD-plant the consortium Comprimo/Bischoff was formed. For the Maasvlakte FGD-project the consortium Tebodin/Bischoff was constituted. The cooperation between Tebodin resp. Comprimo and Bischoff has proven to be a success.

Description of the Process

The process of the FGD-plants in the Borssele and Maasvlakte power station is almost identical and is represented in figure 4.

Downstream of the electrostatic precipitator the flue gas enters a regenerative heat exchanger (GAVO) of Ljungström type. The hot raw gas having a temperature of approx. 120 to 130 °C flows through the GAVO from bottom to top and, via rotating heat storage blocks, conveys a part of its heat to the cold flue gas already cleaned. Then, the raw gas cooled to approx. 70 to 80 oC laterally enters the scrubbing tower.

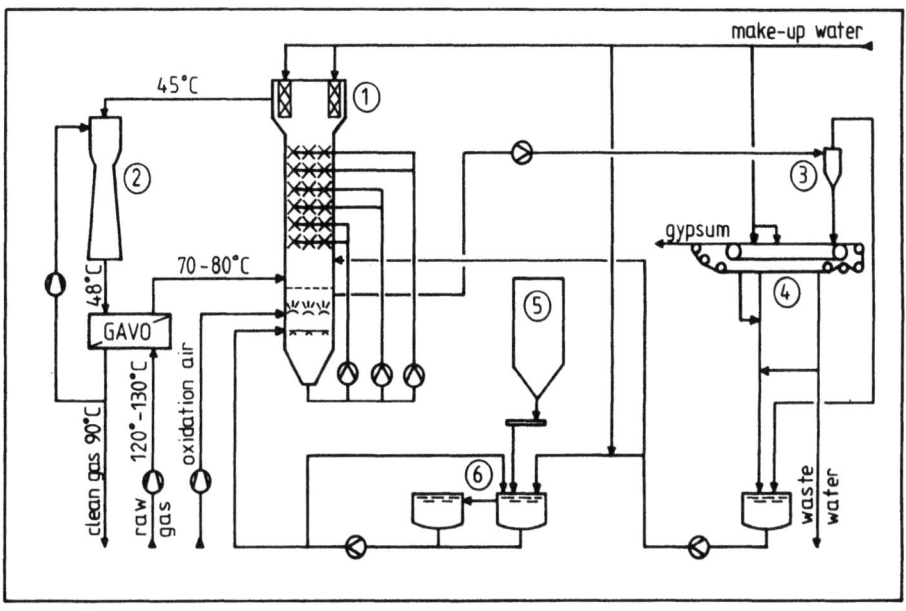

1 scrubber   3 hydrocyclone   5 additive silo
2 mixer   4 vacuum belt filter   6 additive preparation

Figure 4. Flow sheet of the Bischoff-Process

In the scrubber, called absorber, the flue gas flows through the absorption zone formed by six spray levels arranged one above the other. Here, desulfurization is effected by intensive spraying with lime or limestone containing scrubbing slurry. Gaseous sulfur dioxide is removed mainly as calcium hydrogen sulfite dissolved in the scrubbing slurry.

The spray levels are equipped with hollow-cone nozzles operating clogging-free. These sprays of Bischoff design are, with the exception of the upper level, nozzles spraying both upwards and downwards producing a fine droplet spectrum.

In the absorption zone the flue gas is cooled to a saturation temperature of 45 °C. After having left the absorption zone the desulfurized gas passes the annulary arranged mist eliminator stage in the scrubber head, figure 5.

Contrary to the separator surfaces laterally arranged at the gas outlet the annular arrangement offers numerous advantages. With laterally arranged separator surfaces and large scrubber diameters a uniform gas distribution is difficult to achieve due to the small pressure losses in the demister. In this case more liquid droplets are entrained because the lower separation zone is superproportionally loaded. Due to the annular arrangement gas distributors in the Bischoff absorber are not required and a uniform flow is guaranteed. Further advantages of this arrangement are offered by the reduction of the overall height and the better distribution of the weight load.

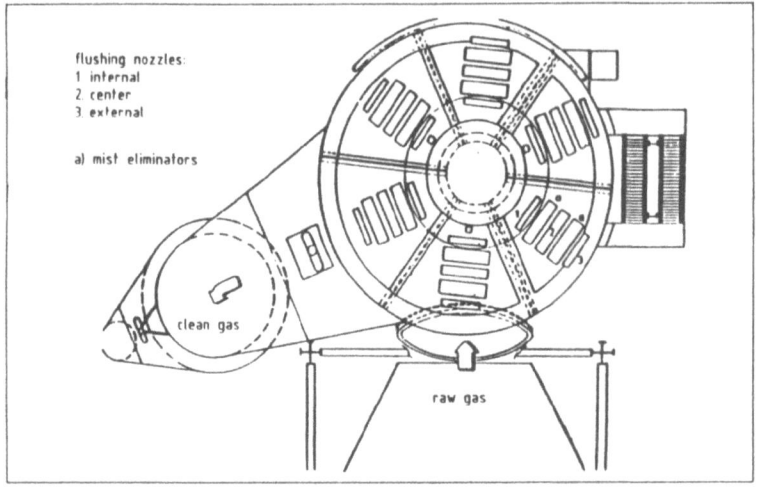

Figure 5. Arrangement of the demister

The mist eliminators are periodically sprayed with flushing water to remove contingent solid deposits. The flushing water also serves to compensate evaporation losses in the scrubbing tower.

Downstream of the scrubber the clean gas enters a venturi-type mixing stage. Here, the residual droplets are evaporated by adding a partial flow of the reheated clean gas. Due to this the heating surfaces of the downstream arranged GAVO are protected against excessive incrustations. Figure 6 shows the scrubbing tower and the gas mixer.

In the heat exchanger the temperature of the clean gas is increased from approx. 48 °C to approx. 90 °C. A small part of the reheated clean gas is fed to the residual droplet evaporation zone via a recirculation bypass. The flue gas is then led to the stack.

The scrubbing slurry is collected in the absorber sump which is divided into oxidation and crystallization zone. A small part of the calcium hydrogen sulfite has already been oxidized in the absoprtion zone due to the oxygen content of the flue gas. In the oxidation zone most of the calcium hydrogen sulfite is oxidized to calcium sulfate (gypsum) at a pH of 4 to 5 by blowing air into the scrubbing slurry. The air is introdueced by an air distribution system consisting of an oxidation tunnel and laterally arranged oxidation hoods, figure 7. The oxidation hoods are shaped as semi-shells equipped with outlet holes. Bischoff has developed this distribution system by long year experiences gained with other oxidation processes, i. e. oxidation of hydrogen sulfide. Due to the special design of the air distribution system the velocity of the slurry between the hoods is raised. So any back-mixing of the slurry between the oxidation zone and the subjacent crystallization zone is avoided. Any back-mixing would result in an pH increase making oxidation more difficult.

Figure 6. Scrubber and gas mixer

Depending upon the produced gypsum quantity a side stream of the slurry is discharged from the oxidation zone and is fed to the gypsum dewatering station.

In the crystallization zone, the fine gypsum particulates produced during oxidation grow to larger crystals that can be well filtrated. Fresh additive, lime resp. limestone slurry, prepared and stored in the additive station is fed into the crystallization zone. The circulation of the slurry from the crystallization zone to the nozzle levels is performed by three pumps.

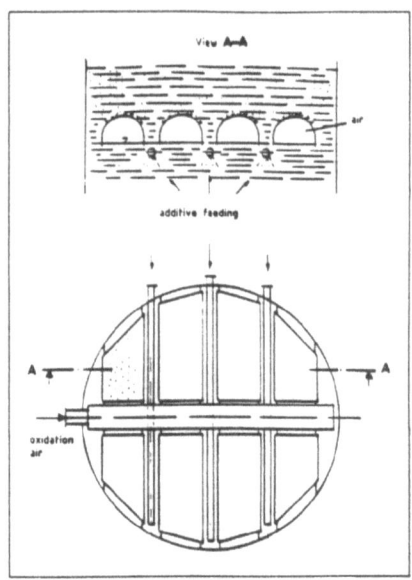

Figure 7. Oxidizer internals and additive feeding

For better understanding figure 8 shows the basic chemistry of the limestone process.

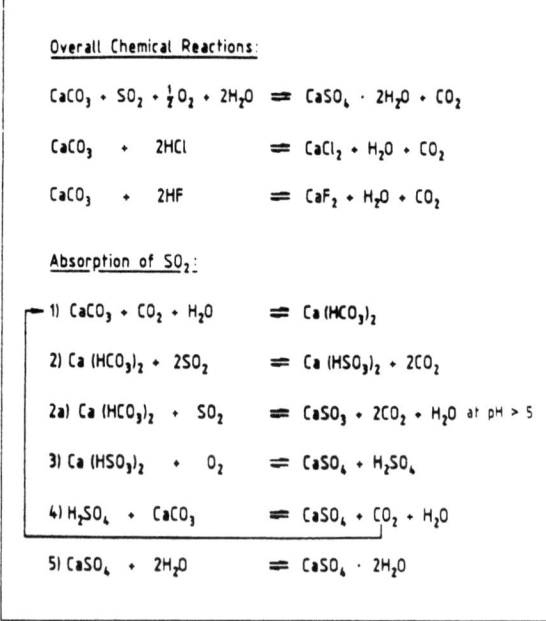

Figure 8. Basic chemistry of the limestone process

Calcium carbonate is in presence of CO_2 dissolved in the scrubbing liquid and calcium hydrogen carbonate is formed. This reacts with SO_2 forming calcium hydrogen sulfite or, at higher pH, calcium sulfite. In the integrated oxidizer the oxygen of the air oxidizes calcium hydrogen sulfite to calcium sulfate. This reaction usually is operated at a pH of 4-5. With the acid formed during the oxidation more carbonate is dissolved.

Following equation 5 calcium sulfate is crystallizing as dihydrate of gypsum according to its low solubility in the scrubbing liquid.

Besides the CO_2-removal hydrogen chloride and hydrogen fluoride are absorbed as well forming dissolved calcium chloride resp. solid calcium fluoride.

The gypsum slurry discharged from the scrubber is thickened in hydrocyclones. The hydrocyclone underflow enriched by solids is fed onto a vacuum belt filter, figure 9.

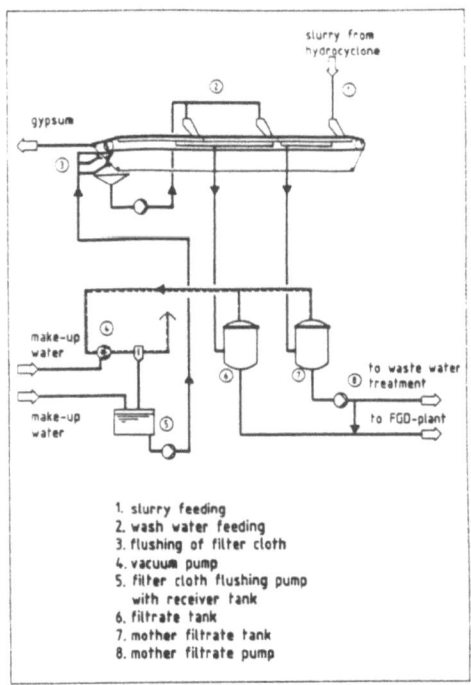

Figure 9. Vacuum belt filter

The gypsum is dewatered to a residual moisture of approx. 10 %. The soluble salts, such as calcium chloride which could trouble subsequent processing of the gypsum are removed from the filter cake by a two-stage water washing process. The gypsum is intermediately kept in a storage building from where it is loaded and transported to the final customer.

Part of the mother filtrate of the vacuum belt filter is pumped to a waste water treatment. The residual filtrate is recycled to the FGD-process together with the overflow of the hydrocyclone.

Waste Water Treatment

In order to maintain the chloride content in a range not affecting the desulfurization efficiency and not causing any material problems the discharge of waste water is required.

The waste water mainly contains calcium and magnesium compounds as well as low quantities of heavy metal ions entering the system via the removed raw gas dust together with the recycled process water and the used additive. A waste water treatment system is required to remove the heavy metal. Bischoff designed the waste water treatment plants for the Borssele and Maasvlakte power station.

When treating the waste water, figure 10, the water is first neutralized with soda lye. In a following step, the still dissolved heavy metal ions are precipitated with sodium sulfide. This process reduces the heavy metal contents to the limit values fixed by authorities.

Further treatment is effected by flocculation and sedimentation with solid separation in a chamber filter press. In a last step, residual solids are removed from the waste water by means of a DynaSand Filter.

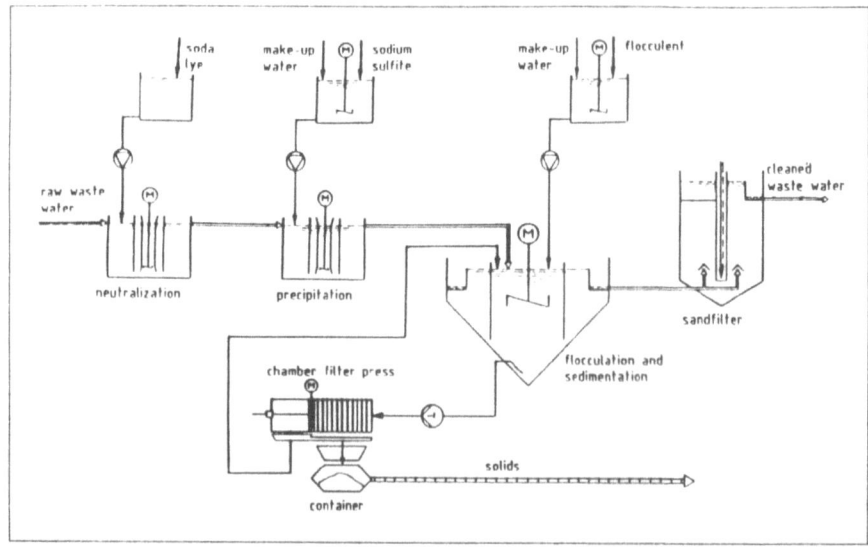

Figure 10. Waste water treatment

Reheating of Flue Gases

In the Borssele and Maasvlakte power station regenerative heat exchangers are applied utilizing the energy of the hot raw gas to reheat the clean gas. This process does not require any extra energy and effects a considerable saving of operation costs.

The regenerative heat exchangers are supplied by Lentjes-BWE, a subsidiary of Lentjes AG which is a parent company of Bischoff.

Raw and clean gases flow vertically through the GAVO in a counter-current, figure 11. Heat storage blocks serve as heat transfer elements consisting of thin profiled plates arranged parallel to the flow direction and mounted on a rotor.

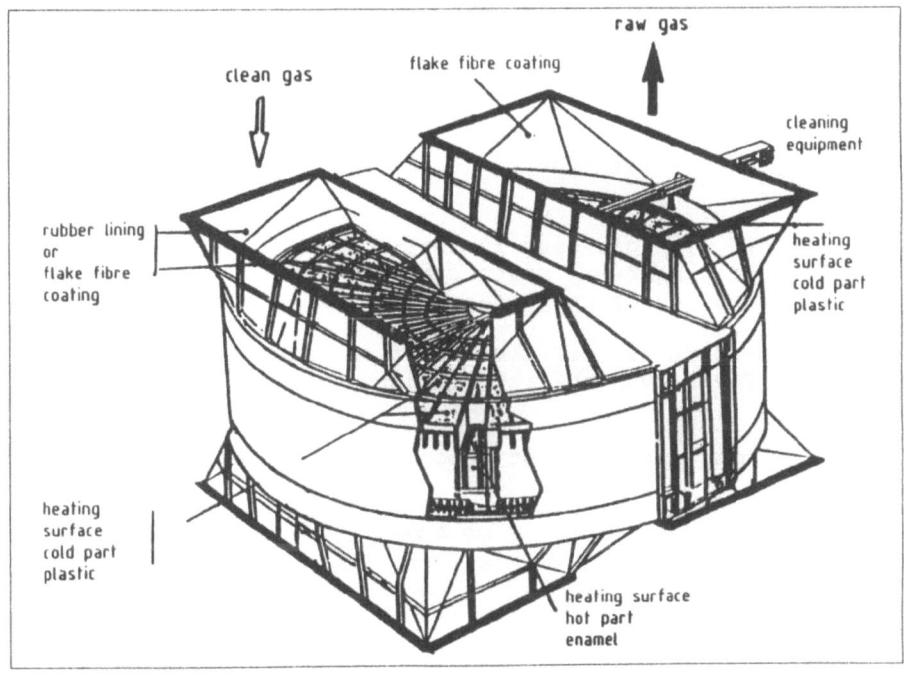

Figure 11. Regenerative heat exchanger "GAVO"

The hot raw gas passes the raw gas side sector from bottom to top and transfers part of the heat to the heat transfer blocks which rotate to the clean gas side sector by reheating the clean gas flowing from top to bottom. On the hot side the heat storage blocks are made of enameled steel. The cold gas side consists of plastic packages. During operation solid deposits can be removed from the heating surfaces by compressed air or high-pressure water.

Materials of construction

The material of flue gas ducts, scrubber, tanks and most of the pipes is steel St 37-2 having an inside coating or rubber lining as corrosion protection. The flue gas ducts on the hot gas side are coated with temperature resistant vinyl resin with inertflake filler. A soft butyl rubber coating is applied to all other plant components exposed to high temperatures, a soft chloroprene rubber coating to those plant components exposed to low temperatures.

The scrubber internals comprise mist eliminators, slurry nozzles and oxidation air distribution system. The mist eliminators are lamella-shaped packages made of polypropylene with talc as filler. The scrubbing suspension nozzles are made of nitridated silicon carbide characterized by high wear resistance. The material of the oxidation hoods is stainless steel 1.4529.

The used pumps are all-metal pumps made of material 1.4464 resp. cast pumps with polyurethane lining having proved to be essentially wear resistant.

Field Experience with the Bischoff Flue Gas Desulfurization Plants

The Bischoff flue gas desulfurization plants in the Wilhelmshaven and Leiningerwerk power stations have operated to the full satisfaction of the plant owners since their start-up.

The plants adapt to fast flue gas flow changes without any problems. Their availability corresponds to that of the boilers.

Table 2 summarizes the most important operation data.

The design of the Wilhelmshaven "FGD 2" based on a total desulfurization efficiency of 80 % considers the addition of hot undesulfurized raw gas for residual droplet evaporation. The high efficiency of the mist eliminators in the scrubber head allowed to do without the raw gas addition thus increasing the total desulfurization efficiency to 90 %.

In the power station of Leiningerwerk a maximum sulfur dioxide concentration in the clean gas of 400 mg/m^3 (stand. dry) is allowed. The values measured since commissioning amount to approx. 200 mg/m^3 (stand. dry). Currently, the minimization of the raw gas supply for residual droplet evaporation is tested.

Like the abovementioned Bischoff plants the Wilhelmshaven "FGD 1a" achieves a scrubbing efficiency of more than 95 % as well. Due to the arrangement of the FGD-fan on the clean gas side the total desulfurization efficiency exceeds 95 %.

Power plant		Wilhelmshaven		Leiningerwerk Zolling
		FGD 2	FGD 1a	
Commission date		1982	1985	1985
Flue gas flow	m_n^3/h	1 500 000	1 300 000	1 500 000
SO_2-content of raw gas	g/m_n^3	3,0	2,8	2,6
Overall desulfurization efficiency	%	> 90	> 95	> 86
Position of FGD-fan		forced draft fan	induced draft fan	forced draft fan
Additive		lime	lime/limestone	limestone
Gypsum quality	%	> 95	> 95	> 95

Table 2. Operation data of the Bischoff flue gas desulfurization plants

Up to now no deposits have occured in the scrubbers. The inside rubber coating of the scrubber is faultless. The slurry nozzles and the mist eliminators do not show any sign of malfunctions. During shut-down the slurry remains in the absorber sump. Despite the sedimentation of more than 100 tons of gypsum circulation and stirring devices are not required. When restarting the plant the scrubbing slurry pumps whirl up the settled gypsum via a flushing system. Afterwards, the pumps are switched over to the nozzle levels. The flushing process operates automatically and lasts a few minutes, after 15 minutes the full desulfurization efficiency is achieved.

Since start-up the regenerative heat exchangers have not shown any failures. Due to the high mist eliminator efficiency only few solid matters settle and are removed by operation flushings. After each flushing process the differential pressure of the initial state is achieved.

The operation of the conveying and feeding devices have been trouble-free up to now.

The gypsum having a residual moisture of approx. 10 % and a purity of more than 95 % is sold to the gypsum or cement industry.

These positive operation results show that the Bischoff flue gas desulfurization process has proven to be a success in practice and is characterized by its high reliability.

THE BIOLOGICAL TREATMENT OF WASTE GASES FROM SMALL URBAN SOURCES
(EMISSION OF VOLATILE ORGANIC COMPOUNDS)

A.J. Dragt[1]), A. Jol[1]), C. van Lith[2]) and S.P.P. Ottengraf[3]

[1]) Department of Energy and Environment, DHV Consulting Engineers,
P.O. Box 85, 3800 AB Amersfoort (the Netherlands)
[2]) ClairTech, P.O. Box 8022, 3503 RA Utrecht (the Netherlands)
[3]) Department of Chemical Engineering, Eindhoven University of
Technology, P.O. Box 513, 5600 MB Eindhoven (the Netherlands)

1. INTRODUCTION

In the last ten years biotechnological methods have increasingly been used
in the treatment of polluted air. The principles of this technique of biofiltration are based on the aerobic conversion of pollutants by microorganisms. At the outset biofiltration was mainly used for odour abatement
in waste water treatment plants. Nowadays there is a clear trend to apply
biofiltration on a much broader scale. One of the most interesting
applications is the degradation of xenobiotic compounds. DHV Consulting
Engineers, Eindhoven University of Technology and Clairtech BV have
developed the so-called Bioton (R), a modern application of biofiltration.
The main advantages of the Bioton are the high possible load, the relatively small space requirements and the large applicability because of the
use of specially isolated strains of micro-organisms. Investigations were
carried out at a laboratory as well as a pilot-plant scale regarding the
degradation of solvents in general.

2. THEORETICAL BACKGROUND

Biofiltration is a technique for the elimination of odour and polluting
compounds in waste gases with the use of aerobic micro-organisms. The
micro-organisms (usually bacteria) are attached to compost, which contains the necessary nutrients. For easily biodegradable components, like
alcohols, ketones and esters, the compost generally contains enough microorganisms suited for biodegradation. The micro-organisms show a tremendous
ability to adapt to the components present in the waste gas. However, to
minimize the adaptation period it is often convenient to use activated
sludge from a sewage treatment plant as an inoculum. Poorly biodegradable
components, such as chlorinated hydrocarbons (e.g. dichloromethane) and
most aromatics (e.g. benzene, toluene, xylene), generally require inoculation with specally cultivated micro-organisms. These micro-organisms
can be isolated in the laboratory from, for example, contaminated soil.
The micro-organisms are able to oxidize the organic and inorganic components concerned into mineral end-products (e.g. CO_2, H_2O, NO_3^- etc.).
Part of the organic components are transferred into new cell material.
The performance of the biofilter is the result of both physical and
micro-biological phenomena, the so-called macro-kinetics of the process.
Some important physical factors are mass transfer, the flow behaviour of
the gas and the average residence time.

The pure rate of the microbiological oxidation is denoted as the micro-kinetics of the process. Only for mono-cultures of micro-organisms the micro-kinetics have been thoroughly investigated. Therefore it is necessary to make simplifying assumptions for a model of the macro-kinetics in biofilters where there usually exist heterogeneous cultures.
Such a model was developed by Ottengraf (lit. 1, 2). The most important characteristics of the model, which proved to be very well in accordance with experimental results, are:
- The macro-kinetics of the elimination process in a biofilter can be described by an absorption process (diffusion) in a wet biolayer, accompanied by a simultaneous biological degradation reaction. The biolayer consists of a small waterfilm around a compost particle.
- The biological degradation reaction of nearly all the components investigated, like alcohols, ketones, esters, aromatics, hydrogen-sulfide, ammonia, follows zeroth order micro reaction kinetics in substrate down to very low (water phase) concentration levels.
- The overall elimination rate of the biofilter is dependent on both the diffusion and reaction rate. Because of the zeroth order of reaction rate the diffusion process will be the rate limiting process below a certain low gas phase concentration. Above this concentration the overall elimination rate is independent of the gas phase concentration.

3. PILOT PLANT TEST

The pilot plant biofilter installation, the Bioton (R), was tested at a lacquer undertaking. During printing and lacquering of wood and plastics solvent vapours were emitted by the ventilation system. The solvents used mainly consisted of toluene, ethylacetate, butylacetate, ethanol, isopropanol and butanol.
The emission of these components occurred discontinuously because the lacquering process was carried out only during a few hours every day. A minor part of the waste gas stream was directly fed to the biofilter pilot plant, while the bulk was led to a buffer installation (carbon adsorber). The pilot plant installation consisted of the following components (fig. 1).:
- carbon adsorber (185 kg)
- gas humidifier
- biofilter (filter bed: height = 1 m, surface area = 5,3 m^2)

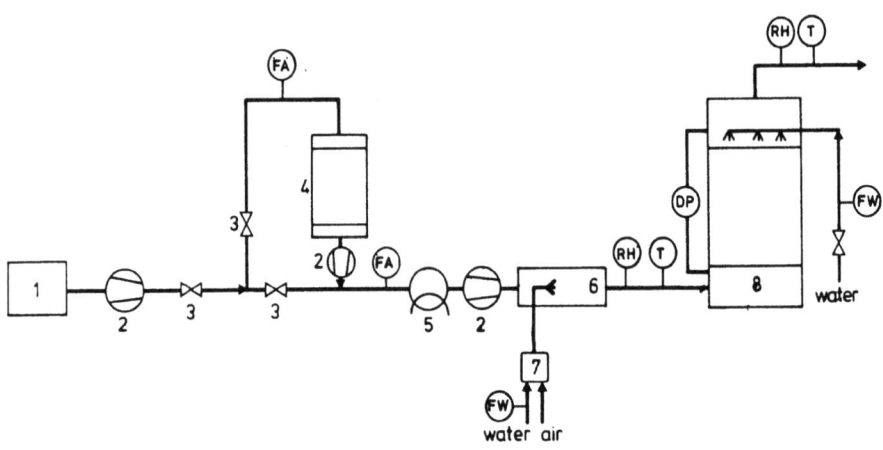

1 = lacquering department
2 = fan
3 = valve
4 = carbon adsorber
5 = heater
6 = sonicore-humidifier
7 = instrumentation

FA = air flow meter (orifice)
RH = relative humidity meter
T = thermometer
DP = open U-tube manometer
FW = fluid flow meter (rotameter)

Figure 1 - <u>Biofiltration pilot installation</u>

About 1000 m³/h waste gas could be treated in the installation. About 650 m³/h of this stream was fed to the carbon adsorber. The purpose of this adsorber was to prevent extremes in the concentration of the components in the waste gas at the inlet of the biofilter.
Adsorption of the solvent vapours on the active carbon occurred during the (short) periods (about 4 hours/day) that lacquering was taking place. Desorption occurred during the rest of the period of operation (about 20 hours/day), while clean air was led through the carbon adsorber. The waste gas stream from the carbon adsorber was combined with the other gas stream of 350 m³/h coming directly from the lacquering department. The total gas stream was then lead to the humidifier. Humidification to 100% relative humidity was necessary to prevent water loss in the biofilter. A water content of 40 to 60 weight percent in the packing material in the compost is essential for a good microbiological activity.

For humidification a so-called sonicore sprayer was used. This sprayer type makes use of pressurized air, by which very small water droplets are produced. The advantage of this system is a good humidifcation with no waste water. The disadvantage is the relatively high cost of operation. The waste gas could be heated before entering the biofilter. This was necessary when the waste gas temperature came below 5°C.
The packing material in the biofilter contained a mixture of compost and inorganic additives, which were necessary to improve the packing structure. Therefore the pressure drop over the biofilter was small even at high loads up to 300-500 $m^3/m^2.h$. These loads are about 10 times higher than the loads used in conventional so-called compost filters.

4. RESULTS

The first two weeks the biofilter was operated at a low load of 100 $m^3/m^2.h$. After this adaptation period measurements showed good elimination of all components. Therefore the load was increased to 200 $m^3/m^2.h$, which was used during the rest of the investigation period. The total period of continuous operation of the pilot installation was about 5 months.
The water content in the packing material decreased slightly during this period, because in practice the relative humidity of the waste gas which was fed to the biofilter was never exactly 100%. However, the water loss was very small and only once, at the end of the investigation period, it was necessary to humidify the filterbed directly.
The pressure drop over the filterbed remained low, about 60 Pa at a load of 200 $m^3/m^2.h$.
The operation of the carbon adsorber was good. Measurements showed that peak-concentrations were reduced with about 65 percent (average).
The elimination of the solvents was measured 12 times. The most important parameter is the elimination capacity of the biofilter, expressed in gram component or organic carbon eliminated per m^3 packing material per hour. The elimination capacity is calculated from the concentrations of the solvents at the inlet and outlet of the biofilter and the load of the biofilter.
In figure 2 the elimination capacity (gram total organic carbon/$m^3.h$). is plotted against the concentration of the total organic carbon in the inlet of the biofilter. The concentration total organic carbon was calculated from the individually measured components.

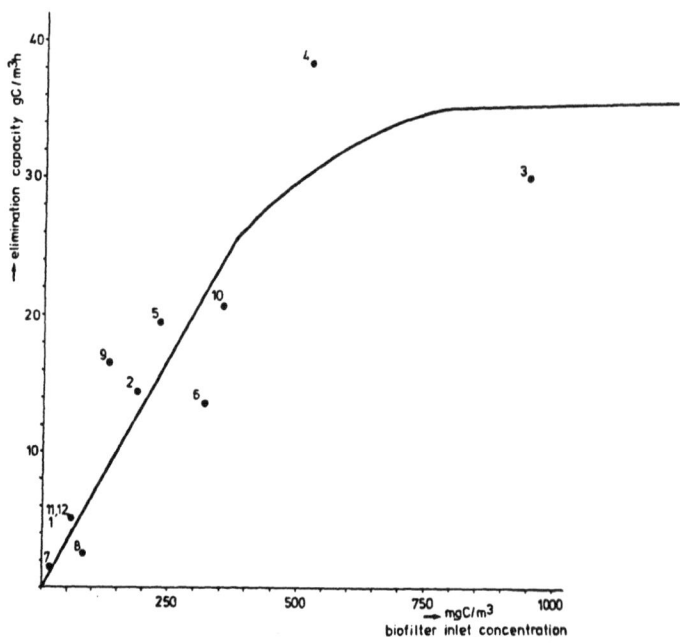

Figure 2 - <u>Elimination capacity versus biofilter inlet gas concentration</u>

From figure 2 it can be concluded, that the elimination capacity is a function of the gas concentration at low organic load of the biofilter. At gas concentrations above 750 mg C/m^3, the elimination capacity is maximal, 35 g $C/m^3.h$.

These results are in good accordance with the previously described theoretical model for the macro-kinetics of a biofilter. Below 750 mg C/m^3 the overall elimination occurs in the diffusion controlled regime. Only in the lower part of the biofilter the elimination capacity is not diffusion-controlled.

With the use of these results it is possible to design a full-scale biofilter for any commercial application in the lacquer industry. Furthermore the optimal operating conditions (adsorption on active carbon, humidification, etc.) found with the pilot installation can be used for a full-scale installation.

5. FULL-SCALE INSTALLATION

From the pilot-biofilter it can be concluded, that biofiltration in combination with a carbon adsorber is a good waste gas abatement technique for discontinuous solvent emissions. These emissions occur at industry which uses paint and lacquer. Usually these undertakings are relatively small and situated in urban surroundings (e.g. car spraying shops). Only relatively cheap techniques can be afforded by these undertakings.

The size of a full-scale installation depends on:
- Amount and solvent concentration of waste gas. These variables depend on the size of the lacquer undertaking.
- Desired emission reduction.

To estimate the costs of biofiltration three representative situations were chosen. The necessary emission reduction was calculated using the interim odour emission standard valid in the Netherlands. According to this standard a concentration of 1 o.u. (= odour unit) per m^3 may not be exceeded at the nearest housing for more than 2% of the year (98-percentile).
The contour around the emission source for this 98-percentile was calculated using a dispersion model. The odour emission was estimated using an odour threshold of 0,1 mg C/m^3 for the sum of all the organic components in the waste gas.
Using this method the calculated conditions for the three situations are given in table 2.

Table 2 - Three representative emission situations for the lacquer industry

	situation 1	situation 2	situation 3
use of solvent (kg/day)	10	50	250
lacquering period (h/day)	2-4	4	12-24
nearest housing (m)	20	50	50
maximum allowable continuous emission (kg/h)	0.17	0.57	0.78

Situation 1, 2 and 3 are representations of respectively a small, medium and large lacquering undertaking.

With the use of table 2 it is possible to design the carbon adsorber and the biofilter for the three situations. Adsorption on the active carbon takes place during lacquering, followed by desorption and loading of the biofilter during the rest of the day. In situation 3 a continuous emission is assumed, which means that two separate carbon adsorbers are necessary. Adsorption during 12 hours is followed by desorption, and loading of the biofilter, during 12 hours for each adsorber. In table 3 the dimensions and total costs are given for each situation. The yearly costs include depreciation and costs of operation.

Table 3 - Dimensions and costs of full-scale biofilter installation for three situations for the lacquer industry

	situation		
	1	2	3
loading period of biofilter (h/day)	20	20	24
removal efficiency of biofilter (%)	65	77	93
load of biofilter (m^3/h)	230	740	3500
height of biofilter (m)	3	3	3
volume of biofilter packing (m^3)	6	46	180
mass of active carbon (kg)	560	2280	10840
total investment (Dfl.)	46,000	151,000	478,000
yearly costs (Dfl.)	9,600	26,600	85,800

6. CONCLUSIONS

1. Biofiltration is applicable as a technique for air pollution control for off gases from small urban sources containing volatile organic components.
2. In the case of an intermittent emission pattern the combination of biofiltration with other accumulation techniques like carbon adsorption is very usefull.
3. The cost of biofiltration is low, compared with other more traditional techniques.
4. Besides the elimination of volatile organic compounds, also odour emissions will be minimized.
5. Biofiltration is a clean technology, the pollutants removed from the gas stream are not drained into other waste streams; they are oxidized to into environmental harmless products or to products that can be neutralized easily.

7. ACKNOWLEGDEMENTS

This research programme has partly been financially supported by the Ministry of Housing, Physical Planning and the Environment in the Netherlands.

REFERENCES

1. Ottengraf, S.P.P. and A.H.C. van den Oever, Kinetics of Organic Compound Removal from Waste Cases with a Biological Filter, Biotechnology and Bioengineering, vol. XXV, 1983, pp. 3089-3102.
2. Ottengraf, S.P.P., J.J.P. Meesters, A.H.C. van den Oever and H.R. Rozema, Biological elimination of volatile xenobiotic compounds in biofilters, Bioprocess Engineering 1 (1986), p. 61-69.
3. Dragt, A.J., S.P.P. Ottengraf and D.M. Zuidam, Biofiltration - a new technology in air pollution, Proceedings of the Seventh World Clean Air Congress, held at Sydney, Australia, August 25-29, 1986, pp. 545-554.

BIOFILTRATION - A RELATIVELY CHEAP AND EFFECTIVE METHOD OF
WASTE GAS TREATMENT

P.G. Paul, Comprimo Engineers & Contractors,
P.O.B. 4129, 1009 AC AMSTERDAM, The Netherlands

F.J. Castelijn, Comprimo B.V. Gas Division,
Vredeweg 3, 1505 HH ZAANDAM, The Netherlands

1. INTRODUCTION

High population density, industrial development and urban civilisation more and more influence the environment and human living space.

For this reason, many governments require technical measures to be taken with respect to waste water and waste air cleaning, noise abatement and waste treatment.

Increasingly severe demands are also issued to limit the emission of air polluting (odorous) compounds. In many cases, process changes are not possible so treatment of the waste gas is the only way to meet these requirements.

2. BIOFILTRATION

Very often, biofiltration is such a suitable technology that combines high effectiveness and reliability with relatively low operation cost.

In the past years, extensive research is executed on biofiltration in the Netherlands. TNO[1] has in cooperation with the VAM[2] carried out research into the effectiveness of several biologically active materials. This has led to the development of a new material called Vamfil, with very favourable performances with respect to biological activity and resistance to flow. Vamfil is a mixture of organic materials (treebark and compost).

1) Applied Scientific Research - a semi government Dutch Institute
2) Dutch Waste Disposal Company

2.1 Biochemical Process

Figure 1 shows the principle of biofiltration.

The carrier in the figure is one of the solid particles of Vamfil. Vamfil is a patented mixture of selected treebark, compost and other nutrients. The pores of the organic carriers are filled with water. In the waterphase, micro-organisms are present, partly free floating in the water and partly attached to the carrier surface.

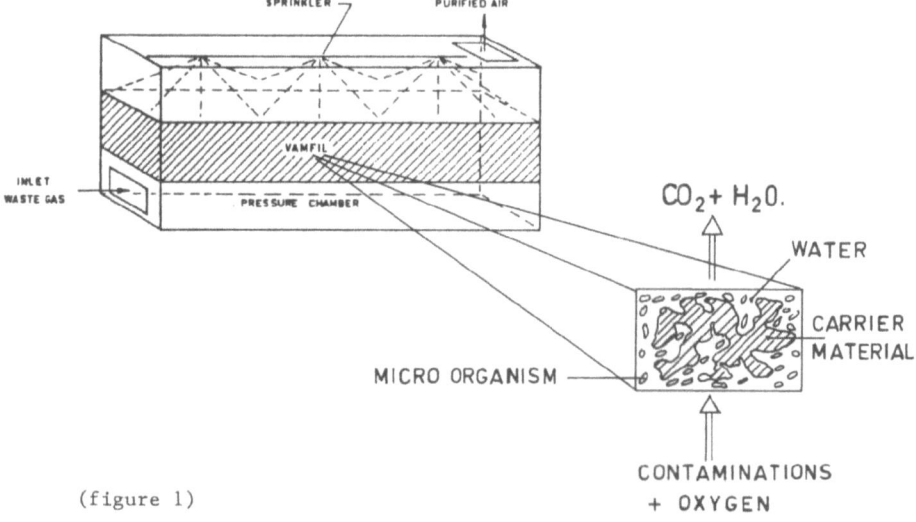

(figure 1)

The mechanism of the biochemical process consists of a combination of adsorption, absorption and biological degradation. By the Van der Waals forces, a part of the pollutants in the waste air, mechanically forced through the Vamfil, is adsorbed by the carrier surface. Another part of the polluting compounds will go from the gas to the waterphase by means of absorption. To maintain the absorption and adsorption capacity of the biofilter, the activity of the aerobe micro-organisms is necessary. The micro-organisms oxydize the contaminations to water, carbondioxide and when applicable sulfate, nitrate, and other components. Carbondioxide is released to the passing air stream.

During this process, the micro-organisms use oxygen and nutrients (out of the Vamfil) e.g. nitrogen, phosphor, kalium calcium, iron and others. The used nutrients are recycled: after dying, the dead micro-organisms are consumed by the others to get the necessary nutrients. In the long run, however, the filling material will be exhausted.

Nutrients stimulate the growth of the mirco-organisms, while toxic compounds can inhibit the growth.

The kinetics of biological elimination reactions can generally be described by the manod equation:

$$\mu = \mu_{max} \frac{S}{K_s + S}$$

where
μ = growth rate
μ max. = maximum growth rate
K_s = Michaelis-Menten constant, the substrate concentration at which the specific growth rate observed is one-half of the maximum value
S = concentration of growth limiting substrate.

It has been proved that easily biodegradable compounds like alcohols, ammonia, esters, ketones and hydrogen sulfide can be sufficiently eliminated from a waste gas in a biofilter. For specific compounds, we have data available of active sludges of waste water treatment plants, which can provide specific micro-organisms. For instance, it has been proved that after inoculation Vamfil material, toluene can be degraded with a maximum break-down capacity of 200 g toluene/ m^3 Vamfil per hour at a temperature of 40°C (1).

2.2 Biobox development
In the near past, mostly open biofilters were used with all the disadvantages of excessive maintenance and almost no process control. With the newly developed filtermaterials and the gained knowledge and experience of the break-down mechanisms in the biochemical process, the modern closed biofilters are rapidly penetrating the air pollution control market.

Comprimo has designed a biofilter installation (BIOBOX) which guarantees optimum process conditions to ensure the effectiveness of the micro-organisms. This biofilter is a container filled with Vamfil (overall surface 13 m^2, biomaterial height depending on application 0,5-1,0 m).

Because of the transportable modular construction of the BIOBOX, the cost of ducting and transport of the waste gas and the necessary area are limited.

TABLE 1: BIOBOX features.

o   High degree of odour and hydrocarbon removal
o   High reliability
o   Low operating and investment costs
o   Easy maintenance
o   Simple inspection and control routines
o   Applicable to waste air with a max. temperature of 40°C.
o   Compact and flexible modular arrangements

The breakdown capacity of easily biodegradable hydrocarbons is approximately 200 g carbon/ m^3 Vamfil per hour or 2.600 g/ BIOBOX module per hour at 40°C. The maximum air capacity of the BIOBOX module is set at 8.000 m^3/ hr. Filter loads higher than 600 m^3/ h per m^2 filter area are technically possible but are economically not advisable. The resulting extra electrical energy costs are more than the extra costs of a larger biofilter.

For biofilter installations where very large volumes of waste gas give reason to install a large filter, special civil constructions are designed.

2.3 Maintenance and inspection

The closed construction of the modern biofilter avoids the growth of plants. Because of this and the mechanical stability of the Vamfil, the periodically rearranging of the filtermedium is not necessary. Long life time is depending on the application: 2 to 5 years or more.

For BIOBOX installations it can be stated that, provided that the process conditions are properly set during the test and commissioning period, the control and inspection procedures are limited to simple routines for which no specially skilled personnel is required. The actual maintenance is reduced to a minimum.

The temperature and the pressure of the feed and effluent streams can be measured and registered just as the relative humidity of the feed gas. Spraying of water above the filterbed as well as the disposal of condenswater can be done automatically. Only a weekly inspection of the filtermaterial is recommended.

2.4 Biofiltration costs
Investment costs per m^3/h waste air are influenced by several factors as can be seen in table 2.

TABLE 2: Factors influencing biofilter investment costs.

cost factors	positive	negative
break-down of components	easy, fast	difficult, slow
pre-treatment waste gas	not necessary	necessary
waste gas quantity	large	small
location of filter	close to source	large distance, roof construction

Taking into account the various factors, the investment costs will vary between hfl. 5-25 per m^3/h waste air.

The operation and maintenance costs consisting of:

o depreciation and interest
o energy cost (absorbed power by the fan)
o water consumption
o replacement Vamfil filter material
o (waste) water cost
o personnel cost

will amount to hfl.0,50-1,50 per 1.000 m^3/h treated waste air.

3. Practical experience examples

In the past biofiltration was developed in and for the waste air treatment of the agricultural sector and sewage treatment plants. The classic biofilters required a large area, space and maintenance but all this was acceptable because of the low investment and operational costs and area was available.

With the present status of biofiltration, available know-how of biotechnology, better process control and the development of modern biofilters, biological air pollution control technology becomes available for an increasing number of industries, among which the chemical and petrochemical industry.

Table 3 shows examples of classic and new application sectors.

3.1 Vamfil experience

On basis of gained knowledge in the Netherlands with respect to biological air treatment, the biological active material Vamfil is available with low pressure drop, good stability and high activity. Vamfil is already used in more than 30 (thirty) operational installations.

3.2 BIOBOX experience examples

3.2.1 Sewage pump plant, Schiedam

Also in the classic application, the BIOBOX has proven its capabilities. The ventilation air of the sewage pump basins caused over 30 years odour nuisance in the adjoining residence area. In a renovation programme, it was decided to use forced ventilation and treat the waste air in a BIOBOX. The waste air contained H_2S and other odour components, whereby the H_2S also corroded the concrete walls and roof of the sewage pump building. Due to limited available space, a BIOBOX was selected which was installed above the pump basins and is satisfactorily cleaning the waste air since.

3.2.2 Polymerisation processindustry, The Netherlands

The available waste air contains an average of 50 mg H_2S/ m^3 and 150 mg CS_2/ m^3. After preliminary tests, a BIOBOX is installed as a first step in the realisation of a large scale installation. Already is proven that the efficiency for H_2S elimination is 99% at high filter loads.

3.2.3 Food industry, The Netherlands

A food producing company has combined waste air flows of several batch processes to one waste air flow, with highly fluctuating odour peak concentrations (60,000-1,100,000 odour units per m^3). The odour elimination efficiency is approx. 95%. The pressure drop over the BIOBOX remained low during more than 12 months of operation (200 Pa at 110 m^3/m^2.h) and the Vamfil obtained a good stable structure.

3.2.4 Chemical industry, West Germany

A chemical industry in West Germany has a waste air, containing tetrahydrofuran, cyclohexanon, methylethylketon and a few other solvents. A BIOBOX is installed to treat the complete air flow. At the submission of this abstract, no operational data were available.

TABLE 3: Biofilter application examples

Classic applications of biofiltration

o   Sewage treatment plants
o   Slaughterhouses
o   Gelatine factories
o   Rendering plants
o   Tabacco industry
o   Sugar industry
o   Blood meal factories
o   Agricultural sector

New applications of modern biofilters

o   Chemical industry
o   Pharmaceutical industry
o   Paint and Ink production
o   Flavour and fragrance production
o   Bulk handling terminals
o   Food industry
o   Synthetic resin production
o   Oil and gas industry
o   Petrochemical industry

REFERENCES

1. Don JA and Feenstra L: Odour abatement through biofiltration. Proceedings Symposium "Characterization and Control of Odoriferous Pollutants in Process Industries". Louvain-la-Neuve, Belgium, April 1984.
2. Don JA and Oosthoek J: Development and Improvement of Filter Materials. TNO report number 85-07544, October 1985 (Dutch).
3. TNO, Dutch patent number 81.04987, 1981 (Dutch).

EXPERIENCE WITH FULL-SCALE BIOPAQ - U.A.S.B.-PLANTS
TREATING VARIOUS TYPES OF EFFLUENT

P.J.F.M. Hack and L.H.A. Habets
PAQUES B.V.
Postbus 52
8560 AB BALK
The Netherlands.

phone : (31)-5140-3441
telex : 46417
fax   : (31)-5140-3342

1. INTRODUCTION

For the past several decades, wastewater treatment in het industrial countries has been dominated by aerobic processes using bacteria that require oxygen to metabolize dissolved organic compounds and convert them into carbon dioxide and settleable solids.

Anaerobic processes - where microbial metabolization of the organic wastes is done in the absence of oxygen - were not feasible until ten years ago. Hydraulic retention times in the anaerobic reactors were considered too long and the entire anaerobic process was regarded as being too sensitive to various extraneous factors. However, since than extensive research has altered the situation radically. It has now become possible and feasible to install well-functioning full-scale anaerobic facilities operation at high loads.

Although good evidence exists today to support this claim, it is still astonishing to see - in spite of high energy prices - that in several countries the wastewater producing industries are still reluctant to adopt the new technology. True, anaerobic techniques cannot fully replace aerobic treatment. A simple aerobic post-treatment phase may still be necessary, especially in situations where a very high effluent quality is required. A combined anaerobic/aerobic treatment plant is more reliable and more cost-effective than aerobic treatment alone. Experience has also shown that it is possible to double treatment capacity through a combined system without an increase in operating costs. This is because unlike aerobic treatment systems, anaerobic treatment does not require aeration, a big consumer of energy and probably the single largest operation expense. Combined with the energy costs savings from the production of biogas, total operating costs are reduced further, thereby offsetting to an even greater extent total capital costs (Habets and Knelissen 1985). The payback period on newly installed UASB reactors for Dutch paper mills has ranged from 1.5 to 2 years.

This paper will give some more background knowledge on the development of the UASB-reactor for industrial use and will mainly deal with recent experience with full-scale and pilot scale operation on effluents from papermills and breweries.

At the moment pilot plants are operating or have been operating under our conduct and supervision in different countries like the Netherlands, Germany, France, Scotland, Austria, Brazil, Finland and Canada. Much experience on different types of wastewater is being gathered and a selection of this is used to produce some figures regarding these experiences. More results concerning paper and also pulp mill effluent have been published elsewhere (Habets, 1986).

2. BIOCHEMICAL PROCESSES

In this paper, only a very short summary of the biochemical processes is given :
The anaerobic treatment of organic wastes and their conversion into biogas and sludge products occurs in four stages:

1. Hydrolysis: Non-soluble organic compounds are hydrolyzed by enzymes excreted from acidifying bacteria. Since the rate of this process is rather slow, it is often regarded as the rate-controlling step for the entire anaerobic treatment.

2. Acid formation: The hydrolyzed compounds are converted into organic acids such as lactic acid, butyric acid, propionic acid, and acetic acid by acid-forming bacteria; as well as into alcohol, hydrogen, and carbon dioxide.

3. Acetogenesis: Organics of the previous step are converted into acetic acid, hydrogen and carbon dioxide.

4. Methanogenesis: Methane-forming bacteria convert the products from the previous step into methane, as follows:
    a) $CH_3COOH \longrightarrow CH_4 + CO_2$
    b) $4H_2 + CO_2 \longrightarrow CH_4 + 2H_2O$

When other hydrogen acceptors like nitrate and sulphate are present, the last mentioned reaction to methane will be replaced partially or completely by other reactions.
Methane bacteria normally loose the competition for hydrogen and if plenty sulphate is present, all hydrogen available will be consumed by sulphate reducing bacteria:

$$4H_2 + H_2SO_4 \longrightarrow H_2S + 4H_2O$$

This is unfavourable for the efficiency of the process (1 mg of sulphide = 2 mg of COD), for biogas quantity and quality and for the effluent odour. Another drawback of the formation of sulphide can be sulphide toxicity.

3. THE U.A.S.B.-REACTOR

Although many experiments with anaerobic treatment have been performed in the past, the real breakthrough only came in the mid 1970's.

At the University for Agriculture at Wageningen, the Netherlands, Lettinga and co-workers developed the "Upflow Anaerobic Sludge Blanket" (UASB) reactor (Lettinga et al 1980).

The UASB reactor is first loaded with a sludge layer consisting of granulated or flocculent anaerobic sludge. Wastewater is then pumped in from evenly distributed nozzles at the bottom of the reactor. The wastewater percolates up through the sludge layer. During this upflow process, the anaerobic bacteria digest the organic material present in the waste and generate a mixture of methane and carbon dioxide (biogas). A small part of the organic matter is used for new cell growth that takes place in the form of granules with extremely good settling ability. The biogas produced by the bacteria is in the form of small bubbles that move upwards through the reactor, providing a natural mixing action. The biogas is removed by three-phase separators at the top of the reactor. A gas-free zone above the collectors allows for the settling of finely dispersed solids, while clarified effluent exits from the top. Granular sludge produced in the reactor can eventually be used to start-up other UASB reactors.

Despite the simplicity of the UASB principle, scaling-up in the early days of its design proved to be rather arduous. However, through many design improvements over the years, successful operation of the reactor has become possible even under somewhat adverse conditions. Excellent results were being achieved with increasingly wider temperature and COD ranges. The number of operating limitations were reduced significantly and the reactor's ability to deal with toxicity of the biomass was found to be at higher levels than reported previously.

By the end of the 1970s, full-scale UASB reactors were already being used for the treatment of wastewater from sugar, starch, french fry, canning, candy, soft drink and alcohol distilling plants. At the beginning of the 1980s, pulp and paper mills, slaughterhouses, and breweries were added to the list.

Since new UASB reactors can be seeded with granular sludge adapted from other full-scale plants, the full organic load can be injected from the very start. Start-up periods of several months - resulting from the very gradual increase of organic loads - are thus reduced to less than a day for the UASB reactor.

Consequently, it is possible to achieve a high removal efficiency within a much shorter time frame than until now was believed.

4. EXPERIENCE WITH FULL-SCALE BIOPAQ - PLANTS

The U.A.S.B.-reactor is being successfully commercialized as the BIOPAQ wastewater treatment system by PAQUES B.V. and her licensees all over the world.

Table 1 gives a review of loadings, temperatures, and results of recently installed BIOPAQ plants. Temperatures of the different waste streams range from 20°C to 40°C.

In general, the concentrations of suspended solids are preferred to be relatively low in proportion to total COD. This due to the reaction time in the BIOPAQ reactor being too short to affect the breakdown of all solids of organic origin. However, higher concentrations are not necessarily harmful and can be tolerated to a high degree. If the biogas production is high enough, the solids are simply washed out again.

Company	Industry	Reactor volume m³	Flow m³/d	Influent COD mg/l	Temperature °C	Volume loading kg COD/ m³d	HRT h	Efficiency of COD removal %	Year installed
Kuibo	french fries	300	380	5000	33-37	6.4	19	70	1981
Fri d'Or	french fries	1300	920	9000	33-37	6.4	34	85	1983
Wheat Starch	starch	2200	840	20000	33-37	7.5	63	85	1983
Roermond Papier	paper	1000	3000	3500	30-40	10.5	8.0	75	1983
Residentie Slachthuis	slaughterhouse	650	1040	4000	33	6.4	15	75	1983
Ruiten Troef	cannery	375	1125	4200	32	8.0	12.5	85	1984
Venco	licorice	50	36	18000	32	13.1	33	90	1984
Bavaria	brewery	1400	6000	1600	20-24	6.8	5.6	80	1984
Celtona	paper	700	2900	1200	20-25	5.0	5.8	60	1984
Industriewater Eerbeek	paper (3 mills)	2200	9600	1300	25	5.7	5.5	70	1985
Davidson	paper	1600	6000	3000	35	11.2	6.4	70	1986

"Table 1. Full-scale BIOPAQ-UASB-reactors"

For low strength and low temperature effluent, the hydraulic retention time becomes the limiting factor. Therefore, applied volumetric loadings at some full-scale-plants are low, while retention times are considerably shorter than in the other cases. The BIOPAQ reactor was especially adapted for operation under short retention time conditions by design of a new three-phase-separator, which features a higher density of gas collection and sludge separation baffles.

The data in Table 1 also show that COD-reduction rates range from 60 to 90 per cent. Biogas production in the mesophilic temperature range (30-35°C) is approximately 0.4 m³/kg of COD-removed. About 80 per cent of this biogas is methane. Hence the energy content is about 90 per cent of the heating value of natural gas. At lower temperatures and at lower COD-values, biogas production is decreased, but the methane content usually remains above 80 per cent.

5. INDUSTRIEWATER EERBEEK B.V. ; AN UNIQUE SITUATION

5.1. HISTORY AND PRESENT SITUATION

In the beginning of the 1960's four papermills in the town of Eerbeek agreed to build a joint biological effluent treatment plant and to share the operating costs in accordance with their share of the total treatment capacity. In the 1970's, the plant comprised a very conventional sequence of processes - primary sedimentation followed by activated sludge biological treatment with mechanical dewatering of sludge.

In 1984, due to increased production of the papermills it became necessary to consider the best way of upgrading the biological treatment capacity. Extensive bench- and pilot-scale tests confirmed that anaerobic pre-treatment before the existing aerobic stage was the most appropriate method of treatment giving the following benefits :
   * relatively low investment cost.
   * reduction in operating costs of the total plant.
   * low space requirements.

A full-scale anaerobic treatment plant was thus constructed in 1985.

5.2. WASTEWATER CARACTERISTICS

The raw wastewater at Industriewater consists of the discharges from three papermills with the characteristics shown in table 2.

Table 2.

RAW WASTEWATER CHARACTERISTICS

Mill	Flow $(m^3 d^{-1})$	Soluble COD $mg.l^{-1}$	$tonne.d^{-1}$	Soluble BOD $mg.l^{-1}$	$tonne.d^{-1}$	Sulphate $mg.SO_4.l^{-1}$
De Hoop	5,500	2,000	11.0	1,100	6.1	130
KNP	5,000	480	2.4	200	1.0	525
Coldenhove	1,500	200	0.3	90	0.14	160

5.3. OPERATING RESULTS

The one year pilot-scale investigation had shown that, despite a low wastewater temperature of 23°C and only moderate COD concentrations (600 - 950 $mg.l^{-1}$), removals of about 63% COD and 76% BOD could be expected. This would more than double the treatment capacity of the existing biological plant. Because of the relatively dilute COD level and low temperature, the full-scale UASB-reactor was designed on a hydraulic loading basis rather than COD loading ($kg.m^{-3}.d^{-1}$).

At the end of 1985, the reactor at Industriewater (see pictures) was innoculated with 250 m^3 of granular sludge from the reactor at Roermond Papier. The reactor was started-up in January 1986 and the results during the first year's operation are summarised in Table 3.

This shows clearly that the full-scale performance is superior to that on a pilot-scale, with COD and BOD removals now having stabilised at about 70% and 80% respectively. The better results on a full-scale are probably caused by the higher inlet COD concentrations and higher temperatures than during the pilot investigation.

Picture 1: 2200 m^3 BIOPAQ-UASB-reactor at Industriewater.

Table 3: Results Industriewater UASB (2200m^3).

Month	Inlet		Outlet		Reactor		Removal efficiency in %	
	COD mg.l^{-1}	BOD mg.l^{-1}	COD mg.l^{-1}	BOD mg.l^{-1}	temp. °C	biogas prod. m^3.d^{-1}	COD	BO$_{D5}$
January	928	478	380	161	26	777	59	66
February	840	403	312	119	27	1.029	63	70
March	907	442	298	102	28	1.867	68	77
April	898	432	242	77	29	-	73	82
May	996	510	279	80	31	-	72	84
June	1.025	522	297	85	32	3.030	71	84
July	976	507	287	88	32	2.517	71	83
August	1.158	630	296	90	30	3.352	74	86
September	941	503	328	115	31	2.629	65	77
October	936	495	302	101	31	2.635	68	80
November	870	466	267	87	29	2.255	69	81
December	1.005	549	320	113	29	2.923	68	79

Remarks

1. Flow rate started at 400 m³ hour⁻¹.
2. 19 March flow increased from 400 to 450 m³ hour⁻¹.
3. 20 November 1986 flow increased from 450 to 480 m³ hour⁻¹.

The composition of the biogas is 75%-80% methane, 18%-23% carbon dioxide and 2.0 - 2.6% hydrogen sulphide. After being washed, the gas is burned in a gas-engine, generating 155kW.
The energy produced by the gas engine has reduced the quantity of power purchased from the public utility from about 180,000 to 80,000 kWh per month . There has occurred a reduction in power consumption down to only 118 kW of the installed aeration power of 227 kW.
A second indentical gas engine is planned to be installed in 1987 in order to use all the biogas generated - this is expected to make the Industriewater plant completely self-suuporting in energy. The production of anaerobic granular surplus sludge is about 50 m³/week.
The reduction in the BOD loading on the activated sludge plant has improved the settleability of the activated sludge solids. Problems with sludge bulking have disappeared now that the sludge volume index (SVI) has been reduced from 200 -300 down to less than 100 ml.g⁻¹ solids. The stable operation of the "polishing" activated sludge plant has permitted the production of a very high quality final effluent (COD = 75, BOD = 5 mg.l⁻¹), the BOD concentration sometimes being lower than the level in the receiving river!

6. BREWERY-EFFLUENT.

In January 1984 the Bavaria brewery (Holland) started to investigate the possibilities of anaerobic pre-treatment of its wastewater by the BIOPAQ-UASB-process. Until that time, the Bavaria brewery treated its wastewater since 4 years in an activated sludge plant, which capacity had become insufficient.
At first, the nature of the wastewater was expected to be not (or less) suitable for anaerobic treatment; especially because of the low temperature (20 - 24°C) and the low COD-concentration (1200 mg/l).
The results of the pilotplant were so good, that the construction of a full-scale plant with the following caracteristics was started immediately (Swinkels et al.1985).

Flow	6000	m³/d
COD-influent	1000-1500	mg/l
Temperature	20 -24	°C
Equalizationtank	1500	m³
UASB-Reactor	1400	m³

The plant was started up with excess granular sludge from other UASB-plants and the design capacity was reached within 2 weeks. The results of the full-scale plant (see picture 2) showed to be corresponding to those of the pilot-plant, and are stable for more than 2 years now:

Load	4.5 - 7	kg COD/m³d
Efficiency on COD	75 - 80	%
Biogasproduction	0.25	m³/kg COD removed

Besides the reduction of energy requirement of the total treatmentplant, other advantages of installing the anaerobic pretreatment are, that the SVI of the existing aerobic system decreased from 300 to 100 ml.gr^{-1} and that the final effluent quality improved to BOD = 5 to 10 mg/l.

The last three years PAQUES continued the research on treating brewery effluent, and in the meantime six more breweries (all over the world) are now installing BIOPAQ-UASB-plants to treat their effluent.

Picture 2: equalizationtank and BIOPAQ-UASB-reactor at Bavaria.

7. CONCLUSIONS

Many full-scale experiences show that anaerobic treatment of effluent is a grown-up technology. Moreover, the application of the BIOPAQ-UASB-system is not restricted the treatment of medium or high concentrated and mesofilic types of effluent, but under conditions, once believed to be sub-optimal, the system performs very well.

8. REFERENCES

1) Habets,L.H.A.;Knelissen,J.H.
   Application of the UASB-Reactor for anaerobic treatment of paper and board mill effluent.
   Water Sci & Tech. 17 (1) 1985.
2) Habets,L.H.A.: Experiences with full-scale and pilot-scale UASB treatment of pulp-paper, and board-mill effluents.
   EWPCA Conference "Anaerobic treatment, a grown-up technology."
   Amsterdam, september 1986.
3) Lettinga.G. et al use of the Upflow Anaerobic Sludge Blanket (UASB) reactor concept for biological wastewater treatment, especially for anaerobic treatment. Biotechn.& Bio eng. 22 (6) 1980.
4) Swinkels, K.Th.M.; Vereijken, T.L.F.M.; Hack,P.J.F.M.
   Anaerobic treatment of wastewater from a combined brewery, malting and softdrink-plant.
   Proceedings 20th EBC-congress, Helsinki, 1985.

WET AIR OXIDATION OF TOXIC WASTEWATER

M.A.G.Vorstman, M.Tels

ABSTRACT

Wet air oxidation of phenol has been studied in a 2 l reactor through which both gas and liquid phases flow continuously. The reactor was used as a stirred cell with a flat gas liquid interface. Reaction temperatures were between 230 and 280°C. Phenol conversions between 80% and 99.6% were found, COD conversions were in between 68% and 96%. Oxygen transfer coefficients were calculated from the experimental data. The experimental results are compared to those of previous workers who mainly used batch reactors. First order reaction rate constants of both phenol and COD conversion prove to be of the same magnitude in both types of operation.

1. INTRODUCTION

Wet air oxidation is a process by which components are oxidized in a liquid water phase by oxygen that is provided by compressed air. The components to be oxidized may be dissolved in the wastewater or may be present as suspended solids. The process must take place at elevated temperatures: 180-330°C in order to obtain a sufficient conversion rate. The application of high pressures is a prerequisite to reach the required process temperatures: a pressure of 20 bars for a process temperature of 180°C and 250 bars to reach the highest reactor temperature. The high pressure also provides a sufficient high partial pressure of oxygen and corresponding oxygen concentration in the liquid. The corrosive conditions and the high pressures that are involved give rise to high investment costs. The costs for the compression of air are also relatively high if the wastewater shows a high value of the Chemical Oxygen Demand (COD). Schulz-Walz and Friedhofen present cost calculations that range from 20 DM/m^3 for 30 m^3/h wastewater of relatively low COD value to 60 DM/m^3 for 5 m^3/h wastewater of COD 80 kg/m^3. Because of these high costs the application of the process is limited to those types of wastewater for which no other alternatives are open. This means that the employment of the process is restricted at one side to wastewaters that contain too little organic material to provide a self-sustaining direct incineration and on the other hand are too toxic to biotreat. Wastewaters with a COD-value over 100-200 kg oxygen/m^3 are cheapest burned directly in an open flame and wastewaters with COD<15 kg/m^3 are preferably biotreated after dilution.

The most wide-spread use of wet air oxidation is in the partial oxidation of sewage sludge. The primary purpose of the treatment of this type of waste, however, is not the oxidation of the sludge but the improvement of the dewatering properties of the material. Further applications are the regeneration of carbon to restore its activity, the treatment of night soil, and various industrial wastewaters originating for instance from paper mills, metallurgical coking, acrylonitrile and petrochemical plants [Perkov, Dietrich]. The major restriction to the application of the process is found in the high costs. A better understanding of the process may result in a lowering of the costs. Reactor modelling asks for kinetic data. The aim of the study presented in this paper is to gain kinetic data both with respect to chemical kinetics and to the influence of the oxygen transfer from gas towards liquid phase. To this purpose experiments have been carried out on a laboratory scale to investigate the kinetics of the oxidation of a model component: i.e. phenol.

Phenol was chosen because it is a component that often occurs in industrial wastewaters. It is very toxic for aquatic life, the 48-hr EC_{50} value based on Daphnia is 23.5 mg/l [Randall] or 6.6 mg/l [Keen]. In biological wastewater treatment phenol concentrations should not exceed 70 mg/l [Katzer] or 200 mg/l [Pruden]. A further consideration was that the reaction kinetic data found by previous workers show large variations, especially in the activation energy of the reaction as can be seen from the next section.

2. PREVIOUS WORK

A number of authors studied the wet air oxidation of phenol. Except for Pruden all workers used batch reactors. In the batch reaction three reaction stadia may be distinguished:
- an induction stage with a low initial reaction rate that increases with time
- a second stage in which the rate of reaction proves to be first order in phenol and the reaction rate constant is far higher than the one in the induction period
- a final stage in which the reaction rate constant is 10-20 times smaller than in the middle stage.

The change from one stage to a successive one may occur quite abruptly [Willms, Helling]. The dependencies of the reaction rate upon oxygen concentration and upon temperature differ from author to author. Table 1 shows the values presented by 5 workers with respect to the order of the reaction in phenol and in oxygen and the energy of activation of the reaction, E_a.

It is generally accepted that the oxidation takes place by a free-radical mechanism. From the table it can be seen that free-radical inhibitors may influence the order of the reaction in oxygen. Both Shibaeva and Sadana found a strong dependency of reaction rate upon the pH-value of the liquid during the induction stage. The maximum initial rate is obtained at pH values of 3.5-4 (measured in the cooled liquid samples) and the rate drops sharply at decreasing pH-values. The increase of reaction rate with increasing pH-value is ascribed to the reaction of the phenolate ion in the initial stage. Sadana also showed that the pH-dependence is only small in the second reaction stage. Phenol conversions are far higher than the COD-conversions because of the formation of intermediate oxidation products. Identified products are polyphenols [Doreau], catechol, (hydro)quinone [Sadana], acetone, acetaldehyde and a number of acids: formic, acetic, maleic, oxalic and succinic [Keen]. At 99.5% conversion of phenol Helling found COD conversions of only 90% and Keen found a TOC removal of 85%. Both authors have shown that like the final reaction stadium of phenol also the COD reduction rate exhibits a very slow final stage at the above values of COD removal. Since the intermediate oxidation products are far less toxic than phenol the

TABLE 1. Summary of activation energies and reaction orders found by previous workers

author	induction stage			second stage			temperature
	activation energy	order phenol	order O_2	activation energy	order phenol	order O_2	°C
Shibaeva	107 kJ/mol	1	1				180-210
Sadana a	109 kJ/mol	1	1	175 kJ/mol	1	0.5	90-180
b				-	1	1	
Helling				20 kJ/mol	1	0-1	185-230
Pruden c				45 kJ/mol	1	1	200-250
Willms	94 kJ/mol	0	1	112 kJ/mol	1	0.5	130-200

a) with use of heterogeneous catalyst
b) with use of heterogeneous catalyst + free radical inhibitors
c) stationary reactor with continuous feed streams

toxicity removal is far larger than the COD-removal and is not as large as the phenol reduction [Randall, Keen].

3. EXPERIMENTAL SET-UP
3.1. Apparatus

The experimental apparatus is given schematically in fig.1. The reactor is a 2 liter autoclave through which both the liquid and the gas flow continuously. Reactor and piping are made of stainless steel. The autoclave is fitted with a 12 blade stirrer in the liquid phase. The stirrer blades are 0.5 cm wide and 3 cm high; the outer stirrer diameter is 3 cm. The gas phase did not contain a stirrer. The liquid feed is transported through an electrically heated preheater into the reactor by means of a piston pump. The feed rate may be choosen by setting the stroke of the piston. The gas feed is withdrawn from gas bottles of compressed air. The gas feed rate is controlled by a mass flow controller. The gas is introduced above the liquid level in the reactor. Both the liquid preheater and the reactor wall heating are provided with temperature control which results in a temperature variation of the liquid in the reactor of less than 2°C. The liquid level in the reactor is held constant. To this aim

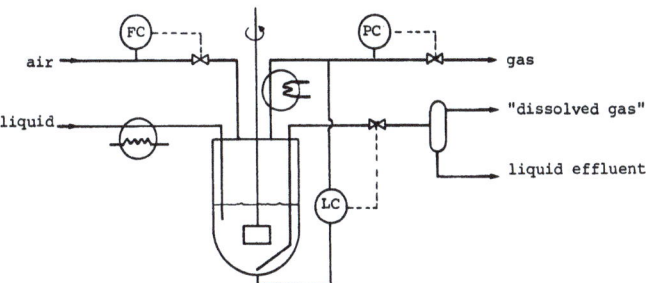

Figure 1. Schematic representation of the continuous reactor

the pressure difference is measured between the bottom of the reactor (liquid side) and the top of the reactor (gas side). This signal is sent to a level controller that acts upon the liquid outlet valve. The liquid is cooled before it passes through the valve. Behind the outlet valve the liquid outlet pressure is atmospheric. Consequently the gasses that were dissolved in the liquid at reactor pressure will escape from the liquid. The mixture of liquid and escaped gasses is separated. The flow rate of the "dissolved gas" is measured by a soap bubble flow meter. The composition of the dissolved gas is determined by injection of a gas sample on a gas chromatograph. The liquid phase that leaves the separator is submitted to measurements of pH, composition and COD. The reactor pressure is regulated by a pressure controller that guides the gas outlet valve. The gas that leaves the reactor passes through a cooler in which water vapour and other volatiles are condensed. The condensate flows back into the reactor. The oxygen content of the gas is determined by an oxygen analyser. The N_2 and CO_2 concentrations of the gas are measured by means of a gas chromatograph.

Gas samples were analysed on a two-column gas chromatographic system equiped with a TCD (thermal conductivity detector). A porapak R column operated at 30°C separates the N_2/O_2 mixture from CO_2. The N_2/O_2 peak is further separated in a molecular sieve 13X column held at 0°C. Carbon dioxide is prevented from entering the molecular sieve column by the operation of a switching valve between the two columns. The phenol concentration is measured on a chromosorb 104 column that is used with temperature programming from 80°C to 240°C. Detection is carried out by means of a FID. COD values of the liquid phase are determined by refluxing with $K_2Cr_2O_7$, H_2SO_4 and silver sulfate catalyst and subsequent titration with ferroammoniumsulfate.

3.2. Experimental procedure

In order to speed up the heating process the reactor is heated under pressure and liquid flow flow conditions thus making use of the feed heater capacity. If the reactor temperature is reached the liquid feed is switched from pure water to distilled water that contains phenol. In the first experiments (1 and 2) the gas phase composition was changed from nitrogen to air after the phenol had entered the reactor. In the later experiments (3,4,5) air was used to pressurize the reactor and consequently the water in the reactor was saturated with air before the phenol feed was started. If the gas outlet oxygen concentration and the pH value of the liquid outlet stream have reached stationary values samples of the liquid and gas and of the "dissolved gas" are analysed. After one hour new samples are taken to confirm that the conditions are really stationary.

3.3. Processing of experimental data

We assume that in our laboratory experiments both the gas phase and the liquid phase are ideally mixed. So the concentrations in the mixed phases are equal to the concentrations in the respective outlet streams. If we assume the reaction rate in phenol, r_p (mol/m^{3}s), to be first order in phenol and n^{th} order in oxygen we may write

$$r_p = k_r c_p c_{ox}^n \tag{1}$$

in which k_r is the reaction rate constant, and c_p and c_{ox} are concentrations in the liquid phase in the reactor of phenol and oxygen respectively in mol/m^3.

The mass balance of phenol over the reactor under stationary conditions is

$$V_1 \, r_p = L \, (C_{p,in} - C_p) \tag{2}$$

in which V_1 = liquid volume in the reactor
$C_{p,in}$ = concentration of phenol in liquid inlet
L = volumetric liquid flow rate

The liquid streams entering and leaving the reactor are practically equal since the phenol concentrations are low and since the water vapour in the outgoing gas is condensed and returned to the reactor. The condensation takes place at reactor pressure and therefore the gas phase will be nearly dry after expansion. The mass balances for O_2 reads for the liquid phase:

$$V_1 \, r_{ox} = N_{ox} + L \, C_{ox} \tag{3}$$

and for the gas phase

$$N_{ox} = G_{in} \, Y_{ox,in} - G_{out} \, Y_{ox} \tag{4}$$

in which r_{ox} is the reaction rate of oxygen, N_{ox} is the number of moles of oxygen transferred from the gas to the liquid phase, G is the gas flow in mol/s and Y_{ox} the gas phase oxygen mol fraction. Subscripts in and out denote the in- and outgoing streams.

The value of C_{ox} may be calculated from the "dissolved gas" composition and the ratio of the flows of "dissolved gas" and liquid. G_{out} will be smaller than G_{in} because oxygen and nitrogen dissolve in the liquid and part of the CO_2 that is produced remains in the liquid phase. The solubility of CO_2, namely, is far higher than that of oxygen and nitrogen. Moreover, the number of moles of CO_2 formed during the reaction is smaller than the number of moles of oxygen that reacts, as can be seen from the reaction equation for a complete oxidation of phenol:

$$C_6H_5OH + 7O_2 \rightarrow 6CO_2 + 3H_2O \tag{5}$$

The equation also makes clear that $r_{ox} \leq 7r_p$. A simple description of r_{ox} like the one presented in eq. (1) for r_p is not available. The oxygen reaction is composed of various parallel reactions of oxygen with phenol and with intermediate products like the ones named in section 2. A balance analogous of eq.(3) can be written for carbon dioxide (r is negative) and nitrogen (r=0). The mass transfer of the gases may be described by an overall liquid mass transfer coefficient $k_{l,i}$, in which i denotes one of the gaseous components O_2, N_2 or CO_2:

$$N_i = k_{l,i} A (C_i^* - C_i) \tag{6}$$

in which C_i^* is the liquid concentration that is in equilibrium with the gas phase and A is the interfacial area.

Equilibrium concentrations were calculated under the assumption of an ideal gas phase from the Henry relation

$$C_i^* = 18 \, p_i / H_i \rho_l$$

TABLE 2. Reaction conditions, compositions

exp.no.		9	10	11	12	13
quantity	units	value	value	value	value	value
reaction conditions						
Temperature	°C	260	279	252	240	230
Pressure	bar	120	120	120	120	110
gas flow	mol/h	6.7	11.8	12.9	8.6	8.6
liquid flow	kg/uur	1.4	1.7	3.3	2.9	5.0
stirrer	r.p.m.	210	190	205	195	300
liquid compositions						
$c_{p,in}$	mmol/kg	20	48	49	29.6	19
$c_{p,out}$	mmol/kg	0.08	1.8	8	0.6	1.4
COD_{in}	mmol/kg	140	336	343	207	133
COD_{out}	mmol/kg	5.84	18.2	111	24	31
c_{ox}	mol/m^3	≃7	≃5	1.5	1.4	1.2
c_{ox}^*	mol/m^3	28	16	23	22	20
$c(CO_2)_{out}$	mmol/kg	-	-	3.2	6.9	-
pH_{out}	-	4.1	4.1	3.0	3.4	3.0
colour		dark	medium	dark	light	medium
suspended matter		little	little	much	none	little
gas composition						
$Y(O_2)_{uit}$	mol%	18.0	14.2	15.5	15.0	16.2
$Y(CO_2)_{uit}$	mol%	1.7	9.3	4.0	3.1	3.3
$Y(N_2)_{uit}$	mol%	-	-	79	80	80
reaction rates and kinetic constants						
r_p	μmol/s	7.7	21.8	37.6	23.3	24.7
r_{COD}	μmol/s	50	150	213	148	144
r_{CO2}	μmol/s	35	-	188	102	119
r_{COD}/r_p	-	6.5	6.9	5.7	6.4	5.8
r_{CO2}/r_p	-	4.5	-	5.0	4.4	4.8
k'_{rp}	s^{-1}	0.94	.012	.005	.039	.018
k'_{rp}/k'_{COD}	-	11	1.5	2.5	6	4.2
$k_{l,ox}$	mm/s	-	-	1.2	0.95	0.85

in which p_i is the partial pressure of component i, H_i is the Henry coefficient for component i and ρ_l is the liquid density at reactor conditions.

Values of H_i for oxygen and nitrogen were taken from [Fernandez] and for carbon dioxide from [Heidemann].

4. RESULTS AND DISCUSSION
4.1. Results

Experimental conditions and measuring results are presented in table 2. In the experiments described here the reactor was used as a so called stirred cell. This means that the stirring rate in the liquid is that low that the liquid level is flat and that no air bubbles will be introduced in the liquid phase. This method allows to work also under conditions in which the oxygen transfer will influence the kinetics of the process. The liquid contents of the reactor amounted 1 liter at reaction conditions. Due to fouling of the pressure transducer the inaccuracy of this value may be as much as 50%. During a run, however, the liquid volume in the reactor was constant. Temperatures have been varied between 280°C and 230°C, and phenol inlet concentrations ranged from 20-50 mol/m^3 (1.9 - 4.7 kg/m^3). The amounts of phenol and COD that are converted show a slightly larger range of variation because also the liquid flow rate was varied between experiments.

4.2. Discussion

The pH of the liquid after separation of the dissolved gas varies between 3.0 and 4.1. As stated before, Sadana showed that the influence of pH on phenol conversion is relatively small in the second reaction stage. Sadana used a heterogeneous catalyst. If the same dependency holds for non catalytic reactions the variation of the pH value between 3.0 and 4.1 will be of minor influence upon the reaction rate. The lowering of the pH value of the liquid during the reaction is the result of the presence not of CO_2 but of organic acid that is formed as an intermediate oxidation product. The following components were identified in the liquid: acetic acid, acetaldehyde, acetone and ethanol. The chromatographic analysis does not provide information about the presence of formic acid or oxalic acid, because these acids are hardly detected by means of a FID. From the low pH-values one may conclude that at least one of these acids has been present in the solution. The amounts of the detected components were seen to increase with increasing COD value. The amount of acetic acid that is produced in experiments 3, 4 and 5 is 15-20 wt.% of the phenol in the feed. This value compares well with the results obtained by Keen in a batch experiment at 230°C. In the experiments at more elevated temperature the fraction acetic acid is far lower.

As can be seen from Table 2 the amount of suspended matter varies within the series of experiments. Except for experiment 1 the trend is that the amount of suspended matter is larger if the phenol concentration in the outlet is larger. Pruden also states that tars were formed in runs with high phenol concentrations. Since the suspended solids are most likely polyphenols this dependency is not unexpected. The rate of formation of polymer will likely be proportional to the phenol concentration. It may be interesting to note that in experiments 1 and 2 that were started under nitrogen atmosphere no tar was produced up until oxygen was admitted to the reactor. It makes clear that oxygen is needed to induce radicals that will give rise to polymerisation. The amount of suspended matter subsequently decreased with time.

The value of the fraction CO_2 in the gas in experiment 2 proves to be unreliable. Oxygen balances based upon the measured oxygen concentrations, flows and COD conversions were calculated from equations 3 and 4. Deviations are large for experiments 1 and 2 and are less than 10% for experiments 3, 4 and 5. For the last 3 experiments oxygen transfer coefficients were calculated by means of equations 4, 6 and 7. Values for $k_{l,ox}$ were found of 1.2, 0.95 and 0.85 mm/s for experiments 3, 4 and 5 respectively. These values agree with the value determined by means of a chemical enhanced reaction in the same reactor. A slight decrease of k_l with decreasing temperature is expected on basis of the decreasing diffusion coefficient of oxygen in the liquid. The expected increase of k_l due to the higher stirring rate in exp. 5, however, is not confirmed experimentally. Mass transfer coefficients for CO_2 that were calculated are lower than those for oxygen. The values however show more variation and seem less reliable. The degree of saturation of the liquid phase with oxygen is small in exps. 3, 4 and 5. This implies that the conversion rate is lowered by the oxygen transfer resistance that was built in by the choice to use a stirred cell. In experiments 1 and 2 oxygen concentration amounts about 30% of the saturation value that is determined by the partial pressure of the oxygen and the temperature in the reactor.

Calculations of the reaction rate constant by means of equations 1 and 2 do not lead to satisfactory results, whatever reaction order in oxygen is chosen. May be this has to be ascribed to a continuation of the oxidation reaction in that part of the reactor outlet system that is still at elevated temperature. The consequence of this continuation would be that both the phenol and oxygen concentration in the reactor are larger than the concentrations measured in the outlet stream. This explanation is not in contrast with the values of the first order reaction constants k'_{rp} that were calculated from the data by means of the relation:

$$k'_{rp} = k_r(c_{ox})^n = r_p/c_p$$

The values of k'_{rp} that are presented in table 2 are of the same magnitude or even larger than those found by Helling or by Keen who found values of 0.01 and 0.006 s^{-1} at equal or higher oxygen concentrations. Slightly higher concentrations in the reactor would improve the accordance between the results of these workers and the present ones. Due to the differences between a batch reactor and a CSTR (continuous stirred tank reactor) higher COD-conversions are found in the CSTR than in a batch reactor at comparable phenol conversion. From the experimental data the ratios may be calculated of the first order reaction rate constants in phenol: $k'_{rp} = r_p/c_p$ and in COD: $k'_{COD} = r_{COD}/COD$. The ratio k'_{rp}/k'_{COD} is presented in table 2. Values vary between 1.45 and 11. From the experiments by Helling values follow for this ratio between 2 and 4 for the reaction in the second stadium and between 6 and 10 for the slow final reaction stadia for both phenol and COD. The data of Helling were obtained at 210 and 230°C so a comparison should only be made with the lowest temperature results presented here in which the ratio varies between 2.5 and 6. Given the uncertainties about the phenol concentrations in the reactor the ratios compare well.

The above results indicate that differences between a continuous operation and batchwise operation are fairly small in the wet air oxidation of phenol as far as COD removal rate and acetic acid production are concerned. This outcome was not predictable at forehand since large differences exist in phenol concentrations. Especially at the start of the batch reaction the concentrations in the batch reactor differ widely from those in a CSTR. Such large differences in concentrations might easily give rise to different reaction paths, for instance in polyphenol production.

In a large scale continuous operation the plug flow reactor is the analogue of the batch reactor. Since mixing lowers the local concentrations in the reactor the rate of conversion will be lowered if mixing occurs. The application of mixing therefore will require a reactor volume that is larger than the volume of a plug flow reactor that provides the same conversion. In industrial scale units partial mixing is applied. Zimpro states that mixing is applied to transport radicals towards the front end of the reactor; this will prevent the induction stadium to occur. The implication of the above reasoning is that the degree of mixing applied in a continuously operated wet air oxidation unit will be of minor influence on COD removal and acetic acid productions provided that the lowering of the concentrations by mixing is compensated by a larger reactor volume.

CONCLUSIONS

The experiments on wet air oxidation of phenol in a continuous operated autoclave show about the same first order reaction rate constants, k'_{rp}, as were reported by authors who worked with batch reactors. The fractions of acetic acid that are produced in the continuous reactor compare well to those found in batch reactors.

ACKNOWLEDGEMENT

This work was financially supported by the Dutch Ministry of VROM in the R&D project on "Wet air oxidation of aqueous waste streams at high pressure and temperature". We are indebted to P.R.Wiers for carrying out the experiments.

REFERENCES

Dietrich, M.J., Randall, T.L. and Canney, P.J., Env.Progr. 4, 171 (1985).
Doreau, G., and Chornet, E., Water Poll.Res.Canada 13, 21-32 (1977/1978).
Fernandez-Prini, R., and Crovetto, R., AIChEJ. 31, 513 (1985).
Friedhofen, G. et al., report BMFT-FB-T 80-132 from Bundesminister für Forschung und Technologie, Fachinformationszentrum Karlsruhe (1980).
Heidemann, R.A. and Prausnitz, J.M., I.E.C.Proc.Des.Dev. 16, 375-381 (1977).
Helling, R.K., Strobel, M.K. and Torres, R.J., report OR NL/MIT-332, Oak Ridge Nat. Laboratory (1981).
Katzer, J.R., Ficke, H.H. and Sadana, A., J.Water Poll.Contr.Fed. 48, 920-933 (1976).
Keen, R. and Baillod, C.R., Water Res. 19, 767-772 (1985).
Perkow, H., Steiner, R. and Vollmüller, H., Chem.Ing.Techn. 52, 943-951 (1980).
Pruden, B.B. and Le, H., Can.J.Chem.Eng. 54, 319-325 (1976).
Randall, T.L., Knopp, P.V., J.Water Poll.Contr.Fed. 52, 2117-2130 (1980).
Sadana, A.J., Ph.D.Thesis, Univ. of Delaware (1975).
Schulz-Walz, A. and Braden, R., Chem.Ing.Tech. 53, 295 (1981).
Shibaeva, L.V., Metelitsa, D.I. and Denisov, E.T., Kinetika i Kataliz, 10, 1020-1025 (1969).
Willms, R.S. et al., Ind.Eng.Chem.Res. 26, 148-154 (1987).

THERMOCHEMICAL TREATMENT OF SOLID WASTE OBTAINED FROM PULP AND PAPER FACTORY WiTH 15 % ACETIC ACID AND NaOH FOR CONVERSION TO CRUDES

TANER, F.; BOZTEPE.H; KiMYONŞEN, Ü.
Ç.Ü., Faculty of Arts and Sciences, Department of Chemistry
ADANA/TURKEY

ABSTRACT

Solid waste obtaiend from wastewater treatment plant of pulp and paper factory; using kraft process, was used for liquefaction studies. Aqueous suspensions of the solid (20 % by weight) have been treated with acetic acid and NaOH (15 % by weight of solid) at the temperatures of 250°C, 300°C, and 350°C in a high pressure autoclave for one hour, starting with one atmosphere initial pressure.

The gas produced during thermochemical treatment was analysed with an Orsat gas analyser. After treatment, the reaction mixtures were separated into solid-oil and aqueous layers. The percentages of acetone extract of solid-oil, the lignin content and calorific values were determined. The solid-oil yield the solid was determined by drying and weighing.

It was found that as operating temperature increases the amount of the acetone extract increases. The calorific values of thes olid-oil phase increase for both acetic acid and NaOH treatment processes. The solid-oil yield of the solid increases with increasing of thermochemical treatment temperature. It can be concluded that the solid waste from pulp and paper industry can be used for the production of crudes.

INTRODUCTION

There have been some attemp to produce liquid hydrocarbons from solid biomass feedstocks in the past 60 years. The basic approach of organic waste conversion is based on the observation (1) that low-rank coals could be hydrogenated with carbon monoxide in the presence of water. Later on, it was shown that cellulose can be converted to liquid materials by high pressure hydrogenation (2) and to a bitumenlike material by treatment with caustic at elevated temperatures and pressures.

The conversion of carbohydrates which are cntained in various plant materials, formed by nature in enormous amounts and with greatest ease into liquid fuels, had been studied over a long period of time (3-7). A reinvestigation and application of this approach to coal liquefaction at the Pittsburgh Energy Research Center showed that the chemistry of the process could be applied to carbohydrates (8). This observation led to a detailed investigation of the mechanism of hyrogenation with carbon monoxide (9-11). Subsequent studies have shown that biomass-derived liquids can be produced under a wide variety of conditions, but analysis of these products suggests that very little hydrocarbon production is taken place (12-16).

Recent studies have now shown that the primary product of biomass liquefaction can be converted with greater success to significant amount of hydrocarbon compounds (17.-29). For direct liquefaction of biomass two processes, BOM and LBL, have been developed (20). In the literature, it has been reported that the sulfiding a cobalt-molybdenum catalyst greatly increased its activity for hydrodexygenation of phenol (21). In our laboratory there are various studies for determining the conditions of the liquefaction of various wastes (23-24). In this study, the solid waste produced at wastewater treatment plant of pulp and paper factory has been used and it was treated under different conditions for conversion to crudes.

MATERIALS AND METHODS

Raw material:

The raw material was obtained from the wastewater treatment plant of the pulp and paper factory, SEKA, constructed at Taşucu, Turkey, where the kraft (sulphate) process is applied. The wastewater treatment plant contains first filtration and the biological treatment in logoons. By may of filtration, about 3000 kg dry solid waste, has been produced. This waste has been piled up and sometime sold commertially.

The waste contains mostly fibers scaping from process. Some properties of this waste is given in Table 1. Throughout the paper this waste will be named as "SEKA waste",

Thermochemical Treatment:

The experiments have been performed by batch mode in a 2L high pressure autoclave (Cook High Pressure, vertical autoclave). In the standard experiments, 100 g of dry solid was slurried in 400 mL of water, containing 15 g NaOH and 15 g acetic acid (15 % by weight of solid). The mixture was put into the autoclave and it was sealed.The autaclave was purged with nitrogen gas by pressurizing the unit to about 50 atm. and then venting the gas. A resistance coiled jacket heated the outoclave to the designated reaction temperature, and the temperatures were kept stable for the specified reaction time of one hour. The mixing was carried out continuously during thermichemical treatment. At the end of the treatment, the autoclave was allowed to cool overnight. The next day the inside pressure was recorded and the gas was analysed by using Orsat gas analyeser, having CO_2 and CO absorbtion bottles. The rest of the gas was vented after controlling it with a flame whether it is flammable or not. Then the autoclave was unsealed to get the reaction mixture out manually. The reaction mixture was filtered through the coarse filter paper to separate solid-oil from aqueous layer. When three phasses (oil, solid-oil, and aqueous) have occured, the oil phase was separated from aqueous layer with a separatory funnel. The solid-oil was washed with water and dried for the determination of solid-oil yield of the waste gravimetrically.

The acetone was used exhaustively to extract the oil from solid-oil phase in a soxhlet unit to determine the percentages of the acetone solubles (defined as oil yield). The calorific values, the lignin content, unsoluble percentages of solid-oil in 72 % H_2SO_4 solution (23), and oil content have been determined. The results are shown in Table 2.

Table 1. Properties of Raw Material

Moisture	: 78.6 % by weight
Ash	: 14.4 % by weight (on a dry basis)
Lignin	: 42.7 % by weight (on maf basis)
Kappa No	: 68.6
Solubility in 1 % NaOH solution: 27 %	
Calorific value (cal/g maf): 4126	

RESULTS AND DISCUSSION

The results given in Table 2, shows that solid-oil yield and oil yield (actone extract) are increasing with increase of operating temperature. These results are contradicting the results of earlier studies (23-25). At 350°C, both with acetic acid and NaOH, third phase which is lighter than the aqueous layers, and separated with separatory funnel, has been obtained. Since the same conditions were applied, to all samples, the production of such an oil phase could be the result of the S present in the raw material. This conclusion can be deducted from the fact that in the combustion gas (flue gas) occured during the determination of calorific values, the amount of SO_2 determined was found to be higher. Presence of sulphide in raw material produced from the Kraft Process waste has to avoid a mechanism resulting from sulfiding catalyst (21). Presence of S (Sulphide) in crude oil, suggests that Sulfide may have such kind of effect for conversion of biomass to petroleum.

As the operating temparature increases. the calorific value of the solid-oil increases. For instance, the calorific values are 8298 cal g^{-1} and 10329 Cal g^{-1} at 250°C and 300°C respectively, for NaOH treatment. When acetic acid was used, it was found that the calorific values were decreasing. In fact it is not the case. Since at 300°C and 350°C, the most of the oil present in solid-oil phase adhered to filter paper. Therefore the calorific value of the solid-oil is not the exact value. For this reason the solid fraction of solid-oil can be used for calorific value determinations. It was not too easy to have homogeneous mixture of solid-oil. But, any way, the calorific value of oil, at 350°C with HAc is also higher than that of lower temperatures.

Lignin content is increasing with increase in operating temperature. Increase in the amount of insoluble fraction in 72 % H_2SO_4 solution, shows that the carbohydrates are converted to biomass-drived fuel. Increase in final pressure, obtained when the autoclave was cooled, shows that the biomass decomposes to lower molecular weight hydrocarbons, such as parafins which are flammable.

Decrease in pH values with increase in operating temperature also suggest that decomposition occurs, and acidic products form.

From all these findings, it can be concluded that acetic acid has about the same effect on the liquefaction of biomass for conversion to crudes. It can be proposed that the sulfide, which is used for paper pulp production, can be effective in liquefaction also.

Table 2. Product characteristics of thermochemical treatment (SEKA Waste 15 % NaOH and Acetic acid, 100 g dry solid, 400 mL H_2O 15 g NaOH and 15 g Acetic acid)

Op. temp °C	S-O % Waste (maf)	Oil yield % S-O (maf)	% lignin of S-O (maf)	Cal. Values of S-O (cal/g)	Pressure Op. (atm)	Pressure final (atm)	Gas Analysis %, dry CO_2	CO	inert[xx]	Aqueous Layers amount mL	pH
250b	36	52.9	87	8298	65	10	38	2	60	390	6.72
300b	38[xxx]	57.9	91	10329	120	10	40	2	58	390	7.17
350b	28[xxxx]	75.3	77	9948	192	10	35	2	63	380	5.25
	x34	98.3	97	12140	–	–	–	–	–		
250a	30	78.9	85	8358	68	10	59	4	37	360	4.80
300a	33	74.5	93	7780	122	22	50	6	44	320	4.83
350a	20	75.2	79	5726	222	22	63	1	36	310	5.25
	x18	99.0	98	12458							

x : Oil seperated with separatory funnel, at 350°C
xx: flammable
xxx: 12.5 % **S** (by weight) determined from the SO_2 content product in calorimeter bomb.
xxxx:11.9 % **S** of solid-oil " " " " "
b: with NaOH   Op.: Operating
a: with Acetic acid
S-O : Solid-oil

LiTERATURES

1. Fischer, F., and H. Schrader Hydrogenation of cool with Carbon Monoxide, Brennstoff-Chemic, V. 2, 1921, pp. 257-261.
2. U.E. Fierz-David, Chem. Ind. (London) 44, 942 (1925).
3. Berl, E., Am. Inst. Mining and Met. Eng. Tech. Publ. No. 920 (1938).
4. Berl, E. Science 99, 309 (1944).
5. Berl, E., Bull. Assoc.Pet. Geologist 24, 186 S (1940).
6. Berl, E., Colliery Guardian 144, 914 (1932).
7. Berl, E., Pro. 3rd Intern. Conf. Bituminous Coal, 2, 820 (1930).
8. Appell,H.R., Wender, I. and Miller, R.D.; Solubilization of Low Rank Coal with Carbon Monoxide and water, Chem. Ind. (London) 47, 1703 (1969).
9. Appel, H.R.; Wender, I.; Miller, R.D.; Conversion of Urban refuse to oil, p. 5. U. S Bur. Mines, Tech. Prog. Rep. 25, May 1970.
10. Appell, H.R.; Fu, Y.C.; Friedman S.;Yavorsky, P.M.; Wender, I. Converting Organic wastes to oil, a replenishable energy source, P. 20. U.S. Bur. Mines Rep. of Investigations. 7560, (1971).
11. Appell, H.R.; Fu, Y.C.; Ullig, E.G.; Steffgen, F.W.; Miller, R.D. Conversion of cllulosic wastes to oil, P. 28. U.S. Bur. Mines, Rep. of Investigations. 8013, (1975).
12. J.A. Knight, D.R. Hurst, and L.W.; Elston, "Wood oil from pyrolysis of Pine Bark-Sawdust Mixture," in Fuels and Energy from Renewable Resources, D.A. Tilman et al., Eds. (Academic, New York, (1977).
13. R.L. Eager, J.F. Mathews, J.M. Pepper, and H.Zohdi, Can. J. Chem., 59, 2191, (1981).
14. D.C. Elliott, "Analysis and Comparison of Products from wood Liqufaction" In proceedings of the International conference on Fundamentals of Thermochemical Biomass Conversion, R. Overend, Ed. (Applied Science Publishers, Barking, England, 1984).
15. C. Roy, B. de Caumia, P.Plante, and H.Menard, "Production of Liquids from Biomass by vacuum Pyrolysis-Development of Data Base for Continuous Proc. in Proceedings of Energy from Biomass and Wastes VII (IGT, Chicago, 1983), P. 1147.
16. D.G.B. Boocock, R.K.M.R. Kallury, and T.T. Tidwell, Anal. Chem., 55, 1689 (1983).
17. H. Heinemonn, Petroleum Refiner, 33 (7), 161; (8), 135 (1954).

18. S. -C. Lin and E.J. Soltes, "Hydrocarbons via Hydrogen Treatment of Pine Pyroytic oil," Presented of 181 st National American Chemical Society Meeting, Atlantic, GA, March 29, (1981)
19. D.C. Elliott and E.G. Baker, "Biomass Liquefaction Producet Analysis and Upgrading," Presented of the 3 rd Canadian Meeting of Biomass Liquefaction Specialists, Sherbrooke. Quebec, September 30, (1983).
20. D.C. Elluott, Precess Development for Direct Liquefaction of Biomass Fuels From Biomass and Wastes, edited by Donald L. Klass (Ann Arbor Science Publishers Inc/The Butterworth Group) Chapter 24 p. 435-450, (1981).
21. D.C. Elliott, Am. Chem. Soc. DIV. Petr. Chem. Prept., 28 (3), 667, (1983).
22. John T. Pfeffer. "Biological Conversion of Biomass to Methane" Final report, Report No. UILU-ENG- 80 - 2009 Junel, 1976 - January 31, 1980, Univ. of Ill., Urbane, Illinois, 61801.
23. Taner, F. and Boztepe, H. "Thermochemical Treatment of sea-grassfor conversion to crudes". Conf. on Res. and App. of Aquatic plants for water treatment and Res. Rec., July 10-24, 1986 Orlando, Florida.
24. Taner, F. "Thermochemical treatment of cottonstalk with NaOH, Int. Conf. On Renew. Energy. Sour. 18-23 May. 1986, Madrid.

OZONATION OF THE AQUEOUS LAYERS OBTAINED FROM THERMOCHEMICAL TREATMENT OF SOLID WASTE WITH 15 % ACETIC ACID AND NaOH

BOZTEPE, H.; TANER, F; KİMYONŞEN, Ü.
Ç.Ü., Arts and Sciences Faculty, Chemistry Deparment,
ADANA/TURKEY

ABSTRACT

Aqueous suspension of the solid waste (20 % by weight) obtained from wastewater treatment plant of pulp and paper factory, SEKA, in Taşucu, Turkey, have been treated with acetic acid and sodium hydroxide (15 % by weight of solid) at 250°C, 300°C, and 350°C in a high pressure autoclave for one hour. The reaction mixture was separated into solid-oil and aqueous layers. The aqueous layers were diluted with water and then subjected to ozonation.

Ozonation has been carried out in an ozonation reactor for 120 min. During ozonation about 20 mL samples were driven out at 20 - minuteintervals. The BOD_5, COD, and pH of the samples were determined.

It was found that pH, COD, and BOD_5 values of the aqueous layers were decreasing with ozonation time. The drop in BOD_5 values is higher than the drop in COD. This shows that the ozonation causes first the oxidation of the biodegradable substances. It can be concluded that ozonation can be applied for wastewater treatment if the organic compounds dissolved are biodegradable.

INTRODUCTION

In recent years there has been interest in the use of ozone as a viable alternative to chlorine in water-, wastewater-treatment. An important reason for this interest is the need to eliminate the formation of trihalomethanes and organo- chlorine compounds when chlorine reacts with naturaly occurring and pollutant organic materials in water sources (1). On the other hand, ozone has high-oxidizing potential, more than 1 1/2 times greater than chlorine, in addition, its effective residual is oxygen. This fact is very important when one is considering effects of the wastewater effluent upon a receiving stream. Unlike chlorine, oxygen forms no toxic compounds if ammonia or phenol is present in the wastewater.

Extensive studies have been made on the reaction of ozone with organic compounds in aqueous medium (2,3). Ozone is a possible candedate because of

be an attractive method for treatment of industrial effluents (4,5). Nebel et al (6) studied several mixer types and evaluated them with respect to phenol reduction efficiency by ozone. Since ozone has a strong affinity for phenol, oxidation of phenol is very rapid. The Chief contribution to total organic carbon (TOC) in potable waters is made by humic substances leached from soils. Despite the interest in the characterization of humic substances, a detailed knowledge of their structure proves elusive; the review of Schnitzer provides a comprehensive text on their known composition and chemistry (7).

The ability of small dosages of ozone to achieve absolute control of all pathogenic and saprophiytic organisms in water is now well-established (8). In the Literature (9, 10, 11), It was reported that the ozone can be used to oxidize organic substances dissolved in water.

Thermochemical treatment of lingo-cellulosic waste for conversion to crudes has been attempted by many researchers over the past 60 years. For direct liquefaction of biomass, two processes, BOM and LBL- have been developed (2). The wastewater produced by these processes is rich in water-soluble compounds forming during the thermochemical treatment of the wastes, mainly ligno-cellulosic in structure. Therefore, the wasterwater has to be treated in some way or another for the removal of the pollutants discharged to wastewater. In this connection, two kinds of experimental studies are carried out in our laboratory. First one is to find the optimum conditions for the use of the aqueous layer for biogas production (13). The second one is to remove the pollutants by ozonation (14).

In this study, the aqueous layers obtained by thermochemical treatment of SEKA solid waste (15) were subjected to ozonation for the removal of water-soluble compounds in wastewater.

MATERIALS AND METHODS

The aqueous layers were obtained from thermochemical treatment of SEKA solid waste (15). After separation of the solid-oil and aqueous layers, the solid-oil phases have been washed to be purified from the watere-soluble compounds. The wash-water and some of the aqueous layers were combined and subjected to ozonation.

Ozonation has been carried out in the system described before (14). Ozone generator (Fischer, Ozone Generator, model 500) was adjusted to produce 10 mg O_3/min. which is kept constant during ozonation. From the ozonation reactor about 2-5 ml of ozonized samples were driven out at 20-minute intervals. Ozonation has continued for 120 minutes. The pH, COD and BOD_5 values of the samples driven out with time were determined using the Standard Methods (16). Observations during the ozonation have been reported (table 1). Ozone Consmuption, COD and BOD_5 concentration with ozonation time were plotted (Figure 1-6).

RESULTS AND DISCUSSION

As seen in Table 1., the color of the aqueous layers are changing from dark to light colors. This shows that the organic coumpounds having the chromophoric groups are oxidizing, and hence the colors diminish. pH's are dropping. This also says that the ozonization produces the acidic groups and therefore pH drops. Forinstance pH of the aqueous layer (250°C, 15 % NaOH) drops from 7,33 to 6.61. For 300°C and 350°, these values are 8.55 to 7.09, and 8.63 to 7.66 respectively (Table 1).

Cumulative O_3 consumption is increasing lineerly. This shows that the various organic compounds resulting from thermochemical treatment can be oxidized with ozone. COD and BOD_5 concentrations are decreasing with ozonation time, except for the aqueous layer obtained at 350°C with 15 % acetic acid (HAC) (Figure 6).

While ozonation of the sample (250°C, 15 % HAC) a white precipitate occured. This says that ozonation convert some organic molecules into nonwater-soluble compounds. Hence the removal of pollutants by oxidation and precipitation is rapid. From this finding it can be said that the wastewater produced in pulp and paper factory which uses thermochemical treatment at max. 190°C, can be treated with ozone to remove the pollutants.

We can conclude that the ozonation first oxidizes the biodegradable compounds in aqueous layers and then the other compounds having chromophoric groups.

Table 1. Changes in pH and color with ozonation time

	15 % NaOH P_{o_3} = 1 atm SEKA					
	250°C		300°C		350°C	
Ozonation time	pH	color	pH	color	pH	color
0	7.33	dark black	8.55	redish brown	8.63	pink
20	7.12	"	7.45	" "	7.59	pale pink
46	6.92	"	7.22	" "	7.57	colorless
60	6.79	brown	7.03	brown	7.49	"
80	6.74	"	7.09	pale brown	7.46	"
100	6.62	pale brown	7.08	yellow	7.66	"
120	6.61	"	7.09	faint yellow	-	

	15 % HAC P_{o_3} = 1 atm SEKA					
	250°C		300°C		350°C	
Ozonation time	pH	color	pH	color	pH	color
0	7.28	dark brown	5.25	pale brown	6.70	yellow
20	6.88	"	5.15	yellow	5.76	pale yellow
46	5.59	pale "	5.03	pale yellow	5.66	colorless
60	6.36	yellow	4.96	faint yellow	5.64	"
80	6.10	pale yellow	4.95	colorless	5.46	"
100	5.85	faint yellow	4.94	"	5.35	"
120	5.80	"	-		-	

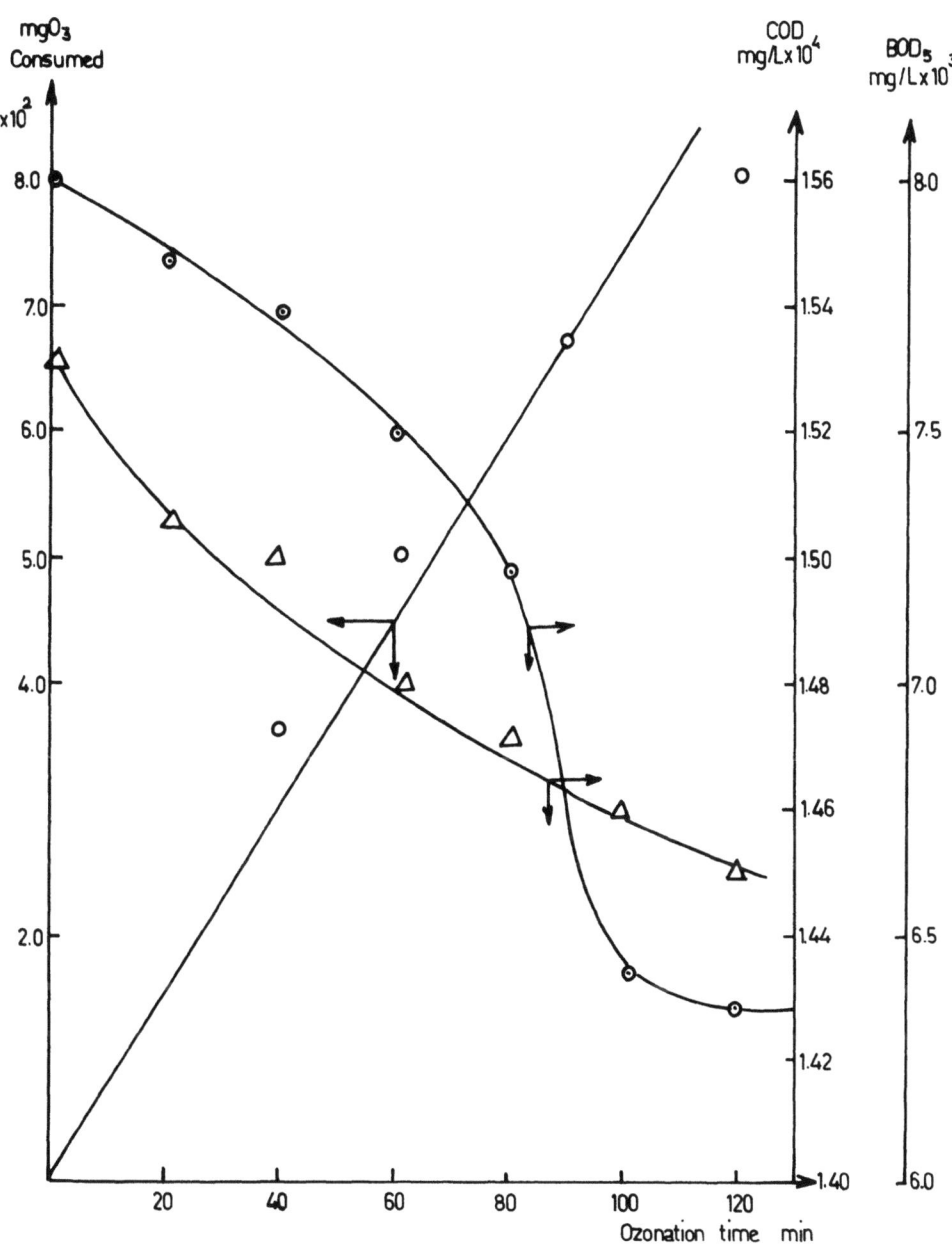

Figure 1. COD, BOD$_5$ and O$_3$ Consumption vs. Ozonation time, SEKA, 15% NaOH, P$_0$=1atm 250°C, ⊙: BOD$_5$, ○: O$_3$ consumed △: COD.

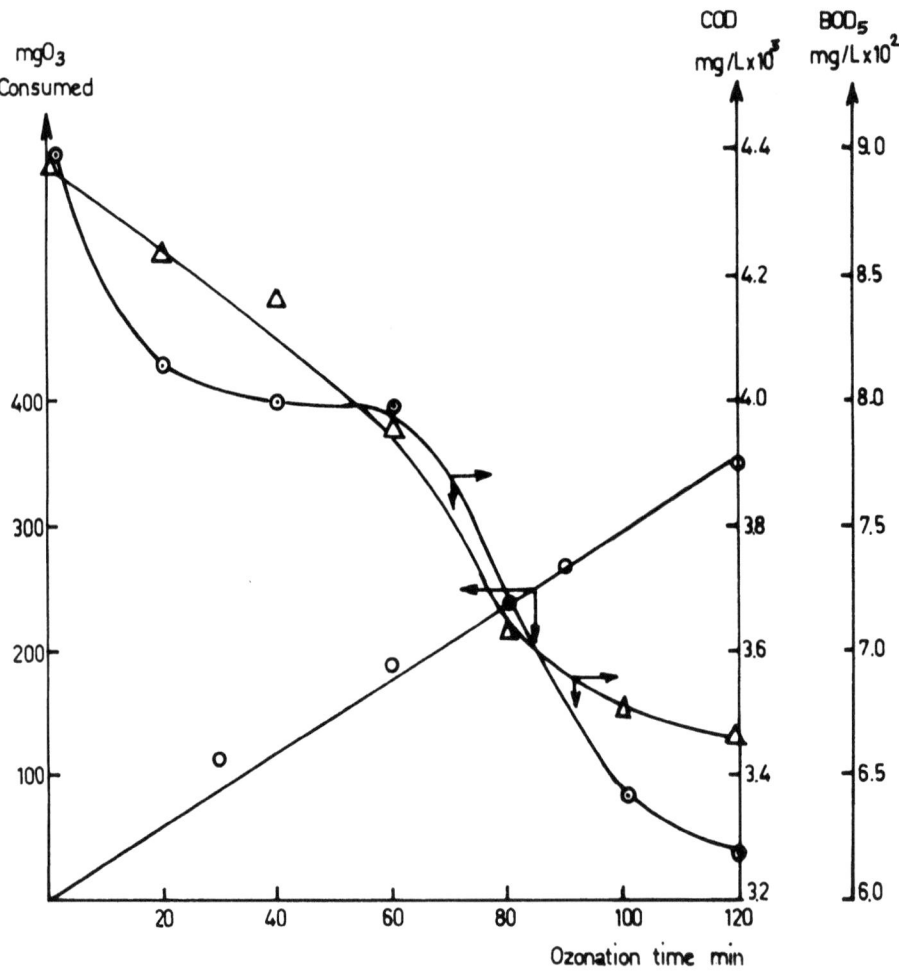

Figure 2. COD, BOD_5 and O_3 Consumption vs. Ozonation time, SEKA, 15 % NaOH, P_0 =1atm, 300°C, ⊙: BOD_5, ○: O_3 consumed △: COD

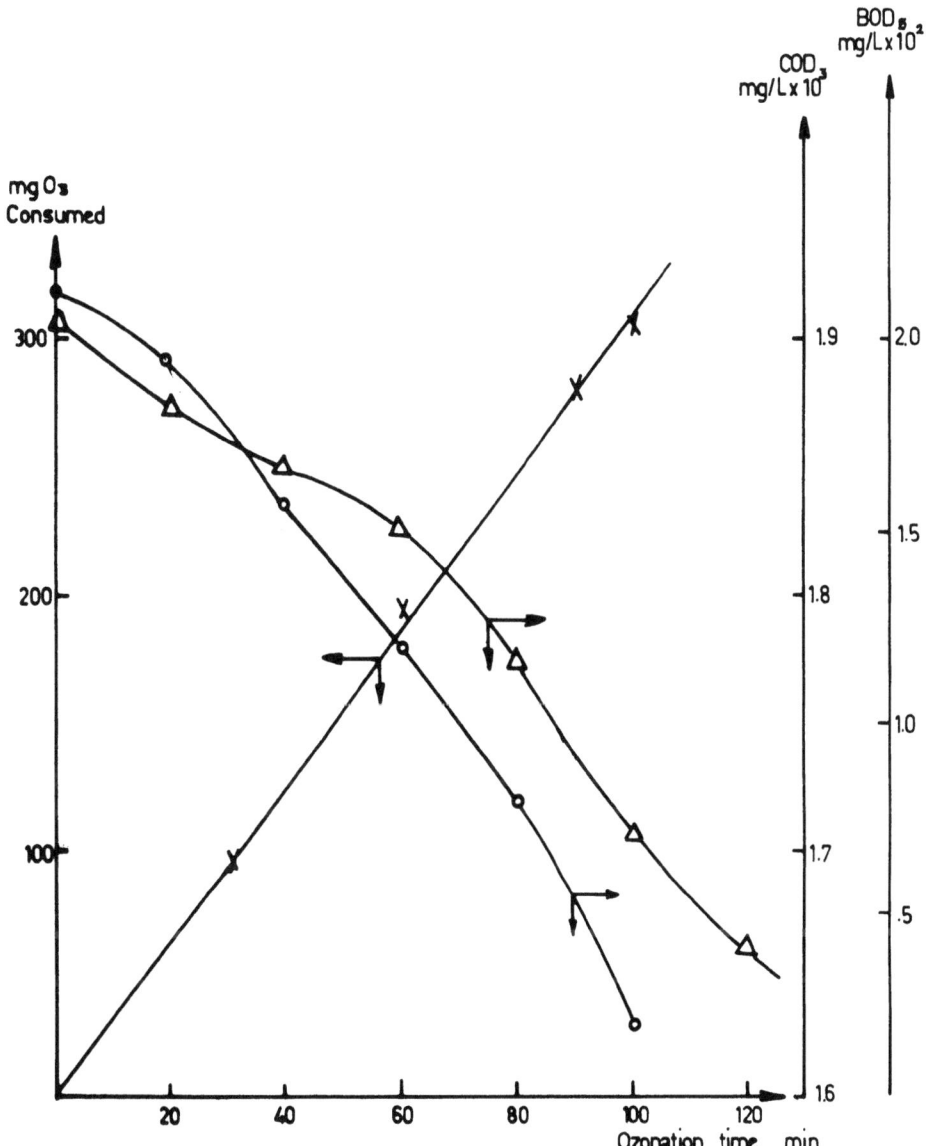

Figure 3. COD, BOD_5 and O_3 Consumption vs. Ozonation time, SEKA, 15% NaOH, P_o = 1atm, 350°C o: BOD_5, x: O_3 consumed
△: COD

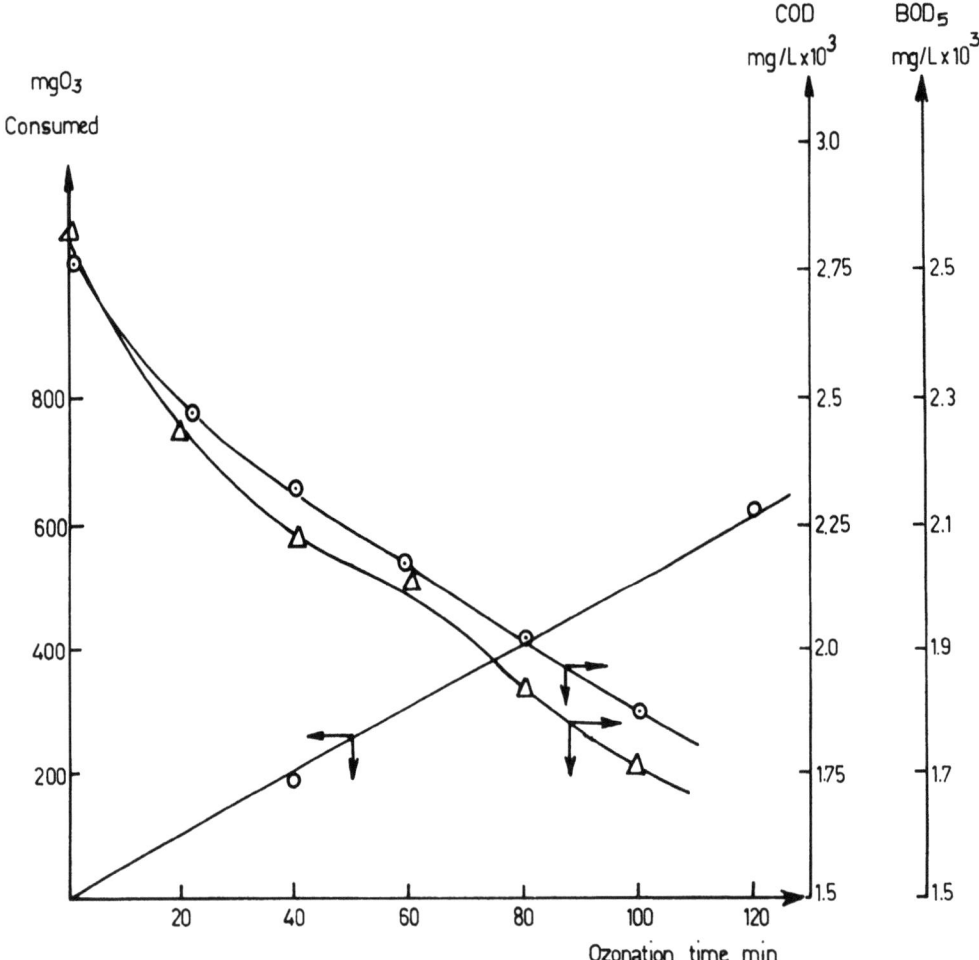

Figure 4. COD, BOD_5 and O_3 Consumption vs. Ozonation time, SEKA, 15 % HAc, P_O =1atm, 250°C, ⊙: BOD_5, o: O_3 consumed △: COD

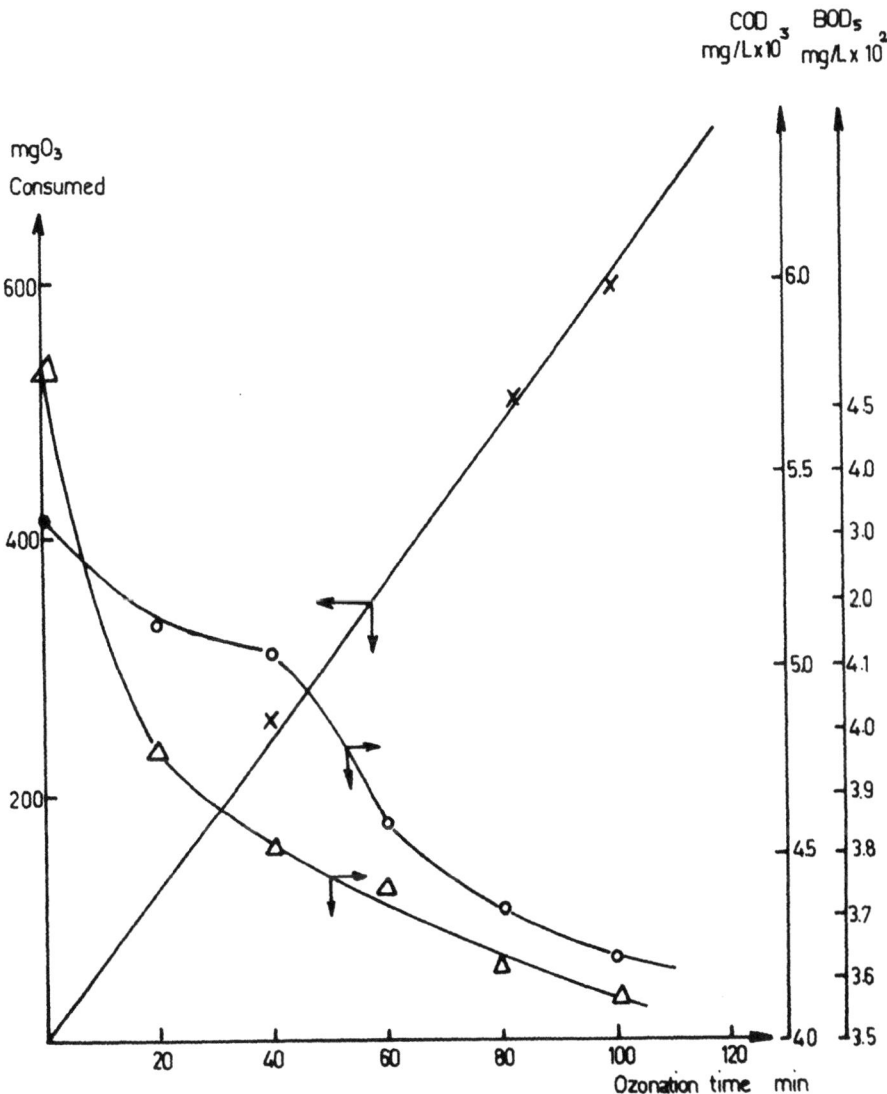

Figure 5. COD, BOD_5 and O_3 Consumption vs. Ozonation time, SEKA, 15% HAc, P_0=1 atm, 300°C, o BOD_5, x:O_3 consumed, Δ: COD

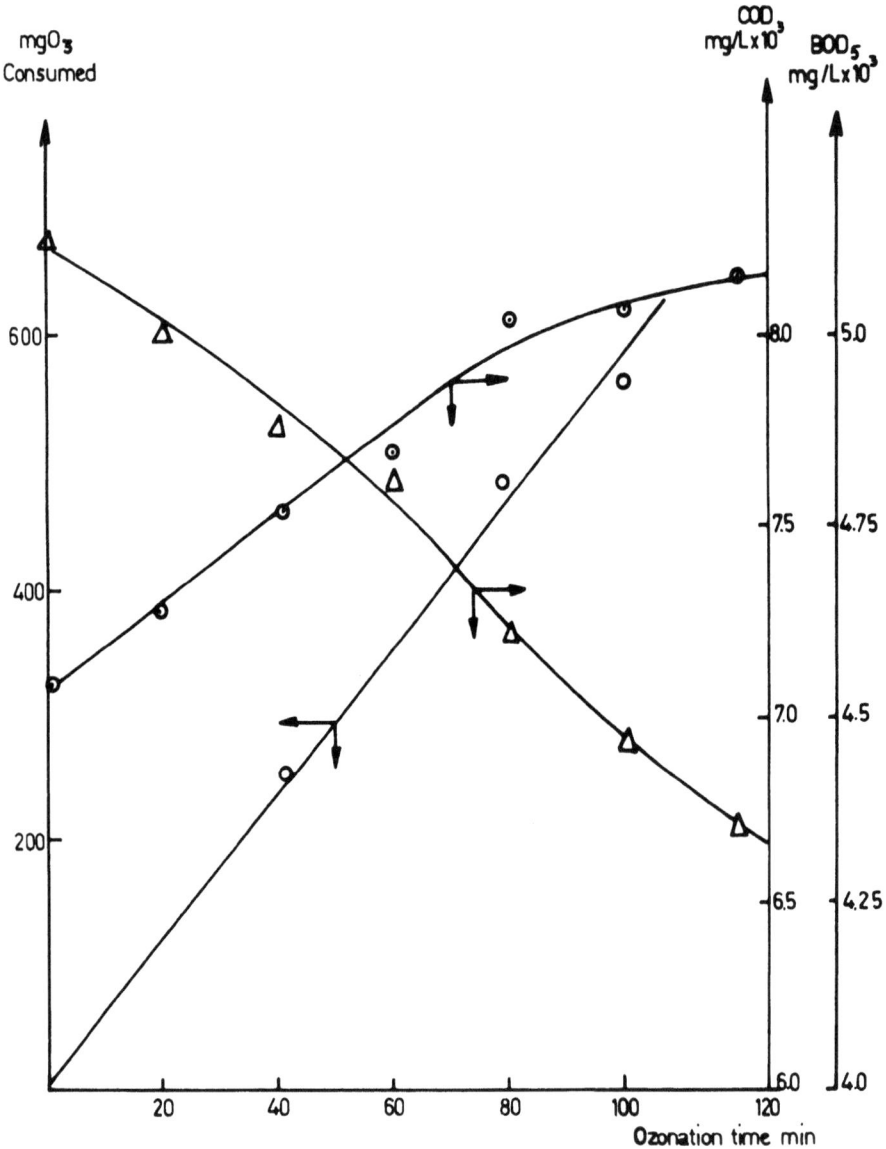

Figure 6. COD, BOD_5 and O_3 Consumption vs. Ozonation time, SEKA, 15% HAc, P_0 = 1atm, 350°C, ⊙: BOD_5, ○: O_3 consumed, △: COD

LITERATURES

1. The Lancet, "Cancer and Chlorinated Water". Lancet, 1142 (1981).
2. P.S. Bailey, "Organic groupings reactive toward ozone. Mechanisms in aqueous media. In Ozone in Water and Wastewater Treatment (Edited by Evans F.L.), pp. 29-59. Ann. Arbor Science, Ann Arbor, Mich. (1972).
3. E. Gilbert, In Ozone/chlorine Dioxide oxidation products of organic Materials (Edited by Rice R.G. and Cotruuol J.A.), pp. 227-242 (1978). International Ozone Institute, Clevland, Ohio.
4. McCarthy J.J.; Smith C.H., A review of Ozone and its application to domestic Wastewater Treatment. J. Am. Wat. Wks Ass. 66, 718-725 (1974).
5. Sliter J.T., Ozone: An alternative to chlorine. J. Wat. Pollut. Control Fed. 46, 4-6 (1974).
6. Nebel, C.; Unangst, P.C., Cottschling, R.D.; An Evalution of Various Mixing Devices for Dispersing Ozone in Water Wtr. Sew. Wks., 120: 4: 6 (Apr. 1973).
7. Schnitzer M., Humic Substances: Chemistry and reactions. Dev. Soil Sci. 8. 1-64 (1978).
8. Weldon C., Harris, "Ozone disinfection" J. Am. Wat. Works ASS. pp. 182-183. March (1972).
9. M.G. Joshi and R.L. Shombough; The kinetics of Ozone - Phenol Reaction in Aqueous Solutions. Water, Res. Vol. 16, pp. 933 to 938, (1982).
10. S.D. Killops, Volatile Ozonation Products of aqueous Humic Material; Wat. Res. Vol. 20. No: 2. pp. 153-165, (1986).
11. P.C. Singer and M.D. Gurol; Dynamics of the Ozonation of Phenol; Wat. Res. Vol. 17, No: 9. pp. 1163-1171, (1983).
12. D.C. Elliott, "Process development for direct liquefaction of Biomass" Fuels from Biomass and wastes Edited by D.L. Klass, (Ann Arbor Sciences Publishers Inc/The Butterworth Group) Chapter 24, pp. 435-450 (1981).
13. Taner, F.; Biodegradation of Aqueous Layers Obtained from Thermochemical Treatment of Cotton Stalk with acetic acid. Int - Conf. on Renew. Energy. Sour. 18-23 May 1986. Madrid.
14. Boztepe, H.; Taner, "F., Ozonation of aqueous Layers obtained from the liguefaction of sea-grass" Conf. on Res. and App. of: Aquatic Plants for Water treatment and Res. Rec. July 10-24 1986, Orlando. Florida.

15. Taner, F.; Boztepe, H., Kimyonşen, Ü., "Thermochemical Treatment of Solid Waste Obtained from Pulp and Paper Factory with 15 % Acetic Acid and NaOH for Conversion to Crudes." Submitted to Sec. Eup. Conf. on Envirn. Thec. 22-26 June 1987 Amsterdam.
16. American Public Helth Assoc., Standart Methords for the examination of water and wastewater, 13th Edition, Washington, D.C. (1971).

CHLOROFF, A NON-DESTRUCTIVE DECHLORINATION PROCESS

P.F. van den OOSTERKAMP
L.J.M.J. BLOMEN
A.S. LAGHATE
H. ten DOESSCHATE

ABSTRACT
Liquid chemical waste streams, like spent lubrication oils, often contain various concentrations of chlorinated hydrocarbon compounds, including toxic substances like chlorinated bifenyls, dioxines and furanes.
Based upon an already existing technology of upgrading waste lube streams, a process has been developed incorporating a hydrodechlorination step which has proven to be very successful in breaking up all chlorine containing hydrocarbons.
Pilot plant experiments have been carried out which revealed that effective removal and destruction took place of the chlorinated compounds, including most toxic compounds like 2,3,7,8-TCDD.
The paper will show some kinetic results as well as some general design information.

1. INTRODUCTION
One of KTI's environmental product lines is concerned with the rerefining of waste lube oil streams. The process is called the KTI RELUBE process and is presently being operated at a number of places worldwide. During operation of the RELUBE process, it appeared that beside the upgrading of the feed stream by hydrogen, also a removal of sulfur- and chlorinated compounds took place.
Based upon this aspect, a second process, the KTI CHLOROFF process has been developed. The CHLOROFF process is quite similar to the RELUBE process and incorporates a catalytic hydrodechlorination step in which chlorine is effectively removed from hydrocarbons and a valuable hydrocarbon stream can be recovered.
Various experiments have been carried out on pilot plant scale with a number of different contaminated feed streams.
Most streams contained polychlorobifenyls (PCB's), polychlorodibenzodioxines (PCDD's) and polychlorodibenzofuranes (PCDF's).

Fig. 1 gives the basic structure of these compounds:

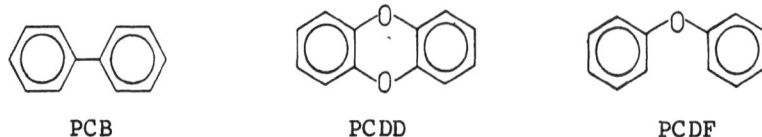

PCB   PCDD   PCDF

FIGURE 1. General chemical structure of PCB's, PCDD's and PCDF's.

Toxicity of the chlorinated compounds strongly depends on the amount and position of the chlorine atoms.
In table 1, some toxicity data are given for PCB's and PCDD's, respectively, based upon LD_{50} values for the Guinea Pig.

TABLE 1. Relation between chemical structure of PCDD's and PCB's and LD_{50} for Guinea Pig.

No.		LD_{50}*
1.	2,3,7,8-tetra-CDD	2
2.	2,3,7,8-tetra-CDF	7
3.	1,2,3,7,8-penta-CDD	3
4.	1,2,3,4,7,8-hexa-CDD	73
5.	3,3',4,4',5,5'-hexachlorbifenyl	223
6.	2,3,3',4,4',5,5'-heptachlorbifenyl	3000
7.	2,3,4,7,8-penta-CDF	10
8.	1,2,3,4,6,7,8-hepta-CDD	600
9.	2,3,4,6,7,8-hexa-CDF	120
10.	2,3,7-tri-CDD	29,444
11.	2,8-di-CDD	300,000

* LD_{50} in µg per kg of body weight.

It should be stressed that toxicity reports vary greatly and toxicity data between animal species vary as well.

2. METHODS AND MATERIALS

Experiments have been carried out with the microflowreactor system, represented in fig.2 and were carried out in the Laboratories of TU Eindhoven.

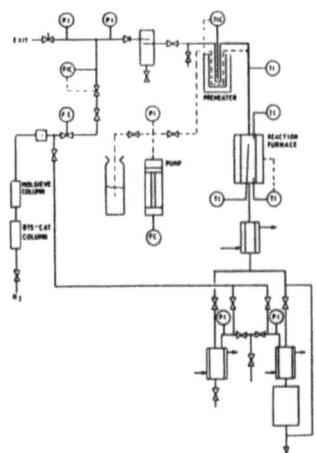

FIGURE 2. The high pressure microflowreactor system for testing of the CHLOROFF process.

The contaminated feed stream is introduced into the system with a high pressure syringe pump. Hydrogen is introduced under pressure and mixed with the feedstream after oxygen and water removal, respectively.

Reaction products are cooled against water and separated in a liquid- and gasphase, respectively. Analyses of reaction products were mostly performed off-line. PCB analysis was performed using High Pressure Liquid Chromatography and GC-MS (Gas Chromatograph-Mass Spectrometer). PCDD and PCDF analyses have been performed by the Toxiclogical Department of Amsterdam University using the method described schematically in figure 3.

- Addition of standard mixture with:
  $2,3,7,8-^{13}C$ TCDD,
  $1,2,3,7,8-^{13}C$ PCDD and
  $1,2,3,6,7,8-^{13}$ HCDF
- Dilution with hexane
- Clean-up procedure via open column chromatography:
  . silica H_2SO_4/NaOH macro column
  . Al_2O_3 - "high aspect" column
  . silica - $AgNO_3$ - Al_2O_3 - "high aspect" column
- Fractionation by reversed phase HPLC
- Analysis with GC-MS (HP-5970-MDS) with a SP 2331 fused silica column (50 m)

FIGURE 3. Principle of chlorinated hydrocarbon analysis (University of Amsterdam).

The CHLOROFF process features a hydrodechlorination step whereby chlorine is catalytically removed from the chlorinated hydrocarbon compound. Reaction takes place in a high pressure trickle-phase reactor at a temperature between 250 and 400°C.

Chlorinated compounds are dechlorinated and partially hydrogenated according to:

$$RCl_n + x \cdot H_2 \longrightarrow RH_xCl_{n-x} + x\ HCl$$

Reaction enthalpy change for this reaction lies between -40 and -90 kJ/kmol. Not only chlorinated compounds are treated in this way, other heterocompounds behave in a similar manner: sulfur is converted to hydrogen sulfide, nitrogen to ammonia and oxygen to water.

For example, with conversion of dioxines, not only dechlorination but also deoxygenation presumably takes place.

Results of conversion of following streams will be presented:
1. Gasoil, contaminated with chlorobenzene and PCB,
2. Waste stream from pesticide production site,
3. Gasoil, contaminated with PCDD's and PCDF's.

3. RESULTS AND DISCUSSION

3.1 Conversion of gasoil contaminated with chlorobenzene and PCB

In table 2, results of the conversion experiments with a contaminated gasoil are presented.

TABLE 2. Conversion experiments with contaminated gasoil.

COMPONENT	UNTREATED SAMPLE mg/kg	TREATED SAMPLE mg/kg	DETECTION LIMIT mg/kg
PCB	530	nd	0.1
Chlorobenzene	0	-	-
PCB	530	nd	0.1
Chlorobenzene	4600	nd	10

As can be concluded from the figures in table 2, concentrations of PCB around 500 ppm are reduced to less than 0.1 ppm, a reduction of more than 99.9%. The same conversion of monochlorobenzene is observed.

3.2 Conversion of waste stream from pesticide production site

In tables 3a and 3b, the conversion of a contaminated waste stream from a pesticide production site is presented.

TABLE 3a. Dechlorination of residual mixture (pesticide production) in the CHLOROFF reactor - Dibenzodioxines and -furanes.

COMPONENT	UNTREATED SAMPLE µg/kg	TREATED SAMPLE µg/kg	DETECTION LIMIT µg/kg
2,3,7,8-tetra-CDD	19	nd	0.02
Sum tetra-CDD (other)	2	nd	0.05
Sum pentra-CDD	7	nd	0.05
Sum hexa-CDD	25	nd	0.05
1,2,3,4,6,7,9-hepta-CDD	20	nd	0.1
1,2,3,4,6,7,7-hepta-CDD	23	nd	0.1
Octa-CDD	64	nd	0.2
2,3,7,8-tetra-CDF	0.1	nd	0.02
Sum tetra-CDF	9	nd	0.05
Sum penta-CDF	8	nd	0.005
Sum hexa-CDF	7	nd	0.05
Sum hepta-CDF	7	nd	0.1
Octa-CDF	12	nd	0.2

TABLE 3b. Dechlorination of residual mixture (pesticide production) in the CHLOROFF reactor - Chloroaromatics and -alicyclics.

COMPONENT	UNTREATED SAMPLE µg/kg	TREATED SAMPLE µg/kg	DETECTION LIMIT µg/kg
2,3,4-trichlorophenol	0.1	nd	0.1
2,3,5-trichlorophenol	0.1	nd	0.1
2,3,6-trichlorophenol	0.8	nd	0.1
2,4,5-trichlorophenol	10.9	nd	0.1
2,4,6-trichlorophenol	0.1	nd	0.1
3,4,5-trichlorophenol	0.1	nd	0.1
2,3,4,5-tetrachlorophenol	0.2	nd	0.1
2,3,4,6-tetrachlorophenol	0.1	nd	0.1
2,3,5,6-tetrachlorophenol	0.3	nd	0.1
Pentachlorophenol (PCP)	0.1	nd	0.1
1,2-dichlorobenzene	7.7	0.6	0.1
1,3-dichlorobenzene } 1,4-dichlorobenzene	10.3	nd	0.1
1,2,3-trichlorobenzene	72.0	nd	0.1
1,2,4-trichlorobenzene	297.0	nd	0.1
1,3,5-trichlorobenzene	14.2	nd	0.1
1,2,3,4-tetrachlorobenzene	76.9	0.1	0.1
1,2,3,5-tetrachlorobenzene } 1,2,4,5-tetrachlorobenzene	53.6	0.1	0.1
Pentachlorobenzene	2.0	nd	0.1
Hexachlorobenzene (HCB)	0.5	nd	0.1
alpha-HCH	0.1	nd	0.1
beta-HCH	0.1	nd	0.1
gamma-HCH	0.1	nd	0.1

As can be concluded from the data in table 3a, PCDD and PCDF concentrations were reduced below detection limits.

Table 3b shows that chlorinated aromatic compounds like chlorophenols and chlorobenzenes can be removed effectively. Aliphatic ring compounds like hexachlorohexane behave in a similar manner. Exact conversion factors could not be established due to the relatively high detection limits of the method of analysis. However, conversion factors are in general more than 98%. Only conversion of 1,2-dichlorobenzene (table 3b) shows a relatively low value of 92.2%.

3.3 Conversion of a contaminated gasoil

For kinetic reasons, experiments have been performed with gasoil, spiked with dibenzodioxine compounds and samples have been processed in the microflowreactor. In table 4, results of a conversion experiment with gasoil, contaminated with OCDD, are illustrated.

TABLE 4. Conversion of gasoil, contaminated with OCDD.

COMPONENT	UNTREATED GASOIL (ppt)	HYDROTREATED GASOIL (ppt) T = 225°C	T = 250°C
Sum tetra-CDD	-	7227	80
Sum penta-CDD	-	5468	80
Sum hexa-CDD	-	3351	70
Sum hepta-CDD	-	1129	70
Octa-CDD	550000	298	199

Besides the elimination of OCDD, which increases from 68% to 99% (based on the sum of concentrations of dioxines) with the small rise in reaction temperature, we observe a redistribution of dioxines with a lower amount of chlorine per molecule. In fact, at the higher temperature, levels of these smaller molecule compounds strongly decrease.

In a second experiment, a gasoil spiked with 2,3,7,8-TCDD, the most notorious dioxine compound, at a concentration of 80000 ppt was processed at different reaction temperatures. See table 5.

TABLE 5. Conversion of gasoil, contaminated with 2,3,7,8-TCDD.

COMPONENT	UNTREATED GASOIL (ppt)	HYDROTREATED GASOIL (ppt) 200°C	225°C	250°C	275°C
2,3,7,8-TCDD	80000	6107	1152	56	40
OCDD	-	75	58	26	20

Also, we observe here some redistribution of chlorine and an increase in conversion with temperature. Note that at or above 250°C, dioxine levels are very low indeed.

As an illustration, fig. 4 shows the conversion of TCDD as a function of the reaction temperature. We observe an almost complete (∼99.9%) elimination of TCDD at a temperature above 250°C.

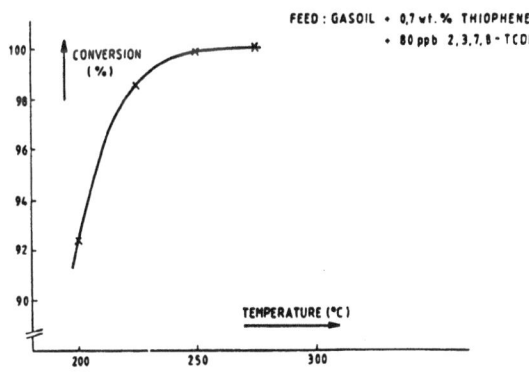

FIGURE 4. Conversion of 2,3,7,8-TCDD as function of reaction temperature.

4. DESCRIPTION OF THE CHLOROFF PROCESS

The CHLOROFF process has been developed from the RELUBE process, which deals with the rerefining of waste lube oil streams.

Both processes are schematically drawn in figure 5.

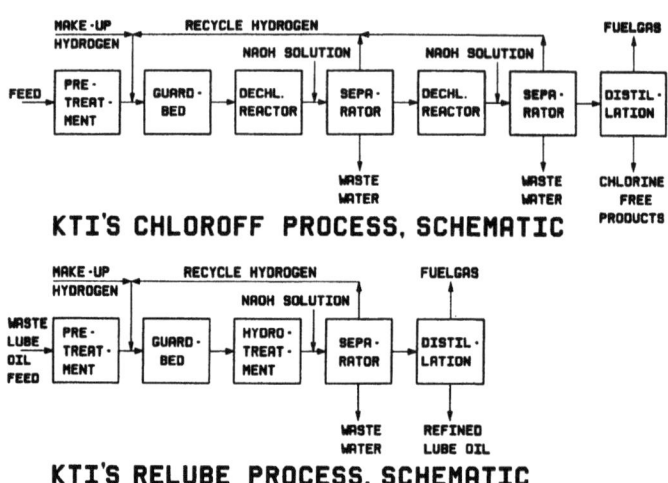

FIGURE 5.

Both processes feature a pretreatment part, a process part and an after-treatment part.

The pretreatment for the RELUBE process mostly consists of dewatering and removal of light ends, gasoil removal and high vacuum distillation.

For the CHLOROFF process, the pretreatment consists mostly only of a filter step where particulates are removed. Both processes apply in the process part a guard bed to remove impurities which might poison the catalyst, a conversion step consisting of a trickle-phase-reactor and a separation step where water is removed. For complete conversion of chlorinated hydrocarbons, two catalytic conversion reactors are foreseen.

Aftertreatment for both processes consists of a distillative step.

Indicative economic evaluation of the process indicates that total operating costs for the CHLOROFF process amount to about 250-300 Dfl/ton for capacities of ca. 10,000 t/yr. Credit of the processes product will generally be in the range of 100-1000 Dfl/ton.

5. CONCLUSIONS

Based upon the information sofar, we can conclude that we developed a catalytic process which is able to eliminate chlorinated hydrocarbons of a wide range of origin with a remarkably high efficiency.

Preliminary characteristic features of the process are:

- Reaction temperature   250 - 300°C
- Reaction pressure   50 - 60 bar

The CHLOROFF process has lots of characteristics of the KTI RELUBE process with which already a lot of experience has been gained on a worldwide scale. The CHLOROFF process is a sophisticated application of the RELUBE process and is therefore not an essentially new process design, but a different application of an existing technique.

The conversion of chlorinated hydrocarbons is going to be done in a closed system - unlike the incineration of chlorinated hydrocarbons. No flue gases, which form a central problem in the incineration process, will be present.

We therefore are of the opinion that the CHLOROFF process can contribute significantly to a solution for treating chlorinated waste stream in a safe and acceptable way from an environmental point of view.

ELECTROCHEMICAL TREATMENT OF ORGANOHALOGENS IN PROCESS WASTE WATERS

D. Schmal, J. van Erkel, A.M.C.P. de Jong, P.J. van Duin
TNO Division of Technology for Society, P.O. Box 217, 2600 AE Delft,
The Netherlands

SUMMARY
In the chemical process industry toxic or non-biodegradable waste waters, often containing halogenated organics, are produced in large quantities. Electrochemical reduction may be a suitable method for the treatment of these waste waters. In order to evaluate its feasibility, research is in progress on a number of representative organohalogens chosen from the EC list of 129 priority compounds (70% of them being chlorinated organics). As electrode material graphite/carbon fibres are used, for their properties make them very suitable for the development of a technical process. In this paper results are presented for eight chlorinated hydrocarbons from the EC-priority list. The experiments performed have shown that it is possible to remove all chlorine atoms from the organic molecules in aqueous solution. This 'dehalogenation' results in a decreased toxicity and an increased biodegradability, thus enabling further biological treatment. Energy consumption and conversion rates are such that a technically and economically viable method for the detoxification of waste waters can be developed.

INTRODUCTION
The chemical process industry produces large quantities of dilute waste waters contaminated with compounds that are toxic or non-biodegradable or both. Two important classes of toxic compounds in these waste streams are toxic metals and halogenated organics. Considerable research effort is being spent on the removal of medium to low concentrations of metals from waste waters, one method finding ever wider application in practice being electrochemical deposition.
By contrast, treatment of dilute (process) waste waters containing halogenated organics has, although being a major problem, received much less attention. The importance of this problem is illustrated by, for example, the fact that the EC list of 129 priority compounds [1] consists for about 70% of chlorinated compounds.
Commercial techniques for the treatment of waste streams containing halogenated organics are scarce. The only method used on a relatively large scale (combustion on land or at sea) is unsuitable for the treatment of diluted waste waters, because of the high cost of transportation, the necessity of adding large quantities of fuel and (when acids, salts or bases are present) the corrosiveness of the liquids. This leads to high costs for this process (>> Dfl. 100/m^3).
Other methods under investigation and sometimes applied are:
- concentrating techniques, such as membrane separation, adsorption on activated carbon or on special resins and stripping (see, e.g., [2]). These methods suffer from the disadvantages that they can be used only in a limited number of cases, and that they produce a concentrated waste that needs to be treated further.
- chemical oxidation techniques, such as treatment with hot humidified air, ozone or other oxidants, UV etc. (see also [2]). Although these techniques can be very useful, they are rarely used for halogenated compounds.

- chemical reduction techniques, such as (catalytic) dehalogenation with hydrogen or other reducing agents. These techniques, currently under investigation, are expected to be suitable particularly for the treatment of concentrated waste streams, e.g. halogenated oils (PCB), solvents etc. (see, e.g., [3, 4]).
- biological techniques using special microorganisms or enzymes. These techniques are at an early stage of development [5, 6]. Investigations into the treatment of contaminated soil are also in progress.

Increasingly strict legislation banning disposal of contaminated waste waters in the environment (sewerage systems, rivers, or the sea) has made it necessary to develop treatment techniques for waste waters containing halogenated organics. One such technique is dehalogenation by electrochemical reduction. Although the reaction is known and has been used for the treatment of (mostly) concentrated halogenated waste compounds [7, 11], its use for the treatment of process waste waters containing halogenated compounds is a novel development. The method is particularly suitable for waste waters containing polar or ionic organochlorine compounds, which are in general difficult to decontaminate by adsorption or stripping.

To investigate its feasibility, a research programme was started some years ago. In 1985 this programme has been aimed at dehalogenation by electrochemical reduction. To ensure a more or less systematic approach we selected eight representative compounds from the EC list of 129 priority compounds [1]. This paper describes the experimental methods we used and some results of experiments we conducted until January 1987.

DEHALOGENATION ELECTROCHEMISTRY AND ITS EFFECT ON TOXICITY
The electrochemical reduction reaction

The overall reaction for the cathodic reduction of halogenated compounds can be written as:

$$R-Cl + H^+ + 2e \rightarrow R-H + Cl^-$$

resulting in the formation of non-halogenated compounds and chloride ions.

In waste waters, generally it will not be possible to prevent evolution of hydrogen as a competing reaction, which decreases current efficiency and, as a result, increases energy consumption. However, at the low concentrations occurring in practice (often less than 100-1000 ppm) energy consumption is not normally a factor of major importance (see Table 1).

TABLE 1. Energy consumptions for the removal of one Cl-atom from a compound at a concentration of 100 ppm, a molecular weight of 250 and a cell voltage of 4V.

Current efficiency, %	Energy consumed, kWh.m^{-3}
100	0.1
10	1
1	10
0.1	100
0.01	1000

Detoxification by dehalogenation

Dehalogenation changes the toxicological properties of waste waters. In spite of isolated exceptions, it generally decreases (aquatic) toxicity and improves biodegradability. Table 2 illustrates this for the example of pentachlorophenol (PCP) and its reduction products. It shows the aquatic toxicity determined with the Microtox method, an instrumental test based on the bioluminescence of specially selected bacteria. Here the EC-50 values represent the concentrations of the compounds in a 2% NaCl solution causing 50% reduction of light emission in 5 minutes.

TABLE 2. EC-50 values of pentachlorophenol and some reduction products.

Compound	EC-50, ppm
pentachlorophenol	0.1-1
tetrachlorophenol	0.1
2,4,6-trichlorophenol	7
2,4-dichlorophenol	4
2-chlorophenol	22
phenol	22-42

Another example is given in Table 3 showing the large influence of the R group on the relative values of biodegradability and accumulation properties. When R is a Cl-atom, the compound is DDT which is only slightly biodegradable and accumulates very much in animal and human tissue. Both examples illustrate the large influence of small changes in molecular structure on the toxic properties and so the usefulness of removing chlorine atoms from the molecule for detoxification.

TABLE 3. Influence of structure on biodegradability and accumulation

R	Biodegradability (relative units)	Accumulation-factor (relative units)
Cl (DDT)	1	15000
OCH_3	60	300
CH_3	500	30
SCH_3	3000	1

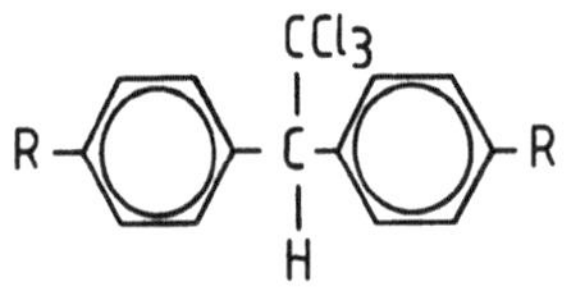

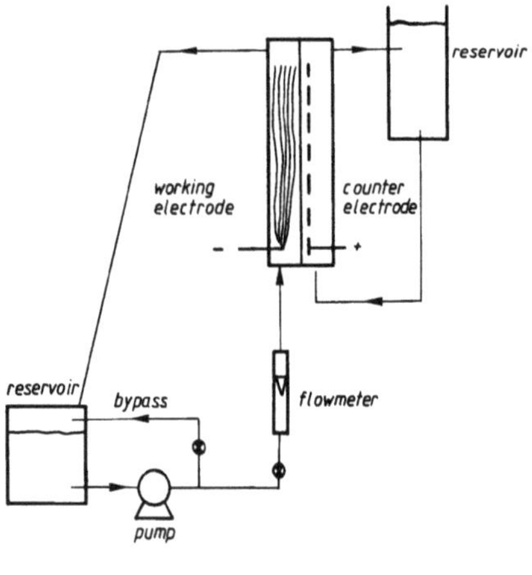

FIGURE 1. Diagram of flow-circuit.

EXPERIMENTAL PART
Electrochemical set-up
Electrode materials and geometry

Carbon/graphite fibres were chosen as an electrode material because of the following reasons:

- Due to the low concentrations of toxic compounds to be converted, electrodes with a high specific surface area will have to be used to make the method economically viable. Carbon fibres, having a diameter of about 10 μm fullfil this requirement (specific surface area of the material 4.10^5 m^{-1}).
- Carbon fiber material is manufactured in growing quantities for the production of strong, light-weight reinforced plastics and so is readily available and expected to become cheaper.
- Electrodes composed of carbon fibre bundles are not very prone to clogging by small solid particles in waste streams.
- Carbon has a relatively high 'overpotential' for hydrogen evolution, which is favourable for the efficiency of the dehalogenation reaction.

Reactor and flow-circuit
The flow-circuit is given in Figure 1. Most of the experiments described below have been conducted in a batch-type recycle mode.
The reactor consists of two compartments. It accepts various fibre configurations (bundles, cloth, felt). The fibres are clamped in the cell with a platinum strip connected to a power source.
Mass-transfer properties of the various carbon fibre configurations used were studied using a model reaction [12, 13].
In the experiments described here, the cathode consisted of a bundle of fibres.

Selected halogenated organic compounds
From the EC list of 129 priority compounds, we selected eight organochlorine compounds (Table 4).

TABLE 4. Selected chlorinated compounds from the EC list of 129 priority compounds [1].

EC list no.	Compound
2	2-amino-4-chlorophenol
30	p-chloronitrobenzene
70	dichlorvos
86	hexachloroethane
102	pentachlorophenol
107	2,4,5-T
111	tetrachloroethene
118	1,2,4-trichlorobenzene

Analytical techniques
Concentrations of starting material, intermediates and products of the reduction reaction were measured using HPLC and GC.
Chloride concentrations were determined by potentiometric titration with silver nitrate with a Metrohm 636 Titroprocessor and a silver indicator electrode.
Aquatic toxicity was determined by the instrumental Microtox method mentioned before. Other standard toxicity tests are available and can be performed routinely, if necessary.

Reactor performance
The performance of the reactor for the various dehalogenation reactions

can be expressed in different ways. We have chosen to base it on two criteria:
- The current efficiency being the part of the electrical current used for the dehalogenation reaction. The figures given are for 90% conversion.
- The normalized space velocity defined as the volume of waste water for which the concentration of the compound to be converted can be reduced 90% per unit of time and of reactor volume [14]. It can be regarded as the inverse of the mean residence time needed for 90% conversion.

To provide insight into the economy of the method, we shall also give the specific energy ($kWh.m^{-3}$) consumed for 90% conversion.

RESULTS
General
All compounds given in Table 4 have been investigated. As an example the results for pentachlorophenol (PCP) will be elucidated somewhat more in detail. This will be followed by a review of the main results for all compounds. Part of the results shown were published before [15, 16].

Pentachlorophenol
This compound is a good example for showing the merits of the method, because it is a polar halogenated non-biodegradable compound. Moreover, its reduction potential is very negative.

Electrolysis at 10A of 1 litre of 50 ppm PCP in 0.1 M sodium sulphate/0.1 M sodium hydroxide solution caused the PCP concentration to decrease as shown in Fig. 2.

After 30 minutes of electrolysis the PCP concentration was below the detection limit of 0.5 ppm (HPLC method). From the chloride content of the solution it followed that five chlorine atoms per molecule of PCP had been removed. The current efficiency for dehalogenation was 1%. Addition of small quantities of certain surface active agents improved the efficiency considerably [17].

During the electrolysis, the toxicity of the solution fell by a factor of 20, i.e. detoxification was virtually complete (Microtox method).

In a separate experiment two litres of the test solution were used for the purpose of determining the intermediates and products formed. Figure 3 shows the formation and decay of tetra-, tri-, di- and monochlorophenols, finally resulting in the formation of phenol and, possibly, monochlorophenols. The experiment shows that the total molar concentration of the phenols remains constant in time.

To get an insight into the effect of the electrode potential on the number of chlorine atoms removed, we conducted tests at various currents. Table 5 shows clearly that large currents (corresponding to more negative potentials) cause more chlorine atoms to be removed. At the same time toxicity decreases (data not shown).

Discussion
In Table 5 the most important results of the comparative tests with the eight organohalogens are reviewed. For the interpretation of these results it should be kept in mind that the conditions of electrolysis were not optimized for the various compounds, as the main purpose was to investigate if and to what extend removal of chlorine atoms of the various compounds is possible.

One of the main conclusions from the experiments with the eight chlori-

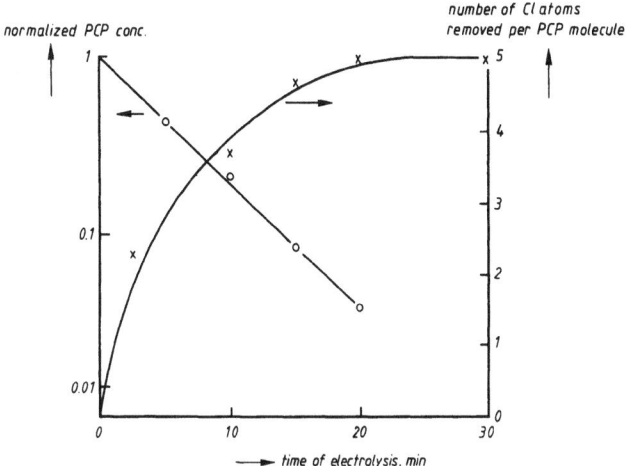

FIGURE 2. Relative PCP concentration and yield of Cl^- ions per PCP molecule during electrolysis:

Structural formula :

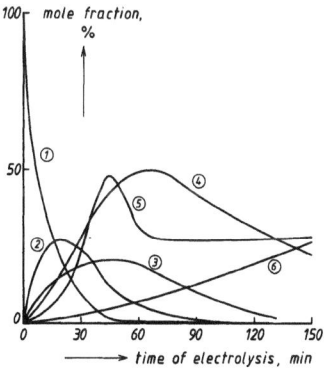

FIGURE 3. Mole fraction of the phenols during electrolysis of 2 l of 50 ppm PCP solution (10 A)
        1. PCP
        2. tetrachlorophenols
        3. trichlorophenols
        4. dichlorophenols
        5. monochlorophenols
        6. phenol

nated organics is that, irrespective of the molecular structure, it is possible to remove all Cl atoms from the organic molecule in aqueous solutions. This generally leads to the formation of less toxic and/or better biodegradable compounds which, if necessary, can be treated further in a conventional biological treatment plant.
The current efficiency is in most experiments of the order of 1%, which is not high, but because of the low concentrations of the compounds to be converted, is an acceptable value for waste water treatment (cf. Table 1). There is an exception (dichlorvos) where a current efficiency of about 75% is found. This is most probably due to the easiness of the reduction of this compound or to the contribution of a non-electrochemical pathway in the decomposition. Whether higher current efficiencies than the 1% mentioned before can be attained at lower current densities is difficult to predict; the experiments with PCP do not give such an indication. Further investigations are necessary to obtain insight in it (e.g. pretreatment of the electrodes).

TABLE 5. Results of the electrolytic dehalogenation experiments with 8 model compounds of the EC priority list.

Compound	Initial concentration ppm	Current A	Final concentration ppm	Number of Cl atoms removed	Current efficiency %	Normalized space velocity $1.1^{-1}.h^{-1}$	Power consumption kWh.m^{-3}
2-amino-4-chlorophenol	100	10	< 0.2	1	0.7	40	60
p-chloronitrobenzene	48	10	< 0.1	1	0.4	50	55
dichlorvos	560	1	< 1	2	75	95	0.8
hexachloroethane	50	10	-	6	1	30	77
pentachlorophenol	50	1	< 0.5	1	3	35	3
		3	< 0.5	3	3	75	8
		5	< 0.5	4	2	45	23
		10	< 0.5	5	2	100	36
2,4,5-T	100	10	< 1	3	1	30	70
tetrachloroethene	74	10	< 0.5	4	3	40	70
1,2,4-trichlorobenzene	39	10	< 0.1	3	0.7	45	70

The normalized space velocity, being in the range of 30-100 $1.1^{-1}.h^{-1}$ is very high, especially when taking into account that the conditions were not optimized so far. A high value of this figure is important from the viewpoint of capital cost. It is expected that after further development a normalized space velocity of 100 (i.e. a treatment capacity of 100 l of waste water per l reactorvolume per hour) should be attainable. The consumption of energy is relatively high at 10 A (about 50 kWh.m^{-3}). At lower currents, however, it decreases very much (see Table 5, pentachlorophenol and dichlorvos). After further development about 10-15 kWh.m^{-3} appears to be a realistic value. This figure has been used for cost calculations. Possibly even lower values will be attainable.

EVALUATION
System definition
The application discussed here is the treatment of toxic process waste waters directly at the source of the process, i.e. before mixing with waste waters from other chemical processes in the same factory. To illustrate this some examples of possible treatment routes are given in Fig. 4.

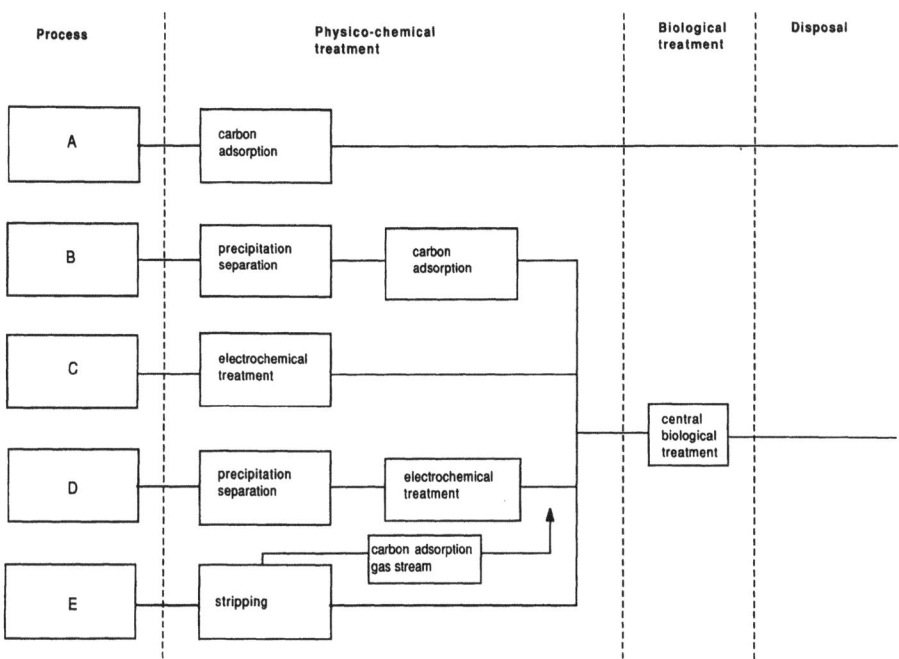

FIGURE 4. Schematic diagram of a factory waste treatment system (example)

Other processes applied to this treatment at the moment are separation of dispersed solids, stripping for removal of volatile compounds and adsorption of dissolved compounds on activated carbon. When the toxic compounds, are volatile the stripped compounds will have to be removed from the waste gas, e.g. by adsorption on activated carbon. Other methods of treatment (e.g. ozone, UV, membrane separation) are also used, but for the time being only to a limited extent.

Cost aspects
The total treatment costs estimated from the experiments carried out so far using a normalized space velocity of 100 $l.l^{-1}.h^{-1}$ and an energy consumption for electrolysis and pumping of 10-20 $kWh.m^{-3}$ are about Dfl. 10 per m^3 of waste water. It must be kept in mind that it is only an order of magnitude at this stage of development.

Comparison with carbon adsorption
The total costs of the electrochemical treatment method will be of the same order of magnitude as those of the currently applied carbon adsorption. Both methods are much cheaper than incineration. Both carbon adsorption and electrochemical treatment will have their advantages, disadvantages and specific applications, some of which are:
- Carbon adsorption is a concentrating technique, leading to a concen-

trated, mostly toxic chemical waste which will have to be treated further or dumped in special depots. Because dumping is getting more and more expensive and possibly will not be allowed anymore in the future, it is expected that the cost of the method will rise.
- The method of carbon adsorption is especially suitable for removal of non-polar compounds, because adsorption of polar compounds from aqueous solutions is less effective, leading to larger apparatus and higher investment costs. The electrochemical method of treatment of aqueous solutions is suitable both for polar and non-polar compounds (the results for these two classes of compounds are similar).

On the basis of the results from the experiments and the cost comparison it can be concluded that further development of the method will lead to an attractive commercial method for the dehalogenation of chlorinated hydrocarbons in process waste streams and will contribute to a further reduction of environmental pollution.

Future investigations

The following aspects deserve further investigation:
- Energy consumption and reaction rates.
  Decrease of energy consumption (lower variable cost) and/or increase in reaction rate (lower investment cost) can be obtained by decrease in current density in combination with measures such as activation of electrodes by coating, pulsing of current, addition of surface active agents, application of more reactive fibre materials and other geometries etc.
- Long time behaviour.
  Tests on long time behaviour of the electrodes (deactivation, corrosion, fouling, dispersed particles) both with model compounds and waste waters.
- Applicability to other types of organic compounds (halogenated or other).
- Semi-technical reactor.
  Testing and further development of the semi-technical reactor using model toxic compounds and waste streams, including in-situ tests for demonstration of the method.

ACKNOWLEDGEMENTS

The research programme described above was financially supported by the European Communities (contract ENV-762-NL(N)), by the Dutch Ministries of Housing, Physical Planning and Environment, Economic Affairs, Transport and Public Works and Agriculture and Fishery (project F60, FEZ 012041) and by Pielkenrood Water Treatment, a Dutch producer of equipment for physicochemical and biological treatment of industrial and other effluents.

REFERENCES

[1] List of 129 priority compounds of the European Communities, Nr. C 176, 1982.
[2] Jørgensen, S.E. (1979), Industrial Waste Water Management. Elsevier Scientific Publishing Company, Amsterdam.
[3] Louw, R., H. Dijks, P. Mulder (1983), Thermal hydro-dechlorination of (poly)chlorinated organic compounds. Chem. and Ind, 1983 3 Oct., 759-760.

[4] Borger, S.K., J. Mc Kenna, J. Karliner, M. Nirsberger (1985), A mild and efficient process for detoxifying polychlorinated biphenyls. Tetrahedron Letters, 26, 3677-3680.
[5] Atlow, S.C., L. Bonadonna-Aporo, A.M. Klibanov (1984), Dephenolization of industrial waste waters catalyzed by polyphenol oxidase. Biotechnol. Bioeng., 26 (1984) 6, 599-603.
[6] Öberg, L.G., K.G. Paul (1985), The transformation of chlorophenols by lacto peroxidase. Biochim. Biophys. Acta, 842 (1985), 30-38.
[7] Swann, S., R. Alkire (1985), Bibliography of electro-organic syntheses 1801-1975. Port City Press Inc., Baltimore.
[8] Farwell, S.O., F.A. Beland, R.D. Geer (1975), Reduction pathways of organohalogen compounds, Part I, Chlorinated benzenes. Electroan. Chem. and Interfac. Electrochem, 61, 303-313.
[9] Harrison, J.M., T.D. Inch, R.G. Williamson (1982), The electrochemical decomposition of TCDD. Chemistry and Industry, 1982, 5 June, 373-374.
[10] Connors, T.F., J.F. Rusling (1983), Removal of chloride from 4-chlorobiphenyl and 4,4'-dichlorobiphenyl by electrocatalytic reduction. J. Electrochem. Soc., 130, 1120-1121.
[11] Connors, T.F., J.F. Rusling (1984), Ultrasonically assisted electrocatalytic dechlorination of polychlorinated biphenyls. Chemosphere, 13, 415-420.
[12] Schmal, D., J. van Erkel, P.J. van Duin (1985), Measurements on mass transfer at fibre electrodes. 36th ISE Meeting, Salamanca, Spain, Sept. 23-28.
[13] Schmal, D., J. van Erkel, P.J. van Duin (1986), Mass transfer at carbon fibre electrodes. J. Appl. Electrochem, 16, 422-430.
[14] Kreysa, G. (1981), Moderne Konzepte und Prozesse zur electrochemischen Abwasserreinigung. Metalloberfläche 35, 211-217.
[15] Erkel, J. van, D. Schmal, P.J. van Duin (1985), Electrochemical detoxification of aqueous waste solutions. 36th ISE Meeting, Salamanca, Spain, Sept., 23-28.
[16] Schmal, D., J. van Erkel, P.J. van Duin (1986), Electrochemical reduction of halogenated compounds in process waste water. Electrochemical Engineering Conference, Loughborough, 21-23 April 1986. The Institution of Chemical Engineers, Symposium Series no. 98, 259-269.
[17] European Patent Application no. 81201302.7. Process for the detoxification of chemical waste material, TNO, 1981.

RECOVERY OF HEAVY METALS BY CRYSTALLIZATION IN THE PELLETREACTOR

M. Schöller   DHV Consulting Engineers
J.C. v. Dijk   DHV Consulting Engineers
D. Wilms   Catholic University Leuven

SUMMARY
Since 1980 DHV Consulting Engineers has been involved in the development of a new system for the recovery of heavy metals (Zn, Ni, Cu, Co, Cd, Mn, Ba, Sr, Pb and Hg) from waste water of the electro-plating industry (plating baths, drag-out baths, pickling baths, passivating baths, rinsing water). Since 1985 the Catholic University of Leuven has been involved in the fundamental research of the crystallization of heavy metal carbonates.
Contrary to conventional waste water treatment methods (hydroxide precipitation) no sludge is produced but pure pellets of metalcarbonate. These pellets can be reused in the electro-plating industry, a.o. for the preparation of a new plating bath.
The plant is very compact, relatively cheap and can be used to treat various baths separately.
Pilot-plant research has yielded very favourable results for Zn, Ni and Cu. The financial possibilities of the system look very bright. In a subsequent project in 1986/87, the implementation of the system will be worked out in cooperation with a partner from the electro-plating industry.
The project is financially supported by the Dutch government.

1. INTRODUCTION
Since the seventies, the Dutch Government has imposed standards for the discharge of heavy metals in industrial waste waters. As a result many factories have started to treat this type of waste water by means of hydroxide precipitation. The sludge produced by this method has no useful application, but must be transported to a hazardous waste landfill: a troublesome and expensive solution.

This article describes a new system by which heavy metals can be removed from waste water without sludge production. Instead of sludge, granular metalcarbonate crystals are produced; these are pure and can be re-used. Considerable economics are made, resulting from the reduced need to dispose of hydroxide sludge and from the lower purchase costs of the metals involved.

The system is a good example of clean technology, avoiding environmental problems and serving to recycle heavy metals. The application of crystallization of metalcarbonates in a pellet reactor has been developed by DHV Consulting Engineers. The crystallization method was developed on the basis of over 15 years of experience with pellet reactors for softening of drinking water (producing calciumcarbonate crystals) [1] and for the removal of

phosphates from waste water (producing calciumphosphate) [2]. The Catholic University of Leuven investigates since 1985 the fundamental aspects of the crystallization of heavy metal carbonates.
The development of this process has been financially supported by the Ministry of Housing, Physical Planning and the Environment and the Minsitry of Economic Affairs; assistance has been provided by the National Institute of Public Health and Environmental Hygiene (RIVM) and the National Institute for Waste Water Treatment (RIZA).

2. SYSTEM DESCRIPTION

The principle of a pellet reactor is shown in Figure 1. The reactor consists of a cylindrical vessel, partially filled with suitable seed material, e.g. filter sand. The fluid velocity in the reactor is so high (40 - 100 m/h) that the grains are kept in suspension so that cementing is prevented. A carbonate solution dosed into the reactor causes the metal-carbonates (e.g. $ZnCO_3$, $NiCO_3$, $CuCO_3$) to crystallize on the seed material. The reaction takes place very quickly, so that only a small reactor volume is needed. A 3 m high reactor with 20 cm diameter can treat a flow of 1.2 m^3/h.

The reaction results in growth of the grains and after some time the larger pellets must be removed from the bed. These metalcarbonates are pure and can be re-used after dissolving the pellets in acid; the carbonate then escapes as CO_2, while the sand core can be used in the reactor again as seed material.

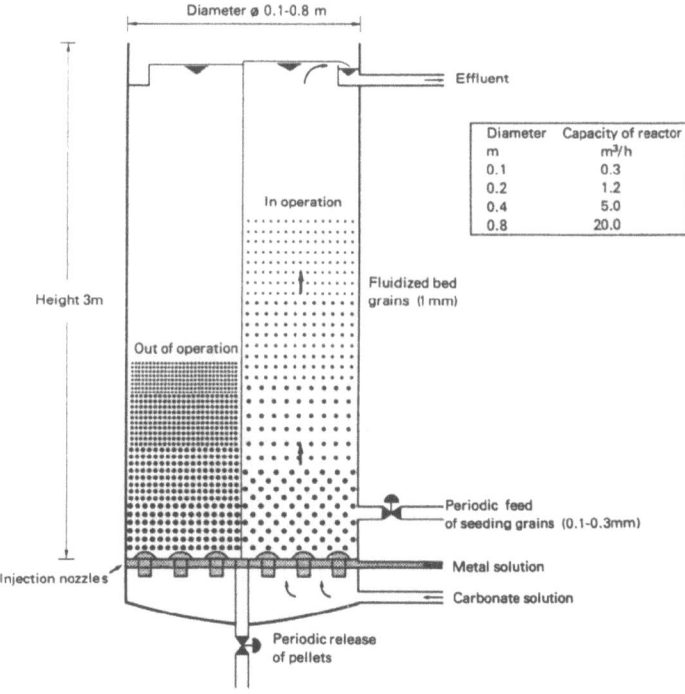

FIGURE 1. Principle of pellet reactor

In principle crystallization of heavy metalcarbonates in a pellet reactor has the following advantages:
- The pellets produced are pure and can be re-used. This provides a saving in the purchase cost of heavy metals.
- The plant is very compact and cheap, and it can treat a relatively large flow. Therefore various baths and flows can be treated separately, allowing better control of the process.
- No chemical sludge is produced whereas the sludge production from a hydroxide precipitation plant if used as further treatment is greatly reduced.

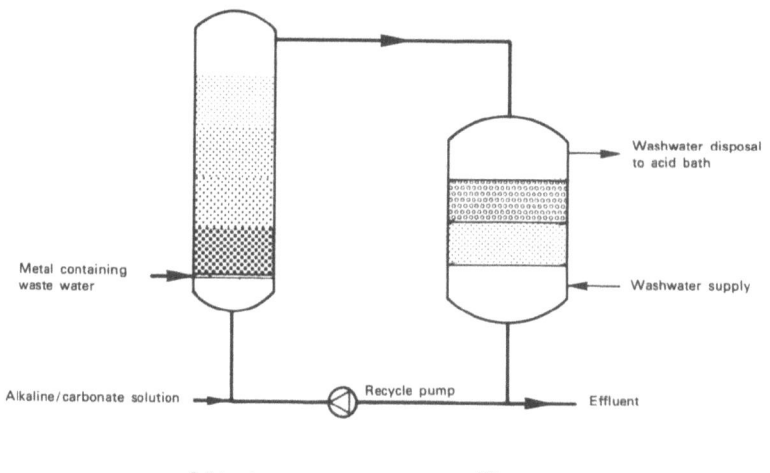

FIGURE 2. System setup of pelletreactor with dual-media filter

In practice, a sturdy plant with very good results can be achieved. The greater part of the heavy metals crystallizes onto the grains. A minor part will remain in the effluent in the form of small suspended particles of metalcarbonate or metalhydroxide (carry-over). These are filtered in the post filter (sand or anthracite-sand).
Waste waters with a metal content of 10 - 100.000 ppm can be treated. In the case of a low metal content no recycle or a small recylce ratio is required while at a high metal concent a big recycle ratio is required.
The effluent from the plant will have one or two destinations, depending on the system selected:
1. Discharge to main sewer or hydroxide precipitation plant. In some cases of Zn and Cu direct discharge may be considered. In other cases if a hydroxide precipitation plant is available, further treatment is indicated. The use of chemicals and the production of sludge by the hydroxide precipitation will be greatly reduced, however.
2. Recirculation to a drag-out bath (process integration). In this case, the carbonate can escape into the air as CO_2, provided there is sufficient acid in the bath.
The backwash water from the sand filter can be returned to an acid drag-out or plating bath where the carry over is dissolved; it can then still be abstracted in the pellet reactor in the succeeding cycle.

In principle the system can treat a number of flows containing metals, such as plating baths, drag-out baths, conversion baths, pickling baths and rinsing water. The heavy metals Zn, Ni, Cu, Co, Cd, Mn, Ba, Sr, Ag, Pb and Hg can in principle be abstracted as metalcarbonate crystals. The metal Cr does not form a carbonate salt and therefore cannot be abstracted in this way.

3. THEORY
   Chemical reactions

Metal (II) ions can form compact carbonate crystals. The carbonate salts of most heavy metals have low solubility, as shown in Table 1. In practice, reaction conditions must be selected to allow the solubility product of the metalcarbonate to be exceeded, with the restriction that the solubility product of the metalhydroxide is not exceeded. Metalhydroxides are not crystalline and form a wet voluminous sludge (as in fact occurs in hydroxide precipitation plants).

This means that the carbonate concentration should be relatively high. For this reason in practice a carbonate solution is recirculated over the pellet reactor. The heavy metals containing waste water is injected in the pellet reactor and enters conditions which are very favourable for the crystallization.

TABLE 1. Thermodynamic constants.

Solubility products of metalcarbonates and -hydroxides
pK_s = negative logarithm of the solubility product

Compound	pKs	Compounds	pKs
$CaCO_3$	8.3	$Ca(OH)_2$	5.0
$NiCO_3$	8.2	$Ni(OH)_2$	14.0
$BaCO_3$	8.3	$Ba(OH)_2$	1.7
$SrCO_3$	9.0	$Sr(OH)_2$	3.9
$ZnCO_3$	9.1	$Zn(OH)_2$	16.0
$CuCO_3$	9.9	$Cu(OH)_2$	19.0
$MnCO_3$	9.3	$Mn(OH)_2$	12.8
$CoCO_3$	10.0	$Co(OH)_2$	14.9
$AgCO_3$	11.1	$Ag(OH)_2$	7.8
$FeCO_3$	10.5	$Fe(OH)_2$	14.0
$PbCO_3$	13.5	$Pb(OH)_2$	15.0
$CdCO_3$	13.7	$Cd(OH)_2$	13.5
$HgCO_3$	16.0	$Hg(OH)_2$	23.7

Pellet reactor

The functioning of a pellet reactor is known from the experiences in softening of drinking water and phosphate removal from waste water.
The primary characteristics are:
- high allowable fluid velocity (40 - 100 m/h);
- no cementing of grains;
- reliable and sturdy operation.

Post filter

The removal of suspended matter from water by antracite-sand filters is a well-known technique; it is used, for instance, for the filtration of carry-over from pellet reactors in drinking-water softening and removal of phosphates from wastewater.
To obtain maximum filtration velocity and suspended matter loading, the most suitable system is a downflow, two-layer (anthracite-sand) pressure filter.

4. PILOT-PLANT RESULTS

Since 1980 DHV has been involved in the development of the system for the recovery of heavy metals from the electro-plating industry.
The system was tested in practice with a pilot-plant consisting of a dry filter, a pellet reactor and a sand filter. Figure 3 shows the pilot-plant. The function of the dry filter was to remove suspended solids and (limited quantities of) iron ions from the process baths, so that this iron would not disturb the crystallization process.
The test programme consisted of trial operation with synthetic solutions, after which a number of waste waters from spent drag-out baths and spent process baths from the electro-plating industry were treated. No or only a small recycle ratio was applied.
Since 1985 Catholic University of Leuven has been involved in the fundamental research of the crystallization of heavy metal carbonates. In a pilot plant consisting of only a pelletreactor and using a high recycle ratio synthetic metal solutions (Zn, Cu) were treated.
The results, which were very satisfactory, are summarized below.

FIGURE 3. Pilot-plant (pellet reactor to the left, post filter to the right)

Iron-removal

Table 2 lists the spent baths included in the programme, the firms which made them available and the composition of each bath. In the low-cyanide bath the cyanide was converted into CO_2 and N_2 by the addition of hypochlorite.

TABLE 2. Composition of baths

Zn-baths	firm	pH	[Zn] (ppm)	[Fe] (ppm)	Other* (ppm)		
1. $ZnCl_2$ drag-out bath on K-basis	AGI	5.5	2,900	190	K	=	4,000
2. $ZnCl_2$ drag-out bath on NH_4-basis	Chromolux	5.6	26,000	± 20,000	NH_4	=	44,000
3. blue passivating bath	Chromolux	1.5	2,300	840	Cr	=	250
4. yellow passivating bath	Duke en Roks	1.4	1,400	370	Cr	=	5,900
5. blue passivating bath	Duke en Roks	2.1	970	< 1	Cr	=	350
6. low cyanide drag-out bath	Holec	13.2	2,100	< 1	CN	=	700
7. alkaline cyanide-free drag-out bath	Holec	12.8	170	< 1	organic complexformer		
Ni-baths	firm	pH	[Ni] (ppm)	[Fe] (ppm)	other* (ppm)		
8. $NiSO_4$/Cl drag-out bath	Holec	6.0	3,200	220	SO_4	=	107,000
9. $NiSO_4$/Cl concentrate bath	Stork Screens	5.9	82,100	4,500	Cl	=	18,000
10. Ni sulfamate concentrate bath	Stork Screens	4.0	62,800	4,500	sulfamate	=	190,000

* The baths also contain shiners, moisteners and other surfactants.

The iron removal was carried out batchwise by means of H_2O_2 at a pH of 6 - 7. The Fe^{3+} flocculates in the form of very poorly soluble $Fe(OH)_3$, which was removed by sedimentation and filtration.
Iron removal by sedimentation and filtration was generally very satisfactory with residual concentrations below 1 mg Fe/l.
The sludge contained in some cases minor amounts of Zn of Ni in addition to Fe.

The necessity of pretreatment by sedimentation and/or filtration was not investigated. It can probably be omitted in cases where the spent baths contain little suspended matter and little iron.

Pellet reactor

The selected system of recirculation of a carbonate solution over the reactor and the sand filter, with injection of synthetic metal solutions or process baths into the reactor, proved to be reliable and robust.
The process conditions and results of the tests are summarized in Table 3.

TABLE 3. Process conditions and results of crystallization

Baths	pH	CT* (mol/l)	Influent (ppm)	Recycle ratio **	After reactor (ppm)	Removal %	After filter (ppm)	Removal %
Zn	7 -9	0,1-0,2	40- 500	0- 1	< 1-140	70 -99,5	< 1-140	70-99,5
Ni	8 -9	0,3-0,5	400-1800	0- 1	320-880	35 -60	60-210	78-90
Cu***	6,5-7,5	0,1-0,2	700-1000	10-15	0,5- 2	99,7-99,95	-	-

*   (CT = CO_2 + HCO_3^- + CO_3^{--}).
**   In the case of Zn and Ni higher efficiencies can be expected if a higher recycle ratio is applied.
*** Synthetic $CuSO_4$ solution.

In general the efficiency can be raised by selecting a rather high pH (figure 4) or applying a high recycle ratio.

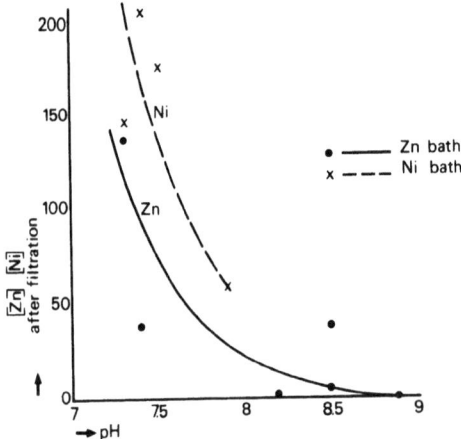

FIGURE 4. Relation between pH and Zn or Ni concentration after filtration

The tests with the synthetic metal solutions and the spent baths gave good results.
The removal of Zn and Ni from the baths is comparable with that obtained with the synthetic solutions and disturbances due to impurities in the baths practically did not occur.

The following conclusions can be drawn:
1. In treating spent Zn-baths hardly any amorphous matter is formed, so that the sand filter is probably superfluous.
   The process efficiency is high and can be influenced by the pH. With a pH of 8.5 - 9 very low effluent concentrations can be achieved (a few milligrams per litre).
2. In treating spent Ni-baths amorphous matter always occurs, so that a sand filter is necessary to increase the efficiency. As known, the backwash water from the sand filter can be returned to the acid bath, where the amorphous matter dissolves again and can be removed in the following cycle in the pellet reactor.
3. In treating synthetic Cu solutions and applying a high recycle ratio very high efficiencies can be achieved. No amorphous matter is formed and the effluent concentration is below 2 ppm.

It should be noted that during this investigation no special attention was paid to achieving the best possible efficiency; therefore an improvement of the results given in Table 3 is likely to be possible with further optimization of pH and CT (particularly for Ni) and applying a higher recycle ratio.

Reuse of produced pellets

The produced pellets are shown in Figure 5.

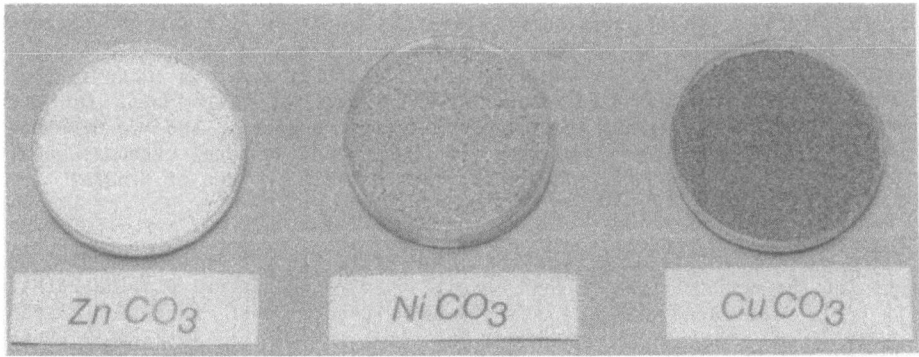

FIGURE 5. - Pellets of $ZnCO_3$, $NiCO_3$, $CuCO_3$

Pellet analyses show that the Zn- and Ni-pellets contain respectively more than 99,7% Zn (rest: Fe, Cr) and 99,8% Ni (rest: Fe). The pellets can be considered as being quite pure. The Zn-pellets contain 70 - 95% CO_3^{2-} and small quantities of other liqands (most probable OH^-). The Ni-pellets contain 50 - 60% CO_3^{2-} and large quantities of other liqands (probably OH^-).

The composition of the produced copper pellets corresponds very well to that of malachite; $Cu_2(OH)_2CO_3$.

The pellets can be reused in the metal industries e.g. to prepare concentrate baths by dissolution of the metal with HCl or H_2SO_4. A Hullcell test was carried out to test the applicability of the produced pellets for these baths. A Hullcell test is a laboratorium test in which a small sample is plated to check the suitability of the bath.
The result was that the produced Zn-pellets can effectively be used to prepare a $ZnCl_2$-concentrate bath.

5. APPLICATION IN PRACTICE

Experience in drinking water treatment and wastewater treatment provides a good basis for the application of pellet reactors in practice. It may also be remarked that scaling up is hardly necessary as the capacity of the pilot-plant is sufficient to treat the wastewater from a 1 m^3 bath in a few hours.

Because of the diversity of the electro-plating industry, application in practice will have to be worked out in further detail. The major aspects of product development will be:
- Further detailing of the application of a pellet reactor for each plant concerned.
- Engineering.
- Re-use of pellets.
- Costs.

The financial possibilities of the system appear very bright.
The investment costs are relatively small, being anticipated at Dfl* 10,000 - 50,000 for plants of 0.5 - 5 m^3/h. Operational costs consist mainly of the cost of chemicals, estimated at Dfl 5 - 10 per m^3. Against these the savings on the purchase of heavy metals can be considerable. At a bath concentration of 5 g/l and a metal price of Dfl 5 per kg these savings amount to Dfl 25 per m^3. Furthermore, there will be lower costs for the operation of the hydroxide precipitation unit (chemicals) and the disposal of hydroxide sludge. The last item greatly depends on local circumstances, but is generally very considerable (approx. Dfl 250 per ton of sludge).

6. FUTURE

The pilot-plant results show that crystallization of metalcarbonates in a pellet reactor can provide good results on treating actual plating baths. The pellets produced are pure and can be re-used amongst others for preparation of new electro-plating baths. The plant is compact, relatively cheap and can be used for various, separate baths and process flows.
Further work is still required to provide an integral solution for specific situations, whereby a pellet reactor can be optimally applied. In a subsequent project in 1986/87 the implementation will be worked out in coopera-

---

*1 US\$ = Dfl 2.0

tion with a partner from the electro-plating industry. This will include further process development (treatment of pickling baths, necessity of pre filtration and post filtration) and product development (application in practice).

After publication of the presented results in the Dutch literature a large number of galvanic industries showed their interest. Therefore DHV has built a laborator scale plant for rapid testing of samples of baths.

The Catholic University of Leuven will continue the fundamental research, with particular emphasis on other heavy metals than Zn, Ni and Cu.

LITERATURE

1. van Ammers, A., van Dijk, J.C., Graveland A., Nuhn. P.A.N.M., State of the art of pellet softening in the Netherlands.
IWSA Specialized Conference on new Technologies in Water Treatment, Amsterdam (1986). Reprinted in Water Supply, vol. 4.

2. van Dijk, J.C., Braakensiek, H. Phosphate removal by crystallization in a fluidized bed.
IAWPRC Conference, Amsterdam (1984). Reported in Water Science and Technology, vol. 17

MSch/Provi/Bur4-33/D

REMOVAL OF ARSENIC FROM WASTE WATERS OF THE LEAD GLASS INDUSTRY

A. KAISER, F. HUTTER, J. KAPPEL, and H. SCHMIDT

1. INTRODUCTION

Arsenic trioxide (As_2O_3) is widely used as refining agent for the production of lead glasses /1/. Because of its toxicity there were attempts to reduce its consumption, which have been successful mainly due to an increasing number of electric melters. The As_2O_3 consumption within the glass industry of the Federal Republic of Germany decreased from about 300 t/a in 1981 to 145 t/a in 1985 /2/. But it was also shown that a complete renunciation of As_2O_3 is not possible up to now /3/. The possible substitution of As_2O_3 by Sb_2O_3 is not recommendable for toxicological and environmental as well as for economical reasons.

The As_2O_3 which is incorporated into the glass matrix during the melting procedure can be leached or dissolved, mainly during finishing procedures like grinding or acid polishing. Therefore, remarkable contents of arsenic in waste waters of lead glass plants are observed: Waste water originating from glass grinding process typically obtains 2 to 4 mg/l As, acid waste waters from chemical polishing even up to some 100 mg/l. In the Federal Republic of Germany no regulations concerning the arsenic content of waste waters of the glass industry exist due to the lack of avoidance. There are efforts, however, to set the limits for the arsenic concentration in waste waters significantly below 1 mg/l. For that reason, it is necessary to remove the arsenic from the waste waters before being drawn off.

Another serious problem arises from attempts to reduce water pollution by recycling the cooling water within the grinding process. The recycling basically offers two major advantages:
- the amount of waste water which has to be drawn off is reduced to about 5 % of the original amount,
- the temperature of the water can be kept at a higher level
  in order to decrease diseases at the glass grinder's arms.
State-of-the-art equipments for the recycling of glass grinding cooling water remove lead ions from that water satisfactorily /4/, but the arsenic content is not affected by that treatment. Therefore, arsenic concentrations in such circulating waters will increase significantly; values up to 10 mg As/l have been measured. A remarkable amount of the cooling water is sprayed as fine aerosols within the working area by high speed rotating grinding wheels. This can lead to health hazards due to their As_2O_3 content.

2. AIM OF THE WORK

Therefore, a technique should be developed with mainly two applications to be investigated: The As_2O_3 removal from recycled water of the grinding process and (within a series of preliminary experiments) the removal of As_2O_3 from other waste waters produced by lead glass plants. The new technology should be applied in a first step in combination with an installed lead precipitation unit, without disturbing its operation. The costs should be bearable to the plants and no secondary environmental problems should arise. The most promising techniques were planned to be evaluated first on laboratory scale. Then, in a pilot scale these techniques should be tested in order to find out the best parameters for the transfer into technical application.

3. STATE-OF-THE-ART METHODS IN REMOVAL OF ARSENIC FROM WASTE WATERS

Almost all papers dealing with the removal of arsenic from aqueous solutions are based on laboratory and a few pilot scale experiments. Mainly two general methods are described:
- adsorption techniques and
- precipitation reactions.

Adsorbents which have been tested for removal of arsenic, e.g. from ground or surface waters are activated charcoal, alumina, clays and materials based on titania /5 - 9/. Additionally, a method is described based on the reduction of the As(III) or As(V) to the oxidation state zero and the simultaneous adsorption at iron or zinc surfaces /10/. Thereby, arsenic contents could be reduced from some 10 mg/l to some 0.01 mg/l. Though it was not sure, that these techniques would work also with glass grinding waste waters, and though it was known that adsorption techniques would be relatively expensive, it was decided to take them into consideration at least within the laboratory scale experiments.

Most publications concerning the removal of arsenic from waste waters deal with precipitation reactions. The following precipitates are reported: arsenic sulphide, calcium arsenate and manganese arsenate /11 - 14/. Earlier publications reported in /15/ describe the precipitations of lead arsenate and magnesium ammonium arsenate. A third group of precipitation reactions is based on the addition of ferric salts to the arsenic containing solutions /16 - 19/.

Most of these reactions are described to be run in a pH range of about 6 to 11. Although it is well known that generally the solubility of As(V) compounds is less than that of As(III) coumpounds /15/, there is no distinction between As(III) and As(V). Starting from different concentrations of As, residual concentrations of 0.5 µg/l to 0.7 mg/l are reported. Therefore, and due to the expected lower costs of precipitation techniques compared to adsorption, these possibilities should be investigated more in detail with respect to their application to glass grinding waste waters.

Only two papers present data from production scale arsenic removal techniques. The first report /11/ deals with the precipitation of As, but also Zn, Cu, Pb, Cd and Hg with sodium sulphide from smelter waste waters. Since this process is optimized for the usual waste water composition of the smelter, a simple transfer to glass plant waste waters cannot be expected. The second method is the so-called "Skorodit-precipitation". It is used in a copper plant and works with ferric salts at pH = 1.9 to 2.0 if chloride ions are in solution or at pH = 2.7 to 2.8 if there are sulphate ions /20/. Grinding waters of the glass industry, however, show a pH in the range of 6 to 8.

In summary no commercially available process could be found from literature study to be easily adapted to the grinding process of lead glass production. Adsorption techniques and mainly precipitation reactions seemed to be promising enough to start laboratory scale experiments. For these experiments genuine waste waters of the lead glass industry were used in order to have all practical parameters included.

4. EXPERIMENTAL RESULTS
4.1. Laboratory scale experiments

The first experiments were done as batch experiments with recycled or non-recycled process waters. The samples contained between 2 and 4 mg/l As and 0.5 to 4.6 mg/l Pb. All analytical data for arsenic were obtained by ICP-AES (inductive coupled plasma - atom emission spectroscopy), which is specific for the element regardless of its oxidation state.

To reduce the arsenic content by adsorption TiO_2-modified silica gels, activated charcoal and alumina were used. Even in the best case only 15 % of the initial amount of arsenic were removed by these methods. No better results could be obtained by a combined reduction/absorption process on the surfaces of iron and zinc metals (table 1). Adsorbent to solution volume ratio and absorption periods were varied in order to make sure that the important range of the adsorption isotherms were covered.

As also can be seen from table 1, most of the precipitation experiments did not yield in a sufficient removal of arsenic within these batch experiments. Addition of sulphide, ammonium, magnesium, manganese, lead or aluminium ions did not reduce the arsenic content remarkably. Variations of concentration of precipitating agent, pH, reaction times, temperatures as well as addition of H_2O_2, hypochlorite or $KMnO_4$ in order to oxidize the As(III) to As(V) did not increase the effect. Therefore these variations are not indicated in table 1.

Favourable results, however, were obtained with iron(III) ions, coprecipitating arsenic with iron hydroxide (presumable as $FeAsO_4$ /20/. The arsenic concentration could be reduced to less than 0.5 mg/l when 50 mg/l iron(III) ions were added (as aqueous solutions of $FeCl_3$, $Fe_2(SO_4)_3$, $FeClSO_4$ or $NH_4Fe(SO_4)_2$) to the batches. The type of Fe(III) salt added did not influence the result. The influence of the amount of Fe(III) ions added to the waste water is shown in figure 1, which also demonstrates that without addition of oxidizing agents much higher amounts of Fe(III) ions are necessary to obtain similar results. The values of As concentration indicated in figure 1 are obtained only when the residence time (time between addition of iron ions and filtration) is kept above 12 minutes. Therefore, in a continuous process a buffer tank which allows a sufficient residence time has to be provided.

TABLE 1. Representative data of laboratory scale batch experiments to remove arsenic from glass grinding waste waters (Initial amount of arsenic: 4.0 mg/l).

Method	% of arsenic removed	Residual amount of arsenic (mg/l)
Adsorption on:		
- SiO_2/TiO_2	< 3	> 3.8
- Activated charcoal	< 3	> 3.8
- Alumina	15	3.4
- Metal surface of:		
- Fe	< 2	> 3.9
- Zn	17	3.3
Precipitation as:		
- As_2S_3, As_2S_5	20	3.2
- NH_4MgAsO_4	< 1	> 3.9
- $Mn_3(AsO_4)_2$	< 1	> 3.9
- $Pb_3(AsO_4)_2$	< 1	> 3.9
- $AlAsO_4$	20	3.2
- $FeAsO_4$	> 85	≥ 0.5

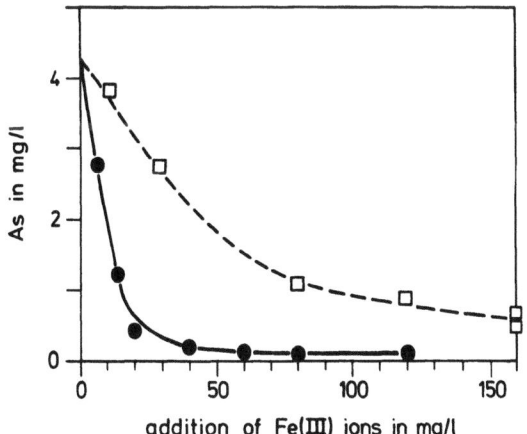

FIGURE 1. Remaining concentration of arsenic after precipitation with Fe(III) ions
●: sample oxidized by H_2O_2, hypochlorite or $KMnO_4$
□: untreated sample.

Since in industrial lead removal procedures CO_2 is injected to precipitate lead ions as $PbCO_3$, analogeous experiments were carried out to find out whether the As precipitation process affects that reaction. No inhibition effect could be seen. All solids suspended in the grinding water (glass particles, $PbCO_3$, $FeAsO_4$, $Fe(OH)_3$, flocculants) could be filtered off together without any difficulties.

Based on these results, it was decided not to investigate the adsorption techniques any longer, and continuous pilot scale experiments with precipitation of arsenic by Fe(III) ions were planned.

4.2. Pilot scale experiments

A pilot scale equipment was designed in which the water supply for ten glass grinder operating positions was recycled with precipitation of lead as $PbCO_3$, flocculation and filtration. The average flow rate was about 400 l/h. Into a buffer container installed upstream the location where CO_2 was added to form lead carbonate, a solution of $FeClSO_4$ and oxidation and neutralization agents were added. With this equipment the influence of different parameters (listed in table 2) was investigated. Figure 2 shows the concentrations of arsenic and lead in the water circuit during periods under different conditions. In periods indicated by "X" no addition of Fe(III) took place. Chloride content (resulting from $FeClSO_4$ addition) was monitored in order to see to which extend its concentration increases.

TABLE 2. Addition of oxidizing and neutralizing agents in the pilot scale experiments (see figure 2). Amount of Fe(III) ions: 0.87 mmoles/l of circulating water.

Period	Addition of H_2O_2 (mmoles/l)	Addition of neutralizing agent (mmoles/l)
A	2.52	3.62 $NaHCO_3$
B	No addition	3.62 $NaHCO_3$
C	2.52	1.35 $Ca(OH)_2$
D	No addition	1.35 $Ca(OH)_2$

As indicated in figure 2 (and proved within the experiments reported in chapter 4.3), oxidation with H_2O_2 was not necessary. The reason for that surprising effect is not yet clear. Perhaps, there is an oxidizing effect when the water gets into contact with oxygen of the air during the recycling. Bubbling oxygen through the batch samples within the laboratory scale experiments did not lead to a similar effect, however. This difference can be due to the fact that the waste waters used in the laboratory scale experiments have aged because of the shipping procedure.

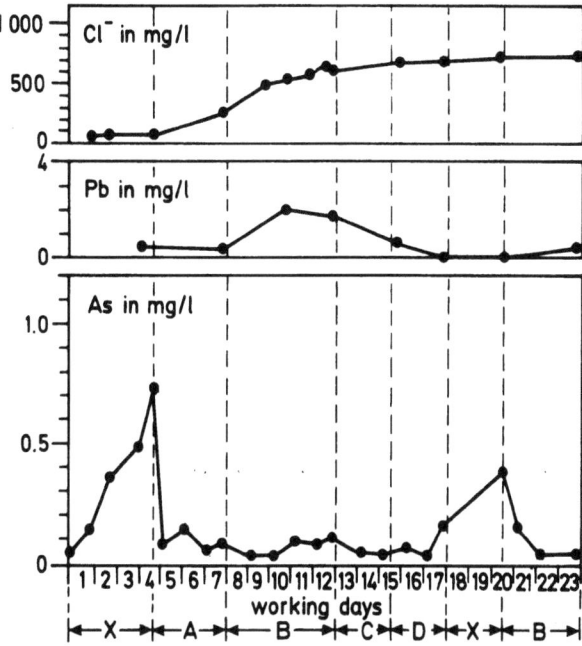

FIGURE 2. Concentrations of arsenic, lead and chloride within the water circuit during the pilot scale experiments. Explanations to periods A to D see table 2; in periods indicated by "X" no Fe(III) was added

Like in the laboratory scale batch experiments, an arsenic concentration of 0.1 mg/l or less could be maintained within these pilot scale tests. Neutralization by a $NaHCO_3$ solution is possible, but using $Ca(OH)_2$ suspension is cheaper and can keep the sulphate concentration at about 1 g/l. This is especially advantageous when $Fe_2(SO_4)_3$ is used instead of $FeClSO_4$ in order to avoid secondary problems by increasing chloride ion concentration. In this case the overall reactions formally can be described by equations (1) - (3):

(1) $\quad Fe_2(SO_4)_3 + 2H_3AsO_4 \longrightarrow 2FeAsO_4\downarrow + 3H_2SO_4$

(2) $\quad Fe_2(SO_4)_3 + 6H_2O \longrightarrow 2Fe(OH)_3\downarrow + 3H_2SO_4$

(3) $\quad H_2SO_4 \quad\quad + Ca(OH)_2 \longrightarrow CaSO_4 \cdot 2H_2O\downarrow$

The $PbCO_3$ precipitation again was not affected by the arsenic removal. Concentrations of far less than 1 mg/l lead ions were maintained. The higher lead concentrations during experiment B were caused by insufficient CO_2 supply. As in the laboratory experiments, all solids suspended in the circulating water could be filtered off without any problems.

4.3. Production scale experiments

Based on the encouraging results of the pilot scale tests an additional experiment with the grinding water of a whole lead crystal glass plant was performed. A flow chart of this system is given in figure 3.

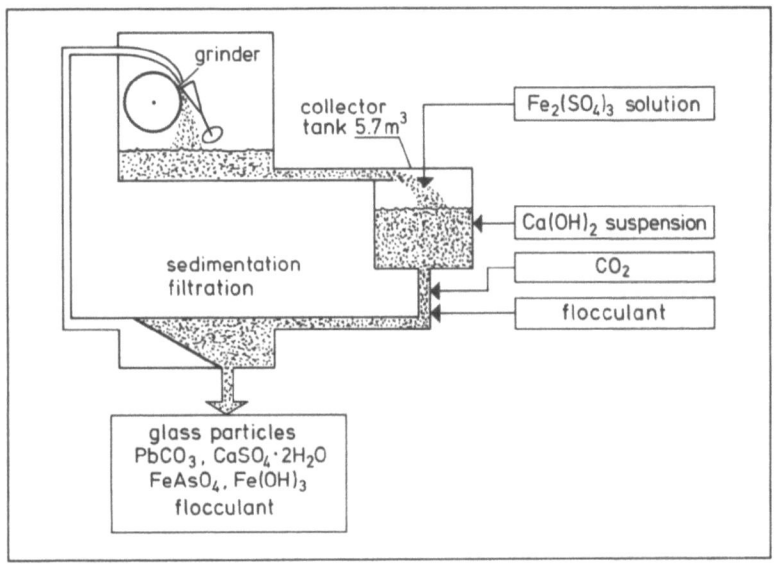

FIGURE 3. Flow chart of the equipment for the production scale experiment.

The flow rate during day shifts was about 16 m^3/h, during night shifts, lower and changing flow rates occurred. Therefore addition of iron(III) ions was not possible and the arsenic concentration increased during night time. Iron(III) ions were added as an aqueous solution of $Fe_2(SO_4)_3$, which was neutralized by the stochiometric amount of a $Ca(OH)_2$ suspension. Figure 4 shows the content of arsenic and lead in the water circuit during this experiment. When iron(III) ions were added, the arsenic concentration decreased to a level of about 0.1 to 0.2 mg/l. The efficiency of arsenic removal could be influenced by the amount of Fe(III) ions added which was varied within the periods A to D (see figure 4). The velocity of lowering the arsenic content of the circulating water depends on its initial concentration. It can be assumed that continuous addition of iron ions controlled by the actual flow rate can stabilize an arsenic level as low as 0.1 mg/l.

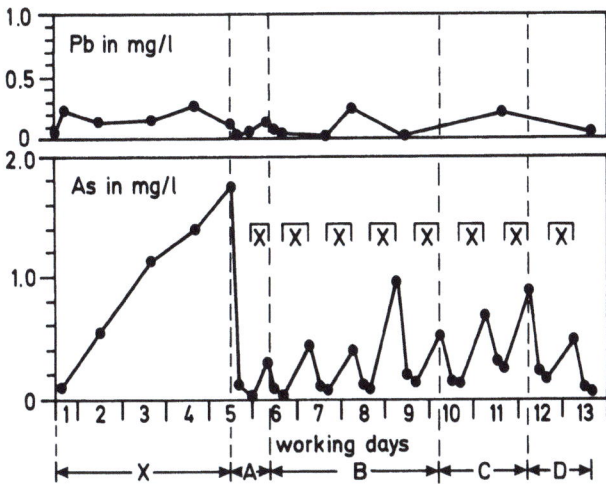

FIGURE 4. Concentration of arsenic and lead during the full scale experiment. Addition of iron ions in mmoles/l within the different periods: A 0.90, B 0,30 C 0.15, D 0.25. In periods indicated by "X" no iron ions were added.

Again, the removal of lead was not affected. Lead ion concentrations could be kept well below 0.5 mg/l. The iron arsenate and hydroxide precipitates sedimented well and could be filtered, as long as not more than about 1 mmole/g iron ions had been added and a reaction time of at least 20 minutes could be provided.

After this period which is documented in figure 4 the whole equipment was operated in the same way another three months whithout problems. After that, some difficulties arose which are mainly due to the lack of controlling and regulating. Therefore in the next step an equipment has to be provided with dosing unit controlled by the flow rate for the iron and Ca(OH)$_2$ addition.

The solid waste filtered off during operation of the arsenic elimination, showed a slightly increased leachability of arsenic compared to the usual grinding sludge. The increase in leachability, however, was not as high as to produce a new quality of waste.

4.4. Removal of arsenic in other waste waters

In a series of additional experiments it was tried to use these excellent results of arsenic removal from glass grinding waste waters also for other arsenic containing waters in glass plants, e.g. from acid polishing process or from the neutralization unit. These experiments failed up to now, however. The reason for that is not yet known, but maybe fluoride containing arsenic compounds produced in the polishing process play an important role. Further research and development activities to overcome these problems are under work.

5. SUMMARY

A new technology was developed to remove arsenic from waste waters of the lead glass grinding process. The method is based on the well known precipitation of arsenic with iron(III) ions. It allows the residual amount of arsenic to be kept at or even below 0.1 mg/l, also when used in combination with state-of-the-art lead removal techniques. The additional costs are low compared to those of the usual water recycling process with lead removal technologies. The process does not create any secondary environmental problems.

6. REFERENCES

/1/ Peters A: Zur Situation des Arsenverbrauchs unter besonderer Berücksichtigung der Glasindustrie. Glastech. Ber. 50 (1977) 328-325.
/2/ Lubisch G: Bundesverband Glasindustrie und Mineralfaserindustrie e.V. Düsseldorf, FRG: private communication (1986).
/3/ Langer A and Scholze H: Untersuchungen zum Ersatz von As_2O_3 als Läutermittel für Kristallgläser. Glastech. Ber. 54 (1981) 223-230.
/4/ Piepho RF: Reinigen von Schleifabwässern in Glasbearbeitungsbetrieben. Sprechsaal 116 (1983) 435-436.
/5/ Mitsubishi, Rayon Co., Ltd., Japan: Granular titanic acid ion-exchanger. Jpn. Kokai Tokyo Koho 82150444, 82.09.17 PAT APP = 8134232, 81.03.10, (1982).
/6/ Sato H, Shigeta S and Uchida H: Inorganic ion exchanger. Can. 1122876, 82.05.04 PAT APP = 336889, 79.10.03, (1982).
/7/ Kochergin VP et al.: Treatment of industrial wastewaters containing arsenic compounds. Deposited Doc (SPSTL 504 khp-D80), (1980).
/8/ Fresenius W and Schneider W: Selektive Entfernung von Fluorid-, Arsenat- und Phosphat-Ionen aus Wasser. WLB, Wasser, Luft und Betrieb 25 (1981) 14-15.
/9/ Watanabe N and Hayakawa O: Adsorption of arsenite and arsenate by soils and charcoal. Kankyo Gijutsu (KAGIDX), (1982), V11 (8) 565-571.
/10/ Blavathnik M et al.: Wastewater purification. Otkrytiya, Isobret., Prom. Obraztsy, Tovarnye Znaki (1980) (23) 130. U.S.S.R. 742,389, 25.06.1980 Appl. 2,388, 919, 01.08.1976.
/11/ Bhattacharyya D, Junrawan AB and Sun G: Precipitation of heavy metals with sodium sulfide: bench-scale and full-scale experimental results. AICHE Symp. Ser. 77 (1981) 31-38.
/12/ Babcock AR and Kuit WJM: Treatment of arsenical effluents. Can. 1111157, 81.10.20 PAT APP = 324552, 79.03.30, (1979).
/13/ Bowers AR, Chin G and Huang CP: Predicting the performance of a lime-neutralization/precipitation process for the treatment of some heavy metal-loaden industrial wastewaters. Ind. Waste, Proc. Mid-Atl., 13th Conf., (1982), 51-62.

/14/ Kuznetsov VL et al.: Removal of arsenic from wastewater by treating with natural pyrolusite. Otkrytiya, Isobret., Prom. Obraztsy Tovarnye Znaki (1982) (25) 103 U.S.S.R. 880999, 81.11.15 PAT APP = 3008113, 80.11.26, (1982).
/15/ Gmelins Handbuch der Anorganischen Chemie, 8. edition vol. 17, pp. 297-300, (1952).
/16/ Krapf NE: Commercial scale removal of arsenite, arsenate, and methane arsonate from ground and surface water. Environ. Perspect. Proc. Arsenic. Symp. $\underline{1981}$ (Sub. 1983) 269-279.
/17/ Hogan JC: A new and simple process for treating heavy metal containing wastewaters. Ann. Tech. Conf. Am. Electroplat. Soc. $\underline{69}$ (1982) 1-18.
/18/ Dyck W and Lieser KW: Coprecipitation of copper, zinc, arsenic, silver, cadmium, and lead with iron hydroxide and iron phosphate. Vom Wasser $\underline{56}$ (1981) 183-189.
/19/ Shannon WT, Owers WR and Rothbaum HP: Pilot scale solids/liquid separation in hot geothermal discharge waters using dissolved air flotation. Geothermics $\underline{II}$ (1982) 43-58.
/20/ Junghanss H, Kudelka H and Dommain K: Verfahren zur Ausfällung und Abtrennung von Arsen aus kupferhaltigen Lösungen. DT 23 42 729 B1 (1975).

7. ACKNOWLEDGEMENTS

The authors want to thank Dr. B. Szillus for helpful discussions and technical assistance, Messrs. R.F. Piepho, Abwassertechnik, Wennigsen (FRG) for their help with the pilot scale equipment, Messr. F.X. Nachtmann, Riedlhütte (FRG) for the cooperation in performing the pilot scale and production scale tests. They also thank the Umweltbundesamt, Berlin (FRG) for financial support.

MERCURY REMOVAL BY ACTIVATED CARBON PROCESS

K.-D. HENNING, K. KELDENICH, K. KNOBLAUCH
BERGBAU-FORSCHUNG GMBH, D-4300 ESSEN 13, FRG

1. INTRODUCTION

In contrast to all other metals, mercury is in the liquid state at room temperature and has a relatively high vapour pressure. As mercury or mercury-contaminated raw materials are used in numerous industrial processes, mercury is often released into waste water and waste air. Mercury emissions mainly originate from the following processes /1/:

- installations for fossil fuel and waste combustion

- alkalichloride-electrolyses according to the mercury-cell process

- battery and catalyst factories

- production of mercury-containing chemicals and fungicides

- production of electric switches, measuring instruments and fluorescent lamps

The inhalation of mercury vapour and of mercury compounds is very dangerous as the inhaled mercury is accumulated and leads to a severe health risk. From waste water mercury compounds are removed by means of precipitation in form of insoluble compounds or by the use of special ion exchangers. For the removal of mercury from product and waste gases "wet processes" (oxidizing scrubbings) or "dry processes" (adsorption processes) are used successfully. For many applications adsorption processes which may be operated with different adsorbents have clear economical and technical advantages. The development of carbon-containing adsorbents for the removal of mercury from waste water and waste air is reported.

2. ADSORBENTS

Activated carbon is a suitable adsorbent for the adsorptive removal of numerous organic and inorganic compounds from the liquid and gaseous phases. In order to operate an economic adsorption process for the removal of mercury from the gas phase, the purification efficiency and the adsorption capacity of the activated carbon have to be increased significantly by means of suitable impregnation. Due to the substances applied at the internal activated carbon surface the separation mechanism changes, too. Mercury is no longer removed by adsorption but by chemisorption.

3. MERCURY REMOVAL FROM THE GAS PHASE

3.1 Test Method

The purification efficiencies of the adsorbents were investigated under dynamic conditions. To this end a model gas (2.2 mg Hg/m^3) flowed through an adsorber filled up to a level of 0.2 m with the adsorbent to be examined. The purification efficiencies were monitored using an atomic absorption spectrometer. The test conditions can be taken from TABLE 1.

TABLE 1: Test conditions

Parameter		Value
Hg content in the raw gas	(mg/m^3)	2.2 ± 0.5
temperature	(K)	298
bed depth	(m)	0.2
flow rate	(m/s)	0.3
contact time	(s)	0.66
analyses method		atomic absorption spectrometry

In order to test the effects of various possible impregnating agents an initial activated carbon (D47/4 1)) was impregnated in various ways (TABLE 2).

TABLE 2: Adsorbents for mercury-removal

Adsorbent 1)	Impregnation
D47/4	none
D47/4, KI	2 % potassium iodide
D47/4, H_2SO_4	8 % sulphuric acid
D47/4, H_2SO_4 + KI	8 % sulphuric acid + 2 % potassium iodide
D47/4, S	11 % sulphur

The mercury elimination rates obtained can be taken from FIGURE 1 as a function of time. Due to the good service life of most of the tested adsorbents the test time was plotted in logarithmic scale.

3.2 Testing of Various Impregnating Agents

The non-impregnated activated carbon D47/4 is basically suited for the adsorptive removal of mercury from waste air streams. This also corresponds to the results obtained by various authors /2/,/3/ for other non-impregnated activated carbons. As in the case of other non-impregnated activated carbons, only a slight dynamic adsorption capacity insufficient for a technical gas scrubbing process could be determinated. The test was terminated after an activated carbon service life of 130 h at a Hg breakthrough of 50 %.

Early work showed (Stock 1934) that an impregnation of the activated carbon with iodine leads to a clear improvement of

---

1) Bergwerksverband GmbH, D-4300 Essen 13, FRG

the mercury adsorption capacity /4/. The first application of iodine-impregnated activated carbon was the use as filter-materials in breathing apparatus in order to provide for a riskless stay in rooms with mercury vapour /5/. Matsuma reports an increase of the Hg adsorption capacity form 1.4 % by weight to 4.8 % by weight if the iodine content is increased from 5 to 20 % by weight. FIGURE 1 clearly shows that purification efficiency and service life of the activated carbon D47/4 is clearly increased by an impregnation with 2 % by weight potassium iodide. The mercury presumably reacts under the catalytic effect of the activated carbon to mercury iodide. The adsorption mechanism at the iodized activated carbon surface has, however, not been clarified /3/.

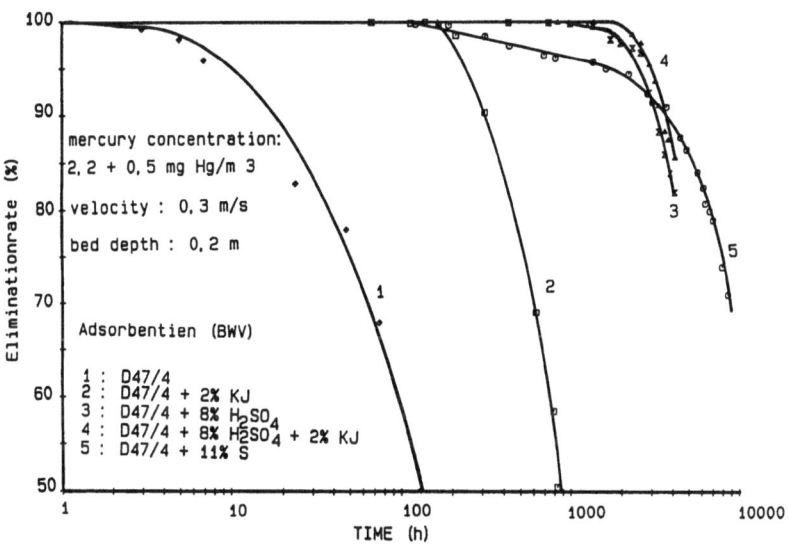

FIGURE 1: Mercury elimination rates various adsorbents

If the activated carbon is impregnated with sulphuric acid /7/ the mercury elimination rate and above all the adsorption capacity is considerably improved (FIG.1). A further improvement of this product can be attained if the activated carbon also contains 0.1 to 5 % by weight of iodide iones in addition to the sulphuric acid (5 to 50 % by weight) /8/. Although the purification efficiencies and adsorption capacities of these products are technically satisfactory acid-impregnated activated carbons are not acceptable for many applications due to corrosion problems.

The suitability of activated carbons impregnated with sulphur was described by Sinha and Walter /9/ in 1972. An impregnation with sulphur yields a product without corrosion problems.

The mercury vapour diffusing into the pore system of the activated carbon reacts to mercury sulphide under the catalytic effect of the activated carbon with the sulphur distributed on the internal surface

$$Hg + S \xrightarrow{\text{activated carbon}} HgS$$

The example of the activated carbon D47/4 impregnated with 11 % sulphur illustrates that Hg elimination rates of 94 % are still measured after a test time of 3000 h.

Chemisorption can take place only as long as the accessible surface of the sulphur distributed on the internal activated carbon surface is covered with a monolayer of mercury. In theory 79 % mercury could be adsorbed stoichiometrically if 11 % sulphur were previously adsorbed. If the sulphur dispersity is assumed with 0.3 to 0.5, mercury loads of 20-35 % by weight can be expected in practical operation.

3. MERCURY REMOVAL FROM WASTE WATER

With the objective to develop an adsorptive process for mercury removal, model examinations are carried out to test the adsorption of mercury onto various activated carbons at pH-values of 4 and 6 /10/.

If the established equilibrium loads at the various Hg residual concentrations are plotted in a double logarithmic coordinate system straight lines will result (FIG.2). Thus, a mathematical evaluation according to the Freundlich equation is possible:

$Q = K \cdot C^n$ Q = amount adsorbed (mg Hg/g AC)
K = constant

C = residual concentration (mg Hg/l)
n = constant

The constants K and n can be taken from TABLE 3.

TABLE 3: Constants of the Freundlich equation

adsorbent	activated carbon type	pH = 4 K	n	pH = 6 K	n
D 55/2	narrow pores	34.5	0.30	9.1	1.45
D 45/2	average pores	23.7	0.34	1.46	1.2
A 35/2	wide pores	17.35	0.34	evaluation impossible	

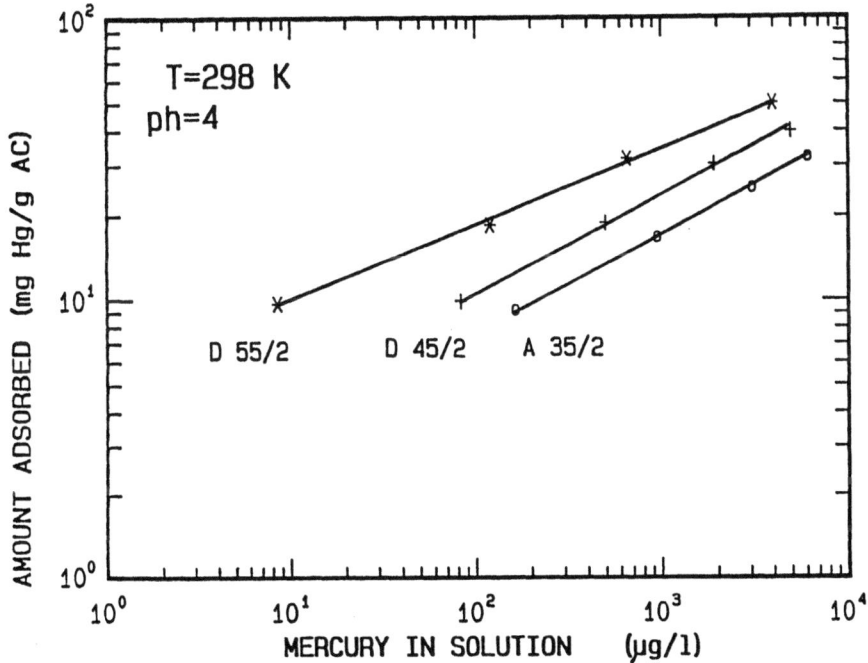

FIGURE 2: Adsorption isotherms of mercury(II)-chloride at various activated carbons

The pH-value has a marked effect on the adsorption equilibria. In accordance with the results of other groups a pH-value of 4 revealed more favourable adsorption equilibria than a pH-value of 6 /10/. In the acidic pH-range there is a relationship between the adsorption equilibria and the pore distributions of the activated carbons. The narrow-pore activated carbon D 55/2 (BWV) has the most favourable adsorption equilibrium (FIG.2). At pH-values above 8 only unsatisfactory adsorption equilibria are obtained. It was found that in this case the use of an activated carbon impregnated with sulphur also leads to more favourable equilibria. The mercury take-up is presumably no longer effected by adsorption, but by chemisorption.

Percolation tests showed that mercury(II)-chloride can be removed from solutions at pH 4 with good purification efficiencies. In FIG.3 the breakthrough curves after 0.6 and 1.2 m bed depth are plotted in relation to the test time.

It can be seen that at pH 4 the activated carbon D45/2 still separates almost 99 % mercury after a service life of 250 h. In the following 250 h the elimination rate decreases to 35 %.

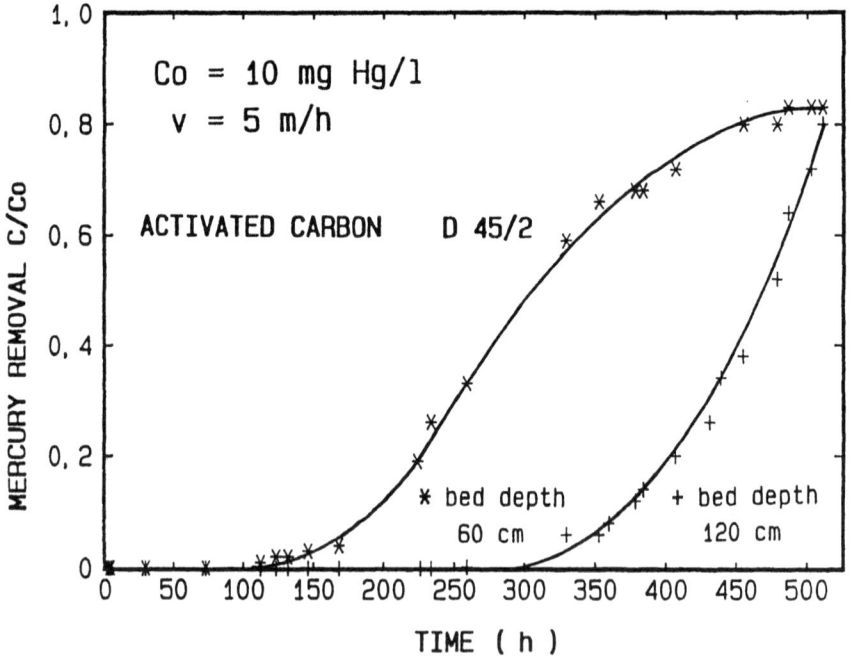

FIGURE 3: Breakthrough curves of mercury(II)-chloride

4. DISCHARGE OF THE ACTIVATED CARBONS LOADED WITH MERCURY

When using activated carbons for the removal of mercury from gases and liquids, activated carbons loaded with mercury are obtained. Thermogravimetric examinations show that mercury(II)-chloride can be desorbed in the temperature range between 300 and 430 °C (FIG.4).

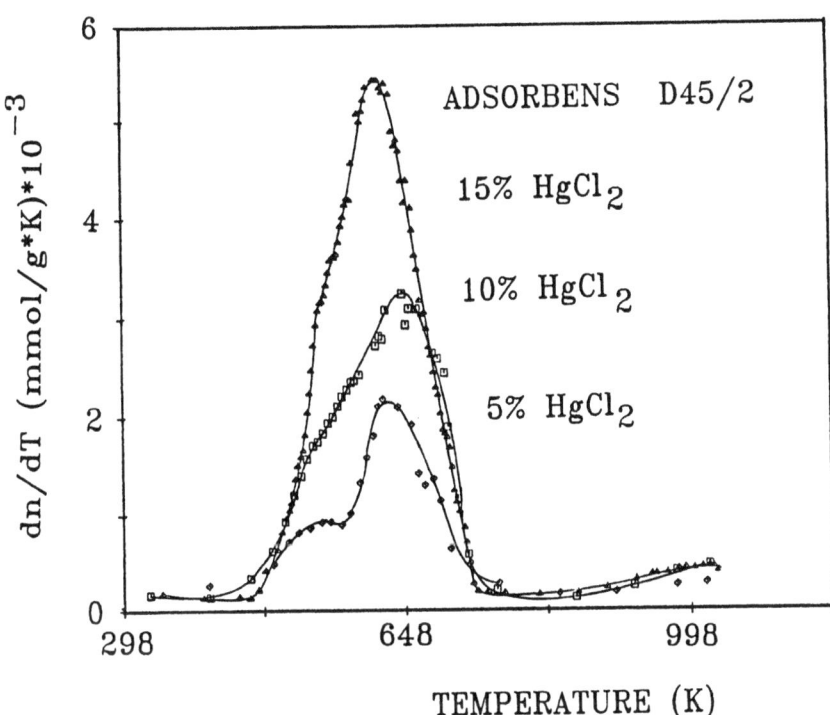

FIGURE 4: Desorption of activated carbons loaded with mercury(II)-chloride

The discharge of activated carbons loaded with mercury is effected by various producers of mercury chemicals recovering the mercury by means of a roasting process. In case of high mercury loads a credit voucher can be obtained.

5. SUMMARY

Due to environmental considerations mercury has to be removed from waste waters and waste gases of various branches of industry. For the removal of mercury, activated carbons often impregnated with potassium iodide, sulphuric acid or sulphur are used. Above all activated carbons impregnated with sulphur have proven their technical value. Good elimination rates and high mercury adsorption capacities are obtained. The loaded activated carbons can be discharged, by mercury-recovery.

6. REFERENCES

/1/ Lohrer, W.; Pahlke, G.:
    Umwelt 5 (1977) S.397/400

/2/ Bergk, K.H.; Wolf, E. und Eckert, S.:
    Z. Chem. 17 (1977) Heft 3, S.85/89

/3/ Matsumara, Y.:
    Atmospheric Environment Vol.8 (1974) p.1321/1327

/4/ Stock, A.:
    Angewandte Chemie, 47. Jahrg. (1934) Nr. 4, S.64

/5/ Pütter, K.E.; Hirsch, M.:
    Angewandte Chemie, 47. Jahrg. (1934) Nr.12, S.184/185

/6/ Revoir, W.H.; Jones, J.A.:
    United States Patent 3,662,523

/7/ Showa Denko, K.K.:
    Offenlegungsschrift 2 358 767

/8/ Krill, H.; Wirth, H.; Rittinger, G.; Hohmann, V.:
    Offenlegungsschrift 2 603 807

/9/ Sinha, R.K.; Walker, P.L.:
    Carbon (1972) Vol.10, p.754/756

/10/ Keldenich, K.:
     Diplomarbeit Universität Essen, 1985

FATE OF SOME TRACE ELEMENTS IN COMBUSTION AND GASIFICATION PROCESSES

Wahab Mojtahedi[1] & Kari Larjava[2]
[1] Laboratory of Fuel Processing and Lubrication Technology
[2] Laboratory of Heating and Ventilating
   Technical Research Centre of Finland
   SF-02150 Espoo, Finland

ABSTRACT

Municipal solid wastes contain many trace elements which are released upon combustion and gasification and have toxic forms at certain concentrations. These pollutants must be removed, if increasingly stricter emission standards are to be met. Four metals, caldmium (Cd), mercury (Hg), lead (Pb), and zinc (Zn), were selected, and thermodynamic equilibrium calculations were carried out to determine which elements volatilise upon combustion and gasification and which condense upon cooling the product gas.

Under combustion conditions mercury, cadmium, and lead seem to be totally volatilised in the temperature range of 700 - 1 000 ^{0}C, but zinc only partially. Sulphates of Cd, Pb and Zn are formed and also solid zinc oxide upon cooling the combustion gases, but mercury does not condense even at 200 ^{0}C.

In a reducing atmosphere, all four metals are volatilised in the temperature range of 700 - 1 000 ^{0}C. Solid sulphides form as the product gas is cooled but also solid chlorides of Cd, Pb, and Zn, at the temperature range of 200 - 300 ^{0}C. Again, mercury is in the vapour phase at 200 ^{0}C, the lowest temperature considered.

PREDICTION OF TRACE ELEMENT DISTRIBUTION BY THERMODYNAMIC EQUILIBRIUM CALCULATIONS

A theoretical distribution of the four heavy metal species in combustion-/gasification products was calculated using the equilibrium calculation program SOLGASMIX. The program, developed by G. Eriksson /1/, has been used in many investigations where a large number of species in various phases are involved. The program uses the method of the minimisation of the total Gibbs free energy of the system. This technique is specially well suited to calculation of the distribution of trace species. The trace elements are present in such small amounts that they do not contribute to the total free energy of the system. Thus, their distribution can be calculated independently from each other and from the major element distributions.

Altogether 100 chemical species were considered, 67 gaseous and 33 condensed species. The thermodynamic data for the species were taken from references /2, 3/.

Composition of the fuel (Table I) was supplied by the Technical Research Centre of Finland, Laboratory of Heating and Ventilating. Most of the projections reported in this paper are applicable to systems defining alternative feedstock compositions.

Table I. Composition of the fuel. Basis: 1 000 moles carbon.

Element	C	H	S	O	Cl	F	N	Cd	Hg	Pb	Zn
Concentration (moles)	1 000	1 600	1	498	5	1	53	10^{-3}	5×10^{-4}	0.02	0.5

RESULTS AND DISCUSSION

Two sets of graphs are plotted: one for the non-metallic compounds (Figure 1) and the other for the gaseous metallic species (Figures 3 and 5). Figures 2 and 4 describe the distribution of the four metals between the different phases present. These graphs must be considered in conjunction with the corresponding graphs for the volatile phase (i.e. Figures 3 and 5). The Y-axis in these graphs (2 and 4) always range from 0 to 1.0. The maximum of 1.0 corresponds to the total amount of the metal concerned in the system (e.g. 0.5 moles for Zn). An empty frame in these graphs indicates that all the metal is in the gaseous state.

ATMOSPHERIC FLUID-BED COMBUSTION

Figures 1 - 3 show the results of the calculations carried out with an air ratio of 1.5 and a total pressure of 1 bar. The air ratio of 1.5 indicates an air input 50 % in excess of the stoichiometric requirement. Figure 1 shows the variation of the partial pressure of the non-metallic combustion products, in the gas phase, with temperature. The major element species deviate slightly from the measured flue gas composition of fluid-bed combustors. Almost all the fluorine (> 99 %) is bound in hydrofluoric acid [HF (g)] and 85 - 99 % of chlorine exits the combustor as hydrochloric acid [HCl (g)]. The only exception is the relatively large percentage of SO_3 in the flue gas, especially towards the lower end of the temperature scale. This is much higher than would be encountered in reality, but it is to be expected in thermodynamic equilibrium calculations. The time element is unlimited in these considerations which in turn will favour more SO_3 formation (by oxidation of SO_2 to SO_3).

The reaction of mercury with fluid-bed combustion gases produces volatile species of mercury: mercury chlorides $HgCl_2$ (g) and HgCl (g), mercury fluorides HgF_2 (g) and HgF (g), mercury monoxide HgO (g), mercury monohydride HgH (g), and monoatomic mercury Hg (g). Whereas mercury dichloride is the favoured trace chlorine compound at the temperature range of 200 - 600 °C, mercury fluorides do not form in significant amounts in this temperature range, Figure 3. (HgF (g) concentration never exceeds 10^{-13} moles). At temperatures above 600 °C Hg (g) concentration and to a lesser extent HgO (g) dominate. The concentration of the latter decreases with increasing temperature above 800 °C, that of former attains a maximum of about 4.9×10^{-4} moles whence it remains constant.

Volatilisation of cadmium would presumably show a similar pattern to that of mercury but due to lack of reliable thermodynamic data the chlorides or fluorides of cadmium were not included in the list of the gaseous trace species. Cadmium is volatilised at a higher temperature than mercury (~ 700 ^0C) and exits the combustor in elemental form, Figure 3. Cadmium sulphate ($CdSO_4$-s) begins to form in the condensed phase below 800 ^0C, the only compound of cadmium in the solid phase, Figure 2.

The volatile compounds of lead formed, under the operating conditions considered, include lead tetrachloride $PbCl_4$ (g), lead dichloride $PbCl_2$ (g), lead difluoride PbF_2 (g), lead oxide PbO (g), lead monohydride PbH (g), and monoatomic lead Pb (g). Over the temperature range of 600 - 1 000 ^0C lead dichloride is the favoured volatile species showing a maximum concentration of 2×10^{-2} moles at about 700 ^0C. At higher temperatures (above 900 ^0C) lead monoxide PbO (g) is the dominant volatile species, Figure 3. Elemental lead concentration becomes significant only at about 1 200 ^0C, but its concentration is much less than that of PbO (g).

The reaction of fluorine with the elements considered does not seem to be thermodynamically favourable. None of the fluorides of the four metals is present in any significant quantity. No zinc fluoride is formed. Zinc behaves in a more or less similar pattern to lead but for the fact that no gaseous zinc oxide is formed. Zinc dichloride $ZnCl_2$ (g) is the dominant volatile species of the metal up to about 1 200 ^0C, Figure 3. At temperatures of interest, relatively small amounts of Zn (g) exists in the gas phase (5 % of Zn (g) compared to 50 % $ZnCl_2$ (g)).

A comparison of the condensed phases of the metals (Figure 2) shows that all three metals (Cd, Pb, and Zn) form sulphates below 700 ^0C. Mercury is an exception in that no compound of mercury is condensed even at 200 ^0C. ZnO-s is the thermodynamically stable compound of zinc at 700 - 1 200 ^0C. Almost 50 % of the metal in the fuel is utilised in ZnO-s at about 900 ^0C (Figure 2). Zinc oxide seems to be thermodynamically unstable below 600 ^0C and is replaced by zinc sulphate in the condensed phase. Sulphates are the only stable compounds found at low temperatures. One would conclude that zinc is only partially volatilised at 700 - 1 000 ^0C temperature range.

ATMOSPHERIC FLUID-BED GASIFICATION

Equilibrium calculations carried out under atmospheric fluid-bed gasification conditions show the metals Cd and Pb to be totally volatilised above 500 ^0C in a reducing atmosphere ($\lambda = 0.5$), Figure 4, zinc at a higher temperature of 800 ^0C. Mercury, on the other hand, is completely in the vapour phase even at 200 ^0C. Partial pressures of gaseous trace compounds are plotted against temperature in Figure 5. Both mercury and cadmium are predominantly in elemental vapour form under atmospheric fluid-bed gasification conditions.

Lead and zinc behave somewhat differently than either of those two metals in that gaseous chlorides of both lead and zinc are thermodynamically stable at AFB gasifier temperatures. Dichlorides $PbCl_2$ (g) and $ZnCl_2$ (g) are the dominant trace species of chlorine. ($PbCl_4$ is never present in significant amounts). $PbCl_2$ (g) forms at relatively lower temperatures (~ 500 ^0C), $ZnCl_2$ (g) at 700 - 800 ^0C. Zinc is mainly in Zn (g) vapour above 800 °C. Whereas significant quantities of PbS (g) (4.5 x 10^{-3} moles) is revealed in the temperature range of 700 - 900 °C in the volatile phase, the concentration of ZnS (g) never reaches 10^{-5} moles. However, monoatomic lead Pb (g) is the dominant gaseous trace species of lead above 800 °C, Figure 5.

Condensed phases of the metals (Figure 4) show sulphides formed as the gasification product gas is cooled. Again, the only exception is Hg which does not condense even at 200 °C. Figure 4 shows that, at 200 - 300 °C, $CdCl_2$-S and $PbCl_2$-S have formed indicating that the sulphides of these two metals are not stable at the lowest end of the temperature scale (below 300 °C). Some calculations were performed under conditions of constant temperature (850 °C) but varying air ratio, λ. Figure 6 shows the results of one of these runs. In this case air ratio variation over a range of 0.3 - 2.0 was considered. The behaviour of Hg and Cd is in contrast to that of Zn and Pb. The former are present in the elemental form in the volatile phase over the entire range of λ considered. The latter two are predominantly in the gaseous chloride form in oxidising atmosphere. Under reducing conditions, Zn (g) and Pb (g) are the dominant trace species of the two metals.

CONCLUSIONS

Thermodynamic equilibrium calculations carried out under atmospheric and pressurised fluid-bed combustion and gasification conditions reveal complete volatilisation of all three heavy metals Cd, Hg, and Pb. Zinc is only partially volatilised under AFB and PFB combustion conditions. In reducing atmosphere (i.e. under gasification conditions) zinc is almost totally volatilised.

Thermodynamic calculations predict the mercury in the fuel to be completely volatilised and exit the incinerator in $HgCl_2$ (g) form, if the temperature is below ~ 600 °C, but in Hg (g), if the flue gas leaves at about 600 °C. Vogg et al /4/ excluded metallic mercury in the crude gas and included only $HgCl_2$ and to a lesser extent HgCl to be present. However, the calculations show mercury to exit a fluid-bed gasifier in Hg (g) form at practically any temperature above 200 °C. This pattern prevails in both atmospheric and pressurised gasification conditions.

One would expect a similar trend with cadmium but the exclusion of cadmium chlorides in the gaseous trace species considered, prevents a similar conclusion to be drawn from our own thermodynamic projections. Others /4/ have concluded that the fuel-bound cadmium volatilises as cadmium chloride in incineration.

LITERATURE CITED

1. Eriksson, G. Thermodynamic studies of high temperature equilibria. Chemica Scripta (1975)8, 100 - 103.
2. Barin, I. & Knacke, O. Thermochemical properties of inorganic substances. Berlin 1973, Springer-Verlag.
3. JANAF thermochemical tables with supplements. Washington, D.C., 1971-1978, National Bureau of Standards of the U.S. Dept of Commerce.
4. Vogg, H. & al. The specific role of cadmium and mercury in municipal solid waste incineration. Waste Management & Research (1986)4, 65 - 74.

FATE OF SOME TRACE ELEMENTS IN COMBUSTION AND GASIFICATION PROCESSES

Wahab Mojtahedi[1] & Kari Larjava[2]
[1] Laboratory of Fuel Processing and Lubrication Technology
[2] Laboratory of Heating and Ventilating
   Technical Research Centre of Finland
   SF-02150 Espoo, Finland

FIGURE TEXTS

Figure 1. Variation of the partial pressure of "non-metallic" compounds with temperature. Low-s coal, λ = 1.3, P = 1 bar.

Figure 2. Variation of the partial pressure of gaseous trace species with temperature. Low-S coal, λ = 1.3, P = 1 bar.

Figure 3. Distribution of the heavy metal species in the gas and condensed phases. Low-S coal, λ = 1.3, P = 1 bar.

Figure 4. Distribution of the heavy metal species in the gas and condensed phases. Low-S coal, λ = 0.5, P = 1 bar.

Figure 5. Variation of the partial pressure of gaseous trace species with temperature. Low-S coal, λ = 0.5, P = 1 bar.

Figure 6. Variation of the partial pressure of gaseous trace species with the air ratio λ. Low-S coal, T = 850 ^{0}C, P = 1 bar.

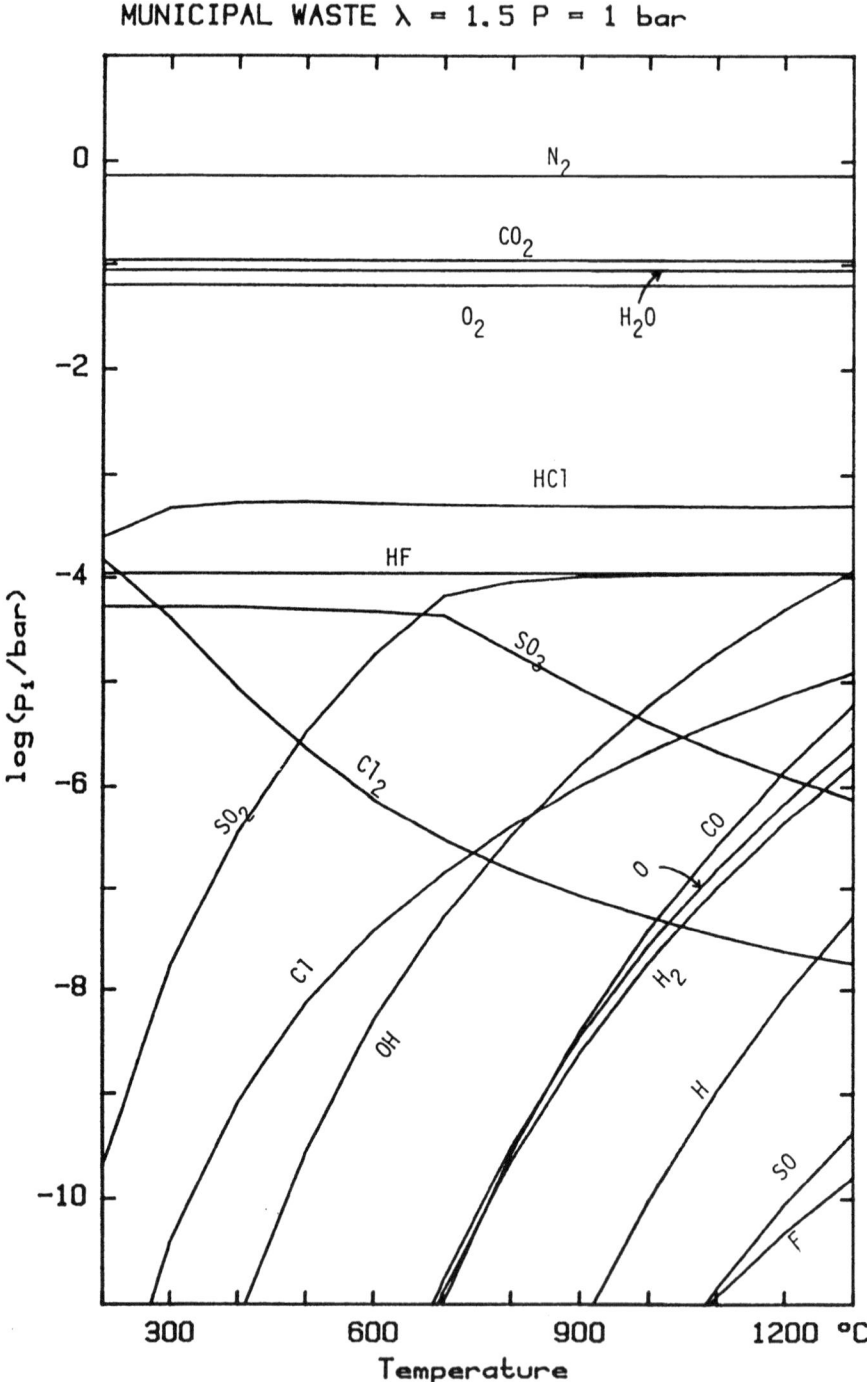

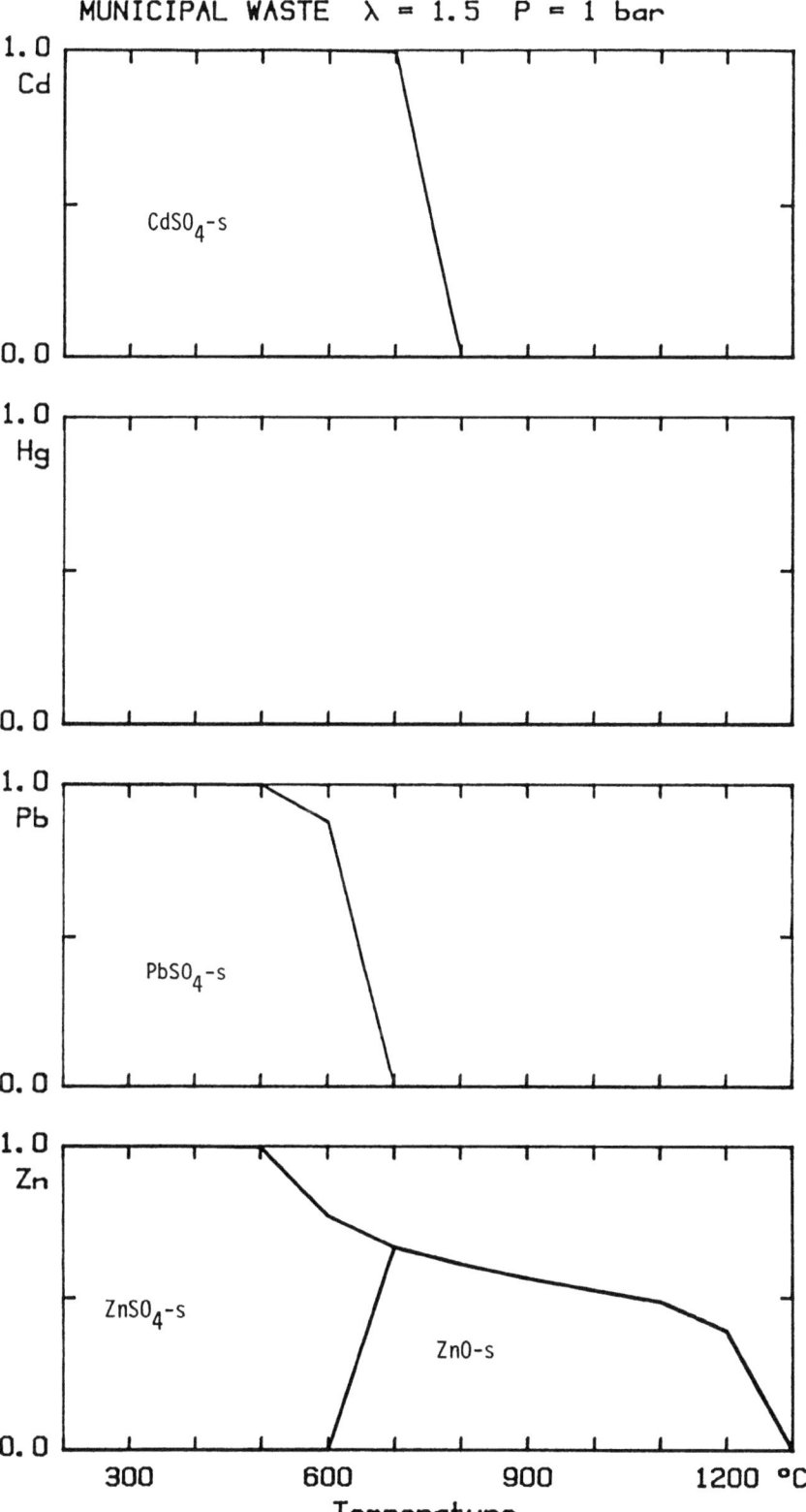

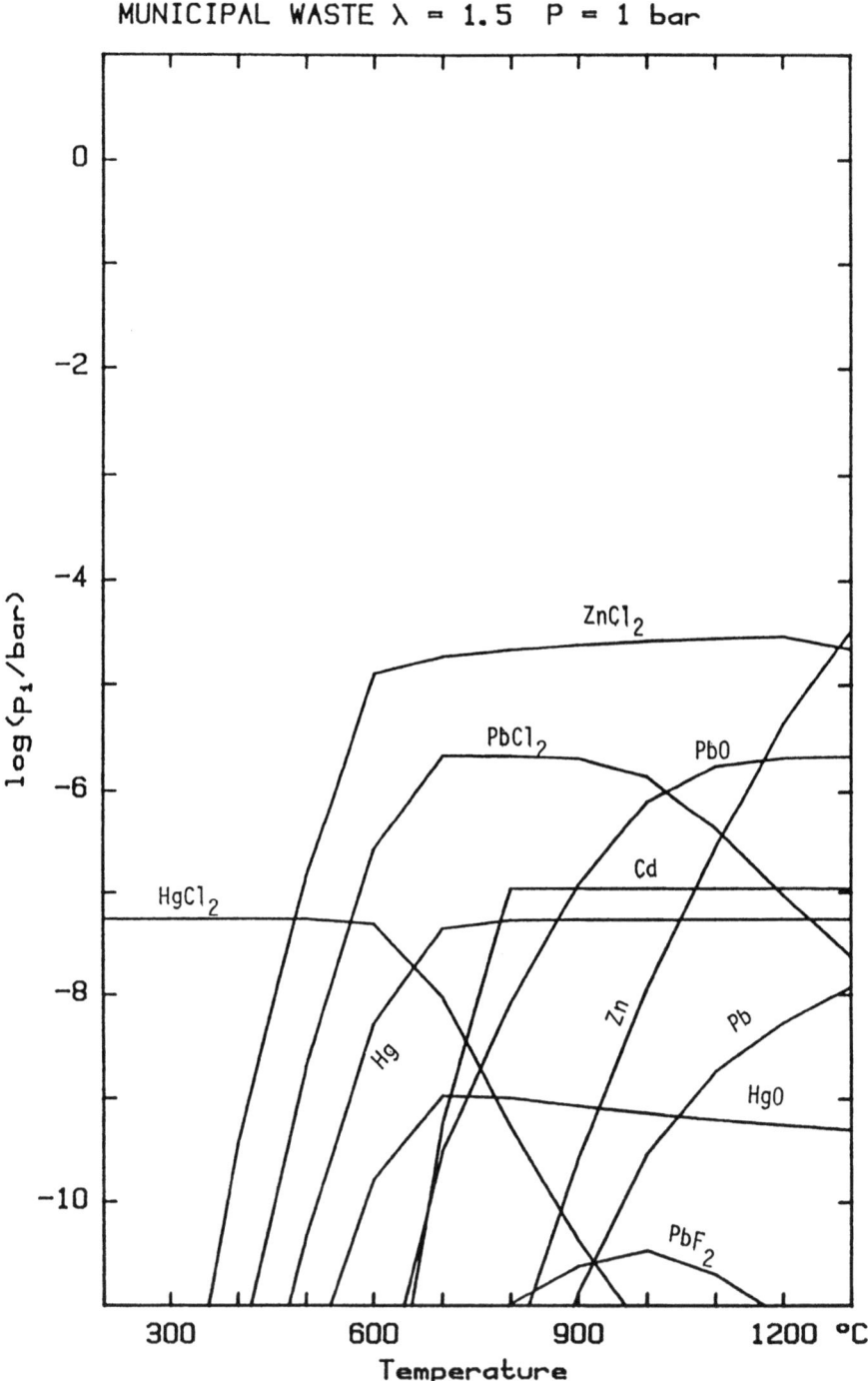

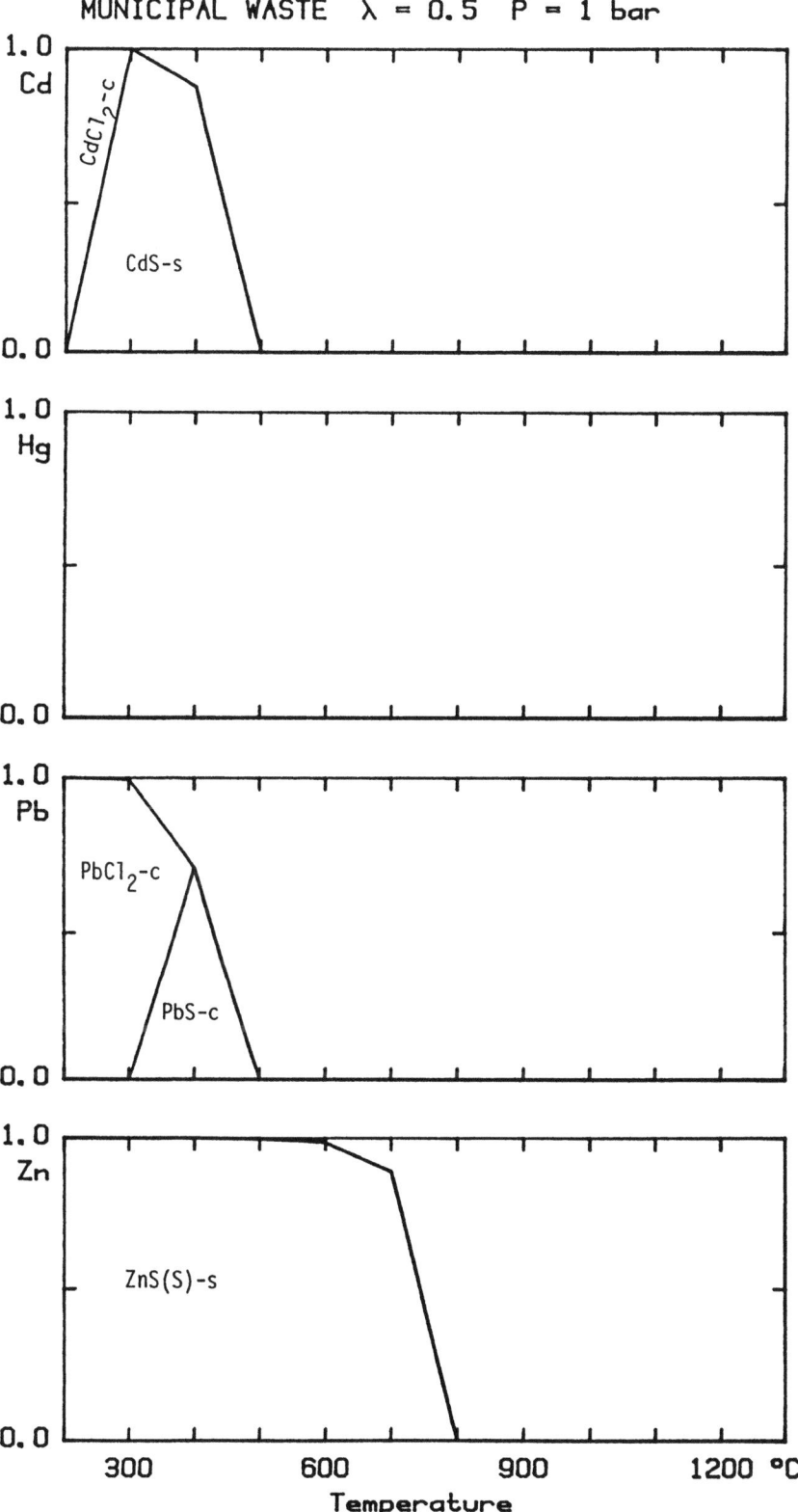

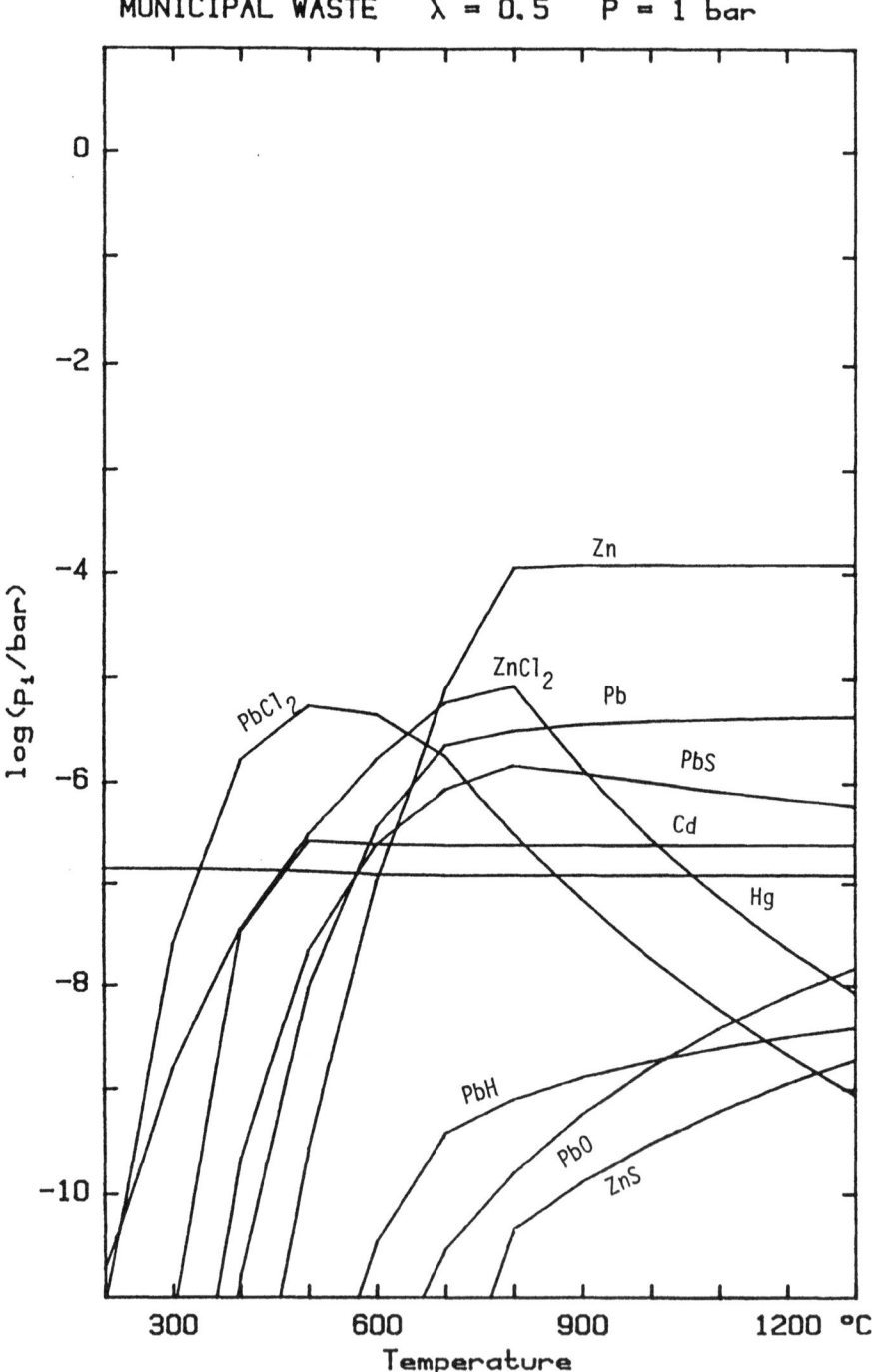

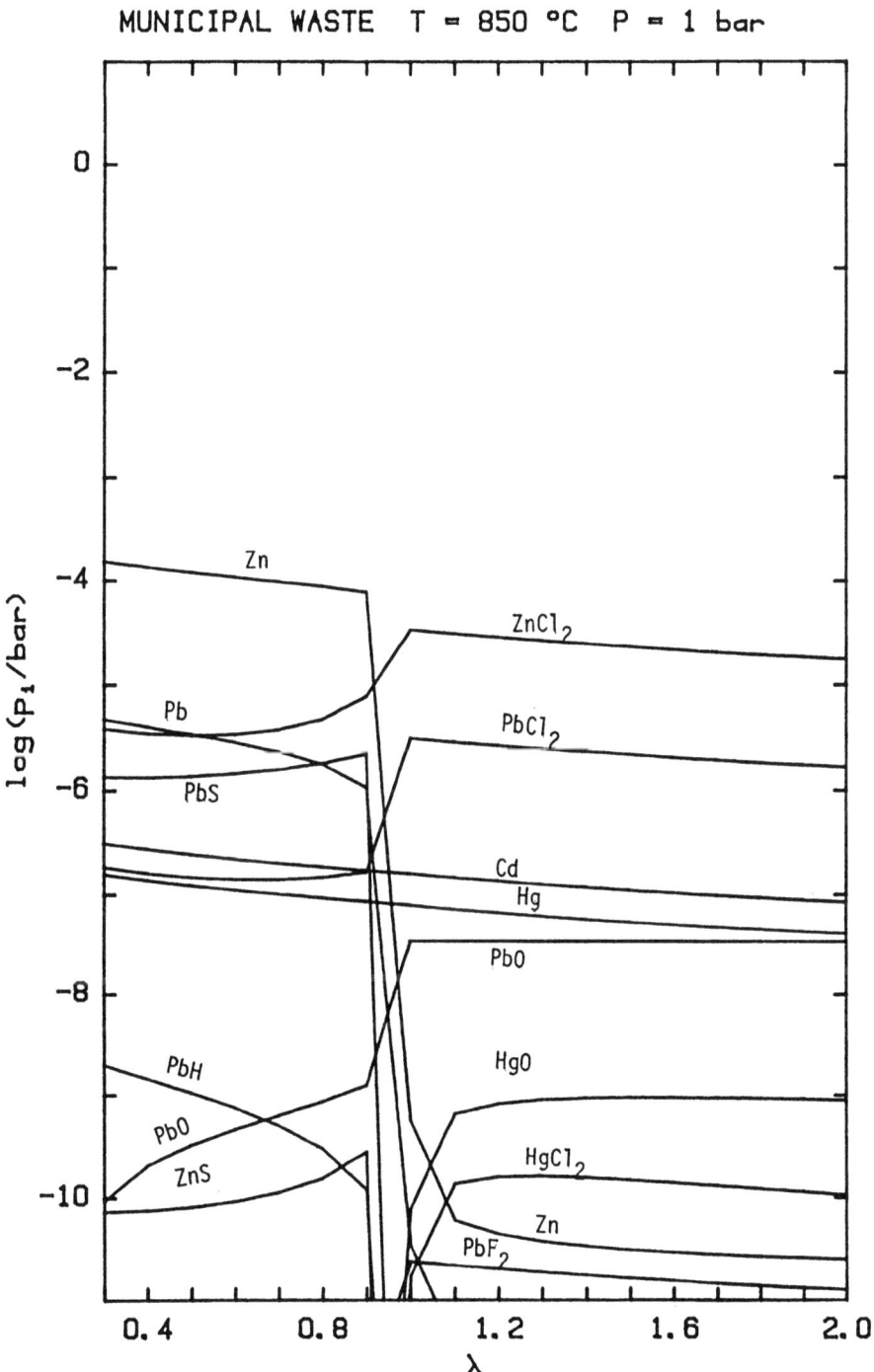

REDUCTION OF THE CADMIUM CONTENT OF WASTE GYPSUM PRODUCED BY THE DSM-PHOSPHORIC ACID PLANT AT PERNIS

R. SPIJKER

DSM RESEARCH - IJMUIDEN LABORATORY

1. INTRODUCTION

Phosphoric acid production at the Pernis site of the DSM-Fertilizer Division is based on the well-known Nissan-H technology (1). Ground phosphate rock is digested with a 75% sulphuric acid solution at 90-100°C and the initially formed calciumsulphate-hemihydrate precipitate is re-crystallized to gypsum by stepwise cooling to a temperature level of 50-60°C. After separation of the gypsum from the phosphoric acid with a rotary tilting pan filter the cake is reslurried in water and discharged into the River Rhine.
From an environmental point of view this is an undesirable situation since about 15-20% of the cadmium input into the system ends up in the solid phase via incorporation in the gypsum crystal lattice.
This contribution deals with the reduction of the amount of cadmium incorporated in the waste gypsum stream by shifting the distribution equilibrium in the direction of the acid phase.

2. EXPERIMENTAL

Preliminary laboratory experiments have shown, that the cadmium uptake of the gypsum phase is strongly dependent on the actual value of the cadmium-calcium ion activity ratio of the acid phase $((a_{Cd}/a_{Ca})_{liquid})$ during the recrystallization stage:

$$\left(\frac{a_{Cd}}{a_{Ca}}\right)_{solid} = 5,8 \times 10^{-4} \left(\frac{a_{Cd}}{a_{Ca}}\right)_{liquid}^{0,83}$$

The cadmium/calcium ion activity ratio in the liquid phase depends, among others, on the amount of excess sulfuric acid present and the temperature to avoid serious disturbances with respect to production rate and process yield. Only slight variations in these parameters are tolerable in the Nissan process.

From published literature it is known, that cadmium ions readily form stable complexes with halides in neutral and acid solutions (2). Therefore, cadmium incorporation in gypsum might be diminished by addition of a sufficient quantity of halide ions prior to the re-crystallization stage of the process.
DSM-research has finally opted for chloride rather than the other halides, mainly because:
- it is a low-priced commodity,
- it is insensitive to chemical oxidation and
- it does not attack rubber-lined process-equipment.

However, there are also some disadvantages connected to the use of chloride:
- an appreciable lower stability of the chloride versus the other halides requires a larger molar concentration in the acid medium (2)
- the increased risk of pitting corrosion of the stainless-steel parts of the plant
- a considerable evoluation of hydrochloric acid during the phosphoric acid vacuum evaporation process and consequently a contamination of the valuable by-product fluosilicic acid with chloride.

Small-scale experiments were performed to quantify the cadmium distribution coëfficiënt $K_d = a_{Cd,solid}/a_{Cd,liquid}$ in pure synthetic phosphoric acid solutions and filter acid form the Pernis-factory with different additions of hydrochloric acid (30% P_2O_5 and 2-4% H_2SO_4). The results are summarized in table 1.

Table 1

HCl-concent of liquid %	Cd-distribution coëfficiënt K_d	Reduction of Cd-incorporation %
synthetic phosphoric acid		
0,0	0,66	-
0,02^5	0,61	8
0,05	0,40	39
0,10	0,32	52
0,25	0,20	70
0,50	0,07	84
1,00	0,02	97
production acid from Pernis-factory		
0,04	0,17	0
0,14	0,12	26
0,29	0,08	52
0,55	0,06	66
1,07	0,04	77

These experimental results indicate a significant lowering of the cadmium content of gypsum in the presence of 0.5% hydrochloric acid.

A recent trial run in the phosphoric acid pilot installation of DSM-research which lasted for one week has confirmed these favorable results. A reduction of 1.8 ppm Cd to 0.7 ppm Cd in dry gypsum was achieved using phosphate rock with 24 ppm Cd under addition of 0.5% Cl (note that the detection limit of the analytical method is 0.5 ppm).
Measurement of the corrosion rate of a number of testprobes in the acid medium during the run has revealed, that both Hastelloy C22 and Sanicro 28 are suitable materials to substitute for AISI-316 in critical areas of the plant.
The problem of the fluosilicic acid contamination with chloride has not been completely resolved at the moment, but further work on this item is planned in the near future.

REFERENCES
1. Fertilizer Science and Technology Series "Phosphates and Phosphoric Acid - Raw Materials, Technology and Economics of the Wet Process", P. Becker, Marcel Dekker Inc (1983), p. 40.
2. Topics in Environmental Health "The Chemistry, Biochemistry and Biology of Cadmium", M. Webb ed., Elsevier/North-Holland Biomedical Press (1979), p. 15.

INVESTIGATIONS OF THE ADSORPTIVE REMOVAL AND RECOVERY OF
HALOCARBONS

K.-D. HENNING, M.SCHÄFER, W.BONGARTZ, K.KNOBLAUCH
BERGBAU-FORSCHUNG, GMBH, D 43 ESSEN 13, FRG

Large amounts of halocarbons (HC) are used in numerous sectors of industry as solving, extraction, purification and degreasing agents. In the Federal Republic of Germany the production of the four most important HCs (dichloromethane, trichloroethane, trichloroethylene and tetrachloroethylene) amounted to 200 000 t in 1982. The resulting emissions polute the environment as follows:

air	60 - 70 %
waste	30 - 40 %
waste water	1 - 3 %

Many HCs are toxically cancerogenic and biologically difficult to degrade and have to be removed from waste gases respectively waste waters due to environmental considerations, legal provisions and their corrosive properties. If the compounds reach the ground water through the waste water comprehensive and costly preparation steps for the recovery of drinking water are required.

Work is carried out in view to the development of adsorption processes for the removal and recovery of HCs from process and landfill gases as well as from waste water using activated carbons (Fig. 1).

Activated Carbon

Activated carbon

Activated carbon is the trade name for highly porous products made from carbon-containing raw materials with a large internal surface of 400 to 1600 m^2/g and a large pore volume of more than 30 cm^3/100 g. Due to the mainly hydrophobic surface properties activated carbon preferably adsorb organic substances and other non-polar compounds from the gaseous and liquid phase.

Removal of halocarbons from gases

The example of the landfill gas cleaning can be used to show that activated carbon is a suitable adsorbent for the removal of HCs from gases. In many cases the utilization of the methane content amounting to 50 - 70 % by volume is severely obstructed by numerous HCs contained in the gas. Fig. 2 shows the adsorption isotherms of various halogenated hydrocarbons contained in the landfill gas obtained with the activated carbon D43/4 (BWV). These adsorption isotherms show that the equilibrium loads of different HCs obtainable at identical concentration may vary by the factor 10. Examinations with real landfill gases revealed that the total chlorine content of approx. 800 mg/m^3 could be reduced to concentrations of less than 25 mg/m^3 by using an activated carbon adsorber. The loaded activated carbons may be regenerated at 120 °C by means of a steam desorption.

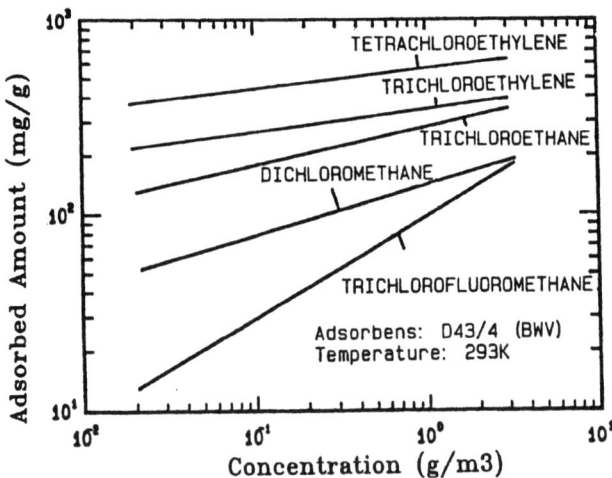

Fig.2: Adsorption isotherms of typical chlorinated hydrocarbons

Removal of halocarbons from waste water

The different adsorption properties of the HCs have to observed also during the adsorption from waste water. Table 1 lists the constants K and n of some HCs in the Freundlich equation.

$$Q = K * c^n$$

Q= adsorbed amount (mg/g)
K= constant
c= concentration (mg/L)
n= constant

Table 1: Constants in the Freundlich equation

Adsorptive	K	n
chlorobenzene	102.2	0.44
trichloroethylene	29.4	0.66
1.2-dichloroethane	5.9	0.67
chloroform	4.8	0.70
dichloromethane	0.8	1.00

Of the examined HCs chlorobenzene has the most favourable adsorption properties.
Work is carried out in view to the development of an activated carbon process for the removal of halocarbons from waste water. To this end flow trials were carried out using 1.2-dichloroethane as model substance. These experiments provide the basis data for the design of adsorbers used for concrete waste water purification. Fig. 3 shows the breakthrough curves of 1.2-dichloroethane after a bed depth of 0.60 m and 1.25m. 1.2-dichloroethane can be removed from the water up to a concentration of less than 1 mg/L.

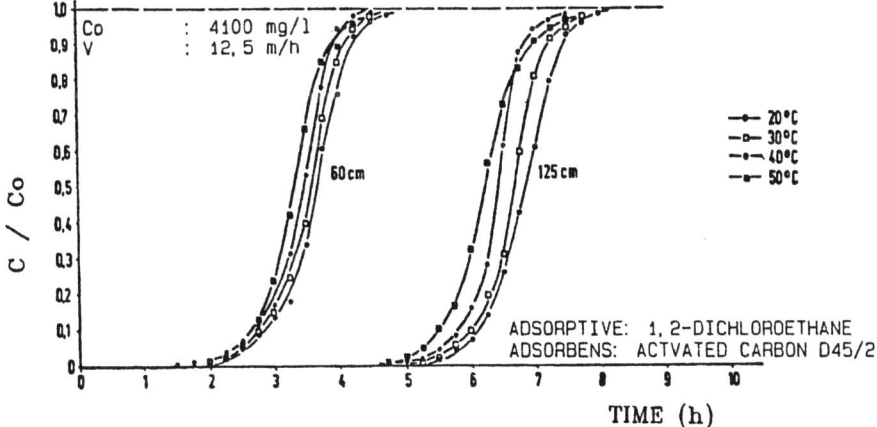

Fig. 3: Break-through curves of 1.2-dichloroethane as a function of temperature

THE USE OF Zoogloae ramigera FOR REMOVAL AND RECOVERY OF Cr^{+6} IONS FROM INDUSTRIAL WASTE WATERS

T.Kutsal and Y.Sağ, Hacettepe University, Chemical Engineering Department, 06532 Beytepe, Ankara, Turkey.

Many aquatic microorganisms can adsorb dissolved heavy metal ions from their surroundings. By making use of their metal uptake properties, microorganisms may be utilized for removal and/or recovery of either harmful or useful metals.

The organism used in this study was Zoogloae ramigera which can be used to remove cupper, cadmium, uranyl and chromium ions from water solutions. At 25°C the optimum adsorption pH of chromium(VI) ion was found 4.0. The adsorption rate increased by increasing initial metal ion concentrations from 25 mg/l to 125 mg/l.

INTRODUCTION

Using microorganisms as biosorbents for heavy metal ions offer a potential alternative to existing methods for detoxification and for recovery of toxic or valuable metals from industrial waste water. Conventional methods for removing dissolved heavy metals include chemical precipitation, chemical oxidation and reduction, ion exchange, filtration, electrochemical measurement and evaporative recovery. Most of them are not practical, economic and efficient (1,2).

The kinetics of metal uptake has been suggested to take place in two stages. The first stage, thought to be physical adsorption at the cell surface, is very rapid and occurs a short time after the microorganism comes into contact with the metal. Microorganism can adsorb metal on to its surfaces from water by many processes.

i) The metal ions could be adsorbed by complexing with negatively charged reaction sites on the cell surfaces (ionic adsorption).

ii) Some microorganisms can synthesize polymers extending from the outer membrane of the cells. These polymers can bind metal ions from the solutions.

iii) Proteins in the cell wall could offer another site for binding. Heavy metals have a strong affinity for protein.

The second stage, related to metabolic activity, is slower and called chemisorption.

MATERIAL AND METHODS

Microorganism and Growth Conditions:

Zoogloae ramigera from United States Department of Agriculture, Agricultural Research Service was used for this study. The cultivation medium had the following composition (as g/l): glucose, 20.0; K_2HPO_4, 1.0; KH_2PO_4, 0.5; NH_4Cl, 0.5; $MgSO_4 \cdot 7H_2O$, 0.5; yeast extract, 1.0; peptone, 1.0. The pH was adjusted to 7.5 with dilute NaOH. The organism was cultivated in a shaker at 25°C. Agitation was 100 rpm.

Preparation of Microorganism for Biosorption:

After the 120 hours of inoculation period, cells were centrifuged and dried at 60°C. For the biosorption studies, 0.2 g of dried biomass was suspended in distilled water to 100 ml. And then adsorption was started in the shaking flasks at 25°C and 100 rpm.

Analysis of Cr(VI) Ion:

Free Cr(VI) in biosorption media was determined spectrophotometrically. The colored complex of chromium(VI) ion with diphenlycarbazid was read at 540 nm (3).

RESULTS AND DISCUSSION

Several mining and metal industries contain undesired amount of Cr(VI) ion. For example, chromium plating industries; chromium concentration may generally reaches up to 555 mg/l. This value is very high according to the water standarts and it must be falled down to the desired value of 0.05 mg/l.

The pH of biosorption media effects the rate and yield of biosorption of metal ions on microbial cells. Generally in lower pH, adsorption rate is rapid. The optimum adsorption pH of Cr(VI) ion was found 4.0. The pH of biosorption media and corresponding maximum adsorbed metal ion concentration and amount of adsorbed metal per unit dried weight of microorganism values are shown on Figure 1.

The adsorption rate of metal ion increased by increasing initial metal ion concentrations to 125 mg/l. The quantity of adsorbed metal per unit dried weight of microorganism increased by increasing initial metal ion concentrations. These values are shown on Figure 2.

REFERENCES

1. Clark, J.W., Viessman, W., Hammer, M.J., Water Supply and Pollution Control, 383-566, Int. Textbook Company, New-York, 1971

2. Khummongkol, G.S., Canterford, C.F., Biotech. Bioeng., 24, 2643-2660, 1982

3. Snell, F.D., Snell, C.T., Colorimetric Methods of Analysis, 267-279, Van Nostrand Co., Canada, 1959

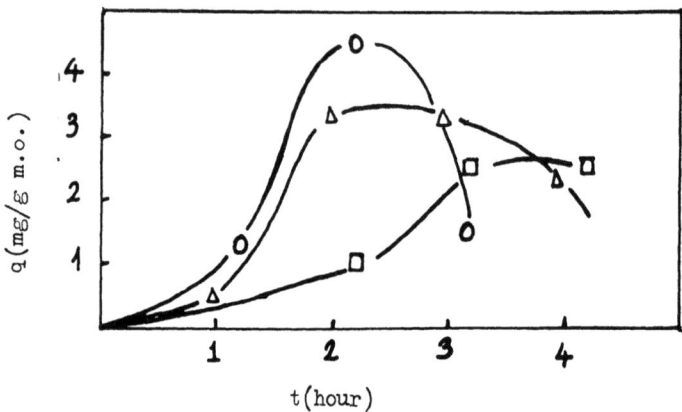

Figure 1. The effect of initial pH on the biosorption of chromium ion (m.o. conc.: 2.0g/l, T:25°C, ▫: pH:3.0; o: pH:4.0, △: pH:5.0).

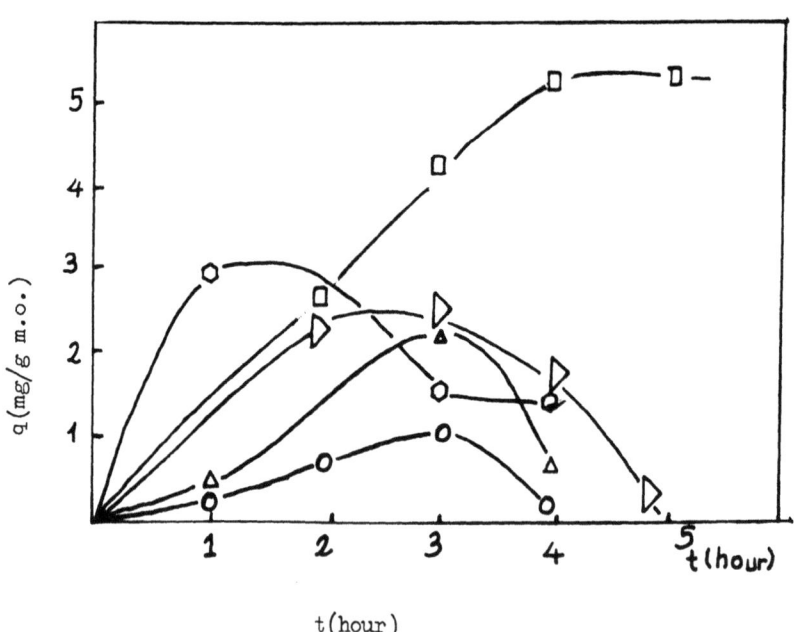

Figure 2. The effect of initial metal ion concentration on the adsorption of chromium ion (m.o. conc.: 2.0g/l, T:25°C, pH:4.0, o: 25ppm, ▷ :50ppm, △: 75ppm, ⬡: 100ppm, ▫:125 ppm).

INACTIVATION OF SEWAGE SLUDGES WITH CALCIUM OXIDE PRIOR TO
AGRICULTURAL USES

T. MARCINKOWSKI
INSTITUTE OF ENVIRONMENT PROTECTION ENGINEERING, TECHNICAL
UNIVERSITY OF WROCŁAW, POLAND

1. INTRODUCTION

Owing to the presence of nutrient substances and microelements (N, P, K), sludges bear a similarity to that of the manure, and may therefore effectively improve the soil structure when applied in appropriate doses. It should be noted, however, that sewage sludge have the inherent disadvantage of containing toxic substances and pathogens which might penetrate soil and contaminate vegetation by persisting there for weeks, months and even years (1). Sludges that are to be used in agriculture, require adequate treatment methods enabling complete (or, at least, partial) kill of pathogens. One of these is inactivation with quicklime.

The quicklime method involves heat produced in the reaction of calcium oxide with water contained in the sludge. The calcium oxide dose required to heat the sludge to the desired temperature may be calculated as follows:

$$x = \frac{0.01 \, Y \, (\text{dry wt} \, C_{\text{dry wt}} + 4.19 \, w)}{1160 / \Delta T - 0.59}$$

The notation is:
x = quicklime mass, kg;
Y = treated sludge mass, kg;
$C_{\text{dry wt}}$ = specific heat of solids content in sludge kJ kg^{-1} K^{-1};
ΔT = increment of temperature, K;
dry wt = kilograms of solids content per kilogram of sludge under deactivation, kg dry wt kg^{-1} sludge
w = kilograms of water per kilogram of sludge under deactivation, kg H_2O kg^{-1} sludge

There are some more factors contributing to the thermal effect: the destructive action of lime to the mixed liquid volatile suspended solids contained in the sludge, or the increased pH level in the sludge-lime mixture (2).

2. AIM AND SCOPE OF THE STUDY

The objective of the study was to answer the following questions: 1. Is quicklime applicable for the thermal-chemical inactivation of sewage sludges? 2. Is the thermal-chemical inactivation method competitive with chemical inactivating procedures? Are inactivated sludges fit for agricultural uses?

The experimental sludges come from five municipal sewage treatment plants operated in industrialized urban agglomerations.

3. EXPERIMENTAL APPARATUS AND PROCEDURES

The experimental apparatus consisted of the following units: a cylindrical tank equipped with an electric stirrer; thermostats separated from one another with foams displaying a heat-transfer coefficient, λ, of 0.0963 kJ m^{-1}h^{-1}K^{-1}, and an electronic temperature recorder coupled with resistance thermometers.

The following procedures were involved: measurements of temperature and duration of the deactivation process; physico-chemical, bacteriological and parasitological analyses prior to, and after completion of the process.

4. RESULTS

4.1. Physicochemical composition

The objective of physicochemical analyses was to obtain information on whether the sludge was fit for agricultural uses. Thus, determinations were carried out for two groups of substances - (1) those acting as fertilizers, i.e., total nitrogen, phosphorus, potassium and calcium, and (2) those of a toxic nature, i.e. chromium, nickel, copper, zinc and lead. Some of the results obtained for digested sludges are listed in Table 1.

TABLE 1. Chemical composition of digested sludge as a function of calcium oxide dose (kg CaO kg^{-1}dry wt).

Substance dry wt. %	Averages for			
	raw sludge		deactivated sludge	
	Plant Z	Plant L	Plant Z	Plant L
CaO dose	-	-	1.32	0.99
Total nitrogen	2.31	2.28	1.07	0.84
Phosphorus (P_2O_5)	3.28	4.00	1.96	1.57
Potassium (K_2O)	0.39	0.37	0.23	0.25
Calcium	3.60	3.63	25.27	32.49
Chromium	0.09	0.01	0.06	0.01
Manganese	0.06	0.01	0.03	0.00
Nickel	0.02	0.01	0.01	0.01
Copper	0.07	0.05	0.04	0.03
Zinc	0.61	0.42	0.28	0.23
Lead	0.03	0.04	0.02	0.02

As shown by these data, sludges treated by the quicklime method are fit for agricultural needs and have no toxic influence on vegetation. Heavy metal concentrations measured in deactivated sludge samples are far below, and fertilizing substance content is almost identical to the data reported in the liliterature (4). The increased calcium content found in the sludge has a favourable effect on the soil (3).

4.2. Thermal treatment

The thermal effect of slacking with the use of sewage sludge was investigated as a function of CaO dose, as well as specific heat and initial moisture content of the sludge involved. The determination of the time-temperature relation and the thermal efficiency of the process made it possible to establish the optimum lime—sludge proportions and, consequently, to minimize CaO consumption. The efficiency of the heating process was found to increase with the increasing dry solids content. When the sludge under deactivation had only been subject to thickening (viz. displayed a water content of about 95%), the process was ineffective and the CaO dose required was very high (approaching as much as 7 kg kg^{-1}dry wt.) . Mechanical dewatering (yielding a moisture content of about 80%) made the required lime dose drop to a level lower than 1 kg kg^{-1}dry wt. Hence, CaO consumption (which is a major factor accounting for the process cost) depends strongly on the initial moisture content of the sludge to be deactivated. It has been found that digested sludges reach the temperature required almost twice as fast as does activated aerobic stabilized sludge of the same initial water content. This should be attributed to the structure of the sludges (which is flocculent in activated sludge, and granular in the remaining two).

The heating efficiencies are gathered in Table 2.

TABLE 2. Thermal efficiency as a function of CaO dose, water content, maximum temperature and time of reaching maximum temperature.

Type of sludge and specific heat kJ $kg^{-1}K^{-1}$	Dry solids, initial wt. %	CaO dose kgCaO kg^{-1}dry wt.	Maximum temperature K	Time of reaching maximum temp. s	Thermal efficiency %
Primary, 1.46	4.26 4.62	4.32 4.62	340 345	32.0×10^3 29.9×10^3	85.0 87.2
Primary 1.39	13.02 13.02	1.28 1.54	339 349	25.6×10^3 20.2×10^3	91.6 93.3
Activated aerobic stabilized 0.93	12.48 12.48	1.60 1.76	328 334	34.2×10^3 29.9×10^3	62.6 66.0
Digested, Imhoff tank 1.09	15.96 15.96	1.10 1.25	338 346	5.1×10^3 4.7×10^3	88.2 90.2
Digested 1.12	21.12 21.12	0.90 0.95	338 342	7.2×10^3 5.4×10^3	77.9 80.3

4.3. Inactivation of sludges

The efficiency of inactivation was established by bacteriological and parasitological analyses performed prior to, and after completion of the process. Bacteriological analyses included the number of mesophilic bacteria and resting spores. Parasitological analyses aimed at determining the number of Ascaris suum ova which had been inoculated before quicklime treatment. CaO doses yielded thermal effects when applied in portions[3] between 0.90 and 7.25 kg CaO kg^{-1} dry solids. No thermal effect was achived with doses varying from 0.15 to 0.60 kg CaO kg^{-1}

dry solids. For comparison, pulverized $Ca(OH)_2$ was applied at doses ranging from 1.0 to 6.0 kg CaO kg^{-1} dry solids. Each sample treated with lime was subject to analyses after 24 h.

The thermal-chemical process yielded the following: in primary sludge (87% H_2O), complete bacterial kill was achieved with a 1.3 kg CaO kg^{-1} dry solids dose at 339 K, and total destruction of Ascaris suum ova with a 0.9 kg CaO kg^{-1} dry solids dose at 322 K; aerobic stabilized sludge (87.5% H_2O) experienced complete bacterial kill at a 1.75 kg CaO kg^{-1} dry solids dose at 334 K, and total Ascaris suum ova destruction at a dose of 1.45 kg CaO kg^{-1} dry solids and 322 K; in digested sludge (79% H_2O), total bacterial kill was achieved at 0.95 kg CaO kg^{-1} dry solids and 342 K, and complete destruction of Ascaris suum ova at 0.70 kg CaO kg^{-1} dry solids at 321 K (Table 3).

TABLE 3. Results of deactivation for sludges.

Type of sludge and initial dry solids wt.%	CaO dose kg CaO kg^{-1} drywt.	Process temperature K	pH of water extract	Bacterial kill	
				mesophilic bacteria	resting spores
Primary 4.62	4.32	340	12.70	100	99.9887
	4.62	345	12.71	100	100
Primary 13.02	1.15	333	12.06	99.99725	99.96539
	1.28	339	12.27	100	100
Aerobic stab. 12.48	1.60	328	12.47	99.90476	98.82353
	1.76	334	12.57	100	100
Digested 15.96	1.10	338	12.16	99.52381	99.8
	1.25	346	12.33	99.48048	100
Digested 21.12	0.90	338	12.50	100	99.90000
	0.95	342	12.52	100	100

The chemical process (low CaO doses and hydrated lime) was not as effective as when involving high doses. The results are shown in Table 4.

TABLE 4. Inactivation of primary sludge (4.9 dry wt. %; Plant L) with low CaO doses (contact time, 20 h).*

CaO dose kg CaO kg^{-1} dry wt.	pH of water extract	Number of bacterial cells in 1 cm^3 of sample			
		prior to lime treatment		after lime treatment	
		mesophilic bacteria	resting spores	mesophilic bacteria	resting spores
0.143	11.92	$1.85 \cdot 10^6$	$2.58 \cdot 10^6$	$6.2 \cdot 10^4$	$3.2 \cdot 10^4$
0.306	11.60	$1.85 \cdot 10^6$	$2.58 \cdot 10^6$	$2.5 \cdot 10^4$	$2.1 \cdot 10^4$
0.612	12.85	$1.85 \cdot 10^6$	$2.58 \cdot 10^6$	$2.2 \cdot 10^4$	$1.85 \cdot 10^4$

*Note: No Ascaris suum ova destruction is found to occur.

Application of $Ca(OH)_2$ failed to yield good effects despite a high lime dose (Table 5). Evident decrease in the number of bacteria and Ascaris suum ova was observed after 7 to 42 days storage. No self-contamination was found to occur during storage of completely inactivated sludges. Variations in the number of pathogens are shown in Table 6.

Neither fungi nor mould were present in sludge samples treated by the lime method.

TABLE 5. Inactivation of primary sludge (10.31 dry wt. %; Plant W) with pulverized hydrated lime (contact time, 24 h).

Dose Ca(OH)$_2$ kg CaO kg^{-1} dry wt.	Bacterial kill, %		Ascaris suum ova in 1 cm^3 of sample
	mesophilic bacteria	resting spores	
1.08	99.88696	95.67568	+++
2.80	99.97392	99.62163	+++ ⊗
6.14	99.994	99.6757	++

+++ approximately 1000,
++ equal to, or less than, 100,
⊗ destruction of single ova

TABLE 6. Variations of pathogens number during storage in the open.⊗

Type of sludge	Dry solids content before and after storage, %	CaO dose kgCaO kg^{-1}dry wt.	Duration of storage τ_s, d	pH of water extract after τ_s	Number of bacteria cells per cu cm before and after storage	
					mesophilic bacteria	resting spores
Primary	22.49 / 45.11	1.15	18	11.40	220 / 10	180 / 10
Aerobic stabilized	30.10 / 58.06	1.28	65	10.43	940 / 60	800 / 0
Digested	36.18 / 61.68	0.71	62	10.33	620 / 0	900 / 0

⊗ No Ascaris suum ova were found to occur after 42 days of storage.

5. ADVANTAGES OF LIME TREATMENT

1. Stabilization of sludges in biological processes and dewatering prior to inactivation or agricultural use may be eliminated (even at increased quicklime consumption), which is of importance to small sewage treatment plants.

2. The sludge-lime mixture meets the sanitary demands made on it. It neither putrifies nor releases objectionable odours; it is easy to dry and disintegrate.

3. Fertilization involving pasteurized sludges accounts for the souring of soils, and this calls for additional manuring with lime. Application of sludge-lime mixtures eliminates the problem.

6. CONCLUSION

1. Sludge-lime mixtures contain valuable nutrient substances

(N, P, Ca), so they may be used as fertilizers for sour soils and bad land.

2. Inactivation is the most effective in sludges dewatered to a dry solids content of over 20%. The sludge-lime mixture meets the sanitary demands made on fertilizers.

3. The inactivation effect is irreversible.

REFERENCES

1. Strauch D.: Mikrobiologische Untersuchungen zur Hygienisierung von Klärschlamm. 1 Mitteilung, 121 /1980/ H. 3, GWF-Wasser/Abwasser
2. Marcinkowski T.: Decontamination of Sewage Sludges with Quicklime. Waste Management and Research /1985/ 3, p. 55 - 64
3. Cebula J.: Kryterium przydatności osadów ściekowych w rolniczym ich wykorzystaniu. Materiały Badawcze. Seria: Gospodarka Wodna i Ochrona Wód. IMGW, Warszawa /1981/
4. Kempa E.S., Marcinkowski T.: Alkalization of Acid Soils with a Hygienized Sludge-Quicklime Mixture. Water Quality Bulletin. Volume 11, No 4 /1986/ p. 200-203 and 222

FLY-ASHES IN SOIL IMPROVEMENT

B.QUANT

POLISH ACADEMY OF SCIENCE, INSTITUTE OF HYDROENGINEERING
GDAŃSK, POLAND

1. INTRODUCTION

Polish energetic industry relies almost entirely upon the hard and brown coals. As a consequence, there are the big problems with disposal of the large quantities of produced fly-ashes. Most of the fly-ashes are stored, and therefore the negative effect of the fly-ash storage on ground water and atmosphere appears. Due to that phenomena, a sealing of the fly-ash waste areas seems to be necessary.

The soil petrification with the solutions of soluble silicates (water-glass solutions), which is called the silicatizations, is one of the well known methods of making the soils impermeable. The silicatization process consits in introducing into a soil, together or separately, the water-glass solution and the coagulant. The electrolytes are usually applied as the coagulants. If a stabilized soil contains the soluble parts which can cause gelation of water-glass, the additions of the coagulants are not required. Such assumption is a base for the elaboration of the methods of soil improvement with the fly-ashes and the water-glass solutions.

2. PRINCIPLE OF METHOD

After mixing of the fly-ashes with the water-glass solutions, which are added in the quantities corresponding to the Proctor optimum moisture content, and then compacting that mixture, a material called the Fly-Ash Composite is obtained. The soluble parts of fly-ash cause the gelation process. The formed silica gel serves as an agent binding the grains of the fly-ash (as a result the strength increases) and as an agent filling the porous (as a result the coefficient of permeability decreases).

3. PHYSICO-MECHANICAL PROPERTIES OF FLY-ASH COMPOSITE

The unconfined compression strength and the coefficient of permeability are the two parameters of the paramount importance for the characterization of the Fly-Ash Composite. The values of these parameters depend on many factors, primarily on a chemical composition of the fly-ash, on a cocentration of the water-glass solution and on a quantity of added solution. The last factor depends on physical properties of the fly-ash and generally correspondes to the Proctor optimum moisture content.

The strength of the Fly-Ash Composite ranges from a fraction to 4-5 MPa and it mainly depends on the chemical composition of the fly-ash. It is proved that the straigth line relation between the strength and the calcium oxide content exists. As an example the results for 10 percent solution are presented in Fig. 1. The influence of the concentration of the water-glass solution on the strength appears less substancial. Its influence, however, appears to be high in a case of the coefficient of permeability. Each increase of the concentration of the water-glass solution decreases the value of the coefficient of permeability (Fig. 2).

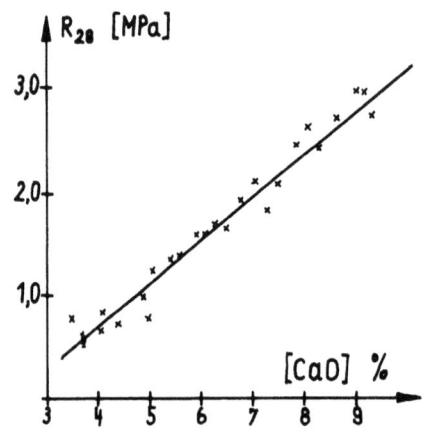

FIGURE 1. The relationship between R_s and CaO.

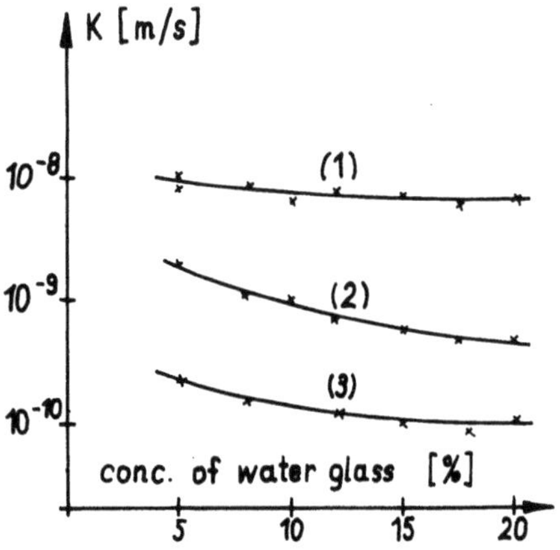

FIGURE 2. The relationship between K and conc. of water-glass for 3 fly-ashes.

When the calcium content in the fly-ash is very low, it is necessary to enrich it by adding the materials with a high calcium content. Different kinds of lime may be used for that purpose, mainly a waste lime - a carbide residue for instance.

From both mechanical and economic points of view, the optimum calcium content ranges from 5 to 7 percent and the concentration of the water-glass solution shouldn't exceed 20 percent per volume of 100 percent water-glass solution (density of 1350 kg/m^3).

Two parameters of the paramount importance for the characterization of the Fly-Ash Composite were mentioned above. A specified engineering case may have special requirements and quite often one must examine also other parameters of the Fly-Ash Composite. A short form of this paper does not permit to present the details. It is only worth to mention, that the tests performed by the author permitted to define the influence of the initial factors on such properties of the composites as the resistance to frost, the resistance to the static and dynamic water activity, the resistance to many external mechanical and chemical agents, the elastic modulus, the rheological properties etc. etc. One should state, that the Fly-Ash Composite exhibits generally sufficient strength and resistance to the most of the external agents.

4. APPLICATION AND TECHNOLOGY

The possibilities of application of the Fly-Ash Composite are very wide. The directions of its applications can be devided into three general groups:

1/ The surface sealing of soils; it concerns first of all the sealing of the bottom and slopes of water and waste reservoirs.

2/ The protection of soils against the errosive activity of water and atmosphere; this group concerns the stabilization of the chanel and riversides, the shores of any lakes or even the seashores.

3/ The stabilization of soils for foundation purposes, which concerns any foundation where the requirements with the reference to strength are not very high. The application of the Fly-Ash Composite as a one of the stratas in road engineering can be also classified to this group.

The costs and the simplicity of a technology are the two most important qualities of the Fly-Ash Composite. For instance, the price of improving of 1 m^2 of soil with the Fly-Ash Composite is 5 do 8 times smaller than in case of other traditional methods.

The practical formation of the Fly-Ash Composite consists of several steps: a transport of fly-ashes, a preparation of a layer of fly-ash on a surface of ground, an eventual addition, dilution and mixing of lime with fly-ash, an addition of water-glass solution and a compaction of the mixture of reagents. Each step can be realized in different ways with the different building equipments. It only depends on an invention of the performer.

The Fly-Ash Composite has been applied several times till now. Among others it has been applied for sealing of the sedimentation basin for the coal-mine sewage (7 ha), for sealing of the storage of the fly-ashes and wastes from the factory of cellulose (8 ha), as a foundation under the floor in an engine room (2 ha) etc. In each case the results were satisfactory.

VEGETATION ESTABLISHMENT ON FLY ASH PONDS BY MEANS OF HYDROSEEDING

Ir. D. DE VLEESCHAUWER & R. IMLER
EBES Communication Department

Fly ash (Pulverized Fuel Ash), a fine powder, is the residual product of the pulverized coal burnt in the furnaces of modern electricity power stations. It constitutes the very finely grained matter that escapes the combustion chamber with the flue gases (photograph 1).
These fly ashes are extracted from the gases by means of electrostatic precipitators before they can disperse in the environment.
The three main elements in fly ash are silicon, aluminium and iron, the combined oxides of which account for 75-95% of the material. Conventional chemical analysis and typical particle size distribution of fly ash, produced by the EBES coal fired power stations, are given in table 1.

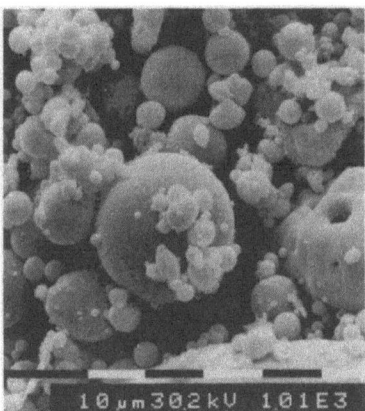

SiO_2	48-53
Al_2O_3	26-30
Fe_2O_3	4-7
TiO_2	1
CaO	2-4
MgO	1-2
Na_2O	0.5-1
K_2O	3-5
SO_3	0.4-0.6
P_2O_5	<0.1
0-2µ	5
2-50µ	80
>50µ	15

Photo 1. Fly ash as seen under the electron microscope

Table 1. Fly ash chemical composition and particle size distribution (as a percentage by weight).

In 1986 fly ash produced by the EBES coal fired plants amounted to about 450 000 tons. About 40% of this quantity was used for different applications, such as cement and concrete production as well as in road construction.

Despite of its increased valorization the remaining part, mostly of lower quality, has to be dumped. At present EBES possesses five authorized dumping sites for fly ash.

Fly ash storage ponds are spontaneously vegetated by mosses, herbs and trees. The vegetation comes about very slowly and is not homogeneous, resulting in a constant threat of dust blowing up. Generally it takes a long time before the landscape is restored. Therefore covering the fly ash pond is mandatory. Traditionally this is done by applying a layer of good quality soil, enabling plants to grow.

Instead of applying this expensive method, another technique has been tried out. This technique, called "hydroseeding", is based upon spraying a slurry of water, fertilizer, seeds, organic compounds and a soil conditioner directly onto the bare fly ash surface (photograph 2).

Photo 2. Demonstration of the hydroseeding technique at the fly ash pond at Mol.

The soil conditioner used in this particular experiment has been developed at the Laboratory for Soil Physics at the State University of Ghent. It consists of polyacrylamide (PAM), a polymer having the property to bind the fine fly ash particles together. Although the polymer film is permeable to water, it prevents evaporation from the soil. After a few months the film desintegrates due to the action of UV light and rain impact. At that time its role is taken over by the grasses which have by then strongly developed and rooted (photograph 3 and 4).

Photo 3. The fly ash pond at Langerbrugge before hydroseeding (tree plantation has already been performed).

Photo 4. The fly ash pond at Langerbrugge some months after the hydroseeding.

The hydroseeding technique has been succesfully applied at the fly ash ponds of Langerbrugge (near Ghent) in 1984 and Mol in 1985. In Mol, besides grass seeds, seeds of herbs, shrubs and trees have been applied. The technique of hydroseeding is supplemented by tree plantation in bare fly ash soil or in treated plant holes. Results so far showed that Poplar and Robinia are most adapted to the growing conditions of fly ash. In the future this ecological friendly technique can also be applied to other fly ash belts.

ATMOSPHERIC PROBING WITH LIDAR

G.J. KUNZ

Physics and Electronics Laboratory TNO
P.O. Box 96864
2509 JG The Hague, NETHERLANDS

INTRODUCTION
Knowledge of the atmospheric vertical structure is of importance in environmental research, meteorology and electro-optic applications. Conventional techniques to obtain information on the vertical structure of atmospheric parameters are either time consuming and expensive (e.g. balloons, aircrafts) or limited in range (acoustic sounders). An alternative remote sensing technique, called LIDAR (LIght Detection And Ranging), provides a way to investigate an atmospheric path of several kilometers with a range resolution of meters within microseconds. Parameters which can be measured with lidar are a.o. backscatter, extinction, wind velocity and wind direction, temperature, pressure and within limits the concentration of some gases. Moreover stackplume structure and behaviour can be measured over ranges of kilometers; a downward probing lidar can provide information on ocean depth and water quality. Each subject requires a specially designed lidar system [1, 2].
This paper gives an impression of the principle of lidar and presents some results of a vertically probing lidar program which was running over a two year period.

PRINCIPLE and RANGE
Lidar is comparable with radar. Laser light is used as a probe instead of microwave radiation. The receiver consists of a telescope (optical antenna) in combination with a photo detector. Lidar signals have to be processed in order to obtain the desired atmospheric quantaties. For homogeneous atmospheres this is a rather simple method but in cases of inhomogeneous atmospheres, more complicated prosessing with additional information is required [3, 4].
The maximum range of lidar in homogeneous atmospheres is on the one hand limited by the amount of atmospheric backscatter and on the other hand limited by the atmospheric extinction. There is a maximum range for each lidar system due to the the proportional relation between extinction and backscatter and due to the geometric attenuation. Transportable lidar systems have ranges in the order of 3 km for normal atmospheres. However for inhomogeneous atmospheres like stackplumes and clouds, this range can be at least a factor of 3 larger.

SOME RESULTS

Extinction profiles resulting from vertical lidar measurements during different atmospheric conditions, are presented in Figure 1. The results are compared with an airborne nephelometer (point visibility meter). Figure 1a shows a homogeneous atmosphere up to 1 km altitude; Figure 1b gives the profile during a ground haze situation and Figure 1c shows the results during a stratus cloud situation.

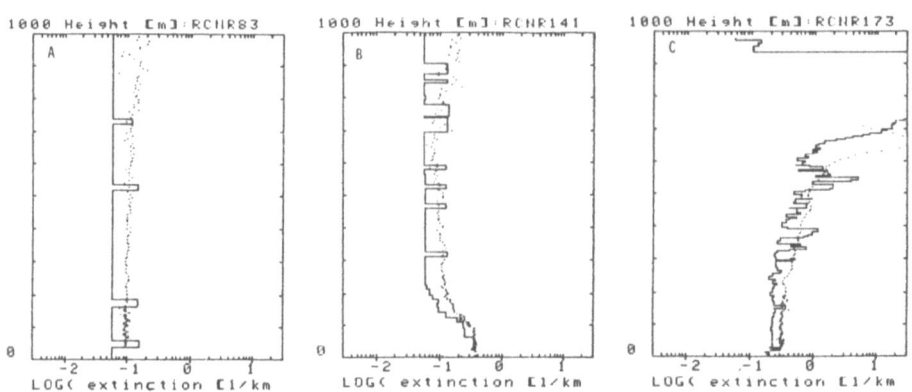

FIGURE 1. Examples of verical extinction profiles measured with lidar under different weather conditions. The solid line represents the airborne nephelometer results.

The diurnal variation in the mixing layer is presented in Figure 2. During a three day period the formation of the mixing layer is clearly visible. The last day a cloud layer decends almost linearly in time.

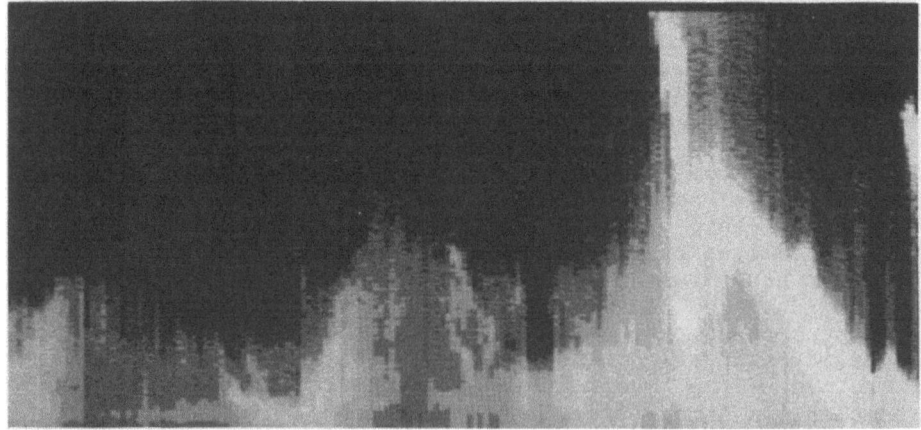

FIGURE 2. Diurnal effect of the mixing layer over a three day period up to 1 km altitude.

Vertical lidar measurements also provide the altitude of the cloud base. An example of the time behaviour of the cloud altitude during a one day period is presented in Figure 3. The gaps mean either that no cloud was detected or that no cloud was present. It appeared that in a large number of cases the increase in the signal strength, due to scattering of the cloud, is compensted or over compensated by the attenuation of the cloud and can therefore not be distinguished from true mixing layer situations [5]. This problem can be met by performing inversion of lidar signals with additional information.

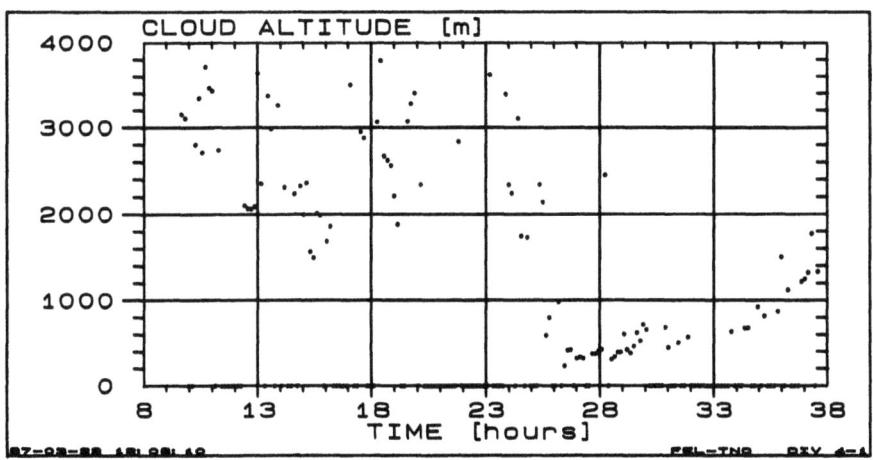

FIGURE 3. Time behaviour of the cloud altitude over a one day period.

CONCLUSION

Lidar provides an efficient method to investigate large areas of the atmosphere within limited time. Not only atmospheric parameters like the mixing layer height can be measured but also clouds and stack plumes can be mapped in three dimensions. Gas concentrations of polluting gases can be measured. Recently we developed a high frequency Nd:YAG scanning lidar system to investigate the dynamical behaviour of the atmosphere on smaller time scales. The system can be used for plume measurements as well as for mixing layer investigations. The usefulness of our calibrated lidar systems has been proven in field trials and aboard ships.

REFERENCES

1. Proceeding 13Th International Laser Radar Conference
   Toronto, Canada, 11-15 August 1986.
2. Ocean Optics VIII, SPIE, Vol. 637
3. Klett, J.D. Appl. Opt. Vol. 20, No. 2, 15 Jan.1981
4. Kunz, G.J. Appl. Opt. Vol. 22, No. 13, 1 July 1983
5. Kunz, G.J. Appl. Opt. Vol. 26, No. 8, 15 April 1987

VT-Biofilter

Ing.Bureau Van Tongeren Beverwijk BV
J.J.W.Bijl

Biotechnology today:

During the past few years biotechnology claims a still
growing part in environmental technologies. This modern
technology utilises the action of micro-organism to solve
major problems in our industries.
Environmental pollution must be prevented.
Today's biotechnology is the answer to these problems.

The working principle of micro-organism:

Purifying air with a biofilter is done by passing the
polluted airstream through a layer of biologically active
material.
Pollutant elements such as odourcomponents and many other
volatile organic and anorganic substances are absorbed by the
filtermaterial. After contact with the micro-organism the
pollutant elements are breaked down, which results in a
clean, odourless airstream leaving the filter.
the conversion of pollutant elements into harmless oxidation
products guarantees a great advantage of this technique in
comparison with other purifying systems.

The advantage of a Biofilter:

The most important characterstic of the Biofilter is the composition of the filtermaterial.
This unique mixture of microbiologically active compost and
environmentally save added components produces highly favourable properties. The result of this composition is a
rich filtermedium in which a high level of micro-organic
life guarantees a maximum of bioactivity. The filtermaterial
allows a large buffer capacity for intermittend loads, an
excellent moisture regulation with optimal pH stabilisation.
A maximum contacttime with the micro-organism is secured,
which results in a maximum cleaning capacity.

Modular Building:

The Biofilter remains adaptable for adjustments in several operating conditions. The modular sections can be linked to one another in series as well as parallel system.
The number of modules depends on the nature and composition of the air to be cleaned.
Different pollutants can be eliminated in various sections of the filter, so a wide cleaning spectrum is possible in the same filter by inoculating specific micro-organism for particular pollutants.
The possibility of combined treatment assures an extremely effective result of cleaning.

Biofilter all-round cleaning:

A wide range of pollutants can be eliminated, such as odour components, alcohols, carboxylic acids, ketones, aldehydes and esters. But also organic conversion products, both aliphatic and aromatic hydrocarbons as well as sulphurous and nitrogenous components can be treated.
Even not essentially biodegradable pollutants, such as dichloormethane and dimetyleformamide can be eliminated through use of specially cultured and in the filter inoculated micro-organism.

Biofilter for industrial purposes:

A wide range of pollutants can be eliminated using the Biofilter. Not only pollutants of natural origin, but also synthetic pollutants can be treated.

The Biofilter is highly effective in removing pollutants:

A Biofilter containing 9 m3 of special prepared and inoculated filter material has been installed to clean process air of a leatherfactory.
The Biofilter is built on only 9 m2 groundarea and cleans approx 6000 m3 processair/hr. The fan needed to feed the Biofilter only requires 7,5 kW. The installation dates of February 1986. After a period of 3 months a double check was made by an independent laboratory to learn the cleaning result of the Biofilter.
3.1 mg/m3 of sulphide entering the Biofilter, was cleaned after passing the filtermaterial to 0.08 mg/m3, which gives a 97,4 % of cleaning efficiency.

A Biofilter built in 3 sections containing 60 m3 of special prepared filtermaterial has been installed for a pharmaceutic Industry in March 1986.

This Biofilter is built on 20 m3 groundarea, but as it is
built in 3 sections, the hight is 7 meters. 2 Sections of the
Biofilter have been inoculated with micro-organism to clean
and eliminate Ethyl-Alcohol and Iso-Propanol.
The topsection has been inoculated with a special micro-
organis to eliminate Dichlorethane.
Before entering the Biofilter, the processair is passed
through a dustfilter. The Biofilter is equipped with 2 fans,
with a total power of 9.5 kW. Approx 5000 m3 of proecessair/hr
is treated in the Biofilter.

After 4 months a double check is made by an independent
laboratory which results are as follows:

	Before treatment:	After treatment:
Ethyl-alcohol	5300 mg/m3	33 mg/m3
Iso-Propanol	4915 mg/m3	<10 mg/m3
Dichlorethane	5210 mg/m3	890 mg/m3

Also in this situation an excellent result of biologically
cleaning is proved. A second filter will be installed against
the end of 1987.

Biofilter conclusion:

The Biofilter requires a relatively low initial investment
for a high operating efficiency. Because of flexible and
compact modular construction it is easy to adapt new operating conditions.
A high removal efficiency of pollutants could excess 99 %.
Low energy consumption caused by low air-resistance over the
filter. However a wide range of polluting components is
eliminated, the filter itself remains clean. And most
important: **NO TOXIC WASTE HAS TO BE HANDLED.**

Close cooperation with the Universities of Wageningen and
Groningen guarantees optimal scientific support and back
up to keep a high grade of knowledge in biotechnological
cleaning systems.
The environmental technology of today offers the possibility
to deal with polluted airflows in a clean and cheaper way
as known and used before.
Polluting problems should not be problems anymore by using
Biofilters.

REMOVAL OF HEAVY METALS AND DIOXINS IN FLUE GASES FROM WASTE TO ENERGY PLANTS

Kurt Carlsson, Product Manager Waste to Energy
FLÄKT INDUSTRI AB, Växjö, Sweden

1. MANY HEAVY METALS ARE VERY POISONOUS
Some heavy metals are essential for living materia e.g. cupper in low concentrations for human beings. On the other hand cupper is a strong poison for lower organisms (and sheep).
Other heavy metals, such as mercury, cadmium and lead, are very poisonous for animals as well as plant life.

2. SOLID WASTE CONTAINS A LOT OF HEAVY METALS
When solid waste is incinerated almost all the mercury is evaporated and also 30 - 60 percent of the cadmium and lead. Downstream the cooling of the flue gas which can be done by boiler or water injection we find most of the evaporated heavy metal as very fine particles - among them cadmium and lead. Mercury and its salts have a very high vapour pressure and exist therefore in gaseous form at normal flue gas temperatures (30 - 230 ^{o}C).
If we compare the specific emissions (in kg/TJ) from combustion of coal and solid waste we will see that solid waste emits much more heavy metals than coal combustion does.

3. MANY OF CHLORINATED HYDROCARBONS ACCUMULATE IN LIVING MATERIA
DDT and PCB have badly hurt some animals - seals in the Baltic have accumulated DDT and PCB and have therefore difficulties to reproduce.
Some years ago it was discovered that salmon contained comparatively high concentrations of dioxins/furans which are a group of poisonous chlorinated hydrocarbons. The symmetrical 2,3,7,8, TCDD is one of the most poisonous elements we know.
Many measurements have shown that waste incinerators produce comparatively high concentrations of dioxins/furans and also other chlorinated hydrocarbons.

4. EFFECTIVE FLUE GAS CLEANING REDUCES EMISSIONS TO A VERY LOW LEVEL
Beside the heavy metals and the chlorinated hydrocarbons the acidifying gases HCl, SO_2 and possibly also NO_x have to be removed. HCl and SO_2 are rather easy to absorb as they react with almost all alkaline dust and/or liquid.
Nitrogenoxides (NO_x) require a specific step for high removal - a specific catalytic reduction stage placed downstream a very good dust collector is today the only efficient way.
To be able to remove particulates, heavy metals and chlorinated hydrocarbons the very best equipment has to be used. Many tests have shown

that the fabric filter is an excellent device in this respect and therefore this equipment should be the base in a modern flue gas cleaning system for at least a new waste incinerator.

Around the fabric filter the complete gas cleaning is built in different ways.

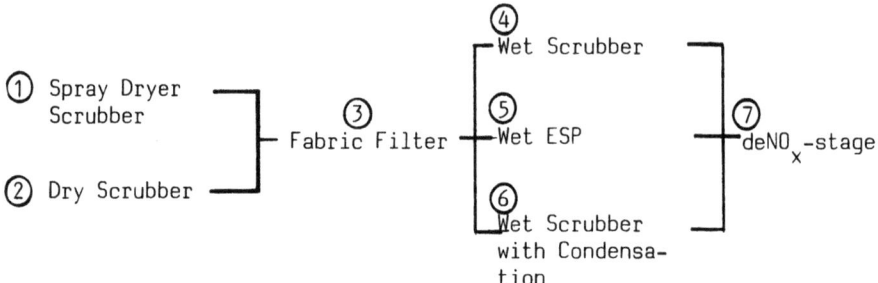

The combinations ① + ③ and ② + ③ have carefully been examined by Environment Canada and Fläkt at a test unit in Quebec City, Canada.

The following results dramatically show the difference in removal efficiency between fabric filter and electrostatic precipitator for dioxins.

TABLE 1. PCDD Concentration (ng/Sm^3 at 8 % O_2) in flue gas and efficiency of removal.

Operating Condition	Dry system			Wet-Dry System		
	110°C	125°C	140°C	200°C	140°C	140°C +Recycle
Inlet (ng/Sm^3)	580	1400	1300	1030	1100	1300
Downstream ESP (ng/Sm^3)	310	570	540	1140	840	1270
Downstream FF (ng/Sm^3)	0.2	ND	ND	6.1	ND	0.4
Efficiency (%)						
With ESP	47	60	57	(11)	24	2
With FF	>99.9	>99.9	>99.9	>99.4	>99.9	>99.9

With the correct temperatures also the HCl/SO_2-removal is excellent at a low excess of lime. Also here the big advantageous of FF can easily be seen.

TABLE 2. HCl and SO_2 concentrations (at 8 % O_2) and collection efficiency

Operating Condition	Dry System				Wet-Dry System	
Flue Gas temp. at FF Inlet	110°C	125°C	140°C	200°C	140°C	140°C +Recycle
Stoichiometric Ratio	1.16	1.03	1.04	1.49	1.19	1.10
Hydrogen Chloride						
Inlet (ppm)	423	464	425	392	366	470
Outlet ESP (ppm)	15	69	129	196	149	152
Outlet FF (ppm)	7	9	29	91	29	42
Eff. with ESP (%)	96	85	73	50	59	69
Eff. with FF (%)	98	98	94	77	92	91
Sulphur Dioxide						
Inlet (ppm)	119	118	99	117	106	106
Outlet ESP (ppm)	24	65	64	103	67	70
Outlet FF (ppm)	4	10	41	83	35	43
Eff. with ESP (%)	80	45	35	11	37	35
Eff. with FF (%)	96	92	58	29	67	60

In some cases it is interesting to use a wet scrubber for HCl/SO_2-absorption as then a cheap alkaline e.g. limestone can be used for the neutralization. Also in this case a fabric filter should be used for dust collection as all heavy metals and chlorinated hydrocarbons then are concentrated to the dry dust which can be stabilized with concrete technology. No extensive water treatment is necessary in this case.

REFERENCES

1. Rosén, Statens Naturvårdsverk: Metallerna i miljön.
2. Kurt Carlsson: Heavy Metals from "Energy from Waste"-plants - Comparison of Gas Cleaning System.
3. Hay, Finkelstein, Klicius. Marentette: Canada's National Incinerator Testing and Evaluation Program Air Pollution Control Technology Assessment.
4. Fichtel: Beurteilung der festen und flüssigen Rückstände aus Rauchgasreinigungsanlagen.

PRODUCTS (DESIGN)

**Second European
Conference on
Environmental Technology**

in the 'European Year
of the Environment'

DESIGNING OF PRODUCTS IN VIEW OF RECYCLING

W. JORDEN

1. SUMMARY

Recycling needs products designed in view of recycling, i.e. that they are prepared for the eventual reuse of the whole product or parts of it and, after the end of the product use, for the reutilization of the materials. From the analysis of the design process and the recycling system, guide lines are derived and explained by examples; the chances in practice are discussed.

2. DESIGN AND RECYCLING

2.1. Circles of information and products

Often the designer feels only like a link in a linear information chain going from the cause (task) over a procedure (design) to the result (product), figure 1 /1 to 4/. Really the result feeds back to the cause, so we have a closed loop. Information feedbacks are caused by defects or by suggestions coming from production, sale or use. So the product is improved and, by and by, optimized for low price and sufficient fulfilment of its function.

But this "optimization" is not sufficient in a higher view of economics. Recycling causes a closed loop of material (dotted lines). The information circle including the designer is closed by analysing the recycling system and gaining new aims and tasks for designing (pointed lines). Further, throwing away or dumping of products causes new problems.

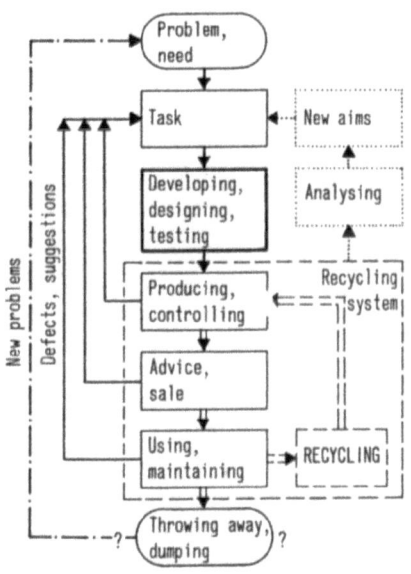

FIGURE 1. Development and life phases of a product; circles of recycling and information.

2.2. Recycling model system

The real flow of material can be simplified into three main circles /1 to 5/, figure 2.

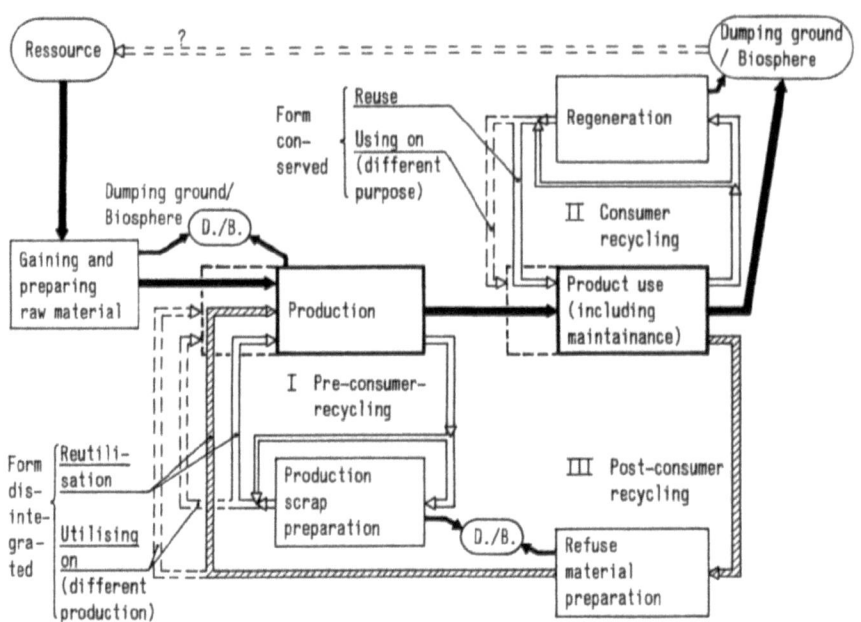

FIGURE 2. Simplified recycling model system

- I Pre-consumer recycling: Recycling of production scrap and refuse.
  o Without preparation, e.g. remelting of casting gates.
  o With preparation, e.g. collecting, cleaning, pressing, and melting chips.
  * For an equal production, called "reutilization".
  * For a different production, called "utilizing-on", e.g. bricks produced from blast furnace slag.
- II Consumer recycling. Recycling during product use means that a product is maintained on a high value level without losing its form or its function ("use").
  o Without regeneration, e.g. glass bottle.
  o With regeneration, e.g. reconditioned engine.
  * For the original purpose, called "reuse".
  * For a different purpose, called "using-on", if no longer suitable for the original purpose (loss of value), e.g. car brake in a test rig.
- III Post-consumer recycling: Recycling of refuse material after the end of the product life; the form is disintegrated, the function is lost ("utilization").
  o Mostly with preparation, e.g. collecting, sorting, separating, etc.
  * For similar products as before, "reutilization", e.g. steel produced from steel scrap.
  * For different products, "utilizing-on", e.g. producing grounds for sport facilities from worn tyres.

Generally circle II should be preferred because here the loss of value is minimal. But often it is a question of economy and ecology whether consumer or post-consumer recycling is more favourable (e.g. reuse of glass bottles or collecting of glass refuse). The decision mainly depends on expense and burdening of the environment within the recycling circles.

All procedures in the circles imply losses of material (and energy) which are to be minimized.

3. GUIDE-LINES FOR DESIGN IN VIEW OF RECYCLING

The following guide-lines have been derived from the processes in the recycling system. Guide-lines are not inflexible rules but recommendations. The normal task of each designer is to find compromises between different demands; recycling gives a new category of design demands and criteria for optimization. The guide-lines are not complete, but they give an idea of what can be done.

3.1. Pre-consumer recycling (circle I)

Circle I is influenced by the designer only in an indirect way. Pre-consumer recycling is to be considered here especially because regaining the material of production scrap or refuse is the most frequent kind of recycling; for these reasons:
- The materials are known.
- They can be collected in a simple, clean and separate way.
- The ways of transportation are simple and short.
- The profit is given and can be calculated.

| 1 | Basic rule for production scrap | Such production processes should be used which cause as little scrap or refuse as possible.

| 2 | Recyclability of scrap | Unavoidable production scrap should be suitable for recycling with as little cost and value loss as possible.

Examples: Steel can be recycled well, thermoplastic material only conditionally, duromere not.

| 3 | Number of materials | The number of different materials should be reduced.

That facilitates collecting and retransporting; furthermore, it saves cost of stocks.

3.2. Post-consumer recycling (circle III)

In the circle III secondary materials are produced resp. mixed with primary materials. The aim is that secondary materials must fulfil the material requirements. The processes in the recycling of refuse material are related to those in circle I; therefore they are regarded subsequently in the following. Conditions for a good function of circle III can be derived from the facts of circle I:
- The materials must be identified.   ⎫ ⎧ Essential
- They must be separated in a clear and simple way.   ⎭ ⎩ problems!
- They must be collected and transported easily.
- Recycling of refuse materials must be economical.

Therefore the general task is designing products which consist of clearly identicable materials which are either compatible or easy to separate.

| 4 | Basic Rule for refuse materials | Every product (in addition to the consideration of all design demands) should be designed in such a way that it is prepared for a process to regain the material after the end of use.

This rule is new but valid in a general sense.

| 5 | One-material-group products | At first it should be aspired to that a product consists only of one recyclable material, at least that all materials of the product belong to the same compatible material group.

An example is a glass bottle. Compatible material groups consist of materials which can be utilized together, e.g. a ball bearing with a cage of steel (but not of brass). Figure 3 is a proposal for a compatible material matrix /5, 6/. It shows which refuse material can be added to which material group unlimited or limited. An overquantity factor specifies that an alloy component exceeds the limit; this material can only be added to such an extent as other materials in the mixture contain a correspondingly less quantity of this component. Such tables should be developed for all usual materials; perhaps combinations of compatible materials may become common by and by.

Materials go with these → Compatible Material Groups	Compatible Material Groups											
Alloy family	Materials of the families (DIN 1725)	AlMn	AlMg	AlMn Mg	AlMg Mn	AlMg Si 0,5	AlMg Si	AlZn Mg	AlCu MgPb	G-Al Si	G-Al Si (Cu)	G-Al SiCu	G-Al Mg
Pure aluminium													
AlMn	(AlMn)			■		●	/	○	○	●	/	○	○	○
AlMg	AlRMg 0,5	■								/	■	■	
	AlRMg 1	○				■				/	○	○	

FIGURE 3. Compatible material matrix (section) for aluminium.

Materials can be part of the concerned Compatible Material Groups
■ unlimited   ○ limited   (overquantity factors see /5/)
■ unlimited if Mg is removed   ◌ very limited
● nearly unlimited   / only in minimal quantities

| 6 | Non-disturbing material combination | If rule 5 cannot be carried out then the materials should not disturb each other during the utilization, i.e. the product should consist of only one recyclable main material and as few as possible other materials which can be removed in the recycling process.

Example: Ball bearing with plastic cage which evaporates and is lost.

| 7 | Separation | If the compatibility of materials is not possible in the whole product then it should be possible to separate the product into compatible resp. non-disturbing groups according to the rules 5 or 6. Groups to be separated should be clearly marked and positioned in the outer areas of the product for easy access and detachment.

Severely disturbing materials e.g. in a normal steel melt are copper (limit $\leq 0,15\%$) and tin ($\leq 0,02\%$). These materials should be separated. Electric motors with their high copper proportion (figure 4 /2/) are concerned as well as high value

materials (e.g. stainless steel elements in a dish washer).
Figure 5 shows the logical procedure for a recycling conform
material selection according to the guide lines no. 5 to 7
/3,4/).

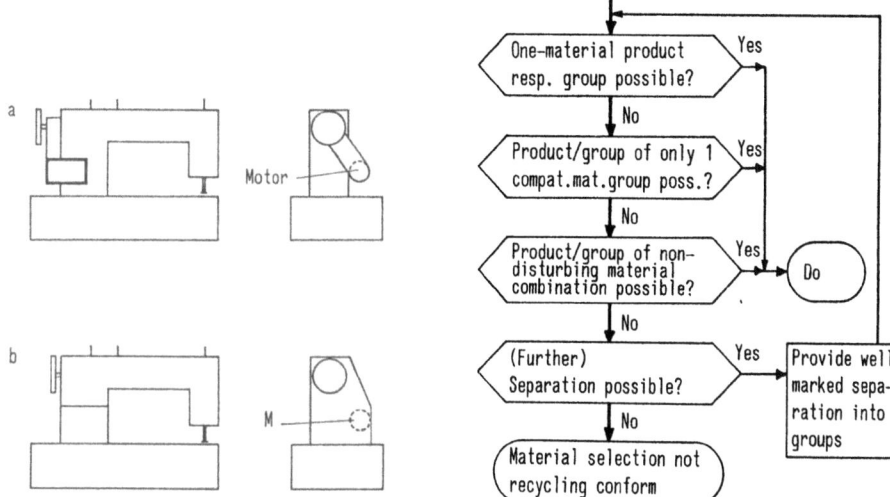

FIGURE 4. Not recycling conform new design of a sewing machine (electric motor M hidden).

FIGURE 5. Logical procedure for a recycling conform material selection.

| 8 | Endangering | Materials resp. devices which may be dangerous in the utilization (e.g. poisonous or explosiv) should be positioned for easy detachment resp. emptying and clearly marked. |

Examples are fuel tanks or refrigeration sets.

| 9 | Undetachable connections | If non-compatible materials are to be connected in an undetachable way, the advantages in the function should be carefully weighed against the disadvantages in the recycling process. |

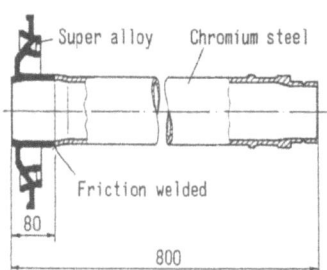
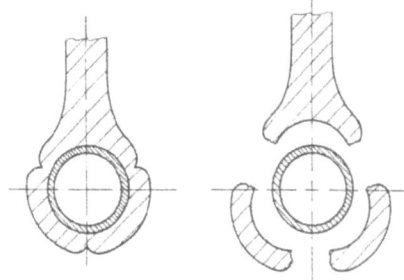

FIGURE 6. Friction welded gas turbine shaft connecting super alloy and steel.

FIGURE 7. Bronze sleeve in a cast iron lever separable in the shredder.

E.g. zinc is one of the best steel protective layers against corrosion, but in the steel melting process it is aggressive against the melting furnace. The friction welding, figure 6 /8/, connects a turbine wheel of super alloy to a shaft of steel, i.e. saves expensive material; but for recycling a separation (cutting) line should be marked. The notches in the lever of cast iron, figure 7 /7/, cause that the bronze sleeve can be separated in the shredder process; but the life endurance of the lever must be high enough.

| 10 | Identification | A clearly visible and durable identification symbol on the product, on groups or parts should state the material or material group and the suitable recycling technology.

For the present it would be sufficient to state the material in a standardized way, e.g. aluminium alloys. If compatible material groups are existing, a symbol like in figure 8 /4/ could code $2.5 \cdot 10^6$ different combinations.

FIGURE 8. Proposal for an identification symbol.

3.3. Guide-lines for consumer recycling (circle II)

In the public consciousness as well as in practice the recycling during use is far less developed than the other recycling circles. But circle II is important because here the product remains on a high value level without losing its function ("function saving recycling", against "material saving recycling" in the circles I and III).

The guide-lines for this circle are on the border between recycling and maintenance. Both of them intend to keep the product as long as possible in the using stage. The repair of a larger product often means that singular units are replaced and recycled. Repair belongs to the normal use; the individuality of the product is saved. On the other hand recycling means a product regeneration in an industry-like manner.

The following guide lines follow the range of procedures in a regeneration factory /9/:
- Disassembling,   - replacing of worn or useless parts,
- cleaning,   - reassembling,
- testing of parts,   - testing of the whole product.

Aims of the design measures are to facilitate the regeneration process and to minimize the expense of new parts.

| 11 | Basic rule for recycling during use | Every product should be suitable for a reuse with as little expenses as possible; therefore regeneration or reconditioning should be as easy as possible.

| 12 | Dis- and reassembly | Every product consisting of several parts should be dis- and reassembled without damage as simply, clearly, and safely as possible.

The snap washer joining, figure 9 /7/, between the ball bearing and the housing is easily to be assembled by the chamfer 2, but not to be diassembled. A second chamfer 3 makes the

disassembly possible.
 In every case disassembly should be possible without any special tools. All parts concerned should be well accessible. A modular structure is to be prefered by which the modules can be dismantled and replaced independently /10/.

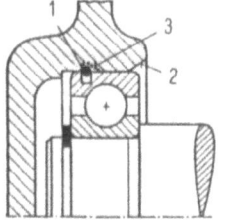

FIGURE 9. Bearing unit. 1 snap washer, 2 and 3 chamfers

| 13 | Connections | Detachable connections must fulfil their function over the whole product life including recycling.

E.g. bolts must neither be stuck by corrosion nor must the thread hole be torn out. In weak materials (e.g. plastic) thread holes should be replaced by through-holes with longer bolts, or by thread inserts.

| 14 | Cleaning | Parts should allow for easy and complete cleaning without any damage.

Cavities and corners should be well accessible, lettering during enough against solvents (casting-in or etching is better than e.g. printing).

| 15 | Standardizing | Parts with the same function should be either absolutely equal or clearly different. Such parts (elements, assembly groups, products) should be standardized in structure, joining dimensions, and materials.

This rule will facilitate the identification and exchangement of parts as well as the identification of materials in circle III; furthermore, it saves cost for storekeeping. Standardizing does not cause narrowing in design (see ball bearings, bolts, etc.)

| 16 | Wear and corrosion | Wear should be minimized. Unavoidable wear is to be limited to special, easily adjustable resp. interchangeable elements. Similar aspects apply to corrosion.

Such elements are well known, e.g. in brakes or tyres. The whereabouts of the worn material are important because it is lost for recycling and often scattered in the environment. Machining allowances and adjustment aids for regeneration are to be taken into consideration. The decision about rolling or sliding friction should depend on the life expectancy (the rolling principle is not always the better one). The tribological knowledge should be transposed into practice in a more consequent way.
 A special example are blades for jet turbines consisting of super alloy /8/. For security reasons they are interchanged after a certain service life time. A HIP procedure (a hot isostatic pressing) after 75% of this time regenerates a beginning material fatigue and prolongs the usable service life time by 50%.

| 17 | Protective layers | Protective layers against corrosion and other destructive influences should be dimensioned for the whole product life; if not possibel, they should be renewable as easily and completely as possible.

That concerns all kinds of protective layers, also on bolts, joints, and other small parts. But we should note that each layer represents a compound material; i.e. we should regard guide-line no. 9.

4. CHANCES OF DESIGN IN VIEW OF RECYCLING

4.1. Economy

The interest of the industry in the recycling conform design is little hitherto because there is the assumption that it would make the products more expensive, at least not cheeper, further that products would live too long, therefore they would become obsolete, and the sale of new products would decrease. These arguments are at least shortsighted concerning the situation of environtment and resources, if not wrong in a general sense, according to the following sentences.

| Cost | Recycling conform products can often be carried out without additional cost, if the designer is aware of the problem.

If a beverage tin e.g. consists of aluminium with a tin plate cover, neither of these materials can be recycled economically, whereas a "tin" only of aluminium is not more expensive, but conforms to recycling.

A positive example is the standard glass bottle for beverages with screw cap. The screw cap closes the bottle ("small" recycling circle II) and protects the bottle thread during the retransportation ("big" recycling circle II). The material of the cap can be regained completely if the "ring" does not separate of the cap.

| Argument for sale | Recycling conform products often are also maintenance conform (service conform).

As the personnel costs are arising, the above is througout an argument for publicity as well as for less maintenance costs.

| New places of employment | The change from "repair" (resp. dumping) to "regeneration" can result in new places of employment in regeneration centres or similar and help the plants to reduce their expensive repair departments.

Examples show that recycling centres can work economically, e. g. for machine tools or office machines, often in contrast to the repair departments.

| Product value | Regenerated secondary products can be of the same value as new products, but considerably cheeper; they are not inferior.

Many parts which are not subject to fatigue or something similar, do not lose their function and their value. If each of the secondary products is tested individually after the reassembly, it is more reliable than new series products which are only tested at random.

| Energy need | Secondary materials often need only a fraction of the energy necessary for primary materials.

E.g. secondary aluminium needs about 4 % of the energy for primary aluminium /7/. That means that recycling unburdens the environment in different ways.

| Government measures | Government measures can advance recycling and recycling conform products.

Such measurements are not popular but sometimes necessary and helpful, e.g. less taxes for recycling conform products or an obligation for the producers to regenerate their products.

4.2. Proceeding in design in view of recycling

The designer solves problems by technical means. Therefore we find the five steps of the general problem solving process in the design process too, figure 10 /1 to 4/, left part. The same procedure is useful for developing the recycling conform design, right part in figure 10.

At first we must realize the problem itself. Then we have to formulate the task. The third step is developing ideas for solutions. In the fourth step we should introduce these ideas into practice, and test them. It is most important to wake up the consciousness of the problem in all men participating in our technical system, especially on the management level. There is a lot of possibilities for the application of the recycling conform design without or with little cost; we must find and pick them out in the fifth step. After that we can prepare and carry out measures.

1. Information (Problem)	1. Recognising the problem
2. Definition (Aim)	2. Formulating the task
3. Creation (Ideas)	3. Drawing up solutions
4. Evaluation (Criticism)	4. Waking up consciousness, estimating possibilities
5. Selection (Solution)	5. Selecting suitable measures
(Realisation)	6. Preparing and realising

FIGURE 10. Problem solving process and measures for the recycling conform design

Looking to the design process, figure 11 /4/, with 7 sections according to VDI 2221 /11/ we find that the design in view of recycling at first is important especially in the first section where the task is to be clarified and specified. Here the demand for "recycling conform" must be part of the list of requirements; otherwise it cannot be carried out. During the next sections we should consider e.g. that the modular structure may facilitate the exchange and the test of moduls or parts (see circle II) and the separation of materials if they are non-compatible (circle III). The last 3 sections contain the proper designing with all necessary details; here the designer must be supported by guide-lines and examples as explained above.

FIGURE 11 (next page). Design process with 7 sections (according to VDI 2221) with the results of the sections and the influence of the recycling.

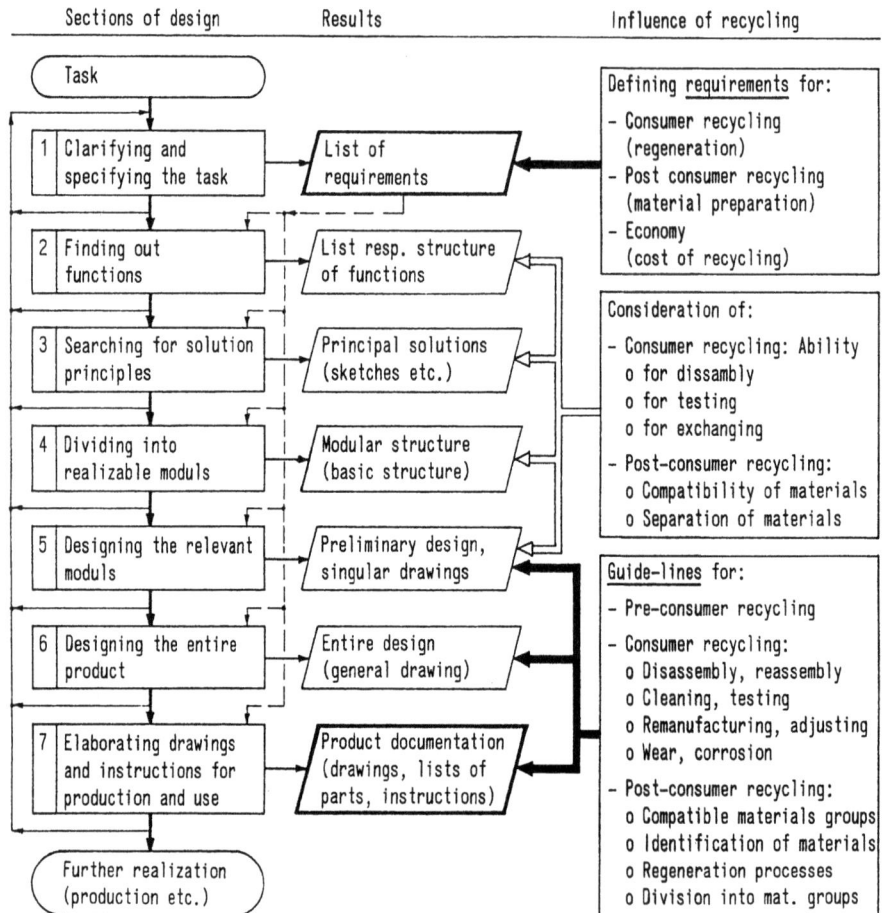

REFERENCES

/1/ Jorden W: Recycling Conform Design. In: Materials Substitution and Recycling. AGARD Conference Proceedings No. 356: Vimeiro, Portugal, 1983.
/2/ Jorden W: Recyclinggerechtes Konstruieren als vordringliche Aufgabe zum Einsparen von Rohstoffen. Maschinenmarkt 89 (1983) No. 61, p. 1406-1409.
/3/ Jorden W: Recyclinggerechtes Konstruieren - Utopie oder Notwendigkeit? Schweizer Maschinenmarkt (1984), p. 23-25 and 32-33.
/4/ Jorden W: Recyclinggerechte Produktgestaltung als Grundlage für ein optimales Stoffrecycling. In: Material and Energy from Refuse (MER 3). Proceedings of the 3rd International Symposium, Antwerpen, Belgium, 1986.
/5/ VDI 2243: Recyclingorientierte Gestaltung technischer Produkte. Düsseldorf 1984.
/6/ Meyer H: Recyclingorientierte Produktgestaltung. Fortschr.-Ber. VDI R. 1 No. 98, Düsseldorf 1983.
/7/ Weege RD: Recyclinggerechtes Konstruieren. VDI, Düsseldorf 1977.
/8/ Huff G, Gräber R, Fröhling R: Measures for Materials Conservation in Aero-Engine Construction. In: Materials Substitution and Recycling. AGARD Conference Proceedings No. 356: Vimeiro, Portugal, 1983.
/9/ Warnecke HJ, Steinhilper R: Instandhaltung, Aufarbeitung, Aufbereitung; Recyclingverfahren und Produktgestaltung. VDI-Z 124 (1982) No. 20, p. 751-758.
/10/ Gehrmann F: Konstruktion und werterhaltendes Recycling niederwertiger technischer Gebrauchsgüter, dargestellt am Beispiel Haushaltskleinmaschinen. Fortschr.-Ber. VDI R. 15 No. 40, Düsseldorf 1986.
/11/ VDI 2221: Methodik zum Entwickeln und Konstruieren technischer Systeme und Produkte. Düsseldorf 1985.

Electrodeposited Aluminium by the SIGAL^R-process. A superior coating
for corrosion protection as an alternative to cadmium and zinc coatings.

H. de Vries

1. Introduction.

The use of approximately 13 million tons of aluminium in the year 1984
in the western world alone shows, that aluminium is a material with
interesting properties. Because of its low density, high ductility, high
electrical and thermal conductivity, good corrosion resistance, non
magnetic behaviour and the possibility of anodizing, aluminium posesses
a wide area of technical applications.

The crude metal aluminium is unlimited in its availility, and there
are no restrictions in its use because of toxicity, as is the case for
example with lead and cadmium. For these reasons numerous attempts have
been made to make aluminium coatings on metals, in order to combine the
good properties of the base metal with the caracteristic properties of
aluminium. Besides plasma spraying, hot dip coating, rolled on coatings,
mechanical plating and PVD-processes, in recent years electrodeposited
aluminium coatings have gained more and more importance.

The electrodeposition of aluminium is only possible from molten salts or
aprotic organic electrolyte systems, as aluminium because of its very
negative standard-potential cannot be deposited out of aqueous solutions.
Although organic electrolyte systems for the electrodeposition of aluminium were described already in 1950, only since a
couple of years electrolytes are available which are suitable for
industrial use.

The breakthrough was realised with the SIGAL^R-process, using a non aqeous
electrolyte which was developed in the SIEMENS R and D centre.
Today the SIGAL^R-process is the only process for electrodepostion of
aluminium on a large industrial scale in operation in the world.
It offers the advantage of a very pure aluminium coating to numerous
applications and industrial branches as galvano-aluminium can be deposited on almost all technically important metals with good adhesion.

PROCESS TECHNOLOGY.

When comparing aluminizing with conventional aqeous plating processes, the parts to be plated pass through the same process steps in both cases:

- Pre-treatment
- Electrodeposition
  and eventually
- Post- treatment.

The parts to be aluminized are pretreated in aqueous solutions. These are matched to the requirements of an adhering aluminium coating, as well as surface protection of the parts by aluminium which is to be effective under various climatic conditions.
As the aluminizing process must take place in a non-aqeous organic medium for electrochemical reasons, the aqueously pretreated material must be subjected to an additional dewatering step beforehand.

The aluminizing process as such is an absolutely clean, ecologically harmless and reliable process step and is performed in an enclosed electroplating cell which is loaded and unloaded through air-locks. The aluminizing electrolyte must remain enclosed, because any contact with atmospheric oxygen and moisture will sooner or later make it useless for electroplating purposes. An eventual post-treatment following the electrodeposition of aluminium would again take place in an aqeous medium.

The electrodeposition proper of the aluminium does not take place from an aqueous solution because of the negative deposition potentional of $E_0 = -1.67$ V but from an aluminium organic complex salt dissolved in toluene. The elctrolyte system has a throwing power comparable with cyanide cadmium electrolyte and can be controlled and even increased considerably by varying the process parameters.

Process-parameters of the electrodeposition of Aluminium.
==

Deposition temperature	100 C_2
Current density	2 A/dm^2
Deposition voltage	5 - 30 V
Specific electric conductivity	$2,5.10^{-1}$ S/cm
Wave shape	Current reversal square wave pulse
Load of electrolyte	0,5 - 0,8 A/l
Rate of deposition	24 um/h
Cathodic current efficiency	100%

Compact anodes of the Raffinal or Super-Raffinal quality grades are
dissolved in the electrolyte when the anode is polarized.
The electrolyte transports the "dissolved" aluminium to the products
which are polarized as cathode.

During this process an aluminium or higher purity than existing in the
anode is deposited on the products. The anode constituents which are
not soluble in the electrolyte are filtered out continuously from the
electrolyte.

The aprotic electrolyte system causes no corrosion whatsoever of the
plating cell, any more then do the reaction products of the
electrolyte with air and moisture. Due to the chemical nature of the
electrolyte it is impossible for hydrogen to be evolved. (Important
for high strength steel).

THE COATING.

Galvano-Aluminium posesses many interesting properties respectivily
property-combinations, which distinguish it favourably from other
elctrodeposited metals. It is very pure and can be deposited without
internal stress on almost every substrate material with conductive
surfaces. The physical and chemical properties are shown in the
following table:

Properties of Galvano-Aluminium (before and after the post-treatment).
==

Purity	< 99,99%
Density	2,67 g/cm^3
Electrical resistivity	3,95 $\mu\Omega/cm$
Thermal conductivity	218 J.s^{-1}.m^{-1}.K^{-1}
Ductility	very high
Elongation of fracture	> 50%
Micro-hardness	21 HV
Micro-hardness anodized layer	490 - 530 HV
Internal stress	15 N/mm^2
Ring-Shear-Test	70 - 90 N/mm^2
(fracture always in Aluminium)	
Layers without pores	from about 8 μm
Coefficient of friction	μ 0,16
(with lubricant)	

An application adepted post-treatment of the galvano-Aluminium coating allows widespread use.

Chromated or phosphated aluminium exceeds the corrosion resistance of other metal-deposits like zinc, nickel and tin. Corossion tests according to DIN or Mill-specs show that galvano-aluminium is at least equal to the toxic cadmium in its behaviour. It also does not form the loose voluminous corrosion products of zinc coatings.

For decorative applications galvano-aluminium coatings can be anodized. Because of the purity of the aluminium the oxidelayers are transparant, very hard and abrasion-resistant and give a good electrical insulation.
They can be coloured weather- and lightresistant in many shades.

Due to the low internal stress and the high ductility the galvano-aluminium coating is suited best to those applications in which the product is subjected to mechanical and thermal loads.

The coating has no columnary structure but it is very dense and pore-free.

Another area for galvano-aluminium is for high temperature applications
until $500^{\circ}C$, whereas cadmium can only be used until $230^{\circ}C$.

Another advantage of the use of the non-aqueous electrolyte is the impossibility for the development of hydrogen. Therefore high-tensile steels can be plated without the danger of hydrogen-embrittlement of the steel. As mentioned before the pre-treatment has to be adepted to the quality-class of the steel substrate.

An independent test-institute in Holland - "The Institute of Metal Technology TNO" - found no deteroriation of the strength of high-tensile steel test-specimens in a fatigue-bending-test according to DIN 50113.

PLANT TECHNOLOGY.

Rack plating.

Fully automatic aluminizing lines are in operation satisfactorily for applying the galvano-aluminizing process on a commercial scale.

Each of these lines comprises a pre-treatment section of conventional design, a dewatering station and the aluminizing section, consisting of the enclosed automatic aluminizing unit with two air-locks and pheripheral equipment such as the electrolyte storage tank and filters.
The automatic aluminizing unit is followed by a conventional post-treatment section. The aluminizing cell with its electrolyte capacity of 15 - 20 m^3 can handle up to 48 racks 500 x 1250 mm simultaneously for coating products with a surface area of 40 to 90 m^2/h with an aluminium layer of 10 - 12 μm.

Barrel plating.

For barrel plating two new installation concepts were developed.
The conventional barrels used in aqeous electroplating electrolytes such as zinc and cadmium, are not attractive for electroplating in non-aqeous
solutions. for economic reasons. The first plant with a capacity of 100 kg/h is in operation in Lelystad.
Production units with a capacity of 600 kg/h will come in operation in the middle of 1988.

ECONOMIC ASPECTS.

Galvano-aluminium is very favourable in cost when compared to its proven quality.
Economically speaking, today the cost level of the plating is below the level of cadmium or copper-nickel coatings with comparable corrosion-resistance. When considering the increasing restrictions on the use of cadmium and the ever more costly disposal of its highly poisonous waste products, galvano-aluminium is the answer to high-quality corrosion protection.

In general galvano-aluminium coatings cannot compete with zinc coatings of comparable thickness in term of price.
However, latest developments in the plating electrolyte as well as the installation technology, have led to a substantial decrease in production costs. The cost-price of the SIGAL coating based on a 90 m^2/h plant has reached a level of 25 - 50% over zinc.

FUTURE ASPECTS.

The ever increasing demand for higher corrosion protection in for example the automotive industry brings the need for other metal coatings than zinc. For environmental reasons, it is no longer desirable or
acceptable to use cadmium for these applications.
Galvano-aluminium is a technical and economical answer to these needs. Now that galvano-aluminium is available on large industrial scale, there is no restriction to implement this superior corrosion protection coating on a wide scale as replacement for cadmium and zinc.

ECONOMIC ASPECTS OF ENVIRONMENTAL PRODUCT DESIGN

GJALT HUPPES
Centre for Environmental Studies
Leiden University
P.O. Box 9318
2300 RA Leiden
31/(0)71-277489

1. INTRODUCTION
Optimizing is the core of micro-economic analysis. The societal rationale for this behavior is that it enhances societal optimality as specified in Pareto-oriented welfare theory. In product design technical and behavioral variables usually can be related to the central value of (in the end) consumer utility or welfare. This welfare is measured practically in money terms. Marginal analysis in welfare theory specifies the individual and societal optimum conditions. Environmental damages do not fit into this framework. Being external effects with public goods charactaristics[1] no monetary measure of (negative) consumer utility or welfare is available. So environmental values are excluded from normal optimum analysis in societal and micro-economic analysis alike. Still at least some optimality conditions, like in minimum cost analysis, would apply it a general measure of environmental damage were available. This measure of effects on environmental values can be constructed for polluting effects of products, using a simplified ecological model. This model converts emissions into 'potential damage'. Addition and subtraction of different types of emissions then becomes possible. In this paper I will explain how this information enables the designer to include invironmental aspects in most types of normal optimum analysis.

2. THE VALUATION OF ENVIRONMENTAL EFFECTS
Specifying all possible chains of cause and effect as induced by design decisions may be possible in principle but not in reality. Correspondingly a designer might specify the chains leading to the consumer utility of his product in principle, but not in reality. If he tries he will learn a lot about relevant factors and mechanisms, the weather, the climate, the education and love affairs of potential buyers, family structures, housing facilities, etc. To some extent this analysis is useful, especially in the creative phases of design. When optimising a product design though these factors are shoveled into a black box and utility is measured in terms of the resulting potential market demand. Of course this demand for a product is the result of all possible uses and their valuation by all potential users. Likewise a full specification and valuation of all relevant environmental effects is impracticable. What can be specified is the <u>potential damage of pollution</u>.[2]

Environmental values may be grouped into three main fields; those on human health, on nature, and on the welfare from material products.
Not all environmental effects emanating from products are relevant. Effects of decisions in product design normally cannot be specified spatially. So all negative environmental effects that exceptionally occur only when the product is made or used in a specific location should not be

also its environmental effects. Its net value is the only justification for negative effects on the environment.

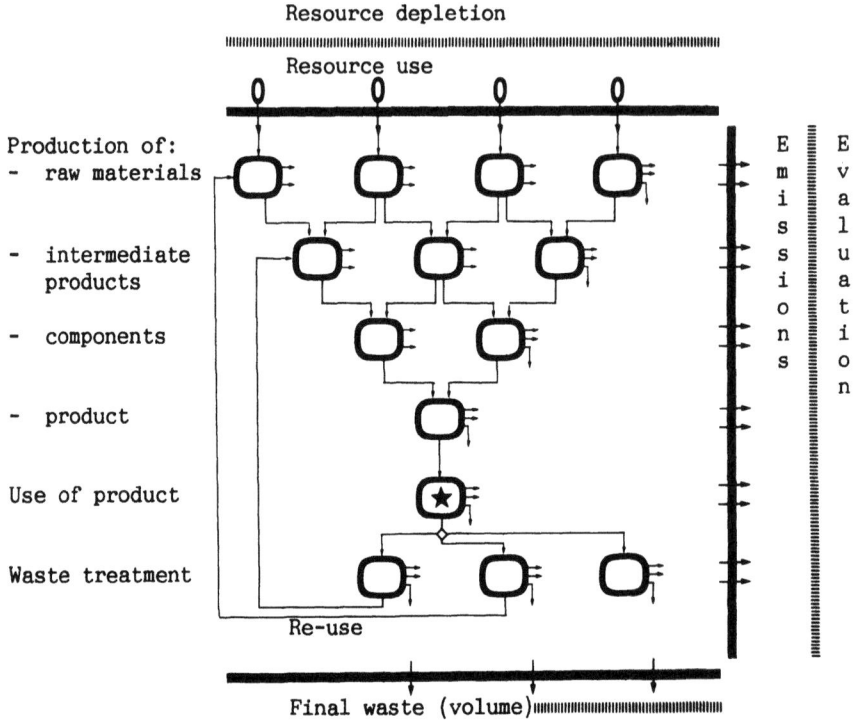

FIGURE 1. Processes and environmental effects of the use of a product.

4. SOCIETAL RATIONALITY.

All methods of emission reduction bring costs to society. For most production-consumption-wastedisposal systems society means the world at large. Every country, firm and consumer tries to confer these costs on others. Societal rationality, be it global or (supra) national, asks for minimizing of costs more or less regardless of their distribution.[8] No central agency can plan these mininum cost points for all production and consumption systems even when given a set of products and processes. Product design by nature cannot be planned at all, but may at best be influenced. What is possible centrally is the sociatal valuation of negative effects on the environment. A unified system for measuring potential damages as described in the preceding paragraphs may guide individuals, firms, private organizations, and governments alike in their search for economically sound environmental improvements. So government planning at the physical level in principle leads to excessive costs, being suboptimal and not being able to plan design. Also physical planning is difficult to implement. The regulations and procedures used cause cumulative costs,[9] totalling to several percent points of the national income within decades.

So why is physical regulation by governments used at all in environmen-

attributed to design decisions. This is the case with local as contrasted with general or diffuse effects on nature. For more practical reasons external welfare effects of design decisions cannot be valued at the moment, not in moneyterms nor in terms of a general physical measure. Loss of some functions of a polluted area is difficult to value in itself and still more diffucult to relate to design decisions. Specific effects of products on human health associated with abnormal use of a product are not the prime responsibility of a designer either. So the main environmental effects that may valued are diffuse effects on human health and diffuse effects on nature as normally caused by pollution.

How can these effects of emissions be evaluated? Generally speaking the damage of an emission is determined by the amount emitted, the chance of (human) exposition by each emitted particle in a certain period of time and the health effect of that exposition. Potential health effects may be quantified using a human ADI (Acceptable Daily Intake) value or the MIC (Maximum Immission Concentration) value as a yardstick for the potential damage of each type of emission.[3] The ADI yardstick results in the number of persons that may be given a just not allowable intake for a day of the given emission. The MIC yardstick as given in quality standards for ambiant air, water and soil leads to the 'amount of air (water, soil) that may be rendered unfit for human use' by a given pollution.[4] Both measures allow addition and subtraction over different types of polluting emissions.

Effects on human health coincide strongly with those on higher animals; they normally even are based on animal experiments. So the 'potential damage on health' is a reasonable indicator of health effects on higher animals[5] as well.

3. ATTRIBUTION OF ENVIRONMENTAL EFFECTS TO PRODUCTS, PARTIAL VERSUS INTEGRAL ANALYSIS

Environmental improvements based on partial analysis of environmental effects, e.g. of 'waste of obsolete products' may lead to perverse, suboptimal results. These only show up when integral systems analysis is applied. Corrosion resistant cadmized metal parts in airplanes e.g. may be replaced by resined carbon fibre parts, which give a relatively harmless waste. The production of this replacement is causing a lot of other types of waste though and the net effect might well be negative. The use of each type of part in the airplane does not generate emissions itself. Thus the integral analysis may show a detoriation instead of the seeming improvement in partial analysis. What is the overall perspective in this type of systems analysis? The central premise is that all human pollution only occurs in activities that are consumptive or are a prerequisite or consequence of that consumption. Material consumption involves the use of a product and this product use is the one and only positive reason for the associated negative things like work and emissions. So products are the focuspoint in this analysis, in contrast to current environmental policy where firms and their processes are focussed on.[6] For any type of product use the integral analysis takes into account the necessary production processes leading to the product, starting with the winning of raw materials; its use; and also the way the obsolate product is disposed of. See figure, for the structure of this lifecycle analysis. Normal market arrangements assure that the benefits of the use to the user of the product outweigh the financial costs of production. The societal costs of emissions are not reflected in the costs for the user[7]. Still the consumptive value of a product should not only outweigh its financial costs but

tal policy? For the sound reason that no other general instruments to combat excessive pollution have been available. Suboptimal inprovements in a very suboptimal state of affairs are better than no improvements at all. When improvements have been effectuated environmental policy may move in two bread directions. Either the instruments of physical regulation may be finetuned, thereby in the end nearing a state of central governmental planning, or more indirect instruments of influencing decisions may be developed. This paper tries to contribute to the latter.

5. MARGINAL ANALYSIS AND OPTIMALITY

In modern welfare theory marginal analysis specifies optimumconditions which assure that the Pareto welfare criterium is satisfied; that nobody's individual welfare (utility) may be increased without reducing the welfare of someone else. These optimum conditions, six in number[10] concerning optimum rates of physical and psychological substition and transformation may or may not be met in reality. One specific set of institutions - perfect markets with purely private goods - would systematically realize the optimum conditions. But, alas, reality is different in some respects. Relevant here is that production and consumption of goods causes pollution of the environment. This pollution may affect the individual welfare of others directly, or indirectly through its technical influence on production processes. Optimal pollution levels may be specified analyzing them as public goods(bads)[11], or by treating them as undepletable external effects[12]. Central in this analysis is the physical charactaristic of pollution that the negative effect on one person or process does not significantly decrease the negative effects on others; that it is undepletable in this sense. For this type of public goods or external effects a Pigouvian tax on emissions may correct the market-mechanism in such a way that Pareto optimality can be restored. The proceeds of the tax should not be payed out as compensation to victims of pollution but should be used for public expenditure or paid out in some lump sum fashion. This theoretical state of affairs which in some form or other has existed since the thirties has not led to much practical environmental policy. It has been impossible to quantify the welfare effects of specific emissions making impossible the specification of the Pigouvian tax[13]. Taxes per unit of emission have been proposed as a means to reach a specified emission level, but nowhere have they been introduced. Emission taxes have gained very limited practical significance, only where they have been used to finance cleaning techniques.[14]

Within the restriction of imperfect knowledge on welfare effects of emissions it is possible to specify necessary conditions for optimality. As a whole these do not yet assure optimality. Assumed is that emissions, though not measurable in welfare terms, can be related to each other in terms of their relative contribution to the unknown welfare effect. So two given sets of emissions defining two states of a public good can be compared to their relative effect on the public good, while the welfare effect of the two states of the public good cannot be compared to the welfare effect of some private good. This means that optimality conditions pertaining to the psychological substitution and transformation conditions for optimality cannot be specified. The physical conditions though do not pose any theoretical problems. They are the minimum cost conditions as specified by Baumol[15] in the context of charges directed at the achievement of some imission standard.

How are decisions in product design related to optimality and institutional arrangements? In this context they are decisions on the joint production of specific outputs; one private product plus some public good (bad). The lack of valuation of external effects and public goods in welfare terms makes it impossible to define an optimum mix of both 'products'. But any given mix should be produced efficiently which means that no reduction in emission should be possible without reducing the product-amount available for private consumption as well.[16] As there are no Pigouvian taxes as an institutional correction of the market - and they cannot be expected in the foreseeable future either - price signals alone cannot give the motivation and guidance for optimal product design. These two elements, motivation and guidance, go together in the market mechanism. Necessary this is not.

Suppose that some incentives do exist giving a motivation to efficiency in producing a product-emission combination. This motivation may be the result of collective arrangements like contests for environmental product design[17], public knowledge of company environmental performance, some information to the public of environmental effects of products,[18] combined with individual ethical motivation of designers, employees and consumers. How then could this motivation be used to approach at least some conditions for Pareto optimality? How could the guidance on which optimal behavior is based be structured and institutionalized? Basically by making available to the designer (and the company and the public) the information on the relative impact of his design decisions on environmental quality. More precisely, what should be known is the effect of a change from one product-emission mix to another, that is the marginal pollution effect of the decision. That effect should be related to the net change in financial costs and benefits of the decision. By systematically choosing the highest cost-effectiveness the minimum cost point is approached. The amount of output in terms of potential emission hazards may first be kept constant in this analysis. However, when it becomes clear that emissions may be reduced at very low or negligible costs the given motivation will lead to some trade off between costs and emissions. This transformation may be optimized at the individual level, as is proposed here, but not at the societal level. The optimality condition for the latter would be that in all situations concerning all firms and households this rate of transformation should be the same. In product design this type of information usually is highly secret, if available at all. What is possible though is that some mechanisms of social control might influence the individual trade off decisions in the direction of some social norm of x cents per reduced unit of potential emission damage.

For society as a whole such a combination of social incentives and informational guidance, however second best they may be compared to the ideal but irreal Pigouvian taxes, may be a major improvement on the physical regulation of mainly production processes and the clean up by governments of polluted waste, sewage, soil, and water.

6. INFORMATION: CONTENT AND FORM

Given some motivation to environmental improvement in product design the guidance is given by the availability of the relevant information. Imagine that knowledge of environmental effects of all possible processes and products and their interrelations is available in huge database with a highly sophisticated thesaurus. How could the poor designer who wants to compare the environmental effects of his two product alternatives retrieve and combine the relevant informational items? Which of the twenty odd possible processes to produce steel should he choose? Which of the many possible ways to handle the diposed of products is relevant? Clearly the designer will be buried under the overload of relevant but unstructured information retrieved.

Economists, when analysing a profoundly interrelated changing reality, make life bearable to themselves by assuming most things constant while varying a few specific variables. This 'ceteris paribus' clause may have theoretical defects, for practical purposes the resulting quantitative approximation may suffice. In a comparable fashion designers of products disregard almost all processes and treat the remaining well chosen set of `typical'processes as constant.[19] Only for those processes the environmental information needs to be specified as the empirical basis for the Database on Environmental Effects of Products (DEEP).[20]

The second way to reduce the information overload is by aggregating related processes. The chain of processes leading to a steel tube as used in a bicycle (winning of iron ore and coal, transport, production of steel, plating, cutting, welding), thereby is seen as one process ('making steel tubes'). It is the cumulation of all emissions of all processes necessary to make the specified amount of tube. Thus not only pro-cesses but also intermediate products, from materials to complex functional parts, become elements in the database. This makes the database more complicated but relieves the user of a lot of specialized work.

The third way to reduce the information load is to give the emission in the aggregated, valued format; in terms of the number of people with a not allowable intake or the volume of environment (water, air, soil) that might be rendered unfit for human use. For overall environmental comparisons this suffices. The technical more complex analytical information may give guidance to environmental improvements in the design.

For very practical reasons the database (DEEP) should be compatible with existing systems for computer aided design. Preferably it is intregrated in these systems. The necessary standardization of terminology and development and expansion of programs may prove impossible between different design fields (small consumer goods, durables, cars , housing, etc.). Then several systems should be developed.[21]

Summing up, the information for optimal environmental design may be made available practically in four steps.

1. The generation of the relevant empirical information on environmental effects of processes. This step cannot be made by individual users and should be financed and organized collectively.
2. The normative agreement on the valuation of the environmental effects in terms of potential of hazards of emissions. This agreement is collective by nature.
3. The structuring of the empirical and normative information in the three named basic formats to make them manageable for designers. The basic formats should be made available collectively with private extensions into specialized fields.
4. The integration of the formatted information in one or more interactive computer programs.

7. MONEY AND VALUES

Societal rationality and that of individual firms generally do not coincide when external effects are concerned. Corporate ethics, private action and public policy can try to repair this regrettable state of affairs by supplying moral, financial, and administrative incentives for pollution reduction. This 'soft' internalization is especially important in product design where government regulation is not feasible. Concerted action on a common empirical and normative basis will enchance optimality. Also there will be a subjectively reasonable sharing of costs as no firms will voluntary take a financial burden that is deemed unreasonable. Increased cost-effectiveness by improved societal optimality implies lower costs to reach an improved environmental quality. Combined with a reasonable sharing of costs environmental product design means a financial improvement for (nearly) all firms and consumers concerned.

REFERENCES

1. See § 5.
2. The parallel goes further. Market research is oriented at specific potential uses. The real uses normally are not analysed. In the same manner environmental valuation is based on some potential chains of effects and the real effects are not analysed.
3. An example of an ecological model is developed by W.T. de Groot en A.J. Murk, 'Risico's van gif; een globale, stapsgewijze integratie van ecologische risicofactoren', Milieu 1986-1, pp 8-14. The chance of intake or immission may be determined as an average for each type of hazardous emission.
4. This latter type of measure has been used in a number of empirical studies:
A.E. Druijff, Milieurelevante produktinformatie, CML, Leiden 1984
Oekobilanzen von Packstoffen, s.n., Bundesamt für Umweltschutz, Bern 1984
M.M.H.E. van den Berg e.a., Potenties van produktbeleid, CML, Leiden 1986
W.G.A. Steeman, 'Milieuaspecten van het ontwerpen', TIO-TUD, Delft 1986.
5. Effects on the abundance of lower animals (e.g. worms) may have profound effects on the population level of higher animals (meadow birds), which is not reflected in this health measure.
6. Empirical possibilities for environmental improvements for a number of products and possible instruments for product directed environmental policy are given in G. Huppes and M.M.H.E. van den Berg, 'Milieuhygienisch produktbeleid ', Milieu 1987-2, forthcoming.
7. Disposal of obsolete products also is included only to a limited extent in the costs of using a product.
8. Redistribution of income can compensate for unwanted distribution all effects of evironmental policy. This possibility of compensation, formulated by Kaldor and Hicks, is the central criterium in cost-benefit analysis.
9. See G. Huppes and H.A. Udo de Haes, Stofstatiegeld voor marktkonform milieubeleid; met een voorbeelduitwerking voor het anti-vermestingsbeleid, CME-notitie nr.6, Leiden 1987.
10. The conditions can be specified in different ways and are based on static analysis. See W.J. Baumol and W.E. Oates, The theory of environmental policy; externalities, public outlays, and the quality of life, Englewood cliffs N.J. 1975; D.M. Winch, Analytical welfare economics, Harmondworth 1971; P. Hennipman, Welvaartstheorie en economische politiek; Alphen a/d Rijn 1977; and F. Hartog, Toegepaste Welvaartsekonomie, Leiden 1973.
11. See winch.
12. See Baumol pp 16-28 or Hennipman pp 180-89 for the different types of and terminology on external effects. External effects can only be specified relative to some defined (real or imaginary) institutional arrangement.
13. Very often is it also practically impossible to measure the physical emission themselves. The emission of NO_x. in the exhaust of a car for example is mainly dependant on driving behavior.
14. The main functioning example is the tax in the Netherlands on oxigen demanding emission to water. See J.Th.A. Bressers, Beleidseffectiviteit en waterkwaliteitsbeleid: een bestuurskundig onderzoek, Thesis 4th imp. Enschede 1984, who analysed the quantitavely very important preventive effect on emissions.

15. See Baumol pp 142-4
16. This is the corollary of the condition that no increase in the production of one product should be possible without decreasing the output of some other product.
17. See the English 'Pollution Abatement Technology Award'.
18. Like the German 'Umweltzeichen' on relatively harmless products.
19. Or an average of several alternative processes.
20. This database can also be used in process development by comparing the environmental performance of a process design to the standard process.
21. The standardization in many fields is taking place now internationally, generally without reckoning with environmental values.

WASTE PREVENTION IN THE ECOLOGICAL BUILDING PROJECT OF DELFT
UNIVERSITY OF TECHNOLOGY

ir. Hans HUBERS

1. INTRODUCTION
A multidisciplinairy workinggroup of 8 departments from 3
faculties of the Delft University of Technology is carrying out
a research on energy saving building and environmental
technology. The preliminary design 'the Egg' is used as
startingpoint. Feasebility-studies on 8 sub-divisions, waste
prevention is one of them, will bring out the necessary changes
in the plan. The starting phase with a duration of allready 5
years showed: a building project is the only possibility for
cooperation of several disciplines. Integrational research (the
second phase) should lead to an appropriate or rather new
design.
The functions of the building are:
- testfacility for multidisciplinairy research
- demonstration of this R & D-work
- office for the Centre for Environmental Technology and
  Environmental Management.
Collaboration with companies, bringing in their knowledge and
experiences in the field of construction and costs, is an
absolute demand. On most of the sub-divisions this is realized.
The financing of project will be beard in equal parts by the
University, the Government and trade and industry. The price of
the Egg is computed at 6 miljon DFl, but still could be reduced
by sharp designing. The research before realization is
estimated at also 6 miljon DFl and the evaluation research at
9 miljon DFl. Those figures are provisionally. A marketing
bureau is consulted for a more definite marketing and study of
possibilities for fundraising.
Initial costs normally make out only one-fifth to one-eighth of
the life-cycle costs of buildings. The calculation of annual
costs give better comparable figures. This has been done for
the Egg and resulted in running costs (all in, ex. tax, 7 %
interest) of 300 DFl/m^2 total ex. ponds. This is at the same
level as a standard office building in the Netherlands. In the
next stage of the project this can be optimized.

2. WASTE PREVENTION AND ENVIRONMENTAL TECHNOLOGY
Allready at the first drafts and through the 'programme of
demands' decisions are being made, that have great influence on
the use of energy and materials for the building, including the
waste. In building (construction and use) 50 % of the world
materials and energy is used. It creates the larger parts of
waste.

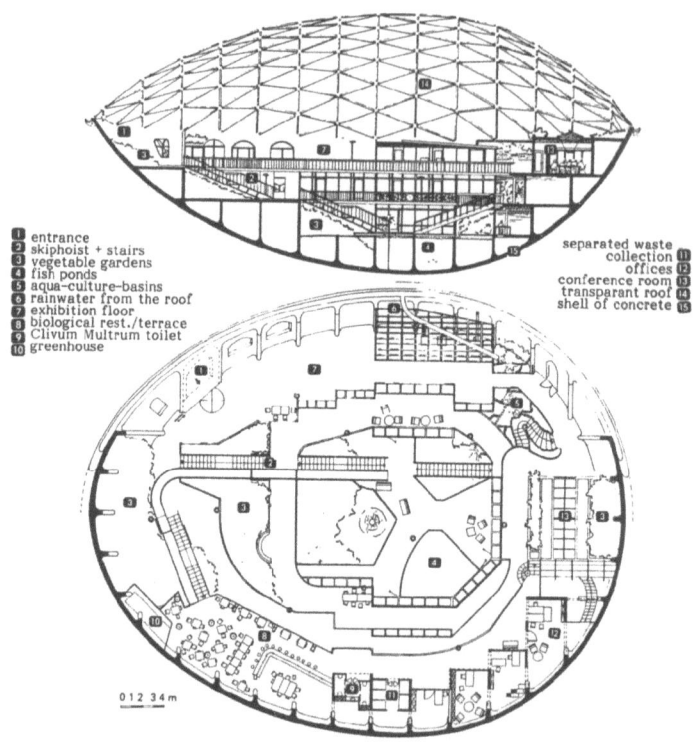

FIGURE 1. Longitudinal section and floorplan

The general idea of the project is that through integration of several activities a better ecological balance is possible. So for instance introduction of fish and vegetable production means low energy and low water costs for this production. At the other hand the garden and fish ponds improve the environment for the users of the building in both psychological and physical way (O_2, H_2O-vapour, dust collection, shade etc.)
The decision to work out a half-underground variant means one can not avoid the use of reinforced concrete, while the 'energy dèmand'of this material (4,5 GJ/m^3), the availability of raw materials and the possibilities for re-use are hardly optimal. Meeting the goals of the project can be done by using minimal reinforcement (60 % of the energy-demand) and prefab elements that could be re-used. This is possible by using a shell. Other energy saving methods could be harnessing fly ash and other waste products in the concrete composition. In any case, the economization in foundation on piles is fundamental in costs and waste.
With an other design there could be used more wood (energy-demand 2,3 GJ/m^3). Of course no tropical wood (10,5 GJ/m^3), because every year the tropical forests are reduced with

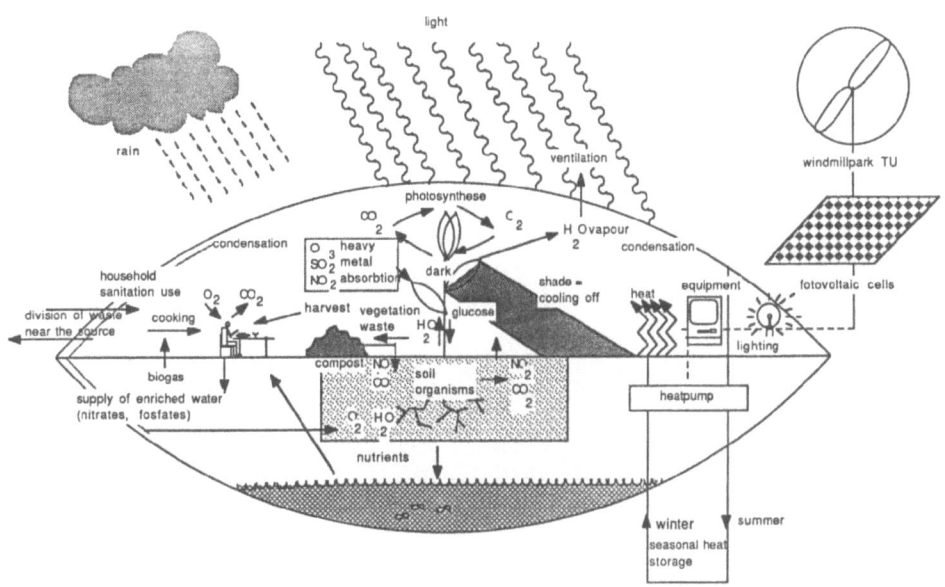

FIGURE 2. Global scheme of the cycles

120.000 square kilometres while they have to produce 32 % of
the oxygen on earth. The Netherlands are for 80 % of their
demand for wood dependant on foureign countries. If there are
not taken any measurements the wordl's wood supply will be
finished in 75 years. At the other hand, an enlargement of 30 %
of 'production forests' will be able to fullfill not only the
world's wood demand but also the world's energy demand.
This is the reason why within the project the possibility of
using Dutch round wood for (dismountable) venerable
constructions is investigated. This can be done with the wire
lacing tool, which gives a good compensation for the negative
effect of contraction cracks. A development of the Faculty of
Civil Engineering of Delft University of Technology. In
Lelystad is made a pilot-shed with such a construction. Also
for other parts of the supporting structure and finishing will
be investigated if materials could be (re-)used in such a way

FIGURE 3. Wire lacing
tool and typical
meeting of struts

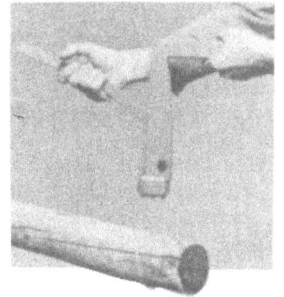
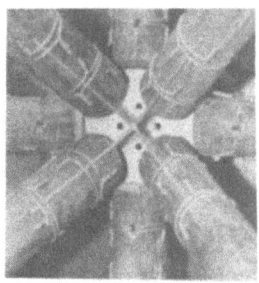

that a minimum of energy and raw material will be spoiled. The
installations are taking care of a pleasant climate without the
use of fossil fuel, due to a design which abelize passive
solar energy and maximum daylight. It can be done with a
combination of photovoltaic cells, a wind-mill, a heatpump and
seasonal heat storage.

3. TREATMENT AND RE-USE OF WASTE
The main objective of waste treatment in the Ecological
Building is: as much re-use of effluent and solid waste as
possible. Valid secondary goals are:
- maximal use of rainwater, together with minimal use of tap-
  water;
- minimizing the use of energy and raw materials;
- economizing on the investments and maitenance costs;
- produced waste should be processed in the building as much as
  possible.
Along with this objective, there are several demands and
limiting conditions which have to be met:
- use of raw materials and products that do not harm
  men and environment;
- the whole should be clean and hygienic;
- stench annoyance is not allowed;
- the necessary behaviour on the toilets by the users should
  not be very different from usual;
- re-use of materials should not be dangerous for health.

To meet the main objective, in the first instance two methods
of waste treatment were considered:
a. biogas installation
b. compost toilets

A biogas installation has to its advantage: it produces gas
which can be used for energy generation. The effluent could be
used for the irrigation of the gardens inside the building. The
hygienic and bacteriological reliability of this kind of
irrigation, especially for consumption crops, shuold be
investigated and guaranteed.
To its disadvantage: the installation should have a constant
temperature of 30 to 35 degrees Celsius. Of course this is
rather negative for energy efficiency. There are some
requirements for the food supply of the methane bacteria in the
installation. One of them is about the carbon/nitrogen rate in
the effluent, which should be at least 16. After analysing all
the expected waste produced in the building, this appeared to
be impossible. Responsible is the, proportionately, large input
of faeces and urine, which contain toomuch nitrogen. In case a
biogas installation is used, there should be a complementary
treatment for this input.
In a compost toilet faeces, urine, kitchen and garden waste are
composted together. To insure the bacteriological reliability a
high temperature is needed during a rather long time. Relevant
literature learned that this can't be guaranteed. Before a
compost toilet could be used for consumption crops, a lingering
period of two to three years would be necessary. This requires

a hughes storage which is hardly available inside the building.
Besides this the produced quantity is much more then needed, so
surplus should be shaken off in one way or an other, for
instance by selling. Also the objective of normal toilet
behaviour con not be realized (toilet flush for instance). Last
but not least, the compost toilet doesn't produce energy. The
conclusion is that a biogas plant is a better solution.
However this means that adjacent treatment of faeces and urine
is necessary. Best methods for this are the trickling filter
and the activated sludge installation. A trickling filter needs
much more space but less energy. In this case there was enough
space.

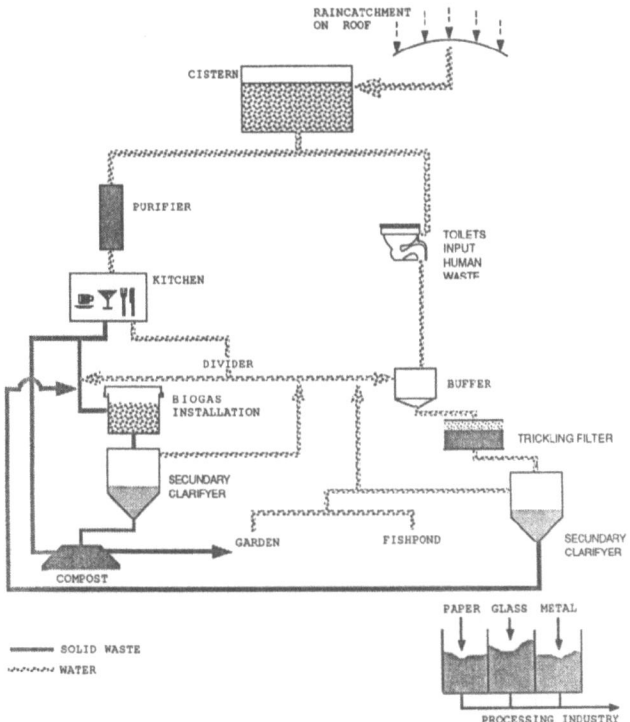

FIGURE 4. Scheme raincatchment and sanitation.

Rainwater from the roof will be collected. In an average year
this will be enough to fullfill the demands. For the use as
toilet flush water no treatment is necessary. However for the
use in the kitchen and for washing hands etc. the water has to
be filtered in a slow sand filter. This needs little energy and
gives a high quality drinking water. It takes very little space
(d = 0,65 m, h = 3,50 m).
Finally the system illustrated in figure 4. has been worked out.
The figure speaks for itself, exept the buffer which need some
explanation. It serves two purposes:
1. the buffering of the input of the trickling filter for
   permanent good conditions for the bacteria;

2. homoganisation of the input, in order to avoid obstruction of the spray installation and other parts of the trickling filter.

The input of the system will be:
- rainwater
- faeces
- urine
- garden and kitchen waste

The output:
- biogas 6-8 m³/day, enough for cooking;
- cleaned water for the garden and fish ponds;
- compost for self sufficient use at the gardens.

Besides this system there will be seperated waste collection of glass, metal, paper and other waste that will be transported to the recycling industry.

installations	trickling filter		activated sludge	
scenario	1	2	1	2
volume (m³)	16,7	8,0	1,71	3,6
h * d or l (m²)	2,5*2,9	2*2,55	1*1,71	2*2,18
sludge production (kg/d)	0,9	0,4	1,05	2,0
energy use (W)	2,3	1,15	24,8	88,4
initial costs (DFl)	4695	2788	4266	4620
energy costs (DFl/year)	5,1	2,55	55,0	202,6

scenario 1: all waste water is treated
scenario 2: only kitchen water and urine

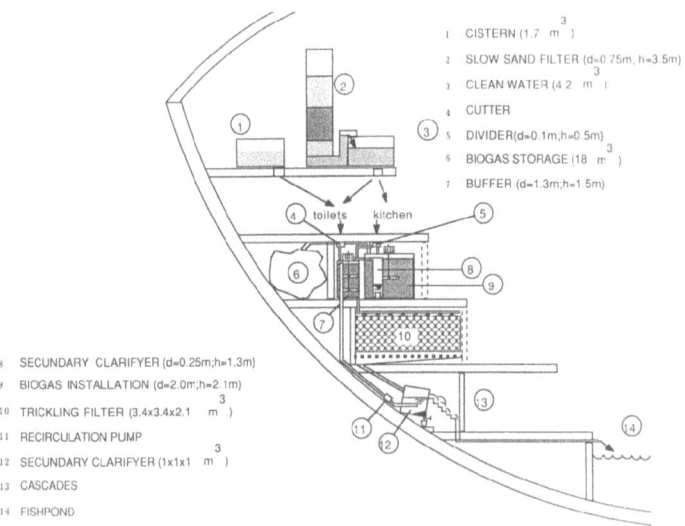

1 CISTERN (1.7 m³)
2 SLOW SAND FILTER (d=0.75m; h=3.5m)
3 CLEAN WATER (4.2 m³)
4 CUTTER
5 DIVIDER (d=0.1m; h=0.5m)
6 BIOGAS STORAGE (18 m³)
7 BUFFER (d=1.3m; h=1.5m)
8 SECONDARY CLARIFYER (d=0.25m; h=1.3m)
9 BIOGAS INSTALLATION (d=2.0m; h=2.1m)
10 TRICKLING FILTER (3.4x3.4x2.1 m³)
11 RECIRCULATION PUMP
12 SECONDARY CLARIFYER (1x1x1 m³)
13 CASCADES
14 FISHPOND

FIGURE 5. Adjustment in the building.

APPLICATION/UTILIZATION OF SALVAGEABLE REFUSE COMPONENTS

Second European
Conference on
Environmental Technology

in the 'European Year
of the Environment'

APPLICATION OF SOME WASTE MATERIALS IN HYDRAULIC ENGINEERING

K.W. Pilarczyk, Rijkswaterstaat, Road and Hydraulic Engineering division,
            P.O. Box 5044, 2600 GA DELFT, The Netherlands
G.J. Laan, Rijkswaterstaat, Road and Hydraulic Engineering division,
            P.O. Box 5044, 2600 GA DELFT, The Netherlands
H. den Adel, Delft Geotechnics, P.O. Box 69, 2600 AB DELFT,
            The Netherlands

1. INTRODUCTION
   Industrial waste products often form a great problem regarding their storage (dumping) and environmental consequences.
In The Netherlands considerable experience has been gained with the application of industrial waste products as alternative materials in hydraulic engineering (i.e. bank and bottom protection, filter constructions, fill material of closure structures).
For the greater part the general applications have been carried out based on practical experience without scientific research into the structural and environmental implications. This kind of research has recently been started, mainly as a result of the growing awareness of environmental impacts of application of industrial waste products.
The research is focussed not only on the commonly used waste materials, but also aims at investigating the usefulness of other residual materials.
Waste materials that we commonly used are minestone, several kinds of slags and silex (a byproduct of the cement industry). A new development concerns the application of the so-called "Euroclay" in dike embankments. Euroclay is formed by consolidation under given conditions of the lightly polluted silt from the Rotterdam harbour area.
The relevant engineering properties of the residual products have already been studied intensively. This has led to the drafting of requirements as to their use, and to the conception of contract specifications for quality control. The ideal is to achieve a situation in which the materials will be supplied under certificate.
A study of the environmental aspects involved has only recently been taken in hand. A first result of this concerns the exclusion of lead slags as material in hydraulic engineering structures.
This study should lead to special requirements with regard to the environment, and to systems for an adequate quality assurance.
This paper will review the existing knowledge and experience on application of waste materials in hydraulic engineering in The Netherlands. Special attention will be paid to define the environmental requirements when using these materials.

2. DUTCH POLICY ON ALTERNATIVE MATERIALS
   The endeavour to increase the use of alternative materials arises from two general public concerns (10):
a. a safe storage and processing of the increasing amounts of waste materials (table 1): this is primarily the responsibility of the Minister of the Environment (waste materials policy).
b. reducing stripping of surface minerals (table 2); this is primarily the responsibility of the Minister of Public Works.

waste materials	amounts in 10^6 ton/year
- building and demolition wastes	6
- dredging sludge	48 (18 dry)
- coal fly-ash	0.6
- coal cinders	0.1
- incineration fly-ash	0.07
- incineration slags	0.65
- steel slags	0.3
- phosphor slags	0.6
- phosphoric acid gypsum	2.0

Table 1: Production of waste materials in the Netherlands

minerals	amounts in 10^6 ton/year
- gravel	15
- sand for industrial purposes	19
- sand for embankment filling	33
- clay	3 to 4
- marl	2

Table 2: Production of surface minerals in the Netherlands

Storage and processing of waste materials, and stripping of surface minerals cause environmental and planological problems, which can be diminished by re-use of waste materials.

Such re-use may also influence, both positively and negatively, factors as energy consumption, employment, labour conditions, and commercial interests (investment in innovative developments). A basic guideline of a policy aimed at stimulating the use of alternative materials in works directed by Rijkswaterstaat, should be that when deciding whether or not to use such materials, not only project-costs, but also the above-mentioned public interests be taken into account. Higher project costs might then in some instances be acceptable. Based on the fore-going, Rijkswaterstaat (RWS) involvement with alternative materials has five approaches:

a. the production of waste materials from construction/maintenance works
b. the possibility to use alternative materials in works directed by RWS
c. the responsibility RWS has for the quality of surface waters; both storage and re-use of waste materials may damage the quality of surface waters
d. the part RWS can play towards innovation; in her policy for awarding research and construction contracts, RWS can control and direct technological renewal.
 For this to work well it is necessary that the needs of both RWS and the industry are clearly formulated at an early stage so that both parties can adjust themselves on time;
e. in close connection with b, c and d, the (co-)responsibility for stimulating, in close cooperation with industry, research institutes, etc., the compilation of general quality requirements, standards and regulations.

From the fore-going it may be concluded that, based on these five approaches, the Dutch strategy is aimed at defining what Rijkswaterstaat can do within its own services to increase the use of alternative materials. Recommendations to other public authorities and industry etc. to promote the use of alternative materials are given, when such is desirable, under the (primary) responsibility of the Minister of Transport and Public Works, in national policy statements regarding borrow pits.

The realisation of the policy mentioned above is the main task of the Road and Hydraulic Engineering Division of Rijkswaterstaat. Many detailed studies on the technological, environmental and practical application have been undertaken in the past 15 years.

Most studies were on the use of concrete- and masonry rubble in road constructions and water works, and as aggregate in concrete; the use of minestone, steel slags and phosphor slags in the same areas; the use of fly-ash and cinders, the use of consolidated dredging sludge ("Euroclay"), recycling of asphalt, the use of incineration slags and the possibility of re-use of gypsum waste (especially phosphoric acid gypsum).

Recently a national project on application of alternative materials has been started at the Dutch Centre for Civil Engineering, Research, Codes and Specifications (C.U.R.) in cooperation with Rijkswaterstaat, research institutes and industrial organizations.

The aim of this project is to formulate generally valid specifications (including environmental aspects) for various civil engineering applications.

3. TECHNOLOGICAL ASPECTS AND DESIGN REQUIREMENTS
3.1. General
Industrial waste materials have a long record of application in Dutch hydraulic engineering construction. Practical experience with the properties of these products and their competitive price compared to natural materials stimulated these applications. Much used materials were granular materials such as minestone, slags from metal production and phosphor production.

In recent years, the environmental aspects of these applications have been studied. Also, the engineering properties of these materials is receiving increasing attention, and their applicability is being increased by optimizing these properties and improving the quality control.

The experience-based introduction of industrial waste materials in hydraulic engineering construction has thus given way to concentrated research into new applications. The increased awareness of the need to re-locate waste materials in a fashion not harmful to the environment, and the limited availability of natural materials has stimulated this research. The research is directed both at obtaining suitable engineering properties and at solving the problems associated with the leaching of harmful matter from e.g. phosphoric acid gypsum, fly-ash, incineration slags and contaminated dredged material. Immobilization of harmful elements in the waste material, and creation of useful engineering elements may be achieved by baking, binding by cement and isolation of the wastes.

This contribution is limited to industrial waste products, with which much experience has been acquired in Dutch hydraulic engineering.

3.2. Alternative materials
Industrial waste materials re-used in hydraulic engineering construction are: minestone, various slags such as LD-slags, phosphor slags, copper slags and silicomanganese slags, silex, "Euroclay" and demolition rubble.
Minestone was generated in large quantities by the coal industry in the Dutch province of Limburg. After this industry came to an end, many of the minestone dumps have been used for landfilling and reclamation of old gravel, sand and clay borrow pits in Limburg. Minestone used by the hydraulic construction industry, about 0.5 million tons annually, has been imported exclusively from the Ruhr area in West Germany and the Zolder area in Belgium, the transport costs being relatively low in these cases.

Part of the slags are produced in the Netherlands. "Hoogovens IJmuiden B.V.", generates about 300,000 tons of LD slags annually (LD is the abbreviation of the Linz-Donawitz steel production process). However, most LD-slags used in hydraulic construction, about 0.2 million tons, are imported from Belgium and the Ruhr area. 0.5 million tons of phosphor slags are generated annually in the production of phosphor by Hoechst near Flushings. Approximately half of this is used in hydraulic construction.
Copper slags and silicomanganese slags are imported from West Germany and Belgium respectively, in amounts of about 35,000 tons each.
Apart from the LD process for steel production, other processes used in the Netherlands generate small amounts of slags which are also used in hydraulic construction.
In the past a little more than 1 million tons of lead slags have been used in construction works. Due to the unacceptable leaching of especially lead, but also copper and zinc, these slags are not used any longer.
Silex is a waste material which results from quarrying marl for the production of cement. Marl is quarried in Belgium and the south of Limburg, in annual amounts of approximately 0.3 and 0.2 million tons respectively.
"Euroclay" is produced from silt dredged in the Rotterdam harbours. Its application has been rather limited up to now. Dredging to maintain the depth of access waterways and harbours produces some 25 million m^3 sludge. Of this, 14 million m^3 is only lightly polluted, and through dewatering and consolidation can be rendered useful as clay for dike construction.
Demolition rubble has long been used in hydraulic engineering. Due to changed construction methods and the development of methods to upgrade rubble to make it suitable for road construction, use of rubble in hydraulic construction is now insignificant.

3.3. Engineering properties

The engineering properties of those industrial waste materials used to date in hydraulical construction are often not inferior and sometimes even superior to those of traditional materials. This is especially true of the good frictional properties of slags due to their roughness and angularity. A number of slags possess a high density making them very suitable to resist waterflow and wave action.
Some of the less favourable properties are the weatherability and low crushing strength of some waste materials (8).
Minestone has a D_{50} (a measure for the theoretical sieve retaining 50 % of the material) up to about 65 mm. Its density varies between 2.4 and 2.6 t/m^3. Minestone is rather sensitive to crushing and weathering. Under water and in deep fills it remains intact. As long as it is not remoulded, even strongly weathered minestone retains a permeability which is at least equal to that of slightly silty sand.
Slag pieces are generally rough, angular and more or less cube-shaped. Gradings up to D_{50} values of about 100 mm are feasible.
The density of LD slags is between 3.1 and 3.4 t/m^3. The basic strength is high. However, individual pieces of slags may crush rather easily due to the presence of cleavage planes, layering and internal stresses.
A small proportion of the slag pieces is unstable due to so-called lime pitting, concentrations of pure calcium.
The iron present in slags can shorten the service life of polypropene fabric, which is used in hydraulic engineering as a filter fabric beneath a layer of stones.

Copper slags have a density of 3.9 t/m³. It may be slightly unstable when the mineral kirschsteinite is present, which swells in water and thus may result in desintegration.
Silex, density 2.6 t/m³, is a strong material. Gradings having a D_{50} value of slightly more than 100 mm can be obtained. Contamination by weak and weatherable limestone can strongly reduce its usefulness.
"Euroclay" is equal in quality to the traditionally applied river clays and marine clays. A proper production process in the so-called "clay factory" is important however. A good quality control both of the mechanical-physical properties and of those properties important from an environmental viewpoint is necessary to stimulate its use and to remove unjustified prejudice. The realisation of a good quality control is presently underway. This is expected to result in the increased application of "Euroclay" and other substances in dike-strengthening projects, for which in the coming years annually a few million m³ is needed.
In general, a good quality control can considerably increase the use of materials produced from industrial wastes. It can especially effectively counter prejudice which has arisen from a justified anxiety regarding environmental effects of re-use of such materials.

3.4. <u>Environmental aspects</u>

The environmental aspects of the use of industrial waste products in hydraulic construction depend not only on their chemical composition, but especially on their leaching behaviour. The Researchgroup Development Standard Leaching Tests Incineration Residue (SOSUV) has developed tests to study this behaviour. These tests concern, amongst others, the determination of the content of anorganic micro contaminants, the determination of the maximum leachability, shaking tests and static tests. These tests allow to compare the leachability of different materials. To decide on the suitability of a given material for use in hydraulic construction, supplementary data are often required, which facilitate the "translation" of laboratory results to in-situ circumstances. The difficulties associated with applying laboratory results to in-situ circumstances have to date prevented the development of unequivocal standards for judging the results of laboratory tests. Models are necessary for this, but these must still be developed. These models must correctly account for the governing parameters such as the particle size, the velocity of waterflow past the particles, the total mass of the material in relation to the quantity of the surface water with which it is in contact, etc.
As such unequivocal standards for the interpretation of laboratory tests are still lacking, some subjectiveness is unavoidable.
In the following the environmental aspects of a number of materials will be treated (5), (6), (7).
Lead slags have been used in large quantities in hydraulic construction in the past. This has now been prohibited due to an unacceptably high level of emission of heavy metals.
Regarding pollution and the leachability of minestone, distinction must be made between recently produced minestone and minestone from dumpheaps, as the latter possibly contain chemical refuse. Moreover, it is unlikely that PCB's which are used in hydraulic fluids, will appear in recently produced minestone as use of oils containing PCB's is prohibited since January 1st, 1986. Minestone from dumpheaps yields on the average twice as much sulphate on leaching as recently produced minestone. Sulphate content varies strongly; in dumpheaps the average is approximately 2000 mg/kg dry matter.

Notwithstanding the above-mentioned differences between recently produced minestone and minestone from dumpheaps, some restrictions in the use of the former will sometimes be necessary.
Leachable PAK content in minestone averages 0.7 mg/kg dry matter. Also minestone may contain leachable arsenic and strontium. Heavy metals pose no problems in minestone.
Application of minestone in stagnant fresh water may result in exceedance of the TMP basic quality norm for sulphate, and plans for any such application must be carefully reviewed. In less sensitive waters (salt water or running water) emission of sulphate from minestone poses no threat to water quality.
The fore-going also applies, though less so, to the other parameters. If PCB's might occur however, use of such minestone is not advised.
LD slags and silicomanganese slags contain small amounts of heavy metals such as zinc, copper, chrome and lead. Emission of these metals in salt and fresh water is quite insifnificant. On the grounds of leaching tests which yielded this result, and on environmental grounds, no restrictions are imposed on the use of these slags in hydraulic construction as yet.
The only possible effect phosphor slags could have on the environment is through the emission of fluoride. These have a high initial level of emission and thus use of fresh phosphor slags in almost stagnant water and in water catchment areas is usually inadvisable. This applies particularly for large quantities of phosphor slags in stagnant and relatively small waters.
Copper slags from two locations have been tested for composition and leaching properties. One of these contained only small amounts of heavy metals and yet was found to leach rather easily, yielding copper, lead and zinc. The other sample, from the Ruhr area had a high copper content and yet its leachability was not worse than that of LD slags. Further research is necessary to confirm whether copperslags from the Ruhr are as acceptable as LD slags.
Silex consists of flintstone, pit gravel and limestone. There are no environmental restrictions against its use in hydraulic engineering.
"Euroclay" is weakly contaminated by a range of materials which might endanger the environment. Further research must be performed into the environmental effects of "Euroclay" in dikes. For the time being its use is permitted by the authorities, as long as it is covered by a toplayer of uncontaminated clay.

4. APPLICATIONS
4.1. General applications
The use of alternative materials such as minestone, slags, silex etc. in Dutch civil engineering is very common and goes back for quite some time. In the scope of reconstruction work which followed the flood disaster of 1953, Dutch civil engineers began to rely more heavily on these materials as a construction material, an their use in hydraulic engineering has increased ever since.
The main reasons for the growing preference for alternative materials in the Netherlands are:
a. shortage of natural rocky materials;
b. greater resistance to current and wave attack as compared to sand and roughly equal to gravel and light sortings of rockfill;
c. availability in relatively large quantities;
d. relatively low cost;
e. growing problem of storage of these materials.

Alternative materials are used in hydraulic engineering in several ways (see Fig. 1). They are often used as retaining-bunds of sand-closure dams (and subsequently as part of a protective dam slope construction) because the slopes of these materials both below and above water level can be built much steeper than sand slopes, thus reducing sand losses during closure. Another application has been as core materials in closure dams, breakwaters and groins. Because of their scour resistance alternative materials are also used for bottom protection and filling scour holes underneath structures and along river banks.

After filling up the holes these materials, if necessary, can be covered by stone-mattresses, blockmats or other protective systems. These alternative materials have been applied on a large scale in the Netherlands as a foundation layer for roads and under slope protection (dikes and riverbanks), in this latter case very often successfully replacing the expensive and difficult to realise (especially under water) traditional granular filters.

Due to this wide applicability of alternative materials there has been an increasing need in recent years for reliable information on the stability criteria of these materials exposed to wave and current action.

Structural design places demands on the mechanical stability of waste materials. Also, specific demands can be formulated with respect to the purpose of the structure; e.g. the protection of the underlaying soil against erosion leads to requirements on the characteristic diameter and density of the material. Other requirements concern the intrinsic properties of the material, e.g. permeability and internal friction. These specific demands originated by the type and load of the structure must be compared with the properties of the available material. In case of disagreement there are two alternatives:
a. modification of a given material or selection of another waste material;
b. modification of design by using standard, though usually more costly materials.

In considering the use of these materials one inevitably comes up against the problems of the expected flow (current) and wave conditions and of the suitability of the various materials under those conditions. In order to solve these problems various laboratory and in situ (prototype) tests have been carried out in recent years. Some of the results of this research will be discussed below.

The hydraulic resistance of loose materials can be related to the magnitude of the critical shearing stress (τ_{cr}) which the current (critical velocity u_{cr}) exerts on the bottom.

The results of the tests indicate that minestone has similar values of τ_{cr} as gravel. However, minestone has a limited mean size (i.e. $D_{50} \approx 0.065$ m) and that means, that the application is limited to infrequent velocities up to 2 m/s but usually not more than 1.5 m/s, and waves usually not higher than 0.5 m.

Contrary to minestone, slags have relatively higher hydraulic stability than gravel or even crushed stone. This is due to two factors: a higher density (depends on type of slags) and a higher internal friction-factor (irregular angular grains). The influence of density is already taken into account in τ_{cr} and can be calculated directly.

In general τ_{cr} (slags) $\approx 1.55\ \tau_{cr}$ (gravel) or u_{cr} (slags) $\approx 1.25\ u_{cr}$ (gravel).

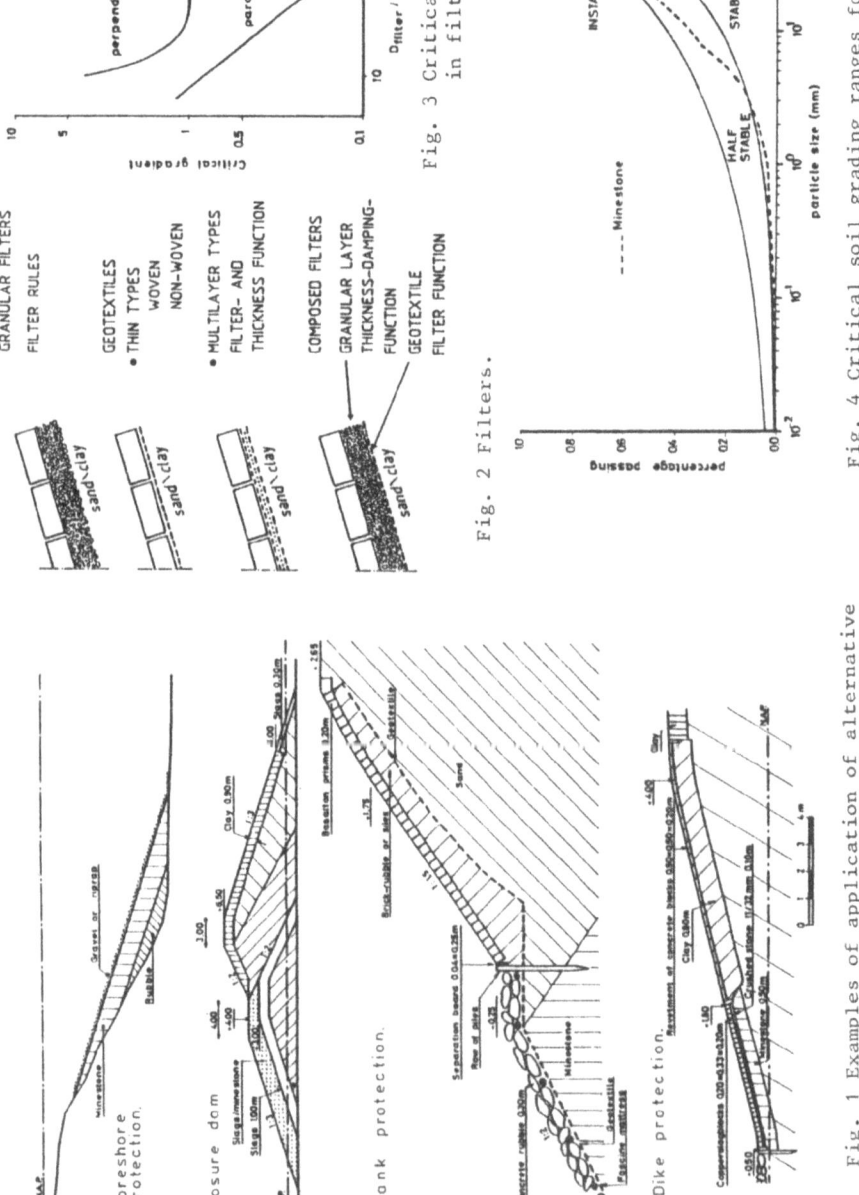

Fig. 3 Critical gradient in filter layers.

Fig. 2 Filters.

Fig. 4 Critical soil grading ranges for internal stability.

Fig. 1 Examples of application of alternative materials in hydraulic engineering.

The laboratory tests have been confirmed by in-situ measurements. As an indication, phosphor-slags and LD(steel)-slags with $D_{50} \approx 0.07$ m have been successfully applied in the Netherlands for velocities up to 2.5-3 m/s and infrequent wave attack up to 1.0 m wave-height (some profile deformation was allowed).
The resistance to current of a heavy clay varies from 1 to 1.5 m/s depending on compaction. However, clay covered by a grass-mat can resist velocities up to 3 m/s.
The recent results on the resistance of clay and grass-mats under current and wave attack are summarized in (9).
More detailed information on the subjects discussed above can be found in (2) and (3), and in various internal reports of Rijkswaterstaat.

4.2. Alternative materials in filter practice

4.2.1. What is a filter?
Granular filters are the most common element in the design of hydraulic structures. However, they are usually expensive and difficult to place (particularly under water) within the requirement limits. An alternative solution consists of a combination of a geotextile (which provides the filter function) and a layer of a certain thickness of graded stone, which attenuates the internal hydraulic loads, see figure 2. A cheaper but equally good solution is to place a thick layer of broadly graded waste products such as minestone, slags, silex, etc. For large hydraulic loads, thicknesses in the order of 0.5 m are required, properly compacted and composed according to internal stability criteria (4). An extensive review of this subject can be found in (1) and (4).
In the design of filters, several factors are involved. Apart from the macroscopic stability (does the material remain in place even under extreme flow conditions?) also the microscopic stability (does the material itself not change?) must be ensured. Both problems are analysed by parts. Filter-theory provides a method to analyse macroscopic stability.

4.2.2. Filter-theory.
A hydraulic filter is composed of one or more layers of different particle sizes. The layer of finest material (base) usually has to be protected against erosion. The filter provides this protection.
There are two methods, based on different principles:
Geometric: By a suitable choice of the characteristic diameter of the filtermaterial, displacement of the fine material is prevented.
The filter acts as a sieve which retains the base material, but lets the water pass. This process is governed by the relative particle sizes in the coarse layer and the base material. Such filters are usually stable.
Hydraulic: By an adequate choice of the particle size of the filter material, the water velocity between the particles is regulated. The water exerts a dragforce on the base particles. Loss of base material is prevented by keeping the dragforce below a critical value. The finer the filter is, the lower the water velocity and the dragforce will be. The obvious choice of very fine filter material however, results in erosion of the filtermaterial itself.
The choice of filter type depends on the magnitude of hydraulic loading, which is often expressed in a gradient (loss of pressure over a certain distance). The capacity of a filter is expressed by its critical gradient, which is the largest gradient in the filter at which base material is not transported. Large hydraulic loads often require geometrically stable filters, but for smaller loads, hydraulic filters are more economical.
Criteria have been developed to simplify the design of granular filters.

These relate particle size and hydraulic load to the critical gradient. The earliest geometric filter criteria date from 1922. Since then, other criteria have constantly been devised for a whole range of special cases. In the early 1970's an overview was published of existing filter criteria and their validity. In many criteria, the influence of certain material parameters is not accounted for. This and other overviews of criteria may be used to develop a design method (4).

Much practical research on filter processes has been done in the Netherlands for the storm surge barrier in the Eastern Scheldt. The results of this research have been published in condensed form (5). Figure 3 illustrates the dependence of the critical gradient on the ratio of the characteristic diameters of filter and base material. It may be seen that the critical gradient is markedly smaller in the lateral direction than in the perpendicular direction.

4.2.3. <u>Internal stability.</u> Another complication arises in well-graded filters with a wide range of particle sizes. This is the case with many waste materials such as minestone and various slags. The filter then cannot retain its own fines when subjected to hydraulic loading. Such a filter is termed 'internally unstable'. Special criteria have been developed to predict internal instability, see figure 4. Generally speaking, materials with a large proportion of fines are potentially unstable internally. Unsorted minestone in particular is prone to large proportions of fines, and should then not be used. However, rejection of minestone in the past probably was not necessary in all cases on hydraulic grounds.

Further research into the hydraulic conditions causing internal instability may increase the applicability of waste materials. For this it is necessary to closely control the production process which yields the waste materials.

4.2.4. **Material stability.** The experience acquired with minestone and steel slags in road construction and their availability have led to their use in hydraulic engineering as well. Their use is not without danger, however. To ensure the long term safety of the structure, the material must not weather or waste. This is particularly important for waste materials such as slags, minestone and silex. These are usually softer than e.g. gravel. Minestone for instance is a kind of claystone and slakes when subjected to alternate cycles of wetting and drying. Should this occur in a filter, the material would eventually become finer, the grading would broaden and the internal macroscopic stability with respect to the other layers would be lost. Thus material instability would trigger, through internal instability, a macroscopic instability. Silex, a waste product from marl mining, can also exhibit such behaviour.

Caution is also necessary with slags. It is necessary to know from which process they originated and how they were cooled. Phosphor slags e.g. can be extremely brittle when cooled too quickly, thus crushing easily during transport and placing. In the presence of water (abundantly present in hydraulic structures) steel slags may induce cementation (hydraulicity). The fines fraction, using water as catalyst, cements the larger particles. While thus preventing washing of the fines, the permeability strongly decreases and the filter no longer functions properly.

4.2.5. <u>Conclusion.</u> Waste materials may certainly be useful in hydraulic engineering construction, if only their specific weaknesses are recognized and effectively dealt with. Some additional research is needed to fill in gaps in existing knowledge. Rijkswaterstaat plans to instigate this research in cooperation with producers of waste materials. In filter design, other than the criteria for traditional stable materials must be adopted. New regulations must therefore be developed. It is expected that, given sufficient funding, material-dependent criteria can be developed within a few years, so that applications of waste materials in hydraulic engineering can be extended further.

4.3. Quality control

Quality control of materials used in hydraulic construction was generally speaking poorly developed up to the recent past. The recent construction of a number of large hydraulic structures has changed this situation.
Most waste products in use in hydraulic construction are now covered by contract requirements regarding mechanical-physical properties and the supervision of these (2), (8).
With waste products one wishes to control not only the mechanical-physical properties, but also those which might affect the environment. This is however hampered by the earlier mentioned lack of unequivocal standards for interpreting the results of leaching tests. Present research is particularly aimed at developing such standards. It is expected this will soon result in requirement standards and descriptions of methods of test for use in contracts. The ideal is to achieve quality control through the delivery of materials under certificate. In such a situation the producer primarily performs the quality control. It is expected such delivery under certificate will greatly increase the extent of quality control of alternative materials in the Netherlands.

5. CONCLUSIONS

Dutch experience shows, and this is supported by results of recent studies, that many alternative (waste) materials are applicable in most circumstances of the hydraulic engineering works. Because of the environmental consequences lead slags should be excluded from the common applications especially in the cases of direct contact with fresh water. In general, application of lead slags should always be supported by detailed environmental studies.
Because of additional properties some types of waste materials produce specific difficulties, which cannot be judged in their full extent yet (i.e. disintegration of minestone at water-air interface or conglomeration of slags). Additional research has to be carried out to quantify these difficulties, to determine the design criteria for different applications and to specify the quality control in order to make the waste materials move generally applicable in civil engineering.
The research on these aspects is still going on in the Netherlands. The final results, including those waste products not yet researched (i.e. fly-ash and phosphoric acid gypsum) can be expected within a few years.

REFERENCES

1. Graauw, A. de, Meulen, T. van der, Does-de Bye M. van der: Design criteria for granular filters. Delft Hydraulics publication 287, January 1983.
2. Laan, G.J., Westen, J.M. van, Batterink, L.: Minestone in Hydraulic Engineering Application, Deterioration and Quality Control. Symposium on the Reclamation, Treatment and Utilisation of Coal Mining Wastes, Durham, England, September 1984.
3. Guide to Concrete Dyke Revetments (1984). Centre for Civil Engineering Research, Codes and Specifications (CUR) and Technical Advisory Committee on Waterdefences, October 1984, The Netherlands.
4. Adel, H. den: Waste products as filter material. Delft Geotechnics report no. CO-272550/27, September 1985 (in Dutch)
5. TAUW Infra Consult B.V.: Composition and leachability of minestone. January 1986 (in Dutch).
6. Luin, A.B. van, Gaastra, D.J.: The leachability of metals and sulphate from basalt and slags. DBW/RIZA - note no. 86.015, April 1986 (in Dutch).
7. Sloot H.A. van der,: Environmental engineering research of phosphor slags. ECN, September 1986 (in Dutch).
8. Laan, G.J.,: Quality and quality control of slags in hydraulic engineering. Rijkswaterstaat DWW, September 1986 (in Dutch).
9. Pilarczyk, K.W.: Dutch Guidelines on Dike Protection, 2nd Intern. Conf. on Coastal and Port Engineering in Developing Countries, Beijing, China, 1987.
10. Rijkswaterstaat: Possibilities of application of alternative materials. Rijkswaterstaat-serie, Report no. 44, 1984 (in Dutch).

THE PROCESSING OF INDUSTRIAL WASTE FOR IMMOBILIZATION AND/OR RECYCLING APPLYING POZZOLANIC REACTIONS

P.D. RADEMAKER
R.B. WIEGERS

1. POZZOLANAS

Pozzolanas are defined as materials which contain constituents which will combine with lime at ordinary temperature in the presence of water to form stable insoluble compounds possessing cementing properties.
Natural pozzolanas are for the most part materials of volcanic origin. The name pozzolana is derived from the Italian town Pozzuoli near the Vesuvius. Other natural pozzolanas are bauxite and diatomaceous earth. Artificial pozzolanas are either produced by heat treatment of natural materials such as clay and shale or are the residues of coal burning such as fly ash. For practical purposes also blastfurnace slag can be described as a pozzolana.
The distinguishing feature of pozzolana is the presence of silica that will react with lime to hydrated calcium silicates, also known as cementous minerals. This reaction goes in a number of steps (ref. 1, 2).
1. The silica usually has siloxane groups on the surface (fig. 1) that will be converted to silanol groups by the hydroxyde.
2. The silanol groups are converted to negatively charged groups.
3. In the presence of lime, the calcium ions are bound to the negative charge on a two to one ratio.
4. After disruption of the surface the silica is amenable to dissolution. A gelatenous structure is formed and monomeric silicates are going into the aqueous phase.
5. The silicates will react with calcium to the hydrated calcium silicates.

$$(CaO)(SiO_2)nH_2O$$

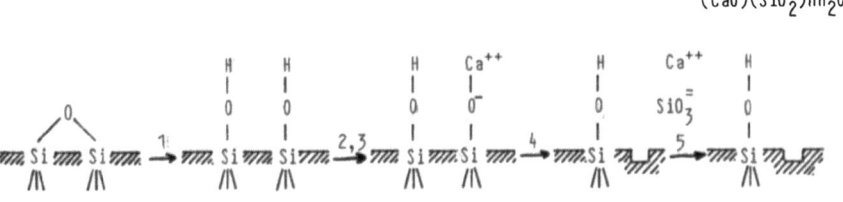

Fig. 1. Mechanism of silica dissolution in lime.

The formation of silica gel by the attack of calcium hydroxyde causes the top layer of a pozzolana grain to swell. Due to osmotic pressure difference an interspace is created and the top layer is peeled off. This process can repeat itself until a grain is totally dissolved (ref.3). Finely divided silica gel will spread into the aqueous phase. This gel will be surrounded by needle-shaped cementous minerals that grow to fill the voids. (fig. 2).

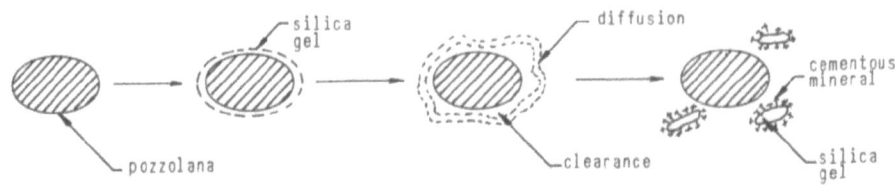

Fig. 2. Dissolution of fly ash particles.

Only a small part of the pozzolana grains will react.
Most grains will remain unaltered. As a result of the adhesion between cementous minerals and through the filling up of void space with the mineral and gelatenous material, the final product is solidified with good compressive strength. At room temperature this hardening process takes several weeks to complete. As in cement, other minerals are formed during that period. In the presence of alumina and sulfur, calcium sulfo- aluminates will contribute to initial strength of the material.
As a result of the pozzolanic reaction, the specific surface area of the material increases. An increase that is comparable to the increase seen during the hardening of cement (ref.4).

2. AARDELITE
The Aardelite process is based on the pozzolanic reaction and was developed to produce a light or medium weight aggregate from fly ash. This aggregate can be used in several concrete applications as substitute for gravel or light weight aggregates.
Figure 3 shows the basic flow diagram of the Aardelite process. Characteristic for the process is the pelletizing of the lime-fly ash mixture and the hardening at elevated temperature (70-100 C) in a humid atmosphere.
Other pozzolanas or latent hydraulic materials can also be used as feedstock for the process. In addition to the pozzolanas and the lime, a certain percentage of inert material or filler can be included in the mixture. This inert material can be bottom ash or sand. If the process is applied for hazardous waste treatment, the filler can also be a waste product, added as a powder or as a slurry. The waste can originate from metallurgical and chemical processes, but can also be the residue from RDF or municipal waste burning systems.

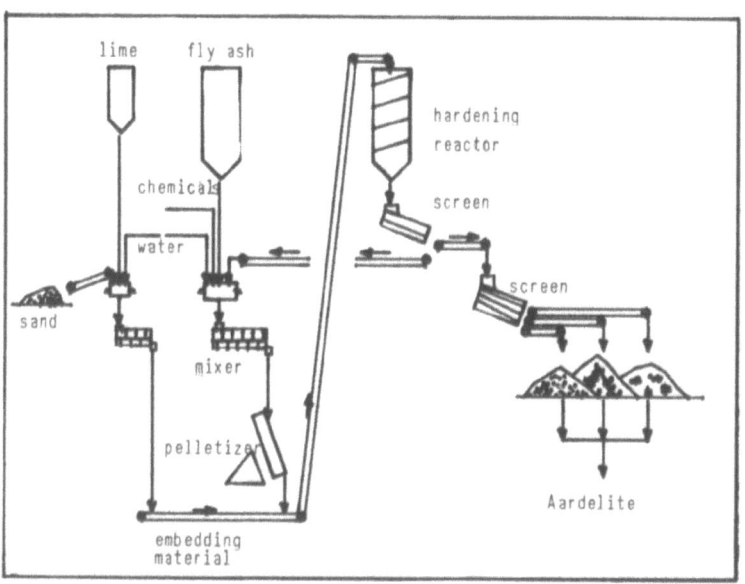

Fig. 3. Aardelite flow diagram.

The solidification of fly ash and of other waste products has the advantage that it is transformed into a product which can be used for various applications. If the composition and characteristics of the final product are such that application is undesirable, the solidification will eliminate dust problems and make the waste easier to handle for safe disposal. The option for disposal may be the least desirable amoung clean technology options. The need for final disposal is however sometimes unavoidable.
Whether the Aardelite process is applied to produce a usable material or it is applied to produce material for safe disposal, the solidification will avoid pollution of the air. The major other question to be resolved is, how well the toxic components are stabilized.

3. STABILIZATION
Stabilization of toxic compounds by pozzolanic reaction is in most cases aimed at heavy metals, although other toxic compounds can be stabilized in such systems as well. This paper will deal with the stabilization of heavy metals only.

The hazardous character of material containing heavy metals is determined by their concentration and the quantity that can be released to the environment.
Stabilization is the treatment where the handling characteristics are improved, the surface area across which transfer to the environment of pollutants takes place is decreased and the solubility of hazardous constituents is limited.
There are a number of methods for stabilization, one of them is the stabilization in pozzolanas. In the U.S. at least eight processes are known and practiced that are on the basis of pozzolanic reactions (ref.5).
This interest in pozzolanic reactions for stabilization purposes is understandable. There is a beneficial effect from pozzolanas on the two elemental steps during release of a metal ion from a solid: dissolution from the solid particles and diffusion through the liquid phase to the surface of the solid where it can be released to the environment. Due to the pozzolanic reaction the dissolution of metals is limited by a number of mechanisms:

3.1. Precipitation and co-precipitation of hydroxydes by an increase of pH.
In the presence of lime, most metals will form insoluble hydroxydes. If hydroxydes of iron or manganese are formed during the pozzolanic reaction, these will adsorb various cationic metals, such as chrome (III), zinc, copper and arsenic (III). (ref.8).
In addition to those reactions the anionic heavy metals will form insoluble calcium salts such as zincates, arsenates and plombates.

3.2. Fixation of metal ions in cementous minerals.
The fixation of metal ions can occur according to 3 different types of mechanisms in analogy to the mechanisms in hardening cement.

3.2.1. It is known that during the hardening of the cement all monovalent metal ions (Li-Cs) are contained in intercrystalline layer of the mono alkali sulphate, which is a derivate of $(CaO)4 \cdot (Al2O3) \cdot x(H2O)$.
It is possible that also other ions are built into the crystalline cement structure during the hardening (ref.17).

3.2.2. A second mechanism of fixation of the metal ions in cement is the formation of an impermeable layer in analogy to the hardening of blast furnace cement (ref.18). This layer is a product of the reaction between calcium ions and silicate ions. This layer makes the cement less porous and therefore the metal ions can no longer be reached by the water (fig.4).

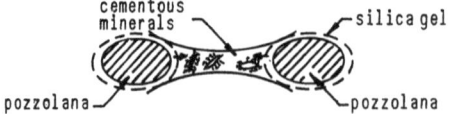

Fig. 4.
Formation of impermeable layers between particles. (ref.18).

3.2.3. The third mechanism is the adsorption of the metal ions on the surface of cementous minerals.
This effect can be influenced by changing the surface charge of the cementous minerals by adding a surface active additive (ref. 19 + 20).
It is known that surface charge (zeta-potential) of cementous minerals can be modified and that it has an effect on the release of metal ions from the solid (fig. 5).

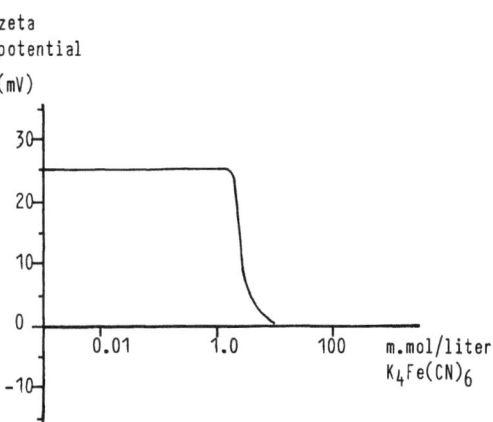

Fig. 5. Modification of zeta potential of cement. (ref. 20)

At high pH metal ions are also chemisorbed on silica, and attach themselves irreversible to the silanol groups (ref.9 + 16).

3.3. Complexation of cationic and anionic heavy metals.
The silicates and the calcium sulfoaluminates have many analogues. A well-known example is the calcium chromoaluminate which is very insoluble. Various heavy metals are known to precipitate in such a form or to be captured into mixed crystals (ref. 6 + 7).

In addition to the lesser degree of solubilization, there is also an effect of pozzolanic material on the diffusion through the solidified material to the surface. Adsorption-desorption phenomena will greatly effect this rate of diffusion, but also porosity and permeability do have an effect. In practice, there is a different rate of diffusion for each element in each different solidified material.

4. LEACHING TESTS
For the evaluation of the ecological effects of a stabilized waste, the release of toxic compounds by the action of natural water on the material in its final application, must be predicted.

A variety of standardized tests have been developed for this purpose. All of them suffer from the fact that longterm behaviour is difficult to simulate. Acidity, redox potential, biological processes and many other conditions will determine the leaching of material from a solid. The leaching mechanism is complex and a myriad of interactions actually takes place. Because of this complexity but also because of analytical problems at the low concentrations in leachate and because of sampling problems, the reproducibility of leaching tests is often very bad. Both from direct experience and from literature, large variations between the results from different laboratories are known. In an EPRI study on the sources of variability it was noticed that a from a standardized EPA leaching of fly ash, the spread in analysis of leachate of 1 sample was (lit.10):

$$0.6 - 110 \text{ ppb for As (average 7 ppb)}$$
$$0.01 - 48 \text{ ppb for Cd (average 0.59 ppb)}$$
$$0.14 - 9 \text{ ppb for Pb (average 1.04)}$$
$$23 - 121 \text{ ppb for Se (average 52)}$$

A leaching test can have two different purposes. It can be carried out to determine how much of the heavy metal is available for leaching, e.a. how much can be solubilised in a certain environment. It can also be devised to simulate longterm behaviour. A clear distinction between the two tests is needed. A third and completely different type of analysis is the one where all solid material is solubilized and where the total concentration of each element is determined.

If the "available" metal is determined it should be realized what it took to make it available. Cationic metals do not get solubilized at high pH, most availability tests are concerned with the availability in "rain water" at a pH of 4 or 5.

In practice it would take 1000 - 5000 years of rainfall at pH 4.5 to neutralize a 10 cm slab of concrete with Aardelite. This assuming that all acid in the rain is used to react with excess lime and cement minerals.

There can be large discrepancy between laboratory results on availability and the practical availability. This even if leaching would go on indefinitely. With groundwater as the leaching medium for example, it can be assumed that the iron from groundwater acts as a scavenger and immobilizes nearly all chromium at the prevailing redox potential and pH (ref.8).

The "availability for leaching" should therefore first of all be considered for comparative reasons. Secondly, it can be used in a model to predict the longterm leaching behaviour.

5. RELEASE
This model to be worked out by the Dutch standardization working group SOSUV is based on the conjecture that the amount of metal released is determined by the rate of diffusion of the metal in that material. This rate of diffusion can be expressed as diffusivity.

In formula (ref. 11).

$$J = f\, S_0\, D^{1/2} (\pi t)^{-1/2}$$

where J = flux of metal ion (m.mol.cm^{-3} . sec^{-1})
  f = leachable fraction
  t = time (sec)
  S = concentration of metal in solidified material (m.mol.cm^{-3})
  D = diffusivity (cm^2 sec^{-1})

The total amount of metal leached from a solid is in first approximation:

$$Q = 2\, f\, S_0 \left(\frac{A}{V}\right) D^{1/2} (\pi t)^{-1/2}$$

where A = outside surface of body (cm^2)
  V = volume of body cm^3
  Q = quantity of metal released per unit volume of the body (m.mol.cm^{-3})

This formula is based on the assumption that diffusion is linear. In reality there will be non-linear gradients for pH, Eh and concentration in the solid. In general this will result in lower values for Q. As can be seen from the ratio A/V in this formula, there is a strong effect of enlarging the dimensions of the solid. But the major factor determining the release of heavy metals is the diffusivity. For metal ions in a pozzolanic bonded material the values for D are in the range of 10^{-8} to 10^{-10} cm^2 sec^{-1} (ref.12).

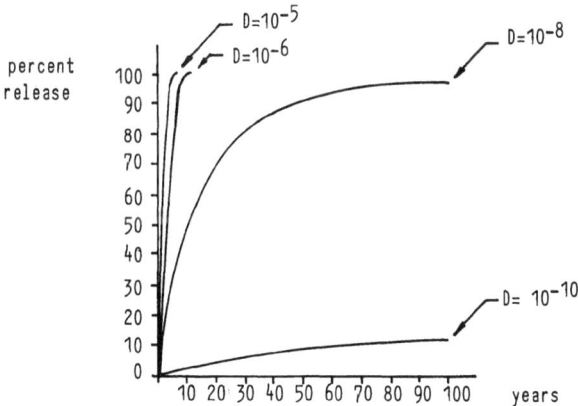

Fig. 6. Release of metal as function of diffusivity.
  Percent of metal released from a semi infinite slab (10 cm) for diffusivities of 10^{-5} - 10^{-10} cm^2.sec^{-1}

Figure 6 gives one example of the calculated release as a function of diffusivity. From this example it can be seen that if a pozzolanic bonded material is used as filler in concrete or in other solidified material the release after 100 years is still only some percentages of the total leachable fraction. The diffusivity in ash is of the order of 10^{-5} cm^2 sec^{-1}, meaning that release is completed in a number of years.

6. EFFECT OF AARDELITE PROCESS ON LEACHABILITY

The practical implication of the foregoing for the Aardelite process is in two applications. First the basic process where fly ash is converted to the light weight aggregate which is used in concrete applications. The second is the process adaptation where other waste products are combined with fly ash to produce aggregates for concrete or other applications and in some instances for the production of material that can be put in longterm storage.

In the Aardelite process various mechanisms can be used to lower the dissolution of heavy metals:

ad 3.1. The lime addition will raise pH and if necessary an excess can be included.

ad 3.2. The formation of cementous materials can be controlled by mixing and curing conditions as to reach specific composition and morphology. The absorption on the cementous minerals can be increased by controlling the surface charge of those minerals.
During the Aardelite process an increase of specific surface area is noticed to at least 4 times the value for fly ash (ref. 14). The specific surface area for Aardelite is at least 12 m2 per gram as determined by the B.E.T. method.
Adsorption on the fly ash particles after addition of lime and formation of silanol groups can be substantial. This also because of the high internal porosity as seen from the difference in Blaine number and B.E.T. specific surface area (0,35 resp. 3.5 m2/gr).

ad 3.3. By mixing fly ash with other selected residues, or by adding small amounts of compounds that can be transformed to insoluble double salts, a co-precipitation can be promoted.

The diffusivity of metals can be effected by modifying the surface charge of the material such that adsorption-desorption phenomena slow-down the diffusion. By hydrofobation of the material the diffusivity can be lowered further. In the Aardelite process the permeability of the final product can be controlled by the lime and other additions.

Mixing components such that a good sizedistribution of original particles is obtained will give a denser packing and this will lower the permeability.

The effect of the solidification in Aardelite on leachability is the most pronounced if wastes with a high level of metal are incorporated in the product. An example of this is the production of pellets with the following composition:

40 % toxic waste, 50 % fly ash, 5 % of another pozzolanic material and 5 % lime.

Both materials were leached according to the SOSUV procedure with L/S = 10.

Element	leachate of waste (ppb)	leachate of waste after solidification in Aardelite process (ppb)
Cd	230.000	< 2
As	3.000.000	10
Cu	220.000	5
Pb	4.600.000	150

Notice that the values for Cu and As are close to those of pure ocean water (3 ppb for both) (ref.15).

For Aardelite on the basis of fly ash only and without specific waste added, the values are of course less spectacular.

From the various studies in the Netherlands it can be concluded that the leachability of metals in fly ash is first of all determined by the form in which it is present. Several elements are mainly present on the surface of the smaller particles (As, Se, Mo, Zn, Cd, V and Sb) (ref. 13). For those elements the leachable fraction is relatively high and their release from pozzolanic bonded material is also relatively high.

If Aardelite is produced from fly ash, the effect on leachability and on actual leached material is different for each element. In the following example fly ash and Aardelite produced from this fly ash, was leached according to the EPA method, which mainly indicates the amount available for leaching. The same sample of Aardelite was then also tested according to the Toxicity Characteristic Leaching Procedure (TCLP-method) which is designed to determine the mobility of contaminants.

I = Concentration in EPA Leachate of fly ash (ppb)
Ia = Percentage extracted from total amount of metal in fly ash
II = Concentration in EPA Leachate of Aardelite (ppb)
III= Immobilization factor (I:II)
IV = Concentration in TCLP Leachate from Aardelite

Element	I ppb	Ia %	II ppb	III I:II	IV ppb
As	885	30	184	4.8	47
Cd	23	22	< 3	> 7.5	5
Cr	235	4	35	6.7	13
Se	26.5	5	6.7	4.0	8.5
Cu	448	4	< 1	> 450	< 1
Zn	359	1.5	2	180	n.d.*
Mg	10800	1.2	338	32	n.d.

* not determined

The enrichment of As and Cd on the fly ash surface makes them more amendable to leaching (column Ia).
By solidifying the fly ash the percentage of leachable metal is lowered by approximately a factor 5 for the trace elements and by a factor 50-500 for other contaminants.
As can be seen from column IV the amount that actually will be leached on medium term (20-50 yrs), for most metals is only a small fraction of the leachable quantity.

7. CONCLUSIONS
The release of toxic components from waste can be reduced substantially by stabilization of the waste in a pozzolanic reaction. The waste can be a pozzolana or can be added as inert material in the mixture.
In the Aardelite process there are several means to control leachability and the rate of leaching by process adaptation or by adding additives. The respons of various metals to the process varies between metals and depends on their chemical nature in the waste product. On the basis of low level waste, a final product can be made which is acceptable for application in the building industry and in road building.

REFERENCES
1. Glaser L.S.D. : Cement and Concrete Research. On the role of calcium in the alkali aggregate raction, Vol. 22, p 321- 331 (1982)
2. Boehm H.P.: Angew Chemie, Intern.Ed, Functional Groups on the surfaces of solids. Vol. 5, No. 6, p 533-544 (1966)
3. Ogawa K.: Cement and Concrete Research, The mechanism of hydration of C3S, Vol. 10, p 683-696 (1980)
4. Lea F.M.: The Chemistry of Cement and Concrete, Edward Arnold Ltd., Publishers, London (1983)
5. Landreth R.E.: Guide to the disposal of chemically stabilized and solidified waste. Environmental laboratory, U.S. Army Engineer Waterways Experiment Station, EPA-IAG-D-4 D569
6. Meric I.P.: Revue des Materiaux de construction, Traitement des boues et eaux residuaires, No. 699, p 79-81, March-April 1976
7. Meric I.P.: Ciments, Betons, Platres, Chaux, La solidification des boues, No. 3/79 718, p 133-136 (1979)
8. Hem J.D.: Geochim. Cosmochim Acta 41, Reactions of metal ions at surfaces of hydrous iron oxide, p 527-538 (1977)
9. Lyklema J.: J.Electroanal, Double layer in SiO2 in presence of bivalent ions, Chem 22, p 1-7 (1969)
10. Eynon B.: A Statistical Comparison of Two Studies on Trace Elements Composition of Coal Ash Leachate, EPRI EA 3181, July 1983
11. Bolt N.: Energie spectrum, Milieu aspecten van vliegastoepassingen, p 226-238, Nov 1985
12. Anthonissen I.A.: RIVM report 3924/141019, March 1984
13. Hanstveit A.O.: Report CL 81/67 TNO-MT, Inleidend onderzoek naar de milieutoxicologische eigenschappen van afvalstoffen van met kolen gestookte centrales, (May 1981).
14. R.S. Dahlin: Paper presented at 2nd Joint Symposium on dry SO2 and Simulataneous SO2/NOx Analysis of LIMB waste management options, control technologies, Raleigh (N.C.), June 1986.
15. Hill M.N.: The Sea, Vol. 2, p 4, Interscience Publishers (1968)
16. Iler Ralph K.: The Chemistry of Silica, p 381, New York: John Wiley and Sons (1979).
17. Dosch W.: Die ein-dimensionale innerkristalline Quellung natuerlicher und synthetischer Schichtkristalle insbesondere von TetraCalciumAluminatHydrat, Habilitationsschrift, Mainz, Jan. 1968.
18. Bakker R.: Ueber die Ursache des erhoehten Widerstandes von Beton mit Hochofenzement gegen die Alkali-Kieselsauerreaktion und den Sulfatangriff Graduation report T.H. Aachen, 1980.
19. Neerhoff A.: Correlation between fracture toughness and zeta potential of cementstone. Report M-81-2 T.U. Eindhoven.
20. Wiegers R.B.: Research to the possibility of stabilizing metal ions in a binder latice, Report M-84-2 T.U. Eindhoven.

UTILIZATION OF INDUSTRIAL WASTE GYPSUM FOR THE MANUFACTURE OF GYPSUM-BONDED PARTICLEBOARDS IN A SEMI-DRY PROCESS

DR. K. LEMPFER

1. Introduction

The environmental control requires a reduction or avoidance of polluant emission. Furthermore, it is called for a utilization of waste - or more precisely - of residues. Both aspects are considered when there are residues as a result of a reduction of polluant emission, especially if the produced quantities imply the utilization. An example herefore is the desulphurization of flue-gases from power plants.

Power plants run with fossil combustibles are emitting considerable quantities of sulphur dioxide which loads the environment, a fact which today is undisputed. Therefore, in the Federal Republic of Germany a limitation of the SO_2 emission was necessary for eco-political reasons. It was realized by the regulation for large combustion plants (13. BimSchV = 13th Federal Emission Protection Regulation). In order to adhere to the required limit value of 400 mg/m³ especially the wet flue-gas desulphurization was successful binding the SO_2 with calcium hydrate or lime. From this process results $CaSO_4$-dihydrate - i.e. gypsum. In the Federal Republic of Germany the amount of industrial gypsum per year from these flue-gas desulphurization plants, the so-called desulpho-gypsum, actually is about 1 million tons. Within the next five years it will increase up to about 3,4 million tons per year and then corresponds to the actual annual demand of the German industry of building materials.

Gypsum is not only a compulsory byproduct of the flue-gas desulphurization but also of the production of phosphoric acid as the so-called phospho-gypsum and of the production of hydrofluoric acid as anhydrite. In the near future worldwide there are expected amounts of 190 millions tons per year of desulpho-gypsum, about 90 million tons per year of phospho-gypsum and 3,5 million tons per year of raw anhydrite.

The utilization of industrial gypsum has become a most important economic object because there is no point in dumping on one hand large quantities of gypsum as waste and to mine on the other hand raw gypsum from natural sources. Due to their genesis, the industrial gypsum disposes of some properties which differ considerably from those of natural gypsum. Therefore the findings and experiences made with natural gypsum cannot easily be applied to industrial gypsum. They have to be reviewed. For example desulpho-gypsum with a moisture of 7 to 10 % is much more humid than a natural gypsum with a mine humidity of about 1 %. This fact complicates the handling and increases the energy demand during dewatering, the calcination, necessary for the preparation of the binder. Impurities resulting

from the manufacturing process, e.g. residual phosphate of phospo-gypsum
have to be removed by washing or floating. If these impurities are of co-
crystalline nature maybe even a costly recrystallization will be necessary.

2. Description of Process

In conventional manufacturing processes the gypsum binder is utilized in a
viscous or liquid state, i.e. it has a considerable amount of excess water.
Due to the morphology and the grain spectrum of the industrial gypsum, the
excess water necessary for a free-flowing consistence is even higher in
comparison to natural gypsum. The strength of gypsum-based building elements is correspondingly low. Furthermore, the thixotropy of such binder
mixtures complicates their processability. The range of the crystal shapes
is exemplified by some scanning electron miscroscope photos of a desulpho-
gypsum, a phospo-gypsum and a natural gypsum (Fig. 1). Especially the
phospho-gypsum binder has a bad flowability due to its lamellar structure.

Fig. 1: Examples of the crystal shape of different gypsum plasters.
Left: desulpho-gypsum; middle: natural gypsum; right: phospho-
gypsum
(SEM-photography, R. Blaschke, Münster)

In the past there has been R&D work carried out searching for a reduction
of the excess water. This was not successful because it was not possible to
attain a homogenous mixture of the binder with the low water quantity aimed
at. This, however, is one of the conditions to obtain a complete hydration
and optimum strength even at a low water level.

The described semi-dry process for the manufacture of gypsum-bonded particleboards allows to reduce the quantity of excess water by up to 70 % in
comparison to the conventional wet process. The water required for the hydration is added to the binding agent via humid wood particles. On contact
of the particles acting as water reservoirs, and gypsum plaster, the water
required for hydration is absorbed by the binder. In contrast to the usually free-flow mixture it develops a pourable mixture consisting of single
particles which can easily be processed (Fig. 2).

Fig. 2: Non-compressed mat for the manufacture of gypsum-bonded particleboard in accordance with the semi-dry process.

Similar to the manufacture of synthetic resin-bonded particleboards, the mix is formed to an endless mat with spreading machines. The board thickness can be increased as desired. There are no waste water problems. After setting of the boards, only about 15 % of excess water have to be dried out (the percentage is 45 to 55 % with conventional methods). There are correspondingly high savings of drying energy compared with gypsum plasterboards and fibreboards manufactured in a wet process.

Compared with synthetic resin-bonded particleboards, energy consumption is further reduced, because the wood particles had not to be dried before being mixed with the binder (in the case of synthetic resin-bonded boards, particles have to be dried from an initial moisture content of between 100 and 30 % down to a final moisture content of between 5 and 2 %). Moreover, the pressing does not need any heat supply (synthetic resin-bonded particleboards require press temperatures of between 170 and 240 °C). Another advantage is that the binder gypsum does not emit any formaldehyde (in contrast to about 80 % of all synthetic resin-bonded particleboards).

The cross section of a homogeneously spread gypsum particleboard shows a very constant distribution of particle furnish and gypsum matrix (Fig. 3). The pourability of the material to be spread also allows, as it is the case with synthetic resin-bonded particleboards, to form a profile varying in structure and density. This allows a specific influence on the board properties.

During the manufacturing process of gypsum-bonded particleboards the spread mat is formed in a press to the required thickness and density. This density is maintained until end of hydration in order to avoid a deterioration in structure due to a spring-back of the particles. This is an essential condition for the utilization of wood particles and for the resulting higher strengths of the gypsum particleboards compared to the gypsum fibreboard.

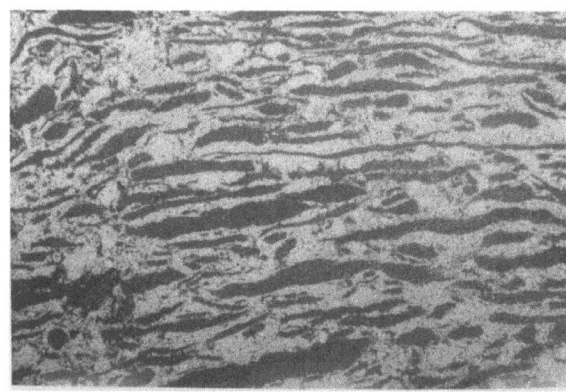

Fig. 3: Cross section of a homogenously spread gypsum-bonded particleboard
(SEM-photography, R. Blaschke, Münster)

3. Properties of gypsum-bonded particleboards in comparison with other panel products

Product properties usually are influenced by different factors, e.g. by the thickness and density of the materials. Therefore table 1 shows ranges of values for characterization instead of single values. The comparison of properties shall be limited to the "bending strength" and the "linear movement".

Table 1: Comparison of some properties of different construction panels

Property		gypsum-bonded particleboard	gypsum plaster-board	gypsum fibre-board	cement-bonded particleboard	resin-bonded particleboard
Density	(kg/m³)	1100...1200	800... 900	1100...1200	1100...1300	650... 750
Modulus of rupture acc. to DIN 52 362	(N/mm²)	6...10	3...8	5...7	9...15	12...24
Modulus of elasticity	(N/mm²)	2000...3500	2000...4000	2500...3500	3000...6000	2000...3500
Tensile strength in board plane	(N/mm²)	2,5...4	1,5...3	1,5...3	4...5	7...10
Tensile strength perpendicular to board plane acc. to DIN 52 364	(N/mm²)	0,3...0,6	0,2...0,3	0,3...0,5	0,4...0,7	0,5...1,0
Swelling (2h) acc. to DIN 52 364	(%)	< 3	< 3	< 3	< 1	< 8
Linear movement (20°C/30% 20°C/85%)	(%)	0,06...0,08	0,03...0,04	0,03...0,05	0,15...0,35	0,3...0,5

The gypsum-bonded particleboard is superior to both gypsum fibreboard and gypsum plasterboard in its modulus of rupture, but weaker than the synthetic resin-bonded and cement-bonded particleboard. The strength in the board's plane is mostly independent of the direction, i.e. it is not determined by the machine direction of the carton, as it is the case with the highly anisotropic gypsum plasterboard.

The linear movement of the gypsum particleboard when exposed to the same climatic conditions, is about 5 times lower than that of the synthetic resin-bonded particleboard. In contrast to conventional particleboards it is possible to hang papers on these boards without the risk of cracks in the area of joints. The linear movement of the other gypsum boards is lower than that of the gypsum-bonded particleboard. But this may be a disadvantage in special cases - as complaints have shown - e.g. when those boards are veneered, namely if the difference between the shrinking movements of the two materials is too great. This may lead to deformations or checks in the veneer. Practical tests have shown that a continuous laying - as it is the case with gypsum plasterboards - is possible, whereas an additional joint band is not necessary.

4. Fields of application

Thanks to their properties which have only been outlined roughly, gypsum-bonded particleboards may be applied in all fields where other gypsum-based boards (such as gypsum plasterboards and gypsum fibreboards) and partly also synthetic resin-bonded particleboards are used, i.e. in the entire interior of buildings (as dry cast, dry cast plaster floor, for interior works in attics, for the renovation of old buildings). Applications in exterior and wet areas are also possible if a modified binder is used. In this case it is of advantage that the linear movement caused by moisture changes is markedly lower than that of the cement-bonded particleboard.

Based on investigations carried out until now a classification of the gypsum-bonded particleboard in the A2 class of building materials according to DIN 4102 (non-combustible) can be achieved. In this case, special interest has alredy been shown for furniture made of gypsum-bonded particleboards, in order to reduce the fire load in hotels and other public buildings and to meet the requirement of having non-combustible building material in the escape areas. The workability of the gypsum-bonded particleboard would also meet the requirements of this kind of application. In comparison to the synthetic resin-bonded particleboard the tool wear from sawing, milling and drilling is less despite higher density and the use of inorganic binder.

Another promising field of application might be the production of panels. Due to numerous open joints, even the use of particleboards belonging to emission class E1 does not always guarantee to fall below the allowable West German limit of 0,1 ppm for formaldehyde concentration in the room air. In this case the formaldehyde-free gypsum-bonded particleboard is of advantage.

5. Industrial Realization

In October 1985, the first industrial plant for the manufacture of gypsum-bonded particleboards was set into operation in Kuopio in Finland. Customer was Saastamoinen Oy, one of the leading manufacturers of wood-based products in Finland. The plant was designed by Bison-Werke in Springe (manufacturing plant) and Salzgitter Industriebau (calcination plant). Generally phospho-gypsum is used. The board size is 1,25 x 3,05 m, the capacity is 125 m³ per day or - related to boards of 10 mm thickness - 12.500 m² per day. In late summer of 1987 a second plant will be set into operation in Norway utilizing phospho-gypsum, too.

For nearly two years, the Finnish plant runs around the clock seven days a week. Thus, it is evident that gypsum-bonded particleboards can be manufactured successfully with the semi-dry process at an industrial scale utilizing residual gypsum.

6. Utilization of gypsum from flue-gas desulphurization

Due to the aforementioned problems with the desulphurization of power plants the utilization of the so-called desulpho-gypsum gains more and more in significance. The WKI therefore recently has tested the applicability of different kinds of desulpho-gypsum for the manufacture of gypsum-bonded particleboards. For these tests different kinds of raw gypsum were used which are residues of all the wet desulphurization processes (on the basis of lime washing) used in the Federal Republic of Germany. Manufacturer of these flue-gas desulphurization plants were:

>Deutsche Babcock Anlagenbau GmbH, Krefeld
>G. Bischoff GmbH, Essen
>Knauf-Research-Cottrel (KRC), Iphofen, Stuttgart
>Saarberg-Hölter-Umwelttechnik (SHU), Saarbrücken
>L. u. C. Steinmüller GmbH, Gummersbach
>Thyssen Engineering GmbH, Essen

The raw gypsum which came from different plants partly run in accordance with the same process, were prepared in a similar way to plaster. With these plasters tests according to DIN 1168 for the assessment of the gypsum properties as well as gypsum-bonded particleboards in accordance with the semi-dry process were made.

The water/binder ratio of the desulpho-gypsum (calculated according to DIN 1168 from the spread weight) varied between w = 0,53 and 0,80. For plaster based on natural gypsum it is approx. w = 0,6. In general the strength of the set plaster prepared of raw desulpho-gypsum was higher than that of natural gypsum. This is not surprising when you consider the partly high portion of inert substances of natural gypsum. The hydration behaviour of the desulpho-gypsum plaster depended essentially on the calcination result. It was possible to influence it similar to natural gypsum plaster by means of additives.

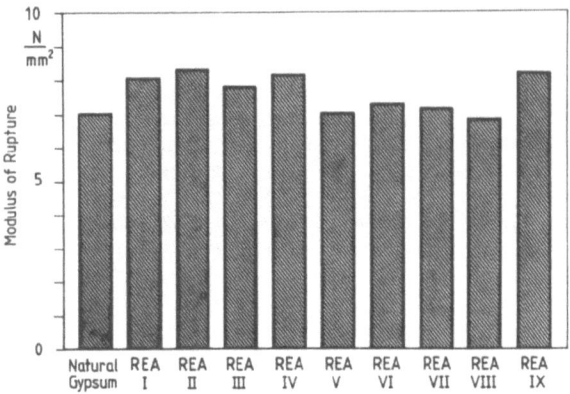

Fig. 4: Bending strength according to DIN 52 563, part 1, of gypsum-bonded particleboards (manufactured in the laboratory) being prepared with different binders (REA = desulpho-gypsum)

The gypsum-bonded particleboards prepared with desulpho-gypsum plaster showed equal or higher strength compared to those prepared with natural gypsum (Fig. 4). Problems during the manufacturing process did not arise. Also the dependence of the bending strength from the density and the particle sort used (e.g. taken from different wood species) is in accordance with the behaviour known from natural gypsum (Fig. 5). Summing up, it may be said that also desulpho-gpysum plaster can be used without any problems in the semi-dry process.

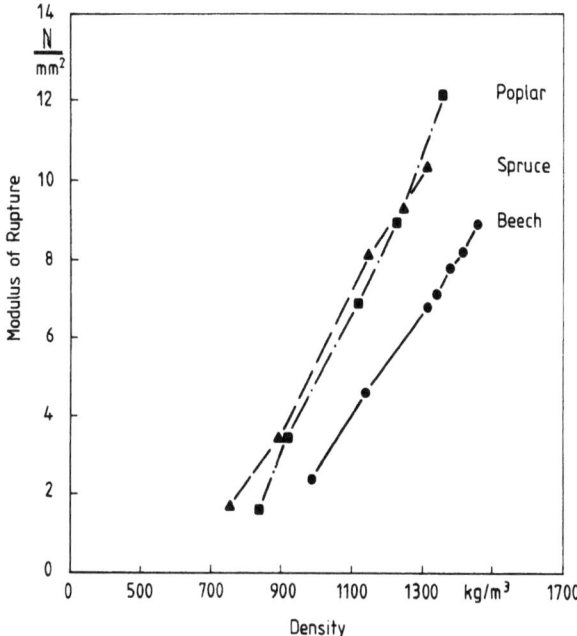

Fig. 5: Bending strength according to DIN 52 563, part 1, of gypsum-bonded particleboards depending on density and wood species using a desulpho-gypsum a binder

7. Outlook

The semi-dry process allows to offer a new product to the market - the gypsum-bonded particleboard. In contrast to conventional wet processes of the gypsum industry this manufacturing process allows an evident reduction of energy consumption. Moreover, by specific change of the recipe and of the substances used, the properties of this new material can be influenced extensively. The industrial plant in Finland using phospho-gypsum shows that with this process industrial gypsum can be utilized successfully. Extensive investigations have shown that also gypsum from flue-gas desulphurization can be used appropriately. Costly preparations as agglomeration and recrystalisation are not necessary. The semi-dry process therefore offers a great advantage in utilizing industrial gypsum.

8. Select biliography

1. BRÜKER, F.-W., SIMATUPANG, M.H.: Mineralgebundene Holzwerkstoffe. In: Willeitner, H., E. Schwab: Holz-Außenverwendung im Hochbau. Verlagsanstalt A. Koch, Stuttgart, 1981, pp. 42-47

2. BÖCKING, G.: Die Herstellung gipsgebundener Spanplatten im Endlosverfahren. Holz als Roh- und Werkstoff Vol. 41, 1983, pp. 427-430

3. HAUG, N.; PIETRZENIUK, H.-J.: Verwertungsmöglichkeiten von Produkten der Abgasentschwefelung. Umwelt Vol. 13 (1983), pp. 442-446

4. HÜBNER, J.E.: The Industrial Production of Gypsum Boards with Reinforcing Wood Flakes. TIZ (vormals Tonindustrie-Zeitung) Vol. 109 (1985) 12, pp. 908-916

5. HÜBNER, J.E.: Gipsplatten mit Holzspanarmierung. Holz als Roh- und Werkstoff Vol. 43 (1985), pp. 433-437

6. KOSSATZ, G., LEMPFER, K.: Zur Herstellung gipsgebundener Spanplatten in einem Halbtrockenverfahren. In: Holz als Roh- und Werkstoff Vol 40, 1982, pp. 333-337

7. KOSSATZ, G., LEMPFER, K., SATTLER, H.: Anorganisch gebundene Holzwerkstoffplatten. In: FESYP-Geschäftsbericht 1982/1983, pp. 98-108

8. KOSSATZ, G., SATTLER, H.: Energieaufwand bei der Herstellung von Spanplatten. Holz als Roh- und Werkstoff Vol. 42 (1984), p. 111

9. KOSSATZ; G., LEMPFER, K.: Gipsspanplatten - ein neuer nicht brennbarer Holzwerkstoff. Mitteilung der Deutschen Gesellschaft für Holzforschung, Issue No. 57/1985, pp. 16-20

10. KOSSATZ, G., LEMPFER, K.: Halbtrockenverfahren zur Herstellung gipsgebundener Spanplatten. TIZ (vormals Tonindustrie-Zeitung) Vol. 109 (1985) 10, pp. 756-759

11. KOSSATZ, G., SATTLER, H.: Verwertung von Industriegipsen. Unveröffentlichte Denkschrift 1986, 28 p.

12. LEMPFER, K.: Einfluß der Rohdichte und anderer Parameter auf die Biegefestigkeit gipsgebundener Spanplatten. Holz als Roh- und Werkstoff Vol. 43 (1985), p. 192

13. PIETRZENIUK, H.-J.: Utilization of Residues of Desulphurization Technologies. TIZ (vormals Tonindustrie-Zeitung) Vol. 110 (1986) 12, pp. 874-876

14. SATTLER, H.: Nomogramm zur Ermittlung der Herstellungsfeuchte bzw. der notwendigen Trocknungsenergie von gipsgebundenen Bauplatten. WKI-Kurzbericht No. 38/1982

15. SCHULZE, H.: Anwendungstechnischer Vergleich zwischen Gipskartonplatten und Spanplatten für den Fertigteil- und Fertighausbau. In: Proceedings of the 36th Session of the Technical Committee of FESYP in Frankfurt, 29-31 March 1977. Gießen 1977, pp. 50-73.

16. SCHWARZ, H.G.: Industrielle Produktion gipsgebundener Spanplatten in einer neuen Anlage in Finnland. Holz als Roh-und Werkstoff Vol. 44 (1986) pp. 385-387

17. SIMATUPANG, M.H., LU, X.X.: Der Einfluß von Holzinhaltsstoffen auf die Erhartung von Stuckgips und bei der Herstellung gipsgebundener Spanplatten. Holz als Roh- und Werkstoff Vol. 43 (1985), pp. 325-331

THE USE OF CONCRETE AND MASONRY WASTE AS AGGREGATES FOR CONCRETE PRODUCTION IN HET NETHERLANDS

Ch.F. HENDRIKS

1. INTRODUCTION

In the course of the past few years the need has become apparent for the use of, besides natural materials, materials such as waste matter and industrial by-products to meet the demand for base materials of the construction industry.
Over the last six years an in-depth analysis was made by CUR in Holland in order to find out whether construction and demolition waste can be recycled for the production of concrete. Detailed information concerning the results is found in (6).
Construction and demolition waste consists for over 80% (m/m) of stony materials, the major components being concrete and masonry waste. Other waste components are: asphalt, glass, rubber, timber, plastics and metals, in volumes ranging from usually less than 1% to a maximum of 5%.
The present paper discusses the use of concrete and masonry waste as aggregate for concrete.

2. DEMOLITION AND PROCESSING TECHNIQUES

A major condition for re-use of building and demolition waste is that, during demolition, part of the timber, and the maximum possible volume of metals are removed. Such a limited degree of preselection can be reached through application of current demolition techniques; however, a requirement is that a great deal of attention is paid to the organisation of the operation.
The process applied for upgrading the waste mainly consists of manual preseparation (to remove e.q. large pieces of timber)and pre-sieving to e.q. 10 mm, to remove granulate impurities. Size reduction is preferably carried out in two stages. In general for pre-crushing a jaw crusher or a hammer crusher is used; the post-crusher is either a jaw, impact, or cone crusher. In between the two crushing stages iron and tin are removed, by means of a magnet.
In this particular stage of the process many plants apply a washing process, to separate from the waste granulates such light materials as timber, paper, and plastics.
Such washing techniques (using an aquamator, or screw washer) are very effective for the purpose of reducing the end product's content of impurities and foreign materials. After the post-crusher the material is sieved to the required fraction size. If a hammer crusher is used for reduction the product will have a rather regular grading.
However, if the objective is to produce mainly coarse aggregates a jaw crusher will yield the best results.
Crusher characteristics are useful tools for the control of the curshing and sieving processes.
The crusher characteristic is the graphic relation between the reduction factor, R (= the ratio between the sieving residue of the fed waste materials and the produced waste granulate) and the sieving residues of the waste granulates on the various sieves.
If for a specific plant the crusher characteristic is known, the grading of the end

product can be forecast on the basis of the size of the fed material.
A technique has been developed for the purpose of determining the cubicity of the waste granulates (index of cubicity) using a vibrating table. The end product must of course be inspected in accordance with the current requirements for the various applications. For all applications it is vital that the granulate is regularly inspected visually, for correct composition, both with regard to the stony main and by-components and to possible impurities.
For the visual inspection a special technique was also developed.
Visual inspection and grading analysis must be carried out with a minimum frequency of once every production day. In view of the possible presence of plaster as an impurity the sulphate concentration must also be measured regularly.

3. COMPOSITION AND PROPERTIES OF WASTE GRANULATES (3) (6)

At a large number of waste processing industries samples have been taken at different times; these samples were inspected to check their composition.
Old cement stone is found mainly in the finer fractions of the concrete waste. Masonry waste has a mortar content of 20% as a maximum.
The content of impurities (other that stony materials) is generally limited to a few % (m/m).
Often this content does not exceed 0.1 to 0.3%. This is particularly so if the waste has been washed during processing. It should be kept in mind, however, that the impurities are mainly relatively light materials, which increases the content in terms of parts by volume. As the density of the old cement stone is less than that of the original aggregates the effect of crushing is, that - with concrete waste granulate - the density of the finer fractions is less then that of the coarser fractions. Table 1 lists some properties of the waste granulates.

TABLE 1. Properties of some waste granulates.

waste granulate	particle density (dry) (kg/m^3)	crushing factor	density of bulk material (kg/m^3)	moisture content (% m/m)	water absorption (% m/m)	loss on ignition (% m/m)
crushed concrete	2150-2450	0,70-0,79	1200	3-6	4-5	5
brick masonry waste	1700-1900	0,60-0,65	950	4	10-14	5
sandlime brick masonry waste	1850-1950	0,60-0,65	1050	8	8-12	7
concrete stone masonry waste	1300-2100	0,50-0,60	650	20	20-25	11

If water can penetrate into alle pores a correlation exists between particle density and the final water absorption (parabolic relation). If, after crushing, the particles show approximately the same angular shape, with approximately the same grading a proper correlation also exists between the particle density and the volumetric mass of the bulk material (linear relation). For both relations regression equations have been computed.

4. PROPERTIES OF CONCRETE PRODUCED WITH WASTE GRANULATE AS THE AGGREGATE

The waste granulates discussed in the previous section were used for the production of concrete mixtures, both for "standard" concrete and for coarse

various concrete mortar plants, on several production days. For crushed concrete granulate produced from 375 kg/m³ of blast furnace cement Class A (slump 100 mm) an average compressive strength/plant was achieved of 41.0 to 50.6 MN/m². With all mixtures with waste granulates the standard deviations between batches at the various plants did not differ significantly from those noted with gravel concrete.

6. MECHANICAL PROPERTIES OF CONCRETE

For the purpose of designing concrete constructions information is required on shrink, creep, tensile strength, modulus of elasticity, and maximum deformations to be taken into account. The applied procedure was developed a few years ago as part of a research programme on light concrete.
The procedure is defined in the Dutch regulations relating to concrete construction.
For the purpose of measuring the properties referred to above, three concrete mixtures, with different types of masonry waste granulate as the coarse aggregate, were compared with concrete made from crushed concrete granulate and with gravel used as an aggregate. These mixtures are optimally comparable in terms of strength and workability.
Strength properties are listed in table 3. This table also shows: maximum stress, the maximum deformations to be taken into account, and the modulus of elasticity, derived from the σ/ε diagram of the concrete prisms (150 x 150 x 600 mm). Concrete prisms of these dimensions were also used for shrink and creep measurements.
Long-term strength was tested by subjecting 28 days old prisms to 0.7 times the average ultimate stres value.
None of thes test specimens of concrete produced from the three different types of masonry waste collapsed during the test period of 48 weeks. Consequently this expensive test was not repeated with concrete specimens made form crushed concrete granulate.
The data listed in table 3 show that, with application of waste granulates, the modulus of elasticity is lower, and that the shrink and creep rates are higher than with gravel concrete. In this respect it should be noted, that shrinking due to desiccation is initially low, through the suppletion of (relatively large volumes of) water from the (porous) waste particles.
The pertinence of the found differences depends on the specific requirements of each individual application; it also relates to the variations in the various properties of gravel concrete, that are obviously being accepted.
Particularly when the latter aspect is concerned detailed analysis is needed.
The frost resistance was determined, after spontaneous saturation of the test unit, on the basis of the front on one face.
Inspection showed that in particular concrete produced with brick masonry waste was damaged. Concrete mixture with gravel and gravel concrete waste did not show any damage. The damage noted presumably relates to the presence of large volumes of water in the brick waste granulate, that has different pore types.
With freezing then the so-called ice chipping phenomenon occurs - in the larger pores ice crystals are formed. This causes a build-up of the pressure, the strength of which is determined by the fineness of the adjacent capillaries. The pressure may cause the cement stone to be chipped off the granulate particles. It should be noted that, in the test, the limestone and argex concrete stone mixtures, contrary to the brick mixture, were only partly saturated with water; the two former mixtures show less damage.
The fact that damage occurs with the test does not necessarily imply that damage must also be expected to occur under normal weather conditions. In this respect a major factor is the water content at the beginning of frost periods. There are reasons for assuming that the degree of saturation with water is at a significantly lower level, in practice, than with the (severe) test. For this reason a laboratory test

concrete (used to make concrete blocks and stone).
Considering the differences in particle density between the various granulates, and in view of the relation between particle size and particle density, all compositions have been computed on the basis of corresponding particle volume percentages.
Gravel concrete was used as the reference mixture. All waste granulates had been sieved to a fraction size of 4 mm; concrete sand was used as a fine aggregate. The final grading must be within the range as defined in the Dutch Specification Standard.
All mixtures had virtually the same slump, i.e. 10 cm.
In the production process that is applied to make concrete mixtures the granulates were previously mixed with the mixing water. This premoistening operation is dictated by the relatively high degree of water absorption of the waste granulates as compared with gravel.
It appears that all mixtures with waste granulates have low compressive strength and a higher degree of shrinkage (40 percent greater on an average) than the reference mixture gravel concrete.
When crushed concrete is used as an aggregate the reduction of the strength is relatively low as compared with gravel concrete. With the use of masonry waste as an aggregate the shrinkage appeared to be less, initially, than with the reference mixture -a higher degree of shrinkage was noted only after several months. Moisture storage inside the granulate will initially protect the cement stone against drying up.
Linear relations were found to exist between the concrete strength and the density. On the basis of the relations referred to earlier, linear relations may be assumed to exist between the particle density and the density of the bulk material, on the one hand and, on the other, between the particle density and the concrete strength (with corresponding grading of the aggregates and the same cement content). The regression equations derived for these relations are shown in table 2. The tabular data can be used to assess the concrete strength under the listed conditions, for various granulates.
Under the given conditions it appears that, depending on the type of waste granulate, the cement demand can be 20 percent higher, as a maximum, than with gravel concrete of corresponding strength. In this respect it should be remembered that the difference in cement demand is lower with reduced concrete strength levels. In addition rounded gravel particles are compared with angular granulates, and it should be noted that, with the use of the latter, the water demand is higher.
With the present study only the coarse fraction was composed of waste granulates. On the basis of the results of research it should be expected that, if the fine fraction is also composed of waste granulates, the reduction in strengh will be greater as compared with similar gravel concrete.

5. CONCRETE BLOCKS AND STONES

It was further investigated to what degree concrete stones, with wast granulate as the aggregate, meet the Dutch NEN 7027 Standard specifications for concrete building blocks and stones. For this purpose a concrete manufacturer was asked to produce four concrete stone series, using as a coarse aggregate: gravel, crushed brick granulate (two quality levels), and crushed concrete granulate.
With the same content (235 kg/m^3 of blast furnace cement Class A) alle mixtures appeared to meet the requirements in terms of shrink and compressive strength. These requirements were met in spite of the fact that, again, the mixtures with waste granulates showed greater shrink and slightly lower compressive strength as compared with the mixtures with gravel as an aggregate.
It was investigated whether, and to what degree variations in quality of the waste granulates supplied by various producers affected the strength of the concrete manufactured from these granulates. The concrete mixtures were produced at

was undertaken, simulating the severe winter condition of 1978/79, as far as temperature and humidity are concerned. It appeared from ultrasonic measurements and visual assessment that no damage occurred. Therefor the conclusion is, when brick waste is used there is a risk of frost resistance being insufficient, however only under the circumstance where the concrete is almost completely saturated with water. Finally, it has appeared that, with the production of the fine fraction with waste granulates, effects similar to those described above for the coarser fraction also occur.

7. PROVISIONAL REGULATION

On the basis of the obtained results a provisional regulation has been published at the end of 1984, that was drafted as a recommendation. The recommendation bears major resemblance to the Dutch Standard for sand and gravel.
The most significant aspects are listed in tablel 4 and 5.

TABLE 2. Regression equations for concrete properties and granulate properties (slump 100 mm; 320 kg of cement/m³).

Equation				r	S_d
D	=	1695 + 0,46 B	$\pm tS_d (1 + 1/n)$	0,990	22,4
D	=	1656 + 0,28 P_d	$\pm tS_d (1 + 1/n)$	0,980	24,1
$T_k 28$	=	14 + 0,0095 P_d	$\pm tS_d (1 + 1/n)$	0,938	1,4
$T_k 28$	=	13,9 + 0,017 B	$\pm tS_d (1 + 1/n)$	0,942	2,0
D	=	Density of concrete (kg/m³)			
B	=	Bulk density of granulates (kg/m³)			
P_d	=	Particle density of granulates (kg/m³)			
$T_k 28$	=	Cube compressive strength of concrete (MN/m²)			
r	=	Correlation coefficient			
t	=	Coefficient (Students distribution)			
S_d	=	Residual variance			

If the aggregate consists of 20 percent, as a maximum, of waste granulate and the remaining part consists of sand and gravel, there is no need to further inspect the mechanical properties of the concrete, as then the effect of the waste granulate(s) is less than the variation that occurs in the various properties of gravel concrete. In case the above percentages are higher the results of the study described in the present publication may be applied (possibly trough interpolation), provided that the type of granulate(s) used matched the concrete quality. If not, additional research can be done into the mechanical properties, insofar as pertinent to the application involved. However, instead of this, the CEB calculation rules for lightweight concrete can be used. It appeared that this gives enough safety for the design, so that, as with normal concrete, it is no longer necessary to determine the mechanical properties experimentally for every construction.

$$E_w = E_g \frac{P_w^2}{2400}$$

$$\psi_w = \psi_g \frac{P_w^2}{2400} \cdot 1,2$$

$$\varepsilon_w = \varepsilon_g \cdot 1,5$$

$$T_w = T_{tg} \cdot (0,30 + 0,70) \cdot \frac{P_w}{2400}$$

E_w = modulus of elasticity of concrete with waste granulates
E_g = modulus of elasticity of concrete with gravel
ψ_w = creepfactor of concrete with waste granulates
ψ_g = creepfactor of concrete with gravel
ε_w = shrinkage factor of concrete with waste granulates
ε_g = shrinkage factor of concrete with gravel
T_{tw} = tensile strength of concrete with waste granulates
T_{tg} = tensile strength of concrete with gravel
P_w = dry density of concrete with waste granulates

TABLE 3. Mechanical properties of some concrete mixtures.

Properties	Gravel	Gravel concrete waste	Brick masonry rubble	Sandlime brick masonry rubble	Argex concrete stone waste	Gravel	Gravel concrete waste
Cement (pc - 8) kg/m³	263	285	311	323	409	370	411
w/c Factor (eff)	0,60	0,63	0,61	0,54	0,42	0,43	0,45
Density (kg/m³)	2333	2243	2135	2125	1796	2344	2267
Compressive strength	30,6	33,1	32,3	32,8	28,1	53,4	57,1
Splitting tensile strength NM/m²	3,31	3,36	2,72	2,82	2,14	4,01	4,09
Maximum stress N/mm²	26,8	27,4	26,2	24,1	> 20,8	40,1	39,2
Maximum deformation %	2,4	2,7	2,5	2,4	1,6	1,9	3,2
Modulus of elasticity M/mm²	30100	27900	20700	20000	16600	37400	30800
Shrink after 52 weeks %	0,37	0,50	0,31	0,47	0,72	0,41	0,64
Elastic deformation after 4 weeks %	0,27	0,34	0,42	0,44	0,41	0,42	0,47
Frost resistance visual after 25 cycles	good	good	moderate poor	fair	fair	good	good
Relatively modulus of elasticity after 25 cycles	0,99		0,78	0,95	1,01		
Carbonation depth mm	4,0		4,5	4,6	3,3		
water penetration mm	20		28	33			

TABLE 4. Main aspects of the provision standard for crushed concrete granulate.

Aspect	Requirements; description
A. Main component	>95 percent crushed granulate, with dry particle density of ≈ 2,100 kg/m³
B. Side component	<5 percent stony materials referred to in table 5
C. Particle catergories	0-4; 4-8; 8-16; 16-32; 4-16; 4-32
D. Grading	If several fractions are used; in accordance with the specification of DIN standard 4136 (4) (narrow range.) For 4-16 and 4-32: in accordance with Sand and Gravel Standard (5) specifications particles < 63 µm; for 0-4 < 4% m/m; for the other categories: <2% m/m; higher percentages are allowed, provided that it has been proven these are not detrimental.
E. Impurities	Soft components, fine substances of organic origin, chlorides, sulphates, carbonates: must meet the requirements of Sand and Gravel (5). Bituminous materials, plastics fire resistant stone, glass, metals, wood, and other light components: total <1% m/m and <1% V/V.
F. Dimensional stability	Expansion phenomena; durability.
G. Staining	Iron and vanadium compounds; for clean concrete: staining index < 20.
H. Retarising, of strength reducing components	Inspection thr. Vicat test — Maximum of 15% difference relative to reference
J. Cubicity	Cubicity index < 70% m/m; flat sections < 30% m/m.
K. Frost resistance	Weathering < 3% m/m.
L. Application area	All classes of non-reinforcement, reinforced, and prestressed concrete.

TABLE 5. Main aspectes of the provisional standard for brick masonry rubble granulate.

Aspect	
A. Main component	<65% m/m crushed material from stones and blocks made from brick and/of cement concrete.
B. Side component	Limestone < 20% m/m; light concrete < 20% m/m; cellular concrete <10% m/m; ceramic materials < 20% m/m; natural stone < 20% m/m; mortar <25% m/m.
C. thr. K	Refer to C thr. K in table 4.
L. Application area	All classes of non-reinforced and reinforced concrete.

8. ECONOMIC ASPECTS

If waste granulates are used as an aggregate for concrete production some savings can be achieved, and extra costs may be incurred, as compared with the use of gravel concrete. Costing was based on current market prices for gravel and waste granulates.

If gravel is replaced by waste granulates the differences in particle density and cement demand must be taken into account. Application of waste granulates further entails extra costs relating to inspection, storage, and production.

These special costs specifically relate to the need for extra storage space (surcharge for angularity), pre-moistening of the waste granulates, extra wear of the plant, and cleaning of the plant with conversion to a different aggregate type.

Another point to be considered is greater compacting energy for mixtures with angular aggregates. If at most 20% of waste granulate is used as aggregate, not a single type of construction will need adjustments. The lower value for the modulus of elasticity may result in a 10% thicker layer at the most if 100% waste granulates is used, compared with gravel concrete.

Cost calculations have shown that at the moment in most cases concrete with waste granulates still is slightly more expensive than gravel concrete.

In this respect it should be noted that the saving on (rising) dumping costs for waste materials has not been accounted for in the present calculation. Furthermore the current market prices are based on sale of waste granulates as unbound stone road base material for road construction, the price/tonne of which is higher than for sand and gravel.

Prospects accordingly are rather favourable, especially when up to a maximum of 20% waste granulates are applied.

9. APPLICATION

Waste granulates 0/40 are already being applied on a large scale as unbound stone roadbase material for road construction.

Current practical experience with concrete containing waste granulate as the aggregate mainly concern projects for the (re)construction of roads, airfields, bridge pathes and walls of houses.

At the Volkel airfield crushed concrete granulate was used for a lean concrete base course and concrete pavement. The mixture applied was crushed concrete granulate 0/40; concrete sand; blast furnace cement, 80 kg/m^3; and crushed concrete granulate 0/31.5; concrete sand; Portland cement A, 350 kg/m^3, respectively. In both cases the compressive strength was found to be 10 to 20% lower relative to similar gravel concrete.

With reconstruction work at Maastricht Airport a mixture was used of crushed concrete granulate 0/15 and 15/30; concrete sand; Portland cement, 380 kg/m^3. The mixture more than amply met the requirements for quality B 37.5.

In the coarse of a field test near the city of Helmond the frost-thaw resistance of cores drilled from the pavement was determined. Crushed concrete granulates and gravel as an aggregate appeared to show no significant differences as far as that aspect was concerned. Concrete with granulates of brick masonry waste was found to have a significantly lower frost-thaw resistance at the laboratory. In practice however, after 4 years still no damage was observed.

In 1982 a minor test was carried out within the framework of the Underground construction in Rotterdam. The results are analogous to the above experiences.

In 1984 a very succesful test was carried out using crushed concrete granulate for the production of concrete to be used for partitions in several houses in Amersfoort. The concrete, having the same cement content as the usual gravel concrete mixture, met the requirement of a characteristic strength of 22.5 MN/m^2. Due to water absorption by the waste granulate the water demand was little higher that with concrete, and compaction required more energy because of the angularity of the material. Several test walls had been cast in advance, special attention being paid to possible cracking due to the higher shrink rate of waste granualte concrete. There was no sign of cracking, however. In 1985 several parts of the Caland windscreen in Rotterdam were constructed with concrete waste granulate.

10. GENERAL CONCLUSIONS

1. The currently applied demolition and processing techniques allow the production of concrete and brick masonry waste that is, in principle, suitable for concrete manufacture.

   That applies in particular to waste granulates used as coars aggregates. Fine fraction waste granulates can contain relatively more weak and/or detrimental components, which presents less of a problem if the waste material is used for road bases than if it is used as an aggregate. For the production and processing of concrete made with waste granulates as an aggregate the usual techniques can be applied.

2. If waste granulates are used as aggregates, it must be expected that the strength is reduced compared with gravel concrete of the same composition. A linear relation exists (with corresponding grading) between the dry particle density, or the dumping weight, of the aggregate material and the concrete's strength (both compressive and splitting strength).

   With gravel concrete, to produce waste granulate concrete with the same strength, the cement demand is higher, up to a maximum of 20% for the higher grade concretes.

3. Crushed concrete granulate can be used, in principle, for all concrete grades, and it is in many respects (virtually) equivalent to gravel. Concrete made from crushed concrete granulate is as frost resistance as gravel concrete. The shrink and creep rates of concrete made from crushed concrete granulate are higher than those of gravel concrete, the modulus of elasticity is lower.

4. Masonry waste granulate can be used, in principle, for an aggregate for concrete production. The frost resistance of concrete made with brick waste granulate, compared to that of gravel concrete, can be lower in case the concrete is completely saturated with water. Its shrink and creep rates are higher than those of gravel concrete, the modulus of elasticity is lower.

5. If the aggregate is composed with at most 20% waste granulates there are no significant differences between the thus poduced concrete and gravel concrete, as regards mechanical properties.

   A mixture of waste granulates with sand and gravel is to be preferred from an economical point of view also.

6. In principle it is possible to produce concrete with waste granulates of a quality as contant as that of concrete made with gravel and sand.

   It is important that the waste granulate producers by further improving their quality control-system, attain the same quality level as with sand and gravel.

   Projects which are initiated by the Government and where concrete made with waste granulates is applied is of great importance because they will inspire confidence with other potential commissioners.

   It should be noted that concrete mortar with waste granulates is supplied with a certificate.

7. The overall conclusion can be drawn that concrete made with waste granulates is a hunderd percent alternative for gravel concrete.

11. LITERATURE

1. Pauw, C. de, Recycling Concrete WTCB, Brussels, 1980 (in French/Dutch)
2. Nixon, P.J., Recycled Concrete as an aggregate for concrete, a review Materials and Structures, nr. 65, September/October 1978, pp 371-378
3. Proceedings of the EDA/RILEM Conference on Recycling, part II, Rotterdam, 1985
4. DIN 4163, Ziegelsplittbeton, Bestimmungen für Herstellung und Verwendung, February 1951
5. NEN 5905, Toeslagmaterialen voor beton, zand en grind, 1985 (in Dutch)
6. CUR-rapport 125 Betonpuingranulaat en metselwerkpuingranulaat als toeslagmateriaal voor beton, September 1986 (in Dutch)

UTILIZATION OF RESIDUES OF DESULPHURIZATION AND DENITRIFICATION TECHNOLOGIES

H.-J. PIETRZENIUK

1. SUMMARY

Large-scale desulphurization plants presently in use in the Federal Republic of Germany produce gypsum, sulphur, sulphur dioxide, sulphuric acid, ammonia sulphate as well as sulphite/sulphite mixtures as residues from dry/quasi-dry processes. Utilization of residues from dry/quasi-dry pollution abatement facilities requires additional treatment processes which involve further costs. The conditions for an orderly, environmentally compatible form to dispose of such residues will be described.

In comparision with dry or quasi-dry processes, wet processes lead to products which can be further utilized. As gypsum counts for the greatest part, its market chances will be discussed in detail. Given a substitution rate of only 50 % of natural gypsum or natural nahydrite presently used and a possible opening of new fields of use, the total volume of the gypsum so obtained can be used primarily by the building industry.

Also in regard to sulphur, sulphur dioxide and sulphuric acid, there are presently no sales problems although it should not be disregarded that the share of sulphur as an unavoidable by-product of other processes in increasing, and that the demand - at least as far as the industrialized nations are concerned - is stagnating as a consequence of utilization processes and avoidance strategies.

In the Federal Republic of Germany, larger amounts of ammonia sulphate are produced only as a by-product which must be sold on international markets. The world market is characterized by considerable capacity expansion in countries having their own natural gas/mineral oil deposists. Competition is aggravating. Prices stagnate or even show a slumping trend.

The denitrification process of "selective catalytic reduction" (SCR) which is expected to become the most common process does not produce any redidues.

2. Useful residues

2.1. Gypsum

As a consequence of the Ordinance on Large Firing Installations, somewhat more than 1.75 million tons SO_2 will have to be precipitated annually in the Federal Rupublic of Germany (from 1995 onwards). The market share of processes with gypsum as their end product presently amounts to approx. 90 % which means that, after 1995, more than 2.4 million tons gypsum are likely to be supplied by coal-fired power station each year, plus another 1.0 million tons of gypsum annually by the brown coal power stations in the Rhine region. About two thirds of this volume, i.e. approx. 2.6 million tons per year, are supposed to be produced already in 1988.

The traditional gypsum industry presently supplies more than 2.7 million tons of gypsum for building plaster, sandwich-type plaster board and building elements per annum. Adding the import/export balance of some 0.5 million tons which plays a role especially for natural gypsum, one obtains a domestic production figure in this area of around 3.2 million tons per annum. These figures must be further increased by 1.4 million tons of natural gypsum and natural anhydrite per annum for the setting control of cements, as well as another 1 million tons of anhydrite contained in mining mortar. This translates into an overall volume of gypsum/anhydrite of 3.6 million tons per annum which is today almost exclusively produced by the German gypsum industry from natural sources.

The major part of natural gypsum undergoes certain pretreatment and, in most cases, is used as a building material for civil engineering purposes after the admixture of various additives. This area counts for 2.5 to 3.0 million tons per annum in the Federal Rpublic of Germany. The following uses are distinguished:

- plaster
- sandwich-type plaster board
- plaster slab partition walls
- plaster floor.

The traditional use of burnt gypsum is plaster. In most cases, plaster consists of so-called multiple-phase gypsum, i.e. mixtures of approx. 25 to 30 % beta-hemihydrate (gypsum) and anhydrite.

DIN 1169 which constitutes the basis for the use of building gypsum is not limited to natural gypsum alone. If gypsum obtained from flue gas desulphurization plants meets the requirements of the standards, no distinction is made between products from natural gypsum and gypsum from desulphurization plants.

The amount of harmful trace elements which might preclude the use of desulphurization gypsum for interior purposes falls within the range of natural gypsum. This also applies to its radioactivity so that there are no objections against the use of desulphurization gypsum as building material. The German gypsum industry has formulated the following additional requirements which desulphurization gypsum must meet:

Humidity	10 %
$CaSO_4 \; 2H_2O$	95 %
MgO	less than or equal to 0.10 %
Chloride	less than or equal to 0.01 %
Sulphite	less than or equal to 0.25 %
pH	5 - 9
Organic constituents	less than or equal to 0.10 %
Color	white
Smell	neutral

Most of these requirements are met by the processes offered, special treatment may be required in some cases. The scope of such processing depends on the crystalline form. Sufficient residence times of the gypsum crystals at low rates of gypsum oversaturation permit the production of the desired coarse-grained gypsum in the absorber; such coarsegrained gypsum is much easier to dehydrate than gypsum having needle-shaped crystals. For other processing stages, too, the use of gypsum having a coarse-grained structure is more favourable. The gypsum so prepared is partly calcined at temperatures of around 140 °C in a calcining furnace. The end product is beta-hemihydrate or plaster.

Another way is to mix gypsum from flue gas desulphurization (FGD) plants with anhydrite obtained from the production of hydrofluoric acid (such anhydrite serving as bonding agent) and to use this mixture in the production of floor pavement substance. Given an area of 40 to 50 million square meters of floor pavement completed annually in the Federal Republic of Germany, and further given a substitution rate of 50 % by the above-described new building material, one obtains an annual demand of some 2 million tons. This figure contains 600.000 to 1 million tons desulphurization gypsum. Such a volume could be marketed additionally without displacing natural gypsum which is primarily used for the production of plaster and sandwich-type plaster board. Such a substitution would primarily affect the market for cement which is produced at high energy input rates and continually rising costs.

Before grinding the cement, gypsum and anhydrite are added to the burnt cement clinker for setting control purposes. According to DIN 1164, the calcium sulphate content of Portland cement is limited to a maximum of 3 % SO_3. Relating to gypsum, this translates into a realistic demand in the order of 5 % which is presently covered almost exclusively by natural gypsum and anhydrite. Requirements concerning the chemical purity of gypsum used in the cement industry have as yet been formulated only as far as DIN 1164 sets forth that the permissible limit percentages of CO_2, MgO, SO_3, Cl in the cement must not be exceeded. As a maximum of 5 % desulphurization gypsum is added, these values are easily adhered to without the need for additional purification processes.

Given a future annual demand for approx. 1.4 million tons of natural gypsum and anhydrite, a realistic figure would be 500.000 to 600.000 tons of desulphurization gypsum per annum as a substitute.

Further development work in the field of a utilization of gypsum from flue gas desulphurization aims at producing from flue gas gypsum - obtained in the form of beta-dihydrate - the more stable alpha-modification.

Uses in coal mining are particularly obvious because of its good early and final strength.

First tests in underground workings have underpinned the suitability of mining mortar made from desulphurization gypsum. Owing to its improved filling rate and a more favourable setting temperature, such a mining mortar is probably superior to conventional mining mortars, especially if it is produced directly in connection with a flue gas desulphurization plant. The demand for mining mortar in the Federal Republic of Germany is in the order of 1.3 million tons per annum, and the trend is increasing. Careful estimates indicate sales of desulphurization gypsum between 600.000 and 700.000 tons per annum.

Another completely new use of gypsum which is becoming more and more important against the background of the present discussion of the dangers inherent in formaldehyde, is the production of emission-free particle board with gypsum binder. Gypsum as a bonding agent is then mixed with wet, undried chips. The pourable mixture can be processed by means of dusting machines. The hydration water for the gypsum is obtained from the wood chips. In comparison with traditional particle board, considerable energy savings can be realized by the omission of the chip drying process and thermal treatment during pressing. This translates into clearly more favourable costs in comparison with both gypsum staff as well as particle board using synthetic resin as bonding agent. Especially as a consequence of the absence of emissions, optimal fields of use are in the entire area of interior construction, e.g. dry floor pavement, interior works in lofts, and modernization of old buildings. Even exterior uses are conceivable with water-proofed gypsum. The total production of wood particle board in the Federal Republic of Germany presently totals some 5.3 million cubic meters per annum. Under the described semi-dry process, approx. 0.8 ton of gypsum is required per cubic meter of particle board. Careful estimates indicate a medium-term gypsum demand between 0.5 and 0.6 million honts per annum, if about 15 % of the present wood particle board could be substituted.

2.2. Sulphur, sulphur dioxide, sulphuric acid

Of the flue gas desulphurization processes used on a large scale in the Federal Republic of Germany, the Wellman-Lord process and the Bergbauforschungs-Uhde process produce alternatively sulphur, SO_2 or sulphuric acid.

Plants which are presently under construction are the Mitte power station of BASF (Wellman-Lord, 30.000 tons of liquid SO_2 per annum), the Buschhaus power station of BKB (Wellman-Lord, 35.000 tons of sulphur per annum at 4.000 operating hours), as well as the power station of the public utility company of Bayreuth (Bergbauforschung/Uhde, 32.000 tons of H_2SO_4 per annum).

These figures are low vis-à-vis the consumption of sulphur on a global level and in the Federal Republic of Germany alone. The world-wide consumption of all forms of sulphur totalled 53.713 million tons in 1983. At the same time, consumption in the Federal Republic of Germany totalled 1.498 million tons per annum. Apart from the distinction between sulphuric acid and non-sulphuric acid products, consumption data is hardly differentiated. This is due to the extreme multitude of sulphur compound products, mainly in the area of specialist chemistry. Sulphuric acid accounts for 80 - 85 % of the total sulphur consumption. Between 1966 and 1981, world-wide sulphuric acid production rose by approx. 73 % to 138 million tons per annum. Expecially high rates of rise are observed in the Soviet Union as a consequence of its expansion of fertilizer production. Production increases in Western Europe and Japan were low or even stagnated because of an only slightly increased demand by the phosphate fertilizer industry. This trend was more than offset by high rates of rise in the Third World since these countries have made the build-up of their own fertilizer industry a predominant task. Especially strong increases in sulphuric acid capacities are observed in those countries which have the possibility of exploiting their own phosphate deposits, e.g. Morocco.

It is interesting to note that the importance of natural sulphur abtained from elemental sulphur ores has been strongly decreasing since the early 1960's in conjunction with increased mineral oil and natural gas consumption. In 1982, its share was down at only 1/3 of the raw sulphur supply indicating that the inflexible percentage obtained in the form of by-products already accounted for 2/3 of the total volume. Owing to a strong expansion of the so-called recuperation sulphur, even Western Europe and Japan have become regions with a sulphur export surplus. Given an increasing rate of avoidance and utilization of sulphur residues, the total consumption of primary raw materials will have to be restricted still further. Against the background of the transition to a more high-grade use of the global coal resources (coal upgrading), the percentage of recuperation sulphur obtained as a by-product will strongly increase. Given the situation, on the other hand, that demand figures will only slightly increase as a consequence of an increasing trend towards the avoidance and utilization of sulphurous residues, it might occur that not even sulphur optained as a by-product can be marketed even if direct sulphur production of elemental deposists would be completely discontinued.

Against the background of the strong expansion of fertilizer production - expecially in Third World countries - such a trend can only represent a long-term development. This is also underpinned by the large number of new sulphur uses, such as sulphur asphalt to substitute bitumen, sulphur concrete, the development of high-sulphur polymers and other products so that an excess sulphur demand can only be expected in the long run.

The present situation as well as the situation which can be expected in the near future is rather marked by a considerable scarcity of the supply of elemental sulphur and thus by increasing prices. In regard to the flue gas desulphurization processes used in the Federal Republic of Germany - such processes producing sulphur, sulphur dioxide or sulphuric acid - this means that there will be a market for such products in any case, apart from cases like PASF where such products are directly used by the company producing them.

2.3. Ammonia sulphate

The Walther process is being used in two blocks of the large Mannheim power station. The by-product abtained is ammonia sulphate which can be used as a fertilizer. Ammonia sulphate can be used as a nitrogen component in the production of multiple-nutrient fertilizer, or directly as a single-nutrient fertilizer in agriculture. It competes with other nitrogen fertilizers. It is not synthetized in the Federal Republic of Germany, it is rather a by-product in the production of synthetic fibres. The quanities abtained cannot be marketed in Germany; they are primarily exported to developing countries. On these markets too worsening competition conditions must be expected as a consequence of the expansion of production capacities in countries having their own natural gas and mineral oil deposists. Prices are likely to stagnate rather than to rise so that - despite of the general possibility of using the ammonia sulphate produced as a fertilizer - its introduction will basically depend on its price and thus affect the overall efficiency of the flue gas scrubbing process.

3. Unusable residues

3.1. Sulphite/sulphite mixtures

With dry/quasi-dry desulphurization, basic sorbents are added in the fire vault and/or after dust removal. Depending on whether precipitation takes place along with the flue

dust or whether flue gas scrubbing takes place after the
dust has been removed, different sulphite/sulphate mixtures
are obtained which may also have a high lime content. Mixtures of such a composition cannot be utilized without prior
treatment and must be dumped. Moreover, landfill uses and
uses in landscape architecture are no utilization in the
strict meaning; such uses rather call for control similar to
dump supervision.

3.2. Dump requirements

Such dumps must be provided above ground water level and
sealed off against the ground water. They must also be sealed off against precipitation, and they have to be provided
with a drainage system to collect gravitational water. The
high cost resulting from these requirements to protect the
environment justify attempts at utilizing the residues. Oxidation of the residues to obtain technical anhydrite is considered the most promising way. For this purpose, the flue
dust should be precipitated beforehand. This anhydrite is
suitable for setting control uses in the cement industry. If
the chloride content has undesired levels, it can be diluted with natural gypsum/anhydrite mixtures.
Other processes under investigation also provide for the
production of building materials.

1. GLASS-RECYCLING

Ir. C.Q.M.Enneking
(N.V.Vereenigde Glasfabrieken)

Of all the glass which comes into the market as packaging material, about 75% is returnable and 25% non-returnable.

Last year 85.000 tons of glass was used for replacement in the returnable circuit, this is higher than normal and is caused by the brewers' changeover to the new style beer bottle.
425.000 tons were necessary last year for the non-returnable circuit; this is a slight drop caused by the fact that the packaging units are now lighter in weight.

The quantity of glass returned to the glass industry is:

        210.000 tons via the bottle bank

        35.000 tons via the industry (eg brewers etc.)

In order to be able to express the success of glass recycling in figures, a number of basic statistics can be given.
In the Netherlands we use the quantity of glass collected via the bottle bank compared with the non-returnable glass which comes on to the market, we then arrive at **49,5%**.
In a European context we take the total quantity of glass collected compared with the quantity of new glass which comes on to the market, this is 48% for the Netherlands.

The following dissertation will only be referring to the 25%, in other words the non-returnable glass, in relation to the three following points:

        2. COLLECTION

        3. THE USE OF THE COLLECTED GLASS

        4. THE PROMOTION OF THE BOTTLE BANK

2. COLLECTION

The spectacular growth of collected glass is shown in figure 1.
The graph shows that there is no longer such rapid growth to be seen; at the present time about 50% of non-returnable glass is collected.
Further growth will only be possible through extra effort.

If we take a look at the situation of the Netherlands in Europe, figure 2, we shall see that the Netherlands in 1985 was way ahead with respect to the quantity of glass collected per head of the population.

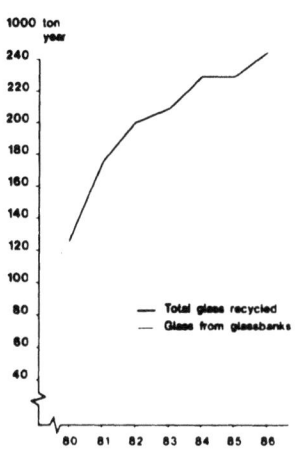

fig. 1

If we convert the Dutch percentage of 53 according to the same methods as used by other countries, the figure of 53 must be changed to 49%.
Even with this corrected figure, we are still at the top in Europe.

EUROPEAN GLASS RECYCLING 1985

Country	Tonnes collected	Share of national consumption
Austria	68,000	38%
Belgium	140,000	42%
Denmark	32,000	19%
France	601,000	26%
Germany	967,000	39%
Great Britain	210,000	12%
Ireland	7,000	7%
Italy	467,000	25%
Netherlands	230,000	53%
Portugal	24,000	10%
Spain	144,000	13%
Switzerland	132,000	46%
* Based on consumption excluding returnables		
TOTAL:	3,022,000	27% average

This high result is due to the fact that practically all Municipal Councils in the Netherlands have installed bottle banks.
The total number of installed bottle banks - including Madurodam with three - amounts to more than 10.000 so that the goal of 1 bottle bank per 2.000 inhabitants has been considerably exceeded.

fig. 2

Dalmijn investigated the relationship between the quantity of glass collected in the bottle bank per inhabitant and the number of bottle banks per inhabitant (figure 3).

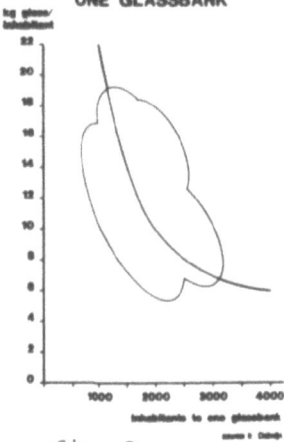

fig. 3

It can be claerly seen, as was indeed expected, that the more bottle banks you install, the higher the amount of glass collected.
We have to make it as easy as possible for the general public.

With respect to collection, there are a number of striking phenomena.
The composition of the glass collected via the bottle bank and the glass thrown away in the dustbin is clearly different (figure 4).
A possible explanation for this is that green is particularly used for wines and spirits.
The advertising campaigns for glass recycling were originally aimed at the collection of drink bottles.

GLASS MIX

	Glassbank	Household waste
Flint	44%	62%
Emerald	42%	21%
Amber/Black	14%	17%

fig. 4

It is also evident that the older people are, the more they participate in glass recycling (figure 5).
The number of non-users of the bottle bank in the category up to 19 years is more than twice as many as in the category 60+.
It would be interesting to find out if this ratio is as large in other environmental situations.

(N.B. A further advertising campaign will be specifically aimed at the even young people).

As a result of the collection of glass, the quantity of glass in domestic waste dropped (figure 6), thereby reducing the risk of injury to the refuse collectors, and the costs of collecting and processing the domestic waste.

On the occasion of installing the 10.000th bottle bank in Zeist, Mr. Rey, director of the Ministry for Housing, Regional Development and the Environment (VROM), referred to a total saving of 60 million guilders for the municipal councils up to September 1985.

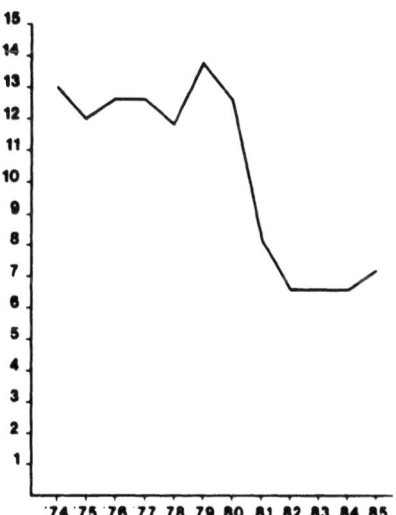

NON USERS OF THE GLASS BANK

Age	Percentage
- 19	35%
20 - 24	27%
25 - 29	22%
30 - 39	21%
40 - 49	21%
50 - 59	17%
60+	16%

Source: Nationaal Rayon onderzoek 1985

fig. 5

fig. 6

3. USE OF COLLECTED GLASS

The glass collected in the bottle banks is taken to the glass processor where it is broken to cullet and cleaned.
In order to avoid unneccessarily high transport costs, there is active import and export with neighbouring countries (figure 7).
Since the glass industry specifies that glass made from recycling cullet must be of the same quality as glass made from raw materials, strict standards are applied to the quality of the supplied; (figure 8), demonstrates how the stone-content in the cullet has dropped over the past years.

In addition to this improvement in the quality of the cullet in order to be able to guarantee the good quality of the bottles the glass industry had to make the necessary investments in equipment to check the bottles produced. In the case of very high percentages of recycling glass, 75% or more, additional investment is needed in the glass furnaces (figure 9).

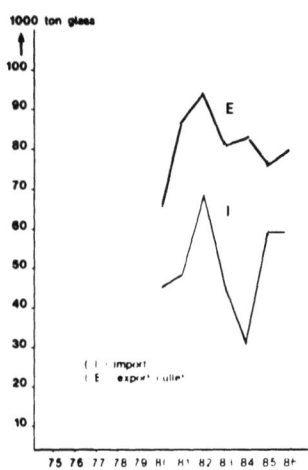

fig. 7

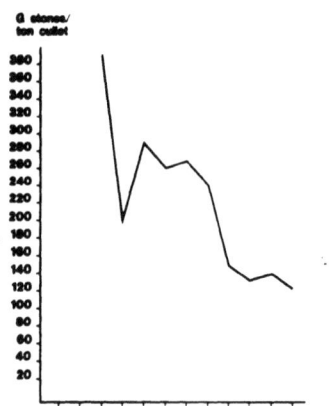

fig. 8

DISADVANTAGES OF RECYCLING CULLET

1) Extra investments for Sorting Machines
2) Corrosion of the furnace by metals in the cullet
3) High cullet levels extra investments on the melting furnace

fig. 9

In addition to these less attractive effects of the use of recycling glass in the glass industry, there is a saving in the use of raw materials and energy consumption. The average saving in energy in the melting furnaces is approximately 2,5% per 10% cullet in the batch.

In the entire production process, including raw materials, the saving can rise to 35% if the percentage of cullet increases from 15% to 100% (figure 10).

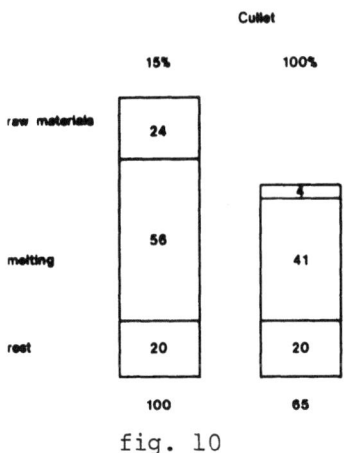

fig. 10

The recycling glass, as collected in the Netherlands, can be processed in coloured glass from 30% in amber up to 80% in green.
Until now, apart from the brown beer campaign, the production capacity in the Netherlands together with the import/export difference is sufficiently large to process the glass which is collected.

Should the supply greatly increase in the future, it will have to be sorted according to colour.
Extra investments will than be necessary.

Research is being carried out with regard to sorting glass according to colour.

4. PROMOTION OF BOTTLE BANKS

A large part of the success of the bottle banks in the Netherlands is a result of the good cooperation between the government and the industry which is closely involved in the collection of glass.
A commission consisting of representatives from:

- the Ministry for Housing, Regional Development and the Environment
- the Ministry of Economic Affairs
- the Ministry of Agriculture
- Trade Commodity Boards
- Central Bureau of Grocery
- Glass Collectors
- Glass Industry

organizes a number of promotional activities each year in close cooperation.
Figure 11 gives a few examples of this cooperation.
The Ministry for Housing, Regional Development and the Environment also publishes a number of brochures concerning the value of glass recycling.

The above-mentioned commission have also designed a logo. The idea is for this to be stuck on the packaging by everyone marketing non-returnable glass. (figure 12).

PROMOTION CAMPAIGNS

Radio Commercials
Television ("Postbus 51")
Advertisements in papers
Stickers on municipal waste collecting cars
Displays and posters in supermarkets shops
Competition "who is making the nicest sticker for a glass-skip"
This year: special promotion for youngsters (primary schools)

fig. 11   fig. 12

As a conclusion for the Netherlands, I should like to say that if glass collection via bottle banks is to succeed, a number of conditions have to be complied with at the very least:

1. high density of bottle banks
2. cooperation with government authorities, etc.
3. promotional activities

As Mr. Rey said on the occasion of installing the 10.000th bottle bank, it is thanks to these activities that:

"THE BOTTLE BANK HAS BECOME PART OF OUR DAILY LIFE, LIKE CLEANING YOUR TEETH EVERY DAY"

References:

1. A.A.J.Cornelissen

    Sorteerproeven met huishoudelijk afval
    Rapportnummer 841613002 RIVM 81 t/m 85

2. W.L.Dalmijn

    Glassrecycling Possibilities and Limitations
    Glass International June 1986

3. N.N. Enquête Glasbak kreeg grote respons
   Recycling Febr. 1985

4. N.N. Van grondstof tot glas
   Recycling Febr. 1985

5. N.N. Glass-Gazette
   Oktober 1986

6. N.N. Glass-Gazette
   April 1986

7. N.N. Nationaal Rayon Onderzoek 1985 Deel I
   PEO-projekt 24.13-030-10

8. Gordon M.Stewart
   How Foreign Cullet usage affects container production
   Glass Industry, December 1985

PHOTOMETRIC SORTING OF CULLET

A. REICHERT
H. HOBERG

1. INTRODUCTION

The municipal solid waste produced per annum in the Federal Republic of Germany contains approximately 2.6 mio tons of waste glass. In the past 4 years the recycling rate of waste glass has increased annually on average by 11 %. It amounted to 890.000 tons in 1985, meaning a share of approx. 34 % of the total mass.

The existing recycling system seems to have proven itself and has largely been accepted by the public.

On the other hand the rising collection rates have recently led to a surplus of waste glass. 76 % of the collected glass is unsorted mixed coloured, which can only be used to any considerable extent in green glass furnaces. Unfortunately the capacity of the green glass production for processing cullets has reached a limit. It is to be anticipated, that it will come to a collapse of the price structure on the waste glass market similar as happened in the waste paper branch. It is clear that counter measures must be enforced in the future to maintain the waste glass recycling system.

However, it is only possible to increase the utilization rate by providing colour sorted glass thus extending the share of cullet in the flint and amber glass furnaces.

In principle there are two ways to that end:

- colour separated collection and
- colour sorting.

For a colour separated collection additional containers and transport facilities or new collecting systems are needed. Experience has shown us besides that the waste glass collected and separated by the consumers does not meet with the specifications of purity required by the glass industry.

For this reason colour sorting seems to be more advisable. This is carried out by hand sorting in all german waste glass treatment plants. However, this is a very expensive method and only includes the share of whole bottles which varies depending on the collection and transport systems, but never amounts to more than 40 %. Therefore a mechanization of the sorting process must be carried out.

Referring to this problem at the Institute of Mineral Processing, Coking and Briquetting a research project is at present being launched to develop a photometric sorter for glass cullet.

2. ECONOMICAL AND TECHNICAL CONDITIONS

The underlying demands to a cullet sorting unit derive from the processed material, the atainable recovery and product quality and the proceeds to be achieved.

Here the main problem of the mechanical colour sorting of waste glass already becomes clear. On the one hand the buyers - the glass producers - demand a very high purity level from the product. The flint cullet admitted to the furnace may not be contaminated with more than 0.005 %; or 50 g/t, coloured glass.

On the other hand the price of the primary raw material for the glass production is relative low. Therefore the proceeds from colour sorted glass of high purity only amount to 130 - 150 DM/t.

The narrow economical margin must be considered in the development of the sorting equipment.

- Complicated processing stages for the pretreatment of the material like washing and drying facilities, requiring expensive waste water treatment systems, must be avoided.

- All machine parts, especially the electronic elements of the detector and the control unit, should be available. They should not need new basic developments, as a series production of cullet sorters is not possible.

- Adjustment to several throughputs should be possible without changing the fundamental construction of the machine.

- The sorter should obtain a high product recovery through an optimal adaptation to the feed.

3. RAW MATERIAL

This adaptation demands exact informations about the characteristics of the raw material. For this reason a lot of samples were taken from different waste glass collecting companies and processing plants.

The grain size distribution of this test material was determined by sieve analysis. The results of these analysis varied depending on the different collecting and transporting systems and on the care at the handling of the waste glass. E. g. the proportion of unbroken bottles varies between 15 and 40 %. The averaged grain size distribution of the cullet is shown in Figure 1.

As the photometric sorting is a separation of individual pieces the mass of the fragments is an important parameter for the throughput of the sorter. Figure 2 shows the mass of pieces in relation to grain size. It becomes clear that with decreasing grain size the number of pieces to be sorted considerably increases. Presupposing the same total mass, e. g. in the size fraction 10 - 20 mm four times as many objects must be examined as in the fraction 20 - 31.5 mm.

size fraction	weight %
- 20	22,2
20 - 31,5	23,0
31,5 - 50	29,5
50 - 63	10,9
63 - 80	9,0
+ 80	5,4

a)

b)

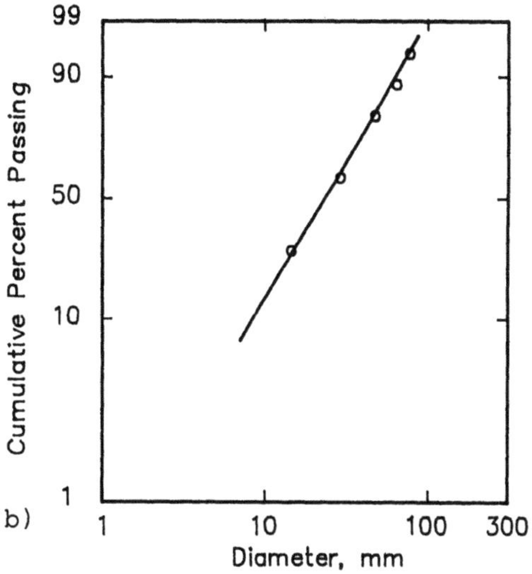

FIGURE 1. a) Medium Grain Size Distribution of Waste Glass Fragments b) RRSB-Diagram of Size Distribution

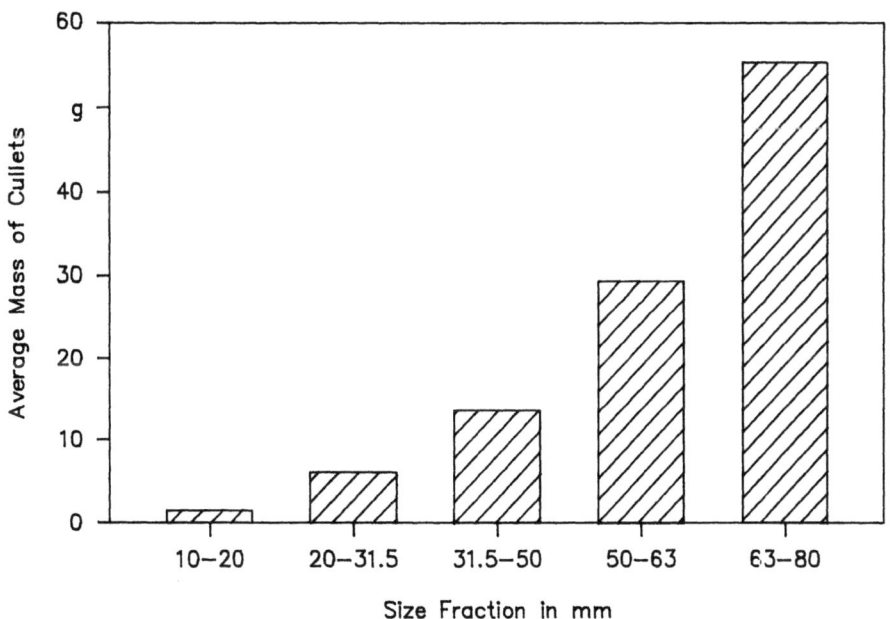

FIGURE 2. Mass of Pieces in Relation to Grain Size

The distribution of colour in the waste glass has also been verified. Figure 3 shows the averaged data. This proportion is similar in all size fractions.

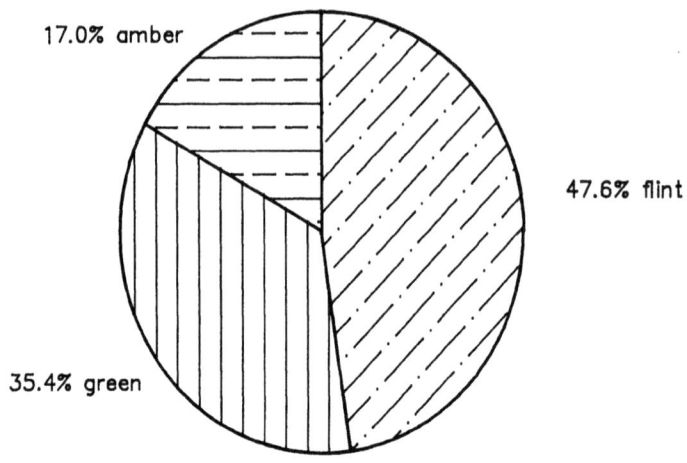

FIGURE 3. Distribution of Colour in Waste Glass

4. STATE OF THE ART

At the end of the sixties experiments were already launched, especially in the United States, to separate waste glass with colour sorters. These machines manufactured by Gunson's Sortex Ltd. were similar to those used in the mineral processing industry and carry out the colour detection by comparing the light reflected from the particle to the reflexion of a coloured background at certain wavelengths. If the particle is lighter than the background the signals of the photocells rise. In the opposite case the controlling processor registers declining diode signals. This measure method presupposes that the cullet is free from soil and labels. Therefore Sortex sorters of this design can only be applied to waste processing plants with wet sorting stages or by interposing a washing unit. But only few waste treatment plants with wet techniques are still in operation.

In Europe the Revalord - BRGM Process with two treatment plants in Nancy and Orleans is the only known process which uses Sortex separators. There the cullet is cleaned in an upward - current -washer before separation. Unfortunately exact informations about these sorting process are not available.

5. PRINCIPLES OF TRANSMISSION MEASUREMENT

The only alternative to separate the cullet by measuring the reflexion is to determine the spectral transmission.

By transmission we mean the relation of the intensities of radiation before and after permeating a substance. It is calculated by the following equation, called the Beer - Lambert Law:

$$T = I_x/I_o = \exp(-aCx)$$

The coefficent a is called the adsorption coefficient, and depends on the material, C is the concentration of the light adsorbing molecules and x is the thickness of the piece.

In the case of glass fragments the transmission is dependent upon the:

- thickness of glass
- kind and concentration of the glass colourants
- kind and intensity of the soil
- surface finish of the fragments.

For clarification of these dependancies extensive spectroscopic tests were conducted. Figure 4 for example shows the spectral transmission in the visible wavelenght range for flint and green glass at constant thickness (2.5 mm) but varying contamination. It can be seen, that the colouring of the glass leads to a characteristic outline of the transmission curve. However, colour-neutral (grey) layers of soil or different glass thickness hardly change the shape of the curve. They only lower the absolute transmission data. For instance with a strongly polluted flint fragment the transmission rate in the entire visible spectral range is lower than with a clear green fragment, but the horizontal course of the graph remains unchanged.

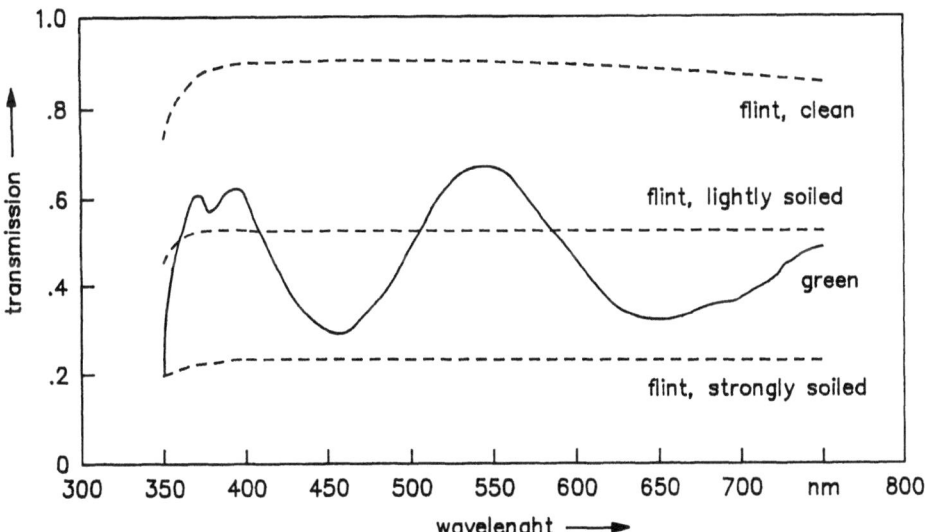

FIGURE 4. Spectral Transmission of Glass

6. FUNCTION OF THE COLOUR DETECTOR

This effect is used for the colour detection. Figure 5 shows a principle sketch of the sensor.

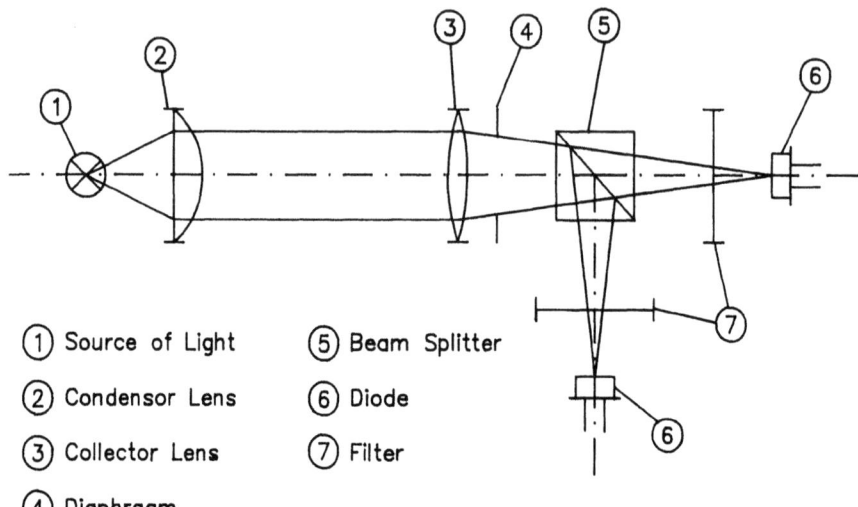

① Source of Light ⑤ Beam Splitter
② Condensor Lens ⑥ Diode
③ Collector Lens ⑦ Filter
④ Diaphragm

FIGURE 5. Principle Sketch of the Colour Sensor

The light from a 150 Watts halogen lamp is bundled and penetrates the fragment. Behind this, the light beam is focused, splitted by a semi-permeable mirror and meets with two photodiodes with different interference filters. The signals of the diodes are amplified, digitalized and evaluated electronically.

The presumption for this evaluation is an assimilation of both measuring signals with empty course of beam. Once that has happened, only two comparison operations are necessary:

- If the signal from one of the diodes is larger than an adjustable minimal value, it must be a transparent particle.

- The difference of the diode signals exceeds a variable level in case the fragment is transparent and coloured. The constant course of the transmission with a flint fragment would cause an equality of the signals.

For this comparison it is necessary that the controller "knows" whether there is an object in front of the sensing point or not. This information is given by an additional infrared-light barrier, which triggers the analysing program of the controller.

The prototyp of the colour detector shown in Figure 6 works according to the measuring method described above. This unit only allows a distinctin between flint and coloured glass. For further separation of amber cullets an additional detector with different interference filters would be necessary. From the technical side the separation of amber and green glass is uncomplicated because of the great difference between the graphs of spectral transmission.

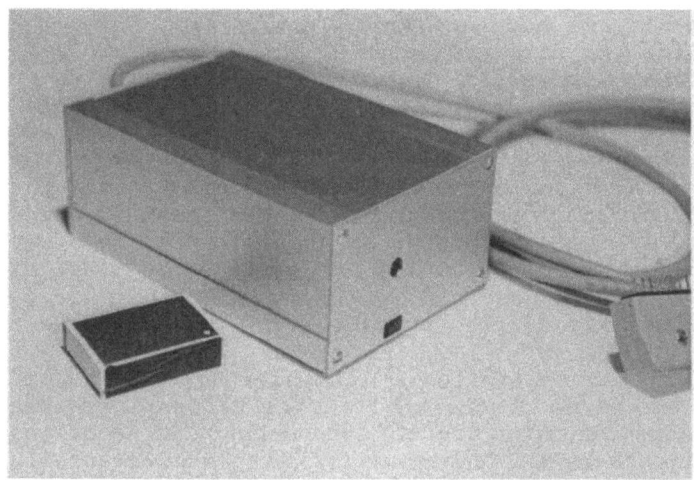

FIGURE 6. Prototype of the Colour Sensor

7. FUNCTION OF THE PHOTOMETRIC SORTER

This detector is the heart of the photometric sorter shown on Figure 7. The function of this instrument is described below:

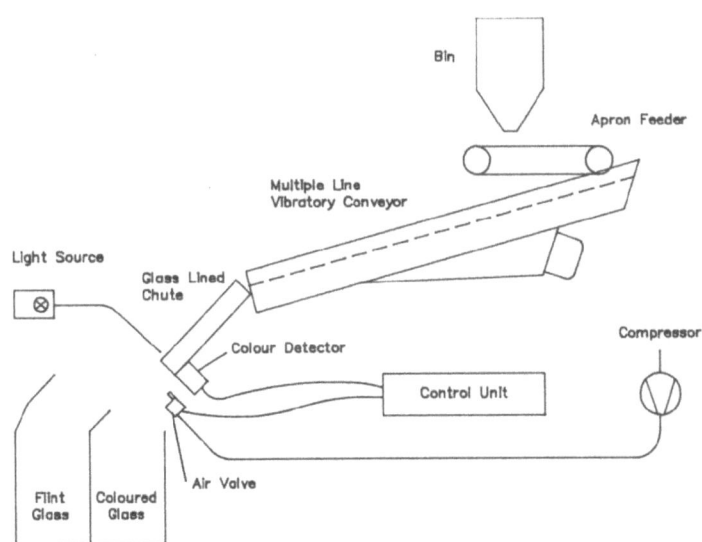

FIGURE 7. Outline of the Sorter

The cullet is charged from the bin by an apron feeder on a multiple line vibratory conveyor. The surface of this conveyor is assembled by a number of chutes that are inclined crossways to the transporting direction. This chutes force the fragments to line up. Underneath each of these channels there's a stronger inclined sluice, lined with glass, on which the fragments are accelerated and pulled apart. At the end of these slide the fragments pass the sensing point in a defined position at a speed of about 3.5 m/s. The white fragments are deflected from their trajectory about 50 mm behind the sensor by an air valve. Flint and coloured glass are caught in separate chambers.

While the fragments are moving above the sensing point, the diode signals are read by the control unit with high frequency and evaluated at the end. The frequency is chosen in a way that objects in a size down to 10 mm are analysed at least at 3 points. This makes it possible to assort also those fragments, which are partly covered by labels.

The input of the sorter varies in the size from 20-63 mm. Fragments + 20 mm are not integrated to the process, because they would cause a distinct drop of throughput. In addition very small pieces increase the amount of misplaced material. The reason for this is, that interference effects caused by refractions at the edges become more effective with small particles than with larger fragments.

Objects that are bigger than 63 mm cause an obstruction on the vibratory chute and disturb the line-up effect.

Basing on the given grain size distribution an a 30 % share of whole bottles, one can say that theoretically 45 % of the entire waste glass can be processed by such a sorting machine.

It is also possible to separate whole bottles by transmission measurement. For that it is only necessary to modify the mechanical part of the machine. A similar sorter for bottles was tested successfully on a dutch wast glass treatment plant.

Up to this moment only a few experiments have been carried out with the demonstration sorter. One can already say, that the separation of flint cullet from waste glass is possible without washing the cullet.

The attainable throughput for one channel is above 350 kg/h. A statement about the purity of the product can't be given at this time. It always stand in an opposite dependency to the flint glass recovery. Therefore this must also be seen from the economic point of view. It can't make sense to bring the purity of the product to a very high level, if that leads to an extreme decrease of output. In that case it is more reasonable to blend the sorted cullet with flint glass from the sorting of whole bottles.

One aim in the future testings shall be to find an optimum concerning this problem.

REFERENCES

1. Cook, R. F.: The Collection and Recycling of Waste Glass (Cullet) in Glass Container Manufacture, Conservation and Recycling, Vol. 1, pp 209-219, Pergamon Press, London, 1976.
2. Dalmijn W. L.: Glass Recycling Prospects and Limitations, Proceedings of the 3rd International Symposium MER, Antwerp, March 1986.
3. Fachverband Hohlglasindustrie e. V.: Reinheitsanforderungen an ofenfertiges Altglas zum Wiedereinschmelzen in Behälterglashütten - Vorläufige Richtlinie, Düsseldorf, Dez. 1977.
4. Fanke, M.: Umweltauswirkungen der Getränkeverpackungen, E. F. Verlag, Berlin 1984.
5. Palumbo, F. J.: Electronic Colour Sorting of Glass from Urban Waste, USBM Solid Waste Research Program, Technical Progress Report 45, Oct. 1971.
6. Stirling, H.: The Recovery of Waste Glass Cullet for Recycling Purposes by Means of Electro-Optical Sorters, Conservation and Recycling, Vol. 1, pp 209-219, Pergamon Press, London, 1976.

THE USE OF OFF-GAS CO_2 IN GREENHOUSES. REMOVAL OF NOx AND ETHYLENE.

C.M. VAN DEN BLEEK, P.J. VAN DEN BERG, A.G. MONTFOORT

DEPARTMENT OF CHEMICAL ENGINEERING
DELFT UNIVERSITY OF TECHNOLOGY
Julianalaan 136, 2628 BL Delft, The Netherlands

SUMMARY
To stimulate crop growth the greenhouse atmosphere is enriched with carbon dioxide. Rich sources of it are available in the off-gases of the utilities of the greenhouse, but their use is limited due to the NOx being present. Selective removal of NOx was so far restricted to NH_3. Based on a theory, presented earlier by the authors, it is made clear that also other nitrogen-containing species can perform the same selective removal. Due to the overall oxidizing circumstances traces of ethylene are removed at the same time. A process is presented to remove NOx and ethylene from the off-gases of gas engines and central heating systems in order to utilize their CO_2 content. A first economical evaluation show good prospects.

1. INTRODUCTION
Crop growth is depending on a combined action of temperature, light and carbon dioxide. During the daytime and periods of artificial exposure carbon dioxide is consumed by the process of photosynthesis. In greenhouses this cannot be supplied sufficiently by only openening the ventilation; the exchange of air between the leaves of the plants is low, so a carbon dioxide deficit will exist. Furthermore it has become common practice for about fifteen to twenty years to enrich the greenhouse atmosphere to a level of about 700-1000 ppm CO_2 to achieve an extra stimulation of growth; in that way an increase in growth of about 20-30% can be reached compared with the open air level of CO_2 of about 340 ppm.
This extra amount of carbon dioxide can be supplied in several ways, depending on the equipment installed at the market gardening. In most cases a gas fired Central Heating System will be present. Originally only meant to produce heat, nowadays its off-gases are used as a carbon dioxide source as well. Up to now the NOx-content of these gases (about 50 ppm) was regarded as harmless to the crop. However the high energy prices have caused a better insulation of the greenhouse, in winter time resulting in an accumulation of NOx above the phytotoxic level of 250 ppb (1,2). Recent investigations (3,4) show, that even at lower concentrations NOx can cause a latent damage (that means not visible, but reducing growth) if at the same time also small amounts of sulphur dioxide and ozon are present. Because it is situated near large industrial activities, such a situation can for instance occur in the South Holland Glasshouse District.
A coming situation is the heat regulation of the greenhouse by means of a Total Energy Installation, an integration of a gas engine and a gas fired central heating installation. By the gas engine heat and electricity are produced; the electricity produced can be used by the gardener himself or fed back to the public grid. Normally the gas engine is designed to produce about 30% of the maximum heat demand of the greenhouse. If more heat is

needed then the central heating installation is put on too. Overall there is a period of about two monthes a year that the gas engine can deliver the heat needed on its own. However, due to its high NOx concentration (700 - 3000 ppm and sometimes even higher) the off-gas of the gas engine cannot be used for CO_2-fertilization. Next to that there is a period of about one month an a half that no heat is needed at all, while still carbon dioxide is needed. To overcome the shortage of carbon dioxide in those three to four monthes, there are three possible options:
- The CH-system is started up either; the heat produced is destroyed; an extra problem is that most of the installations cannot run on a low capacity level (2.5 m^3 natural gas/h/1000 m^2 greenhouse area), which would be high enough to produce the amount of carbon dioxide needed; so extra unnecessary heat is wasted. Nevertheless it is the possibility chosen mostly.
- Another possibility is to use a so called CO_2-gun, a special designed burner to produce a maximum amount of carbon dioxide. Though less heat is wasted an disadvantage is the extra investment needed.
- A last possibility is the use of pure carbon dioxide. Although it is the most cosltly solution, it is avoiding the NOx problem completely.

So it will be clear that there is an overall need for a process to remove NOx from the off-gases from Central Heating Systems and gas engines in order to provide a cheap and safe source of carbon dioxide. It has to be kept in mind that due to the lean firing applied (in order to decrease the amount of NOx in the off-gases as much as possible) there is an increased chance to get traces of not completely burned products in the off-gases, of which ethylene can cause (and has caused already!) the most disastrous effects. So a process to remove traces of ethylene is needed too.

2. REMOVAL OF NOx
2.1. Selective removal

The problem of removing NOx from waste gases is of course already a rather old one in chemical industries where stack gases from power stations and nitric acid plants are well known examples. The problem is tackled in two ways. The most attractive way is to improve the process itself. Lower reaction temperatures are being realized in engines (5) and power stations (6,7), avoiding the NOx-formation from air and better NO_2 absorption in nitric acid plants are proposed (8) and realized (9). The alternative way is off-gas treatment, where the main interest is focussed on the reduction to nitrogen. To save on reductor costs a large part of research has been directed to the reduction of NOx selectively. Only ammonia and its derivates have proved to be capable of converting NOx to N_2 and N_2O without first consuming all the oxygen present. The selective behaviour of NH_3 versus the non-selectivity of the other reducing agents like methane, carbon monoxide etc. was elucidated by a hypothesis proposed by Van den Bleek and Van den Berg (10). The key to the hypothesis was the reaction :

$$2NO + O_2 \longrightarrow 2NO_2 \qquad [1]$$

The particular rôle of this reaction is originating from the fact that it is catalyzed on nearly every surface. Because actually all the catalysts proposed and/or investigated to perform the selective reduction of NOx are consisting of an active function, impregnated or precipitated on a carrier surface, the occurence of this key-reaction cannot be avoided on those carrier surfaces. They showed that this reaction is responsible for the impossibility to reduce NO to N_2 and N_2O as long as oxygen is present. However, they demonstrated that in the case of NH_3 the very same reaction also provides the key to the solution of this problem by promoting a

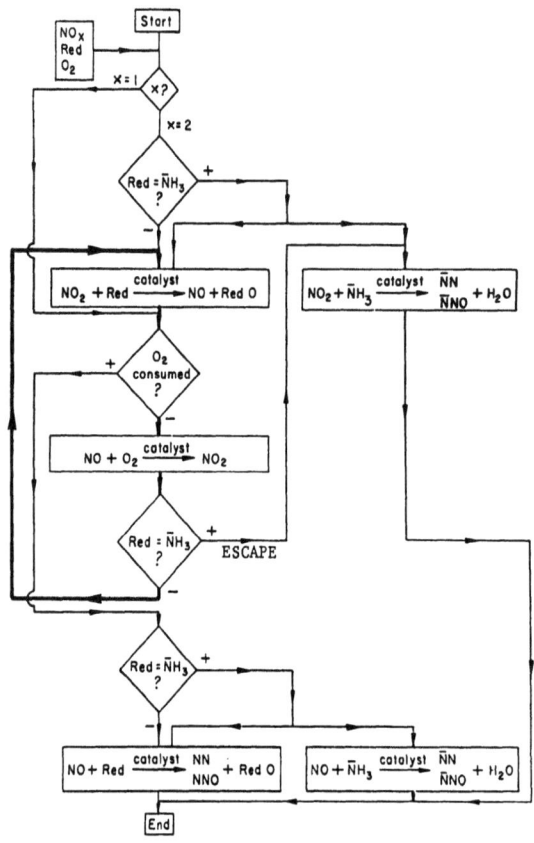

FIGURE 1. General scheme for the reduction of NO_x in the presence of O_2 (10)
N = nitrogen atom belonging to or originating from NO_x
$\bar{N}$ = nitrogen atom belonging to or originating from $\bar{N}H_3$
━ = reductor consuming loop

possible escape to nitrogen-containing complexes which, more or less by coincidence, are converted to N_2 and N_2O. The hypothesis is represented as a model in Figure 1.

The route to escape the reductor consuming loop in fig.1 consists essentially of 'encouraging' NO_2 to take part in another reaction pattern, resulting into valuable, removable and/or harmless products, not necessarily being N_2 and N_2O. A further insight in this route will provide new possibilities to remove NOx selectively. It was shown by Van den Bleek et al.(12), that a proper choice of the reaction partner can result into N_2 and N_2O, but to more valuable products as well. This is important for the removal of NOx from the off-gases from gas engines and CH-installations in the agricultural industry, because the operation of a process based on the handling and storage of liquid ammonia would go beyond the skills of the average market gardener.

2.2. The escape route

The pictured escape route is in essence a nitrosation and/or nitration reaction, both well known and important in other area's of chemistry and chemical technology (11). Nitrosation produces nitroso compounds (X-NO), while nitration results in nitro compounds (X-NO_2), where X can be carbon, sulfur, oxygen, nitrogen etc. If the species which is nitrosated or nitrated (formerly called the reductor) is containing a nitrogen atom itself, these reactions will lead to denitrification. For instance, in the case of a primary amine it was found by Ridd (13) :

$$R.NH_2 \longrightarrow R.NH.NO \longrightarrow R.N{:}N.OH \longrightarrow R.N_2^+ \longrightarrow R^+ + N_2 \quad [2]$$

According to Fridman (14) a similar scheme would be possible for nitration:

$$R.NH_2 \longrightarrow R.NH.NO_2 \longrightarrow R.N{:}N.O_2H \longrightarrow R.N_2O^+ \longrightarrow R^+ + N_2O \quad [3]$$

These reactions make the overall process look like a real selective reduction; de facto it is however a nitrosation/nitration and it is rather a coincidence, that the nitrosation/nitration products decompose into N_2 and N_2O. If R = H, then the reaction products for ammonia are given, just as already overall predicted by the right part of the model given in fig.1 :

$$NH_3 \longrightarrow NH_2NO \longrightarrow HNNOH \longrightarrow HN_2^+ \longrightarrow H^+ + N_2 \quad [4]$$

$$NH_3 \longrightarrow NH_2NO_2 \longrightarrow HNNO_2H \longrightarrow HN_2O^+ \longrightarrow H^+ + N_2O \quad [5]$$

A second possibility may occur if the species which is nitrosated or nitrated does not contain a nitrogen atom itself. Without oxygen SO_2 will reduce NO_2 to NO and NO finally to N_2/N_2O at the same time producing SO_3. This fits perfectly well in the left part of the model demonstrated in Figure 1. If there is no oxygen present the reaction path will directly leave the reductor consuming loop to produce equal amounts of N_2/N_2O and SO_3. In the presence of oxygen, however, this loop cannot be left in that way before all oxygen is completely consumed by the key reaction [1], overall producing an excess of sulfurtrioxide by the repeated reduction of NO_2 to NO. Because SO_2 can also be nitrosated and nitrated, the loop can be escaped as suggested, producing $NOSO_3$, $(NO)_2S_2O_7$ (and $NOHSO_4$ in the presence of water), which are solid, removable and useful products (15). A process for a combined removal of NOx and SOx from stack gases based on these products was presented by Van den Bleek et al.(12).

To check the escape route theory also the denitrification was investigated rather extensively. The experiments will be presented in the next section, resulting in the concepts of two interrelated processes to selectively remove NOx from the off-gases of a gas engine and a central heating installation. These processes will make it possible to utilize the carbon dioxide content of their off-gases to stimulate plant growing processes.

2.3. The removal of ethylene

The presence of ethylene and other not completely burned products in the off-gases of gas engines and CH-systems is due to the low air to fuel ratio applied in order to decrease the NOx-formation from air. Present processes to remove both NOx and ethylene (16) are consisting essentially of two catalytic steps: conversion of NOx to nitrogen in a reductive atmosphere, followed by oxidation of unburned products to CO_2 in an overall oxidizing atmosphere. Because the NOx/urea process presented is operated at overall oxidizing conditions, the conversion of ethylene to CO_2 can be accomplished in the same step.

3. EXPERIMENTS

For the experiments between NOx and the N-containing substances a 12.5 mm ID glass tubular reactor was used, consisting of two bed sections, which could be electrically heated independently (see figure 2). In the top section R_1 the N-containing component was situated, either impregnated (urea and melamine) on an alumina support (average particle diameter 0.72 mm) or as 2-5 mm particles (urea and cow manure). The bottom section R_2 contained a Pd/Al_2O_3 catalyst (typically at 210 °C) to decompose the nitrogen complexes formed.

The feed gas was made up of known flows of helium, helium/oxygen mixture (80/20 mol/mol) and helium/NO mixture (1.2 mol% NO). Helium and the helium/oxygen mixture were dried over molsieves 5A, helium/NO over supported phosphor pentoxide. Dosing of ethylene, carbon monoxide and water vapour (by passing a known flow of helium over a thermostated bed of $FeSO_4.7aq$ particles) was optional. A typical flow consisted of a mixture of

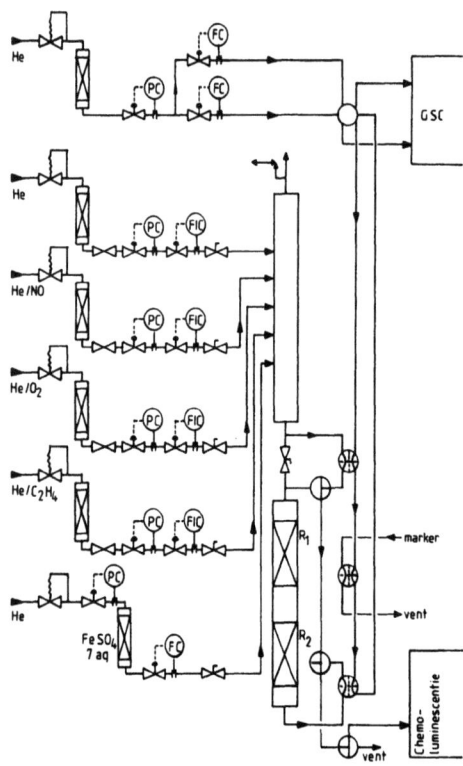

FIGURE 2. Experimental setup

2.7 mol% O_2, 2500 ppm NO and balance helium; a typical space time with respect to the N-containing component was 8.2 seconds.

Feed and product gases were analyzed by gas chromatography (Perkin & Elmer Sigma 2 / Packard 433). CO, CO_2 and N_2O were separated on porapak QS, N_2 and O_2 on molsieves 5A. The NO and NO_2 content of both the feed and the product gas was measured by chemoluminescence (built and developed at the Delft University of Technology). The nitrogen content of urea and melamine on alumina and of the non-decomposed complexes formed, was analyzed by a Kjeldahl procedure on a Technicon Auto-analyzer II. To analyse the non converted ethylene in the reactor outlet, samples of 1 liter off-gas were collected; the ethylene concentration was determined by means of FID detection separately.

Urea on alumina (Ketjen Catalysts, crushed 001-1.5E alumina $0.65 < d_p < 0.80$ mm) was prepared by wet impregnation in a rotating film evaporator. Batches of 10 and 20 wt% on alumina were prepared. Melamine on alumina was prepared in the same way, though only very dilute impregnation solutions could be used because of the low solubility of the melamine. Batches of 2.7 wt% on alumina were made for further use. Dried, pelleted cow manure was available, cylindrical in shape; it was crushed and particles in the range of 2-5 mm were used in the experiments without further treatment.

4. RESULTS AND DISCUSSION
4.1. The removal of NOx

Using only the first reactor (R_1) experiments were carried out over urea (10 wt%) on alumina. In the absence of oxygen the well known S-shaped conversion vs temperature curve was measured. At a space time of 8 seconds a complete conversion of NO was reached at about 105 °C (Fig.3 ◊curve). In the presence of oxygen (2.7%) it was not only possible to convert NO selectively, but the reaction rate was even faster (Fig.4 ◊curve). A hundred percent conversion of NO was now reached at 95 °C. This is completely in agreement with the escape route theory : the most active species for nitration/nitrosation can only be formed in the presence of oxygen (scheme (2)).

However, the production of N_2 and N_2O during both sets of experiments was low and not in agreement with the NO converted, as can be seen in Fig.3 and 4, (+curve) where the conversion based upon the amount of N_2 and N_2O produced is given. White depositions in the glass tubing after the reactor made us suppose that the nitration/nitrosation complexes were formed, but were only partly decomposed to N_2 and N_2O.

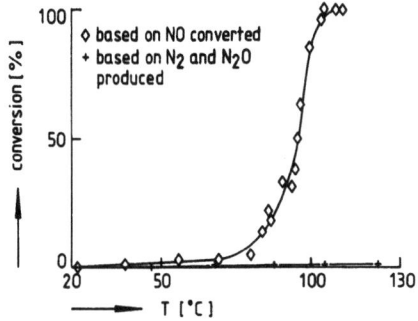
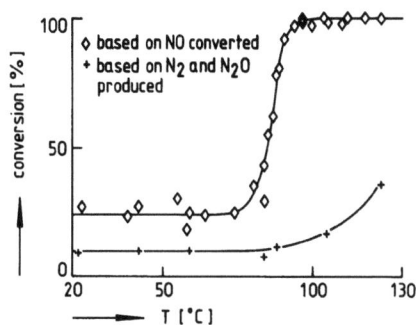

FIGURE 3. NO_x-conversion over urea without oxygen as a function of temperature ($\tau = 8$ s)

FIGURE 4. NO_x-conversion over urea in the presence of oxygen as a function of temperature ($\tau = 8$ s)

So the experiments were repeated with a second reactor (R_2) installed, filled with inert alumina and operated at temperatures up to 400 °C. No increase in N_2 and N_2O production was observed. The inert alumina in the second reactor was replaced by Pd/Al_2O_3 (a well known NOx reduction off-gas catalyst). Operating R_2 at 220 °C and a space time of about 2 seconds the complexes were now completely converted to N_2 and N_2O as is illustrated in Fig.5, where the N_2 production is given as a function of the temperature of the first reactor with and without Pd-catalyst installed.
It is concluded that to perform a selective removal of NOx with urea a second reactor is necessary to decompose the complexes (easily) formed in the first one. Based on these results it was also concluded that at least one of the functions of the catalyst in the well known NOx removal process with NH_3 should be the decomposition of the complexes formed; it is not excluded that this will be the only function.
Though some other catalysts were tested and most promising results were obtained with a cheaper V_2O_5-catalyst, most of the experiments were performed with the Pd/Al_2O_3-catalyst installed in the second reactor and operated at 220 °C.

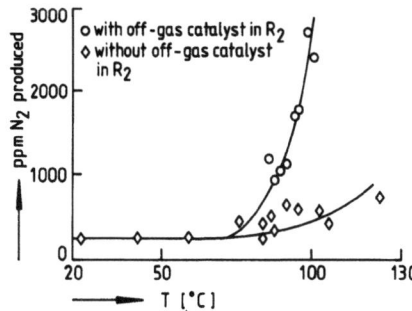
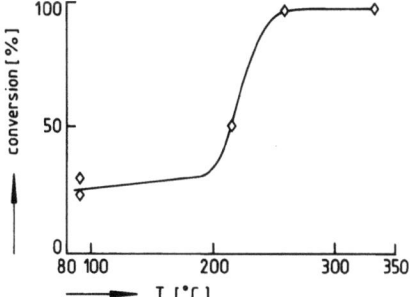

FIGURE 5. N_2 production over urea in the presence of O_2 with and without the Pd-catalyst installed

FIGURE 6. NO_x-conversion over melamine as a function of temperature

At temperatures in the first reactor higher than 110 °C some excess N_2 and N_2O was found due to the thermal decomposition of urea: supported urea can be decomposed at a lower temperature than the prilled one. However, the ammonia formed is oxidized in the second reactor.

Some other N-containing species were tested for the escape route theory. Supported melamine (2.7 wt%) on alumina showed a similar behaviour as urea, only at a higher temperature level (see Fig.6). Complete conversion of NO was possible at about 260 °C, still using a 8 s space time. Even dried pelleted cow manure was tested. Though a selective removal of NOx is possible at very high space times, only 45 % conversion is reached at a space time of 8 s and a temperature of 140 °C. Higher temperatures were not allowed due to an awful smell!

4.2. The removal of ethylene

The conversion of ethylene was measured using only the second reactor (Pd/al_2O_3, space time=2s). The inlet ethylene concentration was kept at 110 ppm. As a function of temperature the S-shaped curve as sketched in Fig.7 was measured. Complete conversion is reached at about 200 °C. So, at the operation conditions typical to decompose NOx/ammonia-complexes, ethylene is converted to CO_2 simultaneously.

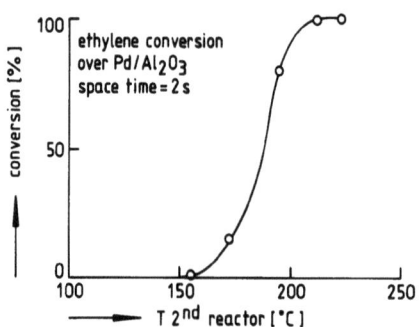

FIGURE 7. Conversion of ethylene as a function of the temperature of the second reactor

5. A PROCESS TO UTILIZE CO_2 FROM OFF-GASES
5.1. Process description

Based on the resultss described above a concept for a process is presented to remove NOx and ethylene from the off-gases of gas engines and CH-systems to utilize their carbon dioxide content. Instead of urea/Al_2O_3 (the alumina has to be reloaded for an economical operation) pure urea prills are suggested based upon preliminary experiments which show comparable results if operated at 100-125 °C using water containing off-gases. The process is based upon a standard greenhouse of 10000 m² surface area and a average height of 3.85 m. Using only a CH-system a 2.3 MW installation is needed. Using a TE-installation a 0.7 MW gas engine will take care of the basic heat load, the remainder being supplied by a 1.6MW CH-system. The off-gases for both situations are specified in Table 1.

TABLE 1. Off-gas specifications.

off-gas	Φv nm³/h	T °C	CO_2 %	NOx ppm	CO ppm	O_2 %	H_2O %
CH-system (on max.cap.)	2700	140	8.7	50	50	1.7	17.3
gas engine	700	500	8.2	700	500	2.7	16.6

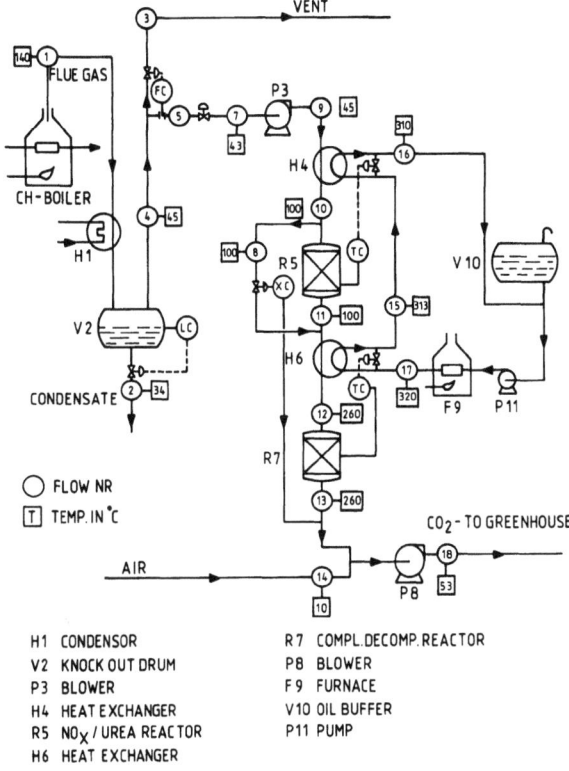

H1 CONDENSOR
V2 KNOCK OUT DRUM
P3 BLOWER
H4 HEAT EXCHANGER
R5 NO_x/UREA REACTOR
H6 HEAT EXCHANGER
R7 COMPL.DECOMP.REACTOR
P8 BLOWER
F9 FURNACE
V10 OIL BUFFER
P11 PUMP

FIGURE 8. Process to utilize CO_2 from CH-system off-gas

The process to clean the CH-system off-gas is outlined in Fig.8. It is connected to the off-gas condensor of an existing installation. The off-gas is partly fed to a first reactor R_5, operated at 100 °C, where NOx is converted with urea to N_2, N_2O and complexes; also some extra urea is decomposed. The complexes are decomposed over a V_2O_5-catalyst at 260 °C to N_2 and N_2O in a second reactor R_7. The bypass flow of the first reactor is mixed up with its outlet in order to achieve complete conversion to N_2 and N_2O in the second reactor. The size of the bypass flow depends on the quality of the outlet of R_2; if the excess decomposition of urea in the first reactor is too low the bypass flow is decreased. Because the unit is not integrated with the CH-system, but connected directly to an existing installation the temperature of the off-gas is too low to be used to heat the reactors R_5 and R_7; so an extra oil circuit is installed. The final outlet NOx concentration of the system is about 2 ppm; ethylene and CO are converted completely; about 120 kg of urea per year is used;

The off-gas from the gas engine of a TE-installation (see Fig.9) at 500 °C is split into a flow (nr.3) which heats the first reactor (R_5) and a secondary flow (nr.2). In this reactor which operates at 120 °C NOx is converted and an excess urea is decomposed. Together with the secondary flow the gas enters the second reactor (R_6) where final conversion is reached at 265 °C on a V_2O_5-catalyst. A concentration of 4 ppm NOx is reached. As with the CH-process fine-tuning is reached by partly bypassing R_5 and complete conversion of ethylene and CO is reached in the second reactor. With this set-up 1640 kg of urea per year is necessary.

5.2. Economical prospects

Most important at this stage of the project is the feasibility of the proposed schemes, according to Fig. 8 (case CH) and Fig. 9 (case TE, where the basic heat load is provided by a gas engine, the remainder by a CH-system). In both cases the supply of CO_2 by means of the unpurified off-gas from a CH-system will be considered as base case, i.e. the case, costs and benefits will be compared with. It will be clear, that in the base case during summertime the CH-system is running only for the CO_2 supply, the produced heat being useless.

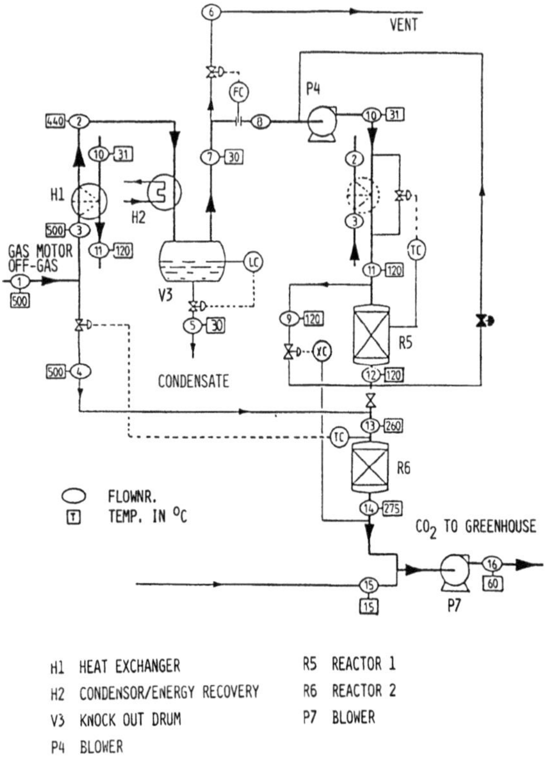

VENT
P4
GAS MOTOR
OFF-GAS
H1
H2
V3
CONDENSATE
R5
R6
CO2 TO GREENHOUSE
P7
FLOWNR.
TEMP. IN °C
H1 HEAT EXCHANGER
H2 CONDENSOR/ENERGY RECOVERY
V3 KNOCK OUT DRUM
P4 BLOWER
R5 REACTOR 1
R6 REACTOR 2
P7 BLOWER

FIGURE 9. Process to utilize CO_2 from gas engine off-gas

5.2.1. Cost. At this stage the investment for the cases can only be guestimated. Mainly due to the additional heating circuit the investment in case CH is app. 20% higher than for case TE, i.e. Dfl. 72000 compared to Dfl. 60000, both for a greenhouse area of 10000 m². Based on a loan of 8% interest, an expected life of 8 years, 4% maintenance costs and insurance inclueded, the total fixed costs (capital charge) will amount to 24% of the investment. The variable cost are mainly power, in case CH at Dfl. 0.16/kwh against at Dfl. 0.11/kwh for the in house generated power in case TE.

5.2.2. Benefits. In both cases the increase of crop output will be equal and is estimated to be 5% due to decreased NOx and 2% due to ethylene elimination. Based on Dfl. 50/m² in the base case, the benefits will be Dfl. 35000/y.
In case TE there will be an additional saving of app. Dfl. 5000, since it will be

TABLE 2. Annual cost/benefits.

all figures in Dfl.	case CH		case TE		case EX	
fixed cost	17300		14400		-	
energy	2200		1400		-	
chemicals	+ 500		+ 1200		+ 55000	
total cost		20000		17000		55000
add.proceeds from crop	35000		35000		35000	
energy savings	+ -		+ 5000		+ 8000	
total profits		− 35000		− 40000		− 43000
surplus		15000		23000		-12000

possible to supply CO_2 during base load heating requirements with cleaned gas engine exhaust instead of CH off-gas.

5.2.3. Conclusion. In Table 2 both cases are summarized, and it is clear, that both are feasible and that the surplus is considerable in relation to the investment, particularly for case TE.
In the table is also given case EX, the supply of CO_2 from an external source. It is clear, that this is no alternative for the base case.

6. CONCLUSIONS

Selective removal of NOx from off-gases is not restricted to NH_3, but also possible with other nitrogen-containing species. The so called selective reduction does not exist, but is de facto a nitration/nitrosation reaction. Because these types of reaction occur under oxidizing circumstances, traces of ethylene can be removed at the same time.
Based upon this insight a process is proposed, using urea, to selectively remove NOx and ethylene from the off-gases of gas engines and CH-systems, in order to utilize their CO_2 content for stimulating crop growth.
An economical operation of such a process seems to be possible, so a further development of this process is desirable.

7. ACKNOWLEDGEMENTS

The authors like to thank P.E. Kleiborn, J.S. Hoornstra and M.A. Schwegler for their contributions during their MSc study in Chem.Eng. and A. Hexspoor, J.B. van Holst and A.A.M. Pruisken for their assistance.

8. REFERENCES

1. Capon, T.M. and Mansfield, T.A., J.Exp.Bot. 27 (1976),1181-1186
2. Capon, T.M. and Mansfield, T.A., J.Exp.Bot. 28 (1977),112-116
3. Wolting, H.G., van Remortel, E.A.M. and van Berkel, N., Air quality in greenhouses with and without CO_2 enrichment; paper presented at the Symp.'Greenhouse climate and its control', Wageningen, 1985, May 19-24
4. Wolting, H.G., van Remortel, E.A.M., Lucht en Omgeving, 1985 (May/June), 58-62
5. NCI, 1985,(nr.20),19 (editorial)
6. Van der Sanden, A.M., Heidweiller, D.J. and Poolman, P.J., Procestechn. 1985,(nr.9),73
7. Flament, G. and Phetan, W., Proc.US-Dutch Int.Symp., Maastricht, 1985, May 24-28,603-621
8. Lefers, J.B., De Boks, F.C., Van den Bleek, C.M., Van den Berg, P.J., Chem.Eng.Sci., 35, (1980),145-153
9. NCI, 1985,(nr.22),15 (editorial)
10. Van den Bleek, C.M. and Van den Berg, P.J., J.Chem.Tech.Biotechnol., 30 (1980),467-475
   A free copy without highly disturbing typing errors can be obtained from the authors.
11. Stedman, G., Adv.Inorg.Chem.Radiochem., 22 (1979),113
12. Van den Bleek, C.M., Montfoort, A.G., Van den Berg, P.J., The selective removal of NOx from stack gases of the process industry; paper presented at the Symp. on 'Recent Advances in the Management of Hazardous and Toxic Wastes in the Process Industries', Vienna, 8-13 March 1987
13. Ridd, J.H., Quart.Rev., 15 (1961),418
14. Fridman, A.L., Ioshin, V.P., Novihov, S.S., Russ.Chem.Rev. 38(1969),640
15. Stopperka, K. Wolf, F. and Süss, G. Z.Anorg.Allg.Chem. 359 (1968),14
16. KLimstra, J. and Wolkotte, B.J., rapport TP/M 85.R.1033, N.V.Nederlandse Gasunie,1985

PRODUCT DEVELOPMENT NEEDS OF WASTE MANAGEMENT

Pekka Vilppunen, M.Sc.(Engineering), Laboratory Manager;
Energy Laboratory, University of Oulu, Finland

1. INTRODUCTION

This project consists mainly of study of the possibilities to produce energy from municipal solid waste in Finland. The experiences in other countries have been taken into account. In addition, the most suitable method to measure quantity and to analyse quality of the municipal refuse has been studied.

The preliminary results show that the separation of combustible and other fractions in the municipal refuse significantly influence to the economical utilization of waste. The sorted municipal refuse enables the preparation of more homogeneous refuse derived fuel (RDF) increasing the possibilities of using the waste in combination with other combustibles in an ordinary burning plant.

2. PURPOSE

The aim of this research is to find out the R&D requirements to increase the recovery of energy from municipal refuse. A partial objective is to develop a separation-handling system to produce refuse derived fuel. This is a two phase separation system: at source segregation and mechanical sorting, followed by a refining stage of the combustible fraction.

The method can be utilized when the recovery of energy and/or materials is pursued by a community.

3. RESEARCH REALIZATION

During the research the following stages in the recovery of energy from waste will be studied:

- the preparation of refuse derived fuel (waste composition, possibilities of mechanical sorting and segregation at source, refining methods...),
- the production of energy (combustion, gasification, anaerobic digestion etc.) and
- the environmental impacts, pollutants and their origin, exhaust gas cleaning systems and equipment.

The realization of the research includes:

- a literature review of the present state of knowledge and foreign experiences in energy recovery methods,
- the technical and economical aspects of the most important handling and energy production methods and
- the definition of R&D requirements.

4. RESULT UTILIZATION

The result can be utilized to achieve economical recovery of energy from municipal refuse.

The utilization increases:
* directly, possibilities to utilize domestic solid fuels (wood and peat) in combination with municipal refuse,
* indirectly, the organic fraction in the collected municipal solid waste by separating at source other components which is then more suitable for energy recovery through anaerobic digestion and
* the degree of recovery of recuperable materials (two phase separation).

The system decreases environmental impacts caused by incineration (more homogeneous fuel - less emissions).

The system facilitates long term planning of the waste management (e.g. diminishes the need of landfills).

The aim of research was to find out R & D requirements to increase utilization of Finnish biogasresources.

The realization of the research included:

* a literature review of the present state of knowledge in Finland,
* the technical and economical aspects of the most important handling and energy production methods and
* the definition of R & D requirements.

R & D requirements of the utilization of the Finnish biogas potential are:

1) **General**

Systematic use of biogas in energy production

* the municipalities which have preconditions to profitable production and use of biogas

2) **Agriculture**

Systematic study of the utilization rate and energy economy as well as the problems and shortcomings of biogas reactors in agriculture.

* the energy and waste management of the small and middle size animal husbandries
* the specific R&D requirements of the biogas reactors in agriculture

3) **Municipalities**

Development of combined digestion of sludges from industrial waste water cleaning plants, organic fraction from household and other digestable organic wastes.
* optimization of the parallel use of anaerobic and aerobic methods
* landfillgas recovery and utilization

4) **Industry**

Development of anaerobic energy production unit which uses:

* pulp and paper industry wastes,
* biosludges from activated sludge plant,
* sedimented pulp and paper industry wastes in the surface waters and
* organic wastes from municipalities and industry in general.

REMEDIATION TECHNIQUES AND WASTE HANDLING

**Second European
Conference on
Environmental Technology**

in the 'European Year
of the Environment'

REVIEW OF SOIL TREATMENT TECHNIQUES IN THE NETHERLANDS

E.R. Soczó, E.J.H. Verhagen, C.W. Versluijs
National Institute of Public Health and Environmental Hygiene (RIVM)
Laboratory voor Waste and Emission Research (LAE)
A. v. Leeuwenhoeklaan 6, P.O. box 1,
3720 BA Bilthoven, The Netherlands

1. ABSTRACT

At the beginning of 1987 ten full-scale thermal and extraction plants were operational in the Netherlands. On the basis of experience with these plants it can be concluded that both types of techniques are suitable for the cleaning of sandy soils contaminated with hydrocarbons (like oil, mono- and polycyclic aromatics) or with cyanides. Thermal plants are better suited for treatment of various types of soil and on the other hand extractive plants are the most feasible for the removal of heavy metals from contaminated soil.

In the last two years there has been an increased interest in the development of biological treatment techniques. Intensive research has been done in this period regarding the improvement of the landfarming method and the development of new techniques c.q. in-situ biorestoration and bioreactors. The first results are promising and research will be continued in this field.

2. INTRODUCTION

In january 1983, the Interim Soil Clean-Up Act was put into force in the Netherlands. On account of the provincial clean-up programmes which are carried out under this act, about 1,600 cases of serious soil contamination will need remedial action during the period up to 1997. For the enforcement of the abovementioned act the development of adequate investigation and clean-up methods was considered to be of prime importance. Research into these methods has been stimulated in various national programmes such as
  o the Research Programme of the Ministry of Housing, Physical Planning and Environment (VROM) on the development of soil clean-up techniques.
  o the Spearhead Programme for Soil Research, which is supported by four ministries.
  o the Innovation-oriented Research Programme Biotechnology within the biological treatment of contaminated soil is indicated as a special priority sub-programme.
The abovementioned programmes and also the Interim Soil Clean-up Act create a good financial framework for the realization and continuing of research in the field of soil treatment techniques.
During the past few years the initial emphasis has been on the development of thermal and extraction techniques. The soil treatment by means of these techniques is carried out in an installation on- or off-site after excavation of the contaminated soil. On the moment ten full-scale plants are available for the treatment of contaminated soil. In the

last two years, however, there has also been an increasing interest in the development of alternatives such as biological treatment techniques, preferably applied in-situ. In the following a short summary is given with regard to the treatment results of operational (thermal and extraction) techniques and the state of the art of research regarding the development of biological treatment methods.

3. THERMAL TECHNIQUES

In principle two techniques can be distinguished namely evaporation of the contaminants in a rotary kiln and combustion of the contaminated soil in a fluid bed incinerator. The evaporation of contaminants takes place in a rotary kiln at a temperature between 200-700° C by means of direct or indirect supply of heat. The vapours of the contaminants are combusted in an afterburner at a temperature between 750-1300° C. In the fluid bed the soil is mixed with fuel and this is combusted at a temperature ranging from 800 tot 1000° C. The room above the fluid bed, also called freeboard, can be seen as the place for afterburning of the gases. After scrubbing flue gases are emitted by the stack. The removed dust may be added to the cleaned soil. The discharged water from the scrubber is neutralized and partly recycled, or used for moisturizing the heated soil.

3.1 Applications

Thermal techniques can be used in case of different types of soil for the removal and combustion of organic contaminants and complex cyanides. These contaminants can be destroyed almost completely by adjusting the temperatures and the residence times. For the afterburner the oxygen content and the mixing of the combustion air with the contaminants are adjusted to this end.
Most of the chlorinated hydrocarbons are stable compounds, so high temperatures are needed to destroy these substances. The formation and emission of ultra-toxic compounds like dioxines and furans have to be avoided as much as possible. Laboratory experiments show that some of the existing thermal installations should be able to clean the soil contaminated with, for instance, lindane (γ-hexachloro-cyclohexane) and drins. It is expected that in 1987 a permission for such experiments on pilot-plant scale will be given.
The thermal treatment leads to a change of the structure of the soil; the higher the temperature the more the organic particles (humus) will be transformed or combusted. However, generally the particle-size of the soil does not change too much, because of the back-mixing of cleaned soil with the dust removed from the flue gases. This has its implications for the re-use of the cleaned soil.

3.2 Treatment results and conclusions

The operational full-scale thermal plants are indicated in Table 1.
[1, 2]
The costs of thermal treatment vary between Dfl 80-190/ton and are depending mainly on both the moisture content of the soil and the type of contaminants. (Dfl 1 is approx. US $ 0.5 (March 1987))
Some treatment results of the plants are summarized in table 2.

Table 1. Operational thermal plants (March 1987)

Name of company	Capacity (ton/year)*
ATM B.V.	60,000**
Boskalis Esdex Bodemsanering B.V.	4,000***
Broerius B.V. bodemsanering	25,000
Ecotechniek B.V. 1st installation:	55,000
2nd installation:	80,000
NBM Bodemsanering B.V.	60,000

*   on the basis of 8 h/day
** this plant is designed for 24 h/day operation
*** the fluid bed combustor of Boskalis Esdex is a pilot plant

Table 2. Some treatment results of operational thermal plants (March 1987) [1, 2]

Type of soil	Contaminants	Concentration mg/kg dry weight	
		Initial	Final
sand	diesel	1000 - 50000	100 - 640
	petrol	1000 - 30000	< 20
sand, < 10% loam	oil	0 - 90000	5 - 10
clay	oil	0 - 1000	< 200
several	oil	0 - 20000	< 300
sand	CN complex	100 - 1000	0 - 1*
sand, > 10% peat	CN complex	200 - 10000	1 - 4*
several	CN	0 - 1000	0 - 7
sand	BTEX	0 - 400	< 1
clay	BTEX	0 - 500	< 1
several	BTEX	0 - 5000	< 1
sand	PCA's	0 - 1000	< 3
sand, > 10% peat	PCA's	700 - 4000	0.1*
sand, > 10% loam	PCA's	100 - 2000	< 0.1*
sand, > 10% loam	PCA's	0 - 8000	< 0.01
several	PCA's	0 - 5000	<10

* thermal treatment by fluid bed combustion

Remarks:
- specifications of the types of oil and the types of cyanides were often not mentioned
- BTEX: benzene, toluene, ethylbenzene, xylenes
- PCA's: polycyclic aromatics
- common measurements of PCA's are: the 6 PCA's of Borneff or the 16 PCA's according to EPA (USA).

On the basis of operational treatment results it can be concluded that:
- about 90-98% of the oil compounds can be removed from the soil
- in the case of cyanides (free or complex) 98% or more of the contaminants can be removed.
- the cleaning efficiencies in the case of BTEX and PCA's are very high namely above 99.5%.

Calculating the efficiency of a cleaning operation generally results in an uncertain value, because in many cases the calculation had to be made from a range and no average concentrations are available.

Summarizing it can be stated that thermal plants are suitable for the cleaning of different types of soil including soils with high contents of humus, peat, loam or clay. The cleaning efficiency varies between 98-99.5% in the most cases. That implies that the "A" standard of Ministry of VROM [3] can be reached in many cases.
In general the thermal techniques are unfit for the treatment of soil contaminated with heavy metals.

4. EXTRACTION TECHNIQUES

Extraction techniques can be classified in two groups being in the first place extraction with an aqueous or organic liquid and in the second place extraction by flotation.
The main principle of the first mentioned technique is an intensive contact between the contaminated soil and the extracting fluid. This can be achieved by means of a scrubber which is the most commonly used technique or otherwise by means of a high pressure water yet (velocity > 200 m/s). During this process the contaminants dissolve and/or disperse into the extraction fluid which is generally water with some chemicals. For instance heavy metals can be dissolved by acidified water (sometimes in combination with complexing agents), cyanides have to be extracted by an alkaline solution to prevent emissions of HCN. Contaminants like hydrocarbons can be removed by adding detergents to the water. Subsequently the treated soil is separated from the fluid and the contaminants are removed by means of a water treatment system. The larger part of the treated extraction fluid can be recycled in the process.
In case of the flotation-process (also called froth-flotation) the contaminants are removed within a foam which is made by forced aeration after addition of specially tailored chemical agents to the water.
The remaining sludge and foam, which contain high concentrations of the contaminants, can be incinerated or decomposed by oxidation or hydrolysis. (Other ways of sludge treatment are under investigation). In many cases however the sludge has to be transported to a controlled waste disposal site.

4.1 Application

The extraction technique is mainly suitable for treatment of contaminated sandy soils. The reasons are:
o the heavy sand-particles facilitate the separation from the extracting fluid
o the relatively small specific area of the sand-particles results in a smaller quantity of adsorbed contaminant.
It will be clear that by the treatment of types of soil other than sand a larger amount of sludge will be produced, which will result in

substantially higher costs. The presence of only small amounts of humus or clay in the sand, however, do not cause a significant problem. The adsorbed contaminants mainly remain in the sludge.
With respect to the types of contaminants it can be concluded that in most cases oil, cyanides, PCA's and heavy metals can be removed sufficiently from sandy soils with extraction techniques as can be seen in the following tables.
After the application of an extraction technique the soil structure is always changed with respect to humus-content and particle-size. The treated soil consists of sand with particle-sizes in a narrow band and can be re-used, for instance, for road-construction or specific building-materials.

4.2 Treatment results and conclusions

The operational full-scale extraction plants are indicated in Table 3.

Table 3. Operational extraction plants (March 1987) [1, 2]

Name of company	Capacity (ton/year)*
BSN B.V.	25,000**
Heidemij Uitvoering B.V.	34,000
Heijmans Milieutechniek B.V.	14,000
HWZ Bodemsanering	27,000
Mosmans Mineraaltechniek B.V.	8,000

* on the basis of 8 h/day
** the new plant of BSN is under construction

The costs of the treatment by means of extraction vary between Dfl 80-200/ton depending mainly on the quantity of small particles in the contaminated soil.
Some treatment results of the plants mentioned above are summarized in table 4.

Table 4. Some treatment results of operational extraction plants (March 1987) [1, 2]

Type of soil	Contaminants	Concentration mg/kg dry weight Initial	Final
sand	oil	500 - 10000	< 100
sand, > 2.5% humus	mineral oil	50000	70
sand, > 10% loam	mineral oil	1500 - 25000	80 - 150
sand	CN	50 - 1000	8 - 25
sand, < 2.5% humus	CN complex	50 - 150	5 - 10
sand, > 10% loam	CN	50 - 1000	10 - 20
loamy sand	CN	250	7*
sand, < 2.5% humus	PCA's (16 EPA)	160 - 290	0.4 - 17
sand, < 10% peat	PCA's (16 EPA)	80 - 190	3 - 9
sand	Ni	15 - 2050	40 - 75
	Cr	105 - 2000	74 - 150
	Cd	1 - 1750	1 - 2.5
	Pb	15 - 2050	42 - 75
sand, > 10% loam	Zn	150 - 1400	90 - 200
	Ni	50 - 900	40 - 75
clayey sand	Pb	300 - 2000	50 - 150*

* Flotation technique

Remarks:
- specifications of the types of oil and the types of cyanides were often not mentioned.
- common measurements of PCA's are: the 16 PCA's according to EPA (USA).

On the basis of treatment results it can be concluded that:
- oil compounds and cyanides can be removed from the soil with efficiency of 90-99%
- PCA's can be removed 94-99%; there is less experience with this type of contaminants
- it is not possible to indicate a reliable range of cleaning efficiency for heavy metals; it is varying up to 95%.

Summarizing it can be stated that by means of extraction techniques an average cleaning efficiency of 95% can be reached in case of sandy soils. That implies that in most cases the concentration after treatment is between the "A" and "B"-values given by the Ministry of VROM. These soils can therefore be applied with some restrictions. In the present situation the extraction technique is the most suitable for the removing of heavy metals. The current extraction techniques are unfit for the treatment of clay or soil with heavy loam or peat content.

5. BIOLOGICAL TREATMENT TECHNIQUES

Biological treatment techniques are at the stage of development. Due to the Research Programme sponsored by the Dutch government (as mentioned in section 2) seven research projects were started in 1985. The majority of these projects will be finished this year, however, some large projects will continue till the end of 1988 [5, 6, 7].
This research project are aiming at:
* The optimization of the environmental and process conditions for landfarming;
* The establishment of the applicability of landfarming for different types of contaminants;
* The development of other biological soil treatment methods such as biorestoration in situ and bioreactors.

In the following a short summary of the main results and conclusions of the research done so far is given for each technique, being in succession landfarming, in situ biorestoration and treatment in bioreactors.

5.1 Research results

Landfarming
In case of soil treatment by means of landfarming the excavated contaminated soil is spread out in a layer of about 40 cm on top ofa drained sand bed. The contaminated soil c.q. sand layer has to be isolated from the subsoil by a plastic layer (e.g. 0.5 mm PVC soil). To stimulate the biodegradation of the contaminants the environmental conditions will be optimized among others by means of adding fertilizers to the soil or regularly mixing of the soil.
In the field of landfarming various projects were carried out during the last two years. In a number of projects the biodegradation of different kinds of oil was investigated and in one project the possibilities for degradation of polycyclic aromatics (PCA's) were studied [6, 8].

Table 5. Degradation of oil products by means of landfarming methods
(mg/kg dry weight total oil concentrations in the soil) [6]

Type of oil	Initial concentration	concentration after 1 growing season*	concentration after 2 growing seasons*
crude oil	35,000	small reduction	**
crude oil	8,000	small reduction	**
gas oil	1,800	400	**
fuel oil	6,800	800	300
mineral oil	1,100	small reduction	400
cutting oil	2,400	800	**

* growing season: the warmest period of the year (4 to 6 months)
** no results available

On the basis of the research results, mentioned in table 5 it can be concluded that, by means of the landfarming method, pollutants such as gas oil, fuel oil and cutting oil can be removed within one "growing season" (4 to 6 months) to a residual concentration of 400 to 800 mg of oil/kg dry weight (see figure 1). During a second growing season a reduction to 200-400 mg of oil/kg dry weight was achieved in some cases. The remaining oil products in the soil are compounds with a low biodegradability.

In case of crude oil the degradation is very slow. Further research will be required before more valid conclusions can be drawn.
In one case, landfarming appeared to be suitable for the removal of polycyclic aromatics (PCA's). Within a period of only two months the initial concentration of 300 mg/kg dry weight decreased to about 100 mg/kg dry weight. From the results it can be concluded that lower PCA's

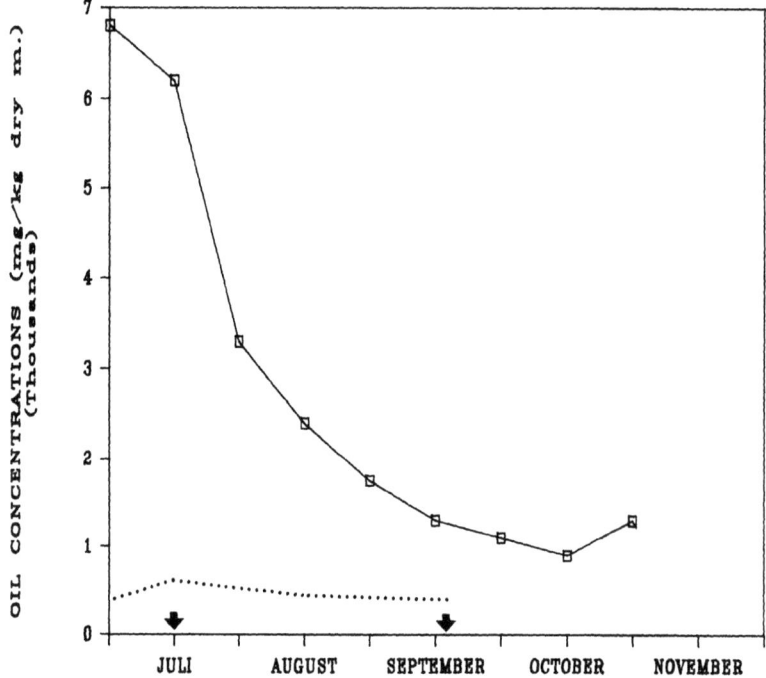

LEGEND:

——— Oil concentration in soil contaminated with fuel oil
..... Oil concentration in underlying land layer
⬇ Time of fertilizer addition

Fig 1: Decrease of oil concentration in contaminated soil as a function of time [6]

such as naphtalene, fluorene and phenanthrene were degraded for 80-100% whereas higher PCA's such as benzo(b)fluoranthene and benzo(a)pyrene were hardly or not at all degraded. After 16 months a final concentration of about 70 mg/kg dry weight was achieved.

In this period much know-how was also obtained regarding the most important environmental and process parameters for biodegradation and among others, about the leaching of pollutants and nutrients [6, 8].

During the last two years, at least three companies have restored different comtaminated sites by means of landfarming and the results for soil contaminated with oil compounds (excluding crude oil) were promising.

<u>Biorestoration in situ</u>

The principle of this technique is indicated in figure 2.

The research regarding in situ techniques is still at an early stage. The possibilities of in situ biorestoration of the deeper layers of oil-contaminated soil has been investigated on laboratory scale in the last year. On the basis of the laboratory research the following conclusions can be drawn [6, 8]:

* The rate of degradation of oil (mainly petrol) in polluted soil is slow. Under laboratory conditions, oxygen diffusion does not constitute a limiting factor. Under field conditions however, oxygen transfer may be limiting.
* The degradation activity is clearly enhanced by the addition of seeding material from a landfarm. The saturation of soil with water also leads to an increased degradation activity.
* The C:N:P ratio has a relatively small influence on the degradation. At increased phosphate concentrations a higher degradation rate was found. This effect is still being investigated.
* The addition of biodegradable detergents does not have any positive effect on the leaching or the availability of petrol for micro-organisms. These results differ from some research data mentioned in literature.

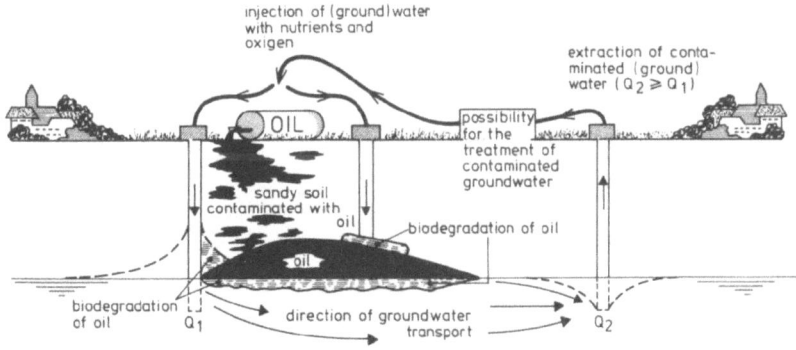

Fig 2: Diagrammatic representation of in situ biological treatment

Further research should be directed towards increasing the knowledge on the use of specially cultivated micro-organisms as seeding material and the application of alternative oxygen sources. During 1987 some companies will apply biorestoration in situ experimentally on selected sites.

Bioreactors
The research into the development of bioreactors has started only two years ago. Generally it can be stated that the development of bioreactors is important for the biological treatment of soil contaminated with substances of low biodegradability (such as halogenated hydrocarbons) and for soils that are generally difficult to treat (such as clay). Some advantages of bioreactors as compared to other techniques are:
* Better possibility for process control;
* Better availability of contaminants for micro-organisms by better homogenization of the soil (e.g. in slurry reactors).
* Possibility for the application of specially cultivated micro-organisms.

Research on the development of bioreactors is carried out at this moment and will be continued in the next years [6, 8].
Based on the first laboratory results the following conclusions can be drawn:
* Biodegradation is strongly enhanced by the addition of selected micro-organisms;
* Biodegradation is better in a soil/water slurry than in the water phase only;
* For the treatment of various types of soils, different process pathways, e.g. reactor systems, have to be developed.

Regarding the degradation of hexachlorocyclohexane (HCH) it can be concluded that:
* the degradation rate of α-HCH is fastest under aerobic conditions during which hardly any intermediates are accumulated.
* β-HCH could not be degraded under any of the three redox conditions (aerobic, methanogenic, denitrifying) were tested.

Field experiments also confirm these laboratory results. Under aerobic conditions a significant degradation of α-HCH was observed within 30 days (see Table 6).

Table 6. Changes in α-HCH concentrations (mg/kg dry weight) after 30 days [6]

Treatment	t(0)	t(30)
wet soil*, aerated	357 ± 8	204 ± 16
wet soil, aerated, sludge	369 ± 14	222 ± 12
slurry**, aerated	424 ± 18	73 ± 5
slurry, sludge, aerated	412 ± 57	83 ± 12
concrete mixer, slurry	343 ± 18	235 ± 19

* wet soil: about 20% water
** slurry : about 30% water

No reduction of β-HCH was noticed. The possibilities of the degradation of β-HCH are studied in a follow-up project.

5.2 Further research

On the basis of provisional research results it can be stated that further investigation into the field of biological soil clean-up will be useful and that it looks promising so far [6, 8]. In the near future research will be done with regard to the following:
* The influence of the physical and chemical soil dynamics on biological soil treatment by landfarming;
* The possibilities of adding and maintaining specially cultivated microorganisms for the different treatment systems;
* The possibilities of biological treatment of soil contaminated with chlorinated hydrocarbons;
* Further development of bioreactor systems;
* The development of simulation models for biological soil treatment techniques in general.

Most of the new projects with regard to the areas mentioned above will be initiated in 1987. These projects will usually be carried out in co-operation with various institutes and universities. Engineering consultants or companies will participate as well at an early stage in those projects in which mainly the development of techniques is studied.
The supervision and co-ordination of the research projects will be in the hands of RIVM/LAE. During the realization of the research programme, an intensive exchange of knowledge and co-operation on a national and international scale will take place.

REFERENCES

1. Verhagen, E.J.H., Versluijs C.W., 1987.
   Soil decontamination by extractive or thermal installations in the Netherlands. Presented on the Danish Seminar Contaminated Soil and Groundwater, Febr. 25-26, 1987, Kolding, Denmark. (Obtainable from RIVM/LAE. Bilthoven, The Netherlands).

2. Handboek Bodemsaneringstechnieken (Handbook for Remedial Action Techniques), Staatsuitgeverij, The Hague, 1983; Revision 1987 (in Dutch).

3. Moen, J.E.T. et al.,
   Soil protection and remedial actions: criteria for decision making and standardization of requirements, p. 441-448, in: Contaminated Soil, 11-15 nov. 1985, Utrecht. ed. Assink, J.W., van de Brink, W.J., Martinus Nijhoff Publishers, Dordrecht, 1986.

4. Hoogendoorn D.,.
   Review of the development of remedial action techniques for soil contamination in the Netherlands, in: Proceedings of the 5th National Conference on Management of Uncontrolled Hazardous Waste Sites; Washington D.C., November 1984, p. 569-575.

5. Soczó, E.R., 1986
   Biotechnologische bodemreinigingstechnieken. Stand van zaken huidig onderzoek en gewenste ontwikkelingen (Biotechnological soil clean-up techniques. State of the art of present research and desired developments). RIVM No. 851105001. National Institute of Public Health and Environmental Hygiene, Bilthoven, The Netherlands, January 1986 (in Dutch).

6. Soczó, E.R., Staps, J.J.M. and Visscher, K., 1986.
   Biotechnologische Bodemsanering. Rapportage van de workshop van 20 en 21 maart 1986 te Bilthoven. (Biological soil clean-up techniques. Report of the workshop in March 20-21, 1986, Bilthoven) RIVM No. 851105002. National Institute of Public Health and Environmental Hygiene, Bilthoven, The Netherlands, August 1986 (in Dutch).

7. Soczó, E.R. and Visscher, K. 1986.
   Biologische Bodenreinigung. UMWELT 7/1986 (in German).

8. Soczó, E.R., 1987.
   Development of biological treatment techniques for contaminated soil in the Netherlands. Presented on the Danish Seminar Contaminated Soil and Groundwater, Febr. 25-26, 1987, Kolding, Denmark. (Obtainable from RIVM/LAE. Bilthoven, The Netherlands).

IN SITU REMEDIAL ACTION TECHNIQUES FOR TREATMENT OF CONTAMINATED SOIL AND GROUNDWATER BY MEANS OF GROUNDWATER EXTRACTION AND INFILTRATION TECHNIQUES

E. DE ZEEUW
De Ruiter Milieutechnologie B.V., Halfweg, The Netherlands.

ABSTRACT
In situ soil and groundwater treatment techniques provide a cheaper alternative for total removal of contaminated soil and treatment on remote site. Some case studies of groundwater clean up and in situ soil treatment are described. The examples show that both solid soil and groundwater can be treated simultaneously by applying groundwater extraction and infiltration techniques. A simple geohydrochemical extraction model describes the clean up process. The addition of oxygen, nutrients or surfactants to the infiltration water may enhance the clean up process.

1. INTRODUCTION
Because of the high costs of remedial action techniques for removal and treatment of contaminated soil, there is a tendency to minimise the amount of soil to be removed. This implies that in situ soil and groundwater treatment will become increasingly important. In this paper some groundwater clean up and in situ soil treatment projects, carried out within The Netherlands by De Ruiter Milieutechnologie B.V., will be described. The examples may provide a brief overview of the applicability of groundwater extraction an infiltration techniques by means of "washing" the soil. First two types of groundwater clean up are described.
When infiltration techniques are applied, the infiltration water is used as transport medium for removing contaminants from the solid soil. In this way both solid soil and groundwater are treated simultaneously (in situ soil treatment). The desorption of contaminants will be described by means of a simple geohydrochemical extraction model. Three cases of in situ soil treatment by infiltration of clean water will be described. Chemicals may be added to the infiltration water to enhance the clean up process.

2. CASE STUDIES OF GROUNDWATER CLEAN UP OPERATIONS
2.1. Removal of a free floating oil layer
Leakage and spill of mineral oil products into the soil may lead to the formation of a free floating oil layer just above the phreatic groundwatertable (CONCAWE, 1981). On a location in Roosendaal a free floating oil layer of gasoline was found over an area of 30 m x 60 m. The thickness of the oil layer as measured in piezometers (∅ 40 mm) ranged up to 1.20 m. In the period April 1985 to November 1986 the spilled oil has been recovered again by two wells with a depth of 5 m (initially only one).
A two pump extraction system was used. Oil and groundwater were pumped separately: the groundwater by two submergable pumps with an average total discharge of 0.5 m3/hr, controlled by level-triggered switches and the oil by a pump controlled by a timer.

The oil was discharged into two storage tanks and the groundwater was disposed into the sewage system via an oil/water-separator. Up to November 1986 a total volume of 12.5 m3 of gasoline has been recovered by this system (figure 1.).

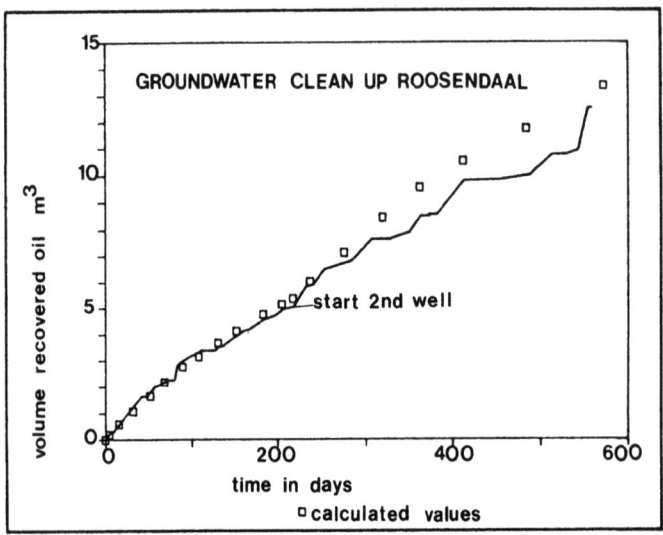

FIGURE 1. Groundwater clean up Roosendaal

The determination of the optimum discharge is a matter of trial and error. If the groundwatertable is lowered too much, the oil layer within the vicinity of the well thins out or becomes discontinuous. Hence, a higher discharge rate does not necessarily lead to an increase in oil yield. Computer simulations indicate that the oil yield of an optimal system mainly depends on the average thickness of the oil layer within the influence range of a well. Figure 1 shows a curve computed by a model in which for each individual well the oil yield-factor (i.e. the volume extracted groundwater divided by the volume of recovered oil in 1/1) is assumed to be inversely correlated to the average thickness of the oil layer within the influence range of the wells. During the pumping period the oil yield-factor of both wells increases from 150-400 l(water)/l(oil) up to over 1000 1/1. A food fit between the calculated and the observed curve is obtained for an initial average thickness of the oil layer of 0.08 m for the first well, an 0.20 m for the second. This is much less than the observed thickness in the piezometers of 0.80 to 1.20 m. Clearly, the observed thicknesses do not agree with the "real" thickness of the oil layer in the soil (CONCAWE, 1981; Umweltbundesamt, 1986). The difference corresponds to a factor 4 to 10. As the clean up process progresses, a further reduction in oil yield of the well is inevitable. In the end, the recovery of oil by a well system becomes uneconomic. In October 1986, the wells were replaced by a horizontal drainage system. Three ditches with a depth of 1.75 m were digged. Drains were placed at 1.5 m below the land-surface, which corresponds to the lowest observed groundwatertable. The lower 0.5 m of the ditches were filled by coarse sand; the rest was filled by the original soil material. The oil yield of the drainage system has been excellent up till now.

In the period December 1986 to Februari 1987 more than 9 m3 of oil has been caught in a storage tank at the surface. The groundwater is now discharged into the sewage system under free gravity flow.

2.2. Treatment of groundwater contaminated by volatile chlorinated organic compounds

On a location near Rolde an unknown amount of chloroform and 1,1,1-trichloroethane was spilled into the soil due to a car accident (Volker, 1986; de Zeeuw, 1986). After an investigation of soil and groundwater, it was decided to isolate the contaminated zone by a steel wall reaching a depth of 3 m below landsurface. At that depth a less pervious loamy layer is present. The groundwater clean up operation was executed in the period October 1984 to April 1985. The groundwater within the isolated zone was extracted by a vertical vacuum drainage system. The discharge varied between 5 and 10 m3/day. Figure 2 shows a plot of the concentration and discharge against time.

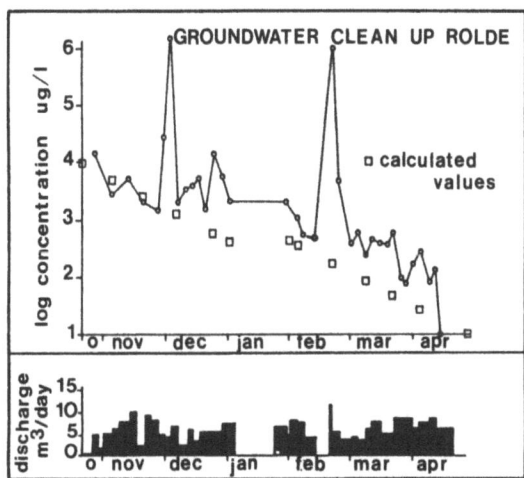

FIGURE 2. Groundwater clean up Rolde

The concentrations of volatile organochlorocompounds (VOCl) gradually decreases from 2,000 - 4,000 ug/l to less than 100 ug/l. Two striking peaks of more then 1,000,000 ug/l show up (December 1984, Februari 1985). These peaks correspond to the extraction of heavily contaminated groundwater from the top of the loamy layer. A part of the contaminants may have sunk down towards the top of the loamy layer as an immiscible phase under the influence of their high specific gravity (1.3 to 1.6 kg/dm3). The investigation of the groundwater has indeed confirmed that extremely high concentrations were present just above the top of the loamy layer. Generally, many extractions are needed to clean up the groundwater, because adsorption/desorption and mixing processes delay the clean up process.
The groundwater clean up at Rolde has been ended after the extraction of 12 to 13 times the original volume of contaminated groundwater.

Another case of a groundwater clean up operation with volatile organo-chlorocompounds has been described earlier by de Zeeuw and Hopman (1987). In that case (in Cuyk) a volume of 150,000 m3 of groundwater was contaminated with tri- and perchloroethylene. From November 1984 up to this moment over 1,000,000 m3 of contaminated groundwater has been extracted using a 18 m deep well (average discharge = 90 m3/hr). Also in this case, adsorption/desorption and mixing processes lead to a prolongation of the clean up process. Meanwhile, a second well has been put into operation and a plan has been made to remove the most heavily contaminated shallow groundwater more quickly by a vacuum drainage system combined with an infiltration system.

3. IN SITU SOIL TREATMENT
3.1. A geohydrochemical extraction model

The number of extractions to clean up a certain volume of contaminated soil (the socalled "extraction-factor") is strongly related to the adsorption/desorption equilibria between solid soil and groundwater. The extraction-factor may lie between 25 and 250 for mineral oil and volatile aromatics, and between 10 and 30 for volatile organochlorocompounds (de Zeeuw and Hopman, 1987). Much higher extraction-factors up to over 10,000 may be needed for the removal of polycyclic aromatic compounds or insecticides. The extraction clean up process is schematically shown in figure 3.

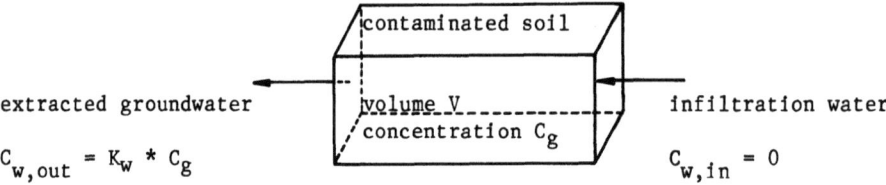

extracted groundwater

$C_{w,out} = K_w * C_g$

infiltration water

$C_{w,in} = 0$

FIGURE 3. The extraction process

After each extraction: $C_{g(n+1)} = C_{g(n)} * \left\{ 1 - \dfrac{K_w * p}{\rho_g} \right\}$ (1)

$C_{w(n+1)} = C_{g(n+1)} * K_w$ (2)

Whereas:
 $C_{g(n)}$ = concentration in solid soil after n extractions (mg/kg dry)
 K_w = equilibrium constant (ug/l)/(mg/kg dry)
 C_w = concentration in groundwater (ug/l)
 p = porosity (dimensionless)
 ρ_g = specific gravity of the soil (kg/m3)

Infiltration water with concentration $C_w = 0$ enters a volume of contaminated soil. The extracted groundwater is then assumed to reach a chemical equilibrium with the solid soil. In case of linear adsoption (constant K_w) the rest concentration in the solid soil after each extraction can be calculated with formula (1). The equilibrium concentration in the extracted groundwater can be calculated with formula (2).

Nonlinear adsorption can be added easily to the model by adjusting the K_w-value after each extraction, with the following general formula:

$$K_{w(n)} = \frac{C_w(n)^M}{C_{g(n)}} \quad (3)$$

Whereas:
  M = constant

The most critical parameter in the model is the K_w-value.
There are several ways to estimate this parameter:
- The K_w-value can be derived from octanol/water partition coefficients (K_{ow}) and distribution coefficients (K_{oc}) known from the literature (Briggs, 1973 and 1981; van der Meijden and Driessen, 1986);
- Laboratory experiments can be carried out in order to simulate the extraction process (shake- and column-experiments);
- The K_w-value can be determined in the field by analyses of groundwater- and solid soil samples taken at representative locations at the same depth intervals;
- The K_w-value can be determined by calibrating a groundwater clean up- or in situ treatment project. In figure 2 the values calculated by the model (project Rolde) show a reasonably good fit in case of a K_w-value of 2,500 (ug/l)/(mg/kg dry). Of course, the model fails to account for the concentration peaks caused by mixing of the extracted groundwater with the more contaminated groundwater from the top of the loamy layer (chapter 2.2.).

3.2. Application of infiltration techniques

Depending on the nature and degree of contamination and on the soil characteristics, an intensive flushing of the soil may be needed to clean up both solid soil and groundwater. Such an intensive flushing can be achieved by the application of infiltration techniques.
Usable infiltration techniques are: - horizontal drains
                                     - infiltration wells
                                     - open systems: ditches, ponds, sprinkling.
Infiltration of water contributes to higher hydraulic gradients and shorter residence times of the water in the soil compartment. It also leads to a local elevation of the groundwatertable. In this way also parts of the unsaturated zone can be cleaned. Other advantages of infiltration are a reduction of the risk of damages caused by subsidence and a reduction of the amount of water, that has to be disposed into a sewage system or surface waters.
Disadvantages are the higher pumping rates, leading to higher energy and water treatment costs. A combination of extraction and infiltration should be designed in such a way, that further spreading out of the contaminants is prevented. This means that some cases infiltration may only be feasible within the centre of the contaminated zone.

3.3. Case studies of in situ soil treatment

Two cases of in situ soil treatment projects in Amsterdam and in Amersfoort have been described earlier in more detail by de Zeeuw and Hopman (1987).
On a former oil depot in Amsterdam an in situ soil treatment has been carried out in the period May to August 1986. Groundwater was extracted from 4 strings of vacuumfilters with a discharge rate of about 4 m3/hr.

The extracted groundwater was purified by an oil/water-separator, followed by a stripping-unit, a sandfilter and an active-carbon filter. After purification the water was infiltrated again through 1.5 m deep ditches filled with coarse sand (see photo 1.). Soil samples were taken before and after the clean up from 0.25 to 1.25 m below landsurface (contaminated, unsaturated zone), on 4 representative locations. About 5 times the volume of contaminated groundwater has been extracted during the clean up operation. A reduction of the average oil content in the soil of 71% was achieved after the clean up, and a reduction of 98% of the volatile aromatics (mainly xylenes).

PHOTO 1. In situ soil treatment in Amsterdam

In the project Amersfoort, both solid soil and groundwater around a petrol station were contaminated with petrol and volatile aromatics. The project had to be carried out simultaneously with the rebuilding of the petrol station. Not all contaminated soil could be removed, because the petrol station and also a garage had to remain in operation. This implied that some of the contaminated soil had to be treated in situ. Column experiments were carried out in order to simulate the clean up of the soil. The actual clean up operation has been started recently. Groundwater is extracted by several 5 to 7 m deep wells with discharge rates ranging from 2 to 8 m3/day. Total discharge equals about 75 m3/day. About 40 m3/day is infiltrated again, after purification, through a central, 1 m deep infiltration drain with a length of 80 m. The drain has been placed such, that the most severely contaminated soil will be flushed the most intensively.

Recently an in situ soil treatment project has been started in Arnhem. In that project the soil in the unsaturated zone up to a depth of 7 m is contaminated with fuel oil. Water is infiltrated - after purification, aeration, and the addition of nutrients - through an infiltration pond. Less pervious layers, which may obstruct infiltration, have been perforated by piezometer pipes (see photo 2.).

PHOTO 2. In situ soil treatment Arnhem

4. CONCLUSIONS AND FUTURE DEVELOPMENTS

In situ soil treatment by means of extraction and infiltration of water is possible and may provide a cheaper alternative for total removal of contaminated soil and treatment on remote site.
Sometimes, eg. below buildings, in situ soil treatment may be the only feasible option. Model calculations and column- and shake experiments have shown that an intensive flushing of the soil is needed. Residence times of the water in the soil compartment of less than one week can be achieved with a properly designed infiltration and extraction system.
In reality, the clean up process is not as simple as in the model discussed in chapter 3.
Column experiments of the project Amersfoort have shown that the flushing of the contaminants out of the soil may take place in different phases:
- water phase (dissolution);
- colloidal phase (adsorbed species transported on small soil particles);
- immiscible phase (eg. a seperate oil phase);
- gas phase (volatilization).

A hydrogeochemical equilibrium may not be reached under field conditions or may be disturbed by mixing processes on micro- or macroscale. In practice, contaminants may also be released by volatilization and microbiological degradation.

In stead of decreasing the residence times of the water in the soil compartment by infiltration, it is also possible to enhance the clean up process by adding chemicals to the infiltration water. Addition of oxygen and nutrients enhances the microbiological activity and the biodegradation of organic contaminants (American Petroleum Institute, 1978, 1982).

It is also possible to influence the adsorption/desorption processes within the soil compartment by means of adding cosolvents or surfactants. Some research has already been carried out in the U.S.A. (eg. American Petroleum Institute, 1979; Ellis et al, 1985). De Ruiter Milieutechnologie B.V. has recently carried out shake experiments in order to test several different types of surfactants on soil samples contaminated with mineral oil. Some surfactants enhance the desorption processes from solid soil to groundwater. However, other surfactants are needed to maintain a good oil/water-suspension.

Problems arise in separating the oil/surfactants/water-phases again, and in the procedures of chemical analyses concerning the distinction between the oil and surfactants. These problems have to be solved, before field application is possible.

REFERENCES

1. American Petroleum Institute 1978. Field application of subsurface bio-degradation of gasoline in a sand formation.
   API Publication no. 4430, Washington.
2. American Petroleum Institute 1979. Underground movement of gasoline on groundwater and enhanced recovery by surfactants.
   API Publication no. 4317, Washington.
3. American Petroleum Institute 1982. Enhancing the microbial degradation of underground gasoline by increasing available oxygen.
   API Publication no. 4428, Washington.
4. Biggs, G.G. 1973. A simple relationship between adsorption of organic chemicals and their octanol/water partition coefficients.
   Proc. 2nd British Insecticide an Fugicide Conference, pp. 83-86.
5. Briggs, G.G. 1981. Theoretical and experimental relationships between soil adsorption, octanol/water partition coefficients, water solubilities, bioconcentration factors and the parachor.
   J. Agric. Food. Chem., Vol. 29, pp. 1050-1059.
6. CONCAWE, 1981. Revised Inland Oil Spill Clean-up Manual.
   Concawe report no. 7/81.
7. Ellis, D.E., J.R. Payne and G.D. McNabb 1985. Treatment of contaminated soils with aqueous surfactants.
   Environmental Protection Agency, U.S.A., report no. EPA/600/2-85/129.
8. Meijden, A.M. van der and A.P.T. Driessen 1986. Betekenis van het sorptie-evenwicht voor de verdeling van organische (micro-)verontreinigingen in de bodem (in Dutch).
   Bodembescherming no. 54, Staatsuitgeverij, Den Haag.
9. Umweltbundesamt 1986. Die wissenschaftlichen Grundlagen zum Verständnis des Verhaltens von Mineralöl im Untergrund (in German).
   Umweltbundesamt, report no. LTwS - nr. 20, Berlin.
10. Volker, P. 1986. Calamiteit met gechloreerde koolwaterstoffen (in Dutch).
    H_2O, Vol. 19, no. 18, pp. 435-438.
11. Zeeuw, E. de 1986. Ervaringen met grondwatersanering en in situ bodemreiniging (in Dutch).
    Land en Water/Milieutechniek, no. 11, pp. 59-63.
12. Zeeuw, E. de and R. Hopman 1987. Interactions between solid soil and groundwater: experiences from the monitoring of groundwater clean up operations and in situ treatment of contaminated soil.
    Contribution to the V.S.G.P.-conference, Noordwijk aan Zee, The Netherlands.

CLEANING SOILS CONTAMINATED WITH HEAVY METALS

J.W. Assink   W.H. Rulkens

MT-TNO, P.O. Box 342, 7300 AH Apeldoorn, The Netherlands

1. INTRODUCTION
It is estimated that 25 to 50% of all contaminated sites in the Netherlands contain too high a concentration of one or more heavy metals (1,2).
Remedial technologies that use a liquid phase to carry off the contaminants are considered the most suitable for removing heavy metals from soil (3). The mechanism for the removal of heavy metals may be based on extraction (dissolving), dispersion (emulsifying) and/or classification (separating certain particles; including flotation).

By order of the Ministry of Housing, Physical Planning and Environment research was carried out to investigate by laboratory experiments the applicability and feasibility of several modifications of the abovementioned methods for cleaning soil contaminated with heavy metals.

2. SOIL SAMPLES AND EXPERIMENTAL APPROACH
2.1 Soil samples
Nine different locations, to a certain extent considered representative for all Dutch locations with heavy metal contamination, were selected; at each location approximately 30 kilos of soil were sampled *at random* at the most affected sites.

The samples were dried at ambient temperature and subsequently homogenized. The fraction larger than 2 mm was sieved off and rejected for the experiments.

Table 1 gives information on the characteristics of the soil samples. These vary from low to high both in humus content and clay content and moreover there are great variations in the origin of contamination, including the form (solid, liquid, sludge) in which the contaminants were dumped in the soil, and the compounds actually present.

2.2 Analyses
The soil samples were analysed according to NEN 6452, 6451, 6453, 6556, 6443, 6457 and 6449. These methods are based on a decomposition in a concentrated acid, which hardly affect the metals in the silicium-matrix. The determination of dissolved metals was carried out with an atomic adsorption spectrometry. All figures given are the mean value of two, and sometimes even four, analysed subsamples.

2.3 Experimental procedures
The soil samples were subjected to several experiments. The most important are briefly described here.

The experiments were carried out batch-wise with 50-150 g soil. Previous to the experiments all samples were scrubbed in order to reach a complete segregation of conglomerates. Scrubbing was mostly carried out by adding 30% (m/m) water and mixing the slurry for 4 hours in a vessel horizontally revolving on its axis. Sometimes the slurry was scrubbed for 30 minutes in an ultrasonic bath as an alternative.

In the first instance all soil samples were subjected to three so-called basic routes:
a) Extraction with hydrochloric acid. The soil was mixed in a jar with an HCl-solution at the desired pH (mostly pH = 1) for approximately 30 minutes. The liquid/soil ratio was 5 : 1 (m/m). The soil was subsequently separated from the liquid by filtration or by centrifugation, combined with filtration of the supernatant. Usually these steps were repeated two times to a total of three extractions. Finally, the soil sample on the filter was rinsed with an excess of demineralized water.

b) Extraction with nitrilo tri-acetic acid (NTA). The soil was mixed in a jar with an NTA-solution at a liquid/soil ratio of 5 : 1 (m/m). Mostly, rather extreme process conditions were chosen: 20 or 100 g Na_3NTA/kg soil, pH = 3-3,5 (adjusted with HCl) and a contact time of 1, 6 or 24 hours (despite these different contact times the removal efficiencies were almost identical). Separation of soil and extractant was realized by sedimentation and filtration of the supernatant.

c) Wet classification by jet-sizing. A sample of 150 g soil was classified in an upflow-column (60 mm inside diameter) with 5 liter 0.1 N NaOH and subsequently 10 liter water. The upflow velocity was 3.3 mm/s, which is sufficient to remove all particles smaller than 60 μm.

These basic routes are relatively simple and supposedly applicable to a wide range of different types of contaminated soil, although the process conditions chosen were relatively extreme.
In order to improve the cleaning results, many additional experiments were carried out; these, however, will not be described in detail. The most important additional experiments were:
- Extraction with acid, preceded by oxidation with H_2O_2, NaClO, $KMnO_4$, air or by thermal treatment.
- Extraction with NaOH.
- Extraction with NaClO (oxidation and complexation by Cl^-).
- Combinations of basic routes, such as wet classification followed by extraction with HCl.

2.4 <u>Evaluation of the experiments</u>
For each sample one or more metals were chosen as "guide metal" to evaluate the experimental results. Main criterium for choosing a guide metal was its relatively high concentration (see table 2).

The extracted soil samples were considered sufficiently clean if the heavy-metal concentrations had been reduced to the A-value of the "Table of Trigger Values" published by the Ministry of Housing, Physical Planning and Environment (4). This table contains A-, B- and C-values, which refer to respectively: a reference value (i.e. the maximum concentration in uncontaminated soils), an indicative value for further site investigation and an indicative value for (investigation of the need for) remedial action. The A-, B- and C-values are given in table 2. From this table it appears that in 13 out of 18 contaminating metals (guide metals) the concentrations have to be reduced by more than 95% to reach the desired A-value.

The experiments were evaluated as to the final concentration of guide metals and their removal efficiency:

$$\text{removal efficiency} = (1 - \frac{\text{final concentration}}{\text{initial concentration}}) \cdot 100\%.$$

3. RESULTS

The results of the most important experiments are presented in the tables 3 to 11. These tables also give a brief survey of the actual conditions during the experiments. In selecting experimental results for this publication, preference was given to those experiments that had the best results or were related to the general discussion on extraction methods.

4. DISCUSSION

Only 2 soil samples (II and VII) could be sufficiently cleaned (that is, below the A-value); these samples had the lowest required removal efficiency: 86% and 68% against 96% or more for the other soil samples.

The samples III and VIII could be cleaned to a final concentration between the A- and B-value while the samples I, IV, V, VI and IX could not be cleaned below the B-value.

In interpreting these results, one should realize that they are based on the best results of over 120 experiments on the one hand and on relatively highly contaminated soil samples on the other hand.

Table 12 summarises the two best treatment methods for each soil sample. It shows that classification, extraction with acid or a combination of these two techniques often give the best results.

This is also demonstrated by table 13 which gives the best experimental results of the three so-called basic treatment routes. Only in three cases does extraction with NTA result in a removal efficiency of over 50%; therefore it is the least effective route.

Extraction with HCl (pH = 1) gave strongly varying results: Cr, Hg and Sn were hardly removed from the soil, while the other metals were removed better or at least comparable to the classification route. Especially Cd and As were well or very well extractable (> 85% removal efficiency). The experimental results indicated that a multiple extraction with HCl gave significantly better results than a single extraction step.

Classification (jet-sizing in an upflow column) resulted in most cases in a moderate to good removal efficiency (60%-90%). It is a relatively simple technique with only one important drawback: a relatively large amount of residual sludge may result from clayey or peaty soils; this residual sludge has to be disposed of properly.

The additional experiments have led to the following conclusions:
- Wet chemical oxidation preceding the extraction with HCl improved the removal efficiency for Cu, Cr and Hg. The results for Pb are sometimes better, sometimes worse; the removal of As is somewhat reduced by this oxidation step.
- Thermal oxidation for 2 hours at 500 °C reduced the metal extractability with HCl in all cases, except mercury. Probably the metals were incorporated, to a certain extent, in the mineral matrix. Mercury was an exception, because it evaporates easily at 500 °C (> 98% removal efficiency).

- Arsenic and chromium (after oxidation to Cr VI) were well extractable with NaOH due to their anionic character.

5. TECHNICO-ECONOMICAL EVALUATION

A rough estimate of cleanup costs for soils contaminated with heavy metals has been made for two methods that are considered most promising:
a. Classification with water or a weak solution of NaOH.
b. Extraction with acid in three steps, in combination with aeration of soil suspension and a separation step in which the very fine particles (< 20 µm) are removed.

Table 14 gives some relevant cost factors for installations based on these two methods. The roughly estimated costs for an installation with a capacity of 20 tonnes of soil per hour (annually approximately 30,000 tonnes) are:
- Classification: Dfl. 95.- to 210.- per tonne of soil (70.- to 110.- per tonne, excluding disposal costs of residual sludge).
- Extraction with acid: Dfl. 180.- to 290.- per tonne of soil.

These cost estimates have an accuracy of ± 40% and exclude costs such as transport of soil, temporary storage, purchase and setting the site for the installation, and analytical control.

The costs for classification are generally lower than those of a three-fold extraction with acid. The difference in costs increases, if less residual sludge is formed (less than 25% (m/m)) and/or if this sludge can be disposed of less expensively (less than Dfl. 250.- per tonne of sludge).

6. CONCLUSIONS

Only two out of nine soil samples could be cleaned sufficiently, that is below the A-value of the "Table of Trigger Values". Despite the large amount of experiments made under strongly varying and sometimes rather extreme conditions, it appeared that it was not possible to clean the other seven soil samples below the A-value.

The results obtained by extraction methods showed strong, variations, depending on the type of contaminant and the type of soil. It was, however, not possible to deduce a clear relation between the type of contaminant, the type of soil, and the removal efficiencies obtained.

Treatment methods that seem to be technically applicable for a majority of contaminated soils are:
- Classification (jet-sizing in an upflow column).
- Extraction with acid.
- Combination of classification and extraction with acid (if desired in combination with an aeration or oxidation step).

Nevertheless it is expected that in many cases neither the A-value nor a removal efficiency of more than 95% will be reached by any of the above methods.

An important advantage of classification over extraction with acid is its relatively low treatment costs. An important disadvantage is the formation of relatively large amounts of residual sludge which have to be disposed of properly. Therefore classification is most suitable for sandy soil.

7. LITERATURE

1. Meerjarenplan bodemsanering - kerngegevens en resultaten van scenario's. Bodembeschermingsreeks 51, Staatsuitgeverij, Den Haag (1986).
2. Informatievoorziening bodemsaneringsprojecten 1986.

Bodembeschermingsreeks 62, Staatsuitgeverij, Den Haag (1987).
3. J.W. Assink
   Extractive methods for soil decontamination; A general survey and review of operational treatment installations in J.W. Assink, W.J. van den Brink (eds), Contaminated Soil, Martinus Nijhoff Publishers, Dordrecht (1986).
4. Leidraad Bodemsanering
   Staatsuitgeverij, Den Haag (1983).
5. J.W. Assink, H.J. van Veen
   Extractieve reiniging van met zware metalen gecontamineerde grond.
   TNO-rapport nr. 85-07553 (1985).

Table 1: General characteristics of the collected soil samples

Sample	Main contaminants (guide metals)	Origin contaminant	Content of organic matter[1] %	pH-KCl[2]	Soil particles distribution (%)		
					< 20 μm	20-60 μm	60-2000 μm
I	Cr	Galvanic	3,0	5,1	6	6	88
II	Pb	Unknown (city centre)	2,9	8,1	12	5	83
III	Pb, Cu, Cd	Galvanic (sludge)	2,9	7,9	6	3	91
IV	Hg	Chlorine electrolysis	2,0	7,9	11	14	75
V	Pb, Zn, As	Pigments?	4,2	7,3	27	8	66
VI	Cd, Cr, Sn	Galvanic	3,1	4,5	7	8	85
VII	As	Pesticide (As_2O_3)	0,9	4,9	1	2	97
VIII	Pb, Cd	Galvanic	11,3	7,0	23	5	72
IX	Cu, As, Hg	Pesticide formulation	5,5	6,9	5	3	92

1) determined by weight loss of dry soil on heating at 500 °C for 2 hours
2) soil liquid ratio = 1 : 5; 1 M KCl-solution

Table 2: Characterization of soil samples: concentrations of heavy metals in the fraction smaller than 2 mm

Sample	Concentrations in mg/kg (dry matter)								
	Cu	Pb	Cr	Cd	Zn	Ni	Sn	As	Hg
I	45	130	<u>12400</u>	<0,1	20	12	-	-	-
II	73	<u>360</u>	24	<1	200	16	-	-	-
III	<u>1160</u>	<u>410</u>	19	<u>23</u>	420	280	-	-	-
IV	45	115	37	0,2	95	12	-	-	101
V	115	<u>870</u>	10	1,8	<u>4800</u>	29	-	<u>73</u>	-
VI	630	620	<u>900</u>	<u>43</u>	270	58	<u>430</u>	-	-
VII	3,3	11	8	<0,5	14	4,2	-	<u>63</u>	-
VIII	360	<u>1590</u>	28	<u>26</u>	480	27	68	-	-
IX	<u>1730</u>	280	27	1,6	1450	20	-	<u>205</u>	<u>3500</u>
A-value	50	50	100	1	200	50	20	20	0,5
B-value	100	150	250	5	500	100	50	30	2
C-value	500	600	800	20	3000	500	300	50	10

- not determined
Guide metals are underlined

Legend to the tables 3-11

A+B	: A followed by B
1*, 3*	: extraction in 1 respectively 3 steps
water	: classification with water instead of 0,1 N NaOH
g/kg	: dosage of NTA or other additives in grams per kilogram dry soil
sl	: amount of sludge (dry matter) separated by classification (% of dry soil sample)
USB	: previous scrubbing in ultrasonic bath
rol	: previous scrubbing in a vessel horizontally revolving on its axis

Table 3: Selected results soil sample I

Guide metal : Cr
Initial concentration (mg/kg) : 12400
A-value (mg/kg) : 100
B-value (mg/kg) : 250

Treatment method	Conditions	Final concentration (mg/kg) Cr	Removal efficiency (%) Cr
Basic routes:			
. HCl-extraction	1*; pH = 1;	12200	1
. NTA-extraction	1*; 100 g/kg	11300	9
. classification	USB; 19% sl	690	94
Other experiments:			
. NaOH-extraction	1*; pH = 13;	10400	16
. classification + oxidation + NaOH-extraction	water; USB; NaOCl 3*; pH = 12	580	95

Table 4: Selected results soil sample II

Guide metal : Pb
Initial concentration (mg/kg) : 360
A-value (mg/kg) : 50
B-value (mg/kg) : 150

Treatment method	Conditions	Final concentration (mg/kg) Pb	Removal efficiency (%) Pb
Basic routes:			
. HCl-extraction	1*; pH = 1;	47	87
. NTA-extraction	1*; 100 g/kg	93	74
. classification	20% sl	96	73
Other experiments:			
. classification + HCl-extration	water; USB; 1*; pH = 1	35	90
. HCl-extraction	3*; pH = 2,5	310	14

Table 5: Selected results soil sample III

```
Guide metal                    :  Pb    Cu    Cd
Initial concentration (mg/kg)  :  410   1160  23
A-value               (mg/kg)  :  50    50    1
B-value               (mg/kg)  :  150   100   5
```

Treatment method	Conditions	Final concentration (mg/kg)			Removal efficiency (%)		
		Pb	Cu	Cd	Pb	Cu	Cd
Basic routes:							
. HCl-extraction	3*; pH = 1; USB	27	76	<1	93	93	>96
. NTA-extraction	1*; 100 g/kg	48	310	<1	88	74	>96
. classification	water; USB; 10% sl	59	520	5,4	86	55	77
Other experiments:							
. NTA-extraction	1*; 20 g/kg	340	1070	12	17	8	48
. classification + HCl-extraction	water; USB 1*; pH = 1	48	96	1,4	88	92	94
. classification + HCl-extraction	water; USB 3*; pH = 1	39	63	0,3	90	95	99

Table 6: Selected results soil sample IV

```
Guide metal                    :  Hg
Initial concentration (mg/kg)  :  101
A-value               (mg/kg)  :  0,5
B-value               (mg/kg)  :  2
```

Treatment method	Conditions	Final concentration (mg/kg)	Removal efficiency (%)
		Hg	Hg
Basic routes:			
. HCl-extraction	1*; pH = 1	101	0
. NTA-extraction	1*; 100 g/kg	98	3
. classification	30% sl	27	73
Other experiments:			
. oxidation + HCl-extraction	aeration 1x; pH = 1	16	84
. NaOCl-extraction ($HgCl_4^{2-}$-complexes)	2*; pH = 6 15 g NaOCl/kg	26	74

Table 7: Selected results soil sample V

Guide metal : Pb Zn As
Initial concentration (mg/kg) : 870 4800 73
A-value (mg/kg) : 50 200 20
B-value (mg/kg) : 150 500 30

Treatment method	Conditions	Final concentration (mg/kg)			Removal efficiency (%)		
		Pb	Zn	As	Pb	Zn	As
Basic routes:							
. HCl-extraction	3*; pH = 1; USB	630	950	63	28	80	14
. NTA-extraction	1*; 100 g/kg	640	2750	71	26	43	3
. classification	USB; 38% sl	360	1850	22	59	62	70
Other experiments:							
. classification with acid	HCl; pH = 1; 40% sl	400	970	44	54	80	40
. extraction with oxidizing acid	0,05 M HNO$_3$ + HCl 3*; pH = 1;	590	770	42	32	84	42

Table 8: Selected results soil sample VI

Guide metal : Cr Cd Sn
Initial concentration (mg/kg) : 900 43 430
A-value (mg/kg) : 100 1 20
B-value (mg/kg) : 250 5 50

Treatment method	Conditions	Final concentration (mg/kg)			Removal efficiency (%)		
		Cr	Cd	Sn	Cr	Cd	Sn
Basic routes:							
. HCl-extraction	3*; pH = 1	720	15	410	20	65	5
. NTA-extraction	1*; 20 g/kg	860	28	410	4	35	5
. classification	USB; 19% sl	110	6,5	80	88	86	81
Other experiments:							
. classification + HCl-extraction	water; USB 1*; pH = 0,5	110	1,6	70	88	96	84
. oxidation + NaOH-extraction	Na OCl 1*; pH = 12	350	12	290	61	70	33

Table 9: Selected results soil sample VII

Guide metal : As
Initial concentration (mg/kg) : 63
A-value (mg/kg) : 20
B-value (mg/kg) : 30

Treatment method	Conditions	Final concentration (mg/kg) As	Removal efficiency (%) As
Basic routes:			
. HCl-extraction	3*; pH = 1	3,3	95
. NTA-extraction	1*; 20 g/kg	8,3	87
. classification	5% sl	2,0	97
Other experiments:			
. classification	water; USB; 4% sl	17	70
. HCl-extraction	1*; pH = 1	6,9	89

Table 10: Selected results soil sample VIII

Guide metal : Pb Cd
Initial concentration (mg/kg): 1590 26
A-value (mg/kg): 50 1
B-value (mg/kg): 150 5

Treatment method	Conditions	Final concentration (mg/kg)		Removal efficiency (%)	
		Pb	Cd	Pb	Cd
Basic routes:					
. HCl-extraction	3*; pH = 1	240	0,5	85	98
. NTA-extraction	1*; 20 g/kg	1070	15	33	42
. classification	water; USB	780	9,7	51	63
Other experiments:					
. classification with acid	HCl; pH = 1; USB; 29% sl	165	<0,3	90	>99
. classification + HCl-extraction	water; USB 3*; pH = 1	120	<0,5	92	>98

Table 11: Selected results soil sample IX

Guide metal   : Hg   As   Cu
Initial concentration (mg/kg): 3500   205   1730
A-value   (mg/kg): 0,5   20   50
B-value   (mg/kg): 2   30   100

Treatment method	Conditions	Final concentration (mg/kg)			Removal efficiency (%)		
		Hg	As	Cu	Hg	As	Cu
Basic routes:							
. HCl-extraction	3*; pH = 1	5400	34	1000	<0	83	42
. NTA-extraction	1*; 20 g/kg	3000	60	1250	14	71	27
. classification	USB; 11% sl	690	48	440	80	77	75
Other experiments:							
. oxidation + HCl-extraction	aeration 1*; pH = 1;	2700	81	780	23	60	55
. oxidation + HCl-extraction	H_2O_2 3*; pH = 1;	2800	102	250	20	50	86
. NaOCl-extraction	2*; pH = 6; 70 + 8 g/kg	330	59	650	91	71	62

Table 12 Best two cleaning methods for each soil sample

Soil sample		Best method	Removal efficiency
I	(Cr)	Classification	94%
		Classification + oxidation + NaOH-extraction	95%
II	(Pb)	HCl-extraction (3*)	87%
		Classification + HCl-extraction	90%
III	(Pb,Cu,Cd)	HCl-extraction (3*)	>93%
		Classification + HCl-extraction	90-99%
IV	(Hg)	Oxidation + HCl-extraction	84%
		NaOCl-extraction	74%
V	(Pb,Zn,As)	Classification	59-70%
		Classification with acid	40-80%
VI	(Cr,Cd,Sn)	Classificationn	81-88%
		Classification + HCl-extraction	84-96%
VII	(As)	Classification	97%
		HCl-extraction (3*)	95%
VIII	(Pb,Cd)	Classification with acid	>90%
		Classification + HCl-extraction	>92%
IX	(Hg,As,Cu)	Classification	75-80%
		NaOCl-extraction	62-91%

Table 13 Best experimental results of the three basic routes

Component (guide metals)	Removal efficiency (%)		
	HCl-extraction	NTA-extraction	Classification
Pb	<u>87</u>; <u>93</u>; 28; <u>85</u>	74; 88; 26; 33	73; 86; <u>59</u>; 51
Cd	>96; 65; <u>98</u>	>96; 35; 42	77; <u>86</u>; 63
Cu	<u>93</u>; 42	74; 27	55; <u>75</u>
Cr	1; 20	9; 4	<u>94</u>; <u>88</u>
As	14; <u>95</u>; <u>83</u>	3; 87; 71	<u>70</u>; <u>97</u>; 77
Hg	0; <0	3; 14	<u>73</u>; <u>80</u>
Zn	<u>80</u>	43	62
Sn	5	5	<u>81</u>

Note: The best results of the 3 basic routes are underlined for each guide metal and each sample.

Table 14: Relevant cost factors for two proposed installations (1984 prices)

	Treatment method	
	extraction with acid	classification
. capacity (tonnes/year)	30,000	30,000
. amount of residual sludge (% m/m)	approx. 10	10-40
. investment (plant cost in 10^6 Dfl) [1]	4-8	2-4
. depreciation and interest (10^3 Dfl/year)	2,500-5,000 [2]	900-1,800 [3]
. chemicals (10^3 Dfl/year)	700	< 200
. energy, water	p.m.	p.m.
. disposal costs residual sludge [4] (10^3 Dfl/year)	750	750-3,000
. labour costs, 5 persons (10^3 Dfl/year)	400	400
. maintenance	300	200
. other costs (overhead, profit, insurance etc.) (10^3 Dfl/year)	1,000	500
. total annual costs (10^3 Dfl/year)	5,500-8,500	2,900-6,100

1) rough estimate, excludes costs for purchase and setting the site.
2) depreciation in approx. 2 years.
3) depreciation in approx. 3 years.
4) approx. Dfl. 250.- per tonne of sludge.

STEAM STRIPPING ORGANIC COMPOUNDS FROM CONTAMINATED WATERS.

F.H.M.M. Langen, P.G. Paul and R. v. Booren
COMPRIMO ENGINEERS & CONTRACTORS, P.O. Box 4129, 1009 AC Amsterdam,
The Netherlands

ABSTRACT
In this paper the technique of treatment by steam stripping is evaluated to determine the extent to which organic priority pollutants can be stripped from contaminated waters. A steam stripper very effectively removes certain organic priority pollutants because they are moderately volatile and their activity coefficients are very large. The benefits and drawbacks of the steam stripping technique for treatment of polluted groundwater are discussed in comparison with two widely used techniques: adsorption on activated carbon and air stripping.

1. INTRODUCTION
   Last years a lot of soil/groundwater polluted sites were discovered in the Netherlands. Based on inventarisations one can expect several hundreds of groundwater reclamations in the next five years (1,2). Besides in groundwater, priority pollutants are also found in industrial waste water, surface water and drinking water. The concern about priority pollutants has intensified the need for gathering data on their occurrence in industrial effluents and for evaluating their treatability.
   Two widely used techniques in the Netherlands for removal of organic pollutants from groundwater and drinking water are adsorption on granular activated carbon (GAC) and air stripping. Experiences from finished clean-up sites have shown the following drawbacks:
- The activated coal costs for treating heavily polluted water by GAC-adsorption are often very high. The saturated carbon has to be transported to a waste treatment facility or has to be regenerated.
- In case of air stripping attention has to be paid to the air emissions. Often the waste air has to be treated by activated carbon or biofiltration.

Many industries use steam stripping to remove organic pollutants from waste water. In refineries the steam stripping technique for removal of hydrogensulfide (H_2S) and ammonia (NH_3) from so called sour water is a normal practice. In this paper the technique of steam stripping is evaluated to which extent organic priority pollutants can be stripped from (ground)water. The benefits, costs and drawbacks are discussed.

2. SCOPE
   A large number of compounds can be present in contaminated (ground)-water (1,2):
   a. heavy metals
   b. cyanides
   c. monocyclic aromatic hydrocarbons incl. phenolics

d. polycyclic aromatic hydrocarbons (pah's)
e. chlorinated hydrocarbons
f. pesticides
g. oil/petrol

From inventories (1,2) three levels of organic pollution can be distinguished. In table 1 these levels are given for a number of frequently found pollutants.

TABLE 1. Levels of concentration of organic pollutants in groundwater (mg/l)

	Extreme	Serious	Weak
aromatics, total	100	10	0.060
. benzene	20	2	0.012
. toluene	20	2	0.012
. xylenes	20	2	0.012
. ethylbenzene	20	2	0.012
. propylbenzene	20	2	0.012
phenolics, total*	10	3	0.12
. phenol	5	1.5	0.06
. cresol	5	1.5	0.06
pah's two rings, total	50	5.50	0.10
. naphthalene	25	2.75	0.05
. methylnaphthalene	25	2.75	0.05
pah's three or more rings, total	10	0.80	0.050
. anthracene	0.073**	0.073	0.017
. phenanthrene	1.6**	0.27	0.017
. pyrene	0.008**	0.008	0.008
chlorinated hydrocarbons, total	1000	10	0.10
. choroform	200	2	0.02
. carbon tetrachloride	200	2	0.02
. 1,2 -dichloro ethane	200	2	0.02
. 1,1,2 - trichloro ethane	200	2	0.02
. 1,2 - dichloropropane	200	2	0.02

* In practice phenolics are specified by the phenol-index, which gives the minimum concentration of total phenolics present, expressed as mg/l phenol.
** Solubility at 20°C

In table 2 present indicative guidelines are given for the quality of groundwater.

TABLE 2. Indicative guidelines for effluent quality (mg/l).

	A-value	B-value	C-value
aromatics, total	0.0010	0.030	0.100
. benzene	0.0002	0.001	0.005
. toluene	0.0005	0.015	0.050
. xylenes	0.0005	0.020	0.060

TABLE 2. (cont'd)

	A-value	B-value	C-value
. ethylbenzene	0.0005	0.020	0.060
. phenolics	0.0005	0.015	0.050
pah's, total	0.00020	0.010	0.040
. naphthalene	0.00020	0.007	0.030
. anthracene	0.00010	0.002	0.010
. phenanthrene	0.00010	0.002	0.010
. pyrene	0.00002	0.001	0.005
Chlorinated hydrocarbons			
. alifatic chloro-alkanes, total	0.001	0.015	0.070
. alifatic chloro-alkanes, individual	0.001	0.010	0.050

A-value: indicative Dutch guideline for no groundwater clean-up
B-value: indicative Dutch guideline for further investigation
C-value: indicative Dutch guideline for groundwater clean-up

For evaluation of the benefits and costs of steam stripping the following flows and duration of treatment are considered:

Flow . low   5 m³/h   Duration . short 0.5 year
     . mean 25 m³/h   . middle 2   year
     . high 100 m³/h   . long   5   year

3. STEAM STRIPPING

The degree of stripping of a component from water depends on the relative volatility of this component, which is defined by the activity coefficient of this component in water and its partial pressure:

$$V_{i,w} = \frac{K_i}{K_w} = \frac{J_i}{J_w} * \frac{Pd,i}{Pd,w} \quad \text{and}$$

$$K_i = \frac{Y_i}{X_i} = J_i * \frac{Pd,i}{Ptot}$$

where
$V_{i,w}$ = relative volatility between component i and water
K_i = vapor-liquid equilibrium ratio of component i
K_w = vapor-liquid equilibrium ratio of water
J_i = activity coefficient of component i
J_w = activity coefficient of water (± 1)
Pd,i = partial pressure of component i
Pd,w = partial pressure of water
Y_i = mole fraction of component i in vapor at equilibrium
X_i = mole fraction of component i in liquid at equilibrium
$Ptot$ = total system pressure

The activity coefficient is a correction factor for the non-ideal behaviour of a component in the liquid phase and depends merely on the liquid composition and the temperature. Since most multicomponent vapor-liquid equilibrium data are often unavailable or unreliable, there are several methods developed for calculating component activity coefficients.

A recent developed method is the UNIFAC-method (5,6,7). UNIFAC stands for UNIversal Functional group Activity Coefficients. With this method one is able to define the vapor-liquid equilibrium on basis of the molecule structure of components in a liquid mixture. The basic idea is that whereas there are thousands of chemical compounds, the number of functional groups which constitute these compounds, is much smaller (p.e. OH, CH_3, C=O, C=C, CCl, CN).
The group contribution method assumes that a physical property of a fluid is the sum of contributions made by the functional groups of the molecules. Predicted coefficients agree well with those obtained from experimental vapor-liquid data. In most typical cases, predicted coefficients at infinite dilution deviate less than 20% from measured results.

4. STEAM STRIPPER EFFICIENCIES

A computer programme was developed to calculate multi-component vapor-liquid equilibrium data and theoretical removal efficiencies in a stripping column using the UNIFAC-method. The degree of stripping of an organic component from water depends on the vapor-liquid equilibrium ratio K, the number of theoretical units (NTU) and the amount of stripping steam:

$$f_i = \frac{S^{n+1} - S}{S^{n+1} - 1}$$

where
f_i = removed fraction of component i
S = stripping factor (= K_i * G/L)
K_i = vapor-liquid equilibrium ratio of component i
G = vapor flow (kmol/h)
L = liquid flow (kmol/h)

The required amount of steam is determined from the heat balance by summing the heat required to bring the feed to ca. 100°C, to vaporize the organic pollutants and to saturate the vapor with steam. The major portion of the steam is required for heating the feed. The heating steam demand can be reduced by exchange of heat between the feed and the effluent stream. The condensed topproducts can be recovered and have to be disposed of. No air pollution or odour nuisance will exist. In table 3 the stripping factors and removal rates are given for a fixed number of theoretical units and a fixed amount of stripping steam.
In table 4 the optimum theoretical removal rates in a steam stripper column are given.

TABLE 3. Theoretical steam stripper efficiencies for extreme polluted groundwater.

Component	activity coefficient	K-value	stripping factor	removal rate (%)
benzene	1173	1993	2.0	99.78
toluene	4977	3534	3.5	99.99
o-xylene	20080	5163	5.2	99.99
ethylbenzene	12720	4208	4.2	99.99
propylbenzene	35460	5781	5.8	99.99
phenol	37	2	0.1	0.21
o-cresol	149	7	0.1	0.64

TABLE 3. (cont'd)

Component	activity coefficient	K-value	stripping factor	removal rate (%)
naphthalene	49260	1244	1.2	95.87
m-naphthalene	192700	1874	1.9	99.67
anthracene	1923000	980	1.0	87.41
phenanthrene	1923000	763	0.8	73.33
pyrene	13230000	15086	14.9	99.99
chloroform	417	1164	1.2	94.52
carbontetrachloride	3458	6224	6.3	99.99
1,2-dichloro ethane	317	491	0.5	48.75
1,1,2-trichloro ethane	1272	820	0.8	78.14
1,2- dichloropropane	995	1062	1.1	91.24

Design steamstripper:
Average temperature in column   106°C
Number of theoretical units   8
Stripping steam quantity 1 kg/ton feed
Level of pollution: extreme (see table 1)

TABLE 4. Optimum theoretical steam stripper removal rates for extreme polluted groundwater

Component	activity coefficient	K-value	NTU	strip. ping factor	stripping steam kg/ton feed	effluent quality	removal rate (%)
benzene	1173	1993	8	4.0	2.0	C-Value	99.98
toluene	4977	3534	8	3.5	1.0	B-Value	99.96
o-xylene	20080	5163	6	5.2	1.0	B-Value	99.99
ethylbenzene	12720	4208	6	4.2	1.0	B-Value	99.98
propylbenzene	35460	5781	6	5.8	1.0	B-Value	99.99
phenol	37	2	20	0.2	95.5	-	20.0
o-cresol	149	7	20	0.6	95.5	-	61.6
naphthalene	49260	1244	8	2.5	2.0	C-Value	99.95
m-naphthalene	192700	1874	8	2.6	1.4	B-Value	99.97
anthracene	1923000	980	8	1.3	1.4	B-Value	97.26
phenanthrene	1923000	763	8	1.7	2.3	C-Value	99.45
pyrene	13230000	15086	4	4.5	0.3	A-Value	99.81
chloroform	417	1164	8	2.3	2.0	C-Value	99.98
carbontetrachloride	3458	6224	8	6.3	1.0	C-Value	99.99
1,2-dichloro ethane	317	491	8	3.9	8.0	C-Value	99.99
1,1,2-trichloro ethane	1272	820	8	3.2	4.0	C-Value	99.99
1,2- dichloropropane	995	1062	8	4.2	4.0	C-Value	99.99

Average temperature in column: 106°C

From these tables it can be seen that most of the considered components can be stripped at high removal rates with a moderate stripping steam requirement. This doesn't apply for phenol and cresol.

In general the total steam demand, using a feed-effluent heat exchanger, amounts, except for the phenolics, to between 10 and 20 kg steam per ton feed. Steam stripping can be carried out effectively in packed columns to treat a large number of organic pollutants from waste waters to concentrations, ranging from 100 pbb to 1 ppb in the effluent:
. monocyclic and polycyclic aromatic hydrocarbons
. chlorinated aromatic hydrocarbons
. apolar halogenated alkanes and alkenes
. poly chlorinated biphenyls (PCB's)
. hexachlorocyclohexanes (HCH's)

Moderately steam strippable components are acoleïn, acrylonitril, nitro aromatic hydrocarbons and some phenolics, alcohols, haloethers, aldehydes and ketones. Slightly strippable components are polar phenolics, chlorinated phenols, nitrophenols, carboxylic acids (humic acids), phthalate esters, benzidines, amines and hydrazines.

5. COMPARISON WITH GAC-ADSORPTION AND AIR STRIPPING

5.1 Steam stripping versus GAC-adsorption

A steam stripper removes very effectively most components considered. However for a number of components the steam stripper doesn't meet the severe effluent requirements (A- and B-values). Therefore a polishing step like GAC-adsorption has to be added. A cost evaluation has to be made for two systems which meet the A-or B-values:
1. GAC-adsorption
2. Steam stripping followed by GAC-adsorption.

The cost of a groundwater clean-up operation is strongly dependent on the groundwater flow to be treated, the type and the concentration of contaminants, the suspended solids and the oil content (primary treatment by a gravity separation technique) and the duration of the operation. In an evaluation only the costs for steam stripping and GAC-adsorption have been taken into account. In general we find that groundwater flows of above 25 m³/h, which contain 100 mg/l or more of good steam strippable components could be treated economically by steam stripping.

5.2 Steam stripping versus air stripping

The following advantages and disadvantages of steam stripping can be mentioned with regard to air stripping.

advantages   : . higher removal rates for moderately volatile organic
                    components.
                  . no air pollution.
disadvantages   : . higher energy requirement (steam)
                  . more complicated equipment

In many publications air stripping treatment costs are given without costs for treatment of the stripped air. The additional costs for air purification can be considerable (Hfl. 0.5-5 per 1000 Nm³ waste air) and have to be taken into account.

REFERENCES

1. VROM/RIZA Purification of groundwater from abandoned gaswork sites (in Dutch Bodemreeks no. 53), August 1985

2. VROM/RIZA Groundwater treatment at soil reclamation (in Dutch), October 1986

3. Provinciale waterstaat Utrecht/Comprimo B.V. Feasibility study groundwater treatment for soil reclamation project Cindu II (in Dutch) July 1984

4. EPA, A screening procedure for toxic and conventional pollutants in surface and groundwater, part 1 revised 1985, EPA-600/6-85/002a

5. Freedeslund A., Jones R.L. and Prausnitz J.M. Group-contribution estimation of activity coefficients in nonideal liquid mixtures. AICHE 21, 1975, 1086

6. Freedeslund A., Gmehling J. and Rasmussen P. Vapor-Liquid equilibria using UNIFAC Elsevier 1977

7. Tiegs D, Gmehling J., Rasmussen P. and Fredenslund A. Vapor-Liquid equilibria by UNIFAC group contribution 4. Revision and Extension. Ind. Eng. Chem. Res. 1987, 26, 159.

EXTRACTIVE CLEANING OF HEAVY METAL CONTAMINATED CLAY SOILS

B.J.W.Tuin, M.M.G. Senden, M.Tels

ABSTRACT
The purpose of our research is to develop extractive techniques that are capable of removing heavy metals from contaminated clay soils. Six different Dutch clays were artificially polluted with five different heavy metals. The polluted clays were subsequently extracted three times with a 0.1 M HCl solution. 90 to 99 % of the Cu, Pb and Zn present could be removed. Nickel proved extractable for 80 to 90 % but chromium(III) was hardly extractable (0-25 %). Cr could be removed to a large extent with a hypochloric acid solution at pH=8. Solid-liquid separation was tested in hydrocyclones. Chemical coagulation and flocculation were investigated. A very good separation was possible with several types of flocculants. This flocculation did not affect the extraction results. Based on the batch extraction and separation results a continuous test installation was built on lab scale. Extraction can be carried out through applying stirred tanks in series or countercurrent extraction in a sieve plate column. Flocculants are used to separate the clay and the extracting agent.

1. INTRODUCTION
Since 1980 many locations in the Netherlands were found to be contaminated with a large variety of pollutants including organics, cyanides and heavy metals. Extraction is a potentially attractive method for removing heavy metal pollutants. Commercial scale extraction plants for decontaminating sandy soils are already being operated in the Netherlands (1). Soils that contain clay cannot easily be treated so far because of their high adsorption capacity and the small size of the clay particles.
The Eindhoven University of Technology and the Netherlands Organization for Applied Scientific Research (TNO) are carrying out a joint research program to develop extractive techniques for clay soils contaminated with heavy metals. Generally, an extraction process consists of three basic parts:
1. extraction of the pollutant from the soil;
2. separation of the cleaned clay from the liquid phase;
3. removing heavy metals from and/or recycling of the extracting agent.

In our research project we concentrate on the first two points, because methods for removing heavy metal ions from watery solutions are well-known from waste water treatment technology (2). We have selected a combined precipitation and flocculation unit to clean the liquid phase in our test installation.
The extraction and separation indicated in points 1 and 2 above were first tested in small scale batch experiments with artificially polluted clay soils: the extraction results are discussed in section 2 and the separation experiments in section 3. Based on the results of these exper-

iments possible treatment schemes were developed for a continuous extraction of metals from clay soils. A laboratory scale continuous installation was then built in which different cleaning routes could be tested. This test installation is described in section 4. Experiments are currently performed in this installation, starting again from artificially polluted clay soils. After these experiments we will investigate the extractive treatment of clay soils from several contaminated sites.

2. BATCH EXTRACTION EXPERIMENTS
2.1. Introduction and experimental methods

Many experimental studies have been performed on the adsorption and desorption (extraction) of heavy metals from soils (3,4,5). It is well-known that metals can be leached out of soil by inorganic and organic acids (5,6,7). We tested the extraction with hydrochloric acid of five heavy metals (Zn(II), Pb(II), Cu(II), Ni(II) and Cr(III)) from six Dutch clay soils. Some properties of these clay soils are given in Table 1. The contaminated clay soils were prepared through adsorption experiments. The clay was contacted with a metal salt solution during 24 hours in a stirred beaker. Thirty contaminated soils were prepared in this way that have metal concentrations of approx. 5000 ppm (mg/kg). A suspension of 2.5 g of this clay soil in 50 ml of 0.1 M HCl solution was then stirred during one hour at room temperature. The pH was kept constant at 1.0 The partially desorbed clay material was centrifuged and dried at the air. Subsequently, the clay was extracted two more times in a 0.1 M HCl solution. These extractions lasted only 30 minutes and were carried out on a shaking machine. Hardly any chromium could be extracted with HCl. Therefore, the Cr contaminated clay soils were also extracted with several other extracting agents including: EDTA, dithionite/citrate mixture,

TABLE 1. Characteristics of six Dutch clay soils

type of clay[††]	Rn	Wa	Ma	Bu	Me	Wi
pH H_2O	7.6	7.6	6.8	7.7	8.0	7.3
pH KCl	7.2	7.2	6.2	7.0	7.3	6.6
Specific density ($10^3 kg.m^{-3}$)	2.60	2.58	2.55	2.59	2.64	2.49
Fraction clay (w% < 2 μm)	42	32	34	36	29	50
d-50 (μm)	3.4	11	9.5	10	21	1.9
Specific surface[†] (m^2/gram)	87	67	81	89	58	101
Organic carbon(%)[Π]	8.1	8.2	6.7	3.8	1.5	6.3
CEC (meq/100 g)[§]	22.8	16.3	21.8	17.6	9.9	22.6
Carbonate content[§§] as $CaCO_3$ (%)	8.3	11.3	0.4	6.8	4.1	0.0

[†] Determined by glycerol adsorption
[Π] Determined by oxidization with dichromate
[§] Determined by $BaCl_2$/Mg adsorption
[§§] Determined volumetrically by the method of Scheibler.
[††] The six clay soils are abbreviated throughout this article as Rn for Rhine clay, Wa for Waal, Ma for Maas, Bu for Burum, Me for Menaldum and Wi for Winsum.

oxalic acid, hydrogen peroxide and sodiumhypochlorite solutions.
The metal concentrations of the supernatant liquids were measured using atomic absorption spectrometry in all cases. The amount of metal desorbed could be calculated from the liquid concentrations. In addition, the metal concentration in the clay was often determined by digesting the dried clay material.

2.2. Results and discussion

The results of the three subsequent desorption experiments are shown in Table 2. Large amounts of Cu, Pb and Zn desorb from the clays in the first desorption step (approx. 80-95 % of total metal initially adsorbed). The percentage of desorbed Cu, Pb or Zn is higher for the clay soils that contain carbonates than for the ones that do not (Winsum and Maas). So it seems that the clays that contain carbonates initially adsorb more metal but that this metal fraction desorbs relatively easily under acidic conditions. When the metal is not adsorbed to the clay surface but is present as a carbonate precipitate it is clear that the metal carbonate will dissolve relatively easily in the HCl solution.

Hardly any chromium(III) desorbs (Table 2) while nickel takes an intermediate position (20-60 % desorption in the first step). Harter (7) and Biddappa (8) also found that the extraction of Ni by means of 1 M or 0.01 M acid solutions was smaller than that of Pb, Cu and Zn. Very low extraction efficiencies for Cr were also measured by Grove and Ellis (9).

A sodiumhypochlorite solution was found to be the best extracting agent for Cr(III) of the ones that were tested. Optimum results were achieved at pH=8 and a ClO^- concentration of approx. 0.072 mol/liter: more than 85% of original present Cr could be removed in three extraction steps of 15 minutes (Table 3). Burum clay showed a different behavior.

The rate of extraction was also measured for several clay-metal combinations: Winsum polluted with Pb and Zn, Rhine polluted with Pb and Ni and Maas polluted with Cu. Generally, the first part of the extraction was rather fast (until about 70-90 % of the end concentration was reached), while the second part was rather slow. The combination Rhine clay polluted with Ni was an exception: the extraction was rather slow over the entire period. It sometimes took four hours before the end concentration was reached. The transition from fast to slow extraction took place after 15 to 30 minutes depending on the metal-clay combination. The second and third extraction showed the same behavior but the transition from fast to slow took place after 5 to 10 minutes.

The chromium extraction with ClO^- showed the same type of time dependence.

From these preliminary experiments, we concluded that it would be worth while to test HCl and ClO^- extraction in continuous stirred tank reactors with mean residence times of approx. 15 to 30 minutes.

3. SEPARATION OF CLAY AND EXTRACTING AGENT

3.1. Introduction and experimental methods

After the metals have been extracted from the clay the resulting suspension must be separated into a solid and a liquid phase. Many methods are known to separate solids from liquids (10). The small size of the clay particles (Table 1) makes simple sedimentation or filtration impracticable. Sedimentation can be accelerated by the use of a centrifugal field (as in a hydrocyclone) or by coagulation or flocculation of the particles to larger aggregates (flocs).

TABLE 2. Results of desorption experiments at pH=1 (in hydrochloric acid)

Metal	Clay	c_{k0} † (ppm)	c_{k3} § (ppm)	% metal removed in extraction no.			
				1 (%)	2 (%)	3 (%)	total (%)
Zn	Rn	5040	140	90	69	12	97
	Wa	5150	241	89	59	10	96
	Me	5220	28	94	85	38	99
	Bu	4670	59	89	84	28	99
	Ma	3800	330	75	63	9	91
	Wi	4170	300	80	60	10	93
Pb	Rn	5050	87	89	79	26	98
	Wa	4960	180	80	74	27	96
	Me	4980	97	91	74	21	98
	Bu	4990	180	86	69	19	96
	Ma	4970	240	75	76	21	95
	Wi	5030	170	85	73	15	96
Cu	Rn	5340	-	93	-	-	-
	Wa	5280	54	93	76	41	99
	Me	5210	210	90	53	15	96
	Bu	5110	160	87	68	21	97
	Ma	3870	390	75	55	9	90
	Wi	4090	110	87	75	15	97
Ni	Rn	4590	1370	18	60	8	70
	Wa	4470	870	34	65	15	81
	Me	4350	790	36	62	25	82
	Bu	4230	560	43	72	18	87
	Ma	3890	690	56	57	7	82
	Wi	3750	340	59	74	16	91
Cr	Rn	5000	4900	-	7	2	2
	Wa	5280	5300	-	3	1.5	-
	Me	5170	5720	-	1.5	1.1	-
	Bu	5250	5080	-	4	1.5	3
	Ma	4000	3110	5	14	5	22
	Wi	3800	2860	3	17	6	25

† c_{k0} is original metal concentration in the clay
§ c_{k3} is metal concentration in the clay after three extractions.

We have tested both of these methods. Hydrocyclones of 25 and 75 mm diameter were fed with clay suspensions of different dry matter content and at different inlet pressures.

Flocculation of the clay was studied in jar tests (10,11,12). Several flocculants and two coagulants ($FeCl_3$ and $Al_2(SO_4)_3$) were used.

The jar test consisted of 2 minutes of rapid stirring (200 rpm) during which the flocculant solution was added, followed by 5 minutes of slow agitation (50 rpm). A sample was taken after 1 minute of sedimentation.

TABLE 3. Results of threefold Cr(III)extraction in 0.072 mol/l hypochlorite solution at pH = 8

Clay	c_{k0} † (ppm)	c_{k3} § (ppm)	% metal removed in extraction no.			
			1 (%)	2 (%)	3 (%)	total (%)
Rn	5000	41	86	77	74	99
Wa	5000	219	80	63	40	96
Me	5400	<30	>99	-	-	>99
Bu	4365	1130	63	24	9	74
Ma	3800	562	69	39	23	85
Wi	4000	215	83	55	26	95

† c_{k0} is original metal concentration in clay
§ c_{k3} is metal concentration in clay after 3rd extraction

The dry matter content of the supernatant was measured by determining the turbidity of the sample in an Elko spectrofotometer. Each flocculant was tested in two suspensions of clay in hydrochloric acid (pH=1 and pH=3). We expected that acidities in this range would occur at the end of the clay extraction.

3.2. Results and discussion

Separation of the clay from the liquid phase by means of a hydrocyclone was not possible. The cut size (10) was in the range of 10 to 20 μm depending on conditions. Although it was not possible to separate the small clay particles, these experiments showed that hydrocyclones could be used to split up the soil suspension in several fractions before the extraction process. This might be very useful when the fraction below 10-30 μm is more strongly contaminated than the coarse fraction of a clay containing soil.

The separation by coagulants was not succesful. The idea was to reduce the charge of the negatively charged clay particles by multivalent ions like Al^{3+} or Fe^{3+}. The particles could then approach each other closely and aggregate. The clay particles may have no net negative but rather a net positive charge in the acidic solution due to H+ adsorption.

However, the addition of flocculants caused a very good separation between the flocculated clay material and the hydrochloric acid solution. The optimum separation was determined by the minimal dry matter content of the supernatant liquid in the jar test. The minimal dosage for an optimum separation was about the same for several types of flocculants.

The optimum dosage was 10-25 mg flocculant/kg clay. The optimum dosage resulted in a dry matter content in the supernatant liquid of less than 14 mg/l (detection limit).

The optimum dosage at pH=3 was slightly higher than that at pH=1. Probably the H^+-ions have already a coagulating effect on the clay. The sedimentation rate without adding flocculant was also higher in a 0.1 M HCl solution than under neutral conditions.

A flocculation unit was designed for the continuous installation. It consists of two stirred cells in series followed by a sedimentation vessel. The flocculant is added in the first cell. The second cell is used for floc growth.

4. TEST INSTALLATION FOR CONTINUOUS EXTRACTION

A laboratory scale continuous installation was built in which different extraction routes could be tested. Two basic process schemes are applied:
1. extraction in two or more stirred reactors each followed by a flocculation unit as described above (see Figure 1);
2. extraction in a countercurrent sieve plate column preceded by a flocculation unit (see Figure 2).

Process scheme 1 is built up from more simple equipment and can be operated more easily than the column process scheme. The second extraction process scheme was chosen because operation in a countercurrent process is expected to be more effective than in stirred tanks in series.

To reach an effective extraction in treatment scheme 1 it is necessary that the extracted metals are not re-adsorbed onto the clay aggregates or complex with the flocculant in the flocculation unit. Experiments in stirred cells and in the installation on Cu-polluted Maas-clay and Pb-polluted Winsum clay showed that the extraction result was not influenced by the subsequent flocculation. These experiments proved also that it was possible to break up the flocs and extract the clay further in a second stirred tank reactor.

It proved possible to flocculate the clay material more than once: flocculation after the second extraction tank caused no problems. The two introductory experiments with the artificially polluted clays resulted in an extraction of approx. 95 % of the original Cu and Pb present. Therefore, further experiments are currently being planned and performed.

Operation in the countercurrent column requires aggregation of the very small clay particles to prevent elutriation of the clay over the top of the column. For this reason a flocculation unit precedes the column and flocculated clay enters at the top. On the other hand, the sedimentation rate of the aggregates is so high that sieve plates are required to increase the residence time of the clay in the column.

The extraction conditions and the hydrodynamics in the column are totally different from those in the stirred tank situation: the polluted clay is present as an aggregate and only a mild mixing is possible or the floc will break up. Extraction of flocculated clay under these conditions was tested in a model compartment of the column with very slow agitation only (20 rpm).

Copper was extracted from Winsum clay for two hours: 72% of the original Cu present could be extracted in this way (a decrease of the metal concentration in the clay from 1940 to 540 ppm). This result shows that further extraction experiments are worth trying in a countercurrent column under the same conditions. The process parameters will be varied to optimize the extraction result of the installation.

Other process routes can of course be used with the installation in its current set-up. It is possible for example to combine the column and one or two of the stirred tank reactors. The column can also be operated without a preceding flocculation: the elutriated clay suspension from the top of the column should then be flocculated. These types of processes will be tested in the future.

The polluted extracting agent from the sedimentation tanks and from the column is currently treated by a combined neutralisation and flocculation. The metal precipitate in the form of a hydroxide sludge remains at the end of the process as chemical waste that must be disposed of. Other ways of treating the extracting agent that may be combined with recycling will be tested in the future.

LITERATURE

1. Assink J.W., van den Brink W.J. (editors), Contaminated soil. Martinus Nijhoff Publishers, 645-667, 1985.
2. Cushnie G.C. Jr., Electroplating wastewater pollution control technology. Noyes Publications, Park Ridge, New Jersey, 1985.
3. Harmsen K., Behaviour of heavy metals in soils. (Thesis Wageningen), Pudoc, 1977.
4. Farrah H., Pickering W.F. The affinity of metal ions for clay surfaces. Chem.Geol. $\underline{28}$, 55-68, 1980.
5. Kiekens L., Adsorption phenomena of heavy metals in soils (Dutch) (Thesis Gent), 1980.
6. Farrah H., Pickering W.F., Extraction of heavy metal ions sorbed on clays. Water, Air, Soil Pollut. $\underline{9}$, 491-498, 1978.
7. Harter R.D., Effect of soil pH on adsorption of lead, copper, zinc and nickel. Soil Sci.Soc.Am.J. $\underline{47}$, 47-51, 1983.
8. Biddappa C.C., Adsorption, desorption, potential and selective distribution of heavy metals in selected soils of Japan. J.Environ. Sci.Health, part B $\underline{B16}$, 511-528, 1981.
9. Grove J.H., Ellis B.G., Extractable chromium as related to soil pH and applied chromium. Soil Sci.Soc.Am.J. $\underline{44}$, 238-242, 1980.
10. Svarovsky L. (Ed.), Solid-liquid separation, Butterworths, London, 1981.
11. Akers R.J., Flocculation, J.Chem.Eng.Services, London, 1975.
12. Packham R.F., The laboratory evaluation of polyelectrolyte flocculants, Br.Polym.J., $\underline{5}$, 305-315, 1972.

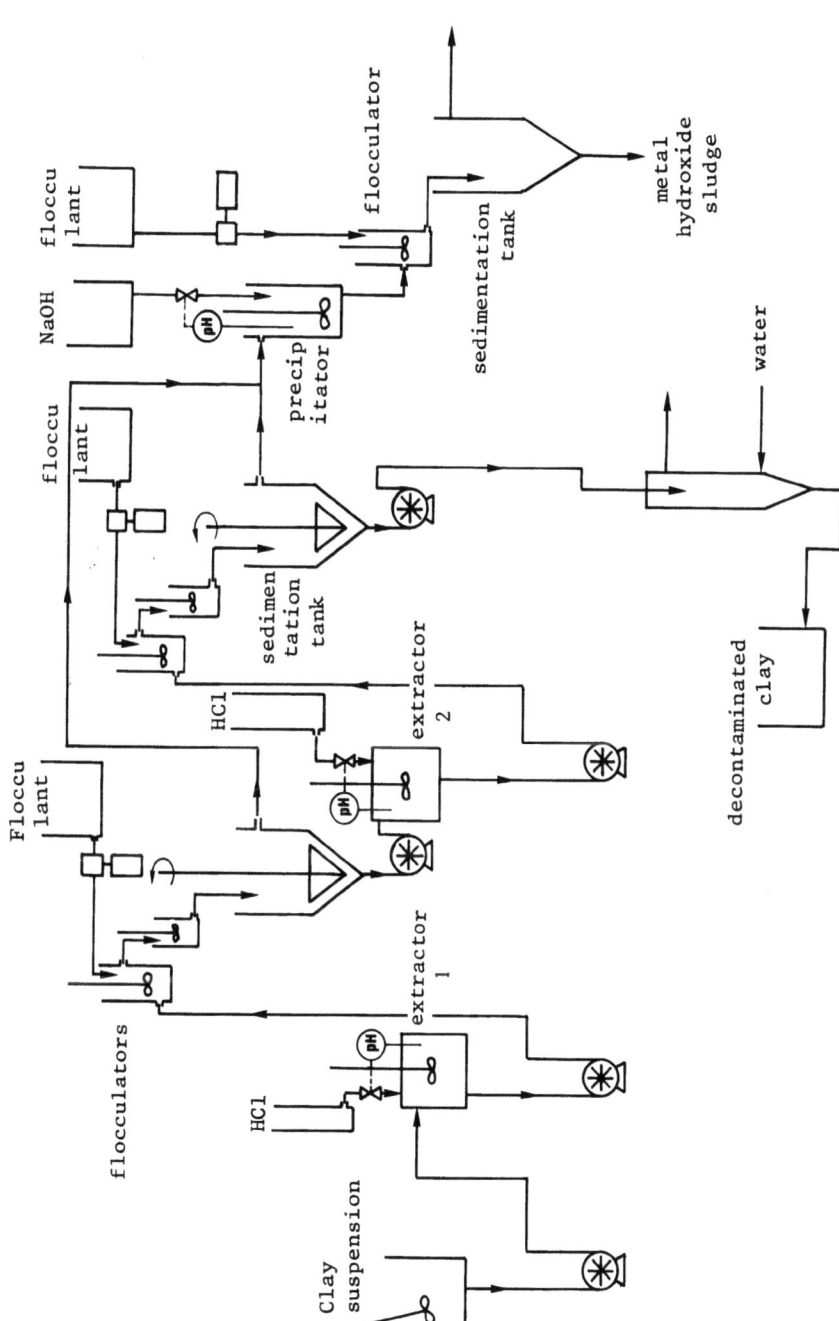

Figure 1. Process scheme with extractors in series

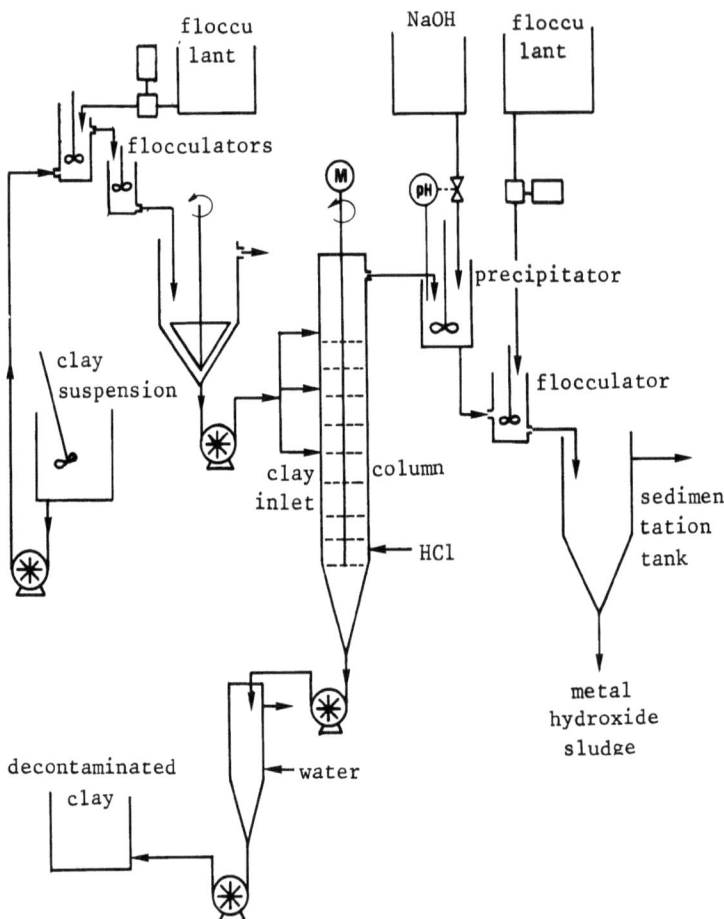

Figure 2. Process scheme with countercurrent column

CHARACTERIZATION AND REMIDIATION OF COKING PLANT SITES IN THE
RUHR TERRITORY

H. Koesters* and H. Spittank**

* ZENIT GmbH, Dohne 54, D-4330 Mülheim a.d. Ruhr
** CLAYTEX CONSULTING, Waidmühlenweg 1, D-5176 Inden

1. INTRODUCTION
ZENIT Zentrum in Nordrhein-Westfalen für Innovation und
Technik GmbH, based in Mülheim a.d. Ruhr, was established for
the promotion of economic activity in technical fields. The
major areas of activity of ZENIT are consultancy services and
the initiation of pilot projects in the areas of computer
science, production engineering and environmental engineering.
  In mid 1985, ZENIT made proposals to the State of North
Rhine-Westphalia concerning pilot projects for the regeneration of industrial waste land in order to promote the urgently
required development of processes for the reclamation of
polluted soil.
  It is now generally accepted that the potential and
limitations of such processes can only be tried and tested in
practice. Until large scale practical tests have been made, we
believe that there is insufficient basis for a public
discussion of the financial requirements and the financial
possibilities for soil reclamation.
  Both from the point of view of technological development
and from the point of view of environmental policies, there is
therefore an urgent need for pilot projects to develop
processes which have previously only been proposed
theoretically or tested on a bench scale to the point where
they can be applied in practice. Such tests would also be
required to assess both the practical usefulness of the
processes and to establish a cost framework. Experience has
shown that the best way of distinguishing between unrealistic
proposals and processes which are effective in practice is to
approach a specific project.

2. AIMS OF THE PILOT PROJECT
  Against this backdrop, the aims of the soil reclamation
pilot projects proposed by ZENIT are as follows:
- to define the objectives of soil reclamation in the light of
  problems posed by specific areas,
- to create a basis for the practical testing and improvement
  of innovative environmental engineering processes on the
  basis of these objectives,
- to test combinations of various possible solutions, and
- to ensure that devolopment and testing are adequately
  financed.
  As these aims indicate, the major objective of the proposed
pilot project is to promote technological development.

Various cities in the Ruhr area have already made available test areas which pose quite varied pollution problems. In view of the importance of disused coking plant sites in the Ruhr area, it was decided to approach a site of this type first.

3. COKING PLANT SITES IN THE RUHR AREA

The area selected for the plant project is the site of a coking plant which was built in the early years of this century. Initially, the capacity of the plant was approx. 86,000 tonnes per year, but this had been increased to 340,000 tonnes per year by 1954. The plant was finally closed down in 1966. The facilities installed were demolished down to 1 m below ground level during the course of the 1970's. The foundations of the plant therefore still remain in the ground. The test area is about 12,000 m^2.

There is probably no need to go into the specific pollution problems encountered at coking plant sites. Various typical pollutants directly connected with the by-products of low temperature carbonization processes are usually encountered (1, 2).

These include:
- various specific heavy metals,
- anorganic products such as ammonium, cyanides, rhodanides and sulphides, resulting from the decomposition of organic material, and
- tar oil.

Now let us turn to the geological situation in the Ruhr area.

Considering the upper strata without the more recent quaternary strata, the overall picture is as follows (3). The Emscher marl deposits represent a major feature of the central Ruhr area. These deposits have a thickness of up to 400 m and consist of a sequence of claystones and sandstones with a high lime content. Due to the high lime and clay contents, the rock has a high ion exchange and buffer capacity, which has a positive effect in reducing the dispersion of pollutants.

The upper 1 or 2 meters of the Emscher Marl formation have weathered to form clayey silt or silty clay and are impermeable to ground water. Below this layer, the clay marlstone may be fissured and bear water down to a depth of 30 to 40 m. As the depth increases, fissures become less common and are finally completely closed. An impermeable layer is then formed. The Emscher marl formation therefore provides a seal between the Cenomanian/Turonian and the Santonian and quaternary water bearing horizons.

In view of this geological situation, the central part of the Ruhr area is of minor importance for water supplies. Both industrial water and drinking water are obtained almost solely from the Ruhr valley in the South, from the Rhine to the West and from the ground water in the North of the area. Up to now potential hazards for ground water and therefore for drinking water supplies have been the main incentive for soil reclamation. However, particularly the municipalities are now showing a great interest in the rehabilitation of industrial waste land in order to provide sites for new industry.

3.1. Test Programme

A major element of the pilot project proposed by ZENIT was to give interested companies an opportunity to have an influence on the analysis of the test area with regard to specific processes.

By their very nature, investigations of polluted soils by public bodies tend to concentrate on risk assessment, i.e. on analyzing the toxicology of the pollutants and the possibility of pollutant escape. Such investigations must also attempt to determine the extent of the polluted area. On the other hand, the procedure proposed by ZENIT, concentrating on technological development, offers a possibility of analyzing a large number of parameters which are relevant from the process engineering point of view, although initially on a rather small area.

The program therefore includes investigations which are relevant both for in situ processes and for on site processes of the thermal, extractive and microbiological types.

The major points of the program are as follows:
- investigation of the geological and hydrogeological situation
- determination of the particle size distribution of the soil
- determination of permeability and ion exchange capacity
- detailed investigation of the anorganic pollution of the soil
- detailed investigation of the organic pollution of the soil
- determination of the biological activity of the soil and the biodegradability of the eluates.

To investigate these points, test holes were drilled in a gridiron network at a spacing of 8 m over a square with 56 m sides. In addition, 3 test holes were drilled to investigate the geological and the hydrogeological situation.

These 3 holes were drilled as dry holes with casings. Drilling was stopped as soon as the firm marl horizon was reached.

To determine the granulometry of the soil, samples were taken at intervals of 0.5 m in these holes and investigated by dry and wet screen analysis.

3.2. Results

3.2.1. The Geological and Hydrogeological Situation.

These investigations were carried out by "Westfälische Berggewerkschaftskasse", Bochum. As a result of its origin the man-made soil deposited on top of the natural soil has very variable granulometryn and ranges from a heterogeneous mixture to coarse material such as broken bricks and stones to clay with stones. The coarse material is highly permeable to water with values in excess of 10^{-4} m/s. As the silt and clay content increases, the permeability decreases to values lower than 1×10^{-7} m/s.

The natural loose quaternary sediments under the made-up soil have a much narrower particle size distribution than the made-up soil itself. These loose sediments consist mainly of silt with a greater or larger proportion of fine sand. The fines portion reaches a maximum value of 10 to 13 per cent by weight. Some of the horizons contain rather more sand, but the maximum sand content does not exceed 30 to 40 per cent. The sand portion consists mainly of fine sand, with a maximum sand concentration of 20 per cent. There is no coarse sand. Due to their fine-grained structure, the quaternary sediments have a very low permeability. The distribution curves indicated k_f values of a maximum of 7×10^{-6} m/s. The more sandy horizons, in which silt and clay fractions predominate, only have permeability values of less than 1×10^{-8} m/s. From a geological point of view, these fine quaternary sediments belong to the loess deposits of the Emscher formation. They also include fluviatile sandy deposits.

As regard granulometry, the weathered lime marl is very similar to the quaternary deposits. It is therefore sometimes difficult to distinguish the two strata. The weathered marl mostly differs from the quaternary strata by its stiffer or tougher consistency. The permeability of the lime marl is also very low and ranges from between 10^{-6} and 10^{-8} m/s in the quaternary loess and loess clay.

The test holes which were drilled down to the natural Emscher did not reveal any water bearing stratum. The strata were only damp with some waterlogging in the lower parts, where water flowed into the test hole. It was therefore not possible to make hydro-tests to determine the permeability of the loose sediments. This result naturally has major consequences for various approaches to the situ reclamation.

3.2.2. <u>Chemical Pollution</u>. Considerable problems arise with the analyses necessary for risk assessment and for the investigation of possible reclamation processes as one is not faced by a single pollutant or a narrowly defined group of pollutants. Tar oil alone represents a mixture of several hundred single compounds. These are aromatic hydrocarbons ranging from benzene, the lightest component, to the high molecular weight constituents of pitch. The variety of compounds concerned is still further increased by alkyl substitution and heterocyclic compounds.

As the individual components of tar oil have very varied toxicological properties, it is clear that environmental and toxicological analyses face very special requirements. In some cases, these requirements are diametrically opposed to demands for practically relevant problem-oriented analyses. Under this aspect, the cost, clarity and practical usefulness of the analyses were in the foreground. On the one hand it was necessary to obtain toxicologically relevant data related to individual compounds in order to assess the possible success of reclamation efforts. On the other hand, it was also necessary to obtain data similar to those used in waste water engineering, which could be used as overall indicators of pollution level and provide the clarity required for first estimates. In this description, we would like to concentrate on the analysis of organic contaminants. Conventional methods were used for the analysis of anorganic ions and heavy metals. After intensive investigation of the various methods available, CLAYTEX CONSULTING decided on the following procedure:

3.2.2.1. Sampling

The samples were mainly taken using core probes. It would be beyond the scope of this paper to go into the problems of representative sampling. It may, however, be sufficient to state that a sufficiently accurate assessment of the type and scope of possible pollution can only be made on the basis of a consideration of the conditions under which the sample was taken together with the results of the analysis. This is particularly the case with samples taken from the edges of polluted areas and from points which are oversaturated with pollutants. Unintentional displacements and freak values may lead to a completely false assessment of the situation.

The samples were transported in appropriately sized glass containers with twist-off caps to prevent the loss of highly volatile components.

3.2.2.2. Extraction

Apart from thermal methods, there are mainly two extraction methods, the Soxhlet process and the ultrasonic process, to separate tar oil components from soil. Preliminary investigations indicated that in view of the substances involved and the structure of the sample material, there would be no disadvantage in using the ultrasonic methods as against the Soxhlet process (4). In addition, the ultrasonic extraction process is much simpler to use and less time-consuming. Only samples which are heavily over-saturated with tar oil may present certain problems. In the case of samples of this type, where the degree of contamination is measured in per cent, a much greater problem is, however, the influence of the sampling method. The solvent used was methylene chloride. The polarity of this solvent gave good results even in the case of more polar components of tar oil, such as the phenols.

After drying, the products obtained by extraction can often be analyzed directly by gas chromatography. In some cases, such as when aliphatic hydrocarbons are also present, further clean-up is required. Products with lower contaminant concentrations have to be concentrated for further analysis.

3.2.2.3. Gas Chromatography

Further analysis was made by gas chromatography using flame ionisation detectors. In addition to the BTX fraction, the material was quantitively analysed for 12 selected polycyclic compounds distributed over the whole boiling point range covered. The selection which was made is by no means binding and can relatively easily be adapted to meet other requirements. In addition, the chromatogram contains a certain amount of other information which may also be useful. In particular, the total FID signal value may with appropriate weighting provide a relatively clear indication of the overall pollution level, as the polycyclic components are indicated almost in proportion to their carbon mass. Special effects such as discrimination etc. must also be taken into account. The value obtained is a direct representation of the overall pollution level.

There is no linear correlation between the overall pollution level and the total concentrations of the individual components analysed. The ratio of the total concentration of the individual compounds analysed to the total pollution level determined by the method described above reaches a maximum value near 30 to 40 per cent.

Overall pollution level data are an essential requirement both for the investigation of many questions arising in connection with possible reclamation processes and for an intensive risk analysis. However, the concentration of individual pollutants such as naphthalene or phenanthrene does not adequately reflect the overall pollution level. A better and clearer indication is given by the weighted total of the FID signals.

Furthermore, a comparison of the overall pollution level with the total concentrations of the individual pollutants identified gives the investigator the opportunity of reacting appropriately to unusual results. If the overall pollution level is relatively high in comparison to the concentrations of the individual polycyclies, the results should be clarified. Our experience shows that there are three major factors which may lead to such a situation and that these factors are of varying ecological and process engineering significance.

Firstly, the sample analysed may also be contaminated by mineral oil. This is normally directly indicated by the chromatogram and the contaminants can be allocated to particular oil fractions.

Secondly, there may be a relatively high concentration of aromatics which were not quantitively analysed, particularly naphthalenes with alkyl substitution such as methyl naphthalenes, dimethyl naphthalenes etc. If such contaminants are indicated over a relatively large area, it may be appropriate to include them in the analysis programme.

Thirdly, there may be a relatively high concentration of phenolic components. It is not so much a question of the phenols which are already contained in the tar oil but rather of these produced by metabolization of aromatic hydrocarbons. It appears that this process only occures under particular soil conditions as this phenomenon has only been observed in exceptional cases. When observed, however, it can occur over larger areas. High concentrations of phenol, alkyl phenols and hydroxy-naphthalenes are of course particularly relevant for ground water pollution as they are both relatively soluble and can promote the solubility of other substances. In such cases, it may only be possible to clarify unusual gas chromatography results by correlation with the results of mass spectroscopy.

All in all, the critical evaluation of the results of gas chromatography using flame ionization detectors can lead to balanced results. The total weighted FID counts may be useful but this parameter cannot be used as the sole criteria for evaluating pollution levels. E.g. in the case of most of the samples, the share of benzpyrene is less than 0.1 per cent; however, there are cases in which this value reaches more than 5 per cent, sometimes with samples where the overall contamination level reaches several thousand mg/kg.

An FID chromatogram using high-resolution capillary columns (in this case 50 m fs - SE 54) can yield a large number of individual data. By means of critical evaluation and weighting, these data can be reduced to give an indication of the overall pollution level taking account of all the data obtained. However, the individual data including the concentration of all the relevant substances should remain available for inclusion in an overall assessment.

3.3. Investigation of the microbiological activity

The hope that a great deal of money can be saved is the reason for much speculations about the cleaning contaminated soil by biological treatment (5). With respect to coking plant sites, however, the results are not very promising. Remembering the data of PAH's all institutions we have asked refused an order to test the microbiological activity. On this account we looked for other samples at surrounding places with much lower contents of PAH's. These samples are now under investigation by TNO, Delft.

4. RECLAMATION PROCESSES

On the basis of these results, the following processes have been proposed by ZENIT after discussion with the various companies interested. As pointed out at the beginning, the main purpose of our proposal is testing new engineering processes and the combinations of various possible solutions.

4.1. Air Extraction (System HARRESS)

The extraction of air from the soil is a frequently practised method of removing volatile organic compounds, particularly chlorinated hydrocarbons, from polluted soil. In view of the results of the analyses made, it is clear that it would not be feasible to reclaim the test area by this method, as organic substances with very high boiling points represent the major source of contamination. However, we believe that there are various points in favour of this process:

Firstly, the extraction of air from the soil would make a contribution to the reduction of acute hazards. The results of the investigations have shown that there is no risk of contamination escaping via the ground water. Air escaping from the soil would therefore represent the major environmental risk. Initial measurements using automatic BTX analyzer have indicated values of up to 1500 mg/m^3. For this reason, action to seal the area has already been discussed. As this would represent a serious obstacle to further reclamation work, we propose that the extraction of air from the soil should first be attempted. We do not believe that this process has already been tried out at a disused coking plant site. In this particular case, this solution would be favoured by the fact that there is no water bearing stratum down to the Emscher formation.

Furthermore, the extraction of soil air would facilitate the subsequent removal of soil for on-site processing, as it would reduce the extent of safety precautions required to protect workers in this area (6). Finally, this process is interesting from a technical point of view as a preliminary stage to solvent extraction. The extraction of air would remove substances which could later form an azeotropic mixture with the solvent and which it would not be easy to separate from the solvent, at least not by distillation.

4.2. Extraction of Pollutants (KRESKEN-WESSLING Process)

It is proposed to use amyl ether or petroleum ether or another substance with a similar low boiling point as the solvent. Such solvents combine the advantage of high efficiency with a high evaporation rate which would mean that pollutants could be removed almost 100 per cent without leaving any solvent residue in the soil. In addition, due to the high evaporation rate of the solvent, it would be possible to operate the process as a closed evaporation - condensation - extraction loop and to concentrate the pollutants by repeatedly using the same solvent.

The reclamation plant consists in principle of two cycles, one carrying the polluted soil and one carrying the solvent. Whereas the polluted soil only passes through the cycle once, the solvent can be recycled almost indefinitely. This offers the possibility of extreme concentration of the pollutants.

The polluted soil is transported to the plant and then passes through an inlet chamber which is operated at a vacuum and is carried to the counter-current extraction chamber by a screw conveyor. The extraction chamber itself operates on the principle of a rotating drum spiral elevator. As the drum rotates, the polluted soil passes upwards through the chamber. Solvent passes downwards and is collected at the bottom of the chamber.

The soil emerging from the top of the extraction chamber still contains solvent and is passed through a drying or degassing section before it emerges through an outlet chamber operated at a vacuum for replacement. The drying and degassing section is connected to the solvent condenser in order to minimize the loss of solvent vapour.

The solvent, now containing pollutant, is fed to a vaporizer in which it is cleaned by destillation before passing to the condenser. In the condenser, the solvent is condensed and then returned to the solvent tank which in turn feeds the extraction chamber. The pollutants removed from the solvent by destillation are fed to the system outlet for disposal.

In the form described above, the extraction plant is designed to prevent both the escape of solvent vapour and the ingress of oxygen into the plant, which could lead to an

explosion risk. The safety problems arising in connection with
the solvent cycle are well known and conventional solutions
are available. The proposed plant, as described, will have a
soil capacity of approx. 10 tonnes per hour at a solvent
consumption of approx. 0.1 tonnes per hour, representing a
pollutant concentration factor of approx. 100:1. In view of
its size and weight, the plant will be suitable for mobile use
although of course it can also be installed on a stationary
basis. The reclamation capacity can also be considerably
increased by connecting several extraction chambers in
parallel.

The only method suitable for the disposal of the highly
concentrated solutions of pollutants collected in the
collection tank of the plant will be extremely high
temperature combustion.

This leads us to the third process proposed.

4.3. Thermal Treatment (DEUTAG-VON ROLL-Process)

A detailed description of the various methods for thermal
treatment was contributed to the NATO-CMS Report (7). The
proposal of DEUTAG/VON ROLL run thus. In the upper section of
a rotary tubular kiln, the soil will be intensively mixed by
the rotation of the tube and ignited by an oil burner with a
capacity of 185 kg/h installed in the end of the kiln at an
average temperature of 800°C which will be sufficient to
obtain the temperatur of 650°C required for the complete
combustion and evaporation of the organic substances in the
soil. Gas velocity in the drum will be limited to ensure that
the soil passes through the drum very slowly. The drum has a
length of 10 m and a diameter of 2 m.

The flue gases from the rotory kiln are then fed to the
combustion chamber. The combustion chamber has an internal
diameter of 2.2 m and a height of 6.5 m. In view of the flue
gas flow rate involved and the temperature of 1200°C and a gas
velocity of approx. 2.2 m/s, the flue gases take approx. 3 s
to pass through the combustion chamber.

The ashes are removed in a dry state and carried by a belt
conveyor to a water bath with a temperature of approx. 70
to 80°C for cooling. The water requirement is approx.
0.5 m^3/h. The hot vapour produced during the cooling process
is extracted and mixed with the preheated combustion air.

The flue gas emerging from the combustion chamber is cooled
in an evaporative cooler to a temperature of approx. 470°C by
the injection of water.

For dedusting by a cloth filter, the temperature of the
flue gas must not be higher than 250°C. For this reason, the
flue gas is cooled by an air/flue gas heat exchanger with the
additional admixture of air before it is passed to the cloth
filter. In order to prevent excessive fouling, a double
cyclone separater to remove solid particles is installed
upstream of the heat exchanger.

The flue gas leaving the heat exchanger, cooled with fresh
air down to 150°C is filtered by a cloth filter which can
reach an efficiency of up to 99 per cent depending on the
solid particle content of the flue gas.

Dust from the evaporative cooler is fed to the water bath by a conveyor chute. Solid particles from the cyclone separator, the heat exchanger and the cloth filter are transported to a mobile dust tray.

5. FINAL REMARKS

Beyond the outlines above two companies proposed extraction processes using water. It is known, however, that these processes are limited by a high content of clay. Unfortunately this is quite typical for the overall Ruhr area.

6. REFERENCES

1. Frank H.G., Collins G., Steinkohlenteer, Springer Verlag, (1979)
2. Saathoff G., Schecker H.-G., Teer, Pech, Teeröl, Bitumen; STAUB Vol. 46 (1986), S. 235
3. Birk F., Coldewey W.G., Altlasten im Rheinisch-Westfälischen Industriegebiet, Modellvorhaben Altlastensanierung/ Bodenregenerierung, ZENIT (1986)
4. Golden C., Sawicki E., Ultrasonic extraction of aromatic hydrocarbons from airborne articles at room temperature, Int. J. Environ. Anal. Chem., 4 (1975), S. 9
5. Eikelboom, D.H., In situ biological treatment of a contaminated soil, Contaminated Soil, 1. International TNO Conference on contaminated soil, Utrecht (1985), S. 686
6. Jansen A., Occupational hygiene during clean up actions of contaminated soil, dito S. 551
7. Rulkens W.H., Assink J.W., van Gemert W.I.Th., On-site processing of contaminated soil, NATO-CMS Report Contaminated land (1985)

IN SITU REMEDIAL ACTION OF CADMIUM POLLUTED SOIL
- progress report of a full-scale sanitation project -

L.G.C.M. Urlings*, A.T. Blonk*, J.A. Woelders*, P.R. Massink**

* TAUW Infra Consult B.V., P.O.Box 479, 7400 AL, Deventer, The Netherlands
** Provinciale Waterstaat Utrecht, P.O. Box 80300, 3508 TH Utrecht, The Netherlands.

INTRODUCTION

Much attention has been focused on the survey and sanitation of Cd-contamination. TAUW Infra Consult investigated a cadmium (Cd)-polluted site of a photopaper producing plant in the central part of The Netherlands, first by order of the industry and later on by order of the Province of Utrecht.

The plant disposed a Cd-containing wastewater into two infiltration ponds in the years 1935-1955. Periodical flooding of these ponds caused also a soil pollution with cadmium in the adjacent dune plot. In the vicinity of the polluted area a groundwater-pumping station is situated.

This paper deals with the survey results of the Cd-pollution in the soil and groundwater and development of an in situ remedial action technic for the polluted soil. Figures 1 and 2 give an impression of the Cd-pollution in the sandy soil in horizontal and vertical profile. Total Cd-content of the soil is estimated 725 kg.

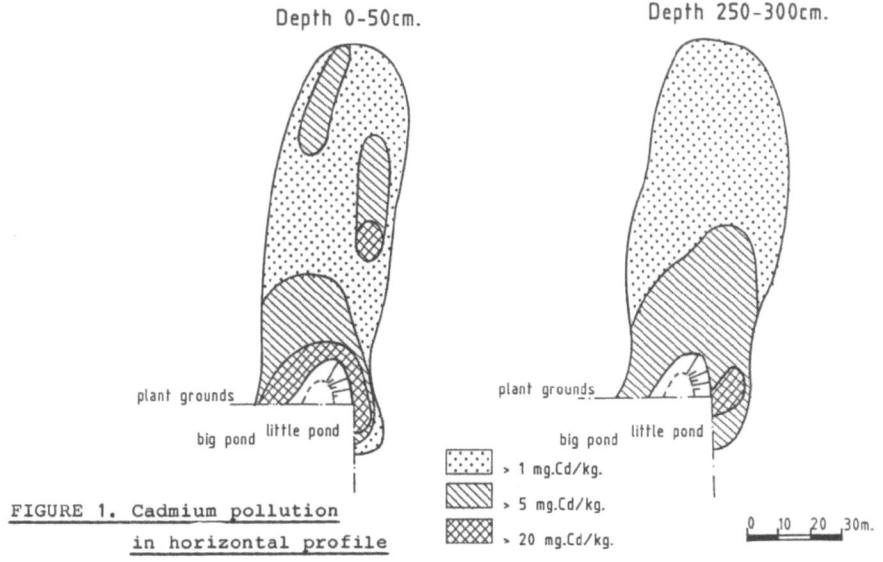

FIGURE 1. Cadmium pollution in horizontal profile

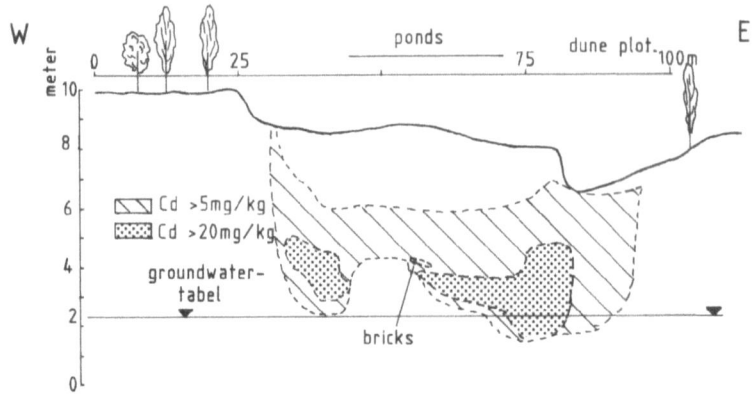

FIGURE 2. Cross section through the two former pounds

Direction of the groundwaterflow changed due to the variation in groundwater withdrawal regime of the drinkingwater pumping station Soestduinen. The Cd-pollution in the groundwater is outlined in figure 3. Besides Cd there is an additional Boron pollution of a different origin.

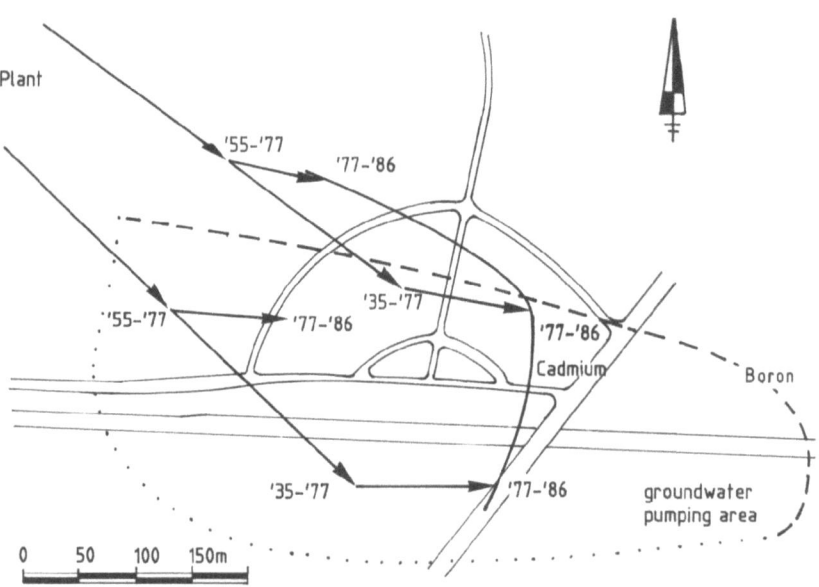

FIGURE 3. Distribution of Cd and Boron in the groundwater in time

CONSIDERATIONS FOR REMEDIAL ACTION

The polluted location is situated within the withdrawal area of the drinking water production facility, and therefore, the Cd-pollution is undesirable. Further, the dune plot on the north boundary of the plant has recreation destination, so exposure to the polluted top soil is undesirable.

REMEDIAL ACTION TECHNICS

Removal of the cadmium pollution out off the environment can be obtained by remedial action of the solid soil on location of the former ponds and the dune plot and by remedial action of the deep groundwater downstreams of the plant (upstreams of the groundwater resource). In this presentation, only the remedial action of the solid soil is outlined. The remedial action on location of the former ponds and the dune plot can be carried out:

- off-site: after excavation the polluted soil, depending on the Cd-concentration, will be disposed on a controlled waste dump or treated in a soil-cleaning installation
- on-site: see above with the difference that the soil cleaning installation is situated on location
- in situ: the polluted sandy soil is cleaned by percolation with acified water (pH = 3.5). The percolate will be pumped into a groundwater-cleaning installation and will be re-infiltrated. The acid water will be recirculated until the Cd-concentration in the percolate reaches constant low value.

IN SITU REMEDIAL ACTION

Preliminary cost estimations indicated that an in situ technic if available would be promising. The benefit can be high as 25% of the total cost of 4 million guilders. For lack of experience with in situ technics in general and with in situ cleaning of heavy metals in particular additional experiments were carried out in the laboratory of TAUW Infra Consult.

Concerning the in situ remedial action three aspects need special attention, because there is no experience available in this matter:
- the Cd-desorption of the polluted soil (soil chemistry)
- the hydrological system of infiltration and withdrawal of water (hydrology)
- the purification of Cd-containing groundwater by low pH (water-treatment).

The laboratory experiments are summarized according to the three topics mentioned above.

Cd- desorption of the soil

A desorption liquid was selected by batch-experiments with polluted soil. By calculation of the Cd-distribution coefficients (C_{soil}/C_{water}) from the batch-experiments (see table 1) a selection was made for the following column-leaching experiment. The water used in the experiments was gained from the groundwater resource of the plant.

TABLE 1. Distribution coefficients (K_d) of batch experiments after 24 h

		groundwater on-site	groundwater 10^{-3} mol HNO_3/l	groundwater 10^{-3} mol HCl/l	groundwater 2,5 g $CaCl_2$/l	groundwater 0,5 g $CaCl_2$/l
soil sample Cd: 20,5 mg/kg d.s.	K_d (dm³/kg) pH Cl⁻ (mg/l)	72 7.0	7,8 4.1	7,8 4.1 110	11 6.5 1300	44 7.2 180
soil sample Cd: 68,5 mg/kg d.s.	K_d (dm³/kg) pH Cl⁻ (mg/l)	102 7.3	3,7 4.4	1,6 3.7 155	5,6 6.5 1350	28 6.9 255

In figure 4 results of the column leaching experiments are compiled.

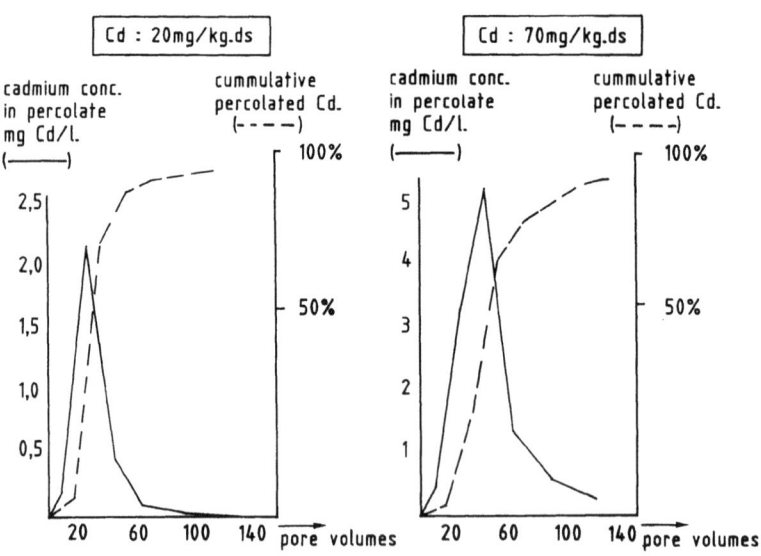

FIGURE 4. Leaching of Cd-polluted soils by 10^{-3} mol HCl/l (pH = 3.5)

The K_d-values for the batch experiments range between 1.6 en 7.8 dm^3/kg and for the column experiments between approximately 2 and 3 dm^3/kg.
Results of the batch experiments can indicate an overestimation of the K_d-value due to the buffer capacity of the sandy soil. The columnn experiments, where approximately 100 pore volumes, are refreshed after 50 days give, a more realistic value for K_d.
Expected quality of the percolate, calculated for a K_d-value of 4 (dm^3/kg) with the assumption of lineair desorption behaviour of Cd, are shown in figure 5 for the fine composing compartments. The hydraulic resistance of the soil is 5 m/day in the calculation.

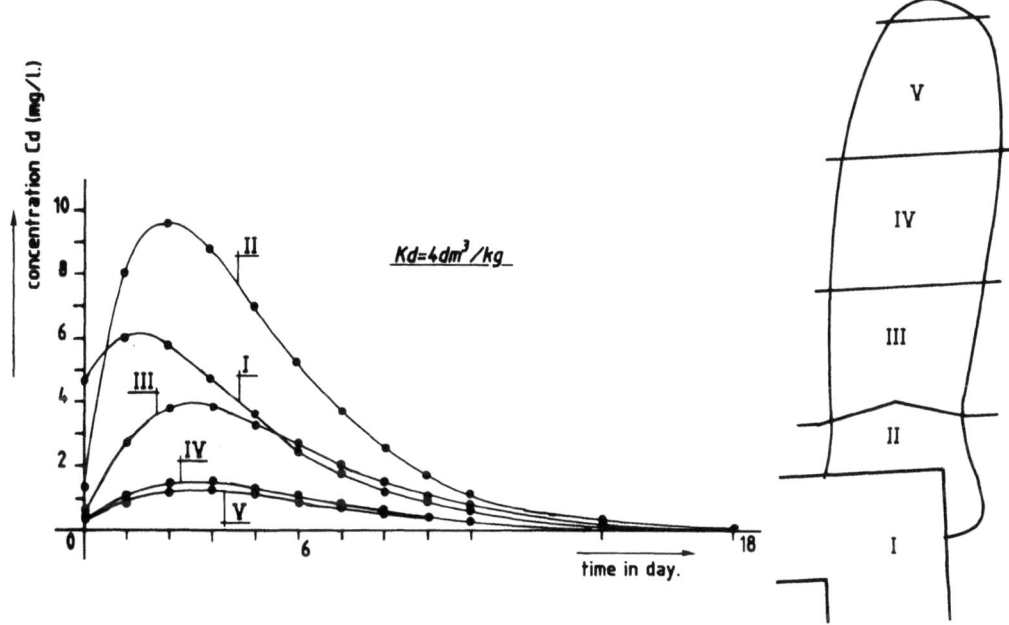

FIGURE 5. Calculated cadmium concentration in percolate in the five separate compartments

The hydrologic system

The total area for remedial action amounts 5500 m^2 (see figure 1). Infiltration capacity of the topsoil is measured in the field and is estimated at 5 m/d (k = 5 m/d, i = 1). Because the capacity of the groundwater treatment installation was limited on 250 m^3/h, the polluted area has to be devided in 6 compartments of approximately 1000 m^2.
Horizontal drains are preferable to vertical deepwells to get strait groundwater flow lines. A cross section of infiltration pond and withdrawal system is outlined in figure 6.

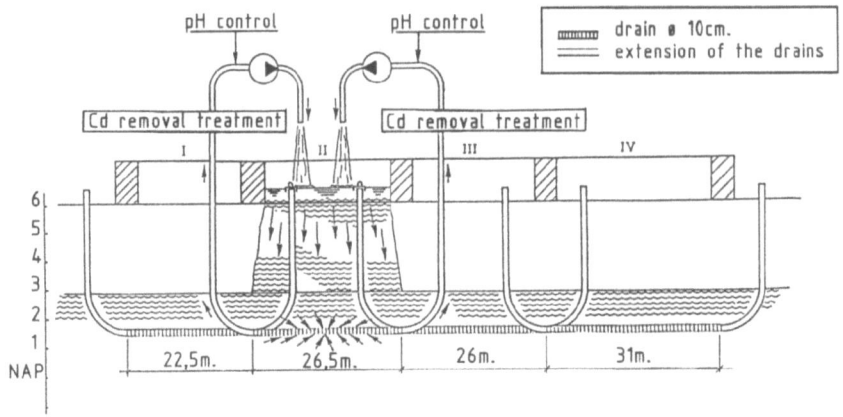

FIGURE 6. Cross section of the infiltration and withdrawal system.

Design and dimension of the infiltration/withdrawal system was carried out by a two-dimensional (vertical) computermodel developed by TAUW Infra Consult. The calculated results are converted to the three-dimensional space. Full recirculation of the infiltrated water can be guaranteed when only a slight quantity of aquifer water is pumped additionally (about 10% of the infiltration capacity).

For a compartment of 1020 m^2, the groundwater flow pattern and accompanying cumulative remaintime distribution are presented in figure 7 and 8.

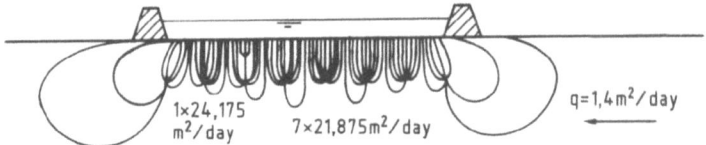

FIGURE 7. Two-dimensional flow pattern of the infiltration water

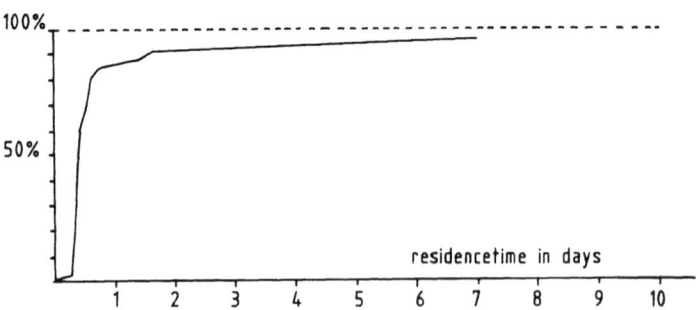

FIGURE 8. Cumulative residence-time distribution of the infiltration water

For the relative transport velocity of Cd in the soil a retardation factor (R) of 20 is estimated. R is based on laboratory experiments which give estimation of the distribution coefficients K_d. By supposing lineair adsorption, R can be calculated: $R = 1 + K_d * \text{density} * \text{porosity}^{-1}$. When the density is 1,5 kg/dm^3 and the porosity is 0.35, R ranges between 14 and 18 for K_d-values of respectively 3 and 4.

The criteria for the in situ remedial action is postulated as follows:
$T_{end} = T_{90} * R$. T_{90} stands for time needed for 90% recirculation of the infiltrated water. T_{90} is about 1,3 - 1,5 days so remedial action time is estimated on 25-30 days for each compartment.

The end of the remedial action will also be determined by the Cd-concentration in the pumped percolate for the separate compartment. In principal, the infiltration with acified water should be continued till the Cd-concentration reaches a constant low value.

The water treatment system

A literature survey was conducted on the removal of Cd from waste water. The three most important treatment technics are:

   precipitation/flocculation
   biosorption
   sorption by resins.

Because the treated percolate is infiltrated again in the polluted soil and waterflow is quite high, viz 250 m^3/h, sorption on resins is preferred.

When precipitation/flocculation technics are used, the salt content in the recirculated water will increase enormously. This is undesirable in the vicinity of a drinkingwater station. Nevertheless, the precipitation unit in such a water treatment system should have oversized dimensions for 250 m^3/h.

Precipitation, based on sulfide, was not feasible because sulfide can cause serious damage to the photopaper-production process in the plant. Biosorption technics seemed not available on this scale and if available the practical execution would be very difficult.

A resin with a quaranteed high-specific Cd-adsorption was not available from various manufactories. A batch experiment was set up to make a selection of the most suitable resin. An artificial mixture of cations was made corresponding with the relative cation composition of the percolate water, but the concentration level was higher in the artificial mixture. The results of the batch experiment (70 ml resin in 700 ml solution) are presented in table 2.

TABLE 2. End concentration of cations in the water phase with different resins after the batch experiments in mg/l

	blanco	GT-73	IRC 718	TP 207	S100	Chel.20
Cd	1,45	<0,01	0,42	0,66	1,25	0,24
Ca	2250	2100	1750	1550	860	1650
Fe	300	190	320	145	170	130
Al	1300	1100	310	560	375	370
Mn	73	55	13	38	55	28
Zn	18	2,55	1,95	1,25	10	1,35
Cu	4,3	0,03	0,27	0,12	2,4	0,16

After batch experiments a column experiment was set up to test the GT-73 resin. The experimental set up is shown in figure 9.

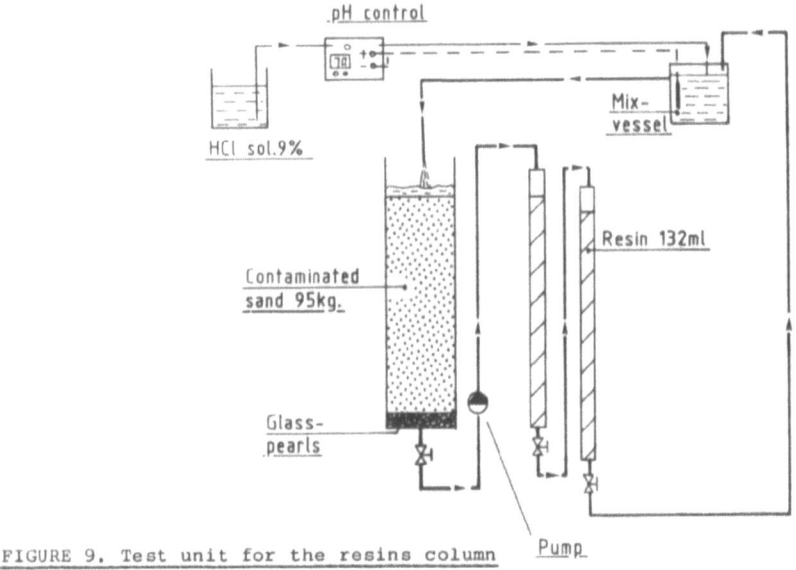

FIGURE 9. Test unit for the resins column

The chemical analyses of influent and effluent of the resin column are given in table 3.

TABLE 3. Chemical analyses of influent and effluent of the resin columns (in mg/l) as a function of refreshment of pore volumes in the polluted soil column.

pore volumes	0.3			3.1			3.9			7.0			9.9			13.8		
	i	e1	e2	i	e1	e2	i	e1	e2	i	e1	e2	i	e1	e2	i	e1	e2
Cd	2.35	0.49	<0.01	13	1.1	<0.01	6.6	0.63	<0.01	2.2	0.96	<0.01	1.4	1.25	<0.01	1.3	1.4	<0.01
Ca	60	58	62	57	57	56	-	-	-	-	-	-	-	-	-	-	-	-
Al	11	9.3	7.2	5.8	5.7	5.7	-	-	-	-	-	-	-	-	-	-	-	-

Table 3 indicates that appearance of cadmium in the effluent of the first column corresponds with a total adsorption onto the second resin column. The calculated load of the first column is 6,6 g Cd/l resin. Loads as high as 34 g Cd/l resin were found in previous experiments with high Cd-influent concentrations.

Reclamation of the Cd-loaded resin is more or less complete when flushed with 4 bedvolumes of 5% HCl.

FULL SCALE DESIGN OF IN SITU REMEDIAL ACTION

The conclusion of the laboratory experiments is that *in situ* remedial action technic can be applied in the present case.

After the evaluation of the three remedial action technics authorities of the Province and Regional Environmental Inspection, the representatives of the drinking water production firm and the photopaper plant all choose for the *in situ* technic.

Because there is less experience with in situ remedial action in general and the application of resins in the treatment of Cd-contaminated groundwater in particular, sanitation of the first of the seven compartments should be accompanied with additional control measurements. Schemes of the watertreatment plant and the infiltration and withdrawal hydraulic watersystem are presented in figure 10 and 11.

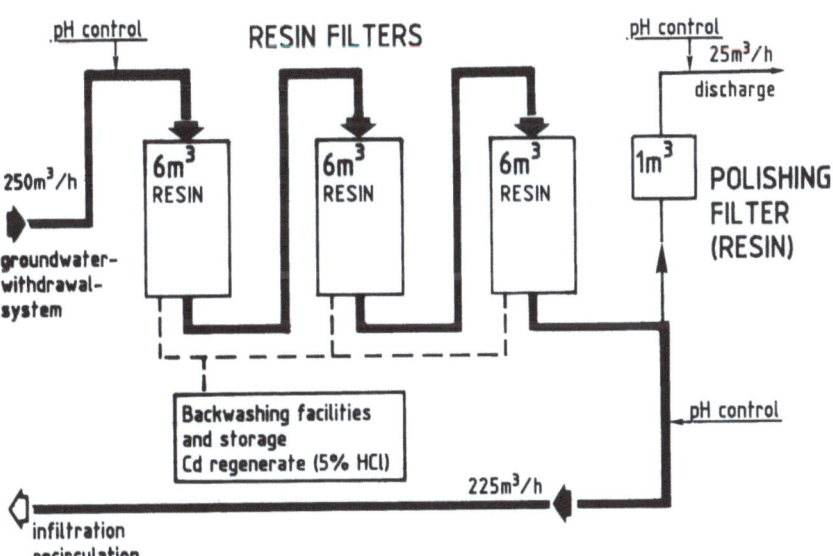

FIGURE 10. Water treatment plant

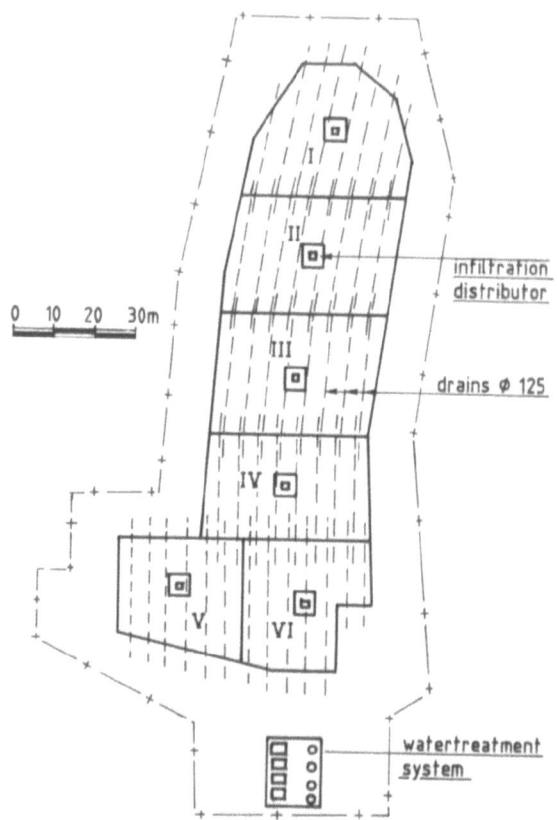

FIGURE 11. Infiltration and withdrawal system

PROGRESS:
The in-situ remedial action will start in summer 1987 and will last for approximately one year.

ACKNOWLEDGEMENT
Preparation of the remedial action project is carried out with B.J. Mathijsen, ing. W. Hagemeijer, ing. V.P. Akkerman (Province of Utrecht), dr.ir. J.F. van Kessel (Regional Inspection Environmental Hygiene Utrecht), ing. J.A.P. Roodhuyzen (Municipality of Soest), drs. W.C.M. Bakkum (WMN), drs. A.C.H. van Peski, and ing. E. van der Steeg (Chemco Soest).

NEED AND AVAILABILITY OF COVER MATERIAL FOR LANDFILL SITES IN SOUTH WALES

E.M. Bridges, A.T. Evans and D.J. Leech

INTRODUCTION

The amount of cover material required by Waste Disposal Authorities is related to the arisings of waste materials. Estimates of the amount of waste produced annually by society vary greatly but in Great Britain it is estimated to be approximately 100 million tonnes. In South West Wales, which consists of the districts of Carmarthen, Ceredigion, Dinefwr, Llanelli, Lliw Valley, Neath, Port Talbot, Preseli, South Pembrokeshire, and Swansea, 'controlled waste' amounts to 528,000 tonnes per annum. Currently, disposal takes place at 17 sites throughout the area and all require a reliable, steady supply of cover material for satisfactory site management.

The present British policy for refuse disposal may be traced back to a Ministry of Health report in 1932 which outlined a number of measures to ensure that domestic refuse was disposed of in as safe a manner as possible. The process of controlled tipping (or sanitary landfill) should conform to the following guidelines: the deposit to be made in layers; no layer to be more than a certain thickness (< 2m); each layer should be covered on all surfaces exposed to the air with at least 20 cm of earth or other suitable substance; animal, fish or organic refuse should be covered by at least 60 cm; no refuse should be left uncovered for more than 24 hours (Cope et al, 1983).

The effect of the Control of Pollution Act, 1974 and EEC Directives has been to re-inforce these earlier guidelines. On-site working practice should reflect the policies outlined in the DoE's Landfill Practices Review. The most common method of landfilling takes place in cells constructed with earth or other inert material; this method is preferred as the wastes are left uncovered for the least amount of time, thus minimising nuisance from smell, flies, vermin and the scattering of lightweight rubbish by wind. The embankments of the cellular method conceal the operation and when a cell is filled it can be rapidly covered, effectively limiting the infiltration of rainfall into the mass of tipped material. The cell walls may be constructed of other inert waste materials, but care must be taken to avoid seepage through the outside walls and into watercourses where contamination of surface waters may occur. When cell walls are made of soil material this represents a loss of space for landfill, thus increasing operational costs.

Definition and concepts of covering systems

A covering system is a sequence of layers of materials placed over domestic or special wastes in an attempt to isolate them from the rest of the environment. It may be necesary to protect aquifers from landfill leachate, to control emissions of gases from the decomposing wastes or it may be politic to re-use a landfill site. In all these cases, wastes must be effectively covered to facilitate efficient after-use of the site.

Parry and Bell (1985) list the primary functions of covering systems under three broad headings: 1) to prevent exposure of harmful contaminants to human, animal or plant life; 2) to sustain growth of vegetation; and 3) to fulfill an engineering role. These three main functions can be subdivided into several secondary functions which are listed here from the surface downwards and not necessarily in order of importance: erosion control; prevention of dust blowing; support of vegetation; limitation of infiltration; control of soil water movement; control of leachate from the landfill; control of gas movement; prevention of an upward migration of toxic constituents through capilliary action; inhibition of root penetration through the barrier; prevention of biological transfer of harmful constituents to the surface; reduction or elimination of harmful surface conditions.

Cover material requirements

The CIPFA Statistical Information Service publishes waste disposal statistics in the form of 'Estimates' and 'Actuals'. In these statistics, cover material is excluded from the tonnages shown unless the material used is waste which the Waste Disposal Authority has a statutory responsibility to accept.

As the reliability of the CIPFA figures has been questioned, letters were sent to the ten Waste Disposal Authorities requesting information on current waste disposal arrangements and usage of cover material in order to confirm them. The data received from the Waste Disposal Authorities in South West Wales are summarised below:

Waste Disposal Authority	Waste collection 1984-85 (tonnes)	Cover material requirements 1986 (tonnes)
Carmarthen	15 816	20 500
Ceredigion	40 000	25 000
Dinefwr	17 500	7 000
Llanelli	42 000	15 000
Lliw	21 000	12 000
Neath	30 100	25 000
Port Talbot	33 790	*
Preseli	20 900	26 500
South Pembrokeshire	16 500	#
Swansea	63 533	85 000

* waste taken to Neath   # waste taken to Preseli

GEOLOGICAL AND PEDOLOGICAL SOURCES OF COVER MATERIALS

Underlying geological materials and soils are the most readily available cover materials. At present they constitute the major source of cover material for the Waste Disposal Authorities, but although ubiquitous may not always be used as availability is limited by good quality agricultural land, built-up areas, National Parks, Areas of Outstanding Natural Beauty or restrictions imposed by Planning Authorities. Sites of Special Scientific Interest and common land also restrict availability of reserves. This review of sources of natural cover materials will consider the geological deposits present in South Wales and the role soil information plays in identifying suitable materials and their distribution.

The geological formations present in South Wales are shown in Figure 1. The rocks of South Wales are mainly Lower and Upper Palaeozoic in age with smaller outcrops of Pre-Cambrian and Mesozoic strata. In addition, unconsolidated Quaternary and Recent deposits are widespread, but of no great thickness. However, in the present context, lithology rather than age is significant, so this is stressed in the following discussion.

Figure 1
General geology map of South Wales:
igneous and metamorphic rocks

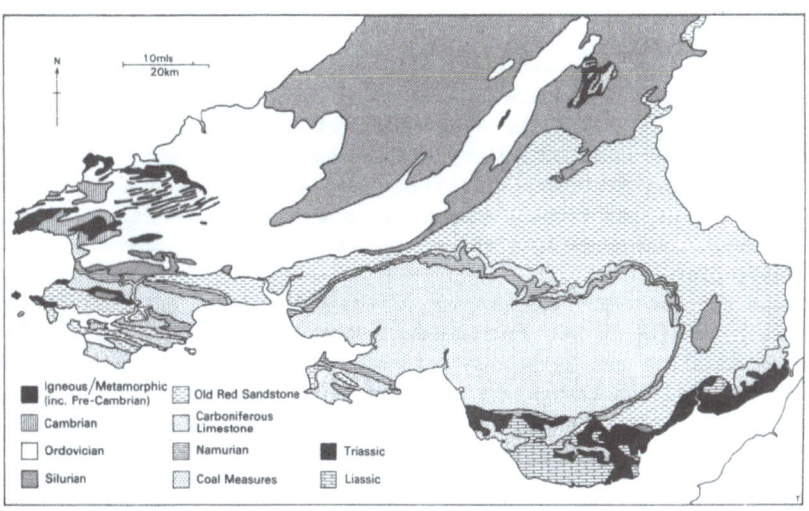

Igneous and metamorphic rocks
These rocks occur in restricted areas of Dyfed and Powys where they are quarried as a source of crushed aggregate, but they are unlikely to be a source of cover material other than as quarry waste. However, where a coarse break layer is required above a toxic waste, dolerite and andesite could be used instead of limestone; granite or rhyolite could be used where base-rich properties are not required.

Arenaceous sedimentary rocks
Sandstones, gritstones and conglomerates are common throughout the Palaeozoic rocks of Wales. Reddish or purple-coloured conglomerates, grits and sandstones of the Cambrian crop out near St David's and Lower Ordovician sandstones occur near Carmarthen. Arenaceous rocks are significant higher in the Old Red sandstone sequence and thin conglomerates cap the highest point of the Brecon Beacons. Conglomerates are also a feature of Devonian rocks in Dyfed, West and Mid Glamorgan.

The Coal Measures are characterised by repeated sequences of sandstone, shale and coal strata, but the upper part of the Coal Measures in South Wales is dominated by the Pennant sandstones which form the plateau which the valleys of South Wales dissect. Triassic sandstones in South Wales are restricted in area.

Many small quarries have been worked in the last century and a few siliceous sandstones continue to be quarried for refractory purposes in the iron and steel industry. Although crushed rock aggregate itself may be too expensive for cover material, there is up to 20 per cent waste resulting from the processing which is an untapped source of cover material.

Argillaceous sedimentary rocks
Shales and mudstones form extensive outcrops of Ordovician and Silurian age in Dyfed and Powys. In some places these argillaceous rocks have been metamorphosed to slate. The lower beds of the Devonian comprise red-coloured mudstones with subsidiary siltstones or sandstones. Grey shales predominate in the Millstone Grit (Namurian) and the Lower Coal Measures, but the extent of the Pennant Sandstones and superficial materials limits their outcrop. Red mudstones are characteristic of the Triassic rocks in the Cardiff district.

Although the mudstones may provide a reasonable cover material, they are difficult to handle in wet conditions. Some Palaeozoic shales were metamorphosed to slate and large quantities of waste are generated by slate quarrying; 50 million tonnes are reported to be available for use as cover material in the Preseli Hills and at Llangynog in Powys. In the coalfield, 50 million tonnes of coal-mining waste from former mines are available for use as cover material.

Limestones
Numerous quarries have exploited the Carboniferous limestone around the coalfield for use as building stone, as a flux in the iron and steel industry, as road metal and as lime and cement. Many quarries on the northern rim of the coalfield have closed but several still operate in the

Vale of Glamorgan. The Liassic limestones are thinly bedded with shale partings and at Aberthaw are exploited for cement manufacture. Reject materials from the quarries, known as limestone 'scalpings' are available from most quarries and are used for cover materials. These are ideal where a calcareous material is desired.

Pleistocene and Recent deposits
Unconsolidated superficial materials lie upon the surface of the solid rocks discussed previously. These include sands, gravels, boulder clay (till), lacustrine and alluvial deposits. The deposits of at least two major periods of glaciation are present; those attributed to the Devensian being the most extensive. Small areas outside the limits of the Devensian glaciation have subdued remnants of an earlier, more widespread, (Wolstonian?) glaciation. Many of these glacial deposits have been affected by periglacial activity and some have been covered by a thin aeolian deposit.

Sands and gravels. The main occurrence in South Wales of coarse-grained glacio-fluvial deposits is in the Vale of Glamorgan with smaller deposits at Swansea and Margam. Elsewhere sand and gravel deposits are confined to narrow strips alongside rivers and the occasional kame feature. These gravels are very poorly graded with a large range of particle sizes and they are often mantled by a layer of finer material of aeolian origin. As a potential cover material these sands and gravels could be a valuable resource in the development of cover systems.

It is estimated that there are 1.7 million tonnes of sand and gravel reserves, with 1.5 million tonnes of marine-dredged sands and gravel obtained from the Bristol Channel annually. Sand extraction takes place in shallow seas over sand banks which lie west of Nash Point, Porthcawl and to a lesser extent Gower. No reliable estimates are available of these marine resources, but it is possible to extract sand and gravel from depths of 25 m, although local dredgers normally operate at 10 m. The marine sands and gravels are less likely be used for cover material as the cost of extraction restricts their use to building aggregate.

Boulder clay (till). As the Pleistocene glaciers slowly moved across the landscape, rock material was eroded, subjected to attrition and eventually deposited as an assorted mixture of boulders and finer material, referred to as boulder clay or till. In the coalfield these superficial materials occupy the mid and lower slopes of many valley sides where they are the site of landsliding activity. Deposits are widespread on the plateau surfaces between the South Wales valleys, often reaching a thickness of several metres. It is impossible to generalise about the composition and physical properties of these materials as they range from fine clays to gravels. Their usefulness as potential cover material will depend on the local variation in composition. The distribution of boulder clay in South Wales is shown in Figure 2. Where these materials are encountered in road reconstruction or other major civil engineering projects they present a good source of cover material.

Figure 2
Distribution of boulder clay in South Wales

Alluvial materials.   The river valleys of South Wales contain deposits of water-sorted material and peat. Near the coast, the river valleys have been infilled with up to 48 m of alluvial or estuarine material in response to the rising sea level of post-glacial times.   In the Teifi valley, particularly around Cardigan, lacustrine silts have accumulated.

Soils
The Soil Survey of England and Wales has recently compiled maps of the soil resources of the whole country (for Wales see Rudeforth et al, 1984). Other information is available in scientific journals (Crampton 1961, 1966; Bridges and Clayden, 1971; Bridges, 1975, 1985).
   Soils are distinguished from each other by the maturity of their profiles, the texture, structure, stone content of the horizons present and the mode of origin of the parent material. It is this last criterion which makes the use of soil maps important in the search for cover material. The 1:250,000 Survey by the Soil Survey of England and Wales gives a good overall impression of the major soil associations. Examination of their soil properties suggests that areas covered by the Arrow, Brickfield, Milford and Wilcocks Associations would appear to be most suitable for use as cover material.

Secondary aggregates as a source of cover material
The use of natural soils and rock for cover material unfortunately necessitates digging up one part of the countryside to re-instate another. Consequently, if suitable alternatives are available these should be seriously considered. The Building Research Establishment has looked into the possibility of using major industrial by-products and waste materials (Gutt et al, 1974). The South Wales Working Party on Aggregrates (1977-83) produced a report in 1977, followed by a supplement in 1979, regional commentaries in 1980 and 1981 and a second supplement in 1983. The most recent document was produced by the British Standards Institute in its series of Guides (BSI 6543: 1985). These surveys report upon the arisings and disposal of waste materials, the location in South Wales is shown in Figure 3.

Figure 3
Sources of secondary aggregates in South Wales
(Gutt et al, 1974; South Wales Working Party, 1977-83; BSI, 1985)

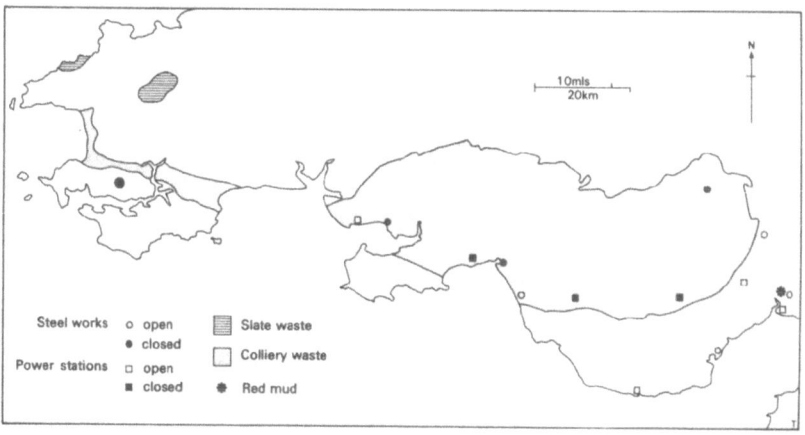

Colliery spoil. Mechanised coal-mining techiques have increased the quantity of unsaleable waste material brought to the surface with the coal. In 1981, the South Wales Working Party on Aggregates recorded that 6 million tonnes of colliery waste was produced each year in South Wales, of which 0.75 million tonnes were sold for constructional fill. Total resources at that date were estimated at 50 million tonnes which are available for use as cover material.

Coal shale waste is not hazardous but the pyrites present may give a strongly acidic weathering product. This material can be used for restoration purposes but precautions must be taken to counter the excessive acidity. However, coal shale, being a mudstone, tends to form a slurry when subjected to traffic and is not favoured by disposal site operators.

Quarry wastes. Specific quarries will have an output of a particular type of stone or aggregate which has been referred to previously but will also create some reject material which is tipped alongside the quarry or in a worked-out area. This waste material may include the overburden of soil and weathered rock; it may be rejected, poor quality stone, or the result of the washing and screening of aggregate. Most of these materials are satisfactory for cover material except for the wastes from mines where metallic ores are won.

Power station ash. Three types of ash are produced from power stations: pulverised fuel ash, furnace bottom ash and clinker. Pulverised fuel ash (pfa) results when pulverised coal is burnt in a stream of air in modern power stations. This ash comprises particles of fine sand size (60 per cent) with a specific gravity of between 1.9 and 2.4, but about 5 per cent of it consists of very light hollow cenospheres with an apparent specific gravity of 0.5. Compared with soil materials, pfa lacks a clay fraction and nitrogen, is alkaline (pH 11-12) and contains soluble salts. Lagooning helps to reduce the soluble salts but an addiditional problem is the presence of between 3 and 250 mg/kg water-soluble boron. It is generally accepted that 4-10 mg/kg boron in soil is slightly phytotoxic, so the presence of boron in pfa may present some problems. However, the content of boron in South Wales coals is lower than elsewhere in Britain. The pfa is difficult to handle, consequently, it is usually made into a slurry and discharged into lagoons. As pfa is a valuable material it could play a greater part in waste material management in the future.

The number of power stations in South Wales has been reduced from eleven in 1970 to six in 1987. The production of ash in the region in 1981 was 0.772 million tonnes and the stockpile of this material is 4.5 million tonnes. Only about 14 per cent of current production is used, which leaves considerable quantities available for use as cover material if required.

Metalliferous slags. The use of metalliferous slags for fill material on construction sites has occurred for many years. Iron and steel-making wastes are widely used in industry and agriculture, but non-ferrous metal wastes contain sufficient toxic metal to make them difficult to use. In 1981, blast furnaces at Llanwern and Port Talbot produced 900,000 tonnes of slag. This was used as roadstone, railway ballast, as a raw material for cement, and as a filter medium for sewage disposal. It has not been used as a cover material. Resources of blast furnace slag in South Wales include the wastes of former iron and steel industries at Merthyr Tydfil and Ebbw Vale.

As far as is known techniques of soil modification by mixing soils of different texture, the addition of dispersants, soil-cement mixtures, soil-asphalt mixtures or the use of synthetic membranes in cover systems has not been used in South Wales. However, sites have been lined by plastic liners to restrict to movement of leachates.

THE DATABASE SYSTEM
A database management system has been developed to maintain and manipulate the information collected through the use of the software package dBASE III, marketed by Ashton Tate of California. dBASE III is designed to be used on an IBM Personal Computer (including IBM PC, Portable, PC XT and PC AT) or an IBM compatible microcomputer. The minimum computer memory required is 256 kb. dBASE III requires MS-DOS or PC-DOS operating systems and the computer should have at least two 360 kb double-sided floppy disc drives. The computer program usually resides on one of the floppy discs, and the database files are saved on the other.

Separate database files were created to store information and data collected from the Waste Disposal Authorities forming the South West Wales Waste area and possible suppliers of cover material. The creation of database files allows the following procedures to be carried out: add records; edit or update records; list and display records and specified fields if necessary; produce report forms; delete records from the file. These operations are easily performed, when using dBASE III, by simply issuing the appropriate dBASE III commands. It was decided to develop a 'menu-driven' system, which means that the person using it need run only one program which controls a series of programs by the user responding to prompts displayed on screen.

CONCLUSIONS
The review of requirements for cover material has revealed that there is a need in the South West Wales region for an amount in excess of 200,000 tonnes per annum. As a source of cover materials, waste rock from the aggregate industry has made an important contribution and in certain circumstances, the aggregates themselves could be used, but the expense of processing the rock makes this unlikely except for specific reasons.

The distribution of Welsh soils is now known with some accuracy, but detailed information on their physical properties is not available. The American Unified Soil Classification System (AMTM, 1964) which utilises particle-size, gradation, plasticity index and liquid limit measurements as criteria appears to have considerable merit when classifying soil materials by using these parameters in the absence of detailed analytical data.

The use of industrial by-products for cover materials is obviously to be encouraged and all authorities have preferential terms for accepting inert wastes which can be utilised. Synthetic cover materials would appear to have distinct possibilities despite their relatively high cost. Considerable experience is available in their use as liners for waste disposal sites but there is less experience of their use in cover systems.

The installation of a cover system should only take place following a thorough investigation and understanding of what it is expected to achieve. The choice of materials for the cover system should begin on site where the properties of soil and subsoil should be investigated. If suitable for use, then the considerable expense of haulage of imported materials is avoided.

BIBLIOGRAPHY

American Society for Testing and Materials 1964 Procedures for Testing Soils. Am. Soc. Test. Mater., Philadelphia.

Bridges, E.M. 1975 Soils in parent materials formed from Devonian rocks in Wales. Welsh Soils Discussion Group Report 15, 73-93.

Bridges, E.M. 1985 Soil survey in Gower. Welsh Soils Discussion Group Report 26 (in press).

Bridges, E.M. and Clayden, B. 1971 Pedology. In: Swansea and its Region (ed. W.G.V. Balchin). British Association for the Advancement of Science, Swansea.

Cope, C.B., Fuller, W.H. and Willets, S.L. 1983 The Scientific Management of Hazardous Waste. Cambridge University Press.

Crampton, C.B. 1961 An interpretation of the micro-mineralogy of certain Glamorgan soils: the influence of ice and wind. Journal of Soil Science 12, 158-171.

Crampton, C.B. 1966 Certain effects of glacial events in the Vale of Glamorgan, South Wales. Journal of Glaciology 6, 261-266.

Gutt, W., Nixon, P.J., Smith, M.A., Harrison, W.H. and Russell, H.D. 1974 A Survey of the Locations, Disposal and Prospective Uses of the Major Industrial By-Products and Waste Materials. Current Paper CP19/74, Building Research Establishment, DoE.

Parry, G.D.R. and Bell, R.M. 1985 Covering systems. In: Contaminated Land: Reclamation and Treatment. NATO Challenges of Modern Society, Volume 8, Chapter 5. Plenum.

Rudeforth, C.C., Hartnup, R., Lea J.W., Thompson, T.R.E. and Wright, P.S. 1984 Soils and their Use in Wales. Bulletin No. 11. Soil Survey, Harpenden.

South Wales Working Party on Aggregates 1977 Interim Report. Mid Glamorgan County Council, Cardiff.

South Wales Working Party on Aggregates 1979 Supplement to Interim Report. Mid Glamorgan County Council, Cardiff.

South Wales Working Party on Aggregates 1980 Regional Commentary Part 1. Mid Glamorgan County Council, Cardiff.

South Wales Working Party on Aggregates 1981 Regional Commentary Part 2. Mid Glamorgan County Council, Cardiff.

South Wales Working Party on Aggregates 1983 1981 Aggregate Minerals Survey. Mid Glamorgan County Council, Cardiff.

DEVELOPMENT OF TECHNOLOGY FOR CONTAMINATED DREDGED MATERIAL REMEDIATION.

H.J. van Veen [1]   A.C. de Waaij [2]

1. INTRODUCTION

Due to the discharge of effluents from industrial and diffuse sources into waterways, very considerable of amounts sediments become contaminated. The contaminants are diverse in nature and very often there is a complex mixture of several contaminants present.
Disposal of dredged contaminated sediments has become very difficult. The reason for this is the sometimes high contaminant content and the uncertainties in predicting the spreading of the contaminants after taking remedial action. Remedial action can be disposal (on land, at sea or in surface water), re-use (possible after immobilisation of the contaminants) and treatment.
For this reason the Dutch government has, among other activities, initiated research into the possibilities for treatment of sediments after dredging. With such treatment we want to achieve a situation in which at least a part of the dredged material can be called 'clean' after the process and can be re-used in any kind of way required.
Until now dredging is mainly carried out for nautical reasons. Now that we know more about the quality of sediments and their effect upon ecosystems, our ideas on dredging and processing sediments in order to save a watersystem are strengthened.
Especially for this second type of dredging projects a well considered national plan must become available before starting the operation. This plan will include a list of locations where dredging is necessary, together with a time schedule and a financial plan. The plan will be ready at the end of 1987, so that probably in 1988 dredging for ecological reasons will be carried out for the first time.
In contrast with other countries the Dutch government does board out activities like dredging and soil/sediment processing. This implies that dredging and sediment processing equipment must be built by private companies. Dredging processing equipment will become available when the government does actually outline processing of sediments in cases with heavy contaminated sediments. This has already been started. There will be an inter-

---

[1] TNO, Netherlands Organization for Applied Scientific Research, P.O. Box 342, 7300 AH Apeldoorn.

[2] Rijkswaterstaat, DBW/RIZA, Institute for Inland Water Management and Waste Water Treatment, P.O. Box 17, 8200 AA Lelystad.

action between governmental outlines and available equipment. Some sediments that are not dredged today, because there is no solution for the sediments after dredging, will be dredged as soon as processing equipment is available that can solve the problem. On the other hand, when there are more assignments that require processing of sediments, more equipments will be built.
It is the intention of this paper to report the results obtained in treating dredged sediments both in the laboratory and in practice.

2. CONTAMINATED SEDIMENTS

The problem of contaminants in sediments in the Netherlands was first recognised in the Rotterdam area. We are now well aware of the fact that there are contaminated sediments all over the Netherlands. Water authorities are investigating sediment qualities with a view to make a survey of sediment qualities in the Netherlands. From the results of the survey it will also be possible to estimate the volume of the market for sediment processing apparatus.
A problem which arises when determining qualities is the definition of criteria involved. This concerns concentration levels to establish the need for sanitation, but also concentration levels, which are permitted to be disposed of.

At the present time no criteria are available either for acceptance or for handling of sediments. Actually, governmental institutes are working out this project. Several interim systems are in use nowadays.
The first criteria that are those used in soil contamination. These comprise a list of concentrations (mg/kg dry matter) of several compounds (heavy metals and organic compounds) that are allowed in soil. There are three levels for each contaminant. The measures that must be taken depend on the level that is exceeded.
In the Rotterdam area a second system is used, this relates allowed contaminant content to particle size distribution. This concept is based on results of ecological studies. This system is being used very often nowadays for sediments.
The concept of relating contaminant content to sediment characteristics is further extended. Not only particle size distribution but also other sediment characteristics, organic carbon content for example, can be taken into account. A draft standardisation system is being presented in 1986. For each contaminant, three guidelines are defined with a 'standard sediment' as a reference point. The quantitative formulation of the sediment classification guidelines will be based on environmental as well as economical considerations.
The systems as used for the time being in soil contamination and in the Rotterdam area are used.

Other criteria requiring formulation are criteria for dredged or processed sediments. Sand separated from sediment can possibly be re-used. Conditions under which this is allowed, for example in terms of contaminant contents must be stated. Another need for criteria lies in rules for the construction of deposits in relation to the contaminant content of disposable sediments. Heavy contaminated sediment must be deposited in better equiped disposal facilities.
Further definition of the two last mentioned criteria will not be given until the sediment classification guidelines have been established.

With this discussion on standardisation still going on, the inventory on sediment qualities is progressing. A first inventory, collecting all data available in 1984, resulted in a list of about 200 locations with heavily contaminated sediments.
The contaminants found vary from heavy metals to oil and PAH, pesticides etc.

The 200 locations are believed to represent a volume of 6 milion m^3 sediment. This is partly an inheritance, which means that once sanitation has taken place, newly sedimentated sediments will be clean. In several other cases there are still contaminating sources present, in again resulting contaminated sediments after a first sediment sanitation. In cases of absence of polluting sources also contaminated sediments can arise in the case of disastrous occurences.

The first survey results show that it is at least worthwhile to investigate the possible ways of treating sediments. The investigations are for the most part initiated by governmental institutions. The results can be used by the industry in building installations.

3. TREATMENT TECHNOLOGY

Treatment means processing the contaminated sediment with the aim:

- To minimize the contamination content in order to make re-use possible and/or
- To minimize the volume of the material that has to be disposed of.

The investigation into the treatment of contaminated sediments started in 1983 with an inventory of techniques to achieve the aim [1]. Nowadays the investigation programs involves testing and upscaling selected techniques. Some of the techniques have been applied on full-scale.

The experimental research on the techniques has resulted in the development of a scheme for the processing of contaminated sediment (figure 1). The process is built up into two stages. First hydrocyclone separation followed secondly by dewatering or decontamination of the slime.

3.1 Hydrocyclones

First the sediment is subjected to hydrocyclone classification. Hydrocyclones separate the sediment in an underflow consisting mainly of sand and an overflow consisting of the slime (fine particles and organic material). It is known that, in the fines and organic material, the contamination content is high compared to the coarse sand fraction. This means that the hydrocyclone separations devides the sediment into a relatively "clean" part (the sand) and one in which the contamination is concentrated (the slime).

Application of hydrocyclones in dredging operations has to cope with some specific dredging conditions such as:

- The dredging capacity is up to 15.000 m^3/h, for inland dredges 50-200 m^3/h. For hydrocyclones, the geometrical proportions is a most determinant factor for the capacity as well as for the cut-size. This means that an

increasing process capacity at a fixed cut-size results in multiplying the number of cyclones used.
A hydrocyclone with a diameter of 25 mm for example has a capacity of about 1 m^3/h and a cut-size of 10 μm. A 250 mm hydrocyclone has a cut-size of abt 45 μm and a capacity of 100 m^3/h. When the dredged sediment has to be classified at a low cut size and a high capacity a great number of cyclones will be required.
- Fluctations in the hydrocyclone feed-properties can be expected due to fluctuations in the composition of the dredged material in several waterways and in one waterway. For instance the particle-size distribution differs from one site to the other.
Hydrocyclones are that sensitive to feed fluctuations, such that the quality of the sand fraction can change strongly when particle-size distribution changes.

The choice of the hydrocyclones or hydrocyclone configuration strongly depends on both aspects, feed capacity and feed fluctuations.

3.2 Dewatering

Dewatering of the contaminated slime fraction results in a volume reduction of the material to be disposed of. Dewatering of sludges is often applied in environmental technology.
Known techniques are:

- sedimentation
- centrifugation
- vacuum-filtration
- belt-press filtration
- press-filter filtration.

To reach high capacity a mechanical dewatering technique is used often. Big mechanical dewatering equipments have a maximum capacity of about 50 m^3/h. So integration of slime dewatering in the dredging operation (50-200 m^3/h) can introduce a capacity problem. Multiplying the number of dewatering apparatus is unattractive, because of the high price and the largeness of these apparatus.

A new research programm is to be started in which the dewatering of the slimes at high capacities plays a major part.

3.3 Decontamination

Decontamination means a separation of the contaminant and sediment particles. There is no universal method to treat all possible contaminants. The possibly applied techniques could be roughly devided as follows:

- techniques for removing heavy metals,
- techniques for removing organic contaminants.

Heavy metal removal from soils and sediments can be achieved by the application of several techniques:

- leaching by acid,
- leaching by complexing agents,

- biological leaching by thiobacilles,
- an electrochemical method.

The results strongly depend on the pollutant involved e.g. leaching of cadmium and zinc is very succesful while the results for mercury are minor. Since the decontamination research is focussed on pollutants of list I of the E.C. cadmium and mercury are more extensively investigated when compared to the other heavy metals.
Based on the experimental research, we suggest that a suitable decontamination method for cadmium is acid leaching/ion exchange. The slime fraction is acidified to pH = 1. This acid suspension is then pumped through four ion exchange colums in series bottom up. The dissolved cadmium is removed from the suspension as a cadmium chloride complex by anion exchange resin. The advantage of anion exchange is that no calcium is attracted to the resin.

Thermal treatment is very succesful for the removal of mercury from sediments and soils. At a temperature of approx. 350° C, the mercury contents in the overflow is reduced from about 27 to < 0,1 mg/kg. A probably suitable technique for thermal treatment of mercurcy polluted sediments is steam-stripping. An advantage of this technique is that the vapour, after thermal treatment, is strongly reduced in volume by cooling. Other types of thermal treatment result in a vapour containing mercury, needing treatment before discharge in the atmosphere.

Techniques for the removal of organic contaminants can be devided in concentration and destruction techniques.

Concentration techniques	destruction techniques
solvent-extraction	incineration
flotation	biodegradation
steam-stripping	hydrolysis
washing	

As mentioned previously herein, our investigation has been directed to list I contaminants. As an example of an organic contaminant Polycyclic Aromatic Hydrocarbons (PAH) is chosen. Techniques which are investigated so far are solvent extraction, washing and biodegradation. It is not yet possible to make a recommandation for the best suitable technique.

4. TREATMENT RESULTS

4.1 Laboratory-scale

Some of the hydrocyclone results are quoted in table 1 and some of the decontamination results in table 2. Both tables give examples of laboratory-scale investigation. Investigations on dredged material from the Rotterdam harbours have shown that the hydrocyclone laboratory-scale results are very favourable comparable to the full scale results.
The hydrocyclone results in table 1 are presented as the efficiency value (E). This is the part of the feed that is separated in the sand fraction. The E value for the dry solids should be high and for the contaminants low. Table 1 shows that for most sediments, hydrocyclonage is rather succesful.

The minor results for sediment 5 are due to the type sediment, this is peat with an organic matter content of about 30%. From table 1 it is obvious that hydrocyclonage is most succesful for sandy sediment as can be expected.

Table 2 gives some decontamination results, and shows that high decontamination reductions are attainable.
Bacterial leaching gives the same results which are comparable with acid leaching for cadmium. Acid leaching is a more simple technique so that in practice this leaching is preferable. For other metals and other sediments however bacterial leaching can be more suitable. The results of leaching by complexing agents and an electrochemical method were less succesful sofar (decontamination reduction for Cd abt 50%).

Oil is extrated from a sediment with gasoline. After extraction the gasoline is separated by centrifugating.
With biodegradation, an oil reduction of 90% has been reached within two weeks for a slime fraction of soil.

PAH is extracted from a sample by using triethylamine as a solvent. This amine is soluble in water below a temperature of approx. 20° C and insoluble at higher temperatures. PAH is exctracted at 15° C and separated with the solvent after the sample is heated to 30° C.

As stated previously herein high Hg reductions are attained only by thermal treatment. Leaching with acid, hypochlorite and bacterial leaching were less succesful (Hg reductions < 10%).

4.2 Full-scale results

It is to be expected in the Netherlands that most of the sediments to be treated are located in inland waterways. Those waterways are dredged by dredgers with a capacity of approx. 50-200 m³/h. A hydrocyclone type used for sediment treatment is the linatex separator. This hydrocyclone is less sensitive to feed fluctuations due to its special construction.

In 1986 two sediment processing plants were in operation in the Netherlands. They both use a linatex separator to remove sand from the sediments. One of the plants removed the sediment from the water with a grabcrane. This plant is further equipped with a belt press to dewater the slime (overflow). The capacity has been about 18 m³ an hour. The water from the press is purified by flocculation/sedimentation.
The second plant is equipped with a suction dredger. The capacity of this dredger is about 100 m³/h. The overflow was dewatered in a basin after addition of flocculant to it. The water from the basin contained about 50 mg dry solids/l.
In both cases it has been possible to remove sand from the dredged sediments. The first plant, which handled fine sediments, removed 18% of the dry matter as sand, the second one 50%. The difference is mainly due to the macrocomposition of the dredged sediments. The main contaminants in the sediments were mineral oil and PAH respectively. Of the mineral oil 0.7% turned up in the sand during execution of the project first mentioned. In the second project about 5% of the PAH turned up in the sand.

Both projects, together with the laboratory results already available showed that application of separation to dredged sediments is very well possible. The costs of the treatment process must be compared with costs of conventional processes, including the construction of larger depots or larger dewatering ponds with many safety precautions against spreading of contaminants. Treatment will not always be economically favourable.

Decontamination of the slime is not yet applied on a full scale-basis. It is still under investigation on laboratory-scale.

5. CONCLUSIONS AND FINAL COMMENTS

- Hydrocyclones are suitable to separate contaminated sediment in a relative clean sand-fraction and a slime-fraction in which the contaminants are concentrated. This process is successful on laboratory-scale and in practice.
- Different contaminants require different decontamination techniques.
- Mercury is succesfully removed by thermal treatment, cadmium by leaching.
- Oil and PAH are removed by solvent extraction with a contamination reductions of 75% and higher.

Commercial application of the technology for contaminated dredged material remediation depends, amongst other matters, to the answers to two questions:

- Which quality criteria have to be reached?
- How big is the market?

Neither is answerable at the moment.

REFERENCE

[1] Methods for the treatment of contaminated dredged sediments.
W.J.Th. van Gemert, J. Quakernaat, H.J. van Veen;
In: Environmental Impact and Management of mine tailings and dredged material.
W. Salomons, N. Foerstner. To be published by Springer-Verlag.

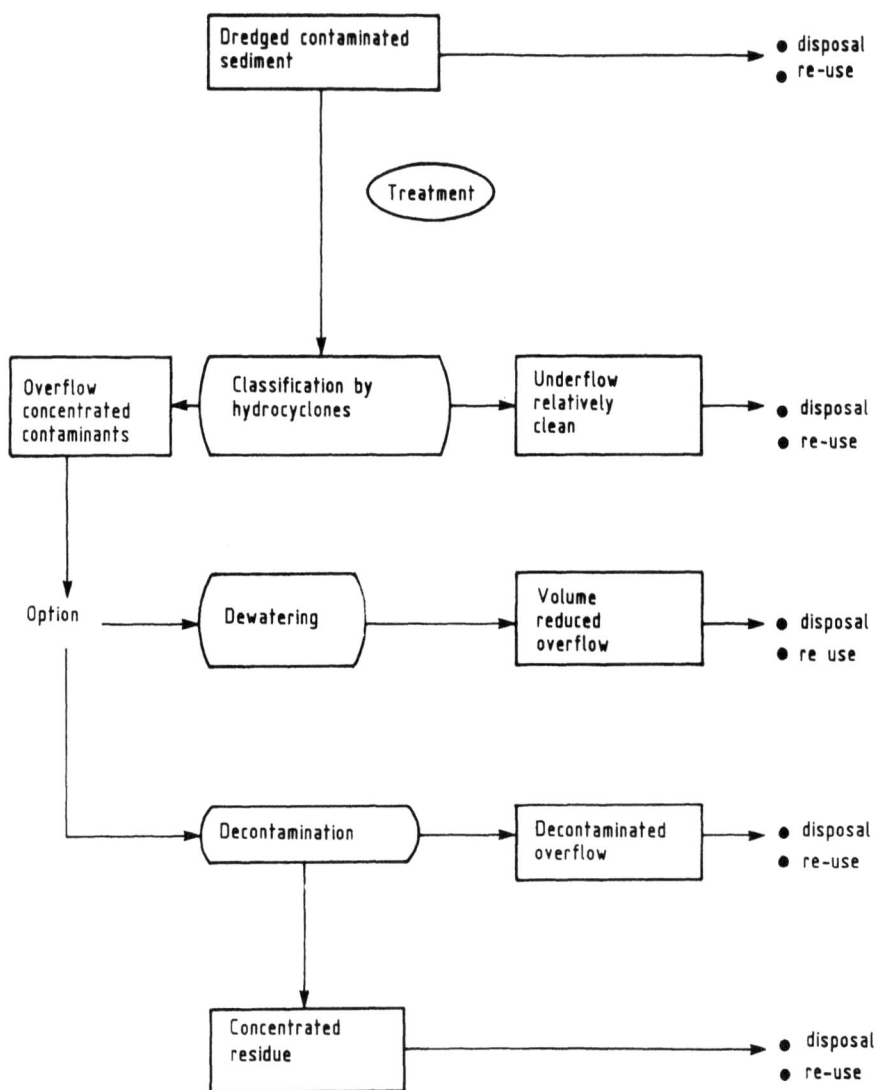

Fig. 1 Scenario for contaminated sediment treatment.

Table 1 Laboratory-scale hydrocyclone results

sediment	type waterway	main contaminant	type sediment	* E.d.s.	* Ex
1	river	cadmium	clay	50	10
2	harbour	oil	clay	60	10
3	river	cadmium	sand	90	15
4	canal	chromium	sand	75	15
5	canal	PAH	peat	40	30
6	river	cadmium	sand	90	30
7	canal	mercury	clay	50	20
8	canal	PCDD/PCDF**	sand	82	10

* E is efficiency = $\dfrac{\text{mass in underflow}}{\text{mass in feed}} \times 100\%$

d.s. is dry solids.
x is contaminant.

** PCDD is polychlorinated dibenzo-p-dioxins,
PCDF is polychlorinated dibenzofurans.

Table 2 Some laboratory-scale decontamination results

overflow of sediment (see table 1)	main contaminant	concentration before decontamination mg/kg d.s.	decontamination technique	contaminant reduction* %
1	Cd	250	acid leaching/ion exchange	90
2	oil	75.000	solvent extraction	85
3	Cd	17	acid leaching/ion exchange	80
5	PAH**	47	solvent extraction	75
7	Hg	27	thermal treatment	95

* reduction is $\left(1 - \dfrac{\text{contaminant content after decontamination}}{\text{contaminant content before decontamination}}\right) \times 100\%$

** PAH are six of Borneff.

RESEARCH ON POLLUTED SEDIMENT

P.J.A. BAAN (DELFT HYDRAULICS)

ABSTRACT

As a result of water pollution in the Netherlands in foregoing years sediments are polluted at many places. The presence of these sediments means a potential threat to the natural environment and functions to be fulfilled such as fisheries. Supply of unpolluted surface water and flushing do not solve this problem completely on short term as pollutants are only slowly released from the sediments.

To get insight in the extent of the problem of polluted sediments an inventory has to be made of the areas where the benefits of removing or cleaning up the sediment exceed the costs of this kind of measures. Delft Hydraulics has executed a first inventory study for the Haringvliet, an area, where polluted sediments have been accumulated for a long time. From this study it turned out that more scientific and technological knowledge has to be gained. Therefore research in this field has started and encompasses among others: i) physical, chemical and biological processes in the water sediment system; ii) costs and benefits of possible measures; iii) technical aspects involved in removing and cleaning up the sediment. A lot of research in this field still has to be done.

1. INTRODUCTION

With respect to water management in the Netherlands the so-called water system approach is followed (1). The approach aims at optimal coordination of the wishes of society with regard to the functions and the functioning of the water systems. A water system is defined as the geographically demarcated, interrelated and functioning whole of surface waters, groundwater, underwater beds, banks and technical infrastructure, including the existing ecosystems and all related physical, chemical and biological features and processes. The demarcation lines of a water system of this kind are determined in the first instance on the grounds of morphological, ecological and functional relationships. The water system approach also may be applied for the problem of polluted sediments.

As a result of water pollution in the Netherlands (and upstream in Germany, France and Switzerland) in foregoing years sediments ate polluted at many places. As ecosystems are affected by the surface water quality as well as by the quality of sediments, the presence of these polluted sediments means a potential threat to the natural environment and functions to be fulfilled such as fisheries. Supply of unpolluted surface water and flushing do not solve this problem completely on short term as pollutants like phosphate, heavy metals, and chlorinated hydrocarbons are only slowly released from the sediments. This means, that sanitation measures at sources of water pollution and waste water treatment are only partly an effective solution to this problem.

This paper deals with the problem of polluted sediments and discusses the results of a first inventory study for the Haringvliet, an area, where polluted sediments have been accumulated for many years.

2. PROBLEM INVESTIGATION

The problem area of polluted sediments is represented schematically in figure 1. Sources of pollution are e.g. waste water discharges of industry and households, atmospheric deposition, inflowing rivers, storm water runoff. The water distribution system (flow direction, flow rates, residence times, sedimentation and erosion patterns) determines the distribution of pollutants through the water system.

Within the water system pollutants are exchanged between the water phase, sediments, and organisms. This exchange and the resulting concentrations in the water phase, the sediments and living organisms determine the effects on functions and interests, which can be related to both water and sediment quality.

Measures to be taken can be directed at water pollution abatement, at preventing distribution of pollutants through the water system, and at compensating functions and interests for damages suffered. After establishing the costs and benefits of the measures the decision making process takes place.

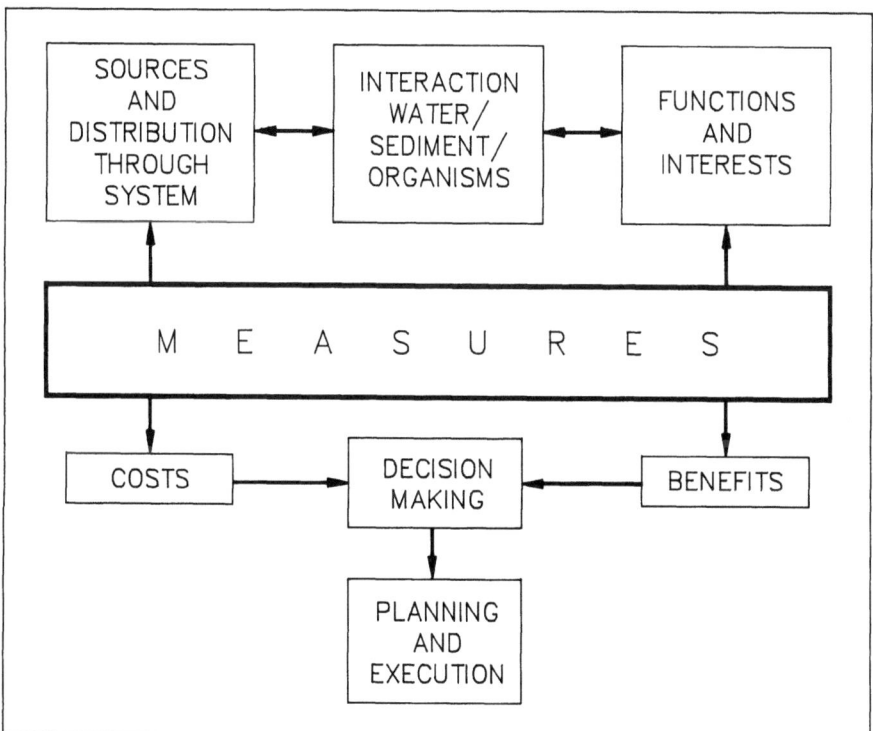

FIGURE 1. Scheme of problem area of polluted sediments

To get insight in the extent of the problem of polluted sediments an inventory has to be made of the areas where the benefits of measures exceed the costs. Delft Hydraulics has executed a first inventory study for the Haringvliet, an area, where sediments are polluted.

The project was aimed primarily at developing a system description and a decision making framework for the problem of polluted sediments, and applying this on the Haringvliet. Secondarily, an overview was wanted of the existing knowledge on this field and the most important flaws in it. The subsequent paragraphs deal with this matter.

3. POLLUTION OF SEDIMENTS

The input of pollutants, including those present in the inflowing streams, to a water system can be considered the gross pollution load. The net pollution load (sedimentation) is obtained by correcting for the pollutants present in the outflowing streams.

3.1. Sources of pollution

Sources of water and sediment pollution comprise:
o inflowing water (including inflowing suspended solids);
o waste water discharged by:
  - industry,
  - households and farmers (including cattle-breeding).
  - ships,
  - sewer systems (effluent of treatment plants);
o inputs from atmosphere;
o inputs due to leaching out and corrosion of bank protection material;
o storm water runoff (containing e.g. manure and pesticides);
o seepage of (polluted) groundwater;
o dumping of waste material (e.g. dredging spoils).

The relative contribution of the sources to the overall pollution depends on the water system under consideration and has to be investigated. After that measures to abate pollution can be analyzed and the effects of the measures assessed. The analysis comprises also the expected development of pollution over time.

3.2. Net pollution load

The physical and chemical water system (flow rates, in- and outflowing quantities of water and suspended solids, including the substances therein) need to be investigated to obtain balances of the pollution load for the sediments in the water system area and to study possibilities to reduce the net pollution load (the sedimentating part).

4. WATER/SEDIMENT INTERACTION

Seven 'compartments' can be distinguished which are in one way or another related to the quality of the sediments:
o surface water;
o suspended solids;
o sediment;
o pore water;
o biotics (organisms);
o groundwater;
o bottom.

The interfaces between some of these compartments are not very clear. E.g. it cannot be stated exactly where the pore water changes into groundwater or where the sediment ends and the bottom begins. For the whole, understanding of the interaction and exchange of substances from one compartment to another this hardly matters. Figure 2 gives an overview of the relations between the compartments. The relations are discussed below.

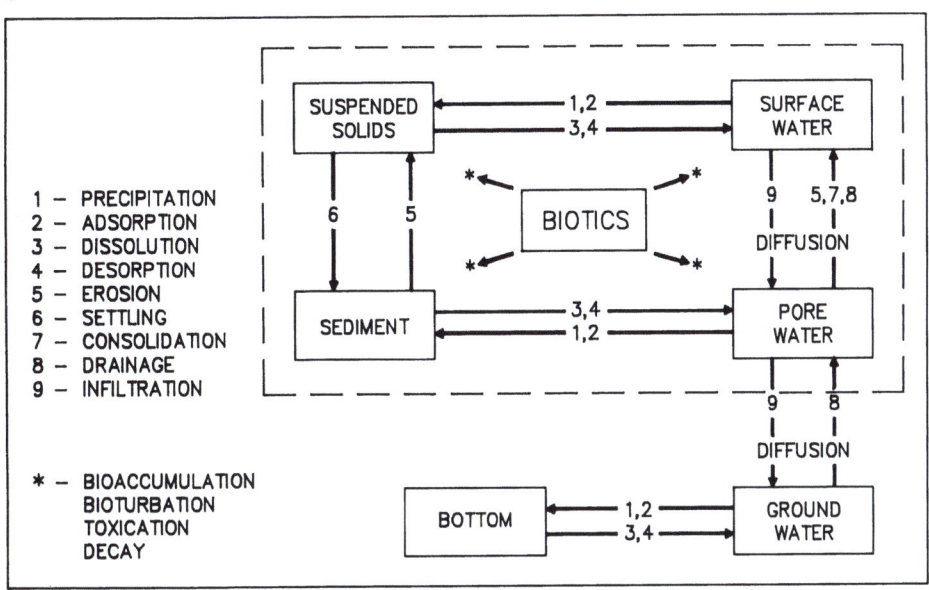

FIGURE 2. Relations between compartments

4.1. Surface water - suspended solids
Precipitation and dissolution together with adsorption and desorption are the most important processes responsible for the interaction. Factors, which influence these processes, differ by substance. E.g. precipitation-dissolution and sorption of heavy metals are affected by acidity and by the presence of complexing agents in the water phase. Organic substances mainly are absorbed on suspended solids; the adsorption capacity is dependent on the composition of the suspended solids.

4.2. Pore water - sediment
The processes that are responsible for the surface water - suspended solids interaction are involved in the pore water - sediment interaction too. Biodegradation of organic matter is an important factor in this case as it increases the presence of complexes in the water phase and changes the composition and adsorption capacity of the sediment.

4.3. Suspended solids - sediment
Sedimentation and erosion are the main processes to be mentioned. Sedimentation occurs at low and erosion at high flow velocities. Wind velocities, waves (also due to shipping) and bioturbation (activities of organisms like grubbing in the upper layer of the sediment) may play an important role in erosion too.

4.4. Surface water - pore water
Processes involved are i) diffusion (exchange of substances at interface of surface water and pore water); ii) consolidation of sediment (pore water is released); iii) erosion (also releases pore water); and iv) infiltration and drainage (results in advective flow through the sediment/bottom phase).

4.5. Pore water - groundwater
Main processes are infiltration and drainage, and diffusion. Infiltration and drainage occur as a result of differences in water level between surface water and groundwater and results in transport of pore water to groundwater or the other way around. Due to diffusion caused by

concentration differences substances are transported through the interface of pore water and groundwater.

4.6. Groundwater - bottom

The interaction between groundwater and bottom is of the same kind as the pore water - sediment interaction. Due to the fact that the physical and chemical composition of the groundwater-bottom system differs from that of the pore water-sediment system, equilibrium concentrations of pollutants in the water phase and the solid phase differ too.

4.7. Biotics - other compartments

Processes worth mentioning are bioaccumulation, toxication, bioturbation and decay of organisms. Bioaccumulation is the phenomenon of strongly increased concentrations of pollutants in organisms compared to those in the environment. Essentially, bioaccumulation is not controlled by the total concentration of pollutants in the environment but by the part that is biologically available.

Toxic effects of pollutants may cause malfunctioning of organisms and may alter their behavior. In the worst case it may cause death.

Bioturbation (moving around of organisms in the top layer of the sediment) may increase the exchange of pollutants between compartments.

Dead organisms constitute suspended solids in the water phase or after sedimentation are added to the sediment.

5. FUNCTION INVOLVED

Important functions and interests related to water and sediment quality are:
o recreation;
o public and industrial water supply;
o agriculture and cattle-breeding;
o fisheries;
o nature conservation;
o navigation;
o mining of sand and gravel.

Considering these functions and interests it must be kept in mind that the effects are not necessarily restricted to the area considered, but that functions and interests in neighbouring areas (as a result of in- and outflowing water and sediment) may be affected too.

5.1. Recreation

Active recreation like swimming and fishing may be affected by water and sediment quality. As a consequence of a bad quality swimming may become unattractive due to possible health effects and fishing may become less attractive (lower abundancy of attractive fish species and deteriorating fish quality). Moreover the appreciation of the landscape (clear water, natural bank overgrowth) may decrease.

5.2. Public and industrial water supply

Treatment costs of input water may increase as a result of deteriorating water (and sediment) quality. If certain pollutants are not removed completely and turn up in the drinking water supplied, health may be affected and the taste may be affected adversely.

5.3. Agriculture and cattle-breeding

The quality of water used for watering of cattle and sprinkling may affect animal and plant health and growth as well as the quality of the final products (meat, vegetables). It concerns surface water used as well as groundwater, as the latter may be polluted due to the above mentioned interactions.

5.4. Fisheries
The quality of water and sediment is an important factor considering the abundance of commercially attractive fish species and the quality of fish.
5.5. Conservation of nature
The conservation of a great diversity of fauna and flora is dependent directly on water and sediment quality. Strong relations exist with neighbouring water areas as well.
5.6. Navigation
Water depth in waterways is an important issue for navigation. As a result problems with the removal and dumping of dredged spoils from harbours and waterways due to a bad sediment quality impose costs on shipping and in this way have impacts.
5.7. Mining of sand and gravel
Potential use of sand and gravel may be affected by sediment quality.

6. POSSIBLE MEASURES
Four kinds of possible measures can be distinguished:
o abatement of water pollution;
o changing water management in the area considered;
o immobilization of pollutants in sediment;
o changing the functions of the water system under consideration.

The effects of the measures on water and sediment quality and as a result thereof on the functions involved have to be investigated. Both single measures as well as combinations of measures have to be studied.
6.1. Water pollution abatement
This comprises all kinds of measures, that force or give water polluting activities incentives to decrease water pollution.
6.2. Changing water management
Water and sediment flows in the water system considered may be altered. As a result sedimentation can be promoted at places wished for and prevented at places where damage is caused. Removing polluted sediment is another possible measure.
6.3. Immobilization of pollutants in sediment
Isolation of polluted sediment in restricted areas by barrages or e.g. plastic foil and (chemical) immobilization of pollutants in sediment can be mentioned as possible measures. Also polluted sediment can be 'diluted' with relative clean sediment or a polluted top layer of sediment can be exchanged with a 'clean' bottom layer.
6.4. Changing the functions of the water system
Measures in this context comprise changing the functions of the water system considered or reallocate them in such a way that adverse effects of a deteriorating water and sediment quality are minimized. Reallocation of functions may occur within the area considered or outside to other water systems.

7. APPLICATION ON HARINGVLIET
7.1. Measures studied
Figure 3 gives an overview of river streams in the Netherlands and the study area of the Haringvliet. Apart from a 'no change' alternative based on pollution conditions remaining unchanged four situations are studied which involved taking single measures:
1. abatement of water pollution;
2. dredging of polluted sediment;
3. decreasing inflow of polluted sediment;

4. disclosure of dam at seaside resulting in exchange of polluted sediment with 'clean' sediment of the North Sea due to tide movement.

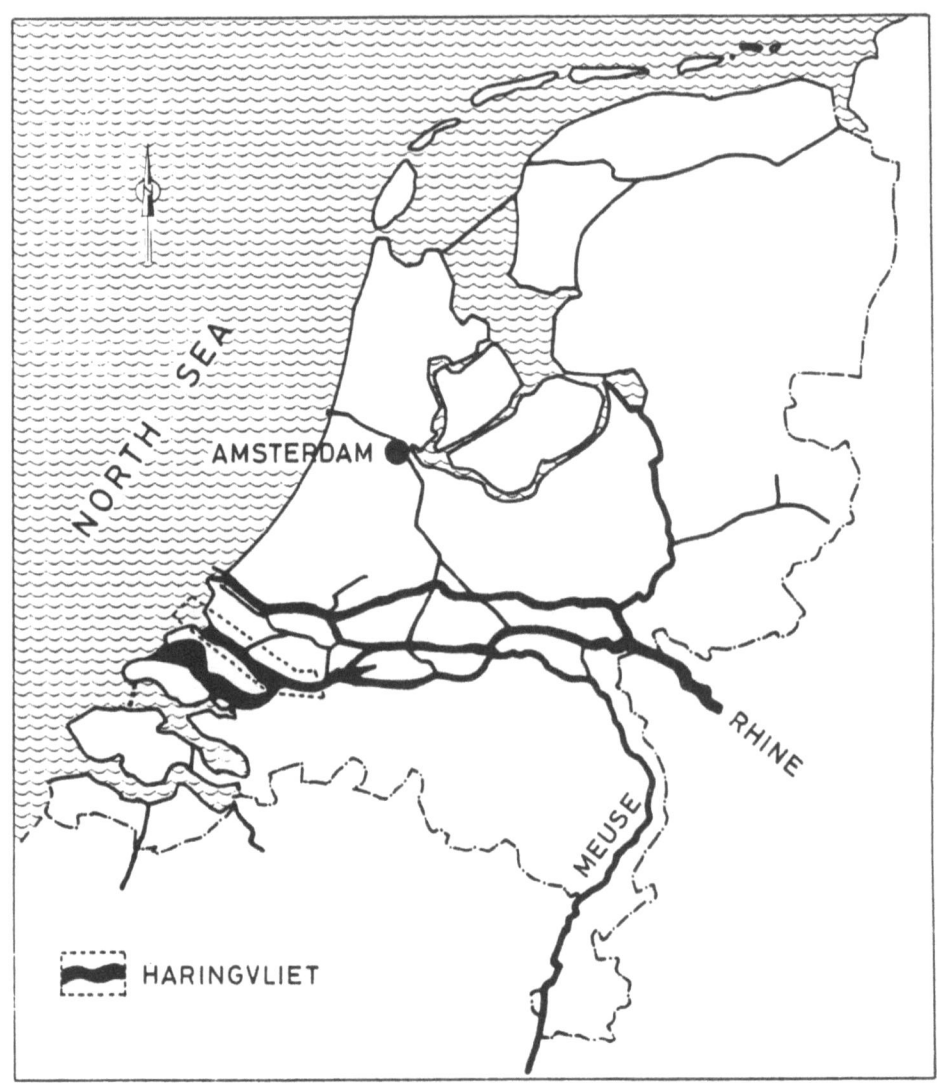

FIGURE 3. River streams in the Netherlands and study area of the Haringvliet

7.2. Effects on functions/interests

The effects of the measures with respect to water and sediment quality and as a result the effects on functions involved were estimated roughly in the following way. Firstly, the change in pollution 'content' of the seven compartments in the water system is estimated relative to the 'no change' situation. Secondly, it is investigated which compartments affect the distinct functions and in what way. Thirdly, the possible benefits are derived from the estimated 'total net present value' of the functions in the area considered. Estimating the effects is troublesome, beset with uncertainties, and as a consequence many assumptions are needed. Due to the limited budget for the study, costs and benefits were not quantified in monetary terms.

An overview of the results for the single measures are given in table 1. Positive effects are presented with one or two plusses (two means strongly positive) and negative effects with one or two minusses. Costs are small for measure 4 (disclosure of dam at seaside). Projected benefits are the largest and most certain for measure 1 (abatement of water pollution). Measures 2 and 3 during execution (dredging activities) have adverse effects on nature and fisheries. Measure 4 affects public water supply adversely (salt intrusion).

The tentative conclusion can be drawn that if a measure should be taken measure 1 looks most promising. Also indications are found that a relative small top layer of sediment is most important for the effects on nature (ecosystems), meaning that abatement of water pollution resulting in sedimentation and formation of a clean top layer over time probably will solve the major problem.

TABLE 1. Some results of analysis for the Haringliet

Effects	Measure			
	1	2	3	4
Costs	--	--	--	-
Benefits	++	+	+/-	+/-
(Un)doubtfulness if benefits are realized	++	+	-	-
Secondary effects				
- within water system		-	-	-
- outside water system				-

7.3. Need for research

In table 2 an overview is presented of the current level of knowledge and the relative importance of the distinct items and processes. A wide gap between relative importance (presented by R) and knowledge (presented by K) indicates that these items earn priority in research. From the table it can be concluded that more scientific and technological knowledge has to be gained. High priority should be given to research on interactions between the compartments in the water system. Also more insight is needed with respect to the costs and benefits (effects on functions) of possible measures. Another field of research encompasses technical and environmental aspects of removing and cleaning up the sediment. Removing sediment by e.g. dredging causes turbidity and release and redistribution of pollutants, which affects nature adversely.

TABLE 2. Level of knowledge and relative importance of items and processes

Level of knowledge/relative importance Items/processes	low	medium	high
Sources of pollution			
Inflowing water			KR
Waste water discharged	KR		
Atmospheric inputs	R	K	
Dumping, dredging		R	K
Interaction			
Surface water-suspended solids	R	K	
Pore water-sediment		K	R
Suspended solids-sediment		K	R
Surface water-pore water		K	R
Pore water-groundwater		K	R
Groundwater-bottom		K	R
Biotics-other compartments	K		R
Effects on functions/interests			
Recreation	KR		
Public and industrial water supply		KR	
Agriculture and cattle-breeding	K	R	
Fisheries	K	R	
Nature conservation	K		R
Navigation		R	K
Mining of sand and gravel	R	K	

REFERENCES

1. Living with water, policy note of the Ministry of Transport and Public Works in the Netherlands.

CONE PENETRATION TESTING IN RELATION TO ENVIRONMENTAL PROBLEMS

J.G. de Gijt
Laboratory Manager and Project Engineer
Fugro B.V., Leidschendam, The Netherlands
G. van Roekel
Project Engineer
Fugro B.V., Arnhem, The Netherlands

ABSTRACT

Pollution of soil and ground water is a significant environmental problem in many parts of the world. An overview is presented of existing cone penetration testing (CPT) techniques related to the study of these type of pollution.
The CPT techniques include sleeve friction measurement to determine the heterogeneity and stratigraphy as well as testing using the piezocone, temperature cone and electrical conductivity cone. For each of these techniques, examples of their use in environmental problems are presented, together with some possible developments for the near future.

INTRODUCTION

Environmental problems, especially soil and ground water pollution, have frequently been encountered all over the world during the last 10 years. A flow diagram is presented (Figure 1) which indicates how such problems may be identified and subsequent remedial measures can be developed. The solution of environmental problems is divided in four main phases. Apart from the historical data and future use of the area, knowledge of the soil type, soil stratigraphy, heterogeneity and geohydrology is essential for estimating the short and long term environmental impact and the determimation of remedial measures.

Traditionally soil and ground water data are obtained by drilling boreholes and installing piezometers, respectively. However, it will be shown that, by using different sensors in the cone penetration test (CPT), the site investigation can be carried out more economically, while at the same time additional environmental information will be collected. The CPT technique is briefly discussed, followed by examples of CPT's with different sensors. Finally, the application and some possible developments for the future are proposed.

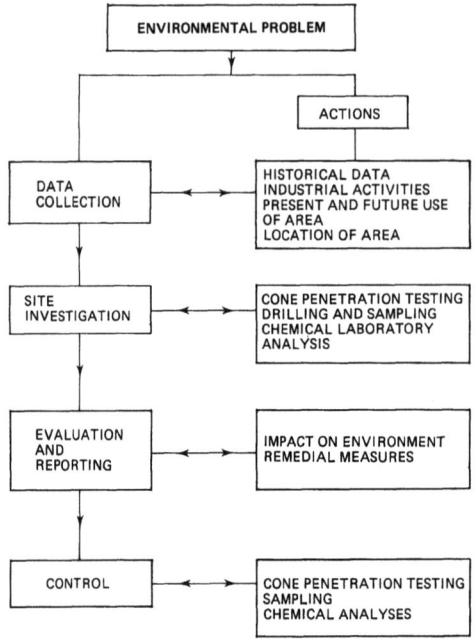

Fig. 1- Approach of environmental problem

CPT OPERATION AND CONE TIPS

The operational technique used to perform CPT's is basically similar for the various types of equipment discussed in this paper.
Detailed descriptions of testing procedures are given by De Ruiter (1981) and Verruyt et al (1982).

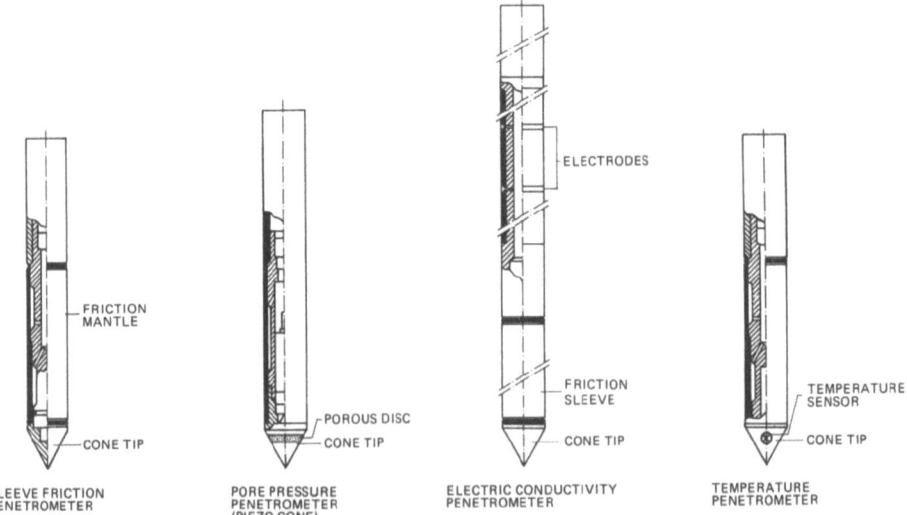

Fig. 2- Cone penetrometers

The standard CPT includes the determination of the cone resistance and the
sleeve friction. In addition, pore pressure, temperature or electrical
conductivity can be measured. If required the deviation of the cone tip from
the vertical can be recorded by means of a built-in inclinometer. Details of
the cones which measure sleeve friction, pore pressure, temperature and
electrical conductivity are shown in Figure 2. The probe dimensions are
generally the same. The only difference is the position and type of the
sensor.

DESCRIPTION OF TYPICAL CPT RESULTS

CPT with measurement of sleeve friction

The results of a CPT with measurement of sleeve friction enables the
computation of the friction ratio, being the ratio of sleeve friction and
cone resistance.
Based on empirical observations various correlations between CPT data and
soil type have been established. One such a correlation for the Fugro cone
is shown in Figure 3 (De Ruiter, 1981).

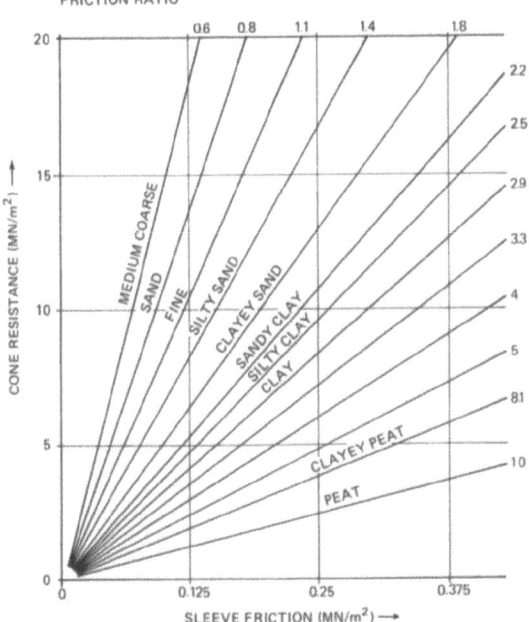

Fig. 3- Relation cone resistance,
sleeve friction and soil type

The results of a CPT, with simultaneous measurement of cone resistance,
sleeve friction and inclination, performed at a waste disposal site is given
in Figure 4.
Very irregular values of cone resistance and sleeve friction were obtained
in the upper 6.5 m. Comparing the measured cone resistance and local
friction with the values given in Figure 3 it appears likely that this

stratum is not of natural origin but is probably related to fill material, which may be potentially suspicious from the environmental point of view. The fill is underlain by Holocene deposits consisting of clay and peat and subsequently followed by Pleistocene sand layers. The boundary between the fill material and the natural ground is distinct. The CPT can be used economically to determine the thickness as well as the area covered by waste material.

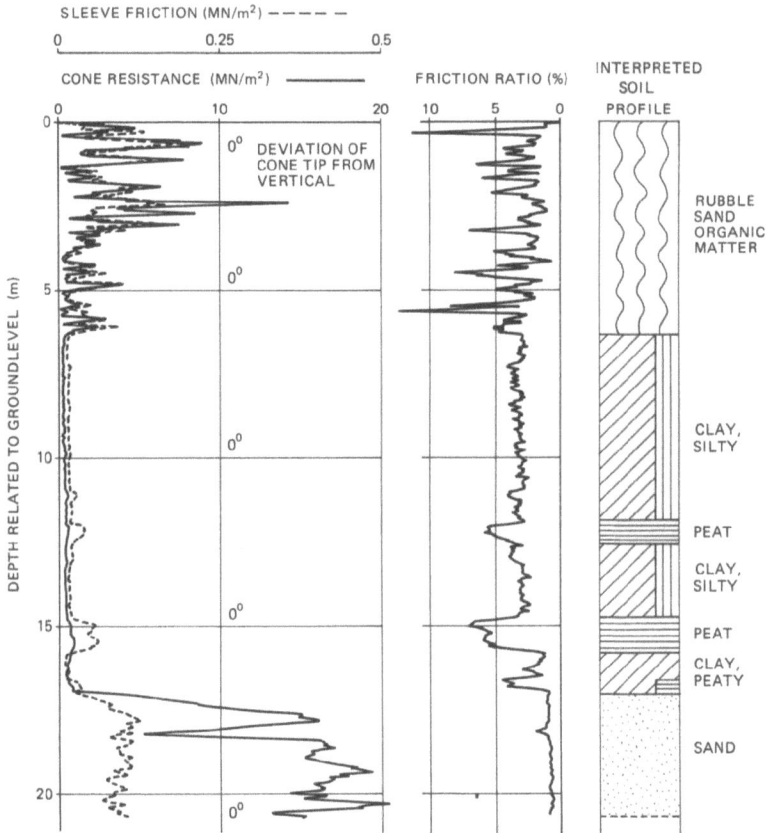

Fig. 4- Cone penetration test in waste disposal site

CPT with measurement of pore pressure

Figure 5 shows an example of a CPT with continuous pore water pressure measurement with depth generated during penetration. The hydrostatic pore pressure is also indicated. The difference between the measured pore water pressure in the cone and the hydrostatic pore water pressure is related to the soil type. In sands low or negative excess pore pressures are recorded; in clays and peats the excess pore pressures are high in relation to the hydrostatic pressure.

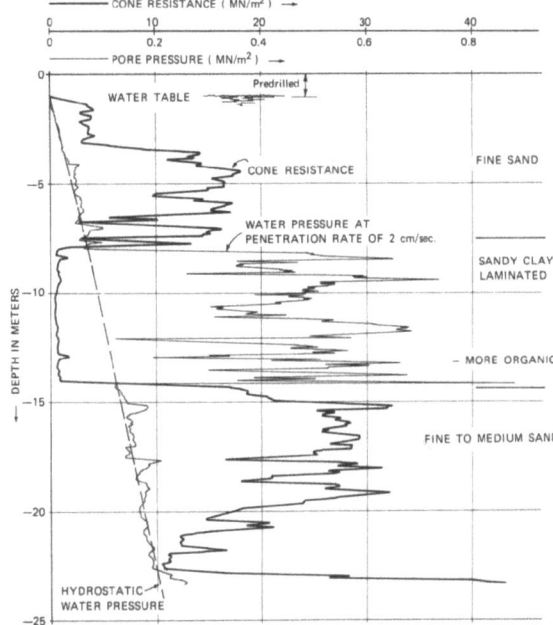

Fig. 5- Piezocone penetration test

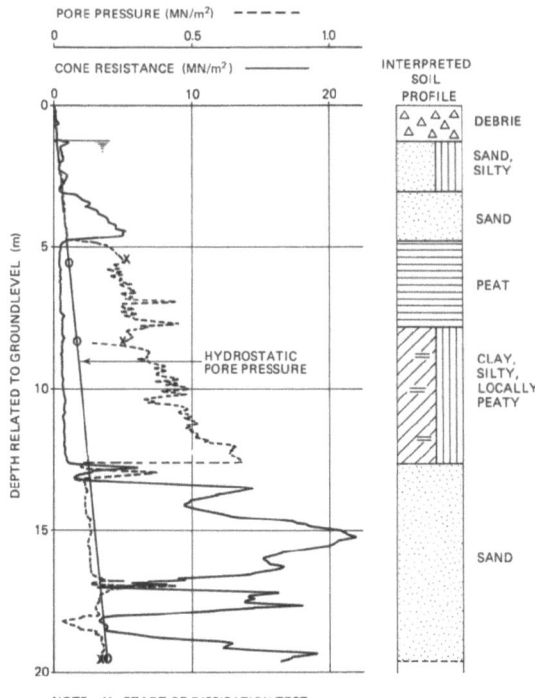

Fig. 6- Piezocone penetration test

If the penetration of the cone tip is halted a dissipation test can be carried out. This consists of observing the decrease of the excess pore pressure with time. This can be interpretated to give an estimate of the soil permeability (Robertson and Campanella, 1983).
The information obtained gives insight in the presence of low and high permeable layers. The static pore pressures give the hydraulic different layers. Both data are essential for environmental statics gradient between. Figure 6 gives an example of CPT with measurement of pore pressure and dissipation tests.

CPT with measurement of temperature

The results of a CPT with measurement of the temperature is shown in Figure 7. This CPT was made at a site where water was infiltrated after it had been used as cooling agent. The soil profile was determined using the cone resistance and the sleeve friction data and the relationships indicated in Figure 2.
For an accurate soil temperature measurement it is necessary to stop penetration because the temperature of the cone tip increases due to friction between cone and soil during penetration. Thus the required depth the CPT is interrupted, and the temperature of the cone tip is allowed to reach equilibrium with the surrounding soil.
Additional information concerning the measurement techniques and applications is given by De Gijt (1983).

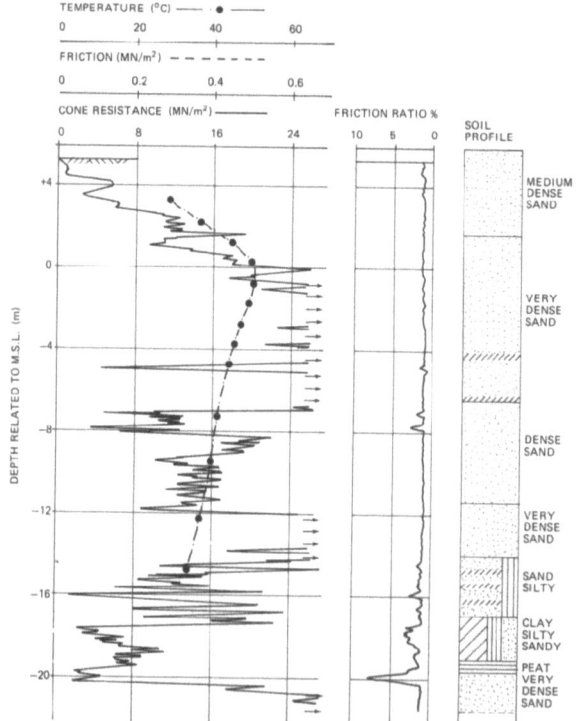

Fig. 7-
Cone penetration test with temperature measurement

The results of the temperature measurement in Figure 7 shows the temperature
at different depths. The results show that a maximum temperature of 50°C was
reached at a depth of 1 m minus mean seawater level (NAP). These high
temperatures are caused by infiltration of water with a temperature around
80°C. The water was used as cooling agent to reduce temperature of steel
plates in a rolling mill.

Under general conditions in Holland the soil temperature will be between 8
and 12°C, up to a depth of 30 m.
In a waste disposal in temperature may be increased due to biological
activity or oxidation processes.
By measuring the temperature inside and outside the waste disposal area an
indication of temperature gradients in the vertical and horizontal
directions is obtained. From these gradients an indication of the flow
pattern of ground water may also be derived. This method always requires
temperature measurements in non-contaminated areas for establishment of base
temperature levels..

CPT with measurement of electrical conductivity

In areas where pollution has occurred the pore fluid might be influenced by
the contamination. One of the properties of the pore fluid which can be
easily be measured in the field as well as in the laboratory is the
electrical conductivity.
The electrical conductivity measured during cone penetration is a function
of the quality of the pore fluid but also of the porosity and the
tortuosity.
To derive the electrical conductivity of the pore fluid alone the influence
of the porosity and tortuosity has to be eliminated. A method for this is
given by Archie (1942). From soil properties a so called formation factor is
established. By multiplying the measured soil conductivity with the
formation factor the pore fluid conductivity is obtained.
In general the higher the conductivity of the pore fluid the higher the ion-
content will be.

The results of two CPT's with measurement of electrical conductivity are
shown on Figure 8. These CPT's were carried out close to an old rubbish dump
in eastern Holland. The soil profile is built up of sandy deposits.
CPT1 was made at the edge of the disposal site. Between 7 meter up to 14
meter below groundlevel a substantial conductivity increase is shown. CPT2
was made at a distance of 100 m downstream. In this CPT it is clear that the
conductivity has decreased and also that the conductive zone is found at a
lower level. The base of this zone is at a 7 m lower level than in CPT1. The
subsequent chemical analysis of a ground water sample from 10 m depth near
CPT1 showed a chloride content of 40 mg/ltr and a sulphate content of 200
mg/ltr, which indicates that the ground water is indeed contaminated.

It is not possible to distinguish the different chemical components present
in the soil using conductivity measurements anone. Interpretation can be
carried out by correlating these CPT measurements with chemical analyses on
samples obtained by drilling and sampling techniques.

In addition to the detection of contaminated soil, this type of CPT can also
be used to determine the interface between fresh and salt water.
The CPT displayed in Figure 9 shows fresh and salt interfaces at respectively 0 and 18.0 meters below Mean Sea Level.

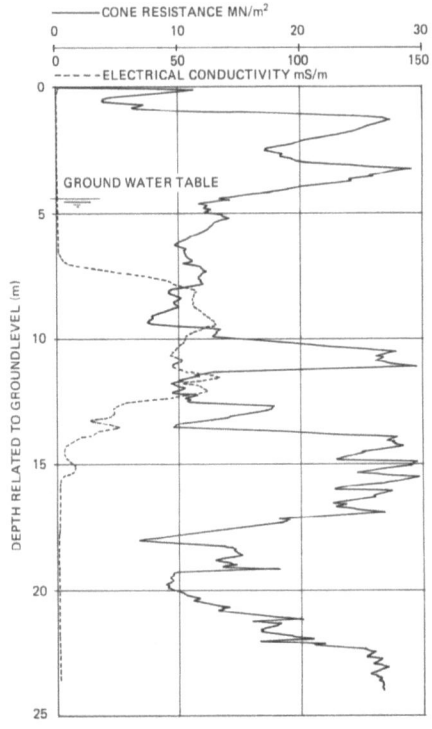

Fig. 8a- Cone penetration test near waste disposal site (Polluted)

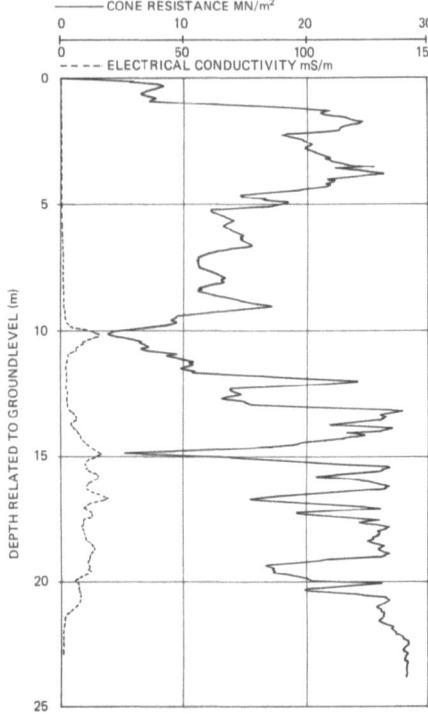

Fig. 8b- (Unpolluted)

APPLICATIONS

The possible applications of the different types of CPT are summarized in Table 1.

CPT with measurement of	De-tailed soil profile	Approximate soil profile	Permeability	Flow patern	Pollution	Fresh/Salt
Sleeve friction	x					
Pore Pressure Dissipation test		x	x	x		
Temperature		x		x	x	
Electrical Conductivity	x			x	x	x

Table 1. Applications of different CPT sensors.

Table 1 shows that the CPT with sleeve friction measurement provides data on the soil profile alone.
The CPT with pore pressure measured provides extra information such as

permeability, hydraulic head and flow patterns. CPT's with measurement of temperature and electrical conductivity give environmental information with impact to pollution. With the latter CPT fresh/salt water interfaces can be established.

Figure 1 has shown that, after the initial phase of data collection, a detailed site investigation has to be carried out.

During this investigation phase the soil and geohydrological data are gathered. This can be economically done by including CPT's in the early stage of the programme.

During the investigation, special emphasis should be given to the detection of layers of relatively high and low permeability. Once the CPT results are available, the engineer can decide more quickly and easily about the number and the location of borings, sampling of soil and piezometer installation for both sampling ground water and monitoring of ground water levels.

Figure 1 shows that, after data collection and reporting recommendations regarding measures to be taken, remedial works may take place. The verification and impact of these works can readily be investigated by making additional control CPT's with different sensors and selected sampling and chemical analysis.

FUTURE DEVELOPMENTS

It may be expected that the number of sensors per cone will increase. At the moment, for onshore purposes, cones with 3 sensors are commonly used (cone resistance, sleeve friction, inclination). It is expected that within a few years cones equipped with up to 5 sensors will be in daily use in soil investigations.
Further developments are expected in the field of chemical component specific sensors which will be built in the cone tip. At the moment it seems attractive and possible to develop sensors for the measurement of acidity and cyanide.
Another field of development concerns the provision of sampling techniques in the cone. Fugro has already developed a cone with a built in screen to sample air from soil pores. This cone was purpose built for sampling in coal stockpilings for power plants.
In addition Fugro is developing a water sample cone. This will have a screen above the tip which allows pore water to enter a chamber from where it is brought to the surface by nitrogen pump.
Some difficulties with the entry of water and cleaning of the system have yet to be overcome. For some specific investigations this technique has a promising future.

CONCLUSIONS

- The cone penetration test provided with additional sensors (e.g. for sleeve friction pore pressure temperature and conductivity) is a fast and economical in-situ testing technique for environmental site investigations. Its importance is still growing.

- The CPT equipment has potential capabilities of being provided with sensors measuring specific chemical components.

- Sampling of soil air during a CPT is already feasible and sampling of ground water is expected to be possible in the near future.

REFERENCES

1. Archie, G.E. (1942)
   The electrical resistivity log as an aid in determining some reservoir characteristics, Transaction American Institute Mineral Metallurgy Engineering 146, pp. 54-62.

2. Verruyt, A., Beringen, F.L. and Leeuw E.A. de, editors (1982)
   Penetration Testing, Second European Symposium on Penetration Testing, Amsterdam, A.A. Balkema.

3. Campanella, R.G. and Robertson, P.K. (1981)
   Applied Cone Research
   ASCE convention St. Louis
   Session 35: Cone Penetration Testing and Experience.

4. Robertson, P.K. and Campanella, R.G. (1983)
   Interpretation of Cone Penetration Tests Part I: Sand
   Canadian Geotechnical Journal 20 pp. 718-733.

5. Robertson, P.K. and Campanella, R.G. (1983)
   Interpretation of Cone penetration Tests Part II: Clay
   Canadian Geotechnical Journal 20 pp. 734-745.

6. De Gijt, J.G. (1983)
   The Measurement of the Soil Temperature with the Temperature Cone.
   Proceedings International Symposium Methods and Instrumentation for the Investigation of Ground water systems, Noordwijkerhout, The Netherlands, pp. 574-584.

RE-INFILTRATION OF WASTE TIP LEACHATE: AN INEXPENSIVE ALTERNATIVE

P.A. de Boks,

1. INTRODUCTION.
In The Netherlands, new waste tips have to be sealed at the bottom to prevent pollution of the groundwater.
By doing this, the highly polluted leachate can easily be collected by means of a drainage system. Before discharging the leachate into the surface water, it has to be treated.
For this treatment several purification systems are available, each system with its own characteristics. Therefore, a case-oriented approach is necessary for the selection of the optimal treatment system.
In the case of the leachate from a waste tip in Zoetermeer, a double re-infiltration system is recommended. In this way the tip itself is used as a bioreactor. The system consists of an anaerobic re-infiltration of the leachate as pre-treatment, and an aerobic re-infiltration of the methanogenic stabilized leachate as post-treatment.

2. CHARACTERISTICS OF THE WASTE TIP 'NOORD-WEST' IN ZOETERMEER.
The regional waste tip 'Noord-West' in Zoetermeer has a total area of 52 hectare. The waste disposal activities started in 1982 and are planned to be continued until 1990.
In principle, the tip is destined only for wastes from building, demolition and agricultural activities. From January 1987 onwards, the pure agricultural wastes, mainly from the greenhouses, have been composted elsewhere.
As a result of these waste sources the tip has, compared to domestic waste tips, the following characteristics:
- a high horizontal and vertical permeability;
- a low water storage capacity;
- low concentration of toxic compounds;
- an easy degradable organic fraction;
- a fairly predictable and rather rapid succession of the acidogenic and methanogenic stages of the stabilization process.

Characteristics of the leachate
In relatively 'young' parts of the waste tip (up to 2 till 3 years) highly polluted 'young' leachate accumulates.

This 'young' leachate is formed, due to the microbial degradation of the organic parts of the waste and the leaching process of the percolating rain water. In this stage the BOD value of the leachate is almost as high as the COD value because of the easily degradable character of the organic wastes.
After about 2 to 3 years, the tip changes from the acidogenic stage into the methanogenic stage. In the methanogenic stage the acid, removed by the methanogenic bacteria, is equal to the acid, produced by the acidogenic and acetogenic bacteria. As a consequence, the COD value of the leachate is now comparatively low. The N-Kj value, however, still increases slightly because of the degradation of the slowly degrading N-containing fraction of the waste. At the same time there is no N-Kj removal by the anaerobic stabilization process. In figure 1 a general course of the COD and the N-Kj development in the leachate is shown.

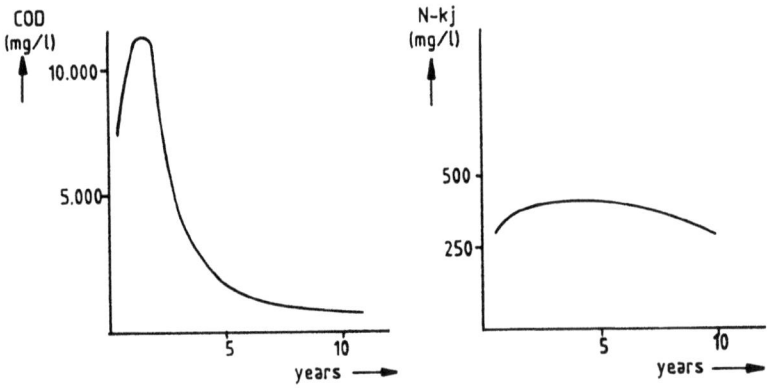

Figure 1: General course COD and the N-Kj development in the leachate.

At this moment about 3/4 of the capacity of the tip is utilized, while the rest is left for new disposal activities. About 2/3 of the disposed area is already in the methanogenic phase and about 1/3 is in the acidogenic phase.
As estimation of the COD and N-Kj load in the leachate is presented in table 1. This estimation is based on:
- the expected disposal acitivities;
- the development of the different stages in the tip;
- the assumptions mentioned in table 1.

Table 1: Estimated BOD and N-Kj load in leachate.

Conditions: Total tipe area Average amount of leachate 'Young' leachate Methanogenic stablized leachate:	: 52 ha. : 5 $m^3.^{-1}.d^{-1}$: 5000 mg/l BOD 300 mg/l N-Kj 50 mg/l BOD 400 mg/l N-Kj		
	1987-1989	1990-1992	1993-1995
* % area not utilized	25	0	0
* % area in acid stage	25	25	0
* % total area in methanogenic stage	50	75	100
* BOD load (t/j)	121	123	5
* N-Kj-load (t/j)	26	37	40
* Oxygen demand (t/j):			
+ Nitrification (4,6 N-Kj)	120	170	184
+ Total demand	241	293	189

3. LEACHATE TREATMENT SYSTEM.
3.1. ANAEROBIC PRE-TREATMENT.

In general, the leachate can be discharged into the municipal waste water collection system, if available, and can subsequently be treated at the treatment works (Lit. 1). However, in the case of a high organic content of the leachate (e.g. 'young' leachate), the liquid disposal charge is very high. Therefore, research has been done to find alternatives. Several treatment systems with a reactor placed on site of the waste tip (e.g. flocculation unit, reversed osmose unit, biorotor, UASB-reactor) have been tested (Lit 2, 3). The treatment costs of those systems are still high (between Dfl. 5,00 and Dfl. 10,00 per m^3); they will only be beneficial for highly polluted leachate. These high treatment costs are mainly caused by the high investment costs, which is from a financial point of view a disadvantage.

For 'young' leachate it can be concluded from these and other (Lit 2, 3, 4) experiments, that the best way to prevent a high liquid disposal charge is anaerobic pre-treatment. For the 'young' leachate in Zoetermeer an anaerobic re-infiltration in a section of the tip in the methanogenic stage is recommended as the anaerobic pre-treatment system.

The advantages of the re-infiltration system in comparison with other treatment systems are:
1. low investement costs;
2. low treatment costs (about Dfl. 0,50 till Dfl. 1,00);
3. an easy to operate technique, so managers of the waste tip can control the system themselves;
4. the treatment takes place underground which camouflages it from the public eye.

Despite these advantages, the system is hardly used, mainly because of flooding of the infiltrated area, due to a bad permeability and clogging processes.
This risk is further minimized by the selected type of re-infiltration system. A scheme of the system is given in figure 2.

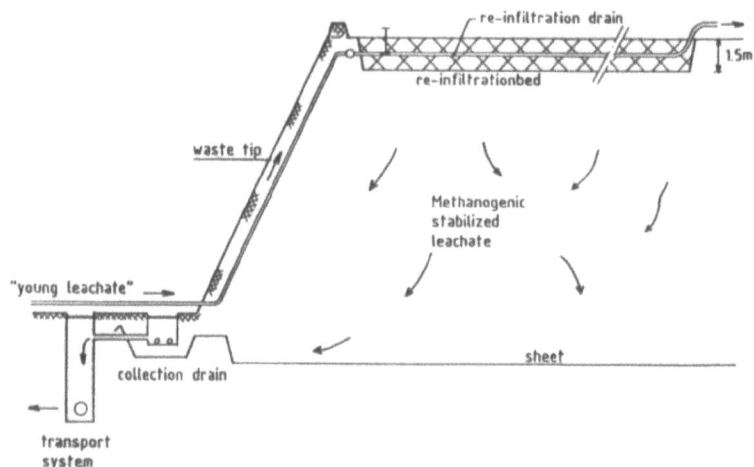

Figure 2: Scheme of the anaerobic re-infiltration system.

The leachate is re-infiltrated in horizontal drains with a depth of 1,5 to 2 m. So the total area for re-infiltration is enlarged and re-infiltration through the side wall of the drain is possible in case of clogging of the infiltration bottom.
Further this system has the advantage that the re-infiltration takes place in a hygienic way, especially in comparison to the sprinkler re-infiltration system. At the moment, experiments with this anaerobic re-infiltration system are carried out in Zoetermeer.

3.2. AEROBIC POST-TREATMENT.
Further removal of the COD and the N-Kj of the methanogenic stabilized leachate can be achieved by an aerobic biological post-treatment. This can be done by aerobic re-infiltration. Special attention has to be paid to the performance of the aerobic re-infiltration, because a succession of processes (nitrification and denitrification) has to take place.
As can be seen from table 1, the oxygen demand is mainly caused by the nitrification of the reduced nitrogen compounds. The nitrification can easily be performed in a modified trickling filter, placed on the tip. Nitrified leachate from the trickling filter can subsequently be re-infiltrated in the waste tip for the denitrification. The required BOD for the denitrification is present in the waste tip or can be added. In this way a two-step re-infiltration system is achieved.

The aerobic re-infiltration system has, in principal, the same benefits as the anaerobic system.
The practical experience, however, is small and is limited to experience with infiltration of household sewage effluent in soil absorption systems and aerobic post-treatment of methanogenic stabilized leachate (Lit. 5). Lab scale experiments with the nitrification and denitrification step are now being performed.

4. CONCLUSIONS.
For the treatment of the leachate from the waste tip 'Noord-West' in Zoetermeer, a two-step re-infiltration system is recommended. A scheme of this two-step system is given in figure 3.

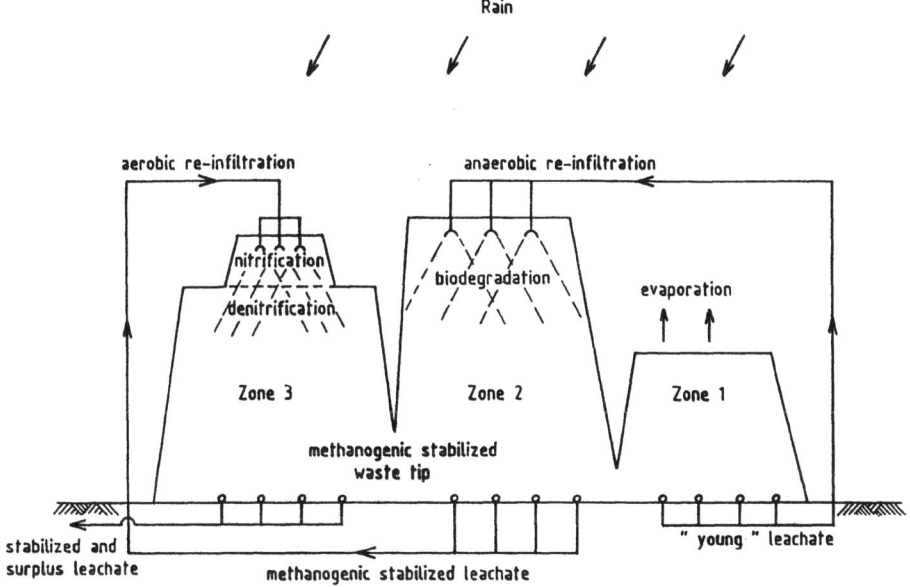

Figure 3: Schematic presentation of the two-step re-infiltration system.

Important factors for the technical feasibility of this system as permeability, clogging processes, water storage, biodegradability, absence toxical compounds pointed to the possibility for a re-infiltration option. The re-infiltration can be performed with simple and cheap materials and the tip itself is used as a bioreactor. Therefore, the re-infiltration is an inexpensive treatment system with low investment risks. The latest experimental results will be presented at the conference.

5. LITERATURE.

(1) Schuk, W.W., James, S.C.
Waste Management and Research, Vol. 4, 265-277 (1986).
(2) Schroeff van der, J.A., Woelders J.A.
Milieutechniek, nr. 8,70,71 (1986).
(3) Mennerich, A.
Veröffentlichungen des Institute für stadtbauwesen, T.U. Braunschweig, Vol. 39, 148-171 (1985).
(4) Mennerich, A., Albers, H.
Anearobic pre-treatment of high concentrated landfill leachates, conference paper Aquatech 1986, Amsterdam, 359-372 (1986).
(5) Schroeff van der, J.A.
Post-treatment of anaerobically stabilized leachate, conference paper Aquatech 1986, Amsterdam, 373-384 (1986).

NITRATE REMOVAL FROM GROUND WATER

J.P. van der HOEK
Agricultural University Wageningen
Department of Water Pollution Control
De Dreijen 12 - 6703 BC Wageningen - The Netherlands

1. INTRODUCTION

In the new E.C. directive relating to the quality of water intended for human consumption the maximum admissable concentration of nitrate in drinking water is decreased from 22.6 mg NO_3^--N/l to 11.3 mg NO_3^--N/l. The guide level is 5.6 mg NO_3^--N/l (1). At the same time, in many European countries an increasing nitrate concentration in ground water is observed. High nitrate concentrations in ground water are a consequence of fertilizer activities in agriculture. Both artificial fertilizers and animal manure cause nitrate problems (2, 3, 4, 5, 6, 7).

In the Netherlands contamination of ground water with nitrate is mainly caused by manure surplusses. These surplusses are a result of an intensification of animal husbandry. As about two-thirds of the drinking water in the Netherlands originates from ground water it is obvious that many problems are expected in coming years (8).

The Netherlands Waterworks Testing and research Institute (KIWA) estimates that 25% of the well fields exploited by the Dutch Waterworks may experience problems, either with nitrate itself or with the reaction products of nitrate reduction (9). The Institute for Land and Water Management Research made some calculations of future nitrate concentrations in ground water in relation to the use of manure as fertilizer in agriculture. It was concluded that with unchanged use of manure in agriculture 16 stations of 166 ground water abstraction stations will exceed the EC standard of 11.3 mg NO_3^--N/l, and 45 stations will exceed the EC standard of 5.6 mg NO_3^--N/l. With the use of manure only 75% of the optimum nitrogen dose, 7 stations will exceed 11.3 mg NO_3^--N/l and 31 stations will exceed 5.6 mg NO_3^--N/l (10).

From these figures it is clear that in the Dutch situation the problem cannot only be solved by restrictions in the use of manure, but that in the near future also nitrate removal processes have to be used to supply water with an acceptable nitrate concentration.

2. TREATMENT PROCESSES

To remove nitrate from ground water several techniques are available. Some of these techniques are summarized in table 1. Only ion exchange and biological denitrification are considered feasible and practical for full-scale treatment of drinking water. However, both these processes have serious disadvantages (12).

Biological denitrification is a process by which nitrate is converted into nitrogen gas by denitrifying bacteria. A direct contact is created between ground water, which is generally free of micro-organisms, and bacteria.

TABLE 1. Nitrate removal techniques (2, 11).

---

ion exchange
biological denitrification
chemical reduction
reverse osmosis
electrodialysis

In the case of heterotrophic denitrification also a carbon source has to be added to the ground water. Both aspects cause a serious risk of a bacteriological contamination of the ground water and to avoid this risk extensive post treatment is necessary (11, 13, 14, 15, 16). Also the production of nitrite, an intermediate product of biological denitrification, is a serious problem. Although the denitrification reactor may be well designed, a relatively high nitrite concentration is often observed in the effluent. Some examples are shown in table 2. The maximum admissable concentration is only 0.1 mg NO_2^-/l (0.03 mg NO_2^--N/l) (1).

Ion exchange is a physical chemical process. With an anion exchange resin nitrate is exchanged for chloride or bicarbonate. A problem is the regeneration procedure of the resin. Normally this is done by using a concentrated NaCl solution (50-100 g/l) at a flow rate of 2-4 BV/h (BV = bed volumes) for a period of 30-45 minutes (17, 18, 19). So, a large excess of salt is necessary for regeneration, while a voluminous brine is produced during regeneration with high nitrate, sulfate and chloride concentrations. Brine disposal can be very difficult. Both aspects cause financial and environmental problems.

3. COMBINED ION EXCHANGE/BIOLOGICAL DENITRIFICATION

The disadvantages of ion exchange and biological denitrification can be avoided by combining both techniques into one process. This process is schematically shown in figure 1. Nitrate is removed from the ground water

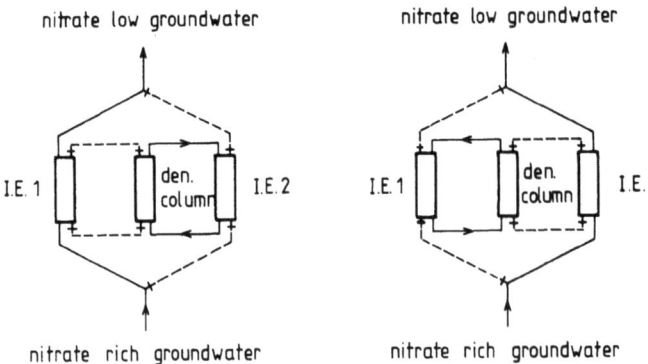

FIGURE 1. Combination of ion exchange and biological denitrification into one process.

TABLE 2. Nitrite production observed in biological denitrification of potable water

I. heterotrophic denitrification for nitrate removal from potable water

reactor type	carbon source	influent NO_3^--N (mg/l)	relative time with effluent NO_2^- > EC directive	max. NO_2^--N in effluent (mg/l)	ref.
.fluidized bed	methanol	14-25	25%	2	20
.fixed bed	acetic acid	18	20%	0.5	21
.immobilized bacteria in a calcium alginate gel	potassium aspartate	22	100%	3	22
.fixed bed with thermoplastic granules	starch di-2 ethylhexyl-phtalat	48-163	100%	8	23
.immobilized bacteria in a calcium alginate gel	ethanol	23	100%	0.09	24

II. autotrophic denitrification for nitrate removal from potable water

reactor type	electron donor	influent NO_3^--N (mg/l)	relative time with effluent NO_2^- > EC directive	max. NO_2^--N in effluent (mg/l)	ref.
.fluidized bed	hydrogen	56	100%	47	25
.fixed bed	hydrogen	25	15%	0.3	26
.fixed bed sulphur/limestone	sulphur	16	35%	0.3	27

by ion exchange but to regenerate a nitrate loaded resin a biological denitrification reactor is used. In the simplest form one ion exchange column (column 1) is used for potable water production while another ion exchange column (column 2) is regenerated by means of a denitrification reactor. When column 1 is exhausted and column 2 is regenerated, column 2 is put into the service mode and column 1 is put into the regeneration mode.

The regeneration process itself is schematically shown in figure 2. It can be carried out with a NaCl solution or NaHCO3 solution as regenerant. The regenerant, for example a NaHCO3 solution, passes over the ion exchanger to exchange nitrate ions for bicarbonate ions. After having passed the ion exchange column the regenerant, now rich in nitrate, is led through a denitrification reactor. In this reactor the denitrifying bacteria convert nitrate into nitrogen gas. The organic carbon source (methanol) which has to be added is converted into bicarbonate, carbonate, water and biomass. Then the regenerant is passed again through the ion exchange column and the denitrification reactor until the ion exchanger has reached a sufficient bicarbonate loading. The regeneration thus takes place in a closed system.

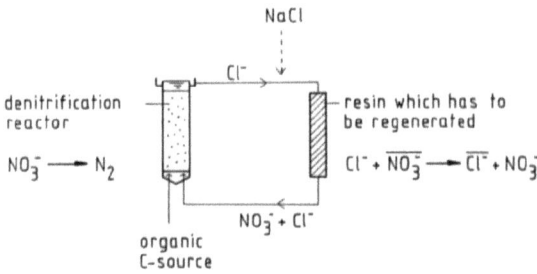

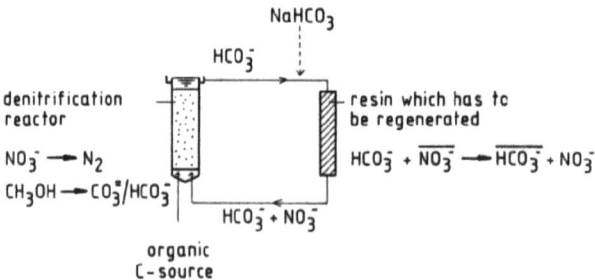

FIGURE 2. Regeneration of a nitrate loaded resin in a closed circuit to the chloride form (above) or bicarbonate form (down).

Compared with the separate techniques the most important advantages of this combined ion exchange/biological denitrification process are:
1. The regeneration is carried out in a closed system by which the production of a voluminous brine can be avoided and regeneration salt requirement can be minimized.
2. There is no risk that nitrite will affect the water quality because the biological denitrification process is not in direct contact with the ground water.
3. There is no direct contact between ground water on the one hand and bacteria and a carbon source on the other. Pollution of the resin by carry-over of suspended material from the denitrification reactor to the ion exchange column is possible, but measures against this can be taken in the regeneration circuit itself, without the need of extensive post treatment.

4. LAB-SCALE PILOT PLANT
4.1. Apparatus

The basic design criteria by which a combined ion exchange/biological dentrification plant for nitrate removal from groundwater can be designed are described in detail in (28, 29, 30). With these criteria a lab-scale pilot plant was built and here the main features will be discussed. Very schematically the pilot plant is shown in figure 3. It consists of three ion exchange columns, filled with resin Duolite A165, a sand filter and an Upflow Sludge Blanket (USB) denitrification reactor. Methanol is used as substrate for the denitrifying bacteria.

Two ion exchange columns are used for potable water production and have a run time of 9 h each. The flow rate through each column is 35 BV/h (downflow). They work 4.5 h out of phase.

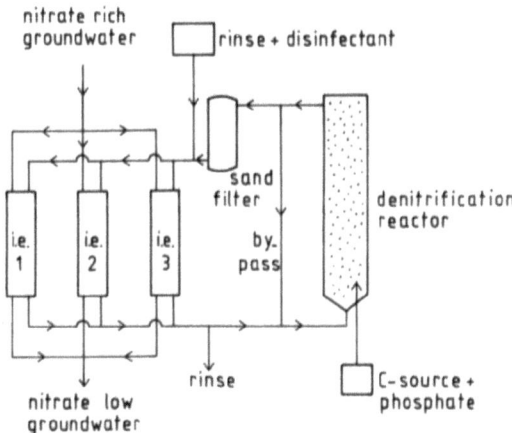

FIGURE 3. Pilot plant for nitrate removal from ground water by the combined ion exchange/biological denitrification process.

The third ion exchange column is connected with the denitrification reactor and is regenerated for 3.5 h. The regeneration flow rate is 10 BV/h (downflow) and the regenerant contains 20-30 g $NaHCO_3$/l. The sand filter in the regeneration circuit is used to prevent contamination of the resins by carry-over of sludge particles from the denitrification reactor. For the next hour this ion exchange column is rinsed with water (downflow) which contains a disinfectant the first 15 minutes. For this purpose peracetic acid (0.075%) can be used. The aim of this disinfection is to eliminate a bacteriological contamination of the resin, which may have occured in the regeneration period of 3.5 h. During rinsing water is recirculated through the denitrification reactor by means of a by-pass. In this way every 4.5 h a regenerated ion exchange column is put into service for nitrate removal from ground water. The pilot plant is controlled by a programmable logic controller. In table 2 the dimensions of the plant and ground water composition during the experimental period are summarized.

TABLE 2. Dimensions of the lab-scale pilot plant and ground water composition.

volume ion exchange column	0.95 l
volume denitrification reactor	5 l
ground water flow rate	65.7 l/h (2 * 32.9 l/h)
regeneration flow rate	9-12 l/h
rinse flow rate	9-12 l/h
ground water composition	
NO_3^--N	19.2 mg/l
SO_4^{2-}	29.5 mg/l
Cl^-	26.1 mg/l
HCO_3^-	98.3 mg/l
pH	7.8

4.2. Results

The nitrate concentration in the treated water is shown in figure 4. All measurements are related to the process-cycle of 4.5 h. During the experimental period three different periods could be distinguished with different denitrification reactor capacities. A higher dentrification reactor capacity results in a better regeneration of the resin, which means that a lower nitrate concentration in the treated water can be reached. During all three periods clearly a sort of break-through profile was visible in the 4.5 h process-cycle. This is caused by the fact that at the start of every 4.5 h process-cycle one ion exchange column is just put into the service mode for water production, while the other is already 4.5 in service. At the end of the 4.5 h process cycle one ion exchange column has been 4.5 h in the service mode and the other 9 h, resulting in a higher nitrate concentration in the treated water. With the used resin Duolite A165, a sulfate selective resin, sulfate was almost completely removed from the water. Only in period I sulfate was present in the treated water ranging from 3.4 to 5.8 mg SO_4^{2-}/l. Chloride concentrations varied between 4.4 and 39.7 mg Cl^-/l. Bicarbonate concentrations in the treated water were always higher than influent concentrations due to a $NaHCO_3$ regenerant. The highest measured concentration was 238 mg HCO_3^-/l. The pH ranged from 7.70 to 8.60.

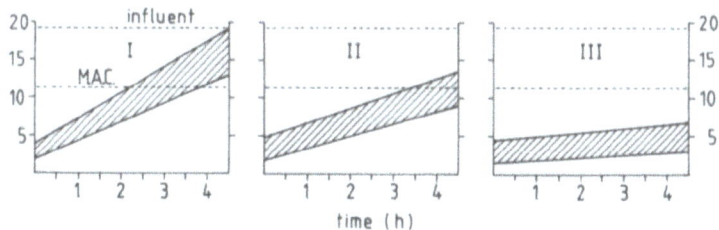

FIGURE 4. Nitrate concentrations in the treated ground water.
I   denitrification reactor capacity 525 mg N/h
II denitrification reactor capacity 625 mg N/h
III denitrification reactor capacity 840 mg N/h
(influent concentration 19.2 mg NO_3^-/l; Maximum Admissable Concentration 11.3 mg NO_3^-/l).

5. DEMONSTRATION PLANT

To study the combined ion exchange/biological denitrification process on a larger scale, a demonstration plant has been built at Montferland (Doetinchem, the Netherlands) with a capacity of 14 m^3/h. The design of the plant is the same as in figure 3. The plant consists of three ion exchange columns, filled with 200 l resin each, an USB denitrification reactor with a working volume of 3.3 m^3, and a sandfilter in the regeneration circuit (sand 1-2 mm). Two ion exchange columns are in the service mode and treat ground water with a flow rate of 7 m^3/h each, and the third ion exchange column is in the regeneration mode. The flow rate in the regeneration circuit is 2 m^3/h. After regeneration the regenerated column is rinsed and disinfected.

Compared with the lab scale pilot plant some changes were made in the design:
- in the demonstration plant a nitrate selective resin, Amberlite IRA 996, is used
- each column is in the service mode for 14 h and in the regeneration mode for 6 h, after which the resin is rinsed upflow for 1 h
- regeneration is carried out with 10.4 g NaCl/l
- disinfection is carried out during the first 45 minutes of rinsing with 0.2% hydrogen peroxide.

6. ECONOMIC ASPECTS OF THE COMBINED PROCESS

The combined ion exchange/biological denitrification process has important financial advantages compared with ion exchange as a result of minimal regeneration salt requirement and brine production. To have an indication of the financial implications of the process some rough cost calculations were made. Three alternatives for nitrate removal from ground water were compared:
I   conventional ion exchange
II   the combined ion exchange/biological denitrification process
III ion exchange with partial regeneration

Biological denitrification was not included for the reasons mentioned in paragraph 2.

Calculations were made for a full-scale plant with the following dimensions: - capacity 75 m³/h = 1800 m³/d = 394200 m³/y
- nitrate removal from 20 mg NO_3^--N/l to 0-3 mg NO_3^--N/l
- sulfate concentration of the ground water 100 mg SO_4^{2-}/l

The calculations were made with several assumptions:
- ion exchange costs were based on data of the Environmental Protection Agency
- construction costs only include manufactured equipment. Housing, excavation and side work, and pipe costs will be approximately equal for the three alternatives. Manufactured equipment is the major part of construction costs
- costs of brine disposal   Hfl 20/m³
  regeneration salt   Hfl 200/ton
  methanol   Hfl 700/m³
  disinfectant peracetic acid 30% Hfl 6500/ton
  energy   Hfl 0.30/kWh
- estimation of the costs of the combined process (II) from the costs of conventional ion exchange (I) with the following assumptions:
  . manufactured equipment 50% more
  . maintenance and labor   20% more
  . energy:energy for pump in regeneration circuit (12 m³/h)
  . brine disposal 5% of brine disposal of conventional ion exchange
  . regeneration salt requirement 10% of salt requirement of conventional ion exchange
  . methanol dosage 2 kg CH_3OH/kg NO_3^--N removed (29)
  . disinfection with 0.075% peracetic acid for 15 minutes
- estimation of the costs of ion exchange with partial regeneration (III) from the costs of conventional ion exchange (I):
  . manufactured equipment 25-30% more due to 25-30% lower resin capacity in connection with partial regeneration
  . regeneration salt requirement 50% less in connection with 150% excess instead of 300% excess
  . brine disposal 50% less

The results are presented in table 3.

ACKNOWLEDGEMENT

These investigations were supported by the Netherlands Technology Foundation (STW); Rossmark-Van Wijk & Boerma Water Treatment LTd.; the Ministry of Housing, Physical Planning and Environment; the Ministry of Economic Affairs; the Water Supply Company "Oostelijk Gelderland" and the Wageningen Agricultural University.

TABLE 3. Cost calculation of three alternatives (75 m³/h)
I   conventional ion exchange
II combined ion exchange/biological denitrification
III ion exchange with partial regeneration

	I	II	III
manufactured equipment	Hfl. 226.000	Hfl. 339.000	Hfl. 290.000
operation and maintenance costs			
. energy	Hfl. 6.480/y	Hfl. 7.340/y	Hfl. 6.480/y
. maintenance and labor	Hfl. 21.700/y	Hfl. 26.040/y	Hfl. 21.700/y
. brine disposal	Hfl. 80.000/y	Hfl. 4.000/y	Hfl. 40.000/y
. regeneration salt	Hfl. 165.000/y	Hfl. 16.500/y	Hfl. 82.600/y
. methanol		Hfl. 13.800/y	
. disinfectant		Hfl. 60.960/y	
	Hfl. 273.180/y +	Hfl. 128.640/y +	Hfl. 150.780/y +

REFERENCES

1. E.C. (1980): Council Directive of 15 July 1980 relating to the quality of water intended for human consumption, 80/778/EEC. Official Journal of the European Community 23: L229, 11-29.
2. Sontheimer H, Rohmann U (1984): Grundwasserbelastung mit Nitrat - Ursachen, Bedeutung, Lösungswege. GWF - Wasser/Abwasser 125: 599-608.
3. Strobel L, König F (1985): Massnahmen in Bayern zur Verringerung der Nitratbelastung des Trinkwassers. GWF - Wasser/Abwasser 126: 199-206.
4. Holtmeier EL (1984): Der Schutz des Grundwassers vor Nitratbelastung. GWF - Wasser/Abwasser 125: 482-487.
5. Furrer OJ, Stauffer W (1986): Stickstoff in der Landwirtschaft. Gas-Wasser-Abwasser 66: 460-472.
6. Marsh TJ (1980): Towards a nitrate balance for England and Wales. Wat. Services 84: 601-606.
7. Richard Y, Leprince A (1982): Pollution par les nitrates: traitements disponibles. Trib. Cebedeau 35: 21-33.
8. Scheltinga HMJ (1985): Nitrate problems in the Netherlands. Proceedings of the congress "Nitrates in Water", SITE 85, Paris, October 22-24, 1985.
9. Beek CGEM van, Kooy D van der, Noordam PC, Schippers JC (1984): Nitrate and drinking water supply. KIWA-Communication 84 (in Dutch).
10. Werkgroep nitraatuitspoeling waterwingebieden (1985): Ground water abstractions and nitrate problems in the Netherlands. Study of alternative measures. Report 12, Institute for Land and Water Management Research, Wageningen (in Dutch).
11. Sorg TJ (1979): Nitrate removal from drinking water. Paper presented at EPA Seminar on Nitrates in Groundwater, Kansas City, Missouri, October 3-4, 1979.
12. Hoek JP van der, Klapwijk A (1985): Nitrate removal from ground water. H_2O 18: 57-62 (in Dutch).
13. Barlog F (1980): Nitrat im Trinkwasser: Ursachen und Problemlösungen. Chem. Rundschau 33: 3, 16.
14. Sontheimer H, Cornel P, Fettig J, Rohmann U (1982): Grundwasserverunreinigung - Bedrohung für die öffentliche Wasserversorgung? GWF - Wasser/Abwasser 123: 521-530.
15. Haberer K (1984): Probleme und Möglichkeiten der Nitrateliminierung bei der Trinkwasseraufbereitung. Gewässerschutz-Wasser-Abwasser 65: 733-752.
16. Leprince A, Richard Y (1982): La bio-technique au service de l'eau de consommation: fiabilité et performance du traitement biologique des nitrates. Aqua Sci. Tech. Rev. 76: 455-462.
17. Guter GA (1982): Removal of nitrates from contaminated water supplies for public use. Report EPA-600/2-82-042, US Environmental Protection Agency.
18. Gauntlett RB (1975): Nitrate removal from water by ion exchange. Wat. Treat. and Exam. 24: 172-193.
19. Deguin A (1982): Elimination des nitrates par échange d'ions dans les eaux potables: Mise en équations du procédé. Trib. Cebedeau 35: 35-41.
20. Hall T, Zabel T (1984): Biological denitrification of potable water - final report to the department of the environment. Report 319 -S/1, Water Research Centre.
21. Frick BR, Richard Y (1985): Ergebnisse und Erfahrungen mit der biologischen Denitrifikation in einem Wasserwerk. Vom Wasser 64: 145-154.

22. Nilsson I, Ohlson S, Häggström L, Molin N, Mosbach K (1980): Denitrification of water using immobilized Pseudomonas denitrificans cells. Eur. J. Appl. Microbiol. Biotechnol. 10: 261-271.
23. Müller WR, Sperandio A (1986): Der Einsatz zweier Kunststoffgranulate für die Denitrifikation in der biologischen Wasseraufbereitung. GWF - Wasser/Abwasser 127: 1-10.
24. Nilsson I, Ohlson S (1982): Immobilized cells in microbial nitrate reduction. Appl. Biochem. Biotechnol. 7: 39-41.
25. Kurt M, Denac M, Dunn IJ, Bourne JR (1984): Denitrification of drinking water using hydrogen in a biological fluidized bed reactor. Proceedings of the third European Congress on Biotechnology, München, September 10-14, 1984, Vol. III: 163-168.
26. Bot, A (1985): Evaluation of two alternatives for nitrate removal from ground water. Report 85-12, Wageningen Agricultural University (in Dutch).
27. Kruithof JC, Paassen JAM van, Hynen WAM, Dierx HAL, Bennekom CA van (1985): Experiences with nitrate removel in the Eastern Netherlands. Proceedings of the congress "Nitrates in Water", SITE 85, Paris, October 22-24, 1985.
28. Hoek JP van der, Klapwijk A (1987): Nitrate removal from ground water. Submitted for publication in Water Research.
29. Hoek JP van der, Latour PJM, Klapwijk A (1987): Denitrification in the presence of high salt concentrations and at high pH levels. Submitted for publication in Appl. Microbiol. Biotechnol.
30. Hoek JP van der, Hoek WF van der, Klapwijk A (1987): Nitrate removal from ground water - use of a nitrate selective resin and a low concentrated regenerant. Submitted for publication in Water, Air, and Soil Pollut.

NITRATES IN GROUNDWATER

A.L. KOWAL, A. POLIK

The ever increasing concentration of nitrates in groundwater is becoming a problem of serious concern. Most of the nitrate loads come from the use of mineral fertilizers. But there are some more sources responsible for the continual increase of groundwater pollution from nitrates, e.g. municipal sewage, dumping sites, feedlots, effluents of septic tanks, etc. Biological fixation of airborne nitrogen in the soil cannot be neglected, either, as a contributing factor.
Shallow Tertiary and Quaternary groundwaters found in agricultural regions experience an evident increase in nitrate concentration due to mineral fertilizers, which ranges from trace amounts to 26 mg N/l. Nitrate concentrations measured in the proglacial Odra river valley varied from 0.2 to 16.0 mg N/l [1]. The lowest measured values, were those occuring in the infiltration wells operated by the waterworks of Wrocław 2 . This is because land use has been eliminated in aquiferous areas, and irrigation involves infiltration of low-nitrate surface water.
Very high concentrations were found to occur in the farm wells of a village belonging to the district of Wrocław. These varied from traces to 60 mg N/l and followed a seasonal pattern. The well displaying trace amounts showed a maximum level which amounted to 2.4 mg N/l. The well with the highest measured nitrate concentration, viz. 60 mg N/l, displayed a minimum which was as high as 20 mg N/l. The contribution of fertilizers was additionally confirmed by the presence of high potassium concentrations varying from 32 to 330 mg K/l [3].
In Hungary a strong relation between the rise of nitrate concentration in groundwater and the ever increasing use of mineral fertilizers has been reported [4]. In one of the districts groundwater pollution from nitrates rose from about 2.5 mg N/l in 1972 to about 12 mg N/l in 1978. In the period of 1950 to 1975 the use of mineral fertilizers in Hungary has become 30 times as high as it was before that date.
The presence of nitrates in groundwater is also becoming a problem in many other countries [5].
Investigations on the removal of nitrates from waterways are carried out in different centres. In terms of WHO standards, the admissible concentration of nitrates in drinking water is 10 mg N/l. The WHO standard is in force in Poland. The health implications of nitrates to infants and adults are commonly known and need no comments.

The methods by which nitrates are removed from waterways fall into two groups: biological and physicochemical. The biological method makes use of denitrification, a process which is well known in the technology of wastewater treatment. In wastewater technology denitrification involves methanol as oxygen acceptor carbon source . But methanol cannot be used in water technology because of the undesirable health implications to human organisms. It must be substituted by another compound which is both harmless to man and easily available to microorganisms. One of the waterworks of France is reported to apply biological denitrification for the degradation of nitrates [6]. In Hungary experiments were run to investigate the problem of nitrate nitrogen removal by lower apathogenic fungi of the Aspergillus genus [7]. The method, however, calls for further investigations in order to determine which of the organics products of fungi metabolism may penetrate the waterway, and how they may affect human health.

In many countries of the world attempts are made to replace biological denitrification by physicochemical methods, e.g. by reverse osmosis, electrodialysis or ion exchange both conventional and selective .

Reverse osmosis and electrodialysis are reported to be effective. In spite of this, both the methods should be regarded as a theoretical rather than practical success in nitrate removal because of the high operating costs and the lack of selectivity. Reverse osmosis or electrodialysis yields a complete removal of cations and anions, but has the disadvantage of contributing to the demineralization of the treated water. Since it is conventional to reduce nitrate concentration to the admissible level only, it seems worthwhile to pass a certain portion of the water through the system of reverse osmosis or electrodialysis and, after passage, mix this portion with the remainder.

Ion exchange is a relatively cheap and simple method. The problem consists in achieving a partial or complete selectivity of exchange in order to decrease the process costs.

NITRATE NITROGEN REMOVAL BY ION EXCHANGE

Nitrate ions may be removed efficiently by ion exchange both on strong base and weak base anion zeolites. But when the nitrate nitrogen removal by ion exchange is for drinking water, the "new" ions must not create health hazards to humans. Needless to say that a selective ion exchange would meet these health requirements, if of course high-selectivity ion exchangers were at hand. The experimental ion exchanger of 1-NMA type [3] displays a good selectivity in relation to NO_3^- ions. However, its high affinity for these ions makes its regeneration a difficult problem.

In a conventional exchange process, nitrate ions are removed from the water by exchanging them for the ions occupying the anion zeolite. Thus, the application of anion zeolites regenerated with soda lye accounts for the passage of hydroxyl ions to the waterway. The result is a drastic change in composition, which makes the water unfit for household supply. To obtain drinking water with no health implications it is advisable to remove nitrate ions by using anion exchangers which work in non-hydroxyl cycles.

REGENERATION OF A STRONG BASE ANION ZEOLITE

The regeneration process aims at providing favourable conditions for the occupation of the anion zeolite centres by active ions which will be exchanged for the ions removed from the water in the course of the cycle. When the removal of nitrates is carried out for drinking water, the following conditions must be fulfilled: a none of the persisting ions should occur at concentrations higher than the admissible level; b the co-ion entering the waterway should neither initiate reactions e.g. precipitation of sediments, liberation of gases nor deteriorate the quality of the water taste, odor, colour ; it must have furthermore, no health implications; c the regeneration procedure should be relatively simple, easy and cheap, and the effluent should create no environmental hazards.

Taking these into account, chloride ions Cl^- and bicarbonate ions HCO_3^- have been selected as appropriate for the regeneration of the anion zeolite.

EFFECT OF SOME IONS ON THE COURSE OF THE NITRATE ION EXCHANGE PROCESS

Natural waters carry predominantly bicarbonate HCO_3^-, chloride Cl^-, and sulphate SO_4^{2-} anions; when the water is contaminated with nitrates, nitrate ions NO_3^- will also be present. In the course of the ion exchange process, the ions contained in the water are competitive with one another, because each of them displays a different affinity for the ion exchanger. As a rule, the ion exchanger has a greater affinity for the counterion of a higher valency. And that is why sulphate ions are competitive with nitrate ions.

To assess the effect of sulphate ion content on the course of nitrate ion exchange, the effective exchange capacity of the anion zeolite was tested as a function of nitrate content and sulphate content. The tests involved a strong base anion zeolite of WOFATIT SBW type made in the GDR , which had been regenerated to the chloride form. The zeolite was fed with water which contained nitrates and sulphates only.

Figure 1 relates the effective nitrate exchange capacity of the WOFATIT SBW anion zeolite to the content of sulphate ions and nitrate ions in the water. As shown by the experimental data, the nitrate exchange capacity of the anion zeolite $E_c^{NO_3^-}$ may be described as follows:

$$E_c^{NO_3^-} = 563.55\ e^{-0.412\ c},\ mol/m^3 \qquad 1$$

where c is the quotient of sulphate ion concentration and nitrate ion concentration.

The correlation coefficient is very high, amounting to r = 0.958. Equation 1 confirms the distinct effect of sulphates on the course of the process. The effect can be deter-

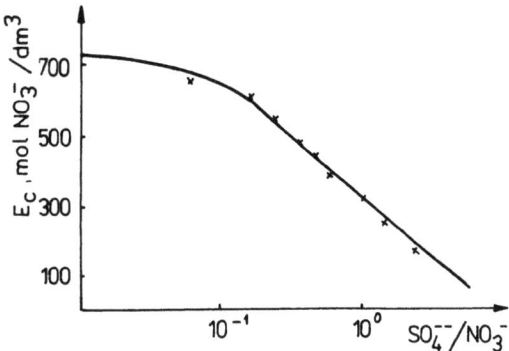

FIGURE 1. Exchange capacity of WOFATIT SBW anion zeolite as a function of nitrate ion and sulphate ion concentrations in raw water.

mined with a good accuracy, when the ion composition of raw water is known.

There is another anion which has an important part in the nitrate ion exchange process - the bicarbonate ion.

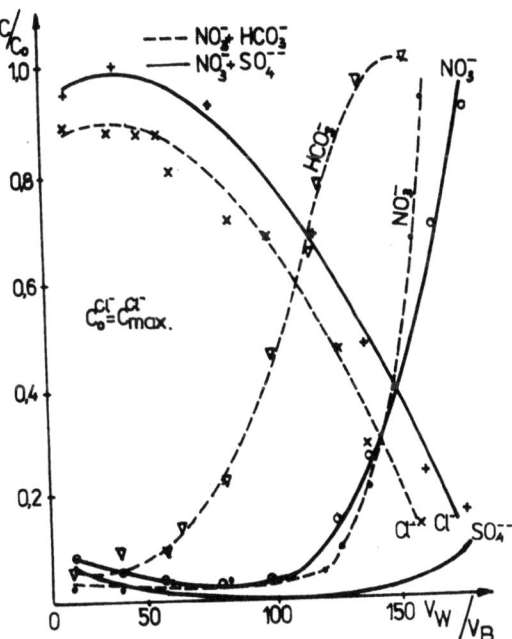

FIGURE 2. Plots of two ion exchange cycles on WOFATIT SBW anion zeolite.

Figure 2 gives the plots for two cycles of ion exchange. The feed for the first cycle contained nitrate ions and sulphate ions, whereas that for the second cycle contained nitrate ions and bicarbonate ions. As shown by these curves, bicarbonate ions have only a slight effect on the course of the process, if at all; they are not competitive with nitrate ions. Of the investigated ions, sulphate ion displays the highest affinity for the ion exchanger of interest. The affinity of nitrate ion and that of bicarbonate ion ranks second and third, respectively. Thus, we can write:

$$A_{SO_4^{2-}} > A_{NO_3^-} > A_{HCO_3^-} \qquad 2$$

NITRATE ION EXCHANGE ON A STRONG BASE ANION ZEOLITE IN THE CHLORIDE CYCLE

The experiments were run on tap water samples with artificial nitrate enrichment, and on WOFATIT SBW. The plots are shown in Figure 3. As it could be expected, the ions of inte-

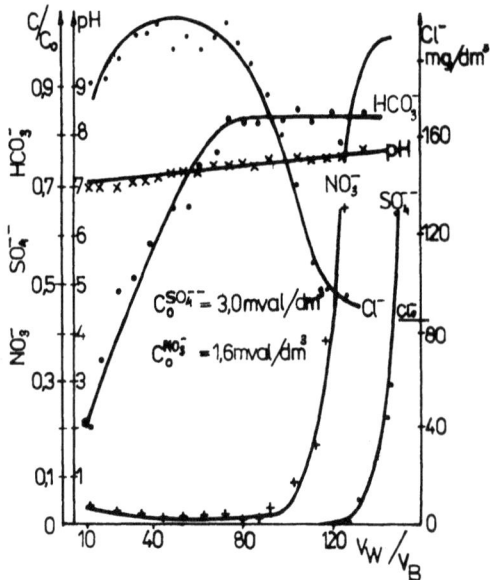

FIGURE 3. Plot of ion exchange on WOFATIT SBW anion zeolite in the chloride cycle.

rest appeared in the effluent in the following sequence: HCO_3^-, NO_3^- and SO_4^{2-}. Up to the moment at which nitrates appeared in the effluent, a water volume V_w of 120 dm^3 with an initial nitrate concentration $C_0^{NO_3}$ of 1.6 mol/m^3 had passed through the anion zeolite bed of a 0.8 dm^3 volume V_1. Thus, the calculated effective nitrate ion exchange capacity of the anion zeolite becomes $E_c^{NO_3} = 240$ mol/m^3. The theoretical exchange capacity calculated in terms of Equ. 1 for c = 1.875 amounts $E_t^{NO_3} = 260$ mol/m^3. On comparing the two values it is obvious that Equ. 1 may be applied in engineering practice, because the difference is below 10%.

No pH variations were found to occur throughout the process, because the chloride cycle does not influence the pH level. When the process was coming to completion, the effluent concentration of chloride ions dropped rapidly. Nitrate ion concentration in the effluent did not increase until the concentration of chlorides had almost reached the initial value. A similar relation was found to occur between nitrate ions and sulphate ions. Sulphate ion concentration began to increase, after nitrate ion concentration had approached its initial value. Knowing these relations, facilitates control of the exchange process. When a single anion zeolite is in use, regeneration must be carried out before its exchange capacity is exhausted. If the number of reactors is greater than one, the bed may be brought to complete exhaustion, but individual reactors should be regenerated after successive cycle.

SUMMARIZING COMMENTS

Shallow groundwaters experience a continual increase of nitrate concentration. If deep groundwater resources are lacking or if there is no possibility to mix waters of various nitrate concentrations, nitrates have to be removed by other methods. One of these is ion exchange on a strong base anion zeolite. Besides the nitrate ion, all of the anions found in the waterway have a more or less distinct influence on the course of the process. This finding may be of considerable help to the designer of the ion exchange system. The chloride cycle exerts no undesirable effect on the quality of the effluent, so it may be of utility for the removal of nitrates by the ion exchange method. The anion exchange zeolite which is to be used for the removal of nitrates, should be subject to conformity certification by Public Health Service.

REFERENCES

1. Roszak, W.: Quaternary groundwater pollution in the Odra Valley, Proc. of Conference. AGH, Kraków, 1986 (in polish).
2. Kowal, A.L., Serwach, A.: Efficiency of slow sand filters and infiltration in the Waterworks of Wrocław. Report of the Inst. of Environ. Prot. Engng., Technical Univ. of Wrocław, 1975 (in polish; unpublished).
3. Kowal, A.L., Leszczyńska, D.: Groundwater pollution. Case study. Report of the Inst. of Environ. Prot. Engng., Technical Univ. of Wrocław, 1986 in polish; unpublished.
4. Kowal, A.L.: Nitrogen compounds as water pollutant. Ochrona Środowiska, PZITS Nr 488/3 29 Wrocław 1986 (in polish).
5. Conrad, J.: Regulation of agriculturally induced nitrate contamination of water in some European countries. IIESS Center, Berlin.
6. Frick, B.R., Richard, Y.: Ergebnisse und Erfahrungen mit der biologischen Denitrification in einem Wasserwerk. Vom Wasser 64, 1985.
7. Szilágyi, F., Major, V.: Nitrates in drinking water - biological treatment. Élet és Tudomány Nr 21, vol. 24, 1985 pp. 660-662 (in hungerian).

INDAVER N.V. - A NEW INDUSTRIAL WASTE TREATMENT PLANT IN BELGIUM

L.M.J. STERCKX, Dr.Ir., General Manager

1. Introduction

As in any other highly industrialised country, generation of industrial waste is comparatively high and diversified in Belgium. Untill now the country has no centralised facilities available to treat the totality of those waste streams appropiately.
In order to provide such installations, Flemish Government and Chemical Industry joined in the foundation of a new company Indaver. The goal of this company is to become a facility for treatment and disposal of industrial, hazardous chemical waste.
As a starting company we would like to build out our relationsship with the outside world and particularly keep all interested parties informed on the progress of the project. Especially on the international scene we do not have the intention to operate in some kind of splendid isolation. Quite on the contrary we feel there are tremendous and growing opportunities for over the border cooperation in many ways.
Consequently it will be a pleasure to give you a short overview of the background, priorities, status and plans of this new company.

2. Background

Indaver N.V. was founded as a private company on the 24 th of october 1985. Besides three governmental organisations, fifteen major chemical companies participate in the initial capital of 533.5 million BF (14 M USD). The prime objective of Indaver is to start up the required facilities in Flanders for treatment and disposal of industrial, hazardous chemical waste.
The clear need for such a centralised plant has been recognised both by government and industry. In this respect the mixed participation of industry and official governmental organisations has been viewed as a major strenghtening factor for the company. Indeed this structure maximises the opportunities for joining all available expertise in the field of waste treatment .

The goal of the company is to become an organisation that provides both a horizontally and vertically integrated structure. Horizontal integration is defined here as the objective the plant has to be able to handle the broader range of problematic industrial waste in Flanders. Through vertical integration the organisation will provide treatment from the point of reception of the waste up to its final disposition including all intermediate steps necessary. This will give direct supervision and control of Indaver on the totality of the treatment and consequently optimally guarantee the quality consistency of the operations to the customer.

In the concept of the company it is important to stress a number of priorities which have been brought forward from the foundation on. First of all a high level of professionalism is aimed at providing safe operation with minimal impact on the environment. Secondly the company will have to be an economically viable unit where waste treatment prices cover operational and investment costs and additionally provide a return to the shareholders. Indaver will thus harmonise ecology and economy, in fact it is the company's point of view that these two priorities are complementary rather than contrary in proper waste treatment. Thirdly the plant will be designed so that it can operate in a very flexible way enabling it to cope with changing quantities and qualities of waste as well as with changing legislation. This might imply operation of pilot plant facilities in order to precisely define the appropriate equipment for some special treatments. Last but not least, the facilities should provide a centralised unit able to guarantee optimal treatment of the larger majority op problematic waste in Flanders.

3. Status

Since the creation of the company a piece of land of 25 hectares has been bought in the North of the industrial port area of Antwerp. The site has been prepared for starting the contruction works. The civil works for the construction of a controlled landfill have progressed to a level of 50 % completion. Essential connections such as electrical power, water and telex/telephone have been realised. Preliminary buildings have been erected for housing the employees and engineering staff during the construction period.

The full management staff of Indaver has been recruited as well as some administrative personnel. The engineering and design of the plant is progressing under the coordination of Belconsulting, a Flemish engineering company. The documents for application of the necessary licences have been prepared. These include a study assessing the overall impact of the plant on environment and safety.

4. Plans

As far as the plans are concerned the plant will essentially consist of three major parts : an incinerator, a set of physico - chemical treatments and a landfill. Firstly there will be an incinerator capable of burning at least 35 000 ton/year of waste. In order to provide high flexibility and versatility it will be a rotary kiln type able of handling solid, liquid and pasteous waste. The intention is to reach average incineration temperatures well above 1000 °C in the after burning chamber. Combined with a retention time of at least 3 seconds the incinerator will achieve complete combustion. Heat generated will be recovered as steam. Thorough gas cleaning including scrubbing and filtering will be the additional features of the installation. The goal is to comply with the highest standards in Europe for this operation.

The second part of the plant will be a set of interconnectable, modular and flexible installations capable of providing a number of physico-chemical techniques such as oxydation, reduction, solidification, immobilisation, breaking of emulsions, neutralisation, and waste oil treatment. The design, operation and further optimisation of these units will be one of the most challenging tasks of this project. In this context it will most likely be necessary to perform laboratory tests and even operate on a pilot-plant scale before making final technology selections.

Finally a controlled landfill operation will be the third part of the project. It will be designed to receive the burned out incinerator residues, the neutralised waste from the physico-chemical units and the directly disposable waste from the customers. A surface of 12 hectares has been foreseen for this activity, annual input is estimated around 50.000 ton/year. The landfill operation will feature triple barrier layers - including a HDPE liner - with two separate draining systems. The infastructure of the plant will include a waste water treatment station involving physical and biological treatment.

From an organisational point of view the plant will be equipped with state of the art information transmission and monitoring systems in a context of maximal automation of operation.

The total investment is estimated in the neighbourhood of 1 500 million BF (39 M USD). Start up of the various units will largely depend on the construction complexity determining the time needed for erection. Accordingly the landfill is to become operational end 1987, the various physico - chemical units will gradually come on stream from mid 1988 on, the incinerator finally is to be started up second half of 1989.

Total personnel is projected to be around 50. They will come on board as the various installatioins are started up.

The total turnover of the organisation when fully on stream is estimated around 1 000 million BF (23 M USD).

5. Conclusion

The background, status and plans of Indaver N.V., the centralised waste treatment facility in Belgium, have been reviewed. The complexity of industrial waste treatment and the many challenging opportunities still open for this new type of industrial activity make international contacts and collaboration mandatory in order to jointly optimise our capabilities and cooperate in any way we can. I thrust this presentation has been a contribution into this direction.

METHODOLOGICAL EVALUATION OF CONTAINMENT STRATEGIES.

C.C.D.F. van Ree, R. Kabos, F.A. Weststrate and M. Loxham.
Delft Geotechnics , P.O. Box 69, 2600 AB Delft, The Netherlands.

1. INTRODUCTION
It is generally accepted that, from an environmental point of view, complete excavation of all contaminants and complete treatment is the best strategy for the reclamation of contaminated land.
However the costs and technical difficulties involved make it practicable for only the most dangerous and urgent cases. For all other solutions, including less rigorous excavation measures, some residual risk of on-going emission from the site has to be accepted.
A similar situation exists concerning soil protection measures. It has long been recognized that a zero emission activity is unattainable and that base line risks exists and will continue to do so.
Both in the area's of land reclamation and soil protection there is a wide range of technologies available of which isolation and barrier techniques, whether used together with in situ remediation methods or not, form the most important category. Selection and design of such complex systems is a field of active research, and there is a growing awareness of the fact that risk analysis techniques can be of major usefulness.
In this study we present a "risk-analysis-technology" methodology, developed by the authors for the Netherlands Environmental Ministry (1).

2. LINE OF APPROACH, RESULTING METHODOLOGY
In view of the Source-Path-Target philosophy, containment means plugging the routes along which the target(s) are threatened.
To develop an adequate containment strategy one has to identify the relevant processes which determine the emissions along these paths. Acceptable limits in terms of emissions and possibilities of occurrence are set and adequate measures to comply with these limits have to be developed.
To make general application of the methodology possible it has a high abstraction level. The methodology can be summarized as shown in figure 1 (next page). It consists of two major sections.
Firstly the design specifications have to be defined for each particular case. Starting with the separate measures which can counteract the identified adverse effects of (potential) soil pollution, these measures are integrated to effective alternatives. Modelling of the soil compartment results in a transfer-function which describes the response of the system (including containment measures) to an input (the source-strength) in terms of an emission (flux) across predefined boundaries. Standards to be met by specific techniques can be specified and designs selected.

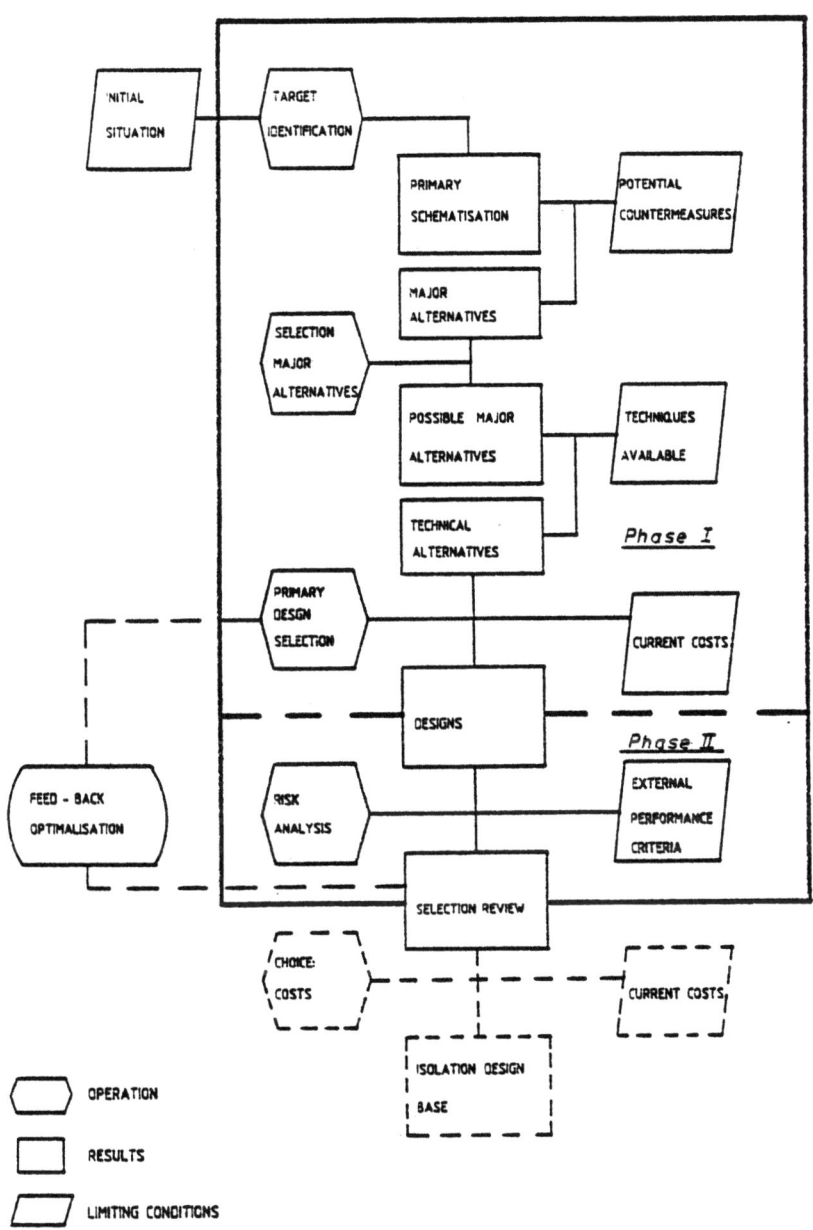

FIGURE 1. Diagram of the methodology's structure.

Once these are selected the second section, a risk-analysis has to be performed. Consequences of uncertainties in the original database (source strength, continuity of natural barriers etc.), uncertainties in the performance of the containment measures resulting from different sources are quantified. The risk-analysis which is based on Monte-Carlo simulation techniques with emission levels as target criteria offers the decision-makers a tool which gives comparisons between seemingly totally different solutions a more rational basis.
Each of the two sections consists of a number of steps which describe the structure of the selection processes using environmental, technical and economic criteria.

3. CONCLUSION

The methodology is a very flexible tool, which can be used both for civil technical containment techniques and geohydrological techniques. The principles also apply to other remedial action techniques and to soil protection measures. It provides a basic tool for taking uncertainties into account in the decision proces.
By using this method before going into the full design stage critical parts of constructions and even faulty designs can be identified.
Successfull design will need a multidisciplinary approach, using systems analysis and risk analysis coupled with knowledge about environmental technology.

REFERENCES

1. Grondmechanica Delft: Risico-analyse isolatie bij bodemverontreiniging. Reeks bodembescherming nr. 61, 5 parts. Staatsuitgeverij, 's-Gravenhage, 1986.

THE USE OF DIAPHRAGM WALLS OF OVER 50 METER DEEP TO STOP SPREADING OF CONTAMINATION.

H.H. v. Breukelen, M.Sc., HASKONING, Royal Dutch Consulting Engineers

1. GENERAL
   The Netherlands have a long history of industrialisation. As a consequence of the operations of a part of the industry, soil contamination has occurred at many incidences.
   As the watertable over about 50 percent of the country is situated only 0,5 to 2,0 meter below grade level, contamination of soil often goes hand in hand with groundwater pollution.
   In such cases the danger exists that the contamination will spread in the surrounding soil and groundwater.
   This has occurred on several sites.

2. SOLUTIONS

2.1 General
   In order to prevent further spreading of the pollution, basically two solutions exist:
   - remove all contaminated soil and groundwater, or
   - stop spreading, using an impermeable wall together with measures to assure a lower watertable inside the area surrounded by the impermeable wall.
   In the Netherlands the second method is named ICC-method, abbreviation that stand for: Isolation, Control and Checking.
   The "Isolation" is achieved using impermeable membranes. The "Control" can be obtained using a lower watertabel and finally the "Checking" has to prove that the system is functioning adequately. Checking includes sampling groundwater and testing of these samples and taking readings of the level of the groundwater table inside and outside the area which has been isolated.

2.2 Requirements
   The isolation requires a maximum of watertightness. For this reason, the surrounding wall has to reach intro a groundlayer of maximum watertightness. When available at the required depth, clay layers of sufficient thickness will perform such a function.
   Since absolute watertightness cannot be achieved we have to allow for measures to assure that leakage which takes place will always be directed from the outside, where the groundwater is clean, to the inside, where the groundwater is polluted.
   Depending upon the quantity of water leakage that may be expected, pumping facilities have to be installed to assure the waterpressure inside the isolated area to be always lower than the pressure outside the wall.

2.3 Site description
The isolation technique described above will be executed in the near future on one or two big sites with an area of some 10 and 6 ha respectively (25 and 15 acres).
In both cases an extensive soil exploration has taken place. On the smaller site a clay layer, sufficiently reliable to take the wall, was found at a depth of some 18 meter below groundlevel. On the bigger site, a clay layer of sufficient magnitude was supposed to be available at a depth of som 45 meters but it became obvious that the thickness on many places was relatively poor and that the depth on which the layer was found reached 45 to 55 meter.

3. DIAPHRAGM WALL
3.1 General
The choise of the diaphragm wall to be used for the bigger site is the result of an evaluation of the different wall types available. This evaluation was based on the assumption that the wall should reach to a depth of 45-50 meter. It became obvious that walls of this length could hardly be executed using one of the other proposed wall types and above all, diaphragm walls have been executed to depths of 75 to 80 meter. When it became clear that the clay layer was even situated at a greater depth than originally envisaged, the choice of diaphragm walls seems even more logical.

3.2 Specific requirements
Besides the proven experience of this walltype, several other points have been examined before the final choice has been made. Among the points studied we mention:
- The system used should not cause more than acceptable nuisance for the people living in the vicinity of the wall under construction (minimum distance from wall to adjacent houses some 35 meters).
- The material of the wall should not or hardly be affected by the polluted groundwater.
- The accuracy of the construction process should be sufficiently developped as to assure the walls to be very near to vertical position. This vertical position should be guaranteed in two perpendicular planes as to assure a minimum wall thickness on the joint between two adjacent wallpanels and to assure that no gap will occur between two adjacent wall panels.
- The wallsystem can be repared, when required, although such repare is expensive.

4. EXECUTION AND CONCLUSIONS
The construction process of the diaphragm walls will be supervised very carefully in order to achieve the quality required.
When for some unforeseen reason the watertightness of the wall should not be 100 percent (also the clay layer will not be 100 percent reliable) the pumping capacity should be increased to achieve sufficient waterpressure from outside directed to the inside of the insulated area.
During the theoretical evaluation of the insulation system, it has already been assumed that a certain porosity will occur and even that wallpanels wil not always reach to the impermeable clay layer.
The allowances thus made have made it clear that the ICC-method chosen, using diaphragm walls can be considered to be a safe and reliable isolation system.

LEACHING OF CYANIDES

W.P. van Oosterom, L.G.C.M. Urlings

TAUW Infra Consult B.V., P.O. Box 479, 7400 AL DEVENTER, The Netherlands

ABSTRACT
The behaviour of cyanides in soils from former gasworks sites has been studied in batch and column experiments for development of (in situ) restoration methods. The effects of organic matter have been determined for soils from two locations. Special attention has been paid to the contribution of microbial degradation on the behaviour of cyanides. Results of the experiments did not support the feasibility of a cleaning technic.

INTRODUCTION
The gas production from charcoal (up till 1950) has led to a number of cases of soil pollution with cyanides. The reclamation of this polluted soil by excavation and incineration will be an expensive operation. The cleaning of cyanide-polluted soils costs about f 250,=/cu m with the present-day technics. Research is needed to develop new and less-expensive technics. First of all, more knowledge is required about the behaviour of cyanides in the soil. In this poster presented research was initiated and conducted by TAUW Infra Consult B.V. Results from batch and column experiments will be presented from two former gas works sites in Tholen and Zutphen, and effects of organic matter additions will be discussed on the behaviour of cyanides.

EXPERIMENTAL

Batch experiments
Four batch experiments are carried out with 190-g soil:
- 1 batch experiment without addition (for both sites)
- 1 batch experiment with dung + compost (for both sites)
The soil was shaken for 24 hours with 1800-ml synthetic rain water (composition: 9.43 mg/l $CaCl_2$, 20.24 mg/l KCl and 4.07 mg/l $NaCl$; pH 5.0). After 24 hours the water phase was filtered over 0.45-μm membrane filters for cyanide analyses.

Column experiments
Five columns have been filled with cyanide-contaminated soil from either Tholen or Zutphen.
- 1 column without addition (for both sites)
- 1 column with dung + compost + granular sludge (for both sites)
- 1 column with dung + compost (Tholen).

The columns have been percolated with synthetic rain water for nearly six months with renewal of the pore volume approximately once a day. All columns were saturated with water (anaerobic) with exception of the last mentioned Tholen column, which was aerated and unsaturated. The percolate water has been analysed on cyanide contents (free cyanide and/or total cyanide) twice a week.

RESULTS
Batch experiments
Results of the batch experiments are given in table 1.

TABLE 1. Results batch experiments

	1 Tholen	2 Tholen+ cow dung (13%) compost (5%)	3 Zutphen	4 Zutphen+ cow dung (13%) compost (5%)
total cyanide content of the soil (mg/kg)	1958	1468	384	289
total cyanide in the water phase ($\mu g/l$)	3900	410	4600	650
free cyanide in the water phase ($\mu g/l$)	36	3	32	7
total cyanide leached (%)	1,9	0,3	11,4	2,2

The total amount of cyanide leached from soils after addition of organic material is lower in comparison with the non-amended soils. This effect is probably due to adsorption of cyanides onto organic matter. The cyanides are leached out better from Zutphen soil, which may contain structurally different cyanides.

Column experiments
Results of the column experiments after six months are given in table 2. Figure 1 shows the distribution of total cyanide in every column after the experiments.

TABLE 2. Results of the column experiments per column

	column 1	column 2	column 3	column 4	column 5
total cyanide content of the soil at the beginning in mg (in %)	7972 (100)	1523 (100)	3625 (100)	751 (100)	4353 (100)
total cyanide content of the soil after the experiment in mg (in %)	6235 (78,2)	537 (35,3)	2601 (71,8)	559 (74,4)	3889 (89,3)
total cyanide leached in mg (in %)	405 (5,1)	806 (52,9)	9 (0,2)	11 (1,5)	429 (9,9)
'rest term' in mg (in %)	1332 (16,7)	180 (11,8)	1015 (28,0)	181 (24,1)	35 (0,8)

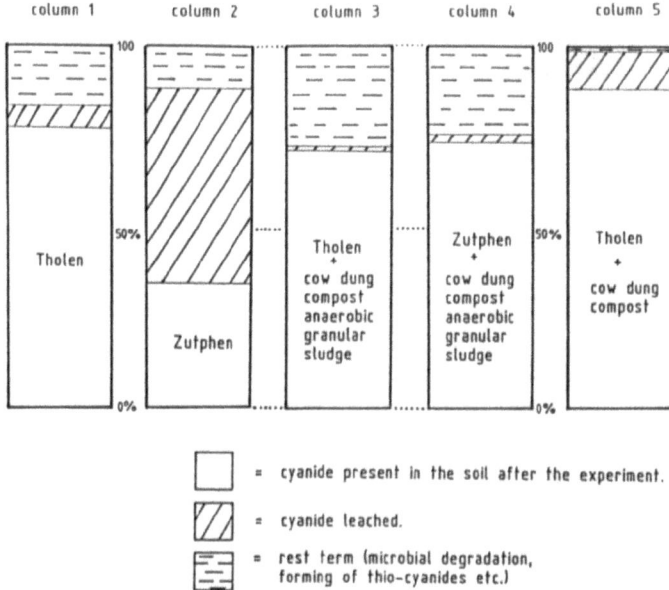

FIGURE 1. Distribution of total cyanide in per cent of the total cyanide content at the beginning of the column experiments

Results of the column experiments (table 2 and figure 1) confirm the ones of the batch experiments in that very little cyanide leached out in presence of organic matter. Besides, the amount of cyanide leached out of Zutphen soil is considerable higher than the amount of cyanide leached out of Tholen soil (table 2). If the 'rest term' is ascribed to microbial degradation, which will be the most favourable case, the half-life time of anaerobic degradation of total cyanide will be at least 1-2 years.
In that case it takes about 10 years to reduce a total cyanide concentration of 500 mg/kg in polluted soils to the background level of 5 mg/kg by a temperature of 20°C.

CONCLUSIONS
The leaching of cyanide out of polluted soils is only partial and depends of the structure of the cyanides present in the soil. Addition of organic matter lowers leaching of the cyanides. The microbial degradation is positively influenced by organic matter, but contribution to the cyanide removal is negligible. Further research is needed. The cyanide leaching was enhanced in the Tholen soil, amended with dung and compost under aerobic conditions. For cyanide removal, this effect was still negligible.
(In situ) restoration technics for cyanide removal by leaching and biostimulation are not feasible under test conditions.

Polymeric flocculants in waste water treatment and in sludge dewatering - technical and economical aspects.

Dr. A. Landscheidt and Dr. J. M. Reuter, Krefeld

Polymeric synthetic flocculants on the basis of polyacrylamide have opened new ways in waste water treatment and in sludge dewatering about 20 years ago. During this time the product development has made many progresses, and the application methods have increased, too.

In municipal and industrial biological sewage treatment and in sludge dewatering preferably cationic polymers are used, whilst for the treatment of mineral solids and metal hydroxides containing waters and slurries anionic polymers will offer best effectiveness. The range of polymers which is available covers the whole area of high cationic load to medium and low cationic load to nonionic types and passes to weak and medium anionic charged polymers up to strong anionic types.

The selection of the optimal polymer (kind and intensity of electrical charge, mole mass) depents from characteristic parameters of the sewage water or of the slurry. Normally the optimal product has to screened by laboratory tests. There are test methods available, by which the effectiveness of the polymeric flocculants is checked as well for clarification and acceleration of settling in water treatment as well as for the dewatering of sludges under different shear and press stressing.

In water purification and clarification processes the solid and colloidal particles are flocculated and coagulated and will be separated from the aqueous phase very quickly and completely by settling or by flotation. The flocculamts here offer the possibility to run the clarifying equipment with very high capacity, and to ensure a clarified water which is free of settleable particles.

Sludge dewatering in centrifuges or sieve belt presses requires the use of polymeric flocculants in order to obtain a satisfying yield of solids and to separate an effluent water which will not cause troubles in the plant when being recycled. Recent developments have confirmed that for sludge dewatering in chamber filter presses, too, the traditional conditioning method with lime and iron or aluminia may be changed to polymer conditioning. By this, important advantages are obtained: No increase in sludge volume by high addition quantities of inorganic materials, no input of materials which will change the natural characteristics of the sludge, no increase in COD or BOD in the dewatered sludge or in the effluent. Furthermore, an important cost saving of more than 30 % is obtained in relation to the inorganic sludge conditioning.

The polymeric flocculants are not hazardous, not toxic materials. The acute oral toxicity has been determined 5 g/kg of rat or mouse.

The polymeric flocculants are adsorbed at the sludge particle surface; they loose their reactivity by the adsorption and will remain in the dewatered sludge. Due to their molecular structure a biological decomposition of these big molecules will not occur. They will remain in the sludge cake like a very small plastic fiber, and no contamination with degradation products is to be expected.

The needed dosage rates of the polymeric flocculants are very small: In clarification systems addition rates are between 1 and 10 ppm, according to the solid load of the water to be clarified. In sludge dewatering systems the addition rates vary between 2 and 6 kg polymer per ton of dry solids in the sludge. These low addition rates confirm the high economic advantages which are offered by polymer conditioning in comparison to other materials or methods.

ON-SITE APPLICATION OF TRICKLING FILTER AND ROTATING BIOCONTACTOR IN TREATMENT OF GROUNDWATER, POLLUTED WITH CHLORINATED HYDROCARBONS

A.L.B.M. van Campen, L.G.C.M. Urlings, B. Bethe.
TAUW Infra Consult B.V., P.O.Box 479, 7400 AL, The Netherlands

1. ABSTRACT

The feasibility of biological treatment of groundwater contaminated with HCH (hexachlorocyclohexane), benzene and monochlorobenzene has been investigated for cost reduction. Loading rates of 1 mg HCH $m^{-2}.hr^{-1}$ and 9,3 mg benzenes $m^{-2}.hr^{-1}$ were applied. Removal efficiencies were alpha-HCH: 40-60%, gamma-HCH 50-80%, delta-HCH: 25-40%, benzene and monochlorobenzene more than 95%.

2. INTRODUCTION

In The Netherlands, the former production of the pesticide lindane (gamma-HCH) has led to a number of cases of soil- and groundwater pollution. The reclamation of the polluted soil and groundwater will be an expensive operation through cleaning by activated carbon adsorption, which is a safe method. Benefits of biological treatment are: total removal of the pollutant by mineralization and cost reduction.

3. MATERIAL AND METHODS

The groundwater first passed an equalisation basin before it was pumped into the on-site installation.
The experiments with the trickling filter lasted five months. The composition of the groundwater during this period (July-December 1986) is given in table 1. The experiment with the rotating biocontactor started in October 1986 and lasted 1½ months. For further experimental information see table 2.

TABLE 1. Concentration of benzene, monochlorobenzene and HCH in the groundwater (n = 25).

Compound	unit	mean	concentration minimum	maximum
benzene	µg/l	1100	530	1800
monochlorobenzene	µg/l	753	380	1050
alpha -HCH	µg/l	70	25	150
beta -HCH	µg/l	6,5	3	10
gamma -HCH	µg/l	150	70	250
delta -HCH	µg/l	110	55	150
epsilon-HCH	µg/l	23	0,5	35

TABLE 2. Experimental conditions

Characteristics	trickling filter	rotating biocontactor
volume (m^3)	0,200	0,075
flow (m^3/h)	0,250	0,200
total surface area (m^2)	40	38
packing medium	filterpack CR50	ribbed discs (D=50 cm)
specific surface area (m^2/m^3)	215	240

4. RESULTS

4.1. Trickling filter

The removal efficiencies for the different HCH-isomeres are given in Figure 1. Adaptation of about 50 days can be recognized in figure 1. Benzene and monochlorobenzene were removed for more than 95%; these compounds could not be analyzed in the effluent of the trickling filter.

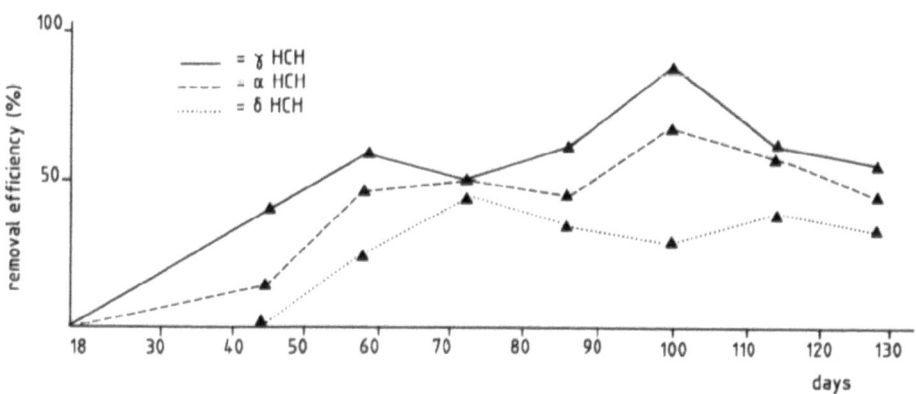

FIGURE 1. Removal effiency of the different HCH-isomers in time (trickling filter)

Oxygen uptake and HCH-biodegradation were measured in the laboratory with some pellets from the trickling filter after 121 days. Bacteriocide NaN_3 was added to discriminate adsorption from biodegradation. The results are presented in figure 2.
The biofilm contained bacteria, ciliata and algae.

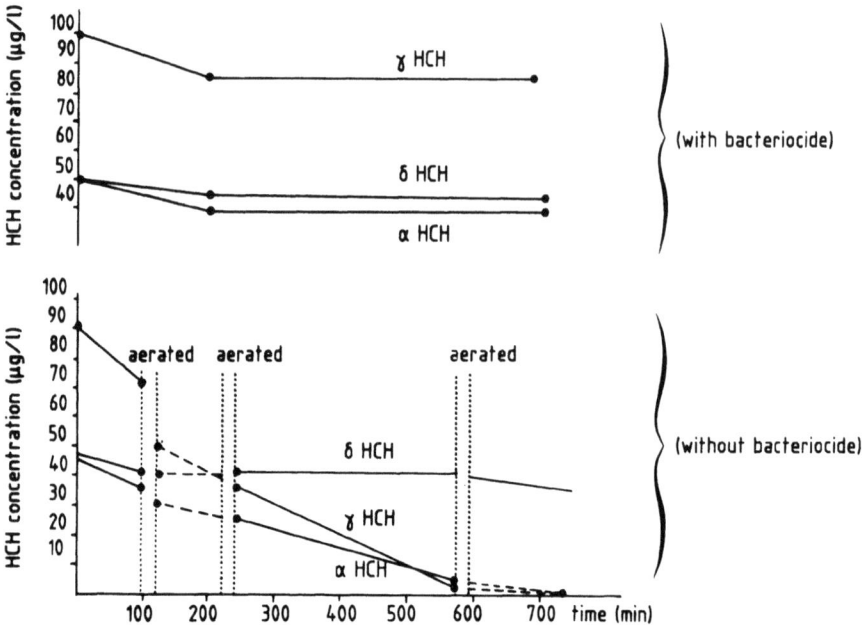
FIGURE 2. HCH-biodegradation-rate experiments with some pellets.

4.2. Rotating biocontactor

The removal efficiencies were alpha-HCH: 85% and gamma-HCH: 92%. The removal rates were respectivily 15.8 and 8.2 mg $m^{-2}.day^{-1}$, which are almost equal to the removal rates in the trickling filter (m^2 = surface area of the packing medium).

4.3. Economic consequences

The costs of activated carbon can be reduced to about 7% by biological pretreatment. In consequence, the total costs of a groundwater purification can be reduced with 30%.

5. CONCLUSIONS

- Biological purification of groundwater, contaminated with HCH, benzene and monochlorobenzene is feasible with removal efficiencies of respectivily more than 50%, 95% and 95%.
- There is an adaptation-phase for HCH of 30 to 50 days.
- The costs of groundwater purification can be reduced 30% by biological pretreatment.

Acknowledgement
This work was supported with a grant by the Institute of Inland Water Management and Waste Water Treatment (DBW/RIZA).

POLLUTION-CONTROLLING EFFECT OF ADSORBING MATERIALS ON THE LEACHING OF POLLUTANTS FROM DREDGING SPOIL DEPOTS

A.W. Grinwis, L.G.C.M. Urlings

TAUW Infra Consult B.V., P.O. Box 479, 7400 AL DEVENTER, The Netherlands

ABSTRACT
The behaviour of organic micro-pollutants from dredging spoil depots on adsorbing materials, such as peat and non-polluted river sediments, was investigated to limit contamination of the depot sites. No 'cocktail-effects' from mixed pollutants on the adsorption rate were found. Also, the effect of oil on the sorption rate was negligible.

INTRODUCTION
The spoil production from regularly dredging of fair ways and docks is quite extensive. Because the spoil is sometimes too contaminated to dump into the sea the (port) authorities have a growing problem in their hands. For storage on land, more knowledge is required to limit and control the out-flow of contaminants from dredging spoil depots.
Earlier literature study (TAUW Infra Consult) suggested the adsorption capacity of peat and non-polluted river sediments for micro-pollutants and heavy metals, based on the content of organic carbon and clay minerals in the sorbent. The adsorption of organic micro-pollutants depends mainly on the organic carbon content and is given by the formula:

$$K_{oc} = \frac{C_{soil}}{C_{water} * f_{oc}} \qquad (dm^3/kg)$$

with: K_{oc} = the partition coefficient of a component between the water and
the organic carbon (dm^3/kg)
C_{soil} = concentration in the sorbent $(\mu g/kg)$
C_{water} = concentration in the water $(\mu g/kg)$
f_{oc} = fraction organic carbon (kg/kg)

This study will focuss on the behaviour of organic pollutants on sorbents (peat, Euroklei) under laboratory conditions with attention to cocktail effects and presence of oil.
This study and the previous literature survey were granted by the Project Group Pollution Controll of MKO (Maintenance Dredging Cost Minimization). Consultants of the project were F. Visser, Rijkswaterstaat North Holland, J.P.J. Nijssen, E.J.E. Wijnen (Public Works Department Rotterdam), R. de Waay and A. Kroes (DBW/RIZA).

MATERIAL AND METHODS

The cocktail containing PCB's, HCH's, PAH's, nonane hexadecane, pentachlorophenol, and hexachlorobenzene was made by enrichment of groundwater from a dredging spoil deposit at Capelle aan den IJssel through a method outlined in figure 1. Three separate enrichments were made. Figure 1 gives a scheme of the enriching method.

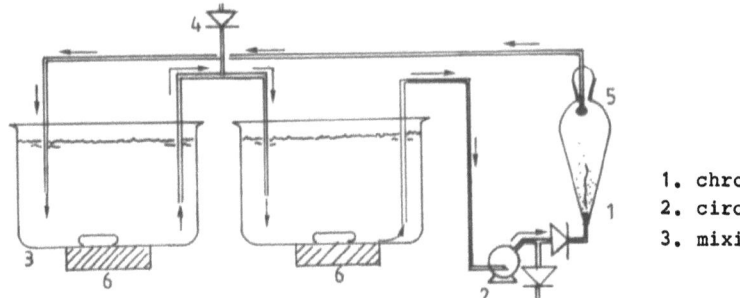

1. chromosorb column
2. circulation pump
3. mixing container

FIGURE 1. Enrichment of groundwater with a chromosorb column.

The adsorption coefficient of each of the organic micro-pollutants was measured with batch experiment under anaerobic conditions. Table 1 gives sorbents, groundwater and solid/liquid ratio used in the experiments.

TABLE 1. Batch processes

S/L ratio * 0,001	2	4	25	0
liquid sorbent	1st 2nd 3rd enriching	1st 2nd 3rd enriching	1st enriching	1st 2nd enriching
peat, Ermelo	x x x	x x x	x	
peat, Klazienaveen	x x x	x x x	x	
riversediment, Euroklei	x x	x x	x	
peat, 0,1% oil		x		
peat, 0,6% oil		x		
blank				x x

RESULTS OF THE BATCH EXPERIMENTS

Results of the batch experiments are presented for gamma-HCH in figure 2.

The mean log Koc for gamma-HCH was 90% of MKO literature value in our experiments; range of the results was small.

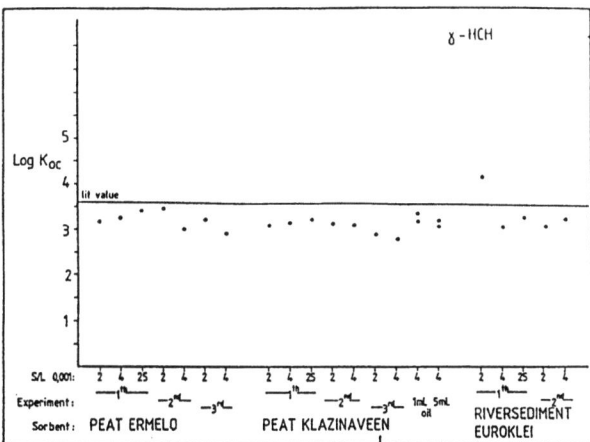

FIGURE 2. Experimental adsorption coefficients of gamma-HCH.

Our statistical reliable log Koc-values for the organics are compared with log Koc-values from the literature in table 2 (Student T-test, alpha = 0.025).

TABLE 2. Experimental log Koc compared with literature values

	experimental log Koc	log Koc Kenaga '80	log Koc Chiou '79	log Koc MKO-R-84-2
nonane	2,7	-	-	3,8
hexadecane	3,0	-	-	-
PCB 28	5,1	-	-	4,2
alpha-HCH	3,0	-	-	3,7
bèta -HCH	3,2	-	-	3,2
gamma-HCH	3,2	-	3,2	3,6
hexachlorobenzene	4,7	3,6	-	5,1

CONCLUSIONS

Experimental log Koc's did not deviate from log Koc's in literature.
The variation in literature values is less than 2 log-units. Thus, no cocktail effects were measured.
Also, the addition of mineral oil did not effect the log Koc-values significantly (Wilcoxon; alpha = 0.0125).
The maximum vertical velocity of pollutants (C/Co= 0.5) is computed in peat at 0.16 cm/year and in river sediment at 0.41 cm/year, in the 'worst case' situation (Koc = 500 dm^3/kg, f.i. nonane). The maximum vertical ground water flux is estimated at 50 mm/year in this case.

REFERENCES

1. Chiou C.T. et al.: Science, 206, 1979, p.831.
2. Kenaga E.E. et al.: STP 707, ASTM, Philadelpia, 1980.
3. MKO-R-84-2: "Het verontreinigingsbeheersend effekt van vastleggende materialen op het transport van verontreinigingen uit baggerspecie depots", Deventer, 1985.

CHARACTERIZATION AND ENVIRONMENTAL EFFECTS OF SLUDGES PRODUCED
BY METALLURGICAL FACTORY.

J.R.DOBOSZ and M.SEBASTIAN
Institute of Envinronment Protection Engineering,Technical
University of Wroclaw,pl.Grunwaldzki 9 , 50-377 WROCLAW POLAND

ABSTRACT

Many of sludges,produced by metallurgical factory,create serious hazards to the natural environment.The storage of liquid and semi-liquid wastes on a specjal dumping ground(the lagoons) is often the final stage of disposal.In this paper,the physicochemical composition of selected wastes by above method is presented. Sludge samples from the lagoon were taken at various points and various depths for the period of two years.The composition of groundwater was determined from samples collected at piezometric bore-holes,driven about 20 m away from the corners of the lagoon.Moveover,heavy metals content in flora growing in the neighbourhood of the lagoon is taken notice.

1.INTRODUCTION.
   Iron and steel works are next to chemical and power industries in polluting the environment.More than a hundred kinds of wastes (both solid and liquid) are produced there.These vary in physicochemical composition,creating - in some instances -technological trouble and environmental hazards.
   The characteristics of those wastes - along with a concept of how to dispose them of - were reported by the autors of this paper in an earlier study [1,2,3].

2.EXPERIMENTAL PROCEDURE AND RESULTS.
   Analises inclued the physicochemical composition of sludges, water extracts,groundwater samples and vegetation samples collected in the immediate vicinity of the lagoons.

2.2.COMPOSITION OF GROUNDWATER AND VEGETATION IN THE IMMEDIATE
    VICINITY OF THE LAGOONS.
   The present disposal method practiced in the steel works of interest involves 100m x 100m x 5m lagoons.This is a system of three units,having walls which are made of earthen structures (with a coping width of 1,5m, a slope inclination of 1:1,5 and a datum of 3,0m above the ground level),and bottoms with no protection against leakage (placed at a depth of 2,0m below the ground level).The surface area of each lagoon amounts to 50x50m. At present,the lagoons are storing an over 26 000 m3 volume of different high-water-content sludges and other wastes (about 10 or so types of wastes materials).The physicochemical composition of sludges which are most frequently found (and in the largest amounts) has been given in Table 1.
The lagoons are also a dumping site for sludges produced by an

TABLE 1

Physicochemical Composition of Sludges Stored in the Lagoons

Type of Slugde	Water content (%)	Organic solids (% dry solids)	SiO_2 + particles insoluble in strong acids (% dry solids)	
from drawing mill, after lime treatment	70.5	5.9	17.3	[1]
from drawing mill, after ammonia teratment	71.0	46.9	0.8	[2]
from scrubbing of flue gases from steel mill	79.4	2.7	8.1	[3]
from electroplating, after neutralization	95.2	19.4	1.7	[4]
from painting rooms of a mechanical plant	84.0	53.9	3.0	[5]
sewage sludge, from a machanical plant, after neutralization	92.7	29.2	3.6	[6]
from grindery of a mechanical plant	38.8	20.6	14.6	[7]
polluted scale	0.8	1.2	9.0	[8]

sludge	[1]	[2]	[3]	[4]	[5]	[6]	[7]	[8]
element	metals (% dry solids)							
Na	0.24	0.10	0.46	1.51	1.06	0.36	0.13	0.19
K	0.07	0.05	0.28	0.01	0.28	0.08	0.09	0.12
Ca	15.03	0.24	4.72	0.76	1.12	5.15	0.58	0.56
Mg	0.04	0.01	4.57	0.12	0.58	0.47	0.02	0.03
Fe	13.98	22.90	41.70	2.00	21.60	10.30	39.50	62.90
Pb	0.02	0.0	0.45	0.25	0.77	0.03	0.0	0.0
Cu	0.01	0.01	0.14	0.90	0.03	0.01	0.02	0.01
Cr	0.23	0.19	0.87	19.80	5.89	0.04	0.14	0.02
Zn	0.01	0.02	1.65	12.10	2.82	0.08	0.03	0.04
Ni	0.05	0.02	0.13	0.05	0.01	0.01	0.02	0.01
Mn	0.08	0.03	0.71	0.06	0.04	0.08	0.18	0.10
Al	0.01	0.01	0.01	0.01	1.00	1.00	n.d.	n.d.
Co	0	0	0	0	0.001	0	n.d.	n.d.
Sn	0	0.001	0.01	0.0001	0	0	n.d.	n.d.
Ti	0.001	0.0001	0.001	0	1.00	0.001	n.d	n.d
Ga	0	0.0001	0.001	0	0	0.0001	n.d.	n.d.
Mo	0	0.00001	0.001	0	0.0001	0	n.d.	n.d.
V	0	0	0.001	0	0	0	n.d.	n.d.
Cd	0	0	0.001	0.1-1.0	0	0	n.d.	n.d.
Ag	0	0	0.01	0.001	0	0	n.d.	n.d.
Ba	0	0	0	0	0.1-1.0	0	n.d.	n.d.

industrial water treatment plant,for sludges from air-chamber wash and a number of waste products,such as spent coelants,cily substances which are unfit for reuse,spent organic solvents or oil emulsions.Physicochemical analises were carried out on sludge samples collected at representative depth and points in two-month intervals for two years (Tab.1).This is an indication that the composition of the the sludges stored is of a highly inhomogeneous naturre.This finding has also been corroborated by the results of water-extract analyses which were carried out in order to determine the water-solubility of the investigated sludges.Here are the variabilities for 14 water-extrakct samples: pH 3.1 to 5.7;colour 10 to 30 g Pt/m3;turbidity 5 to 500 g/m3;dissolved solids 2700 to 12 008 g/m3;mineral dissolved solids 2300 to 6967 g/m3;volatile dissolved solids 400 to 5586 g/m3;total hardness 1321 to 3536 g CaCO3/m3;chlorides 3.0 to 131.0 Cl /m3;sulphates 1695 to 7364 g SO /m3;permanganate COD 2.8 to 320 g O /m3;dichromate COD 54.8 to 410.4 g O /m3;sodium 13.5 to 188.0 g/m3;potassium 5.0 to 21.6 g/m3;calcium 300 to 525.0 g/m3;magnesium 16.0 to 400 g/m3;iron 1.2 to 2000 g/m3 lead 0 to 1.4 g/m3;zinc 1.2 to 38.0 g/m3;nickel 0 to 36.4 g/m3;copper 0 to 1.2 g/m3;chromium 0 to 2.5 g/m3; and manganese 1.2 to 176.0 g/m3.What should be emphasized here is the high acidity level and the relatively high heavy metal concentration.

2.3.COMPOSITION OF GROUNDWATER AND VEGETATION IN THE IMMEDIATE VICINITY OF THE LAGOONS.

As shown these data,the composition of the groundwater coming to the lagoon bottom (bore-hole 2) does not differ from the composition of any other water stream flowing under similar hydrological conditions.However,physicochemical analyses of groundwater samples collected from the remaining bore-holes have revealed increased levels of organics,inorganic substances and heavy metals.This increase should be attributed to the influence of leakage (Table 2).

Analyses of heavy metal concentrations in vegetation samples were carried out for the following species growing i the immediate vicinity of the lagoons: Betula pubescens Ehrh.,Erigeron canadiensis L.,Lepidium rurale L.,Linaria vulgaris L.,Poa annua L.,Filipendula ulmaria max. and Phalaris arundinacea L..The data set obtained for branch-and-bough samples of Betula pubescens Ehrh. shows contamination from sodium 0.05% ; potassium 0.41% ; calcium 1.04% ; magnesium 0.18% iron 2.07% (!); lead 0,01% ; copper 0.005% ; chromium 0.05% ; zinc 0.05% ; nickel 0.01% ; manganese 0.04% ; aluminium 0.01 to 0.1%99 ; titanium 0.001% and molybdenium 0.0001%.These contamination levels are averages from 8 samplings.
Cobalt,tin,galium,vanadium,cadmium,silver and barium were absent.Gras samples (including roots) were subject to physicochemical analyses after remowal of soil remainder and after drying.The presence of the following heavy metals was detected: sodium 0.01 to 0.1% ; potassium 0.01 to 1.0% ; magnesium 0.01 % ; iron 0.01 to 1.0% ; lead 0.001 to 0.01% ; copper 0.0001 to 0.001% ; chromium 0.0001 to 0.1% ; zinc 0.0001% nickel 0.0001 to 0.001% ; manganese 0.001 to 0.1% ; aluminium 0.01 to 1.0% ; titanium 0.001 to 0.1% ; molybdenium 0.0001% ; and vanadium 0.0001 to 0.001%. Silver was found to occur in

trace amounts. Cobalt,tin,galium,cadmium and barium were absent. On comparing the results obtained with the background data well-known from the literature,it becomes obvious that grass-type vegetation groving in the vicinity of the lagoons displays heavy metal concentrations several times as high as those reported elsewhere.Keeping in mind that heavy metals travel from one environmental medium to another, the environmental hazards created by the lagoons call for immediate safety measures.

TABLE 2
Variability of Pollutants in Groundwater Samples from Bore-holes

		Bore-hole 1	2	3	4
Colour	g Pt/m3	140-1000	0-150	10	5-80
Turbidity	g/m3	0-450	0-140	10-20	0-10
pH		4.6-7.5	5.9-7.2	4.4-5.5	2.8-6.0
Dissolved sollids	g/m3	357-2426	123-212	2120-2461	3652-24590
Chlorides	g/m3	22-647	3.0-18.0	56-67	77-940
Sulphates	g/m3	39.9-176.5	42.4-67.6	275-1426	462.3-16052
Ammonia	g/m3	0.2-134.4	0-1.50	88.2-92.4	189-2700
Nitrites	g/m3	0-35.0	0-0.2	0-0.2	trace-0.64
Perm. COD	g/m3	86-275	2-10.8	8.4-9.0	5.6-58.0
Dichr. COD	g/m3	225-1387	0-35.3	66.7-78.4	90.2-354.1
Zinc	g/m3	0.5-5.8	0.5-10.5	0.71-5.80	1.38-68.0
Lead	g/m3	0-2.5	0-2.5	0.2-2.0	0-2.0
Copper	g/m3	0-0.6	0-0.6	0.2-2.0	0-2.0
Nickel	g/m3	0-0.2	0-0.075	0-1.2	0-30.0
Chromium	g/m3	0-0.1	0-16.0	0-0.1	0-1.8
Iron	g/m3	1.55-35.0	0-28.4	1.6-16.5	1.15-228.0
Manganese	g/m3	0.3-2.7	0-0.5	6.4-7.0	6.0-187.5
Cadmium	g/m3	0-0.44	0-0.2	0-0.046	0-1.5
Calcium	g/m3	10-45.4	12-21.5	108.8-238	48-240
Magnesium	g/m3	6-27.5	4-16.8	64-103.7	210-1160
Sodium	g/m3	6.5-128	4.5-62.5	18-23.5	23.2-336.0
Potassium	g/m3	6.5-128	0-9.0	10.2-11.2	10-29.0
Total hardness	g/m3	107-900	62.5-250	1000-1026	1160-6035
Total alkalinity	g/m3	20-500	10-50	10-15	10-15

3.DISCUSSION.

The are two major methods for abating the environmental Impact of sludge disposal-construction of sanitary landfills and reduction of the waste volume at the source of origin.

3.1.SANITARY LANDFILLS.

The structure is protected against leakage by 8-cm-thick concrete plates covered with asphalt carpets having thicknesses which range between 2 and 3 cm.The sludge liquid is passed to a wooden shaft which has a size of 1.0x1.0 m and a height of 5.2m. A 20-cm-thick coke layer surrounds the shaft to prevent penetration of the sludge mass.The sludge liquid is then transported to sewage treatment plants by waste removing trucks.

As soon as the dewatering problem is solved,the sludges

may be stored on the sanitary landfill with a possibility of recultivation.It is therefore necessary to increase the capacity of the site by some 150percent and apply interlayers of slag and a 1.5 m thick soil layer for the recultivation of the top part.A drainage system working in the bottom part will send the sludge liquid to the receiving well.From there, the sludge liquid may by transported to the sewage treatment plant.To prevent atmospheric precipitation from penetrating the interior of the site, it is advisable to form an impermeable layer at a 1 m depht from the recultivated surface and to install a continous drainage system for precipitation water removal to the sewarage.

3.2.REDUCING THE WASTE VOLUME AT THE SOURGE OF ORIGIN.

1.Sludges from Neutralization of Wastewater Produced by the Drawing Mill, the Forging Shop and the Rolling Mill may be May be reduced by modifying the etching technology, i.e. substituting hydrochloric acid for sulphuric acid which is now in use.Sludges precipitated from the neutralization of the etching bath contain iron hydroxide, calcium sulphate and unreacted lime.Application of hydrochloric acid yields iron hydroxide alone.The volume of sludge produced is also reduced.

2.Sludges from Scrubbing owing to their high iron content may be reused, after suitable processing (electromagnetic thickening of iron, dewatering and briqutting),for pig iron melt

3.Sludges from Etching Bath. May be separated from oil emulsion and processed together with scrubbing to yield a valuable blastfurnace charge.

4.Wastes from Painting Rooms are generally disposed of on special sites or, after solidification deposited on sanitary landfills.As hazardous wastes may also be subject to incineration together with other flammable waste x 2 m substances.Prior to the incineration process, hazardous wastes should be dewatered.The process is usually conducted at 1173 to 1253 K and involves a rotary kiln equipped with a scrubber for flue gas treatment.

REFERENCES

1. J.Dobosz et al., Basic Research and Development of a Concept of Liquid and Semiliquid Waste Disposal, Report of the Institute of Environment Protection Engineering Technical University of Wroclaw, 1984 (in Polish).

2. J.Dobosz et al., A Concept of Industrial and Sanitary Waste Disposal for a Steelworks Enterprise in Poland, Report of the Institute of Environment Protection Engineering, Technical University of Wroclaw, 1985 (in Polish).

3. J.Dobosz, M. Sebastian, L.Pekalska: On the advantages of gravitational dewatering as applied to the preparation of industrial sludges for disposal, International Symposium on MER, Antverpen 1986.

RECOVERY OF CHEMICAL WASTES - RESULTS OF A DANISH CASE STUDY
AND IMPLICATIONS FOR OTHER COUNTRIES

Kim Christiansen

In a recent project, carried through by the Technological
Institute and financed by the Technology Advisory Board of
Denmark, the actual and potentiel recovery of chemical wastes
by external recyclers in Denmark was investigated. The results
are assembled in a cataloque comprising data sheets on nearly
20 recyclers of chemical wastes and about 30 catagories of
recoverable chemical waste types depicting sources, fate,
actual and potential recovery evaluated on the basis of literature and knowledge of activities in other countries.

Also a report giving a more qualitative survey of recovery
activities has been published. This report contains surveys
of activities in Denmark and abroad and a comprehensive
bibliography.

Clearly not all data needed to fully complete the data sheets
on the above mentioned recyclers and recoverable chemical
waste types were available, but the present results represent
the most comprehensive and current information material on
external recovery of chemical wastes in Denmark.

In summary the major conclusions were as follows:

o In gathering data concerning activities, the most efficient
  approach was personal contacts with key persons in industry
  and government as identified following screening of obtainable information from the local authorities responsible
  for chemical waste management.

o There is a vast amount of literature dealing with recycling
  possibilities including laboratory research, pilot plant
  testing, on-site industry implementations, and commercialized by external recycling companies, however such information is seldom critically evaluated and updated; The most
  detailed and the latest information was compiled through
  manual search in conference reports and proceedings.

o Danish industry produces approximately 100,000 tonnes of
  chemical wastes every year, excluding waste oils. About
  75% is destructed at Kommunekemi by inceneration or inorganic detoxification followed by landfilling af any residues.
  Only 5% is recycled or reused by companies using wastes from

other industries. Potential wastes being recycled at the source (internal recovery) are not readily identifiable.

o Around 5,000 tonnes of chemical wastes of which organic solvents comprise about 90% is recycled or reused by external contractors for secondary raw materials or by chemical companies as additional feedstocks. Other externally recovered wastes in Denmark are silver from photographic liquid wastes, which is sold at metal exchanges abroad after electrolytic recovery, copper from the electronic print circuit industry and wastewaters which are used as additives to fertilizers af ion-exchange recovery, and ammonium and freon cooling liquors used for ammonia water and plastic foam, respectively af destillative recovery.

o Supplementary to external recycling activities in Denmark an additional 12,000 tonnes of chemical wastes are estimated to be exported. Reclamation of some fraction of these wastes occurs in other countries. Some examples are waste sulphur from pesticide production and waste pickling acids from metal surface treating industries.

o The most promising possibilities for an increase in external recovery of chemical wastes are associated with the following waste categories:

- recovery of iron, zinc, nickel and copper from filter cakes, waste waters and fly ashes and other metal containing wastes from surface plating industries, energy production etc.

- reuse of chromium pickling waste in formulating wood preservatives

- recovery of metallic copper from electric curcuit print industry wastewaters

- recovery of mercury from industrial wastes, batteries and dental wastes

- recycling of solvents from waste products such as paints

These approaches if combined with a suitable waste exchange system could lead to an increase in recovery of 4 to 5 times.

o Opportunities for recovery are often severely limited by financial considerations where a certain monetary gain must be foreseen before investment will be undertaken. Other limitations are the lack of commercialized of full scale tested technologies, fluctuating markets for secondary raw materials and relatively low prices for disposal by means of incineration and landfilling.

Concerning more general and political incentives to increase recovery of chemical wastes the feasability of

- clean technologies/internal recovery
- a selected tax on the most hazardous feedstocks
- inclusion of minimum contents of secondary raw materials in new products and
- a tax on final disposal

are recommended for further investigations.

The project is part of an EEC contract funding a general survey of recovery of hazardous (chemical) wastes in Europe with special reference to transfrontier movements for recycling or final disposal. The EEC-project is finalized mid-1987.

REFERENCES

Resource recovery, reuse and reccyling of constituents of hazardous wastes, Danish case study. Technological Institute, October 1986.

Chemical Engineering Departement
Technological Institute
P.O.Box 141, DK-2630 Taastrup
Denmark
Tel. 02-996611, telex: 334 16 ti dk

WASTE MANAGEMENT
A Comprehensive Service to Industry.

Ing. P.H.M. Meyer zu Schlochtern

Nowadays disposal of industrial (hazardous) waste presents a problem, which almost every industry has to face.
Satisfactory solutions can be found in changes of process, in-house treatment of waste products, alternative raw materials and process-integrated environmental technologies.
But still as long man produces goods, so long waste will be produced as an unwanted by-product with negative value.

To manage the costs which are involved in waste disposal, the waste generator has two options, i.e.:
1) he will do it himself or
2) he will order someone else to do it.

The decision between these two depends on a wide variety of possibilities, for example in house knowledge, complexness, confidentiality, quantities etc., which will not be discussed now.
Options no. 2) is the one which will be presented here.

"Anyone who produces a (hazardous) waste has a moral obligation - as well as a legal duty - to make sure it is disposed of safely."

That's where the waste manager can help. He has the technical expertise to determine the precise nature of the waste and the necessary facilities and competence to dispose of it exactly as demanded by the law, both as it stands now and as it might conceivably develop in the foreseeable future.
Therefore a comprehensive waste management service to the waste generating industry consists of the following activities :

- Consultancy and analytical service.
  A team of highly qualified chemists and trained technicians to provide a comprehensive laboratory and technical back-up service to all areas of its operations, including controlling and checking information received.
  Up-to-date equipment to carry out analyses, details of all wastes tested and treated in a waste data bank and desk research are important tools necessary for the waste manager in advising the customer what to do and what not to do.

WASTE MANAGEMENT
A Comprehensive Service to Industry.

Ing. P.H.M. Meyer zu Schlochtern

- Collection and transport equipment.
  Specialist transportation facilities for the carriage of all types of industrial waste .... tankers, flat beds, skips, demountable containers, vans, trailers, high vac tankers, sewer drain jetters, skip lorries etc. .... are necessary to ensure a safe and reliable waste management service.

- Cleaning and decontamination equipment.
  Full experience in all aspects of industrial cleaning including "gas-freeing" and decontamination of tanks which contained oil, lead solutions, corrosives, tars and other toxic materials is necessary including equipment as high pressure hydro jetting, high pressure (2400 bar) hydro cutting, butter worthing, automatic heat exchangers cleaning devices, etc. to undertake clean-up activities at industrial sites.

- (Pre)-Treatment plants for hazardous waste.
  Roughly waste materials can be divided into three categories :
  1) inorganic
  2) organic
  3) mixtures of 1) and 2).

  The best practical methods of disposal nowadays available are :
  a) incineration
  b) solidification/immobilization
  c) landfill
  d) biological treatment.

  To match 1), 2), 3) with a) b) c) d) it is usefull to operate a transfer/pretreatment plant whereby neutralization, separation, and bulking up activities are operated.

  It's imperative that good commercial relations with other waste treatment centers are necessary to reach a cost effective solution at all time.

- 24-hour emergency response.
  A complete emergency response service, which is on call 24-hours a day, 7 days a week. Spillages, accidents and other emergencies competently handled by expert chemists and technicians to prevent greater damages to man and environment.

WASTE MANAGEMENT
A Comprehensive Service to Industry.

Ing. P.H.M. Meyer zu Schlochtern

- Safety support.
  To optimize constantly safety standards a safety policy is institutionalized into waste management.
  Training of personnel, high class maintenance of equipment, vehicles with ADR (European Road Transport Legislation) licences, individual safety outfit, safety belt, explosion-proof lamps etc. are a few of the many items which are involved in safety thinking and its results.

  Environmental safety standards to the waste manager are as much as important as the ones mentioned above.
  Waste management includes the knowledge of consequences on long term storage, incineration, landfill and biological treatment and what to do with residues obtained from these waste treatment methods.

Conclusion.

Planning of waste generating, evacuation, final treatment and disposal has to start in the earliest stage possible to find the best resolution, environmentally as well as economically spoken.
When the decision to hire the assistance of a waste management service is taken, it is imperative that waste generator and waste manager do have a very close co-operation and interchange information as much as possible.
Only so it will be possible to deliver a comprehensive service to industry and to the public at large.

Oevel, the 18th of March, 1987

Ing. P.H.M. Meyer zu Schlochtern

ATOX WASTE MANAGEMENT

A  Company

Nijverheidsstraat 3
B-2431 Oevel, Belgium.

INCENTIVES/ATTITUDES

**Second European
Conference on
Environmental Technology**

in the 'European Year
of the Environment'

Implementation of energysaving and environmental technologies.

Regina W.Hommes, Johan C.Brezet and Leo W.Baas.

1. Introduction.

Originally the Erasmus University of Rotterdam was a School of Economics. Although since its foundation new faculties in law, sociology, philosophy, business sciences and medicine were added and it became a University in 1973, the main interest in this university is still focused on problems of economics. This main focus can be found also in the new faculties of sociology, law and business sciences.

The city of Rotterdam is located in one of the most polluted areas of Europe. Surrounded by heavy industry, connected with the harbour, it gets its share of air and water pollution. Furthermore, a great deal of its air pollution emanates from the German Belgian an English industry. Still more important is its position at the estuary of the polluted river Rhine. To guarantee the access of tankers, regular dredging is necessary. The slush is so affected by heavy metals and micro pollutants that it can no longer be used for land raising in agriculture and usually also not for laying the foundations of roads. The city of Rotterdam has to put it on land. The first discoveries of chemical soil pollution by illegal dumping to raise the land of new building areas were also made in the neighbourhood of the city of Rotterdam.

It happened that so many students at the university became interested in environmental problems, that within the Faculty of Sociology a new environmental institute came gradually into being. The Erasmus Centre for Environmental Studies (ESM, Erasmus Studiecentrum voor Milieukunde) started its first teaching course in 1984. From the beginning environmental problems were studied in connection with social, juridical, economic and technological aspects. The staff now consists of about ten persons, with completely different disciplinary backgrounds. The research programme is mainly oriented on environmental policy options for government and industry.

Within this programme much attention is directed to the study of the process of adoption and diffusion of energy-saving and environmental technologies. Most projects are subsidized by the city of Rotterdam and by several Ministries.

Although from the beginning the researchers tried to use models to **explain** under which conditions new and cleaner technologies would be adopted, the research projects focus on such different phenomena, that it would be artificial to use only one model. Furthermore, the circumstances under which new technologies are being introduced are so different, that each time again any prediction about the feasibility turns out to be very difficult. Nevertheless, it is possible to use one general model of the field of research, in which all our projects can be placed. This general model gives us the opportunity to draw some general conclusions.

2. Adoption and diffusion model.

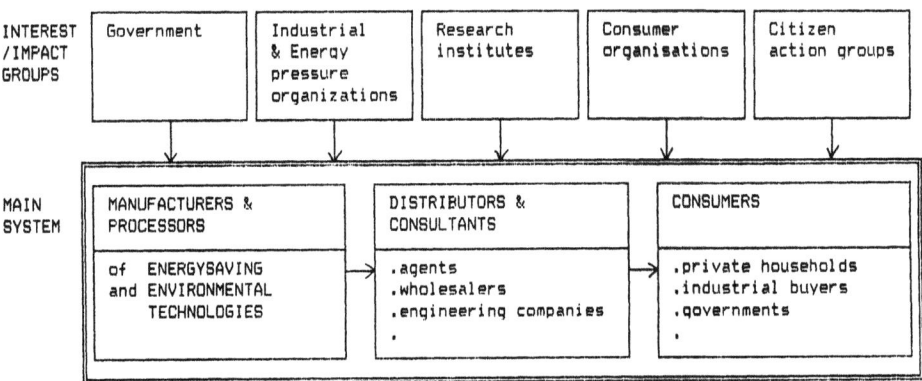

FIGURE 1. Adoption - Diffusion Model

The adoption and diffusion main system involves the major participants who play a role in designing, producing, supplying and consuming energysaving and cleaner technologies. One group of participants consists of manufacturers and processors of these technologies. A second group consists of consultants and distributors, who help to bring the new technologies to their proper markets. On the market a third group, consumers -such as industrial firms or households- are potential buyers and consumers of the energysaving and clean technologies. Thus the core system consists of specialized organisations and individuals who perform certain adoption/diffusion functions. Although the basic direction of new technology flows is shown as moving from the top downwards, it should be kept in mind that the relations and the flows of information and influence are multidirectional.

In addition there is another set of participants who have an interest in the performance of the participants in the main system. Government agencies, industrial and energy pressure organisations, research institutes, consumer organisations and citizen action groups take an interest in evaluating and influencing the participants in the task system. By considering the adoption/ diffusion model as our framework of research, we are able to clarify the main problems and to help develop the most appropriate policy options in the field of implementation of both energysaving and environmental technologies.

In this presentation we want to clarify the design and the results of four different studies, that are either concluded or are already so far under way, that we can present the data. Furthermore we will give a short outline of new projects. After these short presentations we will draw some general conclusions related to the general model. We will conclude the presentation with some remarks about the options for governmental policy.
First we will describe two projects in the field of energysaving technology, followed by two projects in the field of environmental technology.

3. Project 1. The introduction of energy saving heating equipment.

Dutch industry developed in 1981 high-efficiency condensing heating boilers as a means to save energy in households. Because of the probable energy-saving impact of these appliances (15-30% improved energy-efficiency), our Centre became interested in the adoption process of these energy saving boilers. Clarification of the adoption process should also lead to policiy recommendations for the government or government-related organisations as the gas-companies. The study consists of three phases.

First phase. Consumer behaviour.
Brezet and Silvester (Brezet and Silvester, 1984) developed a research design to study the adoption process by consumers of these energy saving central heating systems. This part of the study was supervised by the Section of Economic Psychology of the Faculty of Economics.
Brezet and Silvester developed a consumer decision process model, mainly based on the model of Rogers and Shoemaker (Rogers and Shoemaker, 1971).
They interviewed 84 early adopters of high-efficiency heating systems and 84 buyers of conventional heating equipment. Though the two groups differ in many respects, the main factor to explain the difference in behaviour was their own interest in technical innovations and their own potential to deal with technical problems. Early adopters act independently of retail and service engineers, while buyers of conventional heating systems take their decisions in correspondence with the advices of their retail and service consultants. Obviously, advices of retail and service engineers play an important role in the diffusion of innovative equipment.
Though the study design permitted to compare both groups of respondents in their attitudes towards environment, energy saving and costs of the different types of equipment, all those factors were not so important as the predominant factor of respondents technical interest. Early adopters are of course an interesting group for studies directed towards the first experiences consumers encounter by using innovative appliances, but they form only a small percentage of the population and they are therefore not so interesting for policy options regarding the acceleration of the process of adoption of energy saving central heating systems. As a result of this first phase of the study, logically a second phase was designed to explain the behaviour of retail and service engineers.

Second phase. Behaviour of retail and service engineers.

Though in the case of consumer behaviour it was still possible to make a research design connected with attitudes towards environment and energy saving equipment, in the study of retail and service engineers the study shifted to the market position of the total sector and its possibilities of adopting innovative appliances.
Brezet (1986) developed his own model to discern different factors connected with the advisory process to consumers concerning innovative central heating (CH) appliances. The interviews took place with 100 of the ca. 4000 CH-retail and service engineers in the Netherlands.

Slow adoption turns out to be connected with the following factors:
- characteristics of the new appliances, e.g. perceived high rate of failure and the more complicated service;
- lack of sufficient information, e.g. lack of adequate possibilities for in-service training and updating courses; further more an often insufficient backing-up service from the side of CH-manufacturers and CH-wholesale-dealers;
- market position, e.g. on the one hand an increase of do-it- yourself and non-official installation activities and on the other hand an increase of offical competition by plans of certain energy distribution companies for the leasing and main tenance of CH-appliances.

The study results give the impression that those CH-engineers who, based on careful management, show a relative high level of activity towards groups such as consumers, manufacturers, wholesale dealers and energy companies, experience these problems to a lesser degree and are more apt to adopt innovative appliances.

Third phase. The behaviour of manufacturers and wholesale dealers.

The last part of the study is oriented towards the marketing possibilities, training facilities and backing-up activities of the manufacturers and wholesale daelers. This part of the study is now carried out.

The general conclusion we can draw from this project is that any special attention the government wants to give to the improvement of the adoption of energy-saving heating equipment by households should not only be directed to consumers, but should at the same time be more oriented to the improvement of the intermediate structures of retail and service engineers. Furthermore, this attention can only be effective, if the economic position of retail and service engineers is taken into account.

We must still make a remark about this study. While the study of Brezet was under way, one of our sister institutes , the Centre for Environmental Studies Leiden (Van den Berg a.o.,CML, 1986), became interested in the question if the net result for the environment of high efficiency condensing heating boliers was as effective as was previously thought. Especially the higher emissions of NOx and SO2 was a problem. Certain aspects of their study are still debated and furthermore manufacturers seem to have succeeded meanwhile in improving the new boilers in this respect. But we agree with the CML, as one of our other projects shows as well, that it would be ideal to know the total energetic and environmental impact of a product before studying its adoption process. Unfortunately scientists still cannot measure these aspects on one continuum. This makes it difficult to decide at which moment it is opportune to study the adoption and diffusion process. We did encounter this same problem in our study about the substitution of materials.

4. Project 2. Acceptation of energy-saving appliances by residents.

Silvester does research on the acceptation of energy-saving appliances by residents in six different housing-estates, some newly-built and some rebuilt, while in all cases special attention is paid to install energy-saving appliances. Two of his studies are published (Brezet a.o. 1985 and Silvester, 1986). He works in the tradition of economic psychology (Van Raaij en Verhallen, 1983).
The main blocks of variables to explain the difference in energy use and total residential costs consist of technical and architectural characteristics of the house and its heating system on the one hand and of the energy behaviour of household members on the other. In all the research projects the same model is used and a multivariate analysis is applied.
The sequence of the different research projects permits builders to gain experience from problems that residents encountered by using different energy saving appliances and the studies have a direct practical use. From a scientific point of view the most interesting finding was that technical and architectural characteristics do explain more of the variance in energy use than the behaviour of residents. This fact does not apply for households that have a completely different life style. E.g. the use of energy in households where both partners are working during daytime or are away for long holidays is less than in households where nearly always somebody is at home. But, when a comparison is made between households with the same life style, architectural and technical factors do explain more of the variance than the actual energy saving behaviour of residents, measured e.g.by a high score on the use of the thermostat and the closure of the shutters at night. This theoretical finding could have a high practical value. This is the reason that new studies are carried out in which both blocks of variables, the technical aspects and the residential behaviour are again analyzed. Some technical measurements are carried out by TNO now, to verify whether the different use of energy can be explained by a difference in the quality of the houses. As Silvester is not convinced that the actual energy saving behaviour of residents does not lead to a reduction of energy use, he will develop a new research design where more control factors are used, e.g.

where more attention is paid to the energy behaviour of neighbours in adjoining houses.
The role of government for this project is a very obvious one. As it turns out that several technical appliances have a direct effect on energy use and as the government in the Netherlands has a strong influence on building requirements because most housing estates are subsidized, the government can directly influence the energy use. Apart from that information campaigns on energy saving seem to be most effective, when they are directed to the use of specific energy-saving appliances.

5. Project 3. Transfer of cleaner technologies to small and medium-sized firms.

Another study was directed to the transfer of knowledge of cleaner technology to small and medium-sized firms. As we see a gradual improvement of the environmental impacts of production processes of large factories, in future one of our main problems will be the emissions of small and medium-sized firms. For that reason we became interested in the question how transfer of knowledge of add-on technologies took place from large firms to small industry. The study was carried out in cooperation with the Centre for Research in Business Economics (BEI) of the Faculty of Economics of the Erasmus University.
Frank and Swarte (Frank and Swarte, 1986) studied in an exploratory study the transfer of knowledge concerning anaerobic water treatment, biological filters, scrubbers and machinery for the separation of domestic waste. They interviewed a selection of firms on the market side and on the demand side. Their model consisted of six blocks of variables: A. Clean technology attributes/suppliers; B. Information transfer and circuits; C. Adoption factors/buyers of clean technologies. D. Environmental policy. E. Adoption of clean technologies. F. No adoption or use of other clean technologies.

One of the research results was that small and medium-sized firms usually do not become interested in the environmental impact of their processes of production before one of the governmental agencies puts certain restraints on their activities. In that case especially small firms will find themselves in a difficult situation because of a lack of knowledge of the technologies that could be useful for them and of the ways to get this information on a relatively short term. Though this situation does also occur for medium-sized firms, usually these firms are better equipped with research facilities and laboratories and they are more capable to take the necessary initiatives. Nevertheless also these firms encounter difficulties to find the necessary information. The study shows a very interesting relationship between market parties. Producers of clean technologies prefer a direct relationship with their buyers, but buyers want indepent information about the applicability of different technologies and seek information with intermediate organisations. However, in acting in this way firms face many difficulties to find the right information as the market of intermediate organisations is vast and not well structured. Regarding the role of government, the researchers found that usually firms are not very interested in direct support, but that they saw a general function for government to give support to the improvement of the information transfer. Though this was an exploratory study it seems that one of the main factors to impede a quick adoption of new and cleaner technologies for small and medium-sized firms is the lack of a well structured intermediary infrastructure.

6. Project 4. Substitution of materials.

We started our research with the logical thought that if some materials cause more environmental problems than others it would be interesting to study the process of material substitution and to clarify in which respects government could speed up this process. For this project we worked together with the engineering firm, B & G Consultants (Bureau Fuels and Raw Materials) in Rotterdam. In this case we decided that the engineering firm should start to do research in the environmental impact of some often used materials like iron and aluminium, while our researchers L.W.Baas and R.A.Spoel in the meantime should do research in the different governmental options on the one hand and the adoption process of certain products on the other hand. Though the means of governmental influence are manifold, it took them not much time to make a synopsis of the different governmental means, which they could put on a continuum from direct means like juridical regulation to indirect means like specific campaigns of instruction or general education. More difficult was to trace which governmental measures are more effective under certain circumstances, because this evaluation is dependent on many variables. So they decided to analyse five different products, where different materials could be used. The five products were pylons, greenhouse frames, beverage cans, hard floor-coverings, sea-transport containers.

Then something very interesting occurred. As they got gradually more insight in the use of these five products they became more convinced that to evaluate the usefulness of material substitution and to evaluate which governmental measures could encourage this process it is absolutely necessary to get insight in the whole life cycle of a product from the moment the raw materials are extracted, though to the raw materials processing, the transport of materials, the production and the use of the product and the possibilities for recycling and ultimate disposal. **Only when this whole process is known is it possible to give advices to governmental agencies regarding the use of which materials should be stimulated.**
Especially the life of a certain product and its suitability for recycling and the market possibilities for the recycled materials must be taken into account before the environmental impact of certain materials can be stated.

For this reason, it took B & G more time than we had thought to analyse the environmental aspects of the use of different materials in products and the study is still being carried on.

7. New Projects.

Our Institute started two new research projects this year. In the first project, Clean Technology and Rhine we will try to find out which emissions still take place in the Dutch Rhine basin and we will in cooperation with

several Institutes of the Technical University of Delft and with industry look into the economic and situational constraints for applying new technologies.
Another research project concerns the development of a computerized simulation model for domestic waste, which the city of Rotterdam wants to use to facilitate policy decisions.

8. The role of government and environmental technology.

To evaluate the consequences of our findings for the government we will make use of a policy-options model.
In short this model indicates that for the formulation of an effective and efficient governmental policy the following stages should be considered:

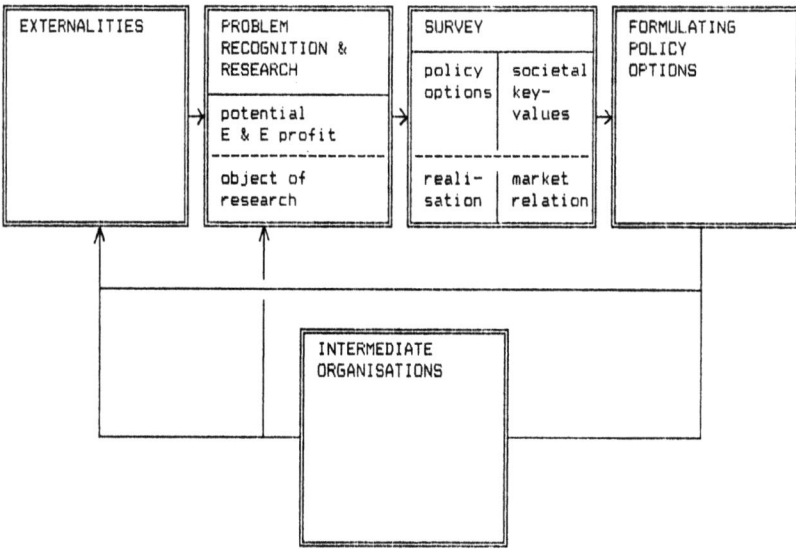

FIGURE 2. Model for policy-options of the government

1. Problem recogniton and research.

Although energy-saving and environmental technology appliances can have a positive effect within a certain sector, very often we still have insufficient insight in their total effect, when the whole system of relations is taken into consideration. Sometimes the net effect of energy-saving appliances and environmental effects is adverse to each other. More systematic research into Energy- and Environment-optima is a precondition for stimulating the most appropriate technologies.

2. Survey of policy options.

The various projects indicate that the degree to which governmental policy is needed to accelerate the application of energy-saving and cleaner technologies varies with the object of research. It is our strong impression that in the field of energy innovations are more naturally adopted by the whole system than in the field of cleaner technologies. In other words: energy-innovations can be stimulated by government mainly by indirect measures, while in the case of clean technology a more direct approach is necessary. Only in the case of the housing estates, government can have a very direct influence by making standard technical requirements for subsidized building projects. Apart from that it can have an indirect influence by stimulating research into new energy saving appliances. The direct approach of government in the environmental field is usually very delicate. Of course, government has the possibility to forbid the use of certain materials like pcb's, but this does not mean it has a direct influence on the use of cleaner technologies. It has though, the option of negotiating with firms to end emissions. Usually it does this by setting a certain time limit for the improvement. At that moment government can influence firms to use new technologies. In our project of transfer of cleaner technology knowledge this seems to be an effective procedure, though very time consuming. Furthermore it should be followed up by the creation of independent centres for environmental technology transfer. Of course government can support the development of these centres. It seems that industry would not be very sympathetic to other types of governmental interference.
In the project of adoption of energy saving heating boilers and in the project of energy saving residential behaviour we got information about the behaviour of consumers of new technology.
In both cases it turned out that other factors were more important for the goal to save energy than the actual behaviour of consumers. This could have an important effect on the attention government wants to pay to consumer information campaigns. Though doubtless consumers should have a general idea of their own contribution towards energy saving, for the government it seems to be more important to improve the infrastructure of transfer of knowledge to intermediate levels between producers and buyers of new appliances. As we have in the Netherlands a long tradition of transfer of knowledge of new discoveries in the agricultural sector, where a whole network of consultants inform farmers effectively about their process of production, we think that for the adoption of environmental technology this example could be useful to explore.

1. Berg,M.M.H.E. van den, D.Schmidt, M. van Koten-Hertogs, G.Huppes, W.T. de Groot, Potenties van produktbeleid, Centrum voor Milieukunde Rijksuniversiteit Leiden, CML-mededelingen, nr.26, 1986.
2. Brezet,J.C. and S.Silvester, De adoptie van hoogrendement ketels en warmtepompen door nederlandse huishoudens, Erasmus Universiteit Rotterdam, 1984
3. Brezet,J.C., A.Frank and S.Silvester, Bewonersonderzoek minimum-energie woningen te Schiedam, Erasmus Universiteit Rotterdam, 1985.
4. Brezet,J.C., CV-innovaties, Erasmus Studiecentrum voor Milieukunde, Publikatiereeks nr.2, 1986.
5. Frank,A. and H.J.J.Swarte, Milieutechnologieën Erasmus Studiecentrum voor Milieukunde, Publikatiereeks nr.1, 1986.
6. Raaij,W.Fred van, and Theo M.M.Verhallen, Patterns of residential energy behavior, Journal of Economic Psychology 4 (1983), Elsevier Science Publishers B.V., North-Holland.
7. Rogers,E.M. and F.F.Shoemaker, Communication of innovations, Free Press, London, New York, 1971.
8. Silvester,S., Minimum-energie woningen, Erasmus Studiecentrum voor Milieukunde, Publikatiereeks nr.3, 1986.

INNOVATIVE STRATEGIES IN THE ENVIRONMENTAL INDUSTRY - AN INTERNATIONAL COMPARISON

Dominique DROUET - Recherche Développement International

1. INTRODUCTION

This paper outlines the major findings of a two year study commissioned by the Groupe de Prospective of the French Ministry of the Environment to compare innovative strategies in the environmental industries of the following countries : Japan, the United States, the Federal Republic of Germany, the United Kingdom, the Netherlands and France(1).The study deals with "add on" techniques applied to industrial and toxic wastes treatment, water and wastewater treatment, air pollution control (stationary sources) as well as instrumentation (2).It is focused on the R & D programs of firms and does not cover research activities of public institutes or universities.

2. MAJOR FIELDS OF R & D

The inventory of new processes under development shows that a majority of R & D programs aims at improving the performance and enlarging the applications of technologies which have already been applied commercialy - sometimes in a very limited manner - in the field of pollution control. Nevertheless a growing minority of projects relies on "new technologies", that is technologies not previously applied to pollution control or derived from recent scientific breakthroughs. This latter trend contributes to a renewal of the technological base of the environmental industry. The technological fields where most of the current R & D efforts are concentrated are the following :

21. Improving and extending known technologies

- Research on empirical selection of natural micro-organisms with usefull characteristics to degrade pollutants in solid wastes, liquid or gaseous effluents. (Incisive manipulations of isolated genes to create artificial organisms is refered to under 22). Simultaneous innovation is focused on improving the processes using such micro-organisms : aerobic/ anaerobic, fixed or fluidized beds, bioreactor design. To various degrees

---

(1) L'innovation dans les industries de l'environnement - Comparaison internationale - D. DROUET (avec la collaboration de S. FAUCHEUX) - RECHERCHE DEVELOPPEMENT INTERNATIONAL - Février 1987 - The full report can be obtained from Recherche Développement International - 37 c3 Avenue Franklin Roosevelt 75008 Paris - France.(An english translation will available shortly).

(2) Technologies not included in the study are : clean technologies, noise abatement technologies, control of air pollution from mobile sources, collection and treatment of municipal wastes.

firms from all countries are engaged in these fields, the main thrust of their programs being on water and waste treatment. West Germany (LINDE, HOECHST, MENZEL...)(3), is globally leading this effort. With regard to fluidized beds and vertical reactors, the Netherlands (GIST BROCADES...) and Japan (KURITA, DENKA...) are also at the forefront. French firms are among the most active developing fixed bed reactors (DEGREMONT, OTV...) as well as techniques for biotreatment of waste air (SAPS...). Dutch firms (CLAIRTECH...) are also investing in odor treatment. Coal desulfurization through biological processes is being mostly investigated by German, American and Japanese firms. Applications to contaminated soils or landfills are primarily being developed by American (DETOX, BIOCLEAN...), Dutch (IWAGO...) and more isolated British companies (BIOTECHNICA...).

- Improving already experienced and sometimes old thermal processes (incineration, fluidized beds, pyrolysis...) to achieve higher destruction ratios for a growing number of pollutants in solid or liquid wastes. Changes in combustion techniques also aim at reducing air pollution. With fluidized bed technology as a prime target, German (LURGI, THYSSEN...) and American firms (G.A. TECHNOLOGIES, MIDLAND ROSS...) are currently the most active in these fields, Japan (TAKUMA, MITSUBISHI HEAVY...) being third. The Netherlands (SMIT OVENS...) and the United Kingdom (COMBUSTION SYSTEMS...) hold an intermediate position while investments of French firms (STEIN, CREUSOT-LOIRE...) in this type of research seems to be lagging behind.

- Improving already tested recycling processes for liquid or solid wastes. Globally, firms in Japan (FUJI FILTER, EBARA...), Germany (DEGUSSA, DORNIER...) and Holland (ESDEX...) are currently leading the innovative effort. Nevertheless in such specific fields as reverse osmosis and electrodyalisis, the United States (ALLIED, IONICS...) and France (EIF ECOLOGIE, SOGEA...) are at the forefront.

- Research on treatment of air pollution from stationary sources (de SOx, de NOx, dust removal...) through experienced methods such as dry or wet systems, catalytic or non-catalytic processes. The following ranking of countries can be proposed according to the current innovative effort of private firms : 1. Germany(LINDE, WALTHER...) and Japan (MITSUBISHI HEAVY, HITACHI...) 2. United States (DORR OLIVER...) 3. Holland (COMPRIMO...) and France (LAB...) 4. United Kingdom (ICI...). With regard to selective catalytic reduction of nitrogen oxides, Japanese firms are way ahead. British firms occupy a better position as far as dust removal is concerned.

- Miniaturizing and improving the reliability and scope of application of existing measuring instruments based on colorimetry, chromatography, spectrography, U.V. spectrophotometry, etc. American (BECKMAN, MERK...), Japanese (HITACHI...), German (DR LANGE...) and British (KENT...) companies heavily invest in these directions,whereas French (ENVIRONNEMENT S.A...) and Dutch (EUROGLAS...) R & D programs appear to be more limited.

22. <u>Applying new technologies to pollution control</u>

- Development through incisive genetic alteration or cell fusions

---

(3) We do not imply that cited firms are the only ones developing important R & D programs. Names of specific companies are given as examples. For a complete outlook with details on research in progress please refer to the full report.

of novel micro-organisms designed to degrade specific toxic pollutants in soil, landfills or liquid effluents. With very few exceptions, research is presently only conducted at the lab scale : American companies (REPLIGEN, GENERAL ELECTRIC, EASTMAN KODAK...) lead this field before their Japanese and German counterparts (especialy those cooperating with the MITI Microbiological Lab and Göttingen University). By comparison Dutch (AKZO..) French (RHONE POULENC...) and British (ICI...) research is less developed.

- Design and production of new types of membranes for filtering industrial effluents or wastewater (hollow fibers for pressurized processes, new materials for ultra or microfiltration). The United States (GENERAL ELECTRIC, ALLIED...), Japan (INTIC, TORAY...) and Germany (ENKA, DORNIER..) dominate this field and are leading in applications to pollution control, whereas other countries pursue more limited efforts. Japan is most involved in developing a new generation of bioreactors incorporating membranes for wastewater treatment.

- Applying advanced thermal processes, such as plasma arcs, laser or electron beams, to detoxify solid wastes, contaminated soils or sludges. U.S. companies (WESTINGHOUSE, J.M. HUBER...) are investigating a large spectrum of applications. R & D programs in other countries are more narrowly focused and sometimes limited (United Kingdom).

- Designing and experimenting instrumentation and automation systems to control pollution abatement processes and/or water and wastewater networks (new technologies integrating sensors, electronic data transfer and storage, with specialized software or expert systems). Firms from all countries have RD & D programs in progress. In the water sector, France (LYONNAISE DES EAUX, FLUTECH...), Great Britain (WESSEX WATER, DYNAMIC LOGIC...) and - to a lesser degree - the Netherlands (PHILIPPS...) are at the forefront. With regards to advanced controls of industrial pollution abatement systems, Germany (SIEMENS...), Japan (YOKOGAWA ELECTROFACT...) and the United States (ROCKWELL...) are investing more in the development of new systems.

- Development of sensors based on new technologies (opto-electronics, microbial biosensors, laser). The United States which dominates the world's instrumentation industry has the largest effort under way, in these new fields (MARTIN MARIETTA, GENERAL ELECTRIC...) before Japan (HITACHI...), Germany (BATTELLE FRANKFURT...) and the United Kingdom (CAMBRIDGE LIFE SCIENCE...). Some French companies (LYONNAISE DES EAUX, REMTECH...) are well positionned for specific developments related to enzyme electrodes and sonic radars. Dutch R & D (HONEYWELL B.V....) sems more limited in these new directions.

3. INNOVATIVE STRATEGIES : GENERAL TRENDS

31. A new factor in international competition

In the seventies, one of the major features characterizing the market for environmental technologies was the need to rapidly and widely install a first generation of pollution control systems in response to the quantitative demand created by newly introduced environmental protection measures. As a general rule, competition among firms was centered on the price of pollution abatement systems designed to achieve primary treatment of effluents or wastes.

In the eighties, innovation in the environmental industry is becoming a key factor to gain a competitive edge on the market. Several phenomena contributed to this change. First of all, in a few leading countries, environmental control regulations are quickly evolving to stricter levels and are extended to new series of pollutants (The 1986 amendment to the American Safe Drinking Water Act or the 1985 german air pollution standards for large boilers are typical examples). Searching for technology adapted to the new standards more and more often imposes to look beyond the mere improvement of existing processes and to develop new ones.

When comparing the present situation with the one prevailing at the end of the seventies, a general change in attitude must also be considered : in most countries, tackling environmental problems can be viewed today as a long lasting national priority, whereas in the early eighties, national agendas - especially in the United States with the first announcements of the Reagan administration - seem to set a more uncertain future. Increased scientific evidence about issues of pollution and human health as well as growing public pressure in key countries (Germany, the United States) are creating this new situation. As a result,more and more industry leaders give a higher priority to pollution control markets when considering long term R & D strategies.

In some countries major corporations - whether already active in the field of environmental technologies or new entrants (such as chemical companies) - are concerned by this important change. Moreover,privatization of operations and management of environmental protection equipement (wastewater treatment plant, waste to energy plants...) in different countries enlarges market perspectives and attracts other firms. As a result increased financial means from the private sector are available for R & D.

Other factors such as the general upward tendancy of research spending in related fields (biotechnologies, microelectronics, etc) also favor innovation in environmental technology. So does the growth of turnkey contracts with operations and financing provided by the equipment supplier, which limit the client's risk and help new processes penetrate the market.

While the trend is for governments to increase their regulatory pressure, thereby creating new markets, it is important to observe that - with the one exception of West Germany - public budgets for environmental research are stagnating and even decreasing in some countries. As a general rule, the currents situation can be summarized as follows: forced by public opinion to introduce stricter environmental protection rules, governments leave it up to the private sector to find ways to adapt. In return, environmental industry leaders seem more and more prone to innovate to take advantage of these new opportunities.

32. <u>Short term vs long term strategies</u>

Today different types of innovative strategies are being developed. For a majority of firms short term strategies prevail. This can be explained by the large number of small businesses in the environmental industry and by the rapid opening of new markets. Nevertheless long term strategies are becoming more important due to the growing involvement of larger groups.

Short term strategies aim at developing pollution control equipment to a commercial stage within one to five years. Upcoming regulations in the domestic market or in a more advanced country are the prime determinant of their targets.

A common situation for a country whose standards lag behing the strictest rules worldwide, is to adopt the most advanced regulations. Technology supplied by domestic market oriented firms within such a country is generaly outperformed by some of the materials developed abroad. As shown by numerous cases, innovation is then mostly introduced by importing foreign processes. This solution can be the quickest way to make up for a technological gap. It does not necesseraly imply a long lasting foreign dependancy, provided that domestic firms have the will to master and go beyond the imported technology.

For some companies, short term R & D priorities are focused on materials which can be operated by the supplier (Privatization of municipal pollution control equipment offer a growing number of such opportunities). R & D budgets can then be amortized both on the sale of the equipment and on subsequent operation.

Longer term innovative strategies come into play when companies have a positive view of environmental markets within a five to fifteen year framework. Related R & D programs always have an international focus. Some companies develop synergies between applications to solid wastes, water and air (Especially when there is a strong investment in biotechnology). Others retain sectorial approaches.

Long term strategies often involve international partnerships. Leading firms worldwide rely on extensive technological information gathering networks. They often have collaborative agreements with universities or public research institutes and try to participate in the most advanced research programs wherever they take place. Consultations between leading domestic companies and governments sometimes help set the pace of enactment of new environmental regulations in order to favor market penetration of domestic innovations. In a further step, governments also play a key role in the international spreading of norms, thereby opening new markets to advanced domestic processes. In Europe, EEC directive negociations reflect this tendancy.

Concerted action between firms and government can either be of the "offensive" type - going a step beyond foreign competition - or "defensive", that is protecting domestic firms through selective authorization procedures of foreign technology. Firms developing long run innovative strategies sometimes also establish new professional associations to counteract lobbying by trade groups who defend conservative technical options.

4. NATIONAL CHARACTERISTICS

Significant differences can be observed between countries with regard to innovation in the environmental industry. Major national characteristics can be summarized as follows.

41. Japan

Japanese environmental protection norms are among the strictest internationally. Although enactment of new regulations has slowed down after the seventies, important new measures are being prepared focused on toxic elements. Close collaboration between government and industry with regard to norms and innovation will most likely result in new progress in the water and waste sectors towards the end of the decade.

Ambitious joint public-private R & D programs (Biofocus, Aquarenaissance 90) combining bioreactor and membrane technologies can put japanese firms at the forefront in these fields and even give them a competitive edge. Japanese companies are among the leaders in air pollution control : as an example their selective catalytic reduction processes for nitrogen oxide reduction from stationary sources are presently being licenced worldwide. Japan also has the third environmental instrumentation industry and is rapidly progressing in biosensors.

In the early seventies the development of the japanese environmental industry was based on foreign licences. Today this dependance is significantly reduced. This transformation was helped by rapid growth in global R & D spending (increasing twice as fast as the United States), the third public budget worldwide for environmental research and high technological levels in related fields such as biotechnology, membranes, microelectronics and instrumentation.

Large diversified corporations are the main R & D players. Their interests cover the entire pollution control field. Most of them are export oriented and this will help the future international distribution of processes under development. When compared with other countries, the structure of the japanese environmental industry presently appears more rigid and less likely to open up to new entrants.

42. The United States

In the first half on the eighties, with all the major environmental laws waiting to be re-enacted, great uncertainty prevailed with regard to regulations. EPA's research budget followed changing trends and globally decreased. In four years three different administrators headed the Agency. In the mid-eighties two trends are emerging : on one hand, Congress is overriding the Reagan Administration's reluctance for stricter environmental standards, and, on the other, public spending for environmental R & D continues to decrease. (Although the US environmental research budget is still the largest worldwide).

Today's perspectives are more tangible for american firms : the 1986 amendments of water and hazardous wastes legislations set new standards which reach and sometimes go beyond the toughest european rules. For air pollution a similar move is likely in the coming years.

As a result european water and waste treatment technology is rapidly progressing on the american market. So do recently some japanese processes for air pollution control. Despite a relatively slow pace of innovation during the last ten years, the US environmental industry retains strong assets : short delays from lab research to commercialization resulting from the best system for university-industry cooperation, global scientific and technical excellence in most fields interesting the future developments of environmental technology.

Different signals indicate that innovation is entering a new phase : chemical companies, biotechnology firms as well as businesses coming from other sectors are investing to develop environmental technologies (Hazardous wastes treatment is today their first target). A major privatization trend also helps market penetration of advanced systems. Some major corporation devise new strategies where the environmental markets have a higher priority.

American firms are also at the forefront in some important fields of research such as the development of novel micro-organisms adapted to specific pollutants, hollow fiber design and manufacturing, advanced thermal processes development and research on new sensors.

43. The Federal Republic of Germany

Germany in the eighties sees several factors coming into play to favor innovation in the environmental industry : enactment of new environmental standards (following a rapid pace in the case of air pollution and a more progressive but durable one in other sectors), dramatic growth of public environmental R & D spendings, major corporations adopting new strategies in the environmental field. The German environmental industry - already well positioned at the end of the seventies - is today strongly reinforced. Moreover, this trend will very likely continue.

The number of new patents filed by German companies during the last years reflects this global effort(4). Nevertheless a difference can be drawn between R & D to improve known processes, with commercial applications in the short run, and research dealing with entirely new technologies. In the first case, German firms are among the leaders in most fields : vertical bioreactors using "natural" micro-organisms, fluidized beds and traditional thermal processes, waste recycling technologies, gas desulfurization and dechloration. In the second one, also quite active, German companies are sometimes behind their american and japanese counterparts (Especially for new sensor technology, membranes or artificial micro-organisms).

Various types of companies contribute to the innovation drive in Germany : small firms, large diversified corporations (some of which are restructuring to focus on environmental markets) and new entrants whose core business is in chemistry, electronics or the nuclear industry. Resulting from close cooperation between government and industry, growing German pressure on her EEC partners can be anticipated in the near future, in order to make them catch up with German regulations. This could further reinforce Germany's leading position in Europe.

44. The United Kingdom

The regulatory trends in the United Kingdom are characterized by slow change and delays in observing EEC directives. As a general rule, the gap with leading countries is increasing since 1980. The share of environmental research in the total public R & D budget is low (1 %, it compares with France and the United States, whereas the Netherlands, Germany and Japan are around or above 2 %). Neither driven by regulatory change, nor supported by public financing, innovation in the environmental industry is globally proceeding at a rather slow pace. This reflects a general trend of British R & D spendings, the growth of which is below OECD's average.

Nevertheless British companies have a good international standing in several fields. In the water sector, R & D in instrumentation control and automation is progressing through public-private collaborative programs. A positive effect is the development of small firms whose products benefit from the demonstration facilities and approval procedures established by these programs. Privatization perspectives in the water industry can also open to new technical options. Because of a large public R & D effort, British combustion technologies - and more specifically fluidized beds - achieved international pre-eminence in the seventies. Today licen-

---

(4) When comparing the total number of patents applied for in France since 1982 for water and air treatment, German firms (754 applications) are ahead of American (610), Japanese (200), British (162) and Dutch (101) companies. The French total (501) includes domestic patents and cannot be directly compared with figures for other countries.

cees operate in several countries.

Innovation in the British environmental industry benefits from research done by large chemical and oil companies in fields like membrane and biotechnology. A tradition of efficient industry-university cooperation and a strong effort in basic biotechnological research led to the start up of new biotechnology firms. Unique in Europe, this phenomenum helped environmental innovation since several new companies selected toxic waste treatment as a primary application.

45. The Netherlands

Dutch environmental standards usually follow the most demanding european regulations and in some cases are even ahead (An example is soil protection since the 1986 bill). The percentage of public R & D spending devoted to environmental research is the highest of the six countries. Today air and waste research are getting a higher priority when compared with water. Two public R & D programs were launched in the eighties to promote environmental biotechnology and fluegas treatment processes.

In the early eighties, Dutch water treatment technologies - biological processes in particular - were at the forefront internationally. Responding to the recent acceleration of R & D programs in Japan and Germany, Dutch companies are presently developing collaborative research with the most advanced institutes worldwide. For deNOx treatment of stationary source, Holland like other countries is buying japanese systems. Dutch fluidized bed technology has a good international standing.

Relying on the country's technological experience in biotechnology and on efficient public-private R & D partnerships, several Dutch companies are ahead in new fields such as biotreatment of odors in wastegases. Dutch R & D activity also covers most processes - whether thermal, chemical or biological - for industrial and toxic waste treatment. On the other hand, investments in the development of environmental instrumentation are weak by international standards.

In the mid-eighties, the general outlook for environmental research is favorable because of sustained public financing, continuous introduction of new pollution control measures and a national policy to promote Dutch environmental industry abroad. Firms involved in R & D programs include not only small businesses and engineering companies but also large conglomerates of the food, chemical or public works industries.

46. France

When compared with more dynamic countries, change in French environmental regulations is occuring at a slower rate since the early eighties. In the water and waste sectors, the gap with leading countries is still limited because French norms were among the most severe ones at the end of the seventies. But when considering air pollution control, France is increasingly lagging behind. Public spending for environmental research has been decreasing in real terms for the last years (As a share of the total public R & D budget, it is among the lowest of the six countries). Whereas in other countries public research priorities are primarily focusing on air pollution, France still devotes the major share of her budget on water and wastes.

A few small companies develop new fluegas treatment processes for stationary sources (dechloration, biological odor removal). Their success on international markets does not make up for a general weakness of french firms due to the small size of the domestic market in this sector. Expanded cooperation with the most advanced institutes and firms abroad allows French water treatment groups to keep in touch with the latest technological developments. However, because of the increasing pace of innovation in this sector, French firms run the risk of loosing their leadership position for specific biological processes.

With regard to waste treatment, some French firms are among international leaders for particular technologies (electrodyalisis). Despite a few exception, innovation in instrumentation generally lags behind competition from the United States, Germany or Japan. On the other hand French water treatment groups and a few small specialized firms are investing heavily to develop control and automation systems.

Because the national context does not provide many incentives for innovation, only firms targeting international markets remain at the forefront of technological developments. With limited domestic perspectives contrasting with the German situation, only a few firms consider environmental technology as a future priority. Outside the water sector, the French environmental industry is not concentrated and attracts only a small number of new entrants (Ciment makers investing in waste treatment are one example). R & D budgets too often remain below the critical mass when compared with top foreign competition. Investigation into new technologies is not broad-based, despite a few bright spots (lasers, enzyme-electrodes.)

5. CONCLUSION

Innovation is becoming a key factor of international competition in the environmental industry. This results in the boosting of R & D programs to improve known processes and develop entirely new ones. The six countries under consideration gather the majority of innovative firms. Nevertheless their contribution to the current R & D effort is uneven.

At one end, the German environmental industry is globally mobilized since the early eighties. Technological research is proceeding today at full speed. For different reasons, the other two main environmental industries, in the United States and Japan, are only at the beginning of a major research drive. In the U.S., following years of uncertainty, the situation becomes favorable to innovation and numerous firms are investing in R & D. Japan, whose research effort increased at a reduced pace in the early eighties, is ready for a new acceleration.

France, which reached a reasonably good international position at the end of the seventies and the United Kingdom - already somewhat behind at that time - do not appear to adequately size up the magnitude of their three main competitor's effort. Despite interesting R & D programs in specific fields, these two countries seem to be loosing ground. On the other hand, the Netherlands are following Germany's tempo : Dutch environmental industry invests in R & D but does not cover the entire field due to the lesser global economic power of the country.

RISK ANALYSIS AS A RATIONAL BASIS FOR SOIL PROTECTION AND CONTAMINATED LAND CLEAN UP.

Dr. M. LOXHAM and Drs. C.C.D.F. VAN REE.

Abstract

In many places the ground and groundwater is contaminated by chemicals arising from past activities in the agricultural, military and industrial sectors. Activity in these sectors will continue and by the nature of things further release of contaminants will occur. Some of the contaminants in the geosphere do or will reduce its use value and will require measures to minimise or counter this. These measures are inherently costly and imperfect and a key problem is the identification of an adequate decision model for their use.
It is proposed that risk analysis models be used for this purpose. These models, based around the source-path-target methodologies, force the explicit quantification of source emission strengths, use scenarios, event chains and final impact assessment at both the end and intermediate ground use-value levels. They form the rational basis of a cost-benifit analysis of the technology addressing both soil clean-up and soil protection problems.
The risk analysis models (and associated event-chain) models can have a determinate or stoichiastic character leading to a requirement for impact norms to be formulated in safety factor or probabilistic terms for the scenarios considered.
The impact of the use of risk analysis techniques in this field, on the design of environmental technology, on the problem of standards and norms and even on environmental policy itself cannot be overestimated.

Introduction

In recent years it has become apparent that the soil and groundwater system is significantly at risk and indeed already severely damaged by a wide range of man-made pollutants. These contaminate the soil eco-system, enter into its internal nutrient cycles and can be taken up, via the groundwater, into the food chain thereby threatening the surface flora and fauna and ultimately man himself.
These dangers are compounded by possibilities of direct human contact with the toxic substances either in the ground itself or by vapour phase diffusion into living-spaces or even by diffusion into piped drinking water supplies.

The extent and severity of the threat cannot be overestimated either in terms of the large volumes of soils involved or the extreme toxicity of many of the pollutants concerned. These include not only petroleum products and inorganic bulk wastes but also heavy metals, cyanides, pesticides, solvents, PCBs and dioxines among other exotic and dangerous organic trace pollutants.

Soil protection measures are being developed and imposed in order to prevent future pollution of the soils and many countries have implimented major soil and groundwater decontamination programs and passed far-going soil protection legislation.

However, to date, these programs are characterised by very high costs and rather less than perfect results. Clear field reclaimation cost of up to $1000,-- per square meter are not uncommon and for a site which is still being utilised, the interuption, dismanteling and reconstruction penalties will increase this figure many fold.

The economics and uncertainties of the clean-up scenarios make it essential that methodologies are developed to ensure that the scarce resources are used in the best practical way to address the right priorities at a minimum of cost. In what follows the technique of risk analysis will be examined in the light of its possible contribution to this problem.

The use of risk analysis techniques

Risk analysis techniques can be used in this field to achieve three objectives:-
         1) priority ranking of sites.
         2) design and evaluation of countermeasures.
         3) absolute "risk to man" calculations.

The requirements on the techniques become heavier on going from one through three in that 1) needs little, 2) a limited and 3) a high degree of exactitude in the emission modelling and dose-effect calculations in a full risk analysis exercise. As at the moment the practical applications are to be found mainly in objectives 1) and 2) they will be dealt with here.

Classical risk analysis methodologies identify three elements in a soil pollution problem:-
- 1) A source characterised by an emission quality and strength.
- 2) A target, usually man, with its associated dose-effect relationship.
- 3) A pathway along which toxic substances can move from source to the target.

In real situations there are a multiplicity of pathways and the emission from the source is a complex mixture of substances which have to be considered both seperately and together in the dose-effect relationship at the target. Further difficulties are associated with the data bases used for setting the parameters in the calculation and in the choice of (potential) exposure scenarios.

This complexity of the real problem has lead to the search for drastic simplifications. Firstly, the toxicological difficulties can be avoided if instead of man as the final target, he were to be replaced by one or several intermediate and substitute targets. This allows the dose-effect relationship to be replaced by flux or concentration condition at an intermediate point on the path.

In practice concentrations are inevitabley chosen as either the trigger or Maximum Allowable Concentration parameters. Drinking water standards are chosen for groundwater or crop production soillimits for the ground. This would seem in turn to suggest that the impact scenario considered for all polluted soils is one of small scale domestic farming with local groundwater winning.

Secondly, the problems associated with the quantification of the path are avoided by assuming it to be zero length and the concentration standards mentioned above are applied to the source itself. The scenario is thus one of a small scale domestic farm built in the polluted site itself.

In terms of initially ranking sites and assigning priorities for urgent action in the worst cases these simplifications have been very succesful as government policy in the United States and The Netherlands has demonstrated.

However the simplifications impose severe restrictions on the technical responses possible at a site. These are necessarily focussed on source elimination and, as in-situ techniques are yet in their infancy, they are, in fact, limited to direct source removal and decontamination. Unfortunately these are precisely the technologies that are the most expensive and, as they require a clear field site, the least applicable to most industrial facilities.

It is thus worthwhile exploring the possibilities and consequences of relaxing the severity of the simplifications discussed above. It is possible to adopt other trigger concentrations at the target and to allow the pathway to take part in the process. This pathway can then be manipulated to achieve the desired safety levels and margins at the new intermediate target. In turn this increases significantly the range of technological responses available some of which may prove to be significantly more cost effective than excavation.

Source - Path - Target techniques

That part of risk analysis technology that applies to the more limited problem scope considered here will be refered to as the Source - Path - Target Methodology. This methodology explicitly does not consider the toxical consequences of the contaminant but takes as given either maximum concentration or flux levels at some given target point in the soil system.

Equally, the methodology does not associate environmental quality parameters with either the source or the path.

Application of the methodology involves the following phases: -

1) Characterisation of the source usually by sampling and chemical analysis.
2) Defining the targets, mostly by check-lists.
3) Quantification of the target MAC values, if need be in cooperation with authorities.
4) Identification of "key" pollutants.
5) Identification of the relevent paths.
6) Calculation of the emission strenght of the source for the paths and keys.
7) Migration modelling of the path.
8) Parameter determination for the migration models either by estimation or by field work.
9) Calculation of impact at target as function of time, coupled with sensitivity analysis and where need be, using probabilistic techniques.
10) If the impact is unacceptable, then the inventorisation of possible counter actions and ranking according to cost and effect.
11) Choice and counter measure and detailed design.

Most of these individual stages naturally overlap each other and experience has shown that it is often more effective to approach the problem in an iterative manner than in a singel pass.

As can be seen the methodology depends strongly on the ability to predict the future effect of the source emission at the target. Fortunately the required, usually computer based, models and codes are readily available both for detailed calculation (migration models) and for the broader overview (system analysis models).

The major limitation in applying this methodology is the cost of the field and laboratory work required. However these costs have usually to be met in any case as part of the site investigation and in fact problem analysis and careful design of the sampling and analysis programs in terms of these specific concepts can lead to significant savings.

Applications

In order to illustrate in more concrete terms some of the uses of the methodology the problem of a (generic but very typical) industrial site will be discussed.

Consider the site shown in figure 1. which illustrates the cooling water basin of a large chemical complex in a low laying esturine region. The geology of the site is quite simple and consists of a single unconfined aquifer directly threatened by any surface discharges. The sands and silts of the aquifer are substantially horizontally bedded and the groundwater table is quite near the soil surface. The groundwater velocity is thought to be 2.5 m/y under the site and the saturated porosity is around 40%.

The cooling water chemistry of the facility is somewhat primitive in that chromate is used as a corrosion inhibitor and a chlorophenol as a biocide. In the cooling water the long term mean concentrations are 10 ppm chromium as Cr and 0.1 ppm phenol.

For reasons beyond this discussion, there has been a leakage from one lower corner of the basin into ground. This discharge has been recently detected but has probably been occuring over the past ten years. The discharge rate has been guessed as being 1000 cubic meters per year.

Simple calculation shows that some 100 kg of chromium and at least 1 kg of chlorophenol have been discharged. Both these substances are recognised as toxic with the Dutch investigation and action trigger levels at 50 and 200 ppb for chromium and 0.1 and 2 ppb for the chlorophenol.

These pollutants will be carried away from the site with the groundwater and could impact on remote targets in an unacceptible way. However for largely legal reasons the intermediate target has been selected at the boundary fence 250 meters and directly downstream from the discharge.

The pathway is simple being from the discharge point to the fence. It is known that chlorophenols can be subject to biodegradation and that they adsorb to a limited extent on clays and soil organic components. It must be assumed that chromates do neither.

The problem and data base sketched above is a typical starting point for a risk analysis study. More information is rarely available at the first pass. Three questions will be addressed:

1) What are the expected concentration developments of chromium and phenol at the boundary fence?
2) Will these give problems and how definite are these conclusions?
3) What technical options are open to mitigate any problems?

In order to answer the first question a mathematical contaminant migration model of the situation has been used. The aquifer was taken to be infinite in extent, two dimensional and homogeneous. Convection of contaminants with the groundwater, the associated hydrodynamic dispersion, and any adsorption or biological decay was appropriately taken into account.
The results of the concentration development as a function of time under the boundary fence are shown for chromium in figure 2. The initial discharge of cooling water for 10 years with a concentration of 10 ppm has been propagated 250 meters to the fence as a peak of 1 ppm passing between 50 and 200 years in the future.

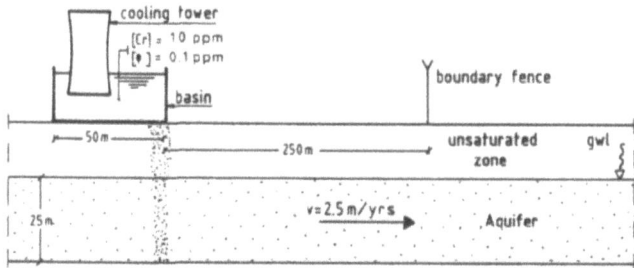

Figure 1. General Problem Sketch

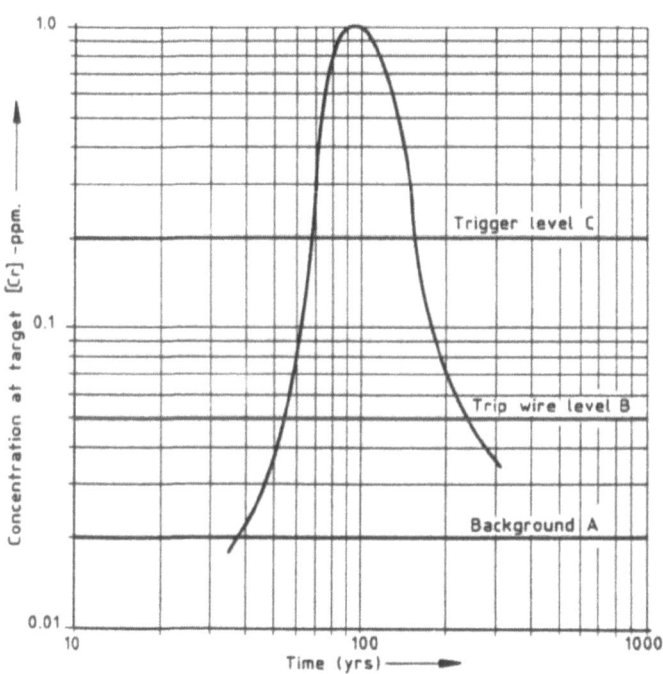

Figure 2. Chromium breakthrough

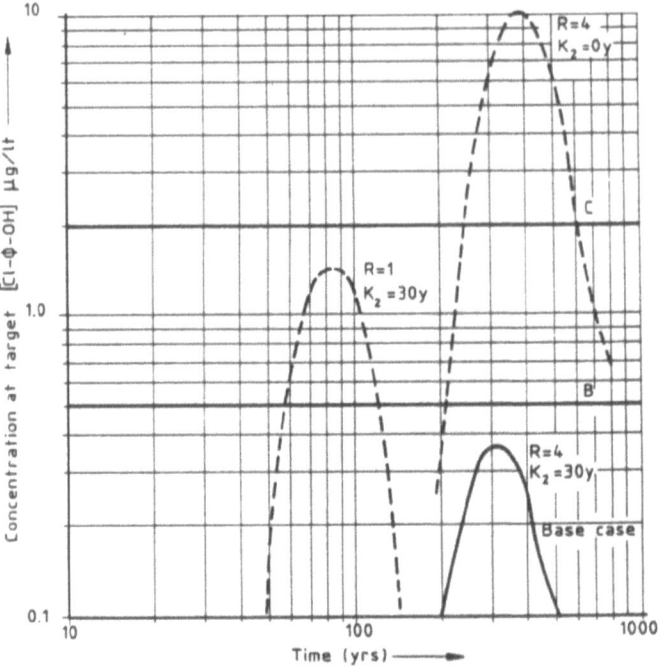

Figure 3. Phenol breakthrough

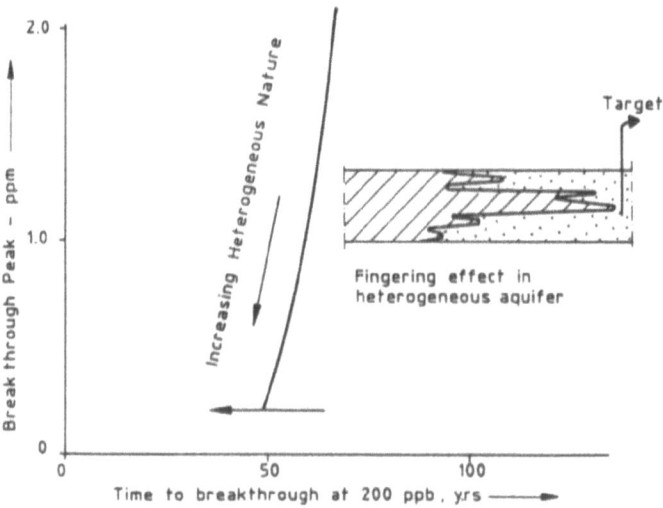

Figure 4. Breakthrough time vrs peak maximum for heterogeneous aquifer case

The corresponding results for the chlorophenol are a peak of 0.4 ppb between 200 and 500 years hence, as shown in figure 3. In this case the adsorption and biological decay along the path have jointly ensured a far greater peak delay and suppression than was the case for chromium. The calculations where these mechanisms were not taken into account are also shown in the figure.

At first the answer to the second question seems self evident. If no countermeasures are taken than there will be a peak discharge of chromium which will exceed the action level by five times between 70 and 150 years in the future. No problems are expected with the phenol where the peak concentration is but a fifth of the action level and this will occur in 300 years.

However these conclusions are bound with uncertainties associated with the discharge scenarios but most especially with the data base used for the modelling exercise. The chlorophenol results, for example, show that whilst the peak falls marginally under the action levels even if no adsorption on the soils of the path takes place, the mechanism of biological decay on the path is of critical importance. The decision between problem or no-problem falls within the bounds of the assumed data base and thus better estimates of this parameter will have to be sought.

The case of chromium illustrates a different aspect. It is to be assumed that there will be a problem but it is highly relevent on which time scale this will occur. The time to the peak is directly related to the groundwater velocity which is uncertain at best. Part of this uncertainty arises because of the difficulties of measuring the overall groundwater velocity on such a small scale but an equally serious problem is associated with the bedded nature of this type (and most) aquifers. This is illustrated in figure 4. where the possibility of fingering is shown where the water and contaminant moves preferentailly along highly permeable beds and will arrive at the target more quickly than otherwise expected. This effect is discounted for in the model as a hydrodynamic dispersion coefficient but this cannot be measured in the field with any degree of certainty. As the dispersion increases, two effects are noted. Firstly the time it takes for the concentrations to exceed a given level decrease but secondly, the peak concentrations also decrease. This relationship for the site in point is shown in figure 4. In any case it can be concluded that if the peak is to be seen as a problem, then it will be at least 50 years before it reaches the target.

Finally, recognising that counter actions will have to be taken to prevent the peak chromium concentrations entering into the environment outside of the boundary fence, several remedial actions will be examined.

These are based around techniques to isolate the polluted soil around the leak and to flush the pollution out from the soil within the isolated zone using forced or natural infiltration. The soluble nature of the pollutants suggest that this is a good choice. Isolation without leaching would simply delay the impact of the release to the environment.

The effect of isolation by constructing a cement-bentonite curtain wall or a steel profile dam on the expected peak discharges across the boundary fence can be coupled with the several design options including the forced leaching rate and the barrier effinciency of the screen. Furthermore allowance has to be made for the finite lifespan of any construction.

All these effects have been modelled using a system analysis program (in this case REFCON of Delft Geotechnics) which downloads aquifer loadings into the contaminant transport model used earlier in the zero option case in the risk analysis. The results are shown for various scenarios in the following table:

Response	Life y	Leach rate m3/y	Screen effic. %	Peak conc. ppb
Zero case		-	-	1079
Requirement		-	-	200
Cement-bent.	40	1000	90	239
"	40	1000	95	169
"	40	2000	90	128
"	100	1000	90	194
Profile	20	2000	40	322
"	20	4000	40	166

Although this table only gives a rough picture of the expected behaviour of the sytem several conclusions can be drawn. In the first place even well built screens with a quite large forced leaching rate (in this case 1000 m^3/y) results in performances with very modest safety factors indeed. Furthermore whilst the leach rate and the screen barrier efficiency are of importance, the ultimate lifetime of the construction would appear to be less so. In other words in this case attention and investement should be devoted to contact efficiency in the leaching and in secure screen construction rather than designing for very long lifespans.

Up to now attention has been devoted to environmental impact expressed in terms of the peak concentrations. However care has to be exercised in this respect. Similar calculations show that the peaks can be suppressed by simple dilution of the source term with a large excess of clean water or similar cosmetic measures. The total flux of contaminant into the ecosystem remains unchanged of course in this case.

Conclusions.

The use of risk analysis techniques to assess the nature and extent of pollution problems has been examined in this study and illustrated by a very simple example. It is possible using these techniques, to predict the expected concentration and flux levels caused by the pollution incident at a remote intermediate target. These predictions can then form the basis of the appropriate choice of countermeasures and their assessment. These techniques can contribute significantly to the overall evaluation of both environmental policies and their consequences.

METHODS FOR ASSESSING THE RISK OF ENVIRONMENTAL CONTAMINATION

C.L. van DEELEN

TNO Netherlands Organization for Applied Scientific Research, Division of Technology for Society, Department of Industrial Safety, P.o. Box 342, 7300 AH Apeldoorn, the Netherlands

SUMMARY

Decisions on environmental issues, either by the authorities or industry, are increasingly made on the basis of a quantification of risks to society (human beings, eco-systems, goods etc.), associated with the release and/or presence of chemicals in the environment. Consequently, there has been a growing demand for a methodology capable to provide a complete picture of the impact of chemicals on the environment.

At present, in many instances, tools are insufficiently available to carry out detailed environmental risk analysis studies. This paper gives a brief survey of the various elements of an environmental risk analysis, i.e. release, transport and fate of chemicals in the environment, exposure and dose-effect relations, and highlights problem areas.

In addition to the above, this paper attempts to point out that a detailed environmental risk analysis is not required in all circumstances. In order to support decision making it may be sufficient to follow a simplified approach. This is illustrated on the basis of two studies related to the risks of soil contamination from industrial activities.

1. INTRODUCTION

Industrialization of our civilization has resulted in the increased use of a wide variety of chemicals. Contamination of the environment may occur at any stage of the life cycle of a chemical, i.e. during manufacturing, storage, use and disposal. After release a chemical will initially be transported within the environmental compartment in which the emission occurs. Depending on its basic properties a released chemical generally will be transformed to a certain extent as a result of physical, chemical and/or biological processes. Moreover, in most cases the chemical will enter other environmental compartments, resulting in subsequent transport and transformation. Regarding the environmental compartments a distinction can be made between abiotic compartments such as air, (ground)water and soil and biotic compartments, e.g. human beings, flora and fauna.

The pattern of exposure of an organism to a chemical largely depends on the fate of the chemical in the environment. Examples of potential exposure routes include inhalation, ingestion via food or drinking water, dermal contact etc. It has been thoroughly recognized by now that exposure to chemicals may have detrimental effects on human beings, animals and vegetation. The extent of these effects and their probability of occurrence are referred to as environmental risks.

Decisions on environmental issues, either by the authorities or industry, are required in order to maintain environmental risks at an acceptable level. Such decisions on environmental issues may involve the reduction or removal of certain emission sources, prohibiting the introduction of new chemicals on the market etc. The instrument of risk analysis has proved to be a powerful support to decision-making in the field of industrial and nuclear safety. In view of the above it is beyond question that risk analysis techniques will be applied to an increasing degree in order to quantify the impact of the release of chemicals on the environment and to evaluate the effects of measures that may be taken.

This paper provides a brief survey of the various elements of an environmental risk analysis study. It is pointed out that at present, in many instances, tools are insufficiently available to enable a detailed quantification of environmental risks to be made. Some of the major problem areas are highlighted in this context. However, it should be realized that a detailed environmental risk analysis is not required in all circumstances in order to facilitate decision-making on environmental issues. This is illustrated on the basis of two studies related to soil contamination carried out by the Department of Industrial Safety TNO.

2. ENVIRONMENTAL RISK ANALYSIS

2.1. General
In this paragraph the major elements of an environmental risk analysis are described. Prior to this the features of an environmental risk analysis are compared with those of a safety risk analysis concerning releases of toxic chemicals.

In a safety risk analysis the following methodology is usually followed:
* identification of undesirable events (=release of toxic chemicals);
* quantification of the consequences of the identified undesirable events (effect and damage analysis);
* quantification of the probability that the identified undesirable events will occur;
* presentation and evaluation of the risk an activity represents.

As for the release of toxic chemicals, a safety risk analysis includes the following characteristics:
* the release results from a technical and/or operational failure; hence the emission is incidental in character;
* the consequences of a release are limited spatially: the location of an emission is properly defined and the distance up to which detrimental effects may occur amounts to a few kilometers at the most;
* the consequences of a release are limited in time, in the sense that only acute detrimental effects following a single exposure are quantified;
* the exposure route is unambiguous: inhalation of a toxic chemical present in the atmosphere;

* owing to the relatively short exposure time, transport of a chemical
  from the atmosphere to other environmental compartments, as well as
  transformation by chemical, physical or biological processes is not
  taken into account.

Environmental risks are predominantly brought on by continuous emissions.
On stating the above it should be realized that there are examples of incidental releases leading to environmental disasters, e.g. Seveso, Chernobyl,
and most recently the fire at the storage premises of Sandoz in Basel.
However, a major distinction between a safety and an environmental risk
analysis is that the latter takes both acute and chronic effects into
account. From the above it will be clear that an environmental risk
analysis is concerned with considerably longer time scales than a safety
risk analysis is. Consequently, transformation of a chemical, as well as
transport to other environmental compartments can play a role of importance
in an environmental risk analysis.
In the event of a calamity the geographic dimensions within which the impact of a released chemical on the environment manifests itself may extend
in orders of magnitude beyond the immediately affected area. An example of
an impact at global level is the attack on the ozon layer as a result of
the presence of carbon dioxide in the stratosphere.
Finally, a safety risk analysis is generally only concerned with damage to
human beings and structures, whereas an environmental risk analysis
attempts to quantify all adverse effects for the environment as a whole.

From the above it will be clear that the quantification of the impact of
exposure of a chemical will be considerably more complicated for an environmental than for a safety risk analysis. Figure 1 provides a schematic
survey of the major elements of an environmental risk analysis. These elements are further discussed below. For a more extensive survey, reference
is made to [1].

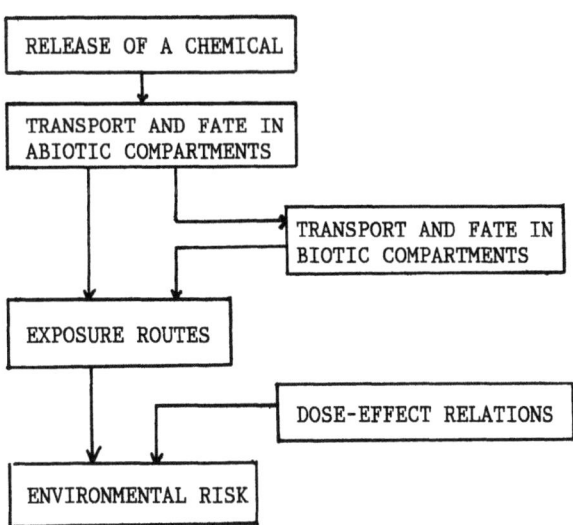

Figure 1. Schematic survey of an environmental risk analysis.

2.2. Release of a chemical

With regard to the release of chemicals into the environment a distinction can be made between (semi-)continuous and incidental emissions. The latter frequently result from calamities or abnormal (process) conditions and can be considered as being highly local emissions (point sources). Additionally, in our society numerous activities produce a (semi-)continuous release of chemicals, e.g. industrial installations, waste disposal sites, agriculture etc. Some of these emission sources will be highly localized, such as waste disposal sites, others will be diffuse, for example agricultural sources. Some of the emissions will show a pattern of diffuse point sources, e.g. domestic waste water, traffic. Irrespective of emission pattern, the identification (where, how?) and quantification (how much?) of emission sources are vital elements in an environmental risk analysis.

An identification of (semi-)continuous sources may be conducted in various ways. In the case of an existing activity, available emission data can be used. When these data are unavailable, identification of emission sources can take place through expert judgement, checklists etc. or through comparison with similar units or activities. The identification of incidental emissions requires systematic analysis of the activity under consideration. For this purpose techniques can be used that are frequently applied in the field of industrial safety studies, in particular hazard and operability studies (HAZOP) [2]. The identification of emission sources should also take into account the environmental compartment in which a chemical is released, as well as its physical and chemical nature, since these factors are important parameters for transport and fate of a chemical in the environment.

The quantification of (semi-)continuous emission sources may be obtained from a calculated mass balance around an operation, or from emission factors [3]. Alternatively, in the case of a local source the released quantity may be derived by sampling and analysing an emission. In this way a vast number of emission data from both industrial and non-industrial sources have been collected in the Netherlands over the last decade [3,4]. As for the quantification of incidental emissions, particularly those resulting from a calamity, it is recommended to make use of models which are employed in risk analysis studies. Risk analysis techniques are also very suitable when estimating the probablity or frequency of incidental releases (see also under 3.).

With regard to the identification and quantification of emission sources it has been concluded from an evaluation of existing techniques that the following aspects require further inverstigation [1]:
* release of chemicals into the environment resulting from use and disposal of products by consumers;
* emissions to soil;
* frequency and quantity of incidental releases.

2.3. Transport and fate of chemicals in the environment

After release a chemical will be distributed into the environment. The distance over which a contamination may spread, the velocity of transport and the concentration pattern in the environment depend upon a large number of factors, such as:

* quantity of released material
* emission pattern (e.g. in which environmental compartment)
* chemical, physical and biological properties of released chemical(s)
* local conditions (e.g. type of soil, level of groundwater, water renewal rate of a lake or estuary etc.).

To date, many models have been developed for the purpose of predicting transport and fate of contaminants in the environment. The complexity of these models varies from simple equilibrium- or transport-models to complicated, dynamic, multi-compartment models. However, the application of most of the models is restricted to specific conditions, and in many cases the quantification by models of transport and fate of a contaminant in the environment is not yet possible. Moreover, the user's requirement regarding accuracy and operability of a model is by definition a conflicting one.

When the course of an approach by model can not be followed, it is still possible to estimate the concentration of contaminants in the environment on the basis of available monitoring data. However, estimating concentrations from monitoring data has some disadvantages:
* monitoring data provide information on the concentration at a specific moment and a specific location; no insight is obtained into fluctuations in time, modifications of (site-specific) parameters etc.;
* unsuitable for new chemicals;
* less suitable for non-homogeneous environmental compartments such as soil;
* time-consuming and expensive

With the exception of models based on the fugacity concept, hardly any other models are available to quantify transport and fate of a contaminant over several environmental compartments. However, the fugacity models represent a limited field of application in that they only quantify the distribution of a contaminant over the environmental compartments, rather than the concentration profile within a compartment. Consequently, these models are most suitable in their application in environmental risk analysis studies at a global or regional level. Moreover, as an initial step the distribution of a contaminant over the environmental compartments can be checked roughly by means of relatively simple fugacity models. After this initial check, specific models can be applied to describe in greater detail transport and fate of a contaminant in a particular environmental compartment.

2.4. Exposure routes

An organism can be exposed in various ways to a contaminant in the environment. An exposure route is made up of the medium in which a contaminant is transported and the manner in which exposure takes place, e.g.
* inhalation of gases or particles;
* ingestion of drinking-water or food;
* dermal contact.

The transport media include, among other things air, food, drinking-water and surface water (recreational facilities).
In an environmental risk analysis a global evaluation has to be made of all potential exposure routes with respect to their relative importance, and subsequently the major ones have to be analysed in greater detail.

Models for the quantification of an exposure dose in particular have been
developed in the field of risk analysis for nuclear activities. These
models require a thorough evaluation with regard to their applicability in
environmental risk analysis. Similar to risk analysis techniques for external safety, the development of models taking into account the frequency
distribution of the level of exposure for a population, requires further
attention. These models are especially required for analyses of site-specific environmental risks. Such exposure models should include behaviour
aspects of (parts of) the population, for example consumer behaviour,
occupation etc..

2.5. Dose-effect relations

An essential step in an environmental analysis constitutes "translating"
into effect the estimated dose to which an organism is exposed. Ideally,
such a translation should be done on the basis of an existing dose-effect
equation relating the fraction of a population suffering from certain effects (e.g. mortality) to the level of exposure. However, such dose-effect
equations are only available for a few chemicals. Moreover, the existing
dose-effect equations nearly all relate to lethal damage, particularly from
carcinogenic chemicals. As for non-carcinogenic chemicals, it is assumed
that they will only cause adverse effects when a specific threshold value
is exceeded. Where these compounds are concerned hardly any dose-effect
equations are available, thus making it impossible to quantify the probability occurrence of detrimental effects with regard to a certain
population. An environmental risk analysis of non-carcinogenic contaminants
will therefore only provide insight into the probability that the estimated
level of exposure will exceed a pre-fixed value, e.g. a no-effect level.

At present knowledge about effects of toxic compounds on non-human organisms is still in its very early stages. Moreover, relatively greater effort
has been put into the subject of toxic effects at species level rather than
into population or eco-system level. In addition to the above there is a
serious lack of knowledge on the effects of simultaneous exposure to different chemicals in general. The same goes for individual sensitivity
towards (combined) exposure to toxic chemicals.

2.6. Evaluation of environmental risk analysis tools

From the above outline it will be clear that, with today's knowledge, a
detailed environmental risk analysis remains out of reach for most conditions. In this context a detailed environmental risk analysis represents
the full quantification of the environmental impact of a chemical release.
On the other hand, it is unrealistic to postpone decisions on environmental
issues till adequate models have been developed.
The next paragraph illustrates how risk analysis techniques can be applied
in order to support decision-making.

3. EXAMPLES OF A SIMPLIFIED APPROACH

3.1. Risk analysis of underground storage tanks

In the Netherlands a large number of underground storage tanks for industrial use are situated at car filling stations. On the basis of available
data it may be stated that leakages from underground storage tanks are

major contributors to cases of soil contamination [5]. For that reason the
Ministry of Housing, Physical Planning and the Environment commissioned TNO
to carry out research into the risk of soil contamination in consequence of
leakage of petrol and diesel oil from underground tanks at filling
stations. Two storage systems have been compared in the study, i.e. a
single-walled tank in accordance with Dutch directives and a double-walled
tank in accordance with German directives. The results of the study have
been reported more extensively in [6] and in full detail in [7]. The risk
assessment concerns the storage tank (including its coating and cathodic
protection), as well as the filling line, filling point and filling procedure for transshipment from the tank car to the tank. The leak detection
system has in addition been considered for the double walled tank. The risk
assessment methodology may be represented schematically as follows:

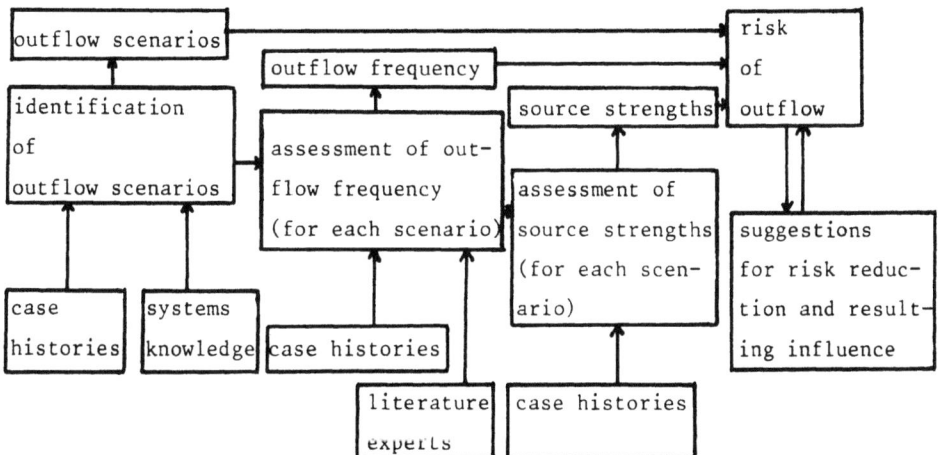

The identification of outflow scenarios has been carried out on the basis
of case histories and a systematic analysis of the storage/filling system.
The following outflow scenarios have been identified in the process:
1) external corrosion
2) internal corrosion
3) outflow during transshipment
4) leakage from mechanical damage
5) leakage from design/construction faults
6) outflow when tanks are being put out of operation

The leakage frequency of oil product has been assessed for each of the
identified outflow scenarios. In assessing the failure frequencies, case
histories - for the greater part - and data from literature - in a few
cases - were made use of. Where possible, the assessment of failure frequencies has taken into consideration (site-)specific parameters such as
aggressiveness of the soil, presence of overfill protection, type of stored
product etc. In case when only insufficient or incomplete information was
available, assumptions and marginal conditions have been formulated. The
obtained failure frequencies were subsequently submitted for checking to
experts in the field of underground storage of mineral oil products.

The velocity at which petrol or diesel oil spills from an underground tank in cases of leakage is dependent on a number of factors: amount and size of the holes, level of groundwater etc. The total outflowing quantity (source strength) however is, apart from these factors, even more determined by the fact that outflow often continues until a certain nuisance level has been reached (e.g. pollution of drinking water, stench nuisance etc.). On the basis of an analysis of case histories of soil contamination the average source strength has been estimated for each of the outflow scenarios.

The combination of outflow frequency and resulting amount of oil product that is released constitutes a measure for the risk of soil contamination resulting from an outflow scenario. Many minor leakages lead in the end to a degree of soil contamination comparable to one major spill, because oil products tend to accumulate in the soil. Therefore it was decided to represent the risk as a product of the frequency of occurrence and the amount of outflow (risk factor). The risk factors provide an insight into the degree to which the various outflow scenarios contribute to the overall risk. Moreover, by totalling the risk factors of individual outflow scenarios, the overall risk of various storage systems to the soil can be compared.

From the study it can be concluded that the double-walled tank is the most reliable of the systems that have been examined. A single-walled tank positioned in a protection zone for public water supply constitutes a somewhat higher risk than a double-walled one. This is caused by the absence of a protection system against overfill and increased leakage frequency from internal and external corrosion. Of the examined systems, a single-walled tank positioned outside a protection zone for public water supply constitutes the greatest risk. The above applies especially to tanks which have not been fitted with cathodic protection. In these cases there is a relatively high leakage frequency from external corrosion caused by the fact that after the tank has been installed, not a single check takes place anymore on the condition of its coating and the presence of a galvanic contact or electrical influence from outside (stray current/interference).

On the basis of the results of the risk analysis the study also provides a number of risk reducing measures, the major ones of which are:
* more stringent rules with regard to the installation for cathodic protection systems;
* periodical checks on the condition of the coating of all tanks, i.e. also those which have not been fitted with cathodic protection;
* increased frequency of inspection of the cathodic protection system for tanks positioned in soils with a high degree of aggressiveness;
* increased frequency of internal inspection, allowing for differentiation between petrol and diesel oil tanks;
* installation of a protection system against overfill.

From the above description it will be clear that no attempt has been made in the study to quantify the effects of a release of oil products, for example in terms of damage to the soil. However, in view of the fact that the storage systems under consideration contain a similar type of product, the simplified approach in this study provides a proper scientific basis for the development of guidelines for underground storage tanks in general. In case when guidelines have to be drawn up for specific conditions the local circumstances will have to be taken into consideration as well and

under those circumstances the released quantity may have to be translated
into environmental impact one way or the other. This aspect is described in
the next paragraph.

3.2. Assessing the risk of soil contamination from industrial activities

The method that has been applied in assessing the risk of soil contamination from underground storage tanks has been further developed into a procedure to quantify the risk of soil contamination from industrial activities in general [8]. The objective of the procedure is to pinpoint, from a wide variety of chemicals, activities, operations and process apparatus, the potential problem areas related to soil contamination from an industrial activity, and to quantify the associated risks. The developed procedure basically consists of 5 major steps:
1) Inventory of chemicals, activities, process apparatus etc. in an industrial activity and identification of potential problem areas regarding soil contamination.
2) Analysis of identified problem areas and definition of scenarios that may actually lead to cases of soil contamination.
3) Quantification of the frequency of occurrence and the released amount of chemicals for each of the defined scenarios.
4) Quantification of a risk factor in relation to the location of an industrial activity.
5) Presentation and evaluation of the risk of soil contamination.

Assessment of the risk of soil contamination with the developed procedure goes one step further than the risk analysis of underground storage tanks does, in the sense that the location of an industrial activity is taken into account. For that purpose risk factors have been derived that constitute a relative measure for the volume of contaminated soil in the case of a leakage. In the procedure the following parameters are taken into account when quantifying the risk factor of a location:
* permeability of the soil, which is related to the natural character of the soil;
* maximum distance over which a contaminant may be transported; this distance is made dependent on the geohydrological conditions and the distance between the source of release and the nearest open water course.

From the above it will be clear that the quantification of a risk factor with regard to surroundings involves a number of significant simplifications. For example, the risk factors have been made independent of type and quantity of released material; processes such as adsorption/desorption, dissolving/depositing and decomposition/conversion are therefore unaccounted for. However, it should be realized that transport and fate of a chemical depends on its chemical, physical and biological properties in relation to the nature of the soil. It will be clear that within the scope of a general applicable procedure it is impracticable to distinguish between all possible combinations of chemicals and types of soil. Moreover, the risk factors have not been developed for the purpose of quantifying the concentration profile of a contaminant in the soil but are only required to provide a relative measure of the potential volume of soil that may be contaminated. For that reason the procedure does not take into account properties of chemicals such as toxicity, tendency to accumulate in biomass etc.

Application of the procedure to an industrial activity provides, in a quantitative way, an insight into the extent to what the various parts of an installation or activity contribute to the risk of soil contamination. On the basis of a quantification of risks it is possible to draw up measures to reduce the risk of soil contamination. Finally, the procedure enables a comparison to be made of the risk of soil contamination from different industrial activities.

With regard to assessing the risk of soil contamination, the procedure has been applied quite successfully in the case of a Dutch company - Chemco Europe NV, Soest - producing photochemicals and photosensitive emulsions. This company is located in the near vicinity of a groundwater pump station. The pumped up groundwater is directly used for public water supply, and any degree of contamination of the groundwater will render it unsuitable for this purpose. For that reason the regional authorities have imposed severe regulations on the company in order to minimize the risk of soil contamination. Implementation of all the provisions imposed in the guidelines would involve a considerable amount of investment. In view of this the company was allowed to introduce the provisions in phases, on the basis of the results of a risk analysis study.
With the procedure developed by the Department of Industrial Safety the potential sources of soil contamination at the company site have been identified and the associated risks quantified. As for the potential sources of leakage, which constitute a relatively high level of risk, recommendations were made regarding measures in order to reduce the risk of soil contamination. In this context it is interesting to note that these risk-reducing provisions differed considerably from those adopted in the regulations.

4. CONCLUSIONS

Risk analysis techniques have proved to be a powerful support to decision-making in the field of industrial and nuclear safety. It has been demonstrated that the quantification of the impact of long-term exposure to chemicals is of a considerable complexity, and that a detailed environmental risk analysis is not quite feasible in a large number of circumstances. This paper gives a brief survey of the areas requiring further development in this respect. Meanwhile, decisions on environmental issues frequently have to be made on the basis of urgency and cannot await the development of models to carry out detailed environmental risk analysis studies. In those circumstances decisions have to be made on the basis of a limited number of data. This paper gives some examples of simplified risk analysis methods related to soil contamination. It is demonstrated that the results of such simplified methods can serve to support decision-making.

REFERENCES

1. C.L. van Deelen and G. Golbach. Programming Note Environmental Risks (in Dutch). TNO report, 1986.
2. Hazard and Operability Study. Why? When? How? Report of the Directorate-General of Labour, 1st edition, 1979.
3. Handbook of Emission Factors, Part 1 - Non-industrial sources. Government Publishing Office, the Hague, 1980.

4. Handbook of Emission Factors, Part 2 - Industrial Sources. Government Publishing Office, the Hague, 1984.
5. Provisional indicative soil program for 1984-1988 (in Dutch).
6. C.L. van Deelen. Assessing the risk of soil contamination in the case of industrial activities; In: Contaminated Soil, editors J.W. Assink and W.J. van den Brink. Martinus Nijhoff Publishers, Dordrecht, 1986.
7. C.L. van Deelen and C.M. Pietersen. Comparative risk analysis of underground storage systems at (car)filling stations (in Dutch). "Bodembeschermingsreeks" no. 45. Government Publishing Office, the Hague, 1985.
8. C.L. van Deelen e.a. A procedure for assessing the risk of soil contamination by industrial activities (in Dutch). To be issued by Government Publishing Office, the Hague.

RISK ANALYSIS AND RISK POLICY IN THE NETHERLANDS

dr. B.J.M. Ale[1], M. Seaman[2]

[1] Ministry of Housing Physical Planning and Environment, The Netherlands
[2] Technica Ltd, London

Abstract

The increasing awareness of the public of the risks associated with the chemical process industry led the Dutch government into the devlopment of an External safety policy. This policy is based on quantification of risk and guideliness for the acceptability of risk expressed in quantitative form.
In support of this policy computer assisted techniques to evaluate the risks of chemical proces plants have been developed.

Introduction

It was in the mid seventies, after the explosion in Flixborough (1) and just after the explosion in the DSM works in Beek, Holland, that the prime minister installed in a new department in the directorate general for the environment. The purpose of this department was to develop a policy directed at the external safety of the population and aimed at the reduction of this risks imposed by the handling of dangerous materials.
At that time risk analysis techniques for chemical plants did not exist and the general line of government policy was aimed at reduction of the risk to zero. The idea that risk cannot be reduced to zero and that finite non zero risks have to be accepted and managed as part of the general environmental policy grew with the year. It was expressed explicitly for the first time in a policy statement regarding the distribution and use of LPG (2) and later in a more general way in the multiyear plan for the environment (3) (1986-1990). One of the prerequisites for risks management is the availability of tools to assess the risks. These have developed from a small set of simple models for the calculation of physical effects into a fully grown system of computer assisted quantified risk analysis.

The Policy

The risk management policy of the Dutch government has been developed from the ideas of W. Rowe (4). It can be described as a five stage cyclic proces.

These five stages are

- identification
- quantification

- decision
- reduction
- control

In the identification phase certain activities come to the attention of the policy makers. This can be the result of systematic search, but also by the occurence of an accident or actions of concerned residents.

In the quantification phase, the following are assessed: what the nature of the hazard is; to what extent people and environment may be harmed; and what the chances are that such an event may come about.
This information is gathered and computed in quantitative form as much as possible.

At the decision stage it is decided whether the risk is acceptable or further reduction is required.
For the acceptability of risk the Dutch government has issued guidelines as to what risks can be accepted and what risks need further consideration .

In these guidelines two measures of risk are used and for each two limits are set. The measures of risk are
- The individual risk. This is defined as the chance that somebody who stays permanently unprotected in a certain spot will get killed as the result of an accident in a certain facility.
- The group risk. This relates to accidents in which multiple fatalities occur, and is defined as the risk of such accidents killing more than a certain number of people. This varies with the number chosen and can best to expected as a graph plotted against that number.

For individual risk there is a limit above which the risk is deemed unacceptable. It is set at 10^{-6}/yr (fig.). There is also a limit below which no further action by the authorities is deemed neccessary. This limit is set at 10^{-8}/yr.
When a risk falls below this latter limit this does not lift the owners responsibility to work as safe as reasonably achievable, it only stops the authorities from further action.

For the group risk the limit of unacceptability is a line passing though 10 people killed at a frequency of 10^{-5}/yr, and at which the frequency decrease with the square of the number killed. The limit of acceptability passes through 10 people killed at 10^{-7}/yr and has the same slope.
In the area between the two limits the responsible authorities should further consider the risk and the feasebility of reducing it.

In the next stage the risk is reduced where necessary or possible and finally the attained level of acceptable risk is maintained by inspection etc.

The method

As mentioned before no policy can be based on quantified risk if the methodology to quantify risks is not readily available.
In the mid seventies the only comprehensive set of models to describe the consequences of accidental releases could be found in the "Vulnerability Model" of the US coast guard. These models were adapted, and in some instances drastically changed as a result of comments by the experts of the Dutch chemical industry, to the Dutch situation and described in the "Yellow Book" of the Directorate General of Labour.
Subsequently an investigation was performed into what could be learned from a complete risk analysis. In this exercise, known as the COVO study (5), six potentially hazardous installations in the Rijnmond area were analyzed. The results were that although the risks could be calculated to produce credible results, the costs were almost prohibitive to perform these kind of analysis on a routine basis.
The major part of these costs were associated with the calculational part of the excercise and it was envisaged that, if these calculation could be automated, such a reduction in cost could be achieved that application of risk analysis for policy purposes would become feasible.

The directorate general for the environment issued a contract to Technica London to develop such computer assisted methods. These methods, now known as the SAFETI package, consist of over 30 programs to assist the risk analyst in doing his job. The SAFETI package os described in the next section.

Computer assisted risk analysis, SAFETI

The Safeti package (6) is a set of computer program's that assist the risk analyst in his analysis. In the first stage of the analysis the plant is modelled as a network of items and pipes. The program contains a comprehensive set of tests to check the validity of the data input. In the next stage common failures like pipe breaks and vessel cracks are generated automatically. Special failures are input by the analyst separately.
The results of this stage is a set of exemplified discrete failures representing all possible failure modes of the installation. This list is further processed though the consequences models producing dispersion distances, flame boundaries etc.
In the final stage, these results are combined with data on population distribution, weather statistics and ignition sources to produce the final risk results, individual risk contours and FN-curves.

Specific decisions

Several decisions have been made and published as official documents based on the above principles. A number of these are described below.

General Electric Plastics

In 1980 General Electric Plastics applied for a permit to extend its plastics facilities in Bergen op Zoom in the south west of the Netherlands. The facility imports chlorine (approx 600 tons/week) and uses phosgene as an intermediate. The inhabitants of the nearest by village, 600 m from the premises, appealed against the granting of the permit on the grounds of the hazards. A risk analysis was performed (This was before the computerised techniques became available and the costs were relatively high, approx f. 250.000,--).
The results showed that the phosgene part of the process did not contribute much to the risk, but that as a result of the risks of the chlorine unloading facility the extension of the plant would bring the 10^{-6} individual risk contour over a large distance into the village. It was therefore decided that the permit for the extension could only be granted if General Electric Plastics agreed to build their own chorine production facility. This would reduce the risk and bring the 10^{-6} contour outside the inhabited part of the village again. General Electric Plastics agreed to this and the permits were granted.

LPG Policy

As a result of the increasing awareness of the risks of the transport and use of LPG the directorate general for the environment ordered TNO to perform a study of the risks of the import, transport, storage and use of LPG in the Netherlands. (7)
This study started in 1979 and was finished in 1982. Subsequently the Dutch government issued several measures to improve the safety of the LPG chain.

- The transport per pipeline proved not to be safer than transport by barge, so pipeline transport was no longer the preferred option.

- About 100 LPG-filling stations had to be closed because the risk at the nearest housing exceeds 10^{-5}/yr (cost 30 million fl. on behalf of government).

- The minimum safety zones around a new station will be at the 10^{-6} in 10^{-7} contour, this being 80 m. (fig. 3)

- Barges will be modified to exclude the dominant contributor ro risk for thins transport made: tearing off the bottom connections of the tank at collisions.

Natural gas pipelines

The Dutch gas company Gasunie operates a nationwide network of natural gas pipelines (1100 km). It appeared to be increasingly difficult to maintain safety zônes around these lines. A risk analysis was perfomed and the minimum zoning requirements were based on the 10^{-6} risk contour as in the table. (fig. 3) (8)

Zoning around DSM

In the south-east of the Netherlands, in the province of Limburg is the main site of one of the larger chemical companies of the Netherlands, DSM. This company produces among other products, over half a million tonnes of ammonia and just under half a million tonnes of acrylonitrile, most of which is further processed to a wide range of fertiliser, polymers and other products. The company employs about 10,000 people on this site. The installations are built on an area that up to 25 years ago housed a large coal mine. The housing developments around the site are still influenced by the previous planning of coal miners housing, which means that residential areas are built in relatively close proximity to the site and thus to the chemical installations.

At this site in 1975 there occurred one of the largest explosions in the history of the process industry. In this accident 14 people were killed and 104 injured. The concern raised by this incident was reinforced by more recent even more damaging incidents in Italy, Mexico and India. In this same period the awareness and concern about the environment in general and especially air pollution, noise, and waste problems led the government of the Netherlands to initiate an environmental action plan for the region.

The objectives of this plan are to, on the one hand, reduce the environmental impact by measures at the site and on the other hand to establish a zoning policy to ensure sufficient separation between the site and the surrounding population. The two determining factors for this zoning exercise are noise, as regulated by the noise abatement law, and risk. The general risk management objective was first laid down in the medium term indicative environmental plan for the Netherlands (3).

The methodology described above was applied to the DSM site. On this site there are some 35 separate installations. A first screening showed that some 10 of these posed a risk outside the side boundary. Each of these was then analysed in detail. The sum total of the resulting risk numbers is given in Figure 4 for the individual risk and the group risk. (This risk analysis, due to the availability by this time the computerized method referred to above could be performed for just over fl. 300.000).

These risks proved to be lower than the authorities had initially (before the study) feared. Nevertheless, there were several problem areas for which meassures had to be taken.

An interesting aspect of the results was the breakdown of risk associated with the various process plant on which quantitative results were prepared. In general, the major production units (ethylene crackers and ammonia plants) were not the dominant feature of the overall risk. Rather the ammonia distribution and storage system of the LPG handling system were more significant. This was not known prior to the study and it is insight of this sort which are often most valuable. (9)

Therefore the results were discussed in a committee in which the Central Government, the Provincial Authorities, the municipal authorities and the company were represented. A major problem in this group was how to handle the uncertainties in the risk quantification.

The results of a risk calculation are probalistic in nature and therefore uncertain. In this particular case this means that althoug the best estimate of the 10^{-6} contour is given as a single line, there is a chance that this line in reality might be somewhere else. In fact it can in principle be anywhere in the hatched area of figure 2. The costs of complying with the general strategy outlined above varies going though this range from nil to about 300 milion guilders (150 US$). It hardly needs explanation that the consequences of being wrong have to carefully evaluated.
It was decided first that the best estimate of the risk resulting from the analysis was to be employed as the basis for any further discussion. It was then decided to take the limits from the environmental plan for new developments, be it housing or industry.
It was than decided that for the existing plants and housing a more lenient approach was to be taken. This means that in areas where the risk is above 10^{-5}/yr and no risk reducing measures can be envisaged, no housing developments will be allowed.
In the zone between 10^{-5} and 10^{-6} only replacement of old houses will be allowed. Furthermore, engineering studies will have to be undertaken to investigate the feasibility of reducing the risk. The costs of these risk reducing measures will be compared with the costs of the zoning requirements and an appropriate course of action set out subsequently. (10)

Conclusion

Risk management has developed into an integrated part of the Dutch environmental and safety policy. It has the support of the parliament. It is based on quantified risk information and quantified risk analysis and has proven to be viable in the decisionmaking context. The directorate general will continue tosupport its development and the expand its application towards general environmental policy.

References

1) Lees, F.P., Loss Prevention in the process industries, Butterworths (1980).
2) Integrale nota LPG, Tweede Kamer der Staten Generaal, vergaderjaar 1983-194, 18233 nrs. 1-2.
3) Multiyear Environmental plan for the Netherlands 1986-1990, Tweede Kamer der Staten Generaal, vergaderjaar 1985-1986, 19204 nrs. 1-2.
4) Rowe W.O., an Anatomy of Risk, Wiley and sons, New York (1977).
5) Report on the COVO study to the Rijnmond Authority, Reidel (1979).
6) The Safety Package Reports I to IV to the Directorate General for the Environment 1983-1986.
7) LPG-Integral study, report to the Dutch Government, TNO, (1983).
8) Proximity requirements around high pressure natural gas transmission lines, Minister of Health and Environmental Hygiëne, the Netherlands (1981).
9) DSM SITE Safety Study. Report to Provinciale Waterstaat Limburg, the Netherlands, Technica, London (1985).
10) Rapport van de Task Force DSM. Directoraat Generaal voor de Milieuhygiëne, March 1987.

dia./pressure	occasional buildings (m)	residential buildings (m)
4"/40 bar	4	4
16"/40 bar	4	20
30"/66 bar	5	30
48"/66 bar	5	50

Converted to individual risk this yields the following results

dia./pressure	occasional buildings (m)	residential buildings (m)
4"/40 bar	$10^{-8} - 8.10^{-7}$	$10^{-8} - 8.10^{-7}$
16"/40 bar	$8.10^{-7} - 7.10^{-6}$	$10^{-7} - 2.10^{-6}$
30"/66 bar	$4.10^{-7} - 3.10^{-6}$	$5.10^{-8} - 2.10^{-6}$
48"/66 bar	$6.10^{-7} - 5.10^{-6}$	$4.10^{-8} - 2.10^{-6}$

Fig 1. Zoning requirements for natural gas lines

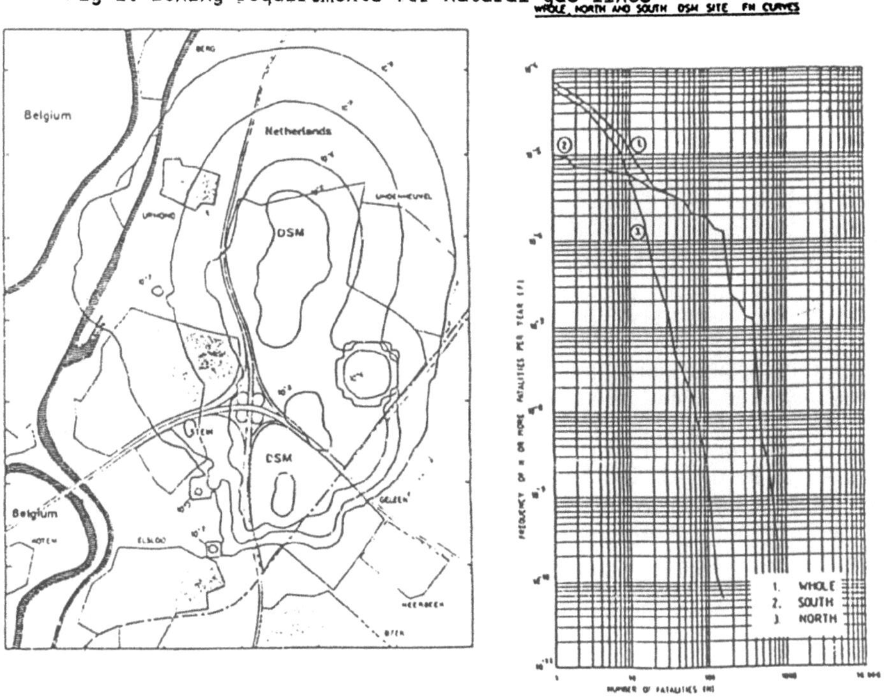

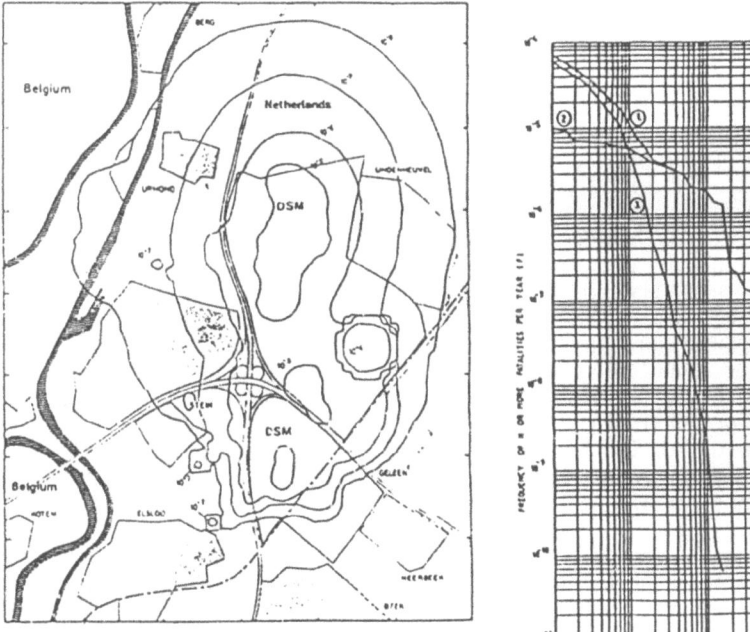

Fig 2. Risk results for the DSM site

GROWTH OF ENVIRONMENTAL TECHNOLOGY AS FORESEEN BY THE
RECENT ADMINISTRATIVE AND LEGISLATIVE DEVELOPMENTS

MÜEZZİNOĞLU, A., ŞENGÜL, F.

1. INTRODUCTION
1.1. Present situation in Turkey
Environmental factors describing the quality of life have shown subtantial and rapid changes for the generations born right after the Second World War. Although these changes are indicators of growth and development in our standards of living., they have also caused several disasterous problems and irreversible disruptions in the physical world we live, in because these changes are realized without due consideration to the environmental quality for the sake of rapid development.
Although Turkey did not fight at the Second World War, our post-war generations are witnessing the same environmental break-down. This is mainly due to demographic factors and rapid industrialization programs being carried out in the country.

1.2. Demographic factors
Turkey covers 778 000 km^2 in Asiatic (Anatolia) and European (Trachia) lands separated by Istanbul and Çanakkale Straits and Marmara Sea. Both sections are in the shape of peninsulas surrounded by seas from three sides. Surroundig seas are Black Sea, Aegean and the Mediterranean Seas. Marmara Sea is like an enlargement of thetwo straits separating the Asian and European lands, and is a Turkish inner sea.
After the fall of the Ottoman Empire, the new Republic of Turkey (founded at 1923) has inherited a broken and burnt-out mainland with a population reduced to a number well below 10 million after decades of tragic wars that took place all over the Near and Middle East. At the 10 th. Anniversary of the foundation of the Turkish Republic a growth in population up to 15 million was heartily blessed and population growth was encouraged by certain social and economic measures. But this high rate of growth in population continued during the decades to come and in spite of several compaigns for population control, a traditional growth rate of 2.16 % in 1985 is still kept, to reach a population outnumbering 52 Millions in 1987. Countrywide proportioning of this population and its dynamics in another problem to be noted. Due to several reasons urbanization rate surpasses the population growth rate. Approximately half of the present population live in settlements having populations above 2000 and are classified as urban population.
Density distribution of the growing population is uneven in the country. Four metropolitan areas having populations between 1.5-6 million cover a big portion of the urban population. Especially these densely populated cities are extremely important environmental problem areas.

1.3. Development methodology
Industrial development has been indicated as the formal means of development for Turkey according to the five year development plans being effective since 1960's. Presently the Fifth Five Year Development Plan covering 1984-1989 period is underway. These plans are made up of several sectors, each complete in itself and also connected with each other; education, energy, industry, transportation, housign, agriculture, health are the main sectors in Turkish development plans.Although enveironment has the capacity to be added to these main sectors, it has purposefully excluded with due claim that "environment" is a "challenging issue" having serious "intersections" with all of the present sectors. Truely enough environment is a challenge in all the planned sectors, thus requiring special attention for the success of these sector plans. But as the environmental consciousness and excitement rose during the last 1.5-2 decades in Turkey although the plans tried to suppress it for the benefit of the planned development.

Today environmental dispute has proven its technological dimension and conscience is rapidly growing among the press and mass media, politicians, industries, scientific and administrative groups, etc. There are two main reasons to this "unplanned rise of environmental issues":

A- New (1982) consitution has put forth a clear view to the environmental matters by suggesting:

(Article 56): "Each citizen has the right to live in a healthy and well-balanced environment.To develop the environment and to protect the environmental pollution are duties given to both the state and the citizens".

In accordance with this new incentive brought by the constitution, a framework-type legislation has been passed in August 1983. Law No. 2872 which has been prepared by the Deputy Minister of the Environment under the Office of the Prime Minister is based on the main principle of "polluter pays".

B- Decisions, suggestions and recommendations drafted at international levels has caused an environmental consciousness to arise among officials and the general public. Especially the works by OECD and Economic Community of Europe are worth to mention here as the firm economical background is not to be overlooked at such international media.Thus the national environmental legislation has got be reinforced, environmental education be better planned, environmental standards be set to invite environmental technology to take part in the development.

1.4. Legal status
Hygiene law (No.1593) which is dated back to 1930's has been in effect to regulate the "environmental health" conditions in Turkey This very powerful law has been regulating the conditions for issuing permits to authorize the industries for operation. These permits are based on the environmental hazards to be probably caused by industries and once the permit is given to an industry it would guarantee itself against any official interrogation for potential environmental conflicts. For this reason obtaining a permit is very difficult and it is known that much more than half of industries near İstanbul is in operation without due authorization by Law No. 1593 and its regulations. Another point to be criticized for these authorizations is that when industries "pass" after convincing the local (or central) health authorities on their pollution control projects to be implemented, they usually don't have to build and operate the treatment facilities in the way indicated in their permits. Another problem area in the procedure is the qualifications of the health authorities. These officials are medical

health personnel to a very big majority and are not qualified to assess the efficiency of the treatment technology. In many cases industries located outside the big cities, the health official they have to face with is the local medical doctor of the nearest town or village. This kind of work is a serious overburden to these medical personnel already under the pressure of public health care in the region and is clearly out of line of their profession.

A more recent development in the field of environmental legislation is related with the Law of Establishment of Metropolitan Municipalities. There are several metropolitan cities in Turkey: İstanbul with approximately 5.5 Million, Ankara with approximately 3.5 Million and İzmir with 2.7 Million population. Another metropolitan area recently established is Adana with approximentely 2.5 Million population. These 4 cities make up about 1/4 of the total population. Metropolitan Law gives the duty related with water supply, sewerage and garbage collection as well as other environmental responsibilities to the Metropoliton Municipalities within their jurisdiction. Thus the responsibilities of Governor's Hygiene Office under Hygiene Law 1593 is now in the Metropolitan Mayor's hands in İstanbul, Ankara and İzmir since 1984. It is interesting to note that very large projects with important environmental implications are being carried out since that time; such as the Golden Horn clean-up; İstanbul water supply, sewarage and marine disposal; İzmir wastewater collection and treatment; Adana metropolitan infrastructure; solid waste management are among these projects. However, environmental technologies and legal techniques reguested by the new Metropolitan system must comply with the old Hygiene Law Nr. 1593.

The most recent developement is the enacted Law of the Environment, Nr. 2872 which has been a milestone in the Turkish Environmental History. This Law is a "framework" type of law setting only the basic principles for the definition, handling and control of environmental problems. Behind these principles and definitions the economic principle of "polluter pays" is notable. Administrative penalties are defined on the basis of this philosophy. Technical details of air, water, soil pollution and noise control as well as the implementation of environmental impact assesment methodologies are left to regulations to be drafted as needed. Among the technical issues to be regulated in accordance with Law Nr. 2872, 10 separate regulations are noted by (Müezzinoğlu, Uslu, 1985). Following is a list of these technical issues to be regulated by amendments. Out of these 10 technical topics to be handled in the form of regulations,

- financial matters are elaborated as "Environmental Fund Regulation" in 1985. This is to collect the fines, donations and several other payments both from the governmet treasury and private bodies and direct these funds for financing research work and allocations necessary for treatment plants,

- air pollution control measures are worked out, as "Regulation for Protection of the Air Quality" dated 2.Nov. 1986. This is in parallel with the German "TA Luft" as modified on 27 Febr.1986.

- noise and vibration control techniques are drafted as "Regulation for Noise Control", enacted in December 1986,

- "Water Quality Control" regulation which will soon be published by the Official Gazette as finalized by a group of researchers of the University in İzmir, among whom the authors of this paper are taking active participation.This regulation follows the basic principles of the German regulations (Wasserhaushaltsgesetz, 1976) in some ways and is in accordance with the basic recommendations of the Economic Commission of Europe. But generally speaking it is not an easy type of legislation and the standars for the industrial effluents and for direct discharges into the receiving water bodies are adjusted according to the bearing capacity of the Turkish economy.

The new "Air Quality Protection" and "Water Quality Control" legislations cover broad spectra of emission/imission constraints and natural/effluent water quality management schemes. Each statement and each figure given in the standards and guidelines tables obviously have a meaning as for the pollution control technology. The basic technological incentives behind the statements, articles and amendments are based on the "best practicable means". This technological definition involves only the "most developed applicable technique which is economically feasible". As Turkey is a rapidly industrializing country with modest resources, the term economical in this basic definition is of utmost importance.

Table 1: List of regulations for the Environment Law Nr.2872 (Müezzinoğlu, Uslu, 1985)

Reference Article Nr.		
	Quality control standards for	
Article 8	Air	Enacted
	Water	About to be enacted
	Soil	To be prepared
Article 9	Environmental protection zones	To be prepared
Article 10	Environmental impact assesment	In preparation
Article 11	Technical guidelines for disposal of wastes into receiving media; principles, efficiencies, etc.	Partly covered by emission (air) and effluent (water) discharge standards (In the form of an amended document to Water Quality Control Regulation)
Article 12	Treatment, waste disposal and management	Partly covered by "water quality control regulation"
Article 13	Production, import, transportation, storage and use of persistent chemicals	Partly covered by an amended document to the "Water Quality Control Regulation"
Article 14	Noise Control	Enacted
Article 15	Inspection, pending, fines, judging and administration	To be summed up in the form of a final worksheet. Partly covered by a special amendment to "Water Quality Control Regulation"
Article 17	Financial matters and "Environment Fund"	Enacted.

2. PROSPECTS FOR THE GROWTH OF ENVIRONMENTAL TECHNOLOGY IN TURKEY

As was mentioned above, Turkey is,
- short of primary energy resources,
- inadequate in basic urban infrastructure such as water supply and sewarage systems, treatment plants, etc.in view of the rapid growth and urbanization rates.

With these and some other indicators, Turkey takes place in the "Developing Country" classification.

Environmental technology is a relatively new dimension in the development history of the country. In has been estimated that with the assumption of all the wastewaters coming from the urban centers and industries are collected and directed to treatment plants,energy requirements of treatment alternatives are shown in Table 2 below.The base year for these projections is 1989 which is the last year of the five-year-plan in effect today.

As can be followed from Table 2, 1989 energy requirement for urban waste water treatment will make up an appreciable percentage of the total energy consumption (and production) projections,if classical treatment technologies will be utilized. If more "natural" methods of lagooning systems or stabilization ponds are preferred, however, energy requirements will be minimized. In case of stabilization pond systems virtually no energy is required.

Tablo 2: Projections of total energy requirements for wastewater treatment for year 1989 (Müezzinoğlu, Uslu, 1986)

Indicator \ Technology	Classical Treatment Methods			"Natural" Treatment	
	Extended Aeration	Activated Sludge	Trickling Filters	Facultative Aerated Lagoons	Stabilization Ponds
Annual Energy requirement per capita, Kwh/cap.	10-18	7-17	3-8	5-7	0
1989 Total energy requirement for domestic wastewater treatment GWh/year	350-1320	240-590	100-280	180-590	0
1989 Total energy requirement for industrial wastewater treatment, GWh/year	580-2200	400-980	180-470	290-980	0
1989 Total energy requirement for wastewater treatment GWh/year	980-3250	640-1570	280-750	470-1570	0
Percentage of wastewater treatment energy in total energy requirement of 1989, %	1.5-5.8	1.8-2.6	4.6-1.2	0.8-2.6	0

Although such non-conventional treatment methods of lagoons and ponds are energy consuming, they require vast land area in order to approach to the same efficiencies offered by the classical treatment systems. Land usage in hectares per unit loading of biochemical oxygen demand per day is calculated from climatological data for Turkey. Results of these calculations are shown in Figure 1 (Soyupak,1973). Four zones of different usability characteristics for stabilization pond systems are to be noted for the same level of efficiency of treatment.

There are two important draw-backs to the widespread use of stabilization ponds in Turkey:
- the land area is either totally non-existent or very expensive near some urban centers,
- industrial effluents or urban wastewaters rich in industrial input might be rather difficult to treat in stabilization ponds because of the toxic effects of some ingradients of industrial wastewaters. That is why separate treatment plants for some industries are to be favored in many cases. The biggest treatment plant feasibility studies have been made for the city of İzmir and Adana (see Fig.1). Both are metropolitan centers located at the most favorable zone (Zone 1) for stabilization pond systems, and feasibility of these treatment systems are very high for these two cities. İzmir sewarage system is under construction for the time being, and as soon as parts of the collection system are finished first stages of stabilization ponds will be established (D.E.Ü., 1985).

İstanbul, which is the biggest city of Turkey,is in the process of erecting a very big sewerage system. A wastewater flow of above a million cubic meters per day is going to be drained with the aid of this system. It is planned that this water will be mechanically pretreated and disposed into the bottom current of the Bosphorus which will carry it away towards the Black Sea.

Water supply, sewerage and treatment plant erection is usually funded by İller Bankası (Bank of Provinces) which is a Government establishment located in Ankara. The Bank also has regional branches to give engineering control service and advice to the municipalities that look for it.Recently, İstanbul, İzmir and Adana Municipalities have seeked international credits and they prepared feasibility reports in order to obtain these funds.

Rest of the urban centers are either in the process of preparing engineering projects or in the erection stage for water supply, drainage and treatment plant systems. These efforts are growing in the form of an "infrastructure campaign" at the municipalities which are concerned. In 1987 budget water supply and wastewater disposal has been shown to be the 6 th. most important sector in the priority list of investments.

For air pollution control, however, a rather pessimistic view is to be cast. Most of the Turkish cities are under a thick smog especially during late autumn and winter. This is due to the very low quality lignites being burned in stoves and central heating systems. Therefore, best method of control of the air pollution is conversion of high sulfur, high ash content and low quality lignites into precleaned lignite briquettes. But neither a widely applicable technology nor the finance exists for the investments to be made for production of hig quality lignites.

3. RESULTS

Turkey is at a very special stage as far as environmental technology is concerned. There is need and also the public awareness about environmental control. With the help of recent legislation such control is being the duty

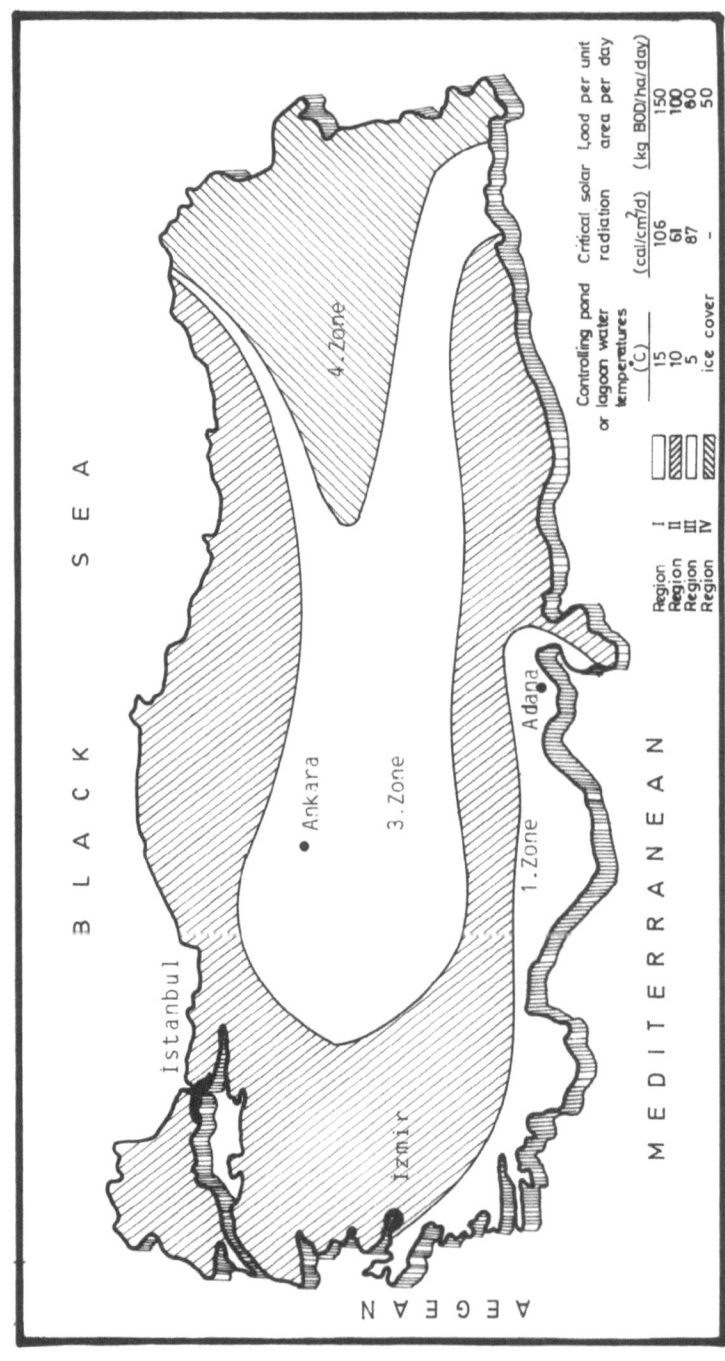

Figure 1: Climatic Zones for Stabilization Ponds in Turkey (Soyupak, 1973)

of the State and the people.

Climatic and economic conditions prevailing in Turkey urges the planners to introduce rather natural methodologies for wastewater treatment. For air quality upkeeping however, lignite preconditioning and processing is a necessity. Such technology is either to be developped or imported into Turkey.

4. REFERENCES

1. 1982 Constitution of the Republic of TURKEY,Article 56:Health Services and Protection of the Environment (in Turkish)
2. Republic of Turkey,Law of the Environment,Nr.2872,passed 9 August 1983,Ankara,TURKEY (in Turkish)
3. Republic of Turkey,Regulation for the Protection of Air Quality, Amendment of Law Nr.2872,passed 2 November 1986,Ankara,TURKEY (in Turkish)
4. Republic of Turkey,"Fifth Five Year Development Plan" 1984,Ankara, TURKEY (in English)
5. Dokuz Eylül University (1985):"İzmir Wastewater Treatment Plant Feasibility Study" Project report prepared for the İller Bankası (in English)
6. Müezzinoğlu,A.,Uslu,O. (1984): "Manpower Requirements according to the Law of Environment (Nr.2872)" paper presented at "IV. Deutsch-Türkisches Symposium für umweltingenieurwesen"11-16 JUNİ 1984,İzmir,TURKEY(in Turkish)
7. Müezzinoğlu,A.,Uslu,O. (1986): "Wastewater Treatment in View of Energy Usage", IV.National Energy Congress, November 1986,İzmir,Turkey.
8. Soyupak,S. (1973): Unpublished Master's Degree Thesis,Middle East Technical University,Ankara,Turkey.

DEVELOPMENT OF APPROPRIATE TECHNOLOGY AND WAY OUT FOR
ENVIRONMENTRAL PROTECTION OF CHINA--A WAY FOR THE
DEVELOPING COUNTRIES LIKE CHINA

Ranjie Hou*

1. INTRODUCTION

Environmental deterioration is still one of major enemys that mankind must struggle with nowdays. The dawn of a decisive victory has not yet appeared in this dramatic struggle even though a vast amount of manpower, material and financial resources has been drained. The present environmental technology is well-developed and has devote much to the solution of the environmental problems. However there are some environmental problems that remain out of man's control still now. Many scientists in industrialised countries are devoting to the development of new environmental technology so as to solve the remained problems. But it is very interesting that the environmental problems which have been solved in industrialised countries appeare again and become even more crititcal in many developing countries. Indeed, some of the developing countries have recognized gravity of the problems and try to control them with the environmental technology that has been proved effective in industrialised countries. Unfortunately their efforts did not achieve a significant result due to lack of fund and skill. Furthermore solution of the environmental problems is, in most of developing countries, often placed on second position even neglected. In those developing countries the experiences and lessons that industrialised countries have suffered from environmental deterioration are often forgoten. But the high living standards of industrialised countries have a permanent attractiveness. Therefore they would rather rise their material living standards than protect their environment. It is understandable to sacrifice environmental quality in oder to alleviate the famine in the developing countries where starvation is still an urgent problem. However environmental deterioration should and can be avoided in the developing countires like China where the food and clothing problems have been solved and the major task of the countries has become how to develop their national economy and rise the living standards as rapidly as possible. It is possible, for this kind of developing countries, to take some time to consider how they can make a balance between deve-

----
* Institute of Environmental Science, Peking Normal University. Working now as a research fellow of Alexander von Humboldt Foundation in Institut für Siedlungswasserwirtschaft, University Karlsruhe, Federal Republic of Germany.

lopment and environmental protection during industrialization. If they don't do this and only follow, step by step, the way that has been trodden by industrialised countries they would find late that they are facing the same troubles, just like that the industrialised countries have been facing in 60s--70s. It has been well known that influence of environmental deterioration is not limited by national boundaries, some time even lead to a global consequence. The efforts of improving environment in industrialised countries will become meaningless, at least will be discounted if the large-scale environmental deterioration in developing countries cann't be avoided. Therefore it is the time now to seek a way out for the environmental protection of developing countries. Is it a right way to transfer the existing environmental technology to developing countries? One should realize the present situation of developing countries before to answer this question. Generally speaking, most of developing countries are lack of either funds and skilled operators. They are rarely willing to spend their limited funds for import of the expensive technology that is considered not relative to their major economic goal. Even they do, the complicated technology cann't still work well due to lack of qualified operators. Therefore a direct transfer of the high technology to developing countries is not considered a right way. Then what is right way? One of such ways is proposed in this paper. That is development and application of appropriate technology. The proposal bases on analysis of the real situation of the developing countries like China. Still, the term of appropriate technology is, from the standpoint of developing countries, clarified; the criteria for selection and evaluation of the appropriate technology are proposed. Finally, the appropriate technology much-needed for environmental protection of the developing countries like China is listed. From that, industrialised countries will be able to find which kind of the existing environmental technology can be transfered as appropriate technology to the developing countries like China directly; which kind of environmental technology should be reformed in oder to suit the special situation of the developing countries like China and which kind of appropriate technology should be developed.

2. PRESENT-DAY CHINA AND ITS ENVIRONMENTAL SITUATION

China is situated in the eastern part of Asia, on the west coast of the Pacific Ocean. It has a vast territory(9.6 million squire km, third largest in the world) and endowed with aboundant natural resources. Also it has a large population(1060 million by the end of 1986) that makes a great pressure on its national economy, but a huge domestic market for industrial and agricultural products, too. By 1986 its per-capita national income is 210 US $, roughly equivalent to 1/40 of the United States', 1/30 of Japan's and 1/10 of the Soviet Union's. Therefore China is still poor in economy and underdeveloped in science and technology even though it has solved its food and clothing ploblems and achieved a fairly high level in some individual scientific or technical fields.

A few years ago a new policy has been put into practice so

as to quicken the speed of development of national economy and change the present poor and backward state of China. In the goals of the new policy the gross annual value of industrial and agricultural production will be quadrupled in two decades--from 192 billion US $ in 1980 to 760 billion US $ in 2000. That means, an average annual growth rate of 7.2% should be achieved from 1981 on. By 2000 the average annual per-capita income will be increased to 800 US $. The result achieved until now shows that the policy is going on smoothly. Table 1 shows the growth of Chinese economy since 1981.

TABLE 1. Growth of Chinese economy since 1981

ITERMS	UNITS	1981	1982	1984	1985	1986
Gross annual industrial & agricultural output value	million yuan	691900	829100	1111900	1326900	1510400
Total agricultural output value	million yuan	172000	248100	346800	375500	394700
Total industrial output value	million yuan	519900	550600	762000	969400	1115700
National income	million yuan	388000	424700	593300	676500	779000
Total import and export volume	million US $	73530	77200	48730	69620	73800

* The figures in TABLE 1 come from the reports of the State Statistical Bureau, except those of 1984, which were calculated from the corresponding values of 1985.

It is believed that China's goals will be achieved if the current policy is continued. It is undoubtful that the rapid growth of economy will bring a lot of benefits to China. However the environmental pressure caused by the growth cann't be neglected, too, especially if a balance between economic development and environmental protection cann't be kept. In fact the present environmental problems, especially pollution problem have become quite critical in some areas of China now. According to the statistics, the waste water and solid wastes discharged only by the large and middle-scale industrial enterprices have reached 80 million tons/day and 1 million tons/day, respectively in whole China. Those figures will increase by 1/3 if small and village-town enterprices that have been rapidly expanded recent years are taken into account. The consequences are that about 19% of all Chinese fresh water resources have been polluted; 20 of annimal species and 354 of plant species are near extinction. The effects of the pollution on the people's health are also significant. Table 2 shows concentration of the suspended solid and its effect in four large Chinese cities.

TABLE 2. Concentration of suspended solid and its effect in four large Chinese cities

	Average value of SS (µg/cub.m)	Maximum value of SS (µg/cub.m)	Number of lost workdays (10000/year)	Number of added death (person/year)
Peking	80	160	2500	850
Shanghai	150	200	4000	1300
Wuhan	170	400	2200	3500
Guangzhou	190	190	2000	1700

In addition to the industrial pollution there are still agricultural pollution and urban problems. Rapid expansion of large cities has made a more critical environmental quality in the urban areas. For example, total amount of urban garbage is 146 million tons per year in China now. Only 40--50% of them are cleaned up and only 1.6% are unharmfulized with proper treatment. The environmental problems caused by agricultural activities are mainly erosion of soil(about 5000 million tons/year), degeneration of grass land(about 0.39 million ha/year) and pollution of pesticides. Total amount of used DDT and HCH has reached 500 thousands tons and 4000 thousands tons, respectively in whole China until they was forbidden in 1983. Table 3 shows the aggregated amount of pesticides in body fat of the persons living in Lake Dongting basin, Hunan Province.

TABLE 3. Aggregated amount of pesticides in body fat of persons living in Lake Dongting basin

The names of prefectures	Number sampled persons	HCH		DDT		Years
		Range	Average	Range	Average	
Changde	81	4.3--58.4	24.3	2.0--42.8	15.9	1980--1981
Yueyang	38	4.6--64.4	24.1	1.1--43.0	14.7	1980--1981
Huxi	47	1.7--35.6	10.4	0.8--35.8	5.9	1980--1981
Chenzhou	51	4.9--47.3	17.6	1.0--32.6	8.7	1980--1981

The environmental problems have brought to Chinese decision-makers' and scientists' attention. A pollution control plan was made by the government departments concerned. In the plan the environmental pollution of China should be brought under control in first five years(1975-1980), and then the problems that caused the environmental pollution should be solved in the following five years(1981-1985). However the goal of the plan was not achieved. Environmental pollution is still expanding in whole China even some improvements have been achieved in a few of large cities or special industrial branches. Ten years are the time that industrialised countries have spent for a significant improvement of their environmental quality. Chinese plan-makers refered to it but forgot the realities of China. Therefore the plan was doomed to failure. Fortunately, the lessons of the failure have been drawn. The principal causes of the failure are that:

(1) The fund for realizing the plan was inadequate. The investment for environmental protection was only 400--500 million yuan RMB(1 US $ = 3.7 yuan RMB) per year during the planned period, which is only 0.3% of G.N.P of China. According to the experiences of industrialised countries it is needed to keep the environmental investment at 0.5--1% of G.N.P. in oder to maintain the environmental quality not deteriorating. In fact this figure should still be higher for China because of the imperfective basic installation. To perfect the basic installation and other public utilities concerned with environmental protection needs even much more money than the environmental investment itself.

(2) The plan didn't take reforming the backward industrial facilities and technologies as an important measure of improving environmental situation. The old and backward industrial facilities and technologies are one of major reasons that cause high pollution in China. For example, in China the average thermo-efficiency of industrial boilers is only 60% and coal consumption of unit G.N.P. is higher with 1--4 times than that of industrialised countries.

(3) The plan didn't obtain an active response from other economic departments of government. It is a usuall case in developing countries that the environmental goal is often disturbed and weakened by economic goal. This was true in China during the planned period.

The lessons drawn from the unsuccessful plan show that China needs a different way which should be suitale to the special situation of the country.

3. DIFFERENT SITUATION NEEDS DIFFERENT WAY OUT

It is at three points that the situation of China is different from the situation industrialised countries used to have in 60s--70s. The first is that industrialised countries have had accumulated enough funds and technologies when high pollution occured in 60s--70s. China is still poor and backward now, and thus has not enough fund for environmental investment. The second is that industrialised countries imported row materials, mainly from developing countries, during the process of their industrialization. China relies on itsself row materials, and thus the form of its environmental deterioration is different. The third is that most of industrialised countries have started transforming from production-type states to welfare-type states when high pollution occured, and thus the policy that put more emphasis on environmental protection was easilly accepted by public. But China is now only on the half way of its development. The enthusiasm that people thirst for a higher living standards is much higher than for a better environmental quality, and thus it is difficult to persuade public and economic departments putting more attention on environmental protection.

From the above analysis one can conclude that China should seek a different way out for its environmental protection. There are two ways that can be taken for China. One is to be enterlly absorbed in economic development no attention to be given to environmental protection. Environmental problems are solved only when a higher level of industrialization has been achieved and enough funds have been accumulated. This way is called "pollution at first, bringing it under control at second". It cann't be considered a successful way even though it has been trodden by industrialised countries and is being followed by most of the newly industrialised countries. There are still many problems remained in unsolved for this way. Indeed a higher growth of economy can be achieved in this way. However the environmental prices for it is very heavy, too. Many of environmental resources will be exhausted and unable to be recovered again. That doesn't accord with the long-term interests of the countries, also the interests of mankind. This way, in our opinion, is a way like "killing hen to get the eggs" and should not be taken by developing contries. Another is the way called "developing economy and protecting environment at same time". This way is obtained from summing up of the experiences and lessons of industrialised countries, and thus is very attractive to developing countries. Many of them, for example China, have attempted to take this way. However there is no successful example until now. That means that it is easy to find a right way, but difficult to march it. A suitable strategy is needed. Then what kind of strategy is applicable for China? We have been enlightened by one of the oldest Chinese philosophic works, Yiging, when we studied the problem. Yiging was written about 3000 years ago. But its idea which advocated harmonization between nature and man has still practical significance at present. Of course we don't want to advocate returning to the original oriental naturalism. We want only promote some kind of combination of modern techology with the old oriental philosophy. We found that the term "appropriate technology" is basicall coincident with our view and thus attempt to promote its development and application in the developing countries like China. We believe that it will supply a way out for environmental protection of China. In our term, the appropriate technology should firstly be in favour of development of a country. Because only an active development but nothing is the way to prosperous and happy for a developing country. At second, the appropriate technology should be favourable to harmonizing and stabilizing the relations between man and nature, man and man. The below criteria can be used for selection and evaluation of the appropriate technology:

(1) It should be adapted to the developing level, social and economic structure and cultural tradition of the country where it will be used.
(2) It should be cheap so that a developing country can afford.
(3) It should be able to make a full use of labour resource which is usually one of the richest resources in developing countries.
(4) It should make a full use of the local and renwable resources as far as possible.
(5) It be geared to needs of small, diversified communities, towns or industrial enterprices.
(6) It should be practically applicable immediately when a technical transfer from industrialised countries is involved.

The appropriate technology can be either existing technology and new development of the existing technology. It should be emphasized that the appropriate technology is not only for developing countries, some of them can also provide solutions for the problems of industrialised countries.

4. THE APPROPRIATE TECHNOLOGY MUCH-NEEDED IN CHINA NOW

To keep environmental quality at an acceptable level a lot of the appropriate technology is needed in China. But the followings are much-needed now:

(1) Appropriate environmental planning technology

Since the reform and open-door policy was carried out the village-town industries have been rapidly developed throughout China, not only in coastal areas but also in inland and rural areas. The development of the village-town industries increased the economic output of China. By 1986 the gross annual production of the village-town industries has reached 330000 million yuan and 33.8% of total industrial output value of China. More important things is that the development of the village-town industries absorbed a quite portion of the labour which were liberated from agricultural work and prevented them from moving to large cities. Total number of the labour working in the vallige-town industries has reached 17 million and 7.5% of total number of workers and staff members of China by 1986. It is undoubtful favourable to mitigating urban problems. However the development of the vallige-town industries will cause new environmental pollution in a more wide scope if the development of them is not rationally controlled. In fact, they have caused critical environmental problems in China. It is agreeable in China that the scale and kind of the vallige-town industries should be controlled. But questions are what kind of the industries should be developed in a special area? how large of its scale should be? in which location of the area should it be constructed? and so on. From viewpoint of environmental protection, to answer those questions one needs the called environmental planning technology, which is based on a rational or optimal distribution of the environmental resources such as natural purificability, environmental capacity of the areas concerned. This technology is especially suitable to the developing countries like China because it can achieve same results with less environmental investment.

(2) Appropriate industrial production technology.

Low efficiency and high consumption of energe and row materials are two of the main problems of Chinese industries. It not only reduces productivity and raise cost but also aggravate environmental pollution. Therefore reforming or renewing the bachward and out-of-date industrial technology is one of the key measures for improvement of the environmental situation of China. But it is impossible to renew all of them with high-technology in view of the present situation of China. Most of them can only be renewed or reformed with appropriate technology at present stage. There is a vast of demand for the appropriate industrial production technology in China, especially in the numerous vallige-town enterprices. A plan called "Spark Plan" is being carried out in China now. Its purpose is to provide technical and financial assisstants for development and application of the technology. However it is insurfficient for the developing countries like China to rely on own strength to do that. Cooperation and surport of industrialised countries are very necessary. We believe that industrialised countries will be interested in doing this and also benefit from doing this.

(3) Appropriate pollution treatment technology

Development and application of appropriate industrial production technology can reduce pollution but not eliminate pollution fundamentally. Therefore a parallel development and application of pollution treatment technology are also necessary. Still, improvement of the present environmental situation of China relys mainly on the pollution treatment technology. One will realize how pressingly the pollution treatment technology is needed in China when a concrete pollution problem, for example water pollution, is concerned. The wastewater produced in China is about 32000 million tons per year. Of them, only 15--20% are treated in different degree. All of the rest are discharged directly into the natural water bodies. With this, an economic loss of about 30000 million yuan per year is caused in China. The development and application of the wastewater treatment technologg will, therefore, obtain not only environmental benefits but also directly economic benefits.

However it is for same reasons that China needs mostly appropriate pollution treatment technology, but not high technology.

5. REMARK AND CONCLUSION

For the environmental protection of the developing countries like China, there exists indeed a way that is different from that industrialised countries have trodden. Human being will greatly benefit if most of the developing countries take the way: "developing economy and protecting environment at same time" during their industrializing their countries. China is the laregst developing country in the world. Its success will surely exert a great influence and encouragement on other developing countries.

In view of the real situation of China at present, the strategy of developing appropriate technology has been suggested. Two things are extramely important for implement of the strategy. One is China's own efforts, it is doing better now. Another is cooperation of industrialised countries. The current policy of China has created a better open-circumstance, which is not only an opportunity for industrialised countries to sell their existing technology to a vast market but also a possibility for joint-development of appropriate technology. Perhaps it is more easily to transfer the technology through the developing countries like China to underdeveloped countries. It is, therefore, obvious that the strategy will benefit both developing countries and industrialised countries, finally mankind.

ACKNOWLEDGEMENT

Author would like to express his thanks to Alexander von Humboldt Foundation for the research subsidization.

REFERENCES

1. Hou Ranjie, Zhu Xuan and Muraoka Kohji, Lake Environment in China----Problem and Effort for Protection, Environmental Research Quartery, No. 57, 1986(in Japanese).
2. Maruyama Motoyoshi, Present State of Chinese Environmental Policies, Environmental Research Quartery, No. 50, 1984(in Japanese).
3. Qi wen, China----A General Survey(Third Edition), Foreign laguages Press, Beijing, 1984.
4. Qu Geping, Environmental Problems and Policy of China, China Environmental Science Press, Beijing, 1984(in Chinese).
5. People's Daily(Overseas Edition), Beijing, 1986, 1987(in Chinese).
6. Su Wenming(editor), Economic Readjustment & Reform, Beijing Review Special Feature Series, Published by Beijing Review, 1982.
7. Su Wenming(editor), Modernization--the Chinese Way, Beijing Review Special Feature Series, Published by Beijing Review, 1982.
8. Technische Universität Berlin, Potential of Appropriate Technology, A Report on Ideas, Research and Actions of IPAT ----Interdisciplinary Projectgroup of Apropriate Technology Berlin, 1979.

SURABAYA INDONESIA: OPTIONS IN SOLID WASTE MANAGEMENT

S.A. VIGIL, CALIFORNIA POLYTECHNIC STATE UNIVERSITY
SAN LUIS OBISPO, CALIFORNIA, U.S.A.

1. INTRODUCTION
The Government of Indonesia requested assistance from the United States Trade and Development Program (TDP) regarding improvements to the solid waste management system of the City of Surabaya. TDP retained the author to prepare a Preliminary Feasibility Study which will enable TDP to make an informed decision on the desirability of funding further engineering work leading to solid waste projects.

Surabaya is the major industrial city of the Province of East Java, Indonesia. It is also the second largest city in Indonesia with a current population of 2.8 million, which is expected to increase to about 3.5 million by the year 2000.

As in many other large cities, the collection and disposal of solid waste has become a critical problem in Surabaya. The current solid waste management system cannot keep pace with the current rate of solid waste generation and will be completely overwhelmed in the near future if significant improvements are not implemented soon.

2. METHOD OF APPROACH
The Preliminary Feasibility Study was conducted in Indonesia during the period August 28 through September 4, 1986. The following procedure was followed:

2.1 Data collection
The primary means of data collection was by interview and site visits. Five Temporary Collection Areas and all of the Final Disposal Sites were visited. Existing solid waste records and reports were reviewed. The U.S. Consulate assisted by translating the Province Solid Waste laws. Translation assistance was also provided by Mr. H. Poeryanto of the Provincial Planning Board.

2.2 Photography
A detailed photographic record was collected by both 35 mm still photography and VHS video. The video tapes shot in-country have been edited into a 20 minute tape. Videotaping of interviews provided error-free transcription which was used in the preparation of this report. The edited tape was also provided to TDP and the City of Surabaya as a supplement to the main report.

2.3 Sorting Study
A field sorting study was performed at one of the Temporary Collection Areas. The study involved manual sorting of the waste into potentially recyclable materials (i.e. glass, paper, metals, etc.) and weighing each component.

3. EXISTING SOLID WASTE MANAGEMENT SYSTEM

3.1 Waste Generation
Current waste production in Surabaya is approximately 5600 cubic meters per day. This is expected to increase to 8500 cubic meters per day by 1992. The projected increase in waste production is caused by both an expected population increase and an increase in waste generation per capita (Reference 1).
Of the current total, approximately 2000 cubic meters per day are not collected (Reference 2). These wastes are either burned or illegally dumped in the streets and into rivers and canals.

3.2 City Solid Waste Management Organization
The city has a total staff of over 1300 workers involved in the solid waste managemnet system. These workers operate hand pushed collection carts, collection trucks, street sweeper trucks, transfer vehicles, and manage the Temporary Collection Areas and Final Disposal Sites. In addition to city workers, a substantial portion of the wastes (2500 cubic meters per day) are transported by private contractors.

3.3 Collection
Solid Waste is collected in the city by both trucks and hand carts. In many parts of the city, hand carts are the most feasible method for collection on the narrow streets.

3.4 Transfer and Transport
After collection, the wastes are taken to one of 20 Temporary Collection Areas, or TCA's, located troughout the city. At these sites, waste is loaded into either steel containers which are later picked up by special container trucks, or reloaded into dump trucks. Several types of TCA's exist:

3.4.1 Off-street yard -The Bukit Barisan TCA is typical of this type. Waste is collected in a walled, concrete floored area away form the street. Wastes are reloaded by hand with pitch forks and baskets into dump trucks for transport to the Final Disposal Site. This process takes up to an hour, greatly reducing the productivity of the truck crew.

3.4.2 On-street container area -At the Penghela Street TCA, waste is loaded into 10 cubic meter containers for truck transport to the Final Disposal Site. Waste is hand loaded and packed into the containers. The containers are parked on the shoulder of a busy street. Heavy traffic is a hindrance to operation of the area and a significant hazard to city workers.

3.4.3 Off-street container area -At the Pirngadi Street TCA, wastes are unloaded from hand carts into 10 cubic meter

containers. The loading area is fenced off from the street.
This removes the loading activity from the street and
improves the cleanliness of the area.

3.4.4 **Off-street market area** -The Keputran TCA is a typical
market area. It services both residential and commercial
customers. At this area, wastes have overflowed the
collection area and spilled over into the street. This poses
a hazard to both city workers and the general public. Again,
the hand loading of trucks reduces productivity.

3.4.5 **On-street "emergency" area** -At the Pandigiling
Emergency TCA, a section of city street has been set aside as
a Temporary Collection Area. This causes health and safety
problems because the waste connot be isolated from the
general public and interferes with the flow of traffic. The
area has been used as an Emergency TCA since 1978.
Fortunately it is not typical of the other 19 TCA's in the
City.

3.5 Problems with Uncollected Waste

Surabaya is criss-crossed by rivers and drainage canals.
In some areas of the City these canals have been used for the
illegal disposal of wastes. These solid wastes float
downstream until they lodge on an obstruction and form banks.
Besides being unsightly, these waste deposits reduce the
carrying capacity of the canals and rivers and cause flooding
during the rainy season. The City is aware of these problems
and has a dredging program to keep the canals clear. A law
has been passed outlawing dumping into the canals. Violators
are subject to a heavy fine.

A related problem exists in regards to the accumulation of
wastes in the drainage ditches. In most cases this is caused
by street litter. In some cases, the filling of drainage
ditches is caused by wind-blown wastes from a TCA, as
sometimes occurs at the Penghela Street TCA. The clogged
ditches allow the collection of standing water and the
subsequent breeding of mosquitoes.

3.6 Final Disposal Sites

After the wastes have been consolidated at the Temporary
Collection Areas, they are transported to Final Disposal
Sites. Solid wastes are spread on the ground at these sites
and compacted with tractors. There are two city owned sites
currently operational and one private site. A new site is
being developed by the City at Asemrowo. It is typical of
conditions in Surabaya in that the site has high groundwater
and is already heavily encroached by housing areas

At the nearby Asemrowo private landfill site, fires are
common, causing hazardous conditions for workers and
generating unhealthy air emissions. Scavenging of recyclable
materials at the landfill is a common, but potentially
dangerous practice. As in the new Asemrowo city landfill
site, this site is surrounded by housing areas.

The main Final Disposal Site for Surabaya is the Keputih
Disposal Site. Over fifty transfer trucks a day deposit
wastes at the site. Operations are conducted from 7:AM to

midnight. Wastes are already over 5 meters deep. Scavenging is conducted at the site by local residents, often at great hazard since visibility is hampered by the piles of waste and the truck and tractor operations. During the rainy season the site is only accessible to heavy duty trucks and crawler tractors.

The site is operated as an open dump. Wastes are uncovered during the operational lifetime of the site. While this makes the wastes available for scavenging, it allows the breeding of disease vectors such as flies and mosquitoes, causes significant air pollution from the emissions of methane and other naturally generated gases, and allows the penetration of rainwater into the landfill. Water percolating through the solid waste causes the formation of leachate, a highly concentrated liquid waste. Leachate can enter the groundwater, contaminating local wells.

3.7 Recycling

Recycling is an integral part of Indonesian life. A prior solid waste study in 1976 (Reference 2) found that up to 25 percent of the waste generated in the test households was recycled before it left the home. This type of recycling, source separation, is the best because it produces the cleanest material.

Additional recycling is also occuring at the Temporary Collection Areas. Typical recycled materials include: plastics, cardboard, tin cans, glass, and mixed paper.

A third type of recycling is taking place at the Final Disposal Sites. Known as scavenging, this is probably the least productive recycling since so much has already been removed from the waste sttream. Nevertheless, scavenging provides a meager living for a considerable older men and women are involved. Due to the operation of tractors and trucks at the disposal sites, scavenging is very hazardous.

3.8 Solid Waste Composition

Prior studies in Surabaya (Reference 3) and other cities in Indonesia (Reference 4) have shown that the solid waste stream is predominantly organic material, 94 and 67 percent respectively. A 353 kilogram sample sorted during this study at the Pandegiling Street TCA confirmed these findings. Over 95 percent of the sample was found to be organic and food wastes. By comparison data from the United States (Reference 5), shows less than 28 percent of the solid waste is organic.

4. SOLID WASTE MANAGEMENT OPTIONS

The current status of solid waste management in Surabaya, Indonesia has been reviewed by means of site visits, interviews, and a review of available translated documents. It is this author's opinion that significant improvements can be made in the solid waste management system in the areas of transfer and transport, final disposal, volume reduction, and energy recovery.

Although these options are site specific to Surabaya, similiar conditions exist at other cities in emerging

midnight. Wastes are already over 5 meters deep. Scavenging is conducted at the site by local residents, often at great hazard since visibility is hampered by the piles of waste and the truck and tractor operations. During the rainy season the site is only accessible to heavy duty trucks and crawler tractors.

The site is operated as an open dump. Wastes are uncovered during the operational lifetime of the site. While this makes the wastes available for scavenging, it allows the breeding of disease vectors such as flies and mosquitoes, causes significant air pollution from the emissions of methane and other naturally generated gases, and allows the penetration of rainwater into the landfill. Water percolating through the solid waste causes the formation of leachate, a highly concentrated liquid waste. Leachate can enter the groundwater, contaminating local wells.

3.7 Recycling

Recycling is an integral part of Indonesian life, A prior solid waste study in 1976 (Reference 2) found that up to 25 percent of the waste generated in the test households was recycled before it left the home. This type of recycling, source separation, is the best because it produces the cleanest material.

Additional recycling is also occuring at the Temporary Collection Areas. Typical recycled materials include: plastics, cardboard, tin cans, glass, and mixed paper.

A third type of recycling is taking place at the Final Disposal Sites. Known as scavenging, this is probably the least productive recycling since so much has already been removed from the waste sttream. Nevertheless, scavenging provides a meager living for a considerable older men and women are involved. Due to the operation of tractors and trucks at the disposal sites, scavenging is very hazardous.

3.8 Solid Waste Composition

Prior studies in Surabaya (Reference 3) and other cities in Indonesia (Reference 4) have shown that the solid waste stream is predominantly organic material, 94 and 67 percent respectively. A 353 kilogram sample sorted during this study at the Pandegiling Street TCA confirmed these findings. Over 95 percent of the sample was found to be organic and food wastes. By comparison data from the United States (Reference 5), shows less than 28 percent of the solid waste is organic.

4. SOLID WASTE MANAGEMENT OPTIONS

The current status of solid waste management in Surabaya, Indonesia has been reviewed by means of site visits, interviews, and a review of available translated documents. It is this author's opinion that significant improvements can be made in the solid waste management system in the areas of transfer and transport, final disposal, volume reduction, and energy recovery.

Although these options are site specific to Surabaya, similiar conditions exist at other cities in emerging

nations. The options do not represent radically new
technology, but rather appropriate technology which can be
easily adapted to Surabaya.

4.1 **Transfer and Transport**
The City Cleaning Section has over 1300 workers. Reference
1 indicates that the City cannot afford to add more wokers to
the payroll. It is therefore important to optimize the
performance of the existing workers. Total mechanization of
the operation is neither desirable nor affordable. Hand
collection of wastes is the only feasible method in many
parts of the city. However, there may be sections of the
city where mini-trucks or scooters could be used. A careful
evaluation of collection routes should be performed to
evaluate this concept.

A major bottleneck in the existing system is the manual
loading of transfer trucks at the Temporary Collection Sites.
Some of the areas are large enough that small front end
loaders could be utilized with no modifications to the
existing area (i.e. the Bukit Barisan TCA). At other areas
redesign is necessary to speed up loading. For example the
sites which use containers are not suitable for loading by
mechanical means. However a ramp system could be devised by
which the hand carts would be unloaded from a platform above
the containers, eliminating the time consuming unloading and
reloading process.

The feasibility of using waste compaction in the transfer
trucks should be investigated as a possible method of
decreasing costs. The weight carrying capacity of the
streets and highways enroute to the Final Disposal Sites
should also be considered.

The optimization of the collection and transfer process
should free up workers who can be reassigned to duties which
still require manual labor. This should allow service to
expand into areas not currently served.

4.2 **Final Disposal**
The current practice of open dumping of solid wastes should
be curtailed for several reasons:
  a. It causes significant air pollution from the emission of
     methane and other hydrocarbons which are natural
     decomposition products of organic wastes. Open fires,
     such as observed at the Asemrowo private disposal site
     are an additional air pollution hazard.
  b. Because of poor compaction, the present practice is
     wasteful of land.
  c. The exposed waste provides a haven for mosquitoes, rats
     and other disease vectors.
  d. The percolation of rainwater through the uncovered waste
     leads to the formation of leachate, an extremely
     concentrated wastewater which can contaminate local
     groundwater.

The technology of choice is the sanitary landfill.
Briefly, an sanitary landfill differs from the current
practice in that the waste is not just piled on the ground,
but rather placed in thin layers, compacted, and covered with

a thin layer of earth every day. This has many benefits including the isolation of the waste from rats and other vermin, control of rainwater infiltration, and control of methane leakage to the atmosphere. A properly designed sanitary landfill can be reclaimed for parks or other public uses after filling. It is also possible to drill wells and recover the methane gas for beneficial use. The ramp and area methods of landfilling would be most appropriate for Surabaya because of the flat terrain.

4.3 Incineration

Reference 1, the Draft Five-Year Solid Waste Plan for Surabaya, recommended the use of incineration for volume reduction. In the opinion of the author incineration is not economically feasible because of the extremely high organic content of Surabaya's waste which has high moisture and low energy contents.

Wastes sampled on September 2, 1986 by the author and in 1984 (Reference 1) had calculated moisture contents of 62.3 and 58.3 percent respectively. Energy content calculations for the same two samples were 2,621 BTU/lb (6,096 kJ/kg) and 2,846 BTU/lb (6,619 kJ/kg) respectively (as discarded basis). A solid waste study in Bandung, a city in West Java, found 66.5 percent organic matter in the waste, with a measured moisture content of 83.5 percent for the combined waste (Reference 4). The study concluded that the waste was not suitable for incineration. By comparison, typical solid waste in the United States has an average moisture content of only 20 to 30 percent and an as discarded energy content of over 4,500 BTU/lb (10,467 kK/kg), Reference 5.

Although it is possible to incinerate almost anything including sewage sludges at 90 percent moisture content, to do so requires the input of additional energy to evaporate the moisture. Typically this energy is in the form of natural gas or fuel oil. The cost of this additional energy was not included in the estimates of Reference 1.

4.4 Composting

The high moisture, high organic content of Surabaya's wastes make them ideal for biological treatment. Two types of treatment are possible: aerobic treatment (composting), and anaerobic treatment (landfill gas recovery). The latter process will be discussed as an energy recovery option.

Composting of Surabaya's solid wastes has already been performed on a commercial scale by a private firm under City franchise. The firm failed about two years ago because it could not market the compost at a profit. The firm used a relatively capital intensive European composting system, the Dano process, which was not cost competitive in the Indonesian market. Twenty thousand tons of the compost have been stockpiled and are still being sold, but production has ceased.

A less costly form of composting, the aerated pile method, has been developed by the U.S. Department of Agriculture Research Laboratory, Beltsville, Maryland. The process

involves building piles of waste over a grid work of
perforated plastic pipes. Air is drawn through the waste
by a small blower connected to the pipes. After composting
is completed, the pipes and blower can be reused.

The process is used in the United States for the composting
of sewage sludge. The liquid sludge is mixed with wood chips
before composting. The high moisture, high organic content
solid waste of Surabaya would be ideal for this method. Any
sales of compost which may result should be regarded as cost
offsets and not profit. Unsold compost can be used to
reclaim closed sanitary landfills and for other City
landscaping and park projects.

4.5: Landfill Gas Recovery

Solid wastes are naturally decomposed by anaerobic bacteria
in the wastes. These bacteria convert organic waste into
carbon dioxide gas and methane gas, about a 50 percent
mixture of each. This process occurs quite slowly, taking up
to 20 years. In contrast, the aerobic composting process is
quite fast, taking place over a period of days.

The mixture of methane and carbon dioxide is known as
landfill gas. Left uncontrolled, it is a serious pollution
and safety hazard. The carbon dioxide gas is heavier than
air and highly soluble in water. It lowers the pH of water
making it acidic. Methane is lighter than air and insoluble
in water. It migrates through the soil or waste along cracks
and fissures. Methane is highly flammable and even explosive
in certain mixtures with air (5 to 15 percent).

Current open dumping practice in Surabaya allows the
landfill gas to escape into the atmosphere or migrate through
the soil, possibly entering adjacent houses which completely
surround all the existing Final Disposal Sites. Landfill gas
also escapes through the top of the open dump into the
atmosphere. This is both a safety and environmental problem
and a waste of a valuable resource.

About 10 years ago, increasing natural gas prices made it
economic to recover landfill gas. The recovered gas can be
used in boilers, engines, and gas turbines. For example Los
Angeles County, California generates 2.8 megawatts of
electricity, enough for 5,600 homes (Reference 6). Landfill
gas is recovered at over 70 other sites in the United States
(Reference 7).

Most of the landfill gas systems in the United States have
been developed under royalty contracts by private developers.
The landfill owner receives a royalty (typically 13 to 20
percent) of gross revenues. In return the developer builds
and operates the system at no cost to the landfill owner.

Some landfill gas projects have been developed and operated
by landfill owners themselves. In this case the owner
receives all of the income but must pay the capital and
operating costs.

A feasibility study should be done to estimate the
potential for gas recovery if sanitary landfills are adapted
in Surabaya. The study should also identify potential
landfill gas developers interested in Surabaya. It may be

possible to offset much of the sanitary landfill costs by a suitable contract with a landfill gas developer. The potential for a landfill gas system owned and operated by the City of Surabaya should also be explored.

5. CONCLUSIONS

A solid waste survey was conducted in the city of Surabaya, Indonesia under the sponsorship of the U.S. Trade and Development Program. The present solid waste management system was found to be overloaded. The following improvements were recommended:
a. The transfer and transport of solid waste should be improved to make more efficient use of existing man power by partial mechanization.
b. The existing open dumps should be converted to sanitary landfills.
c. Composting should be re-evaluated as a volume reduction method.
d. Landfill gas recovery should be investigated a means of controlled gas migration and generating revenues.

REFERENCES

1. Dharmo, Suko. "Draft five Year Plan for Waste in Surabaya Municipality 1987-1992." Surabaya, Indonesia: 1986. (Indonesian).
2. "Dua Ribu Meter Kubik Sampah Tak Terangkut Setiap Hari." ("Two Thousand Cubic Meters of Garbage Undisposable.") Jawa Pos June 16, 1986. (Indonesian).
3. Camp, Dresser, and McKee International, Inc. "Report on Studies for Surabaya Water, Wastewater, Drainage, and Solid Wastes, Volume IV-Program for Solid Waste Management." Boston: 1976.
4. Owens, R.J. "Solid Waste Aspects of Public Health Engineering in Java, Indonesia." In *Practical Waste Management*. Chinchester, England: John Wiley and Sons, 1983.
5. Tchobanoglous, G. et al. Solid Wastes: *Engineering Principles and Management Issues.* New York: McGraw-Hill, 1977.
6. Carry, C.W., et al. "Turbines Produce Energy from L.A. Landfill." *World Wastes* June 1984.
7. "Landfill Gas Survey Update." *Waste Age.* March 1985.

ACKNOWLEDGEMENTS

This project was supported by the U.S. Trade and Development Program, Washington, D.C.. The assistance of Mr. Jack Williamson and Joseph Witczak of TDP and Mr. H. Poeryanto of The Province of Java Timur is gratefully acknowledged. The opinions expressed in this paper are those of the author and not necessarily those of TDP, or the City of Surabaya.

APPLICATION OF THE UASB-PROCESS FOR ANAEROBIC TREATMENT OF MUNICIPAL WASTEWATER UNDER (SUB)TROPICAL CONDITIONS

A.F.M. VAN VELSEN, J.A.W. MAAS

1. INTRODUCTION
Wastewater treatment in developing countries is feasible only when there is a method available which is relatively cheap and easy to operate. This is well recognized these days in many countries, that face an increasing environmental pollution problem.
In general, wastewater is treated in expensive and sophisticated plants. To reduce the treatment costs and to facilitate operation and maintenance, wastewater treatment often is centralized. This, however, requires expensive sewage transport systems.
The combination of sewage collection and conventional treatment has proven to be too costly and too complicated for many developing countries. Plans for these schemes frequently have to be scheduled and be replaced by on-site sanitation schemes or cheaper treatment methods. In looking for cheaper solutions lagooning is considered as the mere alternative. However, lagooning requires extensive land area, which often is not available or too costly. This especially holds for the situation that large quantities of wastewater have to be treated in densely populated urban areas.
In this light we want to focuss at an alternative treatment system: anaerobic treatment in a so-called Upflow Anaerobic Sludge Blanket reactor. This treatment technique has proven to work well, whereas the capital and operational costs are relatively low in comparison to other treatment methods. As a result the UASB-technique offers the opportunity to extend the wastewater treatment capacity within a limited budget.

2. TREATMENT ALTERNATIVES
At present the most extensively applied wastewater treatment systems in developing countries are lagooning and aerobic treatment. These treatment methods will be described briefly. The subject of this presentation, anaerobic sewage treatment, will be presented more in detail.

Lagooning
A conventional wastewater treatment system in developing countries is lagooning. Depending on the oxygen-concentration in the lagoon and the water depth one distinguishes anaerobic, facultative and aerobic or maturation lagoons. In well-designed and well-operated lagoon systems a high degree of purification and pathogen removal can be obtained.

The main obstacles for the implementation of lagoon systems for sewage treatment are
- they require large areas of land. Since the need for wastewater treatment in general is most urgent in densely populated areas, the use of land has to compete with other human activities such as housing, industry and agriculture;
- the costs of lagoon systems largely depend on the local land prices. As a result lagooning may become very expensive;
- the treatment efficiency of lagoons is difficult to control and depends largely on climatological circumstances;
- the presence of large, open surfaces of polluted water may impose hygienic risks and environmental problems;
- in the lagooning system part of the water is lost by evaporation. This is of special importance for arid regions where water is of vital importance;

Aerobic wastewater treatment
With aerobic treatment the organic waste components in sewage are microbiologically mineralized by an enforced supply of air (oxygen) to the water. Through aerobic systems a high degree of treatment efficiency can be achieved. Apart from the oxygen-demanding organic material also nitrogenous components can be oxydized and subsequently eliminated. Furthermore, aerobic processes in general are quite effective in reducing the number of pathogenic organisms. Nevertheless, up to now aerobic wastewater treatment systems have not been successfully implemented on an extensive scale in developing countries because:
- the investment and construction costs are high;
- the construction and instrumentation includes many mechanical parts. Subsequently the maintenance costs are high;
- the running costs are high partly, due to the energy demands for the enforced oxygen supply;
- large amounts of generally unstabilized excess sludge are produced;
- the proper maintenance and control of the treatment plant requires skilled personnel;
- the process is economically very unattractive at small scale application. Therefore, the choice for aerobic sewage treatment will often lead to centralized treatment facilities. This requires expensive serwerage systems.
- although to a lesser extent than with lagooning the installation also requires large areas of land.

Anaerobic wastewater treatment
Organic material can be mineralized under anaerobic conditions as well. In nature this occurs at all places where organic material accumulates and the supply of oxygen is deficient, e.g. in marshes, paddy fields and anaerobic lagoons.
In digestion units, the external conditions during the process can be regulated to speed it up as compared with that occurring in nature. Moreover, the produced biogas is collected and can be used as a fuel. In this way, anaerobic wastewater treatment combines a considerable reduction in the concentration of polluting organic substances with the production of useful energy.

One of the most important conditions to be met for speeding up the process is the maintenance of a high sludge concentration in the reactor. The higher the concentration of viable sludge, the higher will be the conversion capacity of the digester.
The maintenance of a high sludge-retention in the digester is the more important since in methane-producing systems the sludge production is low due to the poor energy profit from these conversion processes.
The retention of bacterial mass in the digester has been, at least until recently, the major obstacle for the application of the process for the treatment of wastewater. This especially holds for low-strength wastewater like domestic sewage.
In the past decades several sludge-retention techniques have been proposed and tested including the anaerobic filter and the anaerobic contact process. Recently a new and appropriate digester technique has been developed at the Agricultural University of Wageningen, the Netherlands, viz. the Upflow Anaerobic Sludge Blanket (UASB) process. The system is shown schematically in Fig. 1. In the UASB-reactor the wastewater is forced upwards through a dense blanket of viable anaerobic sludge. The reactor is equipped in the upper part with a proper system for gas-solid separation and solid sedimentation.

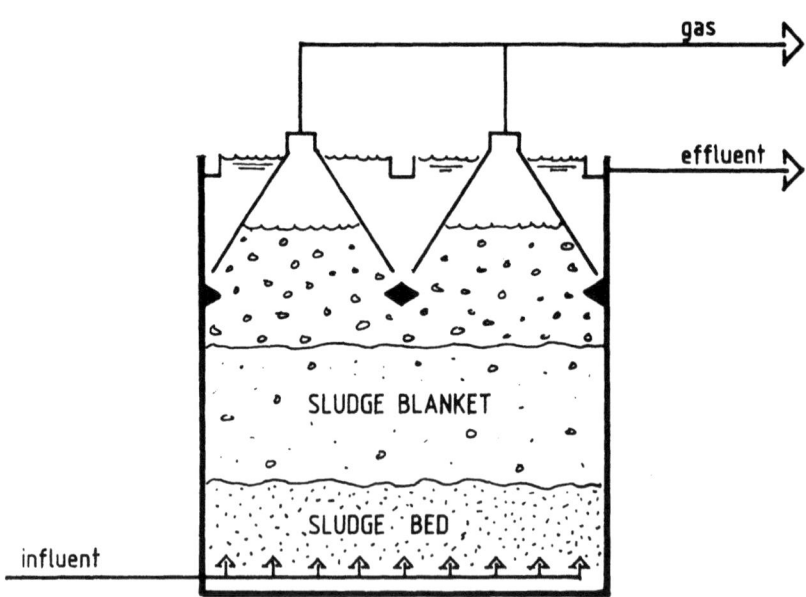

FIGURE 1. UASB-System

The basic ideas underlying the UASB reactor concept are:

* sludge recirculation and mechanical mixing are completely omitted. Due to this the physical conditions are favourable for sludge flocculation, resulting in superior settling characteristics of the digester sludge;
* contact between the sludge particles and the polluting organic substances is achieved (1) by forcing the wastewater through the sludge and (2) by a gentle mixing through the produced biogas. At relatively low gas production rates, e.g. at the treatment of domestic sewage, the biogas- mixing is only poor and a sufficient contact has to be achieved by a uniform spreading of the wastewater over the bottom of the digester;
* washout of sludge-particles released from the sludge blanket can be minimized by creating a quiescent zone within the reactor where the degasified sludge particles can flocculate, settle and return to the digestion compartment.

The UASB-reactor was developed in 1970 at the Agricultural University of Wageningen, in the first instance for the treatment of agro-industrial wastewater flows, such as sugar beet wastewater, potato-processing wastewater etc.
At present more than 50 full-scale UASB-plants are in operation for the treatment of a wide variety of industrial wastewater flows, the largest having a volume of 5,000 m³. In general, these digesters are operated at a temperature of 30-35°C and at a loading rate of 10-15 kg COD (Chemical Oxygen Demand) per m³ reactor volume per day. The treatment efficiency is 70-95% depending on the waste water composition.

TABLE 1   Benefits and limitations of anaerobic wastewater treatment as compared to aerobic treatment

Benefits
- No or little energy requirements
- Production of energy (biogas)
- Low production of excess sludge
- Production of stablized sludge with excellent dewatering characteristics
- Low nutrient requirements
- High loading rates applicable
- Small building area
- Simple in construction and maintenance
- No mechanical parts except a feeding pump
- Low capital and running costs
- Applicable from small to very large scale
- Re-use of plant nutrients possible

Limitations
- Lower treatment efficiencies (BOD, COD, pathogens)
- Relatively long period required for first start-up
- Treatment capacity depends on temperature
- Possible inhibition by specific compounds like chloroform and cyanide.

The main advantages and disadvantages of the UASB treatment process as compared to aerobic processes are summarized in Table 1.

In moderate climates, the main obstacle for application of the anaerobic process for domestic sewage is temperature. The gas production from this low-strength wastewater is so low, that it is insufficient to heat the wastewater significantly thus forcing the process to be carried out at ambient temperatures. The optimum temperature for anaerobic digestion is in the range of 30°C to 35°C, although the process still proceeds at far lower temperatures of 10°C and even below. In general the digester capacity will drop with decreasing temperature as a result of the lower microbial activity.

Fortunately, in many developing countries ambient temperatures are above 15°C. The elevated temperatures make the sewage in these regions amenable to anaerobic treatment. In realizing this, Haskoning and BV the Agricultural University Wageningen investigated the anaerobic treatment of domestic sewage in a 64-m³ pilot plant UASB reactor in Cali, Colombia. The investigations started early 1983 and were funded by the Dutch Government, Directorate General of International Cooperation. The most important conditions and results of this investigation are summarized in Table 2.

TABLE 2   Summary results of anaerobic digestion of domestic sewage in a 64-m³ pilot plant

	Hydraulic retention time (hours)			
	8	6	4	2,4
temperature (°C)	25	25	25	25
BOD-reduction* (%)	82	89	77	73
COD-reduction* (%)	78	78	77	73

gas production (l CH_4/p.e., day)	average 15-20
sludge production (kg TS/p.e., day)	average 0.05
total N-reduction (%)	average 1%
total P-reduction (%)	average 0%

* reduction calculated from raw influent and filtrated effluent

BOD	=	Biological Oxygen Demand
COD	=	Chemical Oxygen Demand
p.e.	=	population equivalent
TS	=	Total Solids

From Table 2 it can be concluded that the results were very promising. Even at hydraulic retention times as low as 2.4 hours the treatment efficiency remains more than 70% for BOD and COD. A BOD-treatment efficiency of 80-90% has to be considered as the maximum for anaerobic treatment of domestic sewage. Apparently the remaining organic material is not anaerobically digestible. In well-working aerobic treatment plants

higher treatment efficiencies are achieved: 90-95% for COD and 95-98% for BOD.
For the interpretation of the results of the Caleplant it is important to note, that the wastewater in Cali is highly septic and relatively rich in inert particles (sand, clay, etc.).

The investigation in Cali also gave strong indications that the reduction of pathogenic organisms in a UASB-reactor is poor (approx. 50%) as compared to that of most aerobic processes.
On the other hand the production of excess sludge is relatively low. The anaerobic sludge is stabilized and has superior settling and dewatering characteristics. Whether or not after dewatering, e.g. on drying beds, the sludge can be used as an organic fertilizer.

The incomplete treatment efficiency and poor pathogen removal efficiency of the UASB-system may necessitate the application of some form of post-treatment. Depending on the effluent quality to be achieved and the local circumstances, proper post-treatment methods include trickling filters, lagoons and sand filtration. As compared to the conventional treatment of raw sewage, post-treatment of anaerobic effluents can be achieved in small and simple aerobic systems. Other post-treatment systems, which combine water treatment with water re-use, are alga/fish ponds and some form of land treatment or crop irrigation.

Anaerobic effluents are particularly suitable for these water re-use systems because all plant nutrients remain in the digester effluent.

3. APPLICATION OF UASB-TECHNIQUE IN DEVELOPING COUNTRIES
The UASB-technique was developed in a period of increasing interest in biotechnology. The progress made in a short period of time can be partly attributed to recent developments in reactor technology and in microbiology. Both these disciplines form the basic elements of biotechnology. From this point of view the UASB-technique can be considered as an element of modern biotechnology.
In spite of the widespread opinion that modern biotechnology is synonymous with sophisticated, high-tech installations, it appears that the UASB-technique can be easily transferred to a wastewater treatment method that fits well in the local conditions of many developing countries.
A UASB-reactor consists merely of an empty pit, provided with a wastewater piping system at the bottom and a gas-solid separator in the upper part. All these elements can be constructed of materials like concrete, steel, plastics, asbestos cement etc. In general these types of construction materials are locally available. Furthermore, local construction practices can be incorporated in the design of a UASB-plant. Though a simple system, it should be stressed that the design of UASB-plants is of utmost importance for a successful process performance. Recent experiences in Colombia learned that an

insufficient knowledge of the basic principles of the UASB-technique often results in inadequate designs and consequent disappointing process performances.

A complete UASB-plant for the treatment of domestic sewage is shown in figure 2. The lay-out of figure 2 represents the full-scale plant which is presently under construction in Cali, Colombia for the treatment of wastewater of approx. 16,000 population equivalents. The plant consists of a coarse screen, pumping facilities (when feeding by gravity is not possible), a sand trap, a UASB-reactor and drying beds for the excess sludge of the UASB-digester. At large treatment plants it can become feasible to convert the biogas produced into electricity. If so, a gasmotor-generator or a dual-fuel gasmotor-generator unit has to be installed.

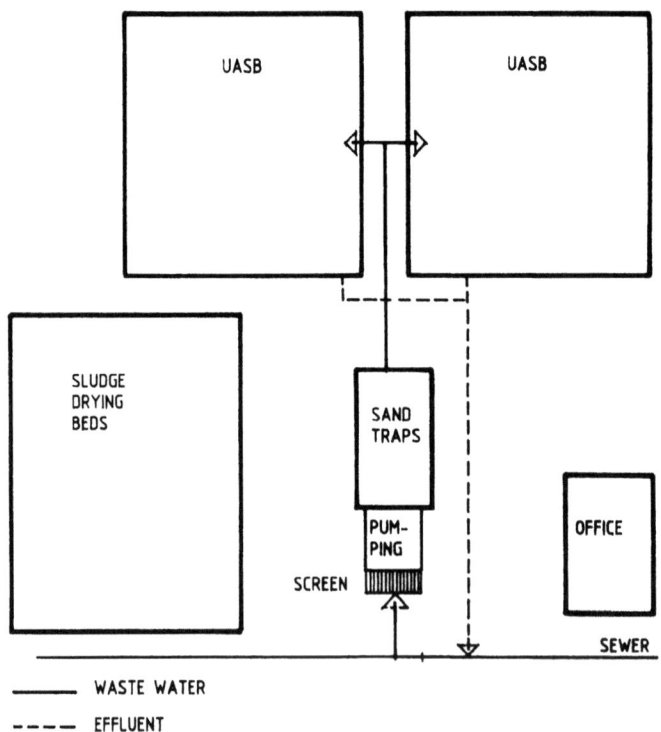

FIGURE 2. Lay-out anaerobic treatment plant for domestic wastewater

As compared to conventional aerobic wastewater treatment plants, the UASB-digestor combines four functions, which in aerobic plants are split up in separate process units, viz. presettling, biological treatment to reduce dissolved organic matter, anaerobic sludge stabilisation and post-settling. In spite of these combined functions, the only operational parameters are the feeding rate (=hydraulic retention time) and the excess sludge discharge.

As a result, operation and maintenance of UASB-treatment plants are quite simple and require relatively low manpower input. Maintenance includes cleaning of the wastewater distribution system, cleaning of the effluent gutters, preventive maintenance of the pumping facilities and sludge discharge and handling. The relatively low financial and manpower input of UASB-plants as well as the small land requirements offers the opportunity to decentralize wastewater treatment facilities. In this way the need for expensive wastewater collection systems can be prevented.

4. COSTS OF DOMESTIC SEWAGE TREATMENT IN A UASB-PLANT

Wastewater treatment costs can be expressed in terms of m³ wastewater treated, in kg BOD-removed or in population equivalents. In this presentation we will present the costs per population equivalent.

Total treatment costs consist of two elements, the capital costs and the costs for operation and maintenance.

As these costs vary widely, depending on the local conditions, the costs of wastewater treatment can not be given in absolute figures. Instead, the costs are related to those for other, conventional treatment systems.

In this way, the relative financial consequences of using the UASB-system for domestic savage treatment become clear, but the actual financial input in any situation cannot be made visible. This input should be calculated on the basis of local prices and conditions.

The global investment costs for domestic wastewater treatment are indicated in figure 3.

The costs in figure 3 hold for the investment costs of activated sludge plants and Pasveer-type plants in the Netherlands (data, RIZA 1985). The costs include pumping station, design, supervision at construction and taxes (19%), but exclude land and the main trunk sewer. The costs for the UASB-plants indicated in figure 3, are calculated by Haskoning on the basis of the general plant design (figure 6), Dutch prices (1985) and the results of the UASB experimental plant in Cali (Table 2). As a result the investment costs for UASB-plants, indicated in figure 3, concern the situation, that the wastewater temperature is approx. 25°C throughout the year. At lower wastewater temperatures, the investment costs per population equivalent will increase.

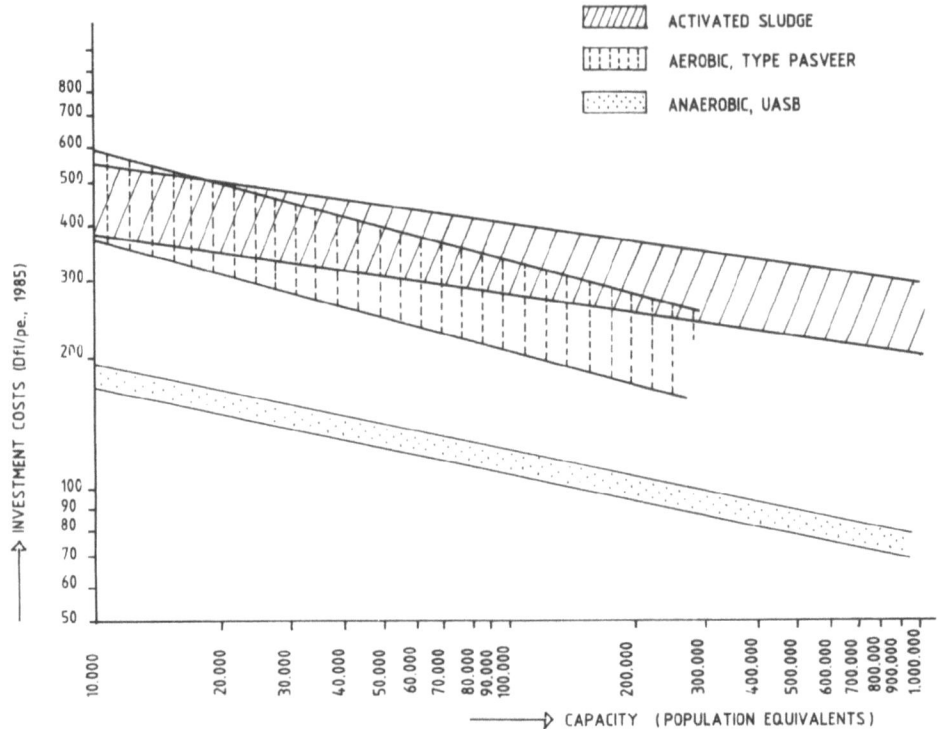

FIGURE 3. Tentative investment costs, in Dfl/p.e., of domestic sewage treatment plants.

The total treatment costs of several wastewater treatment alternatives are related to the costs of the most expensive variant, viz. treatment in a conventional aerobic treatment plant at two different capacities of 16,000 and 135,000 p.e.

The treatment alternatives considered are
a. lagooning system
b. aerobic   system
c. UASB   system

The general flow schemes of these treatment alternatives are presented in figures 4, 5, and 6 respectively.

Since the considerations for process design depend on the plant capacity and specific goals to be achieved, the flow schemes of the treatment plants of 16,000 p.e. differ from those of 135,000 p.e. The differences and starting points of the cost calculations are indicated in Table 3.

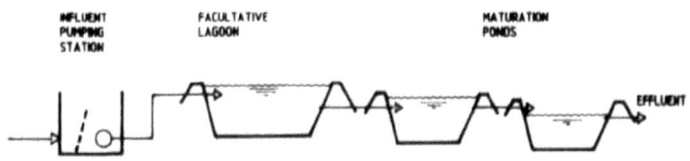

FIGURE 4. Lagooning system

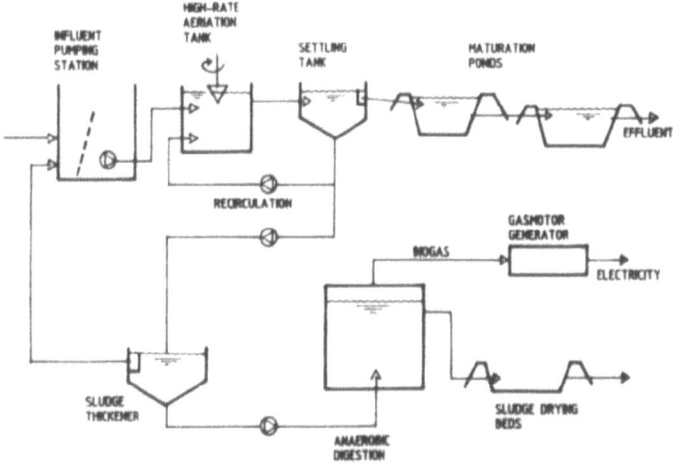

FIGURE 5. Aerobic treatment system (135,000 p.e.)

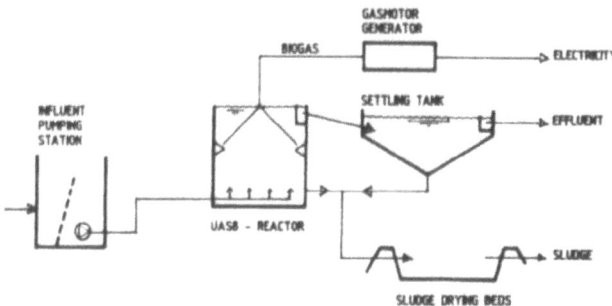

FIGURE 6. UASB system

TABLE 3   Differences in design considerations of treatment alternatives at different capacities

System	Capacity	
	16,000 p.e.	135,000 p.e.
a. Lagooning	see fig. 7	see fig. 7
b. Aerobic		
type	Caroussel	trickling filter
sludge stabilisation	-	anaerobic digestion
electricity generation	-	yes
c. Anaerobic		
post-settling	no	yes
use biogas	no	electricity generation

The tentative treatment costs, as related to the costs of the most expensive variant, viz. aerobic treatment (= 100%) are shown in figures 7 and 8 for treatment capacities of 16,000 p.e. and 135,000 p.e., respectively.

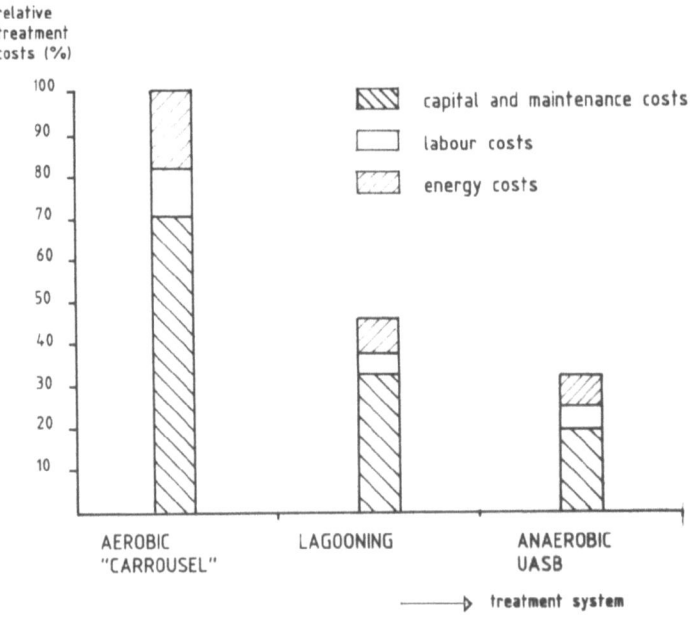

FIGURE 7   Tentative comparison of treatment costs of several sewage treatment systems. Capacity 16,000 p.e. (Aerobic treatment = 100%)

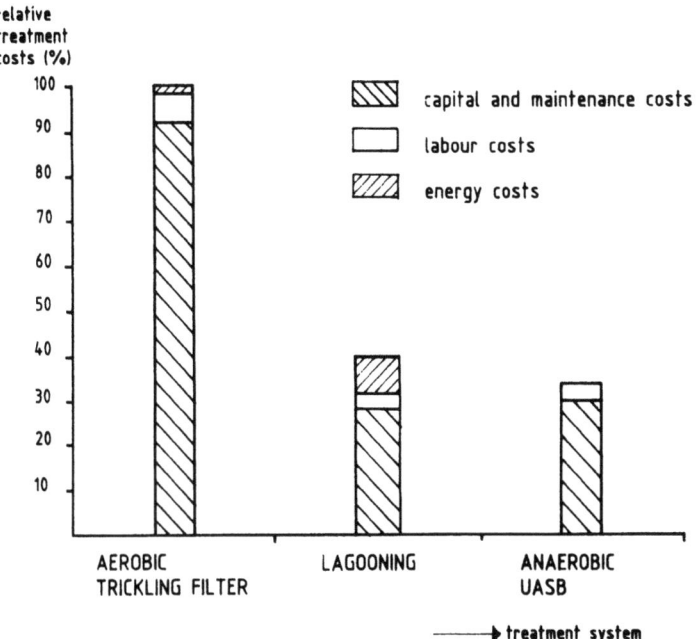

FIGURE 8 Tentative comparison of treatment costs of several sewage treatment systems. Capacity 135,000 p.e. (Aerobic treatment = 100%)

Figures 7 and 8 show that the treatment costs of lagooning and anaerobic treatment are considerably lower than those for aerobic treatment. It should be taken in mind, however, that the treatment efficiency in terms of COD-reduction is somewhat higher for aerobic systems (90-95% COD reduction) than for lagooning and anaerobic treatment (80-85% COD reduction).
Comparing the costs of lagooning and anaerobic treatment, it appears that the capital costs are in the same range. For lagooning these costs strongly depends on the costs for land acquisition, because the lagooning system requires an extensive land area.

In the present calculation the costs of land acquisition are estimated at Dfl 10,- per m².
Figure 8 shows that the operation costs of anaerobic treatment are lower than those of lagooning when the biogas is used for electricity generation. The electrical power generated is sufficient to meet the energy demands of the treatment plant, including pumping of the wastewater. However, electricity generation from biogas is only feasible at relatively large treatment capacities.

5. EPILOGUE

No single wastewater treatment system is so universal that its application is justified under all conditions. The choice for a treatment plant depends on many factors, the most important being water quality standards to be met, the availability of land, the need for water re-use and the climate. Anaerobic digestion in a UASB-system provides a valuable alternative for existing methods because it meets existing demands and offers the opportunity to treat wastewater at relatively low costs.

6. REFERENCES

1. Louwe Kooijmans J., Lettinga G. and Rodriquez Parral.
   The UASB process for domestic wastewater treatment in developing countries.
   Journal of the Institution of Water Engineers and Scientists Vol. 39, no. 5, 1985.

2. Lettinga G., van Velsen A.F.M., Hobma S.W., de Zeeuw W. and Klapwijk A.
   Use of the upflow sludge blanket reactor concept for biological waste water treatment, expecially for anaerobic treatment.
   Biotechnology and Bioengineering Vol. XXII, p. 699-734, 1980.

3. Lettinga G., Roersma R. and Grin P.
   Anaerobic treatment of domestic sewage at ambient temperatures, using a granular bed UASB-reactor.
   Biotechnology and Bioengineering, Vol. XXV, p. 1701-1723, 1983.

WASTEWATER TREATMENT TECHNOLOGY FOR DEVELOPING COUNTRIES

J.G. BRUINS

In recent years BKH Consulting Engineers Bongaerts, Kuyper and Huiswaard has carried out several projects concerning water pollution and wastewater treatment in developing countries.

A selection of these projects is as follows:

- Anaerobic wastewater treatment in the Philippine sugar industry

    A feasibility study was carried out to investigate treatment of waste water from the sugar industry in Upflow Anaerobic Sludge Blanket (UASB) reactors. Quantities and quality of the waste water were determined. The treatability of the waste water in UASB reactors was examined.
    Costs of investment and operations and the benefits of the biogas were calculated.

- Wastewater treatment for leather industry in Sri Lanka

    A wastewater treatment plant for a tannery in Sri Lanka was designed. Implementation of the treatment plant will be combined with improvement of the processes in the tannery, resulting in reduction of the waste production by the processes.

- Water pollution survey by means of aerial photography

    A new time saving technique was developed by BKH Consulting Engineers for water pollution surveys. This technique was applied for a water pollution survey in Colombo, Sri Lanka.
    The survey area was photographed at low altitudes from an ultra light aircraft. From the photographs a wealth of information could be collected on the quality of the water and on the sources of pollution. The sources of pollution were investigated. Hereafter it was possible to assess the situation as to the surface water quality in Colombo and to make plans for improvement.

Further details of the project above are provided in the following sections.

ANAEROBIC WASTEWATER TREATMENT IN THE PHILIPPINE SUGAR INDUSTRY

Introduction

Production of sugar takes a very important place in the Philippine agricultural industry.
The area cropped with sugar cane is nearly 500,000 hectares.
In 1983/1984 about 2,300,000 metric tons of raw sugar were produced by 41 factories.
Important by-products of raw sugar manufacturing are:
- molasses, which is often used for production of alcohol;
- bagasse, which is used as fuel or for manufacturing paper pulp;
- filtercake, which is transformed into an organic fertilizer.

Significant quantities of waste water are produced in the production of raw sugar, refined sugar and alcohol. Usually the waste water is discharged into the nearest river, creek or estuary which often causes untolerable pollution in the receiving water.
A decree of the National Pollution Control Commission (NPCC) indicates that surface water pollution legislation will be inplemented in the near future.
A result of this will be that wastewater treatment will become compulsory for Philippine industries. Some sugar industries already apply wastewater treatment.
Wastewater treatment plants usually consist of aerated lagoons or storage ponds. Aerated systems, however, are often considered to be too expensive, because of the very high energy consumption.
Alternative wastewater treatment methods are currently being investigated and it was decided to undertake studies on anaerobic wastewater treatment.

Studies on anaerobic wastewater treatment

The studies consisted of the following:
- inventory of wastewater production by the Philippine sugar industry;
- treatability study on sugar industry waste water in a laboratory scale upflow reactor;
- design of an anaerobic wastewater treatment installation for a raw sugar mill;
- estimate of investment costs.

Wastewater production by the sugar industry

At a number of sugar factories data were collected and wastewater samples were taken so that estimates could be made of the quantities of waste water that are produced in the sugar industry.

Treatability studies

The application of anaerobic treatment as used in the Upflow Anaerobic Sludge Blanket (UASB) system was investigated in a laboratory scale plant.
The experimental reactor has a volume of 81 l. Experiments were carried out for a period of 20 weeks. The reactor was fed with refinery waste water and with raw sugar mill waste water, respectively.
Composition of these two effluents was as follows:

refinery waste water:
- COD — 5030 mg/l
- total solids — 3450 mg/l
- volatile solids — 2800 mg/l
- alkalinity — 350 mg/l ($CaCO_3$)
- pH — 6.0

raw sugar mill waste water:
- COD — 2040 mg/l
- total solids — 2580 mg/l
- volatile solids — 640 mg/l
- suspended solids — 140 mg/l
- alkalinity — 190 mg/l ($CaCO_3$)
- pH — 7.4

During the experiments the following process parameters were maintained:
- hydraulic loading rate — 0.25 - 0.73 l/l.day
- organic loading rate — 0.9 - 2.6 gCOD/l.day
- pH in reactor — 6.3 - 7.4

Results:

During the first period refinery waste water was fed into the reactor. The total COD removal averaged at 60% at a retention time of 30 hours. During this period the sludge was not very well adapted, and as a result of a substantive wash out of sludge, the sludge concentration in the reactor was quite low.

In the second period raw sugar mill waste water was fed into the reactor. This period was started with fresh unadapted sludge.
Average COD removal was about 80%. Well adapted granular sludge was formed during this period. At the end of the experiment the sludge concentration was 50 g of dry solids per liter at a level of 0.3 m in the reactor.
Production of biogas was about 0.3 l biogas per g COD removed.

From the studies it could be concluded that waste water from raw sugar mills and refineries can be treated effectively in anaerobic reactors, with a COD removal efficiency of approximately 80%.
Granular sludge with good settling properties was formed.
Parallel experiments with anaerobic treatment of distillery waste in an upflow filter system showed a COD removal efficiency of about 60%.

Economic aspects

The costs of anaerobic wastewater treatment in the Philippine sugar industry were estimated. The estimates were based on the quantities of waste water that are discharged by the sugar industries and on the potential biogas production.
It was calculated that a significant percentage of the investment costs is returned, when the biogas can be used efficiently.

Other industries

The study of anaerobic wastewater treatment in the sugar industry had some "spin off" effects for other industries.
Interest in anaerobic wastewater treatment was expressed by the following industries in the Philippines: pulp and paper, soft drinks, canning, coconut processing and leather tanning. Some of these industries are already operating wastewater treatment systems (usually aerated lagoons). Often, however, operation of these wastewater treatment plants has been stopped because of the high energy costs for aeration.
Hence anaerobic wastewater treatment seems could be an extremely feasible method for industry in the Philippines.

WASTEWATER TREATMENT FOR LEATHER INDUSTRY IN SRI LANKA

The tannery of the Ceylon Leather Products Corporation is situated in the North of Colombo, at the left bank of the Kelani River, which is the major river in the Province of Colombo.
Recent surveys on water pollution indicated clearly that the Kelani River receives increasing quantities of effluent from industries of various types.
Hereby the river water quality is deteriorating rapidly and the river is becoming a serious health hazard, since the river water, traditionally, is directly used for all kinds of household purposes.
The Government of Sri Lanka is now implementing laws, that will prohibit the discharge of untreated industrial effluents in the future.
Because of this several industries are now taking measures for reducing their discharge of waste water.
Generally two types of measures are planned:

- measures within the factory that affect the discharge of smaller quantities of waste water, containing less pollutants than before, and
- implementation of wastewater treatment facilities.

The first industry in Colombo that takes measures to reduce the pollutional effect of its wastewater discharge into the Kelani River, is the tannery of the Ceylon Leather Products Corporation.

In this tannery 10 ton of hides, mostly cow and buffalo, per day are processed. A complete sequence of unit processes is carried out in a continuous system.
Processes include soaking, liming, defleshing, deliming, chromium and vegetable tanning, dyeing, fat liquoring and leather finishing. The waste water of all these processes is collected at one point, and it flows directly into the Kelani River.
Recently the production systems in the tannery have been upgraded with technical assistance from UNIDO, whereby consumption of process water was reduced to an acceptable level.

The wastewater treatment system consists of two components:

- system for treatment of chromium containing liquors, by which chromium can be recycled.
- plant for treatment of the combined wastewater flow.

Chromium recycling system
─────────────────────────

The chromium containing liquors of one day production are collected in a sump. Then they are mixed with hydroxide. Hereby chromium hydroxide is formed that settles to the bottom. The following day the supernatans is pumped off and discharged with the other waste water.
The sedimented chromium hydroxide is treated with acid, and may then be reused for the chomium tanning process.
The chromium recycling system is shown schematically below.

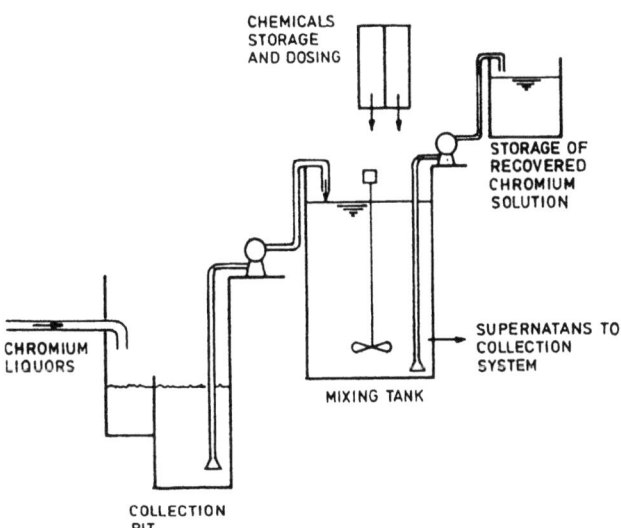

Wastewater treatment system
───────────────────────────

The tannery discharge about 350 m³ waste water per day. Peak flow is about 70 m³/h. The average composition of the waste water is as follows:

BOD_5	1800 mg/l
COD	4900 "
Suspended solids	2600 "
Total solids	10000 "
Sulfides	160 "
Total nitrogen	320 "
pH	2 - 11

At this time two alternative systems for wastewater treatment are being considered.

Alternative 1 - Physical-chemical treatment

In this system the waste water is treated in the following steps:

- screens for removal of coarse materials such as hairs, pieces of hide, etc.
- greasetrap for removal of fatty and oily substances
- equalization tank so that a constant wastewater flow with a constant composition enters the treatment plant
- coagulation tank in which chemicals are added
- flocculation tank in which flows are formed, that contain solids, organics and inorganic pollutants such as sulfur, nitrogen and phosporous
- sedimentation tank in which the flows settle into a sludge
- sludge drying beds in which the sludge from the sedimentation tank is dewatered.

The effluent of the sedimentation tank has a BOD_5 of about 600 mg/l and a suspended solids concentration of about 300 ml/l.
The effluent of the sedimentation tank is further treated in aerated ponds and maturation ponds.

- Aerated pond
  By aeration about 80% of the influent BOD_5 is degraded.
  Power input   = 10 kW
  Retention time = 6 days
  Volume   = 2100 m³
  Depth   = 2.5 m
  Area   = 840 m²

- Maturation pond
  In this pond sedimentation of solids and further degradation of organics take place.
  Area   = 1500 m²
  Depth   = 15 m
  Volume   = 2250 m³
  Retention time = 6.5 days
  Effluent BOD_5 = about 50 mg/l

Alternative 2 - Ponds system

In this system the waste water is treated in the following steps:

- screens
- greasetrap
- anaerobic pond
  Retention time = 5 days
  Depth   = 4 m
  Volume   = 1750 m³

Arrangements are made for desludging of the pond so that continuous operation can take place.
- BOD_5 removal = 50%
- aerated pond
- maturation pond
Aeration and maturation ponds are designed according to the same principle as for alternative 1 although the size of the ponds is larger for alternative 1 because of the higher BOD_5 of the waste water entering the ponds.

System selection

At this stage no final selection for one of the alternatives was made yet. Investment, operation and maintenance costs of alternative 1 are higher, and the operation of alternative 1 is more complicated and is depending on the availability of chemicals.

For alternative 2 more area is needed and this system creates more nuisance for the surroundings because of the development of bad odours.

WATER POLLUTION SURVEY BY MEANS OF AERIAL PHOTOGRAPHY

In order to combat water pollution in an area it is essential to have information about the types and sources of pollution. This information can only be obtained by carrying out extensive surveys to identify the origins of both domestic and industrial sources of waste water.
The discharge and quality of the waste water can then be determined. Such surveys are often time-consuming and expensive and sometimes the results can be unreliable, especially when there is little or no existing data and when some parts of the project area are isolated or inaccessible and can be overlooked. To ensure that coverage is complete an overall assessment survey is essential, prior to commencing detailed surveys. Such an overall survey can form the basis for a water pollution abatement masterplan since detailed surveys and data on wastewater loads are not necessarily required at this stage.
A general survey indicating the quantities of waste water and its origin and quality may, in fact, be sufficient for drawing up an initial masterplan. A new method for this type of survey has been developed by BKH Consulting Engineers and Buwalda BNA of The Netherlands. The method is based on aerial photography taken from an ultra light aircraft. The method was first applied in the recent Colombo Water Pollution Control Survey in Sri Lanka.

An ultra light aircraft is essentially a development of the hang glider. It is fitted with a small power unit and it is very manoeuvrable at low speeds (80 km/h). It is therefore an ideal aerial survey "platform". The aircraft has inherent stability and in terms of safety is "stall resistant". The controls are simple to operate and, with the addition of a second set of rudder pedals, the flight path can be adjusted by the photographer while photographs are being taken.

The aerial photographs can be made with conventional small-format cameras preferably fitted with motordrives, using a selection of filters. Simple stereoscopes are all that is required for the initial photographic interpretation.

The ultra light aerial survey project in Colombo was including the following stages:

The main industrial areas in the project area were identified initially on reconnaissance flights made at 3,000 m. All rivers, lakes, canals and industrial areas were then photographed at an altitude of 900 m, resulting in photographs with a scale of 1 : 14,000. These photographs were studied stereoscopically, resulting in a wealth of environmental data. Differences in the colour of the water gave an impression of the water quality and wastewater discharge points were clearly visible. The structural state of the embankments and the location of garbage dumps and shanty housing areas were clearly shown. Photographs were also taken at an altitude of 300 m to produce slides from which detailed layout maps could be made directly of major pollutors.

Visits were then paid to those industries identified as major pollutors from the aerial photographs and data obtained on production quantities and processes involved. An estimate could then be made of the quantity of waste water, and its quality.
An estimate of the total wastewater load discharged by industries in the project area, could then be made.

Data were also collected on the sewerage system, population and water consumption in the project area from which an estimate of the water pollution of domestic origin could be made.

The survey results will form the basis for an integrated management programme for environmental pollution control.

Ultra light aerial survey has the following advantages:

1. Overall assessment surveys can be made very quickly. Detailed surveys can also be executed in the shortest possible time - 45 km of water courses were surveyed in Colombo in only 15 hours.
2. The manpower input is low. Only two experts are required to execute a survey.
3. Because of the ultra light's manoeuvrability the system is ideal in built-up or other areas where access is difficult. It is also invaluable, for example, for surveying intricate waterway systems, deltas, etc.
4. An ultra light aircraft is relatively inexpensive, the manpower required limited and the cost of the final product is therefore small compared with the cost of conventional ground surveys.
5. The photographs themselves form valuable visual evidence, ideal for increasing public awareness, and for publicity campaigns and for teaching purposes.

The aerial survey in Colombo underlined the urgent need for environmental management, and the findings will form the basis of an integrated management programme for environmental pollution control. The photography clearly identified sources of pollution and areas likely to suffer now or in the future, the intensity of pollution being clearly indicated by the variation in colour.
A very important conclusion to be drawn from the study is that this type of photography provides visual proof of pollution which draws attention immediately to environmental problems. Its visual impact cannot be overstressed because the photographs themselves form a very suitable vehicle for decision makers, and for publicising the urgency of adverse environmental conditions on television, at meetings, in schools and public places. In short, they can be used to educate the public.
As an overall picture of an area is given, the system can be used as an important tool in integrated environmental management, policy preparation, facilitating communication, coordination of the authorities involved and illustrating environmental problems, for example, to financing organisations.

The photographs enabled the consultant to identify areas for the detailed mapping of selected areas and proved to be very useful for identifying basic physical aspects of the canal and river system in the project area. In the absence of base maps, the photographs would also be invaluable for planning sewer systems, drainage networks and the location of wastewater treatment plants, initiating environmental pollution abatement programmes and for monitoring purposes to keep a check on undesirable discharges and waste disposal.

BKH is planning to use an ultra light aircraft for water pollution surveys with a system of water sampling. The ultra light aircraft will be equipped with floats enabling it to take off and land on water.
In this way water sampling will be possible from water courses which are difficult to reach by other means of transport. The time needed for sampling will thus be reduced substantially. Sampling with an ultra light aircraft will be especially useful when, for example, a long watercourse has to be sampled at intervals over its length.

A WASTE REDUCTION PROGRAM AND ASSESSMENT

OF CURRENT STATUS FOR ILLINOIS

David L. Thomas
Daniel D. Kraybill
Gary D. Miller

1. INTRODUCTION

Waste reduction is a national policy in the United States, but it has only very recently gained the support in Congress and in the regulatory agencies to make it a true priority. As Joel Hirschhorn of the Congressional Office of Technology Assessment (OTA) stated at a waste reduction conference in June 1986, "For 20 years we've been saying that waste reduction is a top priority, and we haven't been doing anything about it."(1) Under the Hazardous and Solid Waste Amendments (HSWA) of 1984, Congress declared that it was ". . . the national policy of the United States that, wherever feasible, the generation of hazardous waste is to be reduced or eliminated as expeditiously as possible." The United States Environmental Protection Agency (USEPA)(2) stated that in the broadest sense, HSWA defines waste minimization as any action taken to reduce the volume or toxicity of waste. OTA preferred a more restrictive definition of waste reduction: "in-plant practices that reduce, avoid, or eliminate the generation of hazardous waste so as to reduce risks to health and the environment."(3) Whereas USEPA's waste minimization includes source reduction, treatment and recycling onsite and offsite, OTA considers offsite recycling and treatment as waste management. Although we have generally followed the broader USEPA definition of waste management in this paper, we believe that OTA's distinction is a valid one; it reflects a prioritization of waste reduction first and waste management second.

Geiser(4) stated that 20 to 80 percent of the total hazardous waste streams could be reduced by source (waste) reduction. He went on to state, however, that source reduction is still not fully accepted, either in industrial practice or in public policy debates. "This slowness can generally be attributed to three factors: 1) lack of comprehensive planning to encourage source reduction; 2) lack of institutions to assist industries wanting to treat their toxic by-products; and 3) an absence of capital for process and product changes."(4)

In response to similar concerns about hazardous waste management in Illinois, Governor James R. Thompson and the Illinois legislature created the Hazardous Waste Research and Information Center (HWRIC) in 1984 as part of the state's Chemical Safety Initiative. A brief history of the Center and its programs is described by Thomas, Miller, and Kamin(5). HWRIC was established within the Illinois Department of Energy and Natural Resources (DENR) with a well-defined mission. The new Center would combine research and education; information collection, analysis, and dissemination; and direct technical assistance to industry, agriculture, and communities in a multidisciplinary effort to solve Illinois' hazardous waste problems.

The Center was also charged with specific objectives. Those directly related to waste reduction are:
- Reduce the volume of hazardous wastes generated and the threat they pose to human health and the environment, and
- Help develop and implement a comprehensive hazardous waste management program for Illinois.

This paper describes the basic Center program aimed at reducing the amount of hazardous waste generated in Illinois. The Center's program is meant to supplement and enhance other federal and state waste reduction programs, such as the Illinois Industrial Materials Exchange Service, which is run by the IEPA and the State Chamber of Commerce, and the regulatory waste minimization requirements of both the state and federal governments.

In addition to these ongoing efforts, Illinois industrial data from three sources are also discussed. One source is a survey that asked companies what disposal alternatives they were considering in response to the upcoming (January 1, 1987) restriction in Illinois on landfilling hazardous waste. The second source is information submitted by industries applying for state-sponsored awards, which were given to companies with innovative waste minimization technologies or management strategies. The third source consists of data from waste minimization statements included in the annual hazardous waste generation reports submitted to IEPA for the year 1985.

2. HWRIC'S WASTE REDUCTION PROGRAM

HWRIC, having made waste reduction a priority, is helping Illinois industries reduce the amount of hazardous waste they generate. The following activities are a part of the Center's overall waste reduction program.

- **Provide Technical Assistance** to industry to help them improve general housekeeping; recycle waste when appropriate, either internally or through material exchanges; propose and provide process modifications to reduce hazardous waste generation; detoxify waste; and substitute nonhazardous for hazardous materials.
- **Create a Technology Transfer/Clearinghouse Data Base** on alternative technologies. It will be used by various industries and trade groups to reduce hazardous waste generation.
- **Administer a Matching Grant Program** of up to $100,000 for industry to support modification of existing equipment or processes or to develop new technologies that minimize the generation of hazardous waste.(6)
- **Encourage Waste Reduction** by working through the Governor's office to solicit from industries a description of their waste reduction initiatives. As a part of this effort, the annual Governor's Innovative Waste Reduction Award for industries was initiated in 1986.

. **Conduct Research.** For example, a study was conducted to assess the feasibility of having a central recovery facility for electroplating wastes in the Chicago area. One option from this study that appeared promising was to develop local treatment facilities that would serve a number of similar type generators in an area.(7) Another project is underway to develop in-plant treatment/detruction techniques for certain organic wastes.

Although much waste reduction can be accomplished through better housekeeping, recycling/reuse of materials, product substitution, and process modification, major long-term reductions of waste will often require moderate to major process changes, the use of add-on equipment, and the development of new technology. A long-term research and development program is needed to help industry devise new nonhazardous substitute materials and to evaluate the effectiveness of new equipment or techniques prior to their use. Full implementation of these activities in Illinois will be facilitated by the Center's new Hazardous Materials Laboratory, which is now in the design phase and scheduled to be operational in 1989.

3. HWRIC'S VERY SMALL QUANTITY GENERATOR PROGRAM

HWRIC considers working with very small quantity generators (those that generally produce less than 100 kg of hazardous waste a month) an important component of its waste reduction plan for the state. Although under HSWA the USEPA has recently begun regulating Small Quantity Generators (those who produce 100 to 1000 kg./month of hazardous waste), businesses that produce less than 100 kg./month of hazardous waste are still unregulated for the most part. Besides small businesses, unregulated generators include households, farms, some hospitals, many laboratories (research, industrial, and schools) and other institutions that produce small quantities of hazardous waste. To date, much of this material has gone into municipal landfills, into public treatment works or septic tanks, into storm sewers, or has been poured onto the land. These practices have led to ground-water contamination in some areas. At present the scope and magnitude of the potential problems posed by this group's current waste disposal practices are unknown. Working with this group offers an opportunity to teach people to properly handle and dispose of hazardous wastes. It will also ultimately help reduce the amount of hazardous wastes these businesses generate.

4. ALTERNATIVES TO PRESENT DISPOSAL METHODS

The Hazardous Waste Advisory Council was created by the Illinois General Assembly in 1983 to serve as a body that is to review the implementation of the state's hazardous waste program. As part of its mandate to review the state of the art of alternative technologies it developed a questionnaire that was sent to approximately 9600 individuals in 5800 industrial organizations. The questionnaire was distributed by the Illinois Manufacturers' Association and the Illinois State Chamber of Commerce, and asked questions about waste streams, present disposal methods, and any alternatives being considered to these disposal methods. The objective was to determine what alternatives to landfilling companies were considering in light of the January 1, 1987 Illinois restriction on land disposal of hazardous waste. The returns were received and analyzed by HWRIC and included in the Annual Report of the Illinois Hazardous Waste Advisory Council.(8)

Responses were received from 203 companies (3.5% of those solicited), representing 459 waste streams. However, only the 385 waste streams determined to be hazardous were used for this analysis. Of these waste streams, about 91% were being disposed of offsite. For the 350 waste streams disposed of offsite, 130 (36%) were being incinerated, 97 (28%) landfilled, 81 (25%) recycled and 42 (11%) treated. For waste currently being landfilled, the alternatives being considered included none (62.9%), incineration (16.5%), other treatment (13.4%), recycling (4.1%), and landfilling (3.1%).

A few general statements can be made about the responses reviewed. First, many of the respondents did not understand the questionnaire. Secondly, the large number of respondents who listed "none" under alternatives to landfilling and the even larger number of companies contacted who did not respond at all indicated either that they felt they had no alternative to landfilling or that they had not given much consideration to disposal alternatives. Others apparently were simply reluctant to respond to a government agency or had been saturated with questionnaires and neglected to respond.

Incineration was the most commonly identified viable alternative to landfilling being considered for combustible hazardous wastes. Unfortunately, waste reduction was not indicated as a viable alternative by those companies that responded to the questionaire.

5. PRESENT WASTE REDUCTION EFFORTS IN THE STATE
5.1 Governor's innovative waste reduction award

Industries have been making efforts to reduce their waste and are now required, by both the USEPA and IEPA, to certify that they are doing so. In an effort to learn more about industries' waste reduction accomplishments in the state, to recognize the efforts that have been made, and to encourage other industries to increase their waste reduction activities, HWRIC worked through the Governor's Office to create the Innovative Waste Reduction Award. Industries throughout the state were solicited and asked to apply for the award.

Some 33 applications were received. Other companies called and were interested in applying, but either did not have enough time to apply or could not get approval through their management. It was clear that with a greater lead time, many more companies would have responded.

Company submittals were looked at for a number of different waste reduction and waste management strategies. These strategies and the number of companies employing them include: source segregation or separation (1), process modification (6), chemical substitution or elimination (8), material recovery and recycling (15), treatment (17), material exchange (2), replacement of old and/or installation of modern process equipment (4), and management strategies (17). The latter category includes a number of the processes above, but more specifically refers to corporate plans for waste reduction or introduction of nontoxic processes. Many of the applicants had used two to five of the above strategies. By their estimates, they had reduced their hazardous waste production from between 32 and 100%. The primary wastes being treated were solvents and degreasers, and heavy metals.

Although there are those who do not consider waste treatment a waste reduction technique, this was the most common method mentioned. It generally included concentration or detoxification treatment by the generator or a disposer (either at the generator site or the disposal site). Material recovery and recycling was the next most commonly used method for waste reduction. In some cases this was combined with process modification and treatment; for example, where a company used one waste to treat another waste. Process modification (change in equipment or the way chemicals are handled) and chemical substitution or elimination were the next most frequent methods of reducing hazardous waste. In the latter case some companies were able to replace a hazardous material with one that did not produce a hazardous waste.

5.2 Illinois generators' waste minimization statements

In the State of Illinois, generators of hazardous waste are required to submit annual reports to IEPA summarizing their hazardous waste activities. This report must include basic information on the company, the amounts and types of waste generated, and the ultimate destination and disposal method for the wastes. Reports are generally submitted early in the year for wastes generated during the previous calendar year.

Beginning in 1986 (when data for calendar year 1985 were submitted), large quantity generators (greater than 1000 kg/month) were required to submit statements describing what steps they had taken to reduce the amount of hazardous waste they generated.

We decided to extract data from a statistically significant number of annual reports. Of approximately 1300 generators that were required to submit waste minimization statements, 21% or 275 were chosen at random by computer. Although most of the information was on IEPA computer data tapes each report had to be examined individually to obtain all the desired information. Each waste minimization statement was then placed into one or more of the eight listed catagories (see Table 1).

The data were examined in several ways, as presented in Table 2. Additionally, several predominant waste streams and classes of industries were separated out and examined individually.

Fifty-nine percent of the generators examined responded with a waste minimization statement. Of these, 6.5% simply repeated the wording in the regulation and provided no documentation of their waste reduction efforts. Of the 52.5% that did provide waste minimization statements, most provided a brief explanation (less than one page). This was not

TABLE 1: WASTE REDUCTION AND WASTE MANAGEMENT STRATEGIES.

1. Source Segregation/Sep - Separation of hazardous and non-hazardous constituents of a mixed waste stream into separate streams.

2. Process Modifications - A change in a process or a change in operational parameters of an existing process intended to reduce generation of hazardous waste.

3. Raw Material Substitution - Substitution of ingredients for the purpose of reducing hazardous waste generation.

4. Material Recovery & Recycling - Recovery or recycling of a material for reuse in-house or by others, on site or off site.

5. Material Exchange - Exchange of wastes with another company or individual for their use generally as a raw material.

6. Treatment - Treatment of a waste to either eliminate the exhibited hazardous characteristic or separate and concentrate the hazardous constituent in the waste, on site or off site.

7. New Process Equipment - Purchase of new equipment with at least the partial intention of reducing waste generation.

8. Corporate Strategies - Statements of company policy, descriptions of training plans for employees, descriptions of possible future actions, etc.

TABLE 2: RESULTS OF 275 GENERATOR REPORTS SUBMITTED TO IEPA FOR CALENDAR YEAR 1985.

Survey Sample	Number	Responses	Waste Reduction and Management Strategies (see Table 1)							
			1	2	3	4	5	6	7	8
Total	275	144	7	37	22	59	2	12	20	53
D001*	113	59	5	14	8	32	1	3	4	27
D002*	59	41	2	9	7	15	1	8	8	16
F001*	61	30	2	10	8	12	0	4	5	10
SIC 28**	25	13	1	6	2	8	0	1	3	2
SIC 34**	26	14	1	6	0	7	0	3	6	3
SIC 49**	50	28	2	9	6	14	1	2	5	7

* - Hazardous wastes are specified by a four character numbering system in the United States. Examples are as follows:

D001- Ignitable wastes; Flash point less than 140 Deg. F

D002- Corrosive wastes; pH < 2.0 or > 12.5

F001- Spent halogenated solvents

** - Industries in the United States are given a Standard Industrial Classification (SIC) code for accounting, taxation, and other purposes. Examples are as follows:

SIC 28- Chemicals and Allied Products

SIC 34- Fabricated Metal Products, Except Machinery and Transportation Equipment

SIC 49- Electric, Gas, and Sanitary Services

necessarily undesireable, however, since a simple explanation of actions taken (e.g. "We purchased and are using a solvent still") was often sufficient.

Some companies gave quantified results on waste reduction, but these numbers should be used with caution, as they may reflect changes in production levels or some other effect, rather than true waste reduction. Of the 144 companies that responded (out of 275 generator reports examined), 41% used material recovery and recycling, 37% corporate strategies, 26% process modification, 15% raw material substitution, and 14% new process equipment. A similar group of strategies was used by generators for specific waste streams and for a given group of industries (Table 2). Because an individual company used anywhere from one to five of these strategies for their waste streams, the percentages total more than 100% for any given category.

Extensive details and documentation of a particular company's waste reduction efforts are not usually present in these reports, although there are some exceptions. Since the material in these reports is open for public inspection, inclusion of specific details of a waste reduction plan could provide an advantage to any competitor who chose to examine these files.

Waste minimization statements included with the annual reports must also be viewed in the context of the entire annual report in order to be properly understood. They often refer to data that are in other sections of the report or to the previous years report.

Usually, these type of data will yield general information on large numbers of companies. More meaningful and reliable data will only become available in the next several years as waste reduction programs are more fully implemented by the regulatory agencies and the private sector.

6. DISCUSSION

USEPA (2) listed a number of waste minimization techniques:
1. Recycling (Onsite/Offsite)
    a. Use/Reuse
    b. Reclaim
2. Source Reduction
    a. Product Substitution
    b. Source Control
        -- Good housekeeping practices
        -- Input material modification
        -- Technology modification
3. Treatment

OTA (3) concluded that there were five approaches industry could take to reduce hazardous waste (OTA's definition of waste reduction includes only in-plant practices):
1. Change the raw materials of production,
2. Change production technology and equipment,
3. Improve production operations and procedures,
4. Recycle waste within the plant, and
5. Redesign or reformulate endproducts.

USEPA(2) stated that the potential for future reduction of waste appears to be significant, and that within the next 25 years, aggregate waste generation volume could be reduced an additional 15 to 30 percent by the extension of existing source control techniques. OTA (3) cited a Congressional Budget Office study that indicated a total of 18 percent RCRA hazardous waste reduction nationwide might occur between 1983-1990. However, OTA stated that there are few or no data on the extent of industrial waste reduction, and that it would be more useful to focus on a waste reduction goal, such as 10 percent annually.

OTA(3) concluded that state programs will need to focus their activities on waste reduction if it is to become a significant factor in environmental protection at the state level and if these activities are to be effective in preventing pollution. USEPA(2) stated that they saw their primary role as supporting and encouraging the states in developing their programs. HWRIC has recognized that one of the best places to attack the state's hazardous waste problem is at the source. It is clear that at both the federal and state levels, waste reduction will be given greater priority.

HWRIC is working not only with the regulated generators of hazardous waste, but with very small (unregulated) generators such as households, farms and school laboratories. It is with this latter group that education and information dissemination can have the greatest impact in waste reduction. However, even among the regulated generators we have found a definite need for information and technology transfer.

We conclude that technologies are now available for industries to make substantial reductions in their generation of hazardous waste. Most of the companies that applied for the Governor's Innovative Waste Reduction Award were able to achieve over 50 percent waste reduction or waste minimization through conventional treatment, recycling and reuse, process modification and chemical substitution or elimination. However, it appears that over 50 percent of the generators in the state have not yet begun serious waste reduction and waste management efforts.

Although it is clear at least some generators have reduced a considerable amount of waste, it is not possible to determine what percent reduction is being achieved. This is because of the rather poor response given to our voluntary survey, to unenforced waste reduction requirements of regulatory agencies, and to complicating economic factors. Only by monitoring waste production over several years and by accounting for changes in the associated industrial processes can a more accurate measure of waste reduction be made.

As companies look more closely at the life cycle of chemicals in their plants, including a mass balance or "cradle-to-grave" analysis of the fate of materials, the processes they are using, and the products they are producing, waste reduction methods will be employed more frequently and at an economic savings to the company. This concept is, perhaps, becoming clearer to many companies as they realize the true long-term costs of disposing of their hazardous waste. The potential long term liability and cost for cleanup of land-disposed hazardous wastes are driving, and will continue to drive, many companies to reduce their waste and to find alternatives to land disposal. Information dissemination and education can help speed this process, and they should be a priority of state and federal governments, trade groups and others serving industry.

REFERENCES

1. Aspen System Corporation (Publisher), "Waste Reduction Conference Focuses on Financial, Regulatory Incentives," Hazardous Waste Report, Rockville, NY, 1986, 8-9.

2. USEPA, Minimization of Hazardous Waste, Executive Summary and Fact Sheet, Report to Congress, Washington, D.C. 1986.

3. OTA, Serious Reduction of Hazardous Waste for Pollution Prevention and Industrial Efficiency. Congress of the United States, Office of Technology Assessment, Washington, DC, 1986, 254.

4. Geiser, K., "Source Reduction" in J. Tryens, ed. The Toxics Crisis: What the State Should Do, Conference on Alternative and Local Policies, Washington, DC, 1983.

5. Thomas, D.L., G.D. Miller and J.M. Kamin, Hazardous Waste Management: The Illinois Approach. Chapter 14 in K.B. Levitan, ed., Government Infostructures, Greenwood Press, Westport, CT, 1986, (in press).

6. HWRIC, Annual Report, May 1, 1985 - April 30, 1986, Hazardous Waste Research and Information Center, Department of Energy and Natural Resources (HWRIC 86-008), 1986, 106.

7. Huff, J.E. and L.L. Huff, Feasibility of a Central Recovery Facility for the Metal Finishing Industry in Cook County, ENR Contract No. WR5, 1986, 109.

8. Hazardous Waste Advisory Council, Annual Report, A Report to Governor James R. Thompson and the 84th General Assembly, 1986, 44.

TOWARDS A NETWORK FOR ENVIRONMENTAL TRANSFER

FRANK JOYCE

INTRODUCTION

This short paper presents information on current thinking with regard to a proposed Network for Environmental Technology Transfer. This initiative is being promoted by the Commission of the European Communities within the framework of European Year of the Environment.

ECOTEC Research and Consulting Ltd have been involved with the feasibility and specification work and are currently acting for the CEC to examine the most appropriate mechanisms for its implementation.

NETT : AN INITIATIVE FOR THE EUROPEAN YEAR OF THE ENVIRONMENT

What is NETT ?

NETT is a proposed European Network for Environmental Technology Transfer. The aim of establishing such a network is to help the exchange of know-how between companies and organisations in the rapidly developing field of clean and low-waste technology. It is also concerned with information on more efficient and cost-effective pollution control technologies as a whole. It intends to adopt a "campaigning" approach to stimulating wider adoption of such technologies by firms, and modelled to some extent on the Energy Demonstration Programmes which can be found in the many members states of the European Community. Manufacturers of equipment will be given better information on the markets available for their techologies and new markets which will emerge as the European Community (E.C.) moves to adopt high environmental protection standards as envisaged by the new 'Single Act'.

The Network also intends to bridge the gap between the technological and management or planning aspects of the adoption of clean technologies. This will be done by providing information and advice on programmes which support R&D on the one hand, and on the financial and technical assistance available from the Commission of the EC, national government agencies, and international research and trade associations involved in encouraging this effort, on the other.

Why NETT ?

Within the European Community, there is a growing interest in the promotion of clean and low-waste technologies. The Fourth Environmental Action Programme of the EC is unequivocal in its commitment in this field. Already the Commission of the EC under the ACE Regulation, and the member states through national environmental programmes, implement policies which aim at stimulating the development of clean technologies through financial assistance for R&D and demonstration projects.

The environmental and economic benefits derived from the application of clean technologies have been amply demonstrated through various studies. However, these very studies also reveal that up to now the adoption of such technologies has been rather limited. Industry at large, and small and medium-sized firms in particular are unaware of the recent developments. This lack of awareness also extends to the regulatory agencies concerned. Clearly there is a vital need to bridge this information gap and NETT sets out to do so.

Barriers to clean technologies

In the course of examining the reasons why the adoption of clean and low-waste technologies has not taken off so far, recent OECD work and a study undertaken by ECOTEC have revealed the following barriers in the industrial milieu:

- low level of environmental awareness and training in pollution control; result: a feeling that environmental regulations are an imposed burden;

- unwillingness to consider themselves as waste managers as well as producers;

- a general reluctance to alter production processes and introduce new technologies

- lack of confidence that the application of such technology at their plant will prove technically and economically feasible;

- lack of the technical expertise needed for successful implementation.

- lack of independent advice; result; overdependence on the suppliers of pollution control equipment for such advice;

- lack of access to financial resources to support necessary capital investment to install clean technologies.

These barriers indicate that the efforts to stimulate the development of clean and low-waste technologies by publicly funded or subsidised Research, Development and Demonstration are not enough. The output from such programmes needs to be widely diffused to concerned sectors.

In taking the results of the laboratory to the market place as it were, or in other words, to help convert the results of R&D into marketable products and processes, the EC Commission has a vital role to play. Its success in this field will help create a much needed European technological community.

Another dimension to this whole question relates to market conditions in the EC. Policies concerning the promotion of the internal market form the crux of the issue since it is only at the scale of the European Community as a whole that the market is large enough to make many of these technologies and their products viable. This holds good for pollution control equipment in general and cleaner technologies in particular as the market is strongly influenced by national or regional environmental standards and policies.

The harmonisation of standards and policies among member states cannot be stressed enough here. It holds the key to the development of a dynamic and competitive European pollution control equipment industry.

All this points to the need for a comprehensive programme to encourage stricter standards, R&D promotion, support for demonstration projects focussed upon target segments of an industry, subsidies for investment in clean technologies where necessary stimulation of the pollution control industry, advice and technical support to build up a sound infrastructure at the local level, and most of all, enhanced environmental awareness on the part of industry.

How will NETT work ?

MEMBERSHIP

Membership of NETT is open to a wide variety of organisations such as manufacturers of pollution control equipment, university research departments, consultants, government research laboratories, industrial research associations, users of pollution control products and regulatory authorities such as local and central government organisations.

What services will NETT provide to its members ?

For Manufacturers

* It is intended that NETT campaigns will stimulate awareness amongst industry of the need for and benefits of a range of environmental technologies by means of targeted campaigns at particular industrial sectors

* Provide a mechanism for disseminating information on availability of suitable technologies

* Closer liaison between manufacturers and "users" across the European Community

* Provide market information for suppliers of pollution control equipment.

For Regulatory Authorities.

* At the national and international level it will provide a mechanism for assisting the implementation of environmental protection by organising campaigns to encourage take-up of cleaner technologies, explaining the implication of new regulations to industry and providing feedback on the practicability of new measures

* At the more local level the Network will provide information for regulatory authorities on the most cost-effective pollution control technologies to deal with problems which may arise within their areas of jurisdiction.

For Universities, Research Laboratories and Consultants.

* The Network will provide the opportunity for specialisms to be deployed more widely across the European Community. It will provide a mechanism for improved contacts between the 'research community' and industry, to target R&D resources or specific products for which there is an identifiable market need

* It will offer opportunities for the specialist expertise of consultants, universities and regulatory agencies to be made available to manufacturers and in particular knowledge of local market conditions within the European Community to be made more widely available

* It will act as a focal point for establishing suitable reference plants for new products which may involve joint financing with potential users or regulatory agencies

* It will provide the opportunity to produce at the request of the manufacturer, independent performance monitoring reports to an agreed protocol

* It will assist in a better understanding of the market demand for new products by providing information on 'up coming' regulations affecting particular markets

* Better understanding of R&D priorities in relation to newly emerging demands.

For users of Pollution Control Equipment:

* The Network will provide advice to firms on the most appropriate and cost-effective control measures

* Provision of up to date information on new/emerging cleaner technologies

* It will provide independent validated performance monitoring reports on specific technologies.

How will NETT operate in practice ?

Administration

It is not intended to duplicate the efforts of organisations who are already providing information or advice in this area but rather to invite them to join together to form NETT. In administrative terms it is proposed that national and/or regional groups of organisations are formed which in turn will "federate" to elect an international management board which will be served by a small secretariat. It is envisaged that the legal status will be that of a 'non profit' company. Members will be required to pay subscriptions according to the size of organisation to the national organisations which in turn will provide a 'levy' for the international organisation.

Method of Working

It is envisaged that the NETT will operate in 2 modes - responsive and pro-active.

Responsive

In this mode NETT will provide a mechanism for rapid information exchange between members. A client organisation in one member country may have a 'problem' which requires advice. First point of contact could be a local member of the NETWORK or the national organisation who would suggest a local member based on the character of the inquiry. NETT members would have access to a data base of members skills/ interests/products which he could use to help find the most appropriate solution to the problem.

Pro-active Mode

In many ways the Responsive mode is simply a development of existing practice but the pro-active approach proposed is entirely new to this area although it is being adopted in some member states with regard to promoting energy saving schemes. Working to programmes collectively agreed, NETT will undertake regional campaigns in industrial sectors (e.g. textiles, metal plating) where technologies are available which are both more efficient in environmental terms and more cost effective in relation to capital/operating expenditures. The campaign will use local media -

newpapers, radio, television, to promote workshops for industrialists, breakfast commitment meetings etc. bringing together manufacturers, independent consultants and regulatory agencies together with potential users. Each year two or three sectors will be the subject of the campaign through the E.C. The national/regional organisations may also mount special campaigns relating to local problems but calling upon the expertise of NETT as a whole.

How can you participate ?

It is intended that the NETT should be launched in the autumn of 1987 in the framework of European Year of the Environment. In the meantime national interest groups are being established which will help develop the concept for each national/regional organisation and participate in the establishment of the international organisation. If you wish to know more, you can contact:

ECOTEC Research and Consulting Ltd
5 Rue de la Science
1040 BRUSSELS

or Priory House
18 Steelhouse Lane
BIRMINGHAM
B4 6BJ

More information will be available through the EC stands at the following trade fairs throughout 1987.

RESEARCH AND DEVELOPMENT OF ENVIRONMENTAL BIOTECHNOLOGY IN THE NETHERLANDS

K.Visscher
National Institute of Public Health and Environmental Hygiene

W.H.Rulkens
Netherlands Organization for Applied Scientific Research TNO

1. INTRODUCTION

Biotechnology can play an important role in the development of environmental technology.
For over a century, micro-organisms have been used on an industrial scale for the purification of waste water. Composting of various sorts of waste has also been in use a long time for processing them into useful products.
These processes are aimed in the first place at preventing uncontrolled release of large quantities of oxygen-bonding substances into the environment.
From fundamental microbiology it is known that a great number of substances foreign to the environment can be biologically degraded. Theoretically this opens the possibility of using biotechnological methods for removing the environmental bottlenecks concerning these substances. In this respect techniques for soil purification, air purification using biological filters, waste water purification, and the processing of waste products can be considered.
Recent developments in these areas in the Netherlands show that the possibilities to apply biotechnology in the control of waste flows and emissions into surface water and air, have greatly increased. This applies to improvements of existing processes as well as to the development of new ones. This insight has led to the stimulation of research at universities and institutes as well as in industry.
The Dutch government is prepared in principe to support this type of environmentally and often economically relevant research. It is therefore important that researchers is told where government priorities lie. For this reason a programme for research and development of environmental biotechnolgy has been worked out.

2. ENVIRONMENTAL BIOTECHNOLOGY

In the context of the above mentioned programme, environmental biotechnology is defined as follows:
"Application of biotechnology or biological systems to abate or prevent pollution of water, air, or soil or problems of waste products."

The following categories can be mentioned in this context:
- Added and process integrated provisions under the control of the environmental polluters.
- Provisions applied by others (e.g. collective purification plants).
- Less polluting apparatus, production processes and products.

Biotechnology is defined as follows:
"The integrated use of biochemistry, molecular genetics, microbiology, and process technology to achieve practical applications of the possibilities of micro-organisms, cell cultures or parts of micro-organisms or cells."

Biotechnology can be of great importance for the abatement or prevention of the pollution of water, air, soil or problems of waste products. An important aspect of this is that in biotechnological processes compounds are broken down into products which are part of natural cycles.
This does not mean that biological solutions should always be given preference in the solution of environmental bottlenecks. A choice has to be made between these and solutions based on physical-chemical processes. Economic as well as environmental aspects will have to be taken into account.

3. STIMULATION OF THE DEVELOPMENT OF ENVIRONMENTAL BIOTECHNOLOGY BY THE DUTCH GOVERNMENT

Biotechnology is of great importance to the Dutch society in general and to Dutch industry in particular. For this reason an Innovation-orientated Biotechnology Research Programma is set up. The main objective of this programme is to establish the link between the scientific researchers and industry (chemical, pharmaceutical, food, agricultural and environmental).
The programme consists of four parts, viz. the application of biotechnology in industry agriculture, environment and public health.
The Environmental Biotechnology sub-programme is associated with the activities concerning the development and demonstration of environmental technology of the Ministry of Housing, Physical Planning and Environment. Biotechnology projects are already an essential part of the environmental technology programme of this Ministry. In its environmental technology activities the Ministry of Housing, Physical Planning, and Environment is advised by the Environment and Industry Commission consisting of representatives from industry, research and government. A number of working parties resides under the Environment and Industry Commission. They fulfil a co-ordinating role between the various areas of interest.
An Environmental Biotechnology Working Party has been installed because of the expectation that biotechnology will be of great importance in the development of environmental technology. The most important task of this working party is the co-ordination of the creation and execution of the environment-orientated sub-programme of the Innovation-orientated Biotechnology Research Programme. The financing of this Environmental Biotechnology Research and Development programme is until now covered mainly by existing financial arrangements.
Apart from the Ministry of Housing, Physical Planning and Environment the following ministries are involved: Economic Affairs, Transport, Agriculture and Fisheries, and Education and Science.

4. ENVIRONMENTAL BIOTECHNOLOGY RESEARCH AND DEVELOPMENT PROGRAMME

The most important reasons for setting up the Environmental Biotechnology Research and Development Programme are:
- To establish priorities for researchers, users, and potential users in the development of environmental biotechnology.

- To have an overview for co-ordinating activities and judging research proposals.
- To identify areas of activity to align the work of others.

The bottleneck approach has been chosen for the design of the programme. The environmental bottleneck of the media (water, soil, air, wastes) have been inventorized and it has been established to what extent the development of a biotechnological solution is desirable.
Besides the possibilities of solving environmental bottlenecks by means of environmental biotechnology, process integrated applications of biotechnology in industry have also been considered. One should think here of biotechnological prosesses instead of chemical processes, process modifications, partial flow purification etc.

The inventorizations and evaluations mentioned are a starting point to selecting research areas where further stimulation is desirable.
The programme shows the present state of the art together with recommendations for research based on this. New developments will, however, influence the underlying ideas which will necessitate a regular adjustment to the programme.
A rough classification of the research areas according to the types of contaminants appears to be the most significant. The mutual coherence of the environmental contaminations in the different environmental media is shown as much as possible. Due to their nature, the environmental media (water, air, soil, waste), still have a strong influence on the classification of the research area. As already mentioned however, a too strict classification has not been used, and in this context consideration has been given primarily to the mutual coherence of the contaminations in the several environmental media caused bij one group of compounds. As far as this has been possible, the connection with process integrated biotechnology has also been taken into account.

In the research and development programme, the following clusters of coherent research areas can be distinguished:
a. Removal of micro-biologically degradable organic carbon compounds in waste water, contaminated air and exhaust gases, and contaminated soil.
b. Removal of P-, N-, and S-compounds from waste water and polluted air (including exhaust gases).
c. Removal of heavy metals from waste water.
d. Treatment of wastes and recovery of raw materials from such wastes.
e. Improvement of existing fundamental biotechnological processes and the development of new fundamental biotechnological processes.

These clusters of trends in research have been subdivided into separately described research and development fields. For each research area, the description includes:
a. The environmental problems and bottlenecks for which a solution has yet to be found. Where necessary the relationship between the contaminants in the several environmental media is also given.
b. The research and development trends. Here the direction in which research has to be focused is given in order to develop new and better remedial and other methods of treatment.
c. The current status of the research and development. An indication is given of the type of research considered to be necessary. The following types of research are distinguished: fundamental research,

exploratory research, technological research and research on a pilot plant or demonstration plant scale. This is carried out within the scope of the knowledge of that particular field already present in the Netherlands or elsewhere.
d. The possibilities of research leading to the development of a succesfull process. In assessing these possibilities a very qualitative and brief consideration has been given to the following aspects:
- The effectiveness and costs of the techniques to be developed.
- The amount of the resulting residual material and the emissions of contaminants to water, soil and air.
- The safety of the process in operation
- The extent of the field of application
- The existing non-biotechnological alternative processes available and the processes now in a development stage.
- The savings in energy and raw materials.
- The Dutch and foreign markets for commercialising the technology to be developed and the associated employment prospects resulting from such development.

The results are summarized in three survey tables (table 1, 2 and 3) which have been compiled on basis of the priority assigned to the research field. This priority is mainly based on the Dutch Indicative Multi-Year Programmes in respect of Water, Air, Waste, Soil, Chemical Wastes, and Environmental Protection.

5. PROGRESS OF THE PROGRAMME

With respect to the greater part of the in the programme mentioned research directions, an encouraging progress is made. Some areas of special interest are reviewed:
- Waste water treatment (aerobic and anaerobic)
  Research on anaerobic wastewater treatment in the Netherlands has led to the development of the UASB (Upflow Anaerobic Sludge Blanket) and the Fluidized-bed reactor. At present the research runs along two lines. One line involves the study of fundamental aspects, aimed to further optimization of process and reactor design. The second line aims to adapt both reactors systems for treatment of complex wastewaters such as sewage. Research on the aerobic wastewater treatment is focused on the application of a sludge - on - carrier system for the treatment of sewage and the biodegradation of xenobiotics in polluted groundwater. Much attention is also paid to the microbiological removal of phosphor, nitrogen and sulfur.
- Biological soil treatment techniques.
  In the last two years there has been an increased interest in the development of biological treatment techniques for contaminated soil. Intensive research is now going on regarding the improvement of the landfarming method and the development of new techniques e.g. in-situ biorestauration and bioreactors.
- Biofilters for air cleaning
  Compostfilters to eliminate the hydrogen sulphide odours from sewage treatment plants are used in the Netherlands since 1978. During the last years a lot of research has been carried out to reduce the surface area and to minimize the pressure drop over these filters. A new reactor concept was developed, existing of a vertical column with several horizontal sections containing carrier-material and a mixture

of micro-organisms. For contaminations which are recalcitrant biodegradable, such as chlorinated hydrocarbons, special micro-organisms are cultivated and inoculated in the compost-layer. At the moment the research is also focused on the abatement of the ammonia emissions in agriculture by means of biofilters.

In the near future more attention will be paid to increase basic know how in this field. With the before mentioned developments bottlenecks were met, which were difficult to solve because of a lack in basic know how. Lack in basic know how is also the reason that desirable applications of environmental biotechnology couldn't realized yet. In this context new research projects will be initiated, e.g. on subjects as: the immobilization of micro-organisms, the significance of the application of protozoa, thermophilic digestion, biodegradation of xenobiotics, leaching of heavy metals from waste and the use of selected micro-organisms.

REFERENCES:

Research and development programme environmental biotechnology (in Dutch)
K.Visscher and W.H.Rulkens
Publicatiereeks milieubeheer 85/28
Ministry of Housing, Physical Planning and Environment

Table 1 First priority research area.

Research area	Type of research			
	Study	Basic and explorative research	Semi-technical and pilot-plant research	Research on demonstration plant or practical plant
Removal of easily microbiologically degradable organic carbon compounds from waste water by aerobic treatment		. Immobilisation of sludge	. Reactor development . Concentration and treatment of sludge	. Demonstration plant
Removal of easily microbiologically degradable organic carbon compounds from waste water by anaerobic treatment		. Immobilisation of sludge . Increasing the settlement properties of the sludge formation of granular sludge . Psychrophilic and thermophilic digestion	. Reactor improvement . Research into the practical applicability of the anaerobic process	. Demonstration plant
Removal of non-easily microbiologically degradable organic carbon compounds from waste water by aerobic and anaerobic treatment	. Feasibility study	. Immobilisation of micro-organisms . Selection of micro-organisms	. Reactor development . Research into the practical applicability of the aerobic and anaerobic process	
Removal of non-easily microbiologically degradable organic carbon compounds from air	. Feasibility study	. Several aspects	. Reactor development and optimalisation	. Demonstration plant
Removal of easily microbiologically bio-degradable organic carbon compounds from soil		. Several aspects	. Reactor development (including landfarming bio-extraction)	
Removal of non-easily microbiologically bio-degradable organic carbon compounds from soil	. Feasibility study	. Several aspects	. Reactor development . Research into the applicability of the processes	
Removal of P-compounds from waste water	. Feasibility study	. Several aspects	. Process optimalization	. Demonstration plant
Removal of N-compounds from waste water and air (inclusive off-gases)		. Sludge on a carrier . Removal of NO_x from air (inclusive off-gases)	. Process optimalization	
Removal of S-compounds from waste water and air (inclusive off-gases)		. Several aspects	. Reactor development	. Demonstration plant
Immobilisation of micro-organisms		. Several fundamental aspects	. Reactor development . Process optimalization	
Micro-biological degradation of toxic organic compounds		. Several fundamental aspects	. Research into the practical applicability of the process concerned	

Table 2. Second priority research area.

Research area	Type of research			
	Study	Basic and explorative research	Semi-technical and pilot plant scale research	Research on demonstration plant or practical scale
Production of SCP (Single Cell Protein) from waste water and wastes consisting of easily micro-biologically degradable organic carbon compounds	. Feasibility study	. Several aspects	. Process development	
Treatment of wastes consisting of easily micro-biologically degradable organic carbon compounds		. Several aspects	. Reactor development . Research into the practical applicability of the processes concerned	. Demonstration plant
Treatment of wastes consisting of non-easily microbiologically degradable and toxic organic carbon compounds	. Feasibility study . Comparison with alternative treatment processes	. Basic research into the applicability of the processes concerned	. Process optimalisation	
Application of continuous and recirculation processes	. Feasibility study into the possibilities for application			
Micro-biological chlorination		. Several fundamental aspects		
Possible microbiological alternatives for chemical processes		. Several fundamental aspects		

Table 3. Third priority research area.

Research area	Type of research			
	Study	Basic and explorative research	Semi-technical and pilot plant scale research	Research on demonstration plant or practical scale
Removal of easily microbiologically degradable compounds from air by aerobic processes			. Reactor development . Process optimalisation	. Demonstration plant
Treatment of wastes consisting of non-toxic, organic carbon compounds which are not easily microbiologically degradable		. Biological disclosure of wastes (enzymatically or microbiologically)	. Process optimalisation . Research into the practical applicability	. Demonstration plant
Simultaneous removal of N- and P-compounds from waste water	. Feasibility study into the possibilities for application		. Process optimalisation	. Demonstration plant
Removal of heavy metals from waste water		. Several aspects	. Process optimalization . Pilot plant investigation . Research into the practical applicability of the processes concerned	. Demonstration plant
Possible biotechnological alternatives for the disclosure of raw materials from agricultural origin		. Several fundamental aspects	. Reactor development . Process optimalization . Research into the practical applicability of the processes concerned	. Demonstration plant

AMERICAN NONGOVERNMENTAL ORGANIZATIONS' ROLE IN PROMOTING
WASTE REDUCTION AND CLEAN TECHNOLOGY TRANSFER

Lawrence R. Martin, Director, Hazardous Waste Minimization
The Institute for Local Self-Reliance, Washington, DC

1. INTRODUCTION

The Institute for Local Self-Reliance (ILSR) is a nonprofit NGO. The Hazardous Waste Minimization Project of ILSR was initiated to further an Institute philosophy that resource use should be optimized, and waste reduced, to maximize value.

In the United States, nongovernmental organizations (NGOs) have played an important role in the introduction and growing support for waste reduction and clean technology. NGOs are traditionally viewed as advocacy organizations which engage in the arenas of culture and/or politics to effect a change in social patterns and cultural institutions. Characteristically, they are seen as altruistic and guided by an ethical motivation. They can be nonprofit or incorporated as a business. Many engage in directing change in political and governing bodies; many others focus on giving assistance to special interests or constituencies such as children. Three kinds of activity are pursued by NGOs - assistance, education and the influencing of policy and decision-making; it is often difficult to discern between them.

Historically, U.S. NGOs have operated to affect change in political philosophy (i.e.,the women's suffrage and equal rights movement, labor unions and civil rights organizations). Increasingly, NGOs are debating the value systems needed to support a comfortable and sustainable standard of living. This is seen as a function much like that of the political parties in Europe.

Clean technology is viewed more as a strategy than it is a development of technology. This is because clean technology is typical of traditional chemical and physical processes. The key to the implementation of waste reduction lies in its economic feasibility. For this reason, "clean technology transfer" is practiced as an education in the economics of resource use efficiency and waste, as well as the transfer of data.

2. WASTE REDUCTION AND CLEAN TECHNOLOGY

One component of the work done by ILSR and other NGOs is defining the concept of waste reduction (All future references to waste reduction include clean technology, recycling and any activity resulting in resource conservation and waste minimization) and showing its viability - essentially proving, packaging and selling this concept to government and industry.

The transferability of waste reduction and clean technology between different Standard Industrial Classifications (SICs), and different companies within the same SICs has been identified as a measure of the viability of the hazardous waste reduction strategy. The contribution of waste reduction to the management capacity of a state has often been criticized as being limited in the area of transferability of existing information. ILSR presents evidence that any limitations are the result of inadequate resolve and education.

2.1 Components of Technology Transfer

The issue of technology transfer can best addressed be if it is understood in terms of the component elements, of which technology, proper, is only one. Another is the application of the technology. Thirdly, the strategy for application of the technology is currently seen as the most critical component of the technology transfer elements. The following assertions are based on critical evaluation of U.S. government reports and the review of industry case studies illustrating industry's implementation of clean technology:

2.1.1 Technology. The great majority of "waste reduction technology" is well-known, widely applied for other purposes and easily adaptable; it is in most cases based on standard chemical and physical principles, and therefore easily transferable.

2.1.2 Application. The application of waste reduction technology is an information-intensive endeavor which is custom developed for each installation. Transferability is not the issue in this regard, rather, it is the translation of technology and strategy which is required for the successful application of waste reduction innovations. Individual facilities must take responsibility to translate available information for process-specific applications.

2.1.3 Strategy. The strategies for applying waste reduction technology are new, but increasingly more widely understood and available for study. They constitute the critical information necessary to implement the waste reduction technologies in individual plant applications. Although the respective application of waste reduction modifications are highly site-specific, the technology and strategies for affecting waste reduction are highly transferable between processes.

2.2 Evaluation of Case Studies for Transferability

In recent years numerous case studies have been compiled which illustrate waste minimization practices. The Institute for Local Self-Reliance has prepared and collected case studies of waste minimization innovations wherever they may be found. To date, we have either collected or prepared nearly 400 specific examples of industry's implementation of modification to affect a reduction in the amount of waste generated. ILSR prepared a random sample of the better documented studies and an index of transferability to assess the extent to which the innovations relied on generally-available technology and information. The sample identified 118 modifications from 86 case studies.

INDEX OF TRANSFERABILITY FOR
13 CLASSIFICATIONS OF WASTE MINIMIZATION

CATEGORIES	PERCENT
1. Requires no unusual technology; use of marketed technology; housekeeping improvements; "common sense" management initiatives to maximize process efficiency, reduce waste and turn wastes to profitable resources	45%
2. Materials substitution: either to detoxify the waste stream or permit improved reclamation of a byproduct (i.e., water for solvent based inks; or the substitution of a resource stream with a suitable waste stream	09%
3. Waste exchange -- recycling a waste stream in some other process (the waste generator end of the relationship, as contrasted with the recipient end cited in "2")	01%
4. Evaporation	01%
5. Ion exchange	02%
6. Distillation	05%
7. Adsorption	03%
8. Phase Separation	02%
9. Filtration	02%
10. Electrolytic metal recovery	06%
11. Waste water treatment -- total for a,b & c	06%
a. Closed Loop	02%
b. Precipitation	02%
c. Waste Stream Separation	02%
12. Incineration of waste with heat recovery	07%
13. New innovative technology and/or patent protected	10%

Source: Institute for Local Self-Reliance

The first column lists 13 waste reduction technologies. The second column describes the percentage of case studies which utilizes the respective category to affect a waste reduction. Given that the first 12 classifications are transferable, and that only #13 presents technology which is actually guarded or patent protected, this sampling shows that the 90% of waste reduction modifications cited in three references (Huisingh, et al[1]; Sarokin, et al[2]; United Nations Environment Programme and Economic Council for Europe[3]) are transferable. This finding substantiates that neither the availability of technology nor the availability of strategy pose significant obstacles to increasing the use of waste reduction technologies and strategies. Therefore, application remains as the last element of transferability and as the sticking point for increasing the rate of waste reduction in industry.

If the information is available, but the translating ability is lacking, the obvious conclusion must be that commitment and/or education are the weak links. ILSR identifies the lack of transferability of technology and strategies as a result of little awareness of opportunity, and lack of initiative by the those responsible to take available information and adapt it to specific applications.

3. NONGOVERNMENTAL ORGANIZATIONS' ROLE IN PROMOTING WASTE REDUCTION

NGOs in the U.S. are an important force in the promotion of what little waste reduction policy and technical assistance that exists on the national, state and local levels. In addition, because there is only beginning to be a governmental source of information on waste reduction (excluding work from scholarly NGOs receiving government support), NGOs have had a leadership role in promoting waste reduction among business and industry. The preceding section is an example of work performed for this purpose by ILSR.

Interestingly, the term "Pollution Prevention Pays" (3P), a catch phrase used to popularize waste reduction in the U.S., was coined by Dr. Joseph Ling, a Vice-President of the 3M Corporation, to name 3M's waste reduction program. In the U.S., the 3M Corporation is used by many advocates as a prime example of successful waste reduction. 3M's 3P program has resulted in annually preventing almost 100,000 tons of air pollutants, 150,000 tons of sludge and solid waste, and 1.5 billion gallons of waste water in the time since the program began in 1975. Cumulative first year savings for 3M from the 3P program were $235 million.

Other industrial giants are also beginning to show equally significant results. In some respects, industry is serving as an example to other industry that waste reduction is technically possible and financially desirable. However, nobody in the U.S. who sees the advantage of actively promoting waste reduction is satisfied that business is embracing waste reduction sufficiently quickly, or that the technology transfer and education to industry is adequate. For that reason numerous NGOs have established a variety of programs to encourage industry to adopt waste reduction; often using the success of other businesses to show its feasibility.

3.1. Types of NGOs.

Though typically thought to be structured as a nonprofit, free association, membership organization, NGOs involved in the promotion of clean technology also include institutions of higher education and businesses. NGOs in the U.S. which have been contributing to the promotion of waste reduction and clean technology may be categorized in several ways.
Because each NGO type has a common interest in wider adoption of clean technology and waste reduction strategies there is significant cooperation and collaboration between them. In some cases the respective roles of each NGO type become blurred due to the importance of integrating all aspects of their work, e.g. an advocacy group concerned with environmental quality may promote waste reduction in an educational program, and as a political strategy, but may also offer consulting services - or even market technology as a means for supporting other activities.

3.1.1. Businesses involved with the promotion of clean technology and waste reduction in the U.S. are capitalizing on a growing recognition that waste reduction is a valuable asset to industry. Businesses seeking to build a market for their services or clean technology products engage in promotion of their own product or service, and most any and all activities to promote the adoption of a waste reduction policy nationally and on the corporate level. At present this is a small, but growing, contribution to the transfer of technology by NGOs.

3.1.2. University/College institutions have established programs designed to build waste reduction expertise, to address difficult-to-reduce regional waste streams and assist regional industry reduce waste, and to provide educational opportunities, such as seminars, to professionals. In some cases state university systems have been made key components in a state's waste reduction technical assistance program, or at least receive financial support to provide research or educational support for the state's program.

3.1.3. Advocacy organizations are defined here most broadly; they may or may not be of nonprofit status, they may advocate a position or advocate on behalf of a constituency, and they run the full range between ad-hoc citizen group, to national coalition, to professional service organization. Many environmental organizations, not all qualifying as tax-exempt and nonprofit, include waste reduction as one of many issues they support. Some groups have come to advocate waste reduction very strongly, others only endorse the concept without actively advocating it; often they will continue to advocate pollution control positions. In this respect, some NGOs, like industry and government in the U.S., remain firmly tied to the control of pollution rather than seeking to promote a pre-pollution reduction of waste and adoption of clean technology.

Organizations which advocate waste reduction on behalf of their constituency have been of lesser importance in this regard, however, their role in educating the block of professionals or special interests they represent, such as environmental engineers or chemical workers, is very important. In general, the advocacy organizations codify existing research, and promote waste reduction through social or political channels to institutionalize it as policy.

3.1.4. Nonprofit educational & foundations have performed a crucial service in initially arousing interest around the issue of waste reduction. The nonprofit research and educational NGOs have performed some of the most important work in proving the value of waste reduction in the U.S. It can be said that their pioneering work illustrates the economic and environmental feasibility of waste reduction, and has been instrumental to the growing support of waste reduction by U.S. government and industry. These groups continue to function as supply-houses supporting the adoption of waste reduction. They engage in active educational outreach, organizing and the provision of information to groups seeking to promote waste reduction. Foundations are philanthropic organizations, they have a combined annual budget of $4.3 billion. The money is granted as support to organizations and individuals to perform work judged to be in the public interest, or to suit the respective foundation's requirements, of which environmental issues are only one of many; waste reduction receives only a fraction of this support. Much of the research and advocacy which has occurred on the part of promoting waste reduction has been funded by foundations.

3.2. NGO's Activities.
Generally, NGOs promoting clean technology and waste reduction in the U.S. are involved with one or more of the following activities:
3.2.1. Providing educational material to an identified audience on the opportunities, wisdom or general necessity of clean technology. Audiences include educators, community groups, legislators, the general public, the business community, government environmental regulators and professionals. Educational material may include technical assistance.
3.2.2. Actively seeking to influence policy within the organizations of "audiences" listed in #1 above. This activity is often indistinguishable from providing educational material. However U.S. law regulates the influencing of political bodies by organizations which are tax-exempt and nonprofit. For this reason "educating" and "influencing" are not considered the same even though influencing certainly entails educating.
3.2.3. Offering services to one or more of the referenced audiences for compensation or profit. Increasingly in the U.S., small consulting firms are founded for the purpose of assisting industry to develop an awareness of opportunity for the application of clean technology and waste reduction strategies. Larger established firms are also adopting clean technologies for various applications. Many pre-fabricated pieces of equipment or patented clean technologies are imported to the U.S. from Europe.

3.3 Importance of Nongovernmental Organization Activities
It is postulated here that without the activities of the NGOs very little activity would exist to promote the adoption of waste reduction in the U.S. The publishing, educationing and organizing activities of NGOs were largely responsible for the U.S. Congress requesting a report from the U.S. Environmental Protection Agency (EPA) assessing the feasibility of waste reduction. This 1986 EPA report called largely for more research, for more information gathering from waste generators and for more technical assistance to states. For the most part it assigns primary responsibility to the states. It does not provide any structure to perform the work, nor does it specify any budget. For these reasons it is seen by critics to not provide a waste reduction initiative.

Given this federal response, it is prevalent on NGOs to spur federal and state policy to develop waste reduction initiatives. States generally follow the national lead on "environmental issues", but on this issue they have been more aggressive in exploring options. Several states have already established waste reduction agencies, though budgets are restrained and staffs small. In most cases, NGOs which a clean environment were leading proponents in establishing waste reduction as a cost-effective solution to pollution. In states where no waste reduction initiative exists, local and national NGOs are working to encourage their adoption.

3.4 Coordination and Cooperation Between NGOs and Government

Government programs in waste management are primarily disciplinary and burdensome to industry in the U.S. Unlike Europe, little cooperation in the planning of national environmental quality laws occurs; and government does little to assist industry in the development of clean technology. Government programs focus instead on regulating emissions and pollution control. NGOs seek to facilitate non-adversarial problem-solving between government, industry and the public, and to work with all public sectors to encourage the waste reduction problem-solving approach to preventing pollution.
Several state governments have embraced waste reduction and actively fund its development among industries in the state. Many NGOs work closely with state environmental agencies to advise on strategy, and in some cases have received grants or contracts to research issues pertaining to the development of waste reduction expertise within the state.

In the U.S., lawmakers and their staffs rely heavily on information provided by special interest groups. It is said that if special interests do not promote an issue it is not sufficiently important to consider. Some environmental advocacy groups are well-known by law-making bodies as "the environmental leaders", and have been successful in directing blocks of votes to elected government representatives. Because waste reduction is seen by most as an environmental issue, NGOs' positions on waste reduction are therefore crucial in passing laws to promote it.

4. EXAMPLES OF NGO ACTIVITIES IN PROMOTING WASTE REDUCTION

As was noted earlier, NGOs tend to perform multiple functions, performing education, advocacy, assistance and theory development as an interrelated web of activity.

4.1 The Institute for Local Self-Reliance

ILSR premises that cities could be made considerably more self-reliant, and that both economics and environmentalism should be based on clean and non-exploitive production. The relationship of these two premises is the central political philosophy guiding ILSR's work. ILSR works with governments, other NGOs and industry to provide strategic and technical assistance in the transfer of waste reduction strategies and clean technology. ILSR sees cities as a wealth of opportunity for the development of clean technologies, and for the design of relationships which optimize the use and reuse of local resources. Cities have large internal economies and political authority; therefore the rules developed in cities to govern resource development can be generally applied to the country as a whole. ILSR develops policies and strategies for the introduction of change in socioeconomic and political structures. Promoting waste reduction and clean technology is a key element in strategy because prevailing technologies are a significant force in the development of social structures and cultural relationships.

ILSR seeks to build cooperative relationships with other NGOs, industry and city officials to build local expertise and structures which will enable regions and industry to implement waste reduction and develop waste reduction skills. Our work most recently has been to develop awareness of waste reduction throughout ten midwestern states. Our objective is to form the Midwestern Industrial Waste Reduction Council. The Council will be comprised of government agencies, industrial associations, industries and NGOs. It will operate to share waste reduction information between affiliates, to enhance the visibility of waste reduction concepts to the public and media and to generally provide a clearinghouse on clean technology and waste reduction-related information.

ILSR recognizes the constraints presented by national laws on our regional work. Consequently, ILSR established the U.S. Congressional Working Group on Waste Reduction to examine the waste issues facing the U.S. and the importance of a policy on clean technology. Waste reduction legislation is presently being drafted.

To further support our educational work ILSR researched and published a book of case studies, Proven Profit from Pollution Prevention; the 46 case studies site the technical and economic feasibility for waste reduction innovations and index the studies by industry type, waste reduction strategy and material. The American Management Association, another NGO, has summarized the book, titling it Towards Pollution Free Manufacturing, and made it available to 16,000 of its members. ILSR staff and associates have sponsored or made presentations on waste reduction strategy at over 250 conferences and workshops. ILSR maintains a data base of clean technologies and now totaling nearly 400 examples of waste reduction. ILSR is funded by the Joyce Foundation of Chicago, Illinois.

4.2 The Madison, Wisconsin Audubon Society

This NGO presented a State conference on waste reduction where local industries which had implemented waste reduction received awards from the Governor, and presentations were made on how to implement waste reduction strategies. Planning for the conference included arranging for the co-sponsorship of industrial associations, and the participation of the State government which assured a diverse and large attendance. NGOs in other states have also used this strategy to promote the adoption of waste reduction within their regions. Among them, the Ohio Public Interest Campaign presented the "Governor's Awards" in Ohio; and in Indiana, the People Against Hazardous Landfill Sites and ILSR co-sponsored a waste reduction seminar focusing on the steel industry.

4.3 The League of Women Voters

Among many activities of the League, which is a well-known national political/advocacy organization addressing an array of public issues, they have co-sponsored with Tufts University and the U.S. EPA two conferences bringing together industry,

in the U.S. have participated in these meetings discussing corporate waste reduction programs, national priorities for waste reduction programs, barriers and incentives to waste reduction and waste-reducing strategies.

4.4 INFORM

INFORM studied in detail the waste reduction programs and projects of 29 organic chemical plants in three states. Their findings, entitled Cutting Chemical Wastes, have been widely accepted as an important contribution in determining the information requirements needed to assess the feasibility and implementation of waste reduction. They found:
   Despite the extensive interest and verbal support industry
   and government have given to the concept of waste
   reduction, ...such initiatives still (are) all too rare.
   ... the information on chemical use and discharges of
   wastes to air, land and water was fragmented and piecemeal
   at best.
INFORM's report has been widely referenced by government in support of a position that to tap the potential of waste reduction in the organic chemicals industry requires two "transformations": "changing industries waste-making habits into concrete waste reduction measures; and ...codifying and expanding the fragmented information on total uses and discharges of chemicals, and on those waste reduction practices that are in place."

4.5 National Campaign Against Toxic Hazards

The National Campaign works with a network of community organizers and activists across the U.S. called Citizen Action (CA), and represents CA positions on toxics nationally. State CA groups advocate progressive positions on a wide range of local and national issues, including toxic waste reduction, and operates door-to-door canvases. In addition to general education, conscious raising and advocacy, the National Campaign has prepared model state legislation for introduction by CA groups and others promoting waste reduction. The legislation is comprehensive and professionally prepared.

4.6 Local Government Commission (LGC) in California

The LGC is an association of progressive California local government officials which promotes a wide range of progressive issues in local government. In addition to supporting statements of official policy promoting the goal of waste/source reduction, such has been adopted in Minnesota, Washington, Texas, North Carolina, Michigan and Louisiana, they promote the adoption of local ordinances designed to actively encourage waste reduction by local industries. A model ordinance based on legislation adopted in Contra Costa and Santa Cruz Counties is made available.

4.7 Greenpeace

In addition to a full slate of environmental issues, Greenpeace USA and International address waste reduction as the promise for a pollution-free world. Greenpeace will cooperate with industry and governments, but seeks to do so from a position of strength. To that end, Greenpeace actively engages in non-violent direct action against polluters by publicly embarrassing them with displays such as banners from smokestacks, blocking up wastewater outfalls, and delivering regulated waste streams in 55 gallon drums to the offices of government and corporate leaders.

4.8 The Center For Environmental Management, Tufts University

The Center prides itself on the capacity to conduct research, analyze policy and develop educational programs to address a broad range of issues, and the needs of society, industry and government. In waste reduction, the Center concentrates on improving strategies for waste reduction, analysis of waste classification schemes, techniques for information exchange and facility siting and other management-oriented issues. The Center works closely with the League of Women Voters (see 4.3) to present the inter-industrial conferences on waste reduction and to publish the proceedings of these meetings. It also publishes the journal Hazardous Waste. Other colleges and university systems also perform similar work, such as the University of Pittsburgh and the University of Tennessee.

4.9 Piedmont Waste Exchange, UNCC Urban Institute

The Piedmont Waste Exchange (PWE) is located at the University of North Carolina, Charlotte, Urban Institute and is a nonprofit NGO. One of many regional waste exchanges, the PWE identifies and brings together industrial waste generators, potential waste users, companies seeking waste management services and companies which can provide such services. Increasingly, PWE makes available information on waste reduction provided through North Carolina's excellent state run program on waste reduction.

4.10 Government Institutes, Inc.

This organization is a for-profit NGO which arranges information into conference formats to present to industrial clientele. Largely focusing on environmental regulations of consequence to industry, Government Institutes, like other similar NGOs, present conferences around the country to familiarize their audiences with waste reduction issues, strategies and resources. They have also edited a publication on waste reduction pulling from the research done by NGOs such as ILSR and INFORM.

REFERENCES

1. Huisingh,D., L. Martin, H. Hilger, N. Seldman (1986) Proven Profits From Pollution Prevention, Institute for Local Self-Reliance, Washington, D.C.
2. Sarokin,D., W. Muir, C. Miller, S. Serber (1986) Cutting Chemical Wastes, INFORM, New York, NY
3. United Nations Economic Commission for Europe (1981) Compendium on Low- and Non-Waste Technology

THE WASTE WATER SANITATION PROGRAM OF GIST-BROCADES N.V. LOCATION DELFT

P.A. LOURENS MI Chem E, Coördinator Waste Water Affairs, Gist-brocades N.V.
P.O. Box 1, 2600 MA DELFT, The Netherlands.

SYNOPSIS
Gist-brocades is a major international biotechnology company with its main production facilities in Delft.
Waste water from the various fermentation and chemical production processes is characterized by its high concentration and frequent fluctuation of pollution components. The initial pollution load (1 P.E. = 136 grams of COD) reaches about 0,9 million PE.
Already in the early 70's an ambitious sanitation program has been formulated, which now is nearly completed. In a 15-years period the pollution load is to be reduced to circa 70.000 PE.
This paper will pay attention to the basic strategy, the internal organisation and the external relation with the authorities. The main realised projects to reduce pollution will be discussed; most of them are based on new technologies developed by our R&D-department.
Process-integrated systems for recovery and re-use of byproducts account for about 75% reduction; the remainder is realized in biological (fluidized-bed) anaerobic and aerobic reactor systems.

1. ENVIRONMENTAL PROBLEMS
   Gist-brocades is a major international company with its main production facilities in Delft, and several production locations in Europe and the United States. The main products are yeast and bakery-ingredients, pharmaceutical bulk products like penicillin and cephalosporin, enzymes, pharmaceutical products for human and veterinary use etc.
Fermentation processes i.e. the application of micro-organisms are widely applied. In some cases the micro-organisms are the products (yeast). In other cases micro-organisms produce the desired component (penicillin, enzymes, etc.). Micro-organisms can also convert (cheap) molecules into complex, precious molecules (steroids). Generally more or less complicated recovery and purification processes follow the fermentation process. In Delft the physical-chemical transformation of penicillin into cephalosporin is a major operation.
The biotechnological industry consumes large quantities of agricultural products, like molasses, grain, starch, sugar, proteins etc. as raw materials for the fermentation processes. For the recovery processes several solvents and chemical acids, alkali etc. are indispensable.
The waste water resulting from a complex location like Delft is characterized as extremely varying in flow-rate and composition (BOD, COD, N, sediment, solvents, salts, pH, temperature). The high concentration of biological degradable components is the main point of concern. However heavy metals and "black-list" chemicals are (nearly) absent.

2. THE DUTCH LEGISLATION

The law on pollution of surface waters (W.V.O.) was introduced in 1972. One of the measures to stimulate anti-pollution programs was the so-called discharge tax. The "population-equivalent" PE is based on the average pollution load of an inhabitant. $1 \text{ PE} = \dfrac{Q \cdot (COD + 4,75 N)}{180}$

COD = chemical oxygen demand (mg/l).
N = nitrogen Kjehldahl
In 1986 the formula was changed into: $1 \text{ PE} = \dfrac{Q \cdot (COD + 4,5 N)}{136}$

The discharge tax rose steadily from a few guilders to ƒ 40,-- and more per PE. For Gist-brocades the financial consequenses would be dramatically, in view of the huge pollution loads.
So early in the seventies our sanitation program started to reduce the pollution load. Before giving details on the program, I will make some remarks about the relations with the authorities.
These are very important for the following reasons:
1. The sanitation program for our complicated situation could not make use of standard and simple solutions. New technologies had to be developed and introduced, including marketing for byproducts. These programs are expensive and time consuming.
2. The policy of the national and regional water authorities should be consistent for a long period (10 years or more). Only then the industry may develop optimal solutions. With a variable policy or with crash programs the risk of making wrong choices is large. As an example the per 1-1-1986 modified formula for calculating a PE was mentioned already. The argument that cancelling at the same time the so- called "corrections" nationally had a budget-neutral effect may be true, but the effect for individual industrial dischargers is quite different. Our company was charged with 32% more PE's for the same pollution load.
3. The financing of R&D- and investmentprojects of course is a point of main concern. In The Netherlands there are many stimulation programs and subsidies. At least 3 ministries are involved in environmental programs: Ministries of Housing and the Environment, Economic Affairs, Transport and Public Works. It is a time consuming task to find the optimal way of financing. The advantages of subsidies do not need an explanation; disadvantages may not be forgotten however (for example: rather long period of waiting before decision; obligation to commercialize in the MVT-program).
To conclude this subject, Gist-brocades realized that intensive and open relations with all authorities involved, was and still is of main importance. At different levels, including top management, appropriate contacts on a regular basis take place.

3. THE GIST-BROCADES ORGANIZATION

The steadily increasing pressure to reduce emissions to the environment is not just a Dutch problem. In most industrialized countries where Gist-brocades operates plants, restrictions became sharper and sharper. However, the production departments do not have enough facilities to tackle these problems (waste water, emisions, noise, external safety, soil pollution, etc.). The foundation of a Corporate Staff Department Environmental Affairs, Energy Conservation and Process Automation (in 1982) reflects the need for a coördinated and well-balanced policy on these special subjects. The Staff Department reports directly to the Board of Management, prepares advises, coördinates, stimulates.
For each of the four main fields of interest a coördinator is responsable: waste water, dry environmental affairs, energy conservation and process automation.

Within the divisions and the main locations the environmental engineers are responsable for the implementation in the production departments.

4. BASIC STRATEGY

The following items are always evaluated in the given order, when environmental problems have to be solved:
1. prevention of pollution by process-integrated measures (if possible)
2. purification and re-use of spent materials
3. valorisation of spent-materials to useful by-products
4. added-on equipment to treat and purify emissions.

Examples for category 1. are: efficient dry dust filters in stead of wet air washers, change of toxic benzene into toluene; for category 2.: destillation of solvents. More recently Gist-brocades developed special processes to purify (to a high degree) ammonia, tri-ethylamine and pyridine from very complex effluent streams. For category 3. examples are: concentration in multiple-effect evaporators of yeast-effluent to a protein rich liquor (Vevomix^R) that is used in the cattle-feed industry. Anorganic crystals contain enough potassion to be used as a low grade fertilizer; for category 4.: anaerobic and aerobic treatment of waste waters, sedimentation and surplus-sludge treatment.

The total effect of the projects in the second and third category account for $\pm$ 75% of the PE-elimination! Some by-products have a considerable (market)value. The physical projects operate stable and with high efficiency. It needs however a continuous marketing effort to sell these byproducts in the appropriate markets. Constant quality and attractive price settings are important.

5. THE SANITATION PROGRAMS

As already mentioned Gist-brocades could not use simple and proven technologies. In the next sheet the gross pollution load, expressed in PE's is indicated as for the situation of 1980. The main production processes yeast/alcohol, penicillin and others (mainly chemical conversions) each account for a high percentage. The so-called "source projects" to recover byproducts are indicated. A lot of efforts was required within the company to develop these technologies. Even after performing intensive laboratory and pilot scale experiments, often problems arise in full scale practice. These may be due to a variety of reasons, like corrosion, mechanical damages, unexpected scaling factors, medium or long term deterioration of products or changing marketing demands etc. The R&D departments therefore stay involved with trouble-shooting and optimation programs for several years after the initial start-up.

By far the largest development project is the biological waste water treatment system, based on the principe of "micro-organisms attaching to heavy granules". The principle is that a biofilm of active micro-organisms will develop on the surface of heavy granules (like sand). The micro-organisms then may be kept easily in the reactor without the use of external sedimentation tanks. A high biomass concentration may be obtained, thus high conversion rates and low hydraulic residence time may be realized. The first full scale application was the two stage anaerobic fluid bed project that started up in 1984. In 1985 a duplication was constructed. Now the whole waste water flow ($\pm$ 350 m^3/hr) with $\pm$ 300.000 PE pollution load is treated. Biogas production = 5.000 - 10.000 m^3/day.

A disadvantage of anaerobic treatment is that sulphates are reduced to H_2S and $S^=$. This noxious and bad smelling gas requires accurate safety measures and closed equipment. Our fluid bed system has been designed according to very strict rules.

To increase the effluent quality an aerobic treatment is applied, the so-called Biox-process. In a single reactor, also based on the principle of biomass-attached-to-granules, sulphides are oxidised to sulphate and ammonia to nitrate-iones.

After the 3 sequential biological steps sedimentation of suspended solids is performed. The obtained sludge of ± 10% dry solids will be stabilized and used as a low-grade fertilizer. Experiments are being prepared now.

When all projects operate to their specifications we will eliminate over 90% of the inital pollution load of 900.000 PE and discharge only ± 70.000 PE. It is realistic to say that anaerobic treatment with biogas production and sludge application in agriculture also are examples of useful byproducts!

6. THE COSTS

The total investments in the period of 15-years waste water treatment at the Site Delft are calculated at ca. ƒ 110.000.000. Subsidies and financial stimulation programs (like WIR) accounted for ca. ƒ 40.000.000.

The research expenses are estimated at ca. ƒ 25.000.000 for the 15 years period. Subsidies for R&D are ca. ƒ 5.000.000.

The operational costs for the projects and the discharge-costs for effluent are not included in these figures.

These data indicate that the sanitation program has a heavy impact on the concern activities and financial results. For several years within our R&D- and Engineering Departments environmental affairs formed a relatively large share of their workload.

We hope that 1987 will be the year that the sanitation program can be terminated. We have to prove now on full-scale practice that with the available installations the target (less than 100.000 PE) can be met and kept as low.

ENVIRONMENTAL TECHNOLOGY AND ECOLOGIZATION

H. MARINOV

Technologies' increasing role in society life is a 20th century characteristic feature. It is being explained by the rapid acceleration and the systematic fundamental science achievements in new technical arrangements and technological processes. All contemporary science results have brought to deep changes in production field. There appeared new productions in the industry and others disappeared.

The problem for searching principally new ways for initial raw material processing in order not only 5-6% of it useful substance to be used arose. That is an index for irrationality in using natural resources. Rationality requires consequent and full resource utilization and bringing its wastes to minimum. The idea for resource economical and clean technologies, known in literature as low and non-waste technologies was created in that way.

In connection with that in 1973 Academician M.V.Keldish, opening the Soviet Union Academy of Sciences session on the problems of biosphere rational use and protection, showed the need for rational technological principles to guarantee biosphere maximum preservation and not admitting violation of the necessary balance (1). Academician S.S.Schwartz (1975) formulated the technology essence. "One has to include his production processes in the substance normal rotation and energy in biosphere" (1). That means a gradual transition to closing the cycles to be realized. N.N.Colossovsky's rational idea for power output cycles at contemporary conditions should be further developed in the theory for nature-anthropogenic rotation of substances, energy, factory products and wastes. Changing the linear character of production cycles in a rotation (non-linear) one will guarantee continuity, breaking down and decomposing of final product and reproduction in the system of biotic-economic rotation (2). In connection with that serious discussions about the future of substance energetic rotation arise. N.Reimers and N.Fedorenko think that two rotations will practically exist isolated from nature environment:

1. Natural biogeochemical rotation.
2. Anthropogenic reutilizational substance turnover (3).

According the same authors in the nearest stage of development the necessity for creating a new type of productions, oriented not to a narrow economic specialization but to a wide and ecologic profitableness will appear (3).

At the same time a process of the kind would lead to a rapid product nomenclature extending to dividing of narrow branch point of view, because the production complex will not be able to occupy a boundless big spatial volume. The technological production complex will also seriously be changed as far as it aims will be not only economical but ecologic economical as well. (3).

Undoubtedly the question is about non-waste technological production complexes, which, according to both other authors and us, will allow the co-evolution between the two mentioned kinds of rotation and the realization between production and natural processes to be realized. "Combining between the objects of human economic activity, the inhabited and nature environment in an united system will be done" (H.Marinov), developing according to peculiar laws which are not enough known to us (4).

According to the same authors and the written by us before (5) at the basis of each non-waste technological production complex lies the rotation of substances and energy (Kolosovsky energetic chains). Analogically with the nature biogenic rotation of substances and energy the rotation in technological production complex may be called technogenic and will be different from the self-regulating biogenic one mainly in its conscious forming and regulating by man (4).

Now in geochemical respect the industrial production has the nature of an open linear system. In one of the sides a stream of natural components (ores, coal, petroleum, water, air etc.) enters it, on the other - a two-sided stream goes out:
1. ready production + waste matters (dust, semi-manufactured goods, slags, gas etc.) which comprises 96% of the initial substance raw material mass as an entire stream. The contemporary human civilization in the person of scientific-technical progress could not remain indifferent. That is why the rational and careful attitude to the drawn on a large scale natural resources is being personified by the so called non-waste technologies. They aim at achieving maximum natural substances use as raw materials for different productions. The question is about reducing both the volume of the objects taken off nature and the minimum restoring to the environment of their transformed equivalent.

On the basis of definitions and classifications in the sphere of non-waste technologies one may put those which aim at a considerable decrease of tested natural objects for production needs and diminishing the volume of the generated substances, which dirties the environment.

In order to determine the essence of production influence over the environment, some authors use the term 'technologic wastes' (6). These are labour products, created in the production process, which are not used for satisfying social needs and whose presence in the environment leads to its qualitative characteristics deterioration. According to their substance form, the technologic wastes may be insufficiently used objects of labour, worn out means of labour and substances formed as a result of physical (mechanical) wearing out, corrosion and other means of labour. According to the way of their forming they may appear as production wastes, accompanying products, losses from raw materials, energy, ready production etc.

The production wastes are very undesirable result of production. That is why the main direction for speeding up the scientific-technical progress is connected with the need of diminishing the wastes - the by-products are inevitable from complex raw materials revision.

The consumption wastes are also an inevitable element of society vital activity.

Society exists only for granting of certain needs, i.e. for material welfare use and the objects of consumption that have lost their ability of that quality become wastes of consumption.

The transmission from polluting productions to ecologically clean ones is being realized through not allowing the forming of certain kinds technological wastes and using others. That is being gradually done on the strength of social economical and technical reasons.

But ecologically pure production supposes utilization of technological wastes. In connection with that we have to point out that talking of the concept 'technological wastes' the concept 'non-waste technology' is not very correct. We have to search the main reason in the circumstance that the production is an open system and the exchange of substances and energy between production and environment are unremoved. On principle a complete use of labour objects could be reached but it is impossible for the physical wearing out of labour means to be taken away which is one of the essential sources for forming technological wastes, an inevitable product of interaction between environment and production.

Speaking of non-waste technologies usually they have in mind ecologically pure technologies, i.g. such technologies in which the standards for environment quality are fulfilled. Most of the contemporary technologies are accompanied with a lot of wastes. As far as it is difficult to suppose that the newly projected technologies will be ecologically pure the concept 'technologies with little wastes' is being accepted. The introducing of that concept is not well grounded. First, there is no criterion for carrying away this or that technology to those with little wastes.

Second, the technological process may give a significant quantity of wastes but compared with the competing technologies it pollutes the environment to a less degree. It is expediently for us to think that such technologies are ecologically pro-gressive.

In this manner, the ecologically pure technology is the main task for scientific-technical progress in the field of environment protecting. That is why they think with right that the 'non-waste technology' is a narrow concept and it does not completely answers the scientific-technical progress tasks in the nature protecting activity of environment. The term 'non-waste production' would be more suitable for the aim. It may be defined as complex production, processing the objects of labour with complete use of valuable components and useful properties, enduring the standard environment qualities (6).

Since we have accepted technology as a totality of means and methods for extraction, treatment and processing of raw materials, materials and semi-manufactured goods we could agree with some authors that the concepts 'non-waste production' and 'non-waste technologies' are identical in meaning.

As the same author says the non-waste technologies conception is conditional to some extent. Taking the non-waste technologies as an ideal model for production, which in most cases cannot be realized as a whole, it is quite natural to accept as nearer to reality of developing processes the term ecological (pure) technologies. According to prof. G.D.Haralamkovich they are the technologies which may be 'registered', coordinated and harmonized with the environment inhabited by man without unfavourably influencing the nature (7).

Creating of non-waste productions in the form of new technological production complexes like a most perspective field in production technological schemes projecting obliges us to make an assessment of the especially harmful and polluting productions by the index of specific production ecology (8).

According to prof. D.W.Pearce (9) the longevity and severity of the current recession in the international economy has undoubtedly served to reduce both the rate of economic throughput (matter and energy) and the rate of increase in the quantities of waste requiring disposal in the industrialised economies. Individual plants have been forced to husband their resources more efficiently and have also been faced with a reduction in aggregate demand in the market place. It is likely that an industry suffering from over-capacity problems will be less inclined to introduce cleaner processes because of low rates of investment in new plants.

On the other hand, the recessionary effects are not uniformly negative as far as LNWTs are concerned. The aluminium industry, for example, has been forced to reduce operating costs and also to diversify into higher quality products (such as thinner packaging foils). Both these trends have encouraged the introduction of LNWT such as the FILD aluminium degassing process (reviewed below). Much seems to depend, however, on the exact circumstances and technological requirements that particular industries face.

In these circumstances it seems to us necessary to set down the basis for the social evaluation of LNWTs: i.e. to seek an answer to the question of how to decide whether a new technology, ostensibly embodying low waste characteristics, is socially worth while.

LNWT AND MATERIAL BALANCE

According to the same author LNWTs seek to reduce the demands made on such natural resources (many of which are unpriced) in the production of goods and commodities for capital investment and final consumption purposes. The essence of LNWT is, therefore, to reduce materials and energy inputs into any technological process for the manufacture or creation of goods, and to reduce the waste outputs from that process. LNWT is thus 'environmentally benign' and is best illustrated by a simple materials balance model (see Fig.1).

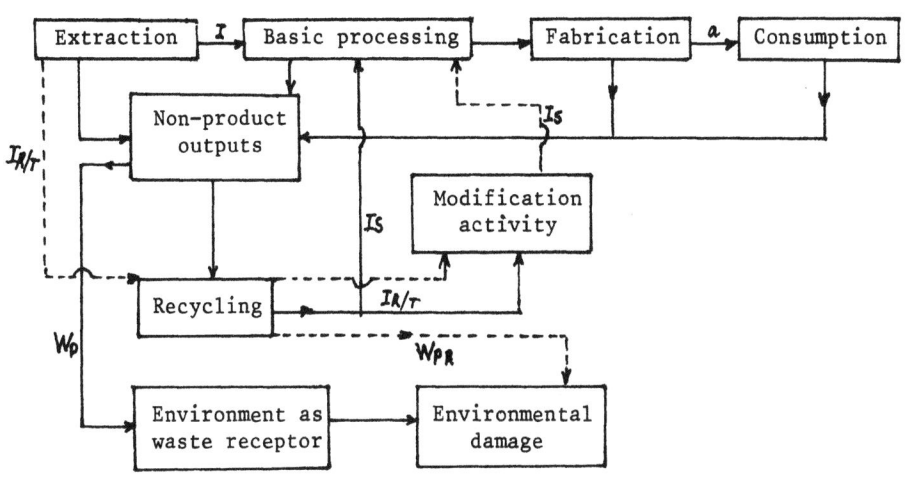

Fig.1. Simplified materials balance, where: I = primary material and energy inputs; I_s = secondary (recycled) inputs; $I_{R/T}$ = primary inputs for recycling and/or modification processes; W_p = residuals requiring disposal; W_{pr} = residuals generated during treatment and/or recycling processes; Q = final product output.

During the transformation of low entropy inputs (I) into outputs of goods (Q) in the production process, a range of non-product outputs (NPOs) are also produced. These jointly produced NPOs (waste materials, gases and energy) are not demanded or required by the final consumer. Procedures for recovering these by-products result in the recycling of at least some part of NPOs for reuse in the production process or for direct sale to some other production activity. Typical examples of such flows are so-called 'home scrap' flows within the paper and board or metals basic processing plants, and 'direct use of scrap' or 'prompt' scrap flows from fabricating plants back to basic processing units in the same industries. These types of recycling loops can often be achieved with a minimum of additional material and energy inputs. But recycling is not a costless activity and even reuse systems within a basic processing unit, such as the introduction of closed cycle water circuits in paper and board mills, can involve latent costs. Thus the technically simple, and low input, process of installing new tanks, piping and plumbing facilities necessary for a closed water circuit system can, after a time lag, cause end product quality deterioration, increased corrosion and slime build-up and odour problems.

THE ECOLOGIZATION OF PRODUCTION AND TECHNOLOGIES

The ecologization of production and technologies was born by the contemporary economical-ecological development which changes the biosphere into noosphere. According to Vernadsky's studies the environment of man is gaining new qualities. The man's role as a geological and geochemical factor is being increased. The living substance mass and its organization begins to dominate as a regulating power. But at the same time changing the biosphere man with his economic activity begins to play an unstabilizing role with respect to the ecological balance.

The increased scales of different substances drawing are in measures that do not correspond to its ecological possibilities for their natural reproduction. At the same time machines and technical means which are incompatible with the character and ecological features of the corresponding environment are being used for substances drawing from earth entrails.

The wastes' mass is quickly increasing. They have to be taken as an independent element of the productive forces. That is why the ecological aim of production now is not to diminish the environment pollution degree but to change the technologies in such a way that pollution should be avoided as a whole.

The changes in technological way of production concerns all fields of entire social life. The ecologized production will develop on the basis of overcoming the contradictions between the natural production conditions and the existing productive forces. Ecologization itself may be defined as a totality of methods and means for reaching a rational use of nature by <u>transferring</u> the structures of consumption, protection and re-<u>production</u> of nature resources in <u>strictly keeping with</u> the ecological principles for biosphere <u>functioning</u>.

The reproduction of nature environment as a specific stage of means of life production is a qualitatively new element of ecologized production.

The second characteristic is connected with nature forces use like <u>primary</u> not like transformed ones.

The third characteristic of ecologized production is its complexity.

The fourth ecologized production characteristic feature is that it is being realized through <u>closed</u> production cycles. The main ecologized production <u>task and</u> peculiarity is the transforming of terminal linear production process into a non-terminal nature production cycle.

Non-waste technologies become an engineer ideal.

Low waste, non-waste and other progressive technologies and closed cycles of water use are being transformed into a main structure defining element of ecologized production.

Ecologized production will gradually get across a qualitatively new energetic basis: solar energy and cosmos mastering; increasing of transformational possibilities of autothrophic organisms for solar energy processing.

REFERENCES

1. Metodologicheskie osnovy ratzionalnogo prirodopolzovania. Vladivostok, 1985.
2. Bannikov NG et al: Ohrana prirodi. Moskva, 1985.
3. Fedorenko NP, Reimers HF: Ecologia i economika, evoliutzia, vzaimootnoshenie, ot Economii prirodi do Bolshoi ecologii. In: Filosofskie problemi globalnoi ecologii. Moskva, 1983.
4. Iagodin GA, VA Zaitzev, SE Mokarov, U Istokov: Novoi nauki. In: Priroda i chelovek, No.7, 1984.
5. Marinov H: Nauchno-tehnicheski osnovi na ecologizatziyata i obshtestvenoto proizvodstvo i problemite na teritorialnata organizatzia. In: Ecologizatzia na obshtestvenoto proizvodstvo. Svieko 81, Svishtov, 1981.
6. Economicheskie problemi ispolzovania promishlenih othodov, Kiev, 1983.
7. Gussev AN: Economicheskie problemi bezothodnih proizvodstvah. In: Izvestia AN SSSR. Seria Economicheskaya No.5, 1985.
8. Haralamkovich GD: Ecologicheskaya tehnologia - novaya oblast issledovanii. In: Vestnik Vishaya shkola, No.5, 1979.
9. Pearce DW, RK Turner: The Economic Evaluation of Law and Non-waste Technologies. - Resources and Conservation, 11(1984), Elsevier Science Publishers B.V., Amsterdam.

OPTIMIZATION OF ENVIRONMENTAL POLICY BY INTEGRATED ENVIRONMENTAL RESEARCH

J.S.A. LANGERWERF

INTRODUCTION
Until recently, environmental research has mainly been confined to separate scientific disciplines and to studies of the grossest forms of pollution where it is relatively easy to establish correlations between sources of contaminants and the effects, and to formulate the rather evident remedies.
Because growing emphasis is being laid on the more subtle effects of chronic exposure of biological systems to low concentrations of contaminants problems become more and more complex and remedies for such effects are fare less evident. Moreover, current environmental technology not seldom provides only partial solutions or even replaces one environmental problem with another.
In the face of a welter of alternatives policy makers are confronted with decisions which have to be based on the magnitudes of parameters of widely diverging accuracy. A number of these parameters frequently fall into entirely different spheres, making direct comparison of e.g. ecological and social and economic effects impossible.
In this field of decision-making, the usually less accurate results of ecotoxicological research frequently have to compete with the far better definable technological and financial consequences of policy alternatives.
Important forces in society ask on the one hand for zero emissions by industry to protect environment rigorously and for undisturbed industrial activity to enhance economic development on the other hand.
An optimum compromise prescribes a well-balanced contribution of the different sciences and a well-informed society; all relevant effects have to be taken into consideration and the financial and social sacrifices society is ready to make for protecting the environment should be put to maximum effect.

ECOTOXICOLOGY
The contribution of ecotoxicology to policy making is still very limited. Although the significance of effects on development and function of ecosystems is paramount, hardly any contribution to the setting of environmental standards has so far been made.
At the moment, ecotoxicology, a young and complex conglomerate of a variety of sciences, is still unable to cover the wide field of biological interactions in ecosystems and the influence of the great variety of possible chemical stress factors on these interactions.
Promising developments are concerned with the reduction of the complexity of problems with only minor losses of relevance, as exemplified by the search for key biological processes in ecosystems and the influences of

different classes of contaminants on it.

Evidently many ecotoxicologists are very concerned about effects of contaminants in the environment. Instead of mixing up science and politics, ecotoxicologists should confine themselves to putting the results of their work in politically understandable terms, and leave decisions to the policy-makers.

The task of translating ecological effects into terms the man in the street can understand should not be underestimated. Even when ecotoxicology is able to describe relevant effects quantitatively, the perception of these effects by society will remain doubtful. Contrary to human safety studies where a rather clear perception of acceptable death risks already exists, in environmental risk analyses the formulation of upper limits of acceptable environmental risks seem hardly feasible. By their very nature different biological species and systems will ask for different limits, moreover the appreciation of these species and systems may vary enormously from one individual member of society to another. The level at which protection of the environment is appreciated individually will depend first on the extent to which the sacrifices mentioned influence daily life and secondly on the personal perception of the importance of attainable reductions in environmental effects. In other words the gain in environmental quality must compensate clearly for the social and financial sacrifices.

This situation, as well as the lack of insight into processes in stressed ecosystems, make ecotoxicologists loth to put their questionable - though still their best - data at the disposal of policy-makers.

INTEGRATED ENVIRONMENTAL RESEARCH

To proceed at this very moment with the solution of environmental problems, policy-makers must be made aware of the margin of error in their conclusions as a result of the uncertainties in the various parameters on which their conclusions are based.

Apart from positively influencing the relation between scientists and policy-makers such assessment of inaccuracies offers the advantage of a possibility to single out by sensitivity analysis those disciplines whose contribution to the weakness of conclusions is greatest. Such an exercise in its turn enables one to formulate unambiguously priorities for the development of the various environmental sciences concerned.

To realize the said maximum output of the financial, social and economic sacrifices of society, an integrated approach to environmental problems is needed. Such an approach evidently contributes strongly to the conservation of the environment, by setting priorities in the right order. A subject of major concern in integrated environmental research is the relation between sacrifices and their effect.

If the environmental sciences are able to produce such quantitative relations, policy-makers can do their work properly by selecting those areas where the social, economic and environmental needs of society match. The sacrifices mentioned comprise social and economical consequences of alternatives of human activities as well as less severe changes in technologies or remedial activities. Some sacrifices can be expressed explicitly in financial terms. Where personal perception plays an important role, the input of the social sciences is essential.

Usually there is no direct causal link between reduction of emissions and a decrease of environmental effects. The evaluation of effects is a largely statistical excercise involving quantitative information about emissions, their dispersion and conversion, and the presence within

the radius of action of the source of vulnerable biological species which play a key-role in ecosystems.
Moreover mixtures of contaminants may be so composed as to evoke antagonistic or synergistic effects.
If a full knowledge of possible biological effects is assumed the probabilistics of the emission and fate of contaminants and the effects of exposure of vulnerable systems, can be dealt with adequately by the methods of risk analysis together with modelling of dispersion and conversion of contaminants.
Thus, integrated environmental research requires the assembly of expertise in:
- process emissions, dispersion and conversion of contaminants in air, water and soil.

- detection of stressed locations and stress factors, and quantification of these factors by application of physical, chemical and biological measuring techniques in the in- and outdoor environment.

- dose/effect relations on the different levels of biological integration: (sub)cellular, tissue, individual organisms, population, community and (model) ecosystem as well as the deterioration of abiotic systems resulting from exposure to external stress factors.

- risk evaluations based on dose/effect relations and on figures of emitted contaminants, their contribution to local concentrations and effects on the (biological) systems locally present.

- removal of contaminants by (forced) conversion by means of microbiological, (physico)chemical and technological processes.

- prevention of emissions by application of cleaner technologies involving adjustment of: product-choice; product design; choice of raw materials and energy consumption.

- evaluation of the impact of implemented environmental policies and applied remedial and preventive technology on the environment and on the social and economic aspects of life.

Efficient integrated environmental research requires close cooperation between a great number of experts. Here one is inevitable confronted with the availability of the essential expertise, with interface problems and with the tendency of research projects to burst their seams, making them poorly manageable and hardly marketable.
Availability of expertise for integrated environmental research is not a problem within the TNO organization. To overcome the other problems mentioned we have recently formed a project group engaged in integrated environmental research. In this project group the wide range of environmental expertise that has been built up within TNO over the last decade is represented.
In this way we have been able to replace an unwieldy and rigid interactive construction with a set of smaller fields that can be more easily surveyed by individual experts and put into action for specific purposes.
    At regular meetings the group exchanges views on integrated environmental research as a whole and prepares and conducts research projects in specific fields.

At the moment the group comprises the following expertise:
    environmental biology and biological modelling
    (eco)toxicology
    environmental chemistry and physics
    dispersion- and conversion models
    environmental technology
    system dynamics
    risk-analyses and -management
    sociology
    management science.

In case gaps in expertise arise, other scientific disciplines within TNO or from outside are fitted in. By this form of integrated environmental research and essential developments of the sciences involved TNO helps to create a sound scientific basis for governments and industries to make environmental policy decisions.

NATIONAL ENVIRONMENTAL CENTRE
Outline summary:
Objectives, structure, function and procedures.

A.W. Veenman

Introduction.

In mid-1986, several people involved in the environmental sector initiated the establishment of the National Environmental Centre (NMC), under the motto:

' For enterpreneuring Holland, through enterpreneuring Holland'.

In March 1987, the National Environmental Centre is expected to be formally established.

The goals of the NMC follow the findings of the Environmental Production Plan (Nijman Committee). The goals also coincide with the conclusions of Broers en Partners, who, acting under orders of the Nijman Committee, performed a feasibility study on the NMC.
Within the NMC, both the supplier and consumer sides of the environmental market are equally represented. There is also space reserved for intermediary organisations who focus on small and middle-sized businesses and who function as a bridge between consumers and producers of environmental advice, services and equipment, by providing services. Regional aspects are also covered by these organizations.

The initiators have chosen for a small and efficient organization. The tasks of the centre are to be performed only by Members of the NMC, with a clear distinction between the centre's functions and the functions to be delegated to participants on the basis of a contract.
The centre's functions include serving as the general addressing body for the government on environmental aspects, and initiating and coordinating activities. The functions to be delegated include information and communication activities, as well as stimulation of the development and application of environmental technology. These ,functions will operate together to form a complete service, and this will be explicitly shown in the contracts for the delegated functions.

Objectives.

The objectives of the NMC are:

* to stimulate the development and application of environmental technology, also in the economical sense, for the purpose of a better environment, in the interests of the total environmental sector;

* to optimize efficiency of communication of knowledge and information supplied to the environmental sector;

* to optimize communication and coordination within the environmental sector, on behalf of business organizations and institutions.

Structure of the NMC.

The NMC consists of a Central Organization, the Bureau NMC and its Members. The Central Organization and the Bureau are housed within the NMC and the Members are separate legal personalities who participate in the NMC.

The functions of the NMC are divided into two categories:

1. Functions performed by the Central Organization supported by the Bureau NMC:

* identification strengthening -the Central Organization functions as the central address, formulating common viewpoints and visions and pointing out aspects of environmental concern to others;

* initiating and coordinating -the Central Organization is responsible for the necessary negotiations with respect to initiation, stimulation and execution of the delegated functions.

2. Functions to be delegated to the members of the NMC.

The above described functions are characterized by their specific interests in the development and stimulation of activities (project-approach).

Development and stimulation projects can cover the following areas:

* environmental technology;
* communication within the environmental sector;
* information structures;
* structured export promotion;
* structured services to the business sectors concerning procedures, etc.

Composition of the organs.

Only existing or future foundations, organizations and corporations from the environmental sector will be considered for membership in the NMC.
Members of the NMC can represent the following sectors:

* businesses;
* municipal consumers;
* intermediary institutions and organizations;
* investigative and research institutions and universities;
* private persons.

Each member will have a seat on the General Board which decides on admission of members upon recommendation of the Daily Board in accordance with the articles on admission criteria.
These articles, which define three groups of members - the consumer, the supplier and the intermediary - determine that each group is to be equally represented on the Daily Board.

Membership in the NMC without participation in the delegated functions is possible, however, in accordance with the NMC's articles, this kind of membership can be restricted.

Financial aspects.

The Members are to support the financing of the NMC by contribution. Of the yearly funds contributed, 60% is to be raised by the Members themselves and paid to the NMC; each of the Members contributing the same amount.

The remaining 40% is to be raised by the Members through a surtax on their services.

For each participating business, the maximum yearly charge including surtax will be HFL 2.500.--.

The NMC will charge commission for the delegation of tasks which are offered to the NMC by the government.

Working procedures of the Central Organization.

The General Board will determine yearly policy plans with an attached budget for the delivery of activities. The board is also responsible for the performance of these activities as well as for coordinating the initiation, stimulation and execution of the delegated activities.

In addition to defining the boundaries of the delegated activities, the delegation contract states the extent of the General Board's control over the performing activities.

In principle the organization prefers to delegate the activities within a chosen category to one member, however, if necessary, activities can be distributed amongst several members.

Activities can only be delegated to participants directly represented by at least one member on the General Board.

As stated in the delegation agreement, bureaus, institutions, and organizations to whom activities are delegated, form an integrated part of the NMC. Bureaus however, need to operate under the surveillance of institutions or foundations.

Activities of the main organ.

Within the framework of delegating activities, it is the task of the Central Organization to indicate which projects are necessary and to then supervise their actual start-up activities.

In this context, the General Organization evaluates:

* the plans for estimating expenses and income;
* the extent of the activities within each project;
* guarantees for execution of the project.

Further, the Central Organization advices financing or subsidizing institutions. The Bureau NMC serves as a support function to the Central Organization.

Delegated activities.

The activities necessary for the performance of projects are delegated to members.

The following projects are in preparation:

* Development of a meeting point within the environmental sector.
  Through this, the NMC hopes to provide an instrument for improving communication and information-gathering within the sector, with the end result of strengthening identification of the environmental sector.

* Development of a service centre for individual businesses in the field of environmental technology.
  The service centre is supplier-oriented and aims to educate individual businesses about the environmental sector.

* Development of an information network.
  The goal of this project is to provide a structured information system to transfer knowledge on developments in the field of environmental technology, including products and services, for commerce and industry.

* Stimulation and development of environmental technology.
  The goal of this project is to create a framework for stimulating the development and production of environmental goods and services, including re-use and waste technology, in order to promote export activities.

  Stimulation is achieved by encouraging interested participants in a project to gather confidentially to form a cooperative alliance. Participants in the cooperation can be, in accordance to the project's requirements, its initiators, consumers, suppliers, research institutions, and financiers.

* Export stimulation.
  The aim of this project is to promote exportation of equipment, techniques and services, which can be offered by commercial markets.

* Increasing the supply of information to the different sectors. This means informing and instructing consumers of environmental equipment and services.

Plan of approach.

The main premise of the NMC is to establish the necessity of its own existence, which has resulted in the choice for a phased approach. Phase one, the starting phase, is followed by the growing phase. With this approach, success is essential from the start.

It is also important that good structures for communication are set up for the different interested participants.

It should be pointed out that the Central Organization and the Bureau NMC will only function as initiators for both the starting and the growing phases. It is therefore, not the intention of the NMC to become a performing body itself.

ANNEX 1: INITIATORS AND FOUNDERS OF THE NMC.

Drs. P. Berculo, on behalf of the Utrecht Environmental Centre.

Drs. A. Boone, on behalf of the Chamber of Commerce, Amersfoort.

Dr. L.A. Clarenburg, Member of the Board of the Foundation for the Realization of an Industrial Centre for Environmental and Waste Technology (ICMAT).

Dr. ir. P.J. Huiswaard, Member of the Order of Dutch Consulting Engineers (ONRI).

Ir. W.H. Knoll, Chairman of the Foundation to Realize a Centre for Technology of Interior and Exterior Environment.

Dr. ir. N. van Lookeren Campagne, Member of the Environmental Hygiene and Spatial Planning of the Dutch Entrepreneurs League and the Dutch Christian Employers League.

Ir. A. Nijdam, Chairman of the Dutch Union of Processed Soil Cleaning Industries (NVPG).

H.F. Peze, MI. Chem. E, Chairman of the Foundation for the Realization of an Industrial Centre for Environmental and Waste Technology (ICMAT).

Drs. W.A. Telkamp, Director of the Union of the Chamber of Commerce and Industries, NV Databank.

Dr. ir. A.W. Veenman, Chairman of the Union of Suppliers of Environmental Equipment and Services (VLM).

Ir. J. Voskamp, Head Environmental Control, Member of the Environmental Hygiene and Spatial Planning of the Dutch Entrepreneurs League and the Dutch Christian Employers League.

Environmental Care Programme: Exploring the market for environmental technology.

E.J. VLES (Association of the Dutch Chemical Industry)

With the beginning of universal environmental awareness, not yet two decades ago, governments all over the world have initiated legislation that aimed at the reduction of pollution load at the main concentration points of emission. The industry, and in particular the chemical industry, has been early identified as a main source, if not as the main culprit, of environmental disturbance.

Moreover, each single compartment of environmental impact was treated as an isolated entity: water, air, noise, waste were alle considered well-defined and un-related issues. Laws and regulations were poured out over the industry at an ever increasing speed. In all industrialised countries, the decade of the seventies saw a speedy generation of obligations which confronted the industry.

The industry itself had at first to adapt itself to this new phenomenon of environment and to overcome its initial shock to a changed culture - up till then it was, by society, considered only as the bringer of wealth and well-being, by virtue of its growth which was believed to be unlimited. When it did respond, it did so with rather massive investments, initially almost exclusively with so-called add-on techniques, mainly in the field of effluent treatment.

Let me give you an example:

Discharge of untreated effluent in i.e. by the Dutch chemical industry

	1973	1978	1983
	4.10^6	$2.3.10^6$	$1.3.10^6$
investments (cumulative)	30.10^6	210.10^6	500.10^6

For various reasons, this example is of interest for our further considerations.

First of all, installations for the aerobic degradation of the biological
oxygen demand in effluent were realised, as pointed out, as add-on techniques.
Only few companies had the vision and the time allowed to them by the authorities to scrutinize at depth their in-house water management. Those who
did, and still do, managed to reduce substantially the volume of effluent to
be treated. In an earlier paper during this conference you have been given
a very noteworthy example. The result of this is, by the way, that public
effluent treatment plants in this country now start to suffer from under-
occupancy.

Secondly, the units installed were all variations on the theme of the well-
known aeration of active-sludge systems. The amount of energy required
- certainly for those entities who could not afford a very large aeration
area for this purpose, expressed per unit of BOD resp. COD removed - is
substantial.

Energy requirement of effluent treatment $^{x)}$
(k Wh/kg BOD reduction)

aerobic	anaerobic
0,35 à 0,50	0,10 à 0,20

x) excluding energy benefit of biogas production

With the evolution of energy prices this has become a rather costly affair,
while the amount of energy to be generated for this purpose creates, in
itself, new air pollution problems! Finally, the treatment units were designed
solely for BOD/COD removal. Since 1970, however, we have become aware of a
multitude of important environmental parameters in industrial effluent, and,
for that matter, also in domestic effluent.

Obvious as these considerations might appear at the present day, they were
not so a decade ago. Within industrial enterprises, environmental matters
were delegated to the odd specialist in the field-specialist in the sense
that he was asked to devote part of his time to these matters. Consequently,
management failed to become aware of the magnitude of environmental matters
in general.

It were however, other causes, and not so much the need for an integrated technological and research effort, that led to a marked change in this situation.

In the early eighties governments became more and more aware of the fact that legislation in itself was not enough to guarantee an improvement of the environment. On the one hand discrepancies between rules and practice - be it legal or illegal - became apparent. On the other hand, discontinuous releases to the environment - mainly as a result of operational failures - proved to have very grave impacts, as highlighted by some major accidents. Governments were thus induced to tighter control set-ups, but even in an optimal situation, these improved controls could, and can, ensure only the compliance of entities - be it industrial or non-industrial - with the set rules regarding continuous or semi-continuous releases. It can not monitor industrial activities itself, nor prevent occasional releases - that can only be done from within.

A new approach was developed in the United States of America, where, as is well known, liability aspects and the impact of court procedures on legislative matters are more pronounced than in Europe. It was notably this aspect of industrial liability with regard to environmental matters that initiated the development of environmental auditing. Since then, a wide experience with these and similar management techniques concerning environmental matters has been acquired.

To be perfectly clear, environmental auditing techniques have not been set up primarily in order to assess the market for technological improvements or, in general, for new research and development activities. It was clearly done in order to improve on compliance survey techniques, or, in the words of the Environmental Protection Agency in the USA, to come to a 'higher level of compliance'. According to a recent policy document of EPA, environmental audits are meant to:

verify compliance
        evaluate effectiveness of environmental management
        assess risks

It is, of course, the second of these goals that relates environmental auditing to the main subject of this conference.

At a somewhat later date, similar developments took place in the Netherlands. Here it was not so much the liability of a single, individual company but, in general the observation that enforcement and control of environmental measures were badly lacking, as highlighted in a nation-wide scandal involving chemical wastes. This triggered off new developments. Starting with studies of various models of the US set-up of environmental auditing the view became manifest that such a management tool had to be set in a much wider framework of industrial environmental policies in order to be efficient and effective. After a year of study, and after fruitful exchanges of thought with the authorities concerned, the Federation of Netherland Industries formulated its policy, and coined the conception of Environmental Care in Industry. In a publication dated may 1986 the Federation laid down the principle and programme of this policy   2)

Let me sketch shortly the path that leads from compliance survey to this company-integrated approach.

Initially, it is of course by governmental bodies that the environmental requirements under which plants are allowed to operate are set. These requirements, which are by themselves the 'translation' of government policies, have to be maintained, for which registration of data is indispensable. An evaluation of these registrations by the company itself is the first step to any internal maintenance system

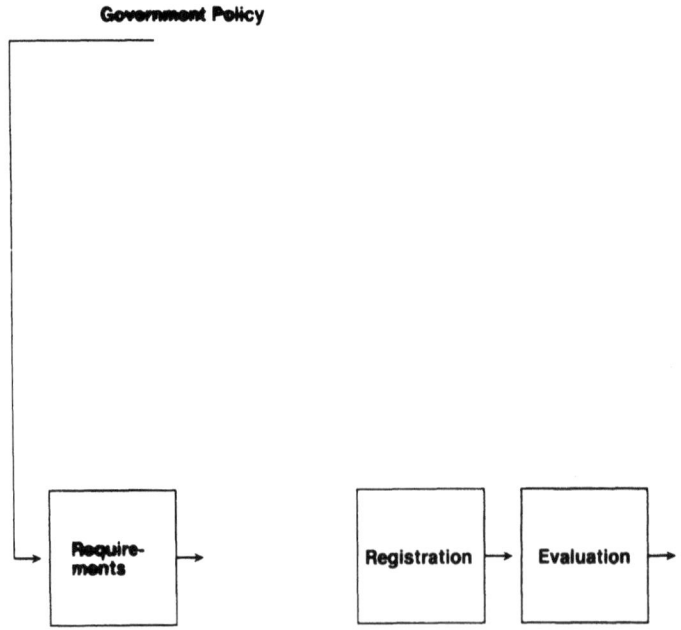

Simple as this picture seems, it must be noted that registration of data
is sometimes extremely complicated. Permits tend to include maximum tolerated
levels for which adequate continuous measuring devices are not available
yet. The more a company will systematically approach the issue of environmental controls, the better it will be able to state its needs for improved
measuring equipment and adequate sampling policies , a clear link with the
main issue of this conference.

This picture is of course highly incomplete. On the one hand, the authorities
will not only set the appropriate requirements but also indicate by what
technical means they have to be realised.

Furthermore, an evaluation by the company is, in itself, not sufficient if,
for example in the case of noted descrepancies between desired and actual
data, no actions are planned. The scheme will so be extended as follows

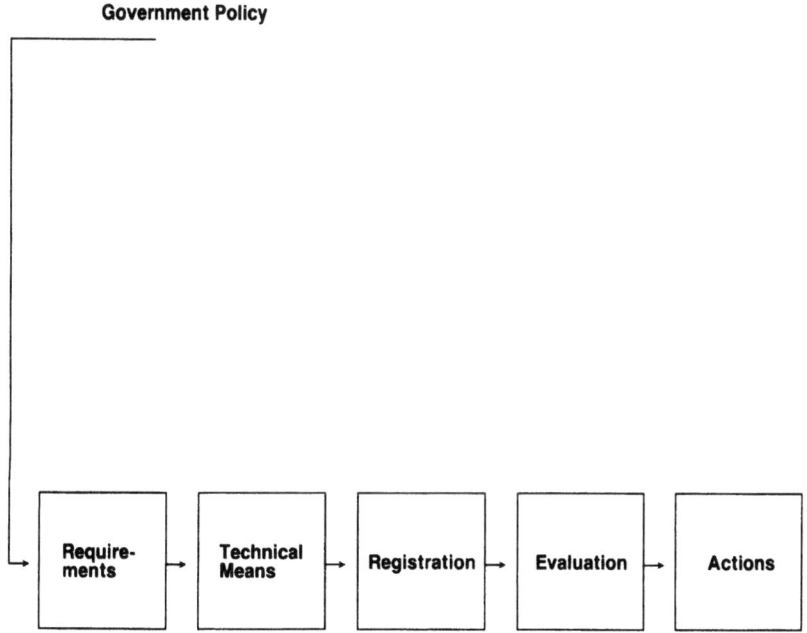

Again, this is not the way the complete picture should look like.
On the one hand, outside policies will only be viable within a company if the
company management supports them. The least a management can do is to
state clearly that it will fairly implement all requirements imposed. In most

cases, however the company, for very good reasons of its own, will add certain internal conditions and rules, for example in order to guarantee its good practices and good name. In such a case, management will also have to be informed of the outcome of a compliance survey or, more effectively, on the progress made with set actions derived from such a survey. That brings us the following scheme

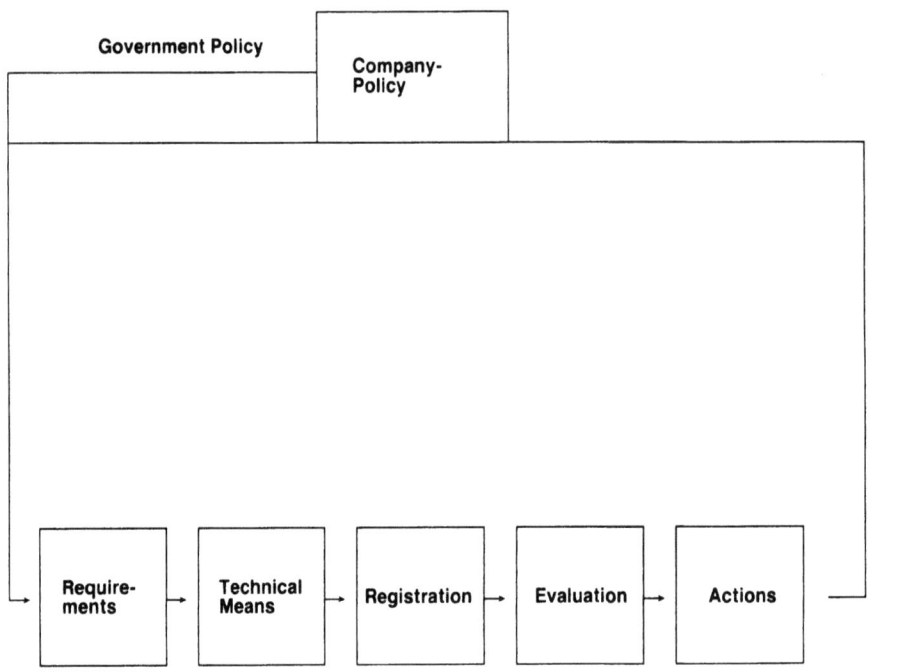

Up till now, we have only considered those operations within a company that can directly create an environmental impact, in other words, production. Limiting the company policy to this would however create frictions within the company and, moreover, fail to ascertain that other activities within the company will not simultaneously pursue conflicting goals.
Therefore, a company-integrated environmental care structure should look like this.

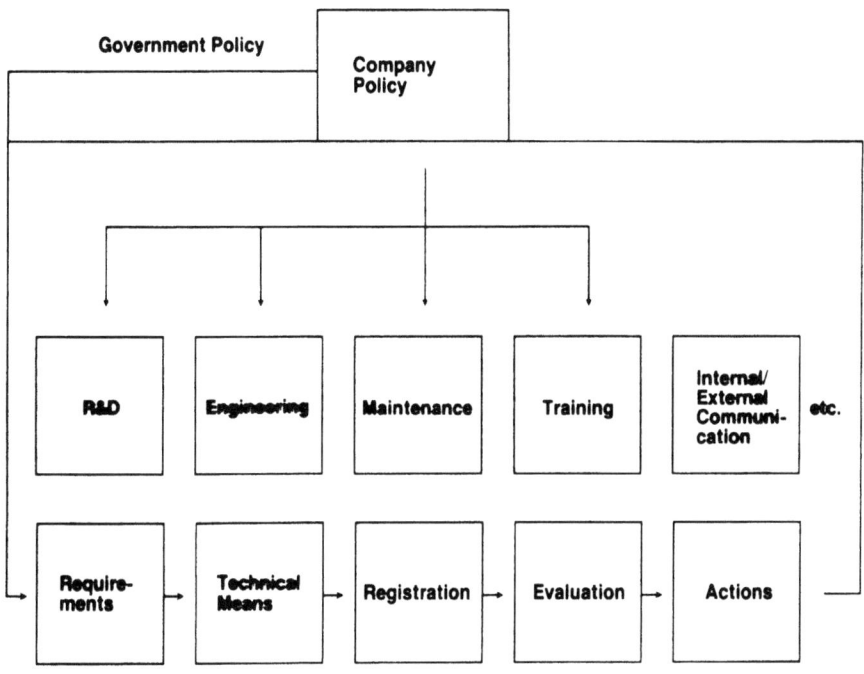

This aproach has gained wide support within the Dutch industry. The benefits to be derived can be summarised as follows:

- insight into the actual environmental situation of the company
- optimalisation of environmental techniques installed
- view on future requirements, monitoring R & D engineering effort
- planning of future environmental activities
- enabling companies to discuss requirements with authorities on equal footing
- improve communication on environmental matters.

The benefits will in first instance become apparent when, as a result of research and engineering, new production facilities are devised and installed. Here compliance with new regulations is obvious, although it still requires a fair amount of information and vision in order to incorporate in new products and processes the requirements of, say, the next decade.

Within the companies, an integrated environmental policy will, amongst others, result in standard procedures by which research and development, as also engineering activities, are to be checked on a regular basis in order to comply with environmental and safety goals. As far as research is concerned, this will mainly result in the evaluation of product and process alternatives.

A more fundamental insight in, amongst others, relations between molecular structures and toxicological respectively ecotoxicological properties is needed, but this is well outside the scope of this conference. In the field of engineering, design parameters are to be based also on recent insight or, even better, understanding of future requirements, resulting in demands for lower energy consumption, reduction of chemical waste, of process and cooling water, and abatement of air pollution and noise.

In short, a company which has integrated its environmental policy in line with the Environmental Care Programme, will be an experienced partner for innovation in the field of environmental technology.

As indicated in one of the organograms, Environmental Care equally touches upon other disciplines within the company, as for example Organisation and Personnal Development. Here again, cooperation with outside experts can prove to be very fruitful.

So, also in this respect, an integrated approach to the matter will form an incentive for new developments.

It is sometimes said in this respect the Pollution Prevention Pays. In the absolute sense of the word, that is not true, Prevention of pollution, like all human efforts geared at installing order where chaos has reigned, is a costly affair, according, amongst others, to the rules of thermodynamics.

This is illustrated by the increased costs for modern process equipment versus its traditional types. Increase of energy consumption for environmental measures is another noteworthy item.

In some cases, prevention of pollution has gone hand in hand with savings on a.o. raw materials. However, those savings should have been realised anyway by any good management, and legal measures have only focused attention on potential savings where attention was already due.

An integral approach to prevention of pollution does, however, make an optimalisation, or better a minimizing of costs possible.

In comparing for example costs of prevention or abatement of emissions with duties payable, in paying for more training and communication in order to reduce costs linked with poor house-keeping, companies can arrive at an optimalisation of capital lay-out and manhours invested. It must be borne in mind, however, that requirements are not static, so that such a company effort of optimalisation is an on-going affair. As an example, new requirements aimed at the protection of the soil, are now coming into force in the Netherlands. These will have a marked effect on construction and installation costs of process equipment and product storage. And, more important, it must be stated clearly that even with the most sophisticated management techniques on environmental matters and with the optimal use of pollution prevention techniques, requirements can reach a point where any company will cease to be competitive, be it with companies abroad, or be it with alternative products on the marketplace, and be forced out of business.

There are various ways in which companies start to set up their Environmental Care Programme, as became clear from recent interviews carried out by the Association of the Dutch Chemical Industry among member-companies.
We have reviewed these in a recent paper.
The most vital, and also the most common items of an overall programme are:

- policy statements by management
- detailed function analyses
- formal registration policies
- environmental planning
- training
- environmental audits
- communication policies on environmental matters

The sheet shows that the initial environmental audit, developed a decade ago, forms only a part of the total Environmental Care set-up, although it will prove to be a very vital one.
The audit, when properly executed, will give management a clear view however of where the company stands, at what the real needs in respect to environmental management are.
In order to be a successful instrument, the following general elements will have to be included:

- explicit topmanagement support
- audit independent of other activities
- adequate audit staffing
- explicit audit programme objectives
- analyses and interpretation of information
- written reports
- quality assurance procedures included.

Execution of company audits that do include the above mentioned elements is a quite serious affair.
Notwithstanding an avalanche of information and literature on the subject, many companies will look for outside help, at least for the set-up and execution of their first environmental audit. In the Netherlands, this has led to an increase of interest in the subject shown by auditors and organisation advisers.
A clear distinction must, however, be made between outside councillors who will act on behalf of and in collaboration with the company, and certified auditors, who, at the request of the authorities, should verify the data, submitted by a company to the authorities. The latter idea, with which politicians have toyed, has been rejected both in the USA and in the Netherlands on both principal and practical grounds.

The question again rises: what good can Environmental Auditing, as part of the Environmental Care Programme, do to the advancement of environmental technology?

The answer lies in the fact that it will include a systematic evaluation of pollution abatement units installed including both their reliability, - as illustrated by frequency of failure - and their cost/effectiveness. We have to realise that the first units were installed more than 10 years ago, so that not only are vital parts already fully depreciated, but also overtaken by newer technology. Effluent treatment, as showed in the beginning of this short talk, underwent a drastic technological change when anaerobic breakdown of BOD was introduced. But also the treatment of excess active sludge has been improved over the last years.

Moreover, solutions which, a decade ago, were either in their experimental stage -like biologic air treatment - or only applicable to very large set-ups, like floating roofs and certain stripping equipment, now became available for an ever increasing amount of production units.

Only a systematic, well-documented, frequently executed analysis as Environmental Auditing is, can identify the need or possibility to improve on the ongoing practice of the company's day to day operation, and thus explore the market for environmental technology.

INNOVATION AND POLLUTION ABATEMENT

D. WIERSMA AND T. PULLES

1. INTRODUCTION
Efficient abatement of pollution means the use of environmental technology at minimum costs. In the long run this efficiency goal can be reached by technological progress which can provide new pollution abatement technologies with higher productivity or with lower abatement costs. The process of development of new environmental technologies consists of two stages:
- research and development (innovation);
- introduction of the new technology (diffusion).
In both stages scarce resources and development time are needed. If the development of the new technology is successful, a flow of benefits (savings of abatement costs) after introduction is possible. Acceleration of the research and development program causes in most cases additional development costs, but can also result in an earlier flow of benefits (1).

This paper deals with the impact of shortening the innovation period on the expected benefits of a polluting firm. Attention is paid to the influence of different environmental policies on the allocation of resources of the innovating firm. First a brief outline is given of the economic theory of allocation of innovation resources to be made by an innovating firm. Next the theory is elaborated for the case of a polluting firm which is forced to pollution abatement by government policy.

2. ECONOMIC THEORY OF INNOVATION
The conditions of optimal allocation of innovation resources of an individual firm are elaborated by Scherer (1). Following this approach we assume that firms are profit maximizers. According to this premise the innovating firm will allocate the resources of innovation in such a way as to maximize the surplus of expected benefits over research and development costs. Research and development effort has two dimensions: time and costs which can be substituted within certain limits. Accelerating the development is costly for three reasons. First the probability of making errors is higher when development stages overlap each other. Second, parallel experimental approaches may be necessary to deal with uncertainty of the results. Third, there are diminishing returns when scientific research labour is increasing for a given project. For the cost-time trade-off relation of innovation we use a function, specified by Mansfield (2), (3).

$$C = V \cdot \exp\{\frac{\varphi}{t/\alpha - 1}\} \qquad (2.1)$$

when C is the expected research and development (R+D) costs, t the expected development time, V the minimum possible R+D cost, α the minimum development time and φ a dimensionless constant. Mansfield (ed) has estimated the value of φ for 20 innovation projects in the USA. On the basis of these estimates the average value is 0.1 (standard deviation: 0.2).

The expected benefits of research and development consists of cost reductions after introduction of a new production technology or additional benefits after introduction of a new product in the market. Total expected benefits are a decreasing function of development time for two reasons. First, an earlier introduction date allows the firm to tap the benefits over a longer time span. Second, earlier introduction can improve the competitive position of the firm, which can result in additional revenues. For simplicity we postulate the following expected benefit function:

$$W = W_0 - R \cdot C (t-\alpha) \qquad (2.2)$$

where W is total expected benefit of the new technology, t the expected development time, R the expected benefit per unit of time and per unit of investment C in new technology. W_0 is total expected benefits when development time is at the minimum level ($t=\alpha$). The profit maximization problem of the innovating firm is to choose a development time t* for which W - C gives a maximum. Setting the derivative of this expression to zero and solving for t*, we obtain:

$$t^* = \alpha + \sqrt{\frac{\alpha\varphi}{R}} \qquad (2.3)$$

Figure 1 represents this relation graphically.

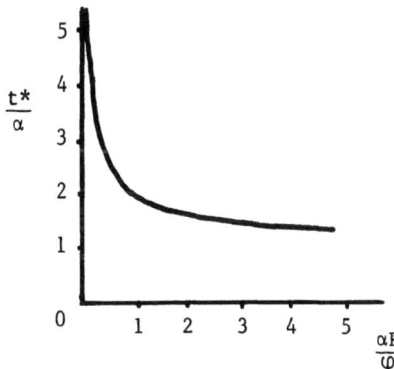

FIGURE 1. The relation of development time and expected benefit.

3. INNOVATION AND POLLUTION ABATEMENT

From equation 2.3 one can conclude that the square of t* decreases at increasing R. When a firm produces pollution as a by-product of market activity, the expected benefits of a new technology may consist of a mixture of additional sales revenues, reductions of production costs and probably changes of pollution abatement costs. The decision to start the development of pollution abatement technology depends of the direction of the technical progress and of the incentives of the environmental authorities. To illustrate these impacts we will distinguish two different situations.

First we assume a situation of technical progress which results in cost reductions of production and an increase of pollution. When environmental authorities have imposed a pollution standard, the firm is forced to increase of pollution abatement. In the case of a pollution tax the tax payments will be increased. In both cases the expected benefits of the research and development are lower than without environmental policy. As a result of this the pace of technical progress will be lower. When technological progress, however, results in jointly reductions of production costs and savings of abatement costs, the situation is different. We shall illustrate this point with a simple graphical figure. In figure 2 F gives the marginal pollution abatement cost function of an existing abatement technology. We measure along the quantity axis, the pollution abated. Maximum pollution abatement in E is corresponding with zero pollution. Suppose the government wish to reduce the firms pollution to QE. Using a standard of Q, the firm is forced to pollution abatement of OQ. Total pollution abatement cost is the area OQA. The government can get the same result by imposing a pollution charge of P_T. In this case the pollution costs for the firm are the area OQA plus the tax to be paid on the rest of the emission (area QABE). Now we suppose that research and development results in a decline of the marginal pollution abatement cost function to F'. Under unchanged conditions the research and development of pollution abatement technology results in savings of abatement costs. With the standard of Q, the savings of abatement costs are the area OAC. With the tax P_T, pollution abatement of the firm will increase to Q' and the savings of abatement costs increase to area OAD. Since the expected benefits in the latter case are higher, the acceleration of innovation shall be faster.

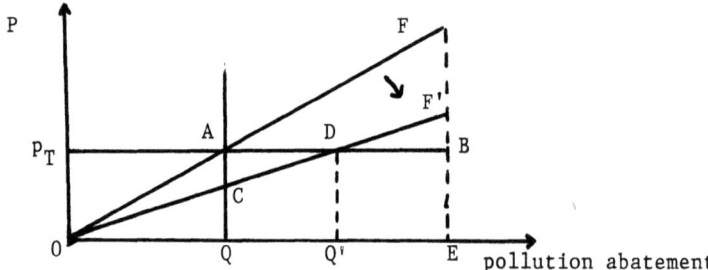

FIGURE 2. Pollution abatement costs and technological progress.

4. CONCLUDING REMARKS

Technical progress of pollution abatement technology can be a means to contribute in achieving environmental goals. Because of the cumulative effects of some pollutants, the rate of this technological progress is important. If the results of technical progress of production and pollution abatement works out in the same direction a pollution tax results in a faster pace of innovation than a quantity restriction.

REFERENCES

1. Sherer FM: Industrial Structure and Economic Performance. Chicago: Rand McNally, 1980.
2. Mansfield E(ed): Research and Innovation in the Modern Corporation. New York: Norton, 1977.
3. Mansfield E(ed): Technology Transfer, Productivity, and Economic Policy. New York: Norton, 1982.

A MICRO-COMPUTER PACKAGE FOR CALCULATING THE COST OF WASTE MANAGEMENT OPTIONS

F. ETEMAD AND D.J. LEECH

1. INTRODUCTION

From time to time, a Waste Disposal Authority needs to choose between available options for the disposal of domestic and commercial waste. This was the case, for example, in 1984/85, when all 37 W.D.A.'s in Wales were required to submit to the Welsh Office, waste disposal plans for the forthcoming ten years. Generally the available options are limited to controlled landfill of unprocessed waste; the landfill of baled waste; and various combinations of metal extraction, energy recovery and the landfilling of unrecovered material.

It has generally been the case that landfilling is the cheapest way of dealing with waste, but the decreasing availability of landfill sites and the increasing standard to which landfill sites must be built and managed are increasing the cost. This increased cost of landfill makes waste reclamation more attractive.

Because any waste disposal system involves considerable capital investment - probably several million pounds - it is important that the system is chosen carefully. In Wales, the options seem to be determined by the availability, size and geography of the landfill sites, whether planning permission to landfill demands baling to meet site management standards, the market for recovered energy and the markets for dirty or de-tinned scrap iron.

Before choosing, it is necessary to know the cost of each of the options available. Currently, it is most important to determine the cost of landfill, partly because this is the most commonly used method of waste disposal, but also because, however waste is reclaimed or disposed, there will always be a residue which must be landfilled.

Choosing from the available waste management options is not simply a question of choosing the cheapest option. In each case, there must be some measure of optimization so that, for example, if the costs of landfill are being predicted, it will be necessary to choose optimal values of the parameters of landfill (phase size; cell size; height of tip; degree of compaction; etc.) as far as is reasonable.

We thus have a need to make several calculations of the cost of each option. We also have a need to determine and store the costs of the equipment required for any particular waste management option.

2. A PROGRAM TO CALCULATE COSTS

It is clearly desirable to provide the officer responsible for choosing a waste disposal option with a computer program which will make the calculations quickly and with acceptable accuracy. It will also be desirable for the program to record first costs and predicted operating costs of equipment (such as balers, compactors, etc.).

Because the officer is unlikely to be expert in the use of computers it is desirable that any program be user friendly.

The program which has been written will calculate the cost of disposing of a tonne of rubbish, allowing for
(i) the civil, mechanical and electrical engineering and labour costs of building and operating a landfill site, and taking account of site life, phase sizes and cell sizes;

(ii) the civil, mechanical and electrical engineering and labour costs of building and operating a baling plant;
(iii) the civil, mechanical and electrical engineering and labour costs of extracting ferrous metal from waste;
(iv) the civil, mechanical and electrical engineering and labour costs of de-tinning recovered ferrous metal;
(v) the civil, mechanical and electrical engineering and labour costs of converting suitable waste to fuel floc or fuel pellets.

In the case of options (iii), (iv) and (v), opportunity costs of landfilling will be taken into account. It is possible to calculate the costs of combinations of options and, in most cases, it will be necessary to do so (for example, if waste is converted to fuel, there will be a residue of inconvertible waste, which must be landfilled). In all options, due allowance will be made for the cost and availability of capital and the regulations which govern waste disposal (control of leachate, daily cover, etc.).

3. THE USE OF THE PROGRAM

The program will run on an IBM PC using floppy discs, although more flexibility is available if an IBM XT or AT is used because hard disc storage permits easier combinations of options.
(i) The user is first asked to choose an option from

```
CONSOLIDATED    LANFILL
BALING   AND   LANDFILLING
FUEL   ENERGY   RECOVERY
FERROUS   EXTRACTION
```

(ii) The program will calculate waste arisings from population data or will accept the user's estimate.
(iii) According to the option chosen, the user will be provided with a check list of the equipment required, the land available, the power required, the labour required, the equipment availability, the spares requirement, the cost of capital, the lives of the project, sub projects and equipment.

Where the user has not chosen a parameter, the program will suggest a value although this facility is limited at present. As an example, the program currently contains information on the costs of only one baler but if and when a user supplies information on a different baler, that information will be stored and become available to later users.
(iv) The program will discount the equipment cost over its life (or lives, if mechanical and civil engineering systems and the site have different lives) and determine the cost of each tonne of waste that is dealt with.
(v) The program will calculate the operating times of the plant, in terms of shifts per day, days per week, weeks per year, permissible down time and plant capacity. The user will be able to change from the arrangement recommended by the computer, if he wishes.
(vi) The program will calculate the labour costs as a function of wages paid, hours worked, etc. The user will be able to change from the arrangement recommended by the computer.
(vii) The program will calculate running costs as a function of rent, power, spares, indirect material costs, down time costs, etc. Again the user may change from the displayed arrangement.
(viii) The program will display the capital and running costs of the combinations of options that have been chosen.
(ix) The program will calculate the sensitivity of cost per tonne to such parameters as cost of capital,

4. SOME TYPICAL COMPUTER OUTPUTS

```
                    CONTROLLED    LANDFILL    BY   BALE   DISCHARGE
                    ----------------------------------------
                    SELECTION   OF  THE  BALER  AND  OPERATING   TIME
                    ----------------------------------------
RECOMMENDED      ARRANGEMENT:

1-   NO.  STANDARD   TYPE   BALER   35.00   T/HR                    1
2-   NO.  WEEKS   PER   YEAR                                       50
3-   NO.  DAYS   PER   WEEK                                         5
4-   NO.  SHIFTS   PER   DAY                                        1
5-   NO.  HOURS   PER   SHIFT                                       8
6-   EXTRA   HOURS   AFTER   SHIFT                                  0
7-   DOWN   TIME   RATE                                         10.00%
8-   MAX   CAPACITY   OF   THE   PLANT                         63000.00
9-   AVE.   TONNAGE   OF   WASTE   RECEIVED   (NO   FERROUS)   57072.74

IF  THE  ANALYSIS  IS  ACCEPTABLE  TYE  "Y".  TO  CARRY  OUT  YOUR  OWN
ANALYSIS  WITH  ANY  ALTERATION  TYPE  "N".
```

```
                              ANALYSIS   RESULTS
                              ----------------
CAPITAL   COSTS:

1-   CIVIL   ENG.         644000.00      3-   ELECTRICAL                45900.00
2-   MECHANICAL           115500.00      4-   MECHANICAL(FE.   EX.)    116950.00
TOTAL   CAPITAL   COSTS                                                922350.00
DEPRECIATED   CIVIL.       75644.00      DEPRECIATED   MECH.&EL.        45300.18

RUNNING   COSTS   :

1-   ANNUAL   WAGES        45783.88      3-   PLANT   COSTS   (FE)       3100.00
2-   PLANT   COSTS        119025.28
TOTAL   RUNNING   COSTS                                                167909.16
ANNUAL   REVENUE   :  (FERROUS   METAL)                                 14888.54

TOTAL   ANNUAL   COSTS   (DEPRECIATION   &  RUNNING)                   273964.80
COSTS/TONNE   (PLANT)          4.42      COSTS/TONNE   (LANDFILL)           5.80

TOTAL   COSTS   PER   TONNE   OF   WASTE                                   10.22

TO   CONTINUE   TYPE   "C"
/-C
```

5. DISCUSSION

Because the program is user friendly and takes only a few minutes to determine the cost of a waste disposal scheme or combination of schemes, the decision maker may explore many options before choosing one for the Waste Disposal Authority to adopt. It is also quick and easy to explore the sensitivity of the cost per tonne of waste disposed to any input parameter (such as cost of tin, cost of cover material, selling price of fuel). The investigations required to write the program suggest that not only are the costs of landfill considerably greater than many W.D.A.'s had supposed, but also many reclamation schemes that had been applauded do not meet their specifications.

HAZARDOUS WASTE MANAGEMENT ON THE REGIONAL, MACROREGIONAL AND NATIONAL LEVELS

R.SZPADT

INSTITUTE OF ENVIRONMENT PROTECTION ENGINEERING, TECHNICAL UNIVERSITY OF WROCŁAW, POLAND

1. INTRODUCTION
Present state of hazardous waste management in Poland presents many hazards for environment and also for health and life of population. Considerable part of waste is disposed in an uncontrolled manner in the environment by storage on inadapted dumps, municipal landfills and waste ponds.
Numerous research works and studies financed by the government have been undertaken in order to find optimal technical and organizational solutions of hazardous waste management. Up to the present there are worked out (1,2) :
- principles of waste classification with respect to their harmfulness for environment,
- detailed characteristics of composition and physico-chemical and technological properties and their harmfulness for environment of several hundred of significant hazardous wastes generated in the national economy,
- balances of hazardous wastes for some industrial branches /in national level/ and macroregions,
- general instructions for programming, organization and designing macroregional systems and facilities for hazardous waste management,
- preliminary technological and organizational conceptions of hazardous waste management for two macroregions.
Started works are continued, but difficult economic situation of the country does not promote rapid implementation of technical solutions.

2. PLANNING OF HAZARDOUS WASTE MANAGEMENT
Due to complexity and importance of hazardous waste utilization and disposal it is considered that it is necessary to work out a general conception of these waste management in the whole country (2,3).
There are two basic approaches to solving this problem:
1 - working out a conception for the whole country initially, taking no account of administrative and economic borders /provinces/, determining the type, capacity and location of hazardous waste management stations and range of their operation and then separate the tasks arising from realization of this conception for each province,
2 - working out a conception for particular provinces and macroregions and then for the whole country on the basis of

province and macroregion conceptions. The object of co-ordination at national level is optimization of solutions, possible limitation of station numbers by forming in reasonable cases supraregional /and even national/ stations of treatment and disposal of hazardous wastes, elimination of ineffective stations and waste transportation, determining organizational forms and principles for the whole system.

At the present moment there are being carried out works on macroregional plans of hazardous waste management. Macroregion comprises several basic economic and administrative regions /provinces/ and is a unit only for planning purposes.

In accordance with the accepted assumptions macroregional plans of hazardous waste management should contain the conception of system organization on the level of a macroregion as a whole and also determine methods and technical solutions of control and recording of generation, transport, reloading, storage, treatment, utilization and disposal of every waste from all sources. This will make possible a rational management of hazardous wastes with undertaking necessary precautions and control.

Within macroregional plans various organizational solutions are admissible, they must however form a coherent system. There are possible in general the following variants of solution:
- building of central station for treatment and disposal of hazardous wastes from all sources in the whole macroregion and comprising all necessary technological processes /in the intensively industrialized macroregions there can be 2-3 macroregional stations/,
- separate stations of hazardous waste management /incineration plants, special landfills, etc./ for very large sources of waste generation and one central station for wastes from remaining sources,
- common stations of waste management for plants of the same branch located very closely one from the other,
- building of separate stations of treatment and disposal of specific groups of wastes for the whole macroregion /e.g. central station for treatment of electroplating sludges and the like/,
- utilization of efficient but not fully loaded equipment /e.g. incineration stations/ in large plants for disposal of wastes from small and medium size plants,
- building of transportable installations for treatment of wastes from plants of the whole macroregion /e.g. installation for solidification of waste, splitting of oil emulsions or dewatering of sludges/,
- building of regional /province/ networks of stations for storage, preliminary treatment and transfer of wastes which will be an intermediate link between waste producers and the macroregional station of treatment and disposal,
- building of supraregional /on national level/ stations of treatment of strongly toxic wastes generated in relatively small quantities /e.g. for regeneration, reuse and disposal of spent hardening salts, catalysts or active carbon/.

Built stations must be equipped with all necessary installations protecting environment against secondary pollution connected with waste treatment.

The following initial assumptions necessary for rational solving of problems occuring during working out macroregional and regional conceptions of hazardous waste management have been accepted:
- first of all it is necessary to bring to ordered state water and wastewater, sludge and waste management in particular industrial plants. Macroregional or regional station cannot be loaded with treatment and disposal of wastes which can be easily reclamed and reused at the place of their generation or are generated due to uneconomical management of plants,
- macroregional and regional stations should service first of all small and medium sized plants, generating relatively small quantities of the same kind of waste. Large plants can organize disposal of their own wastes.
- treatment and disposal of specific waste generated only in one plant in the given region should be performed by this plant, especially in case when these waste cannot be treated together with other waste,
- if the quantity of given kind of waste in one plant is decidedly /many times/ larger than in not numerous small plants generating the same waste, it is recommended that this plant should organize its treatment and accept waste from remaining plants /against payment/,
- general principle of hazardous waste management is reclaiming of secondary raw materials and their utilization. Only those wastes should be disposed for which there is no possibility or technology of reuse.

In planning and designing of regional and macroregional stations it is necessary to secure appreciable flexibility of their operation, arising among others from large irregularity of waste generation, their heterogeneity, etc.

Selection of detailed solutions for particular macroregions depends on kind of industry, wastes, local conditions, but it can be preliminarily assumed that in each macroregion it is necessary to build several regional stations for storage and pretreatment of waste, several utilization, treatment and disposal stations in industrial plants and at least one central facility of hazardous waste treatment and disposal. Regional stations for waste storage and pretreatment should comprise with their operation particular provinces or their parts, whereas the central facility should be situated as far as possible in the centre of waste generation in given macroregion.

It follows from the rewiev of proposed solutions that the hazardous waste management is a many level one - from local level of industrial plants and particular towns, through regional /province/ solutions of waste storage and pretreatment stations, macroregional stations of waste treatment and disposal up to central /national level/ stations of utilization and treatment of some specific wastes. This complicity of the problem, quantitative and qualitative variety of wastes and necessary technological processes of their treatment and disposal and also social and economic factors are the cause of many difficulties in planning and implementation of optimal solution.

3. DIFFICULTIES ARISING IN PLANNING AND IMPLEMENTING OF SOLUTIONS

Till now attempts of implementation of regional /province/ systems of municipal waste management in Poland in most cases ended in failure, and regional plans worked out with great effort and optimized economically were not realized.

Causes of this state are as follows:
- regional authorities did not undertake real co-ordination of waste management in their regions and did not initiate implementation of worked out plans,
- attempts of location of regional landfills for several administrative units /town, commune/ were met with firm opposition of territorial self-governments, which did not allow the possibility of storage of wastes from other regions on territory under their administration,
- municipal and industrial enterprises were against realization of joint stations which require close co-operation during their erection and exploitation. There prevailed principle that own waste dump is better even if it is flimsy than a joint station of several partners,
- production of equipment for transfer stations which as intermediate objects between waste generators and stations of disposal are indispensable elements of regional systems, was not undertaken,
- production of equipment for landfills and compost plants was not undertaken,
- economic factors, all municipal managements in Poland are lacking finances and are undercapitalized.

Therefore, no radical change in municipal waste management is possible and regionalization problem cannot extend beyond the sphere of theoretical considerations and conceptions.

All these factors have caused return at present to building small local waste dumps for particular towns, communes and even villages. Often it is the only lawful sanction for coming into being or existing wild waste dumps. It means many years regress and it increases potential hazards for environment.

There is a danger that strong decentralization tendencies will increase also in solutions of hazardous waste management problems, which would be absolutely unacceptable due to range and scale of potential hazards by unsuitable management of the waste.

Increasing general pollution and degradation of environment in Poland, and also examples of local ecological disasters have caused considerable alarm and at the same time increase Polish public interest in problems of environment protection and in necessity of undertaking appropriate technical and organizational solutions of wastewater and gases treatment and waste management in order to protect people health. However, if in the case of gases or wastewater treatment there are solutions of local character and they are accepted by population then solutions of waste management and hazardous waste in particular exceed plant or commune scope and this creates additional problems.

Numerous bad experiences with landfilling of municipal

and industrial waste in the past and often also at present and connected with it arduousness of waste dumps for environment caused that villagers very unfavourable see waste disposal stations on their territories.

Efforts of industrial plants to obtain grounds for not troublesome waste often last many years and are connected with additional costly services for given village or community inhabitants.

Also authorities of particular provinces are not favourably disposed towards co-operation with neighbouring provinces, especially if arising from optimization location of macroregional facility for disposal of hazardous wastes is on their terrain, what is connected with necessity of accepting for disposal wastes from outside given province.

So often, proper in assumption solution encounter difficulties in realization which are inherent mainly in psychical sphere, in lack of understanding needs and necessity of co-operation. Without explaining and overcoming these resistances it will not be possible to implement any solution in larger than local scope, and this would mean lack of possibility for rational management of hazardous wastes.

There are however few examples of joint initiative undertaken by groups of industrial plants which decided to build joint station for treatment of spent oil emulsions or hazardous waste landfills and also incineration of wastes from various sources in one incinerating plant.

It seems that only in this way of gradual bringing closer different standpoints it may be possible to reach finally optimal solution on the regional and macroregional levels.

More effective influence of implemented economic reform on projects of waste management undertaken by industrial plants is expected.

REFERENCES

1. Bartoszewski K, Szpadt R: Hazardous Waste Management in the Polish Chemical Industry. Proceedings of the MER-3, Antwerp, 1986
2. Kempa E, Szpadt R: Criteria for Evaluating the Environmental Impacts of Industrial Wastes and Possibilities of Their Utilization. Institute of Environment Protection Engineering, Wrocław Technical University, Internal Report No. 49/85, Wrocław, 1985, /unpublished - in Polish/
3. Wilson DC, Parker CJ: Hazardous Waste Management - the Way Forward. Proceedings of the MER-3, Antwerp, 1986

SITING OF INDUSTRIES AND STATUS OF WETLAND ECOSYSTEM ON THE INDIAN COAST

DR. RAJESHWARI MAHALINGAM, DR. KAMALA GOPINATHAN, MINISTRY OF ENVIRONMENT & FORESTS PROJECT, DEPARTMENT OF CHEMICAL ENGINEERING, I.I.T., MADRAS-600 036, INDIA

The Environment (Protection) Act, 1986 of the Government of India is an act of Parliament to provide for the protection and improvement of environment and the prevention of hazards to human beings, living creatures, plant and property. The Government has the power for prohibitions and restrictions in siting of Industries and the carrying on of processes and operations in different areas. Persons carrying on industrial operations etc. are not to allow emissions or discharge of environmental pollutants in excess of the standards as may be prescribed. Persons handling hazardous substances are required to comply with procedural safeguards.

However, developing country like India is at present not in a position to assess the cumulative effect of a number of Industries at a single or adjacent locations, which ultimately would cause significant damage to the environment and the ecological features of a productive ecosystem. Thermal power generation, Fertiliser, Petrochemical and Chemical Industries are site specific which depends on the process technology. They are required to obtain environmental clearance for siting only from the end of 1984.

The industries which are already in existence especially on the main coast make the wetland ecosystem unproductive. Wetland ecosystem comprises mangroves as woody coastal formations, seagrass beds as submerged macrophytes and coral reefs as carbonate skeleton of the seafloor animals and algae. They are recognised to be the critical and linked habitat in fisheries production. The industries such as Tuticorin Thermal Power Plant; Southern Petrochemical Industries Corporation Limited, Tuticorin Alkali Chemicals, Dharangadhara Chemicals, Gulf Olefins; and Tuticorin Salt and Marine Chemicals are located in Tuticorin, situated on the main coast of South India in the Gulf of Mannar. The seafloor in this area is utilised for effluent disposal and the exploitation of coral reefs for calcium carbide in the acetylene industries. Large areas of the mangroves have been converted for salt pond construction. Various activities of the industries using the seafloor for dumping hazardous chemicals, solid wastes etc. are a threat to the environment in the region as a whole. There is an urgent need for synchronisation of policies to monitor the standards of the effluents to protect the environment and to integrate the technology for environment conservation.

List of addresses of first-named authors

Ale, B.J.M.
Ministry of Housing, Physical Planning and the Environment, P.O. Box 450, 2260 MB Leidschendam. The Netherlands.

Arikol, M.
Chemical Engineering Department, Bogazici University, P.K. 2, Bebek, Istanbul. Turkey.

Assink, J.W.
TNO Division of Technology for Society, P.O. Box 342, 7300 AH Apeldoorn. The Netherlands.

Ataer, Ö.E.
Gazi University, Faculty of Engineering and Architecture, Mechanical Engineering Department, Maltepe, Ankara. Turkey.

Baan, P.J.A.
Delft Hydraulics, P.O. Box 177, 2600 MH Delft. The Netherlands.

van den Bleek, C.M.
Department of Chemical Engineering, Delft University of Technology, P.O. Box 5045, 2600 GA Delft. The Netherlands.

Blomen, L.J.M.J.
KTI, P.O. Box 86, 2700 AB Zoetermeer. The Netherlands.

de Boks, P.A.
IWACO, P.O. Box 183, 3000 AD Rotterdam. The Netherlands.

Boztepe, H.
Ç.Ü. Faculty of Arts and Sciences, Department of Chemistry, Adana. Turkey.

van Breukelen, H.H.
HASKONING Royal Dutch Consulting Engineers, P.O. Box 151, 6500 AD Nijmegen. The Netherlands.

Bridges, E.M.
University College of Swansea, Department of Geography, Singleton Park, Swansea SA2 8PP. Great Britain.

Bruins, J.G.
BKH Consulting Engineers, P.O. Box 93224, 2509 AE The Hague. The Netherlands.

Bijl, J.J.W.
Van Tongeren B.V., P.O. Box 34, 1940 AA Beverwijk. The Netherlands.

Calis, G.H.M.
DSM Research B.V. (PT/KG), P.O. Box 18, 6160 MD Geleen. The Netherlands.

van Campen, A.L.B.M.
TAUW Infra Consult B.V., P.O. Box 479, 7400 AL Deventer. The Netherlands.

Carlsson, K.
Fläkt Industri AB, Kvarnvägen, 35187 Växjö. Sweden.

Christiansen, K.
Technological Institute, P.O. Box 141, 2630 Tåstrup. Denmark.

van Deelen, C.L.
TNO Division of Technology for Society, P.O. Box 342, 7300 AH Apeldoorn. The Netherlands.

Dobosz, J.R.
Institute of Environment Protection Engineering, Technical University of Wroclaw, pl. Grunwaldzki 9, 50-377 Wroclaw. Poland.

Don, J.A.
TNO Division of Technology for Society, P.O. Box 342, 7300 AH Apeldoorn. The Netherlands.

Dragt, A.J.
DHV Consulting Engineers, P.O. Box 85, 3800 AB Amersfoort. The Netherlands.

Drouet, D.
Recherche Développement International, 37 c3 Avenue Franklin Roosevelt, 75008 Paris. France.

Durmaz, A.
Gazi University, Faculty of Engineering and Architecture, Mechanical Engineering Department, Maltepe, Ankara. Turkey.

Enneking, C.Q.M.
N.V. Vereenigde Glasfabrieken, P.O. Box 46, 3100 AA Schiedam. The Netherlands.

Etemad, F.
University College of Swansea, Department of Management Science, Swansea SA2 8PP. Great Britain.

Grinwis, A.W.
TAUW Infra Consult B.V., P.O. Box 479, 7400 AL Deventer. The Netherlands.

de Gijt, J.G.
Fugro B.V., P.O. Box 63, 2260 AB Leidschendam. The Netherlands.

Hack, P.J.F.M.
Paques B.V., P.O. Box 52, 8560 AB Balk. The Netherlands.

Harmsen, L.W.F.
Gist-Brocades N.V., P.O. Box 1, 2600 MA Delft. The Netherlands.

Hawkins, R.G.P.
Cleanaway Limited, The Drive, Warley, Brentwood, Essex CM13 3BE. Great Britain.

Hendriks, Ch.F.
CUR, P.O. Box 420, 2800 AK Gouda. The Netherlands.

Henning, K.-D.
Bergbau-Forschung GmbH, P.O. Box 130140, 4300 Essen 13. Fed. Rep. of Germany.

van der Hoek, J.P.
Agricultural University Wageningen, Department of Water Pollution Control, De Dreijen 12, 6703 BC Wageningen. The Netherlands.

Hommes, R.W.
Erasmus Centre for Environmental Studies (ESM), Erasmus University, P.O. Box 1738, 3000 DR Rotterdam. The Netherlands.

Hou, R.
Institute of Environmental Science, Peking Normal University, Peking. China.
Present address: Institut für Siedlungswasserwirtschaft, Universität Karlsruhe, Am Fasanengarten, 7500 Karlsruhe 1. Fed. Rep. of Germany.

Hubers, H.
CICAT (Centre for International Cooperation and Appropriate Technology), P.O. Box 5048, 2600 GA Delft. The Netherlands.

Huppes, G.
Centre for Environmental Studies, Leiden University, P.O. Box 9318, 2300 RA Leiden. The Netherlands.

Jorden, W.
Gesamthochschule Paderborn, Laboratorium für Konstruktionslehre, FB 10, Pohlweg 47/49, 4790 Paderborn. Fed. Rep. of Germany.

Joyce, F.
ECOTEC Research and Consulting Ltd., Priory House, 18 Steelhouse Lane, Birmingham B4 6BJ. Great Britain.

Kaiser, A.
Fraunhofer-Institut für Silicatforschung, Neunerplatz 2, 8700 Würzburg. Fed. Rep. of Germany.

Koesters, H.
Zenit GmbH, Dohne 54, 4330 Mülheim a.d. Ruhr. Fed. Rep. of Germany.

Kowal, A.L.
Institute of Environment Protection Engineering, Technical
University of Wroclaw, pl. Grunwaldzki 9, 50-377 Wroclaw. Poland.

Kroon, J.A.
DSM Research B.V. (PT/KG), P.O. Box 18, 6160 MD Geleen. The
Netherlands.

Kunz, G.J.
TNO Physics and Electronics Laboratory, P.O. Box 96864, 2509 JG
The Hague. The Netherlands.

Kutsal, T.
Hacettepe University, Chemical Engineering Department, 06532
Beytepe, Ankara. Turkey.

Landscheidt, A.
Chemische Fabrik Stockhausen GmbH, P.O. Box 570, 4150 Krefeld 1.
Fed. Rep. of Germany.

Langen, F.H.M.M.
Comprimo Engineers & Contractors, P.O. Box 4129, 1009 AC
Amsterdam. The Netherlands.

Langerwerf, J.S.A.
TNO Division of Technology for Society, P.O. Box 217, 2600 AE
Delft. The Netherlands.

Lempfer, K.
Fraunhofer-Institut für Holzforschung, Bienroder Weg 54E, 3300
Braunschweig. Fed. Rep. of Germany.

Lourens, P.A.
Gist-Brocades N.V., P.O. Box 1, 2600 MA Delft. The Netherlands.

Loxham, M.
Delft Geotechnics, P.O. Box 69, 2600 AB Delft. The Netherlands.

Ludwig, H.
Fichtner Consulting Engineers, P.O. Box 572, 7000 Stuttgart 1.
Fed. Rep. of Germany.

Mahalingam, R.
Indian Institute of Technology, Department of Chemical
Engineering, Madras 600 036. India.

Marcinkowski, T.
Institute of Environment Protection Engineering, Technical
University of Wroclaw, pl. Grunwaldzki 9, 50-377 Wroclaw. Poland.

Marinov, H.
ul. Ami Boué 51, 1606 Sofia. Bulgaria.

Martin, L.R.
Hazardous Waste Minimization, The Institute for Local
Self-Reliance, 2425 18th Street NW, Washington D.C. 20009. USA.

Meyer zu Schlochtern, P.H.M.
Atox Waste Management, Smet Jet N.V., Nijverheidsstraat 3, 2431
Oevel (Westerlo). Belgium.

Mojtahedi, W.
Laboratory of Fuel Processing and Lubrication Technology,
Technical Research Centre of Finland, Vuorimiehentie 5, 02150
Espoo 15. Finland.

Müezzinoglu, A.
Dokuz Eylül University, Engineering and Architecture Faculty,
Environmental Engineering Department, Bornova, Izmir. Turkey.

Müller, K.
Ministry of the Environment, National Agency of Environmental
Protection, 29 Strandgade, 1401 Copenhagen K. Denmark.

Neumann, U.
UHDE GmbH, P.O. Box 262, 4600 Dortmund 1. Fed. Rep. of Germany.

Nitschke, E.
UHDE GmbH, P.O. Box 262, 4600 Dortmund 1. Fed. Rep. of Germany.

van den Oosterkamp, P.F.
KTI, P.O. Box 86, 2700 AB Zoetermeer. The Netherlands.

van Oosterom, W.P.
TAUW Infra Consult B.V., P.O. Box 479, 7400 AL Deventer. The
Netherlands.

de Oude, N.T.
Procter & Gamble European Technical Center, Temselaan 100, 1820
Strombeek-Bever. Belgium.

Paul, P.G.
Comprimo Engineers & Contractors, P.O. Box 4129, 1009 AC
Amsterdam. The Netherlands.

Pietrzeniuk, H.-J.
Umweltbundesamt, Bismarckplatz 1, 1000 Berlin 33. Fed. Rep. of
Germany.

Pilarczyk, K.W.
Rijkswaterstaat, Road and Hydraulic Engineering Division, P.O.
Box 5044, 2600 GA Delft. The Netherlands.

Quakernaat, J.
TNO Division of Technology for Society, P.O. Box 342, 7300 AH
Apeldoorn. The Netherlands.

Quant, B.
Polish Academy of Science, Institute of Hydroengineering, ul.
Cystersów 11, 80-953 Gdansk. Poland.

Rademaker, P.D.
Aardelite Holding B.V., P.O. Box 301, 8070 AH Nunspeet. The
Netherlands.

van Ree, C.C.D.F.
Delft Geotechnics, P.O. Box 69, 2600 AB Delft. The Netherlands.

Reichert, A.
Institut für Aufbereitung, Kokerei und Brikettierung der RWTH
Aachen, Wüllnerstrasse 2, 5100 Aachen. Fed. Rep. of Germany.

Risse, T.
Gottfried Bischoff GmbH & Co. KG, P.O. Box 100533, 4300 Essen 1.
Fed. Rep. of Germany.

van Roosmalen, G.R.E.M.
'Waste Management', KU Tilburg - TU Eindhoven, P.O. Box 513, BG
3.16, 5600 MB Eindhoven. The Netherlands.

Schmal, D.
TNO Division of Technology for Society, P.O. Box 217, 2600 AE
Delft. The Netherlands.

Schöller, M.
DHV Consulting Engineers, P.O. Box 85, 3800 AB Amersfoort. The
Netherlands.

van der Sluis, S.
Department of Chemical Engineering, Delft University of
Technology, De Vries van Heystplantsoen 2, 2628 RZ Delft. The
Netherlands.

Soczó, E.R.
National Institute of Public Health and Environmental Hygiene
(RIVM), Laboratory for Waste and Emission Research, P.O. Box 1,
3720 BA Bilthoven. The Netherlands.

Spijker, R.
DSM Research, IJmuiden Laboratory, P.O. Box 463, 1970 AL
IJmuiden. The Netherlands.

Sterckx, L.M.J.
Indaver N.V., Poldervlietweg, 2030 Antwerpen. Belgium.

Szpadt, R.
Institute of Environment Protection Engineering, Technical
University of Wroclaw, pl. Grunwaldzki 9, 50-377 Wroclaw. Poland.

Taner, F.
C.Ü. Faculty of Arts and Sciences, Department of Chemistry,
Adana. Turkey.

Thomas, D.L.
Hazardous Waste Research and Information Center, 1808 Woodfield Drive, Savoy, Illinois 61874. USA.

Tize, R.
Seghers Engineering n.v., Genèvestraat 10, 1140 Brussel. Belgium.

Tjioe, T.T.
Delft University of Technology, De Vries van Heystplantsoen 2, 2628 RZ Delft. The Netherlands.

Tol, L.A.J.
ESTS B.V., P.O. Box 10.000, 1970 CA IJmuiden. The Netherlands.

Tuin, B.J.W.
University of Technology Eindhoven, Department T/TF, P.O. Box 513, 5600 MB Eindhoven. The Netherlands.

Urlings, L.G.C.M.
TAUW Infra Consult B.V., P.O. Box 479, 7400 AL Deventer. The Netherlands.

van Veen, H.J.
TNO Division of Technology for Society, P.O. Box 342, 7300 AH Apeldoorn. The Netherlands.

Veenman, A.W.
Bureau Op den Kamp, Veursestraatweg 74, 2265 CE Leidschendam. The Netherlands.

van Velsen, A.F.M.
Haskoning, P.O. Box 151, 6500 AD Nijmegen. The Netherlands.

Verbeek, A.
TNO Division of Technology for Society, P.O. Box 342, 7300 AH Apeldoorn. The Netherlands.

Vigil, S.A.
California Polytechnic State University, San Luis Obispo, California 93407. USA.

Vilppunen, P.
University of Oulu, Faculty of Technology, Energy Laboratory, 90570 Oulu. Finland.

Visscher, K.
National Institute of Public Health and Environmental Hygiene (RIVM), P.O. Box 1, 3720 BA Bilthoven. The Netherlands.

de Vleeschauwer, D.
EBES N.V., Mechelsesteenweg 271, 2018 Antwerpen. Belgium.

Vles, E.J.
Association of the Dutch Chemical Industry (VNCI), P.O. Box 443, 2260 AK Leidschendam. The Netherlands.

Vorstman, M.A.G.
University of Technology Eindhoven, Department T/TF, P.O. Box 513, 5600 MB Eindhoven. The Netherlands.

de Vries, H.
HGA Galvano-Aluminium B.V., Loodsweg 1, 8243 PH Lelystad. The Netherlands.

Walpot, J.I.
TNO Division of Technology for Society, P.O. Box 342, 7300 AH Apeldoorn. The Netherlands.

Wiersma, D.
Faculteit der Economische Wetenschappen, Universiteitscomplex Paddepoel, P.O. Box 800, 9700 AV Groningen. The Netherlands.

Witkamp, G.J.
Delft University of Technology, De Vries van Heystplantsoen 2, 2628 RZ Delft. The Netherlands.

de Zeeuw, E.
De Ruiter Milieutechnologie b.v., P.O. Box 14, 1160 AA Zwanenburg. The Netherlands.

MIX
Papier aus verantwortungsvollen Quellen
Paper from responsible sources
FSC® C105338

If you have any concerns about our products,
you can contact us on
ProductSafety@springernature.com

In case Publisher is established outside the EU,
the EU authorized representative is:
Springer Nature Customer Service Center GmbH
Europaplatz 3, 69115 Heidelberg, Germany

Printed by Libri Plureos GmbH
in Hamburg, Germany